Dipak Dutta

Structures with Hollow Sections

Dipak Dutta

Structures with Hollow Sections

Dipl.-Ing. Dipak Dutta
Marggrafstraße 13
D-40878 Ratingen

This book contains 549 figures and 125 tables

Die Deutsche Bibliothek – CIP-Cataloguing-in-Publication-Data
A catalogue record for this publication is available from Die Deutsche Bibliothek

ISBN 3-433-01458-2

© 2002 Ernst & Sohn Verlag für Architektur und technische Wissenschaften GmbH, Berlin

All rights reserved, especially those of translation into other languages. No part of this book shall be reproduced in any form – i.e. by photocopying, microphotography, or any other process – or be rendered or translated into a language useable by machines, especially data processing machines, without the written permission of the publisher.

Typesetting: ProSatz Rolf Unger, Weinheim
Printing: betz-druck GmbH, Darmstadt
Binding: Großbuchbinderei J. Schäffer GmbH & Co. KG, Grünstadt

Printed in Germany

Dedicated to my beloved wife Helma

Preface

In order to answer to the question why a book specially dealing with the structural applications of circular and rectangular (also square) hollow sections in steel has to be written, the following two essential points have to be underlined:

1. Although circular hollow sections had been in structural use since last century, the industrial manufacture of rectangular and square hollow sections started first at the end of the fifties (Stewarts and Lloyds in England, 1959). The hollow section is, so to say, the youngest member of the family of the structural steel profiles. Not only for rectangular (square) hollow sections but also for circular hollow sections, the stringent necessity existed even in the seventies for the technical knowledge in practically all sectors e.g. static and fatigue loads and resistance, stability (flexural, lateral-torsional and local buckling), composite constructions with concrete, protection against corrosion or fire, wind loads, fabrication, assembly, erection and transport. Extensive research works were required to make up for the competitiveness against other profiles (I, L, U etc.) as well as other materials (concrete, timber) with regard to the technique and the economy. The steel constructors needed the exact knowledge about how to construct structures with hollow sections, which are technically secured and economical at the same time. The professors in the universities and the teachers in the polytechnical schools and colleges could very little impart to the students regarding the application technology of hollow sections. This is documented by the fact that the hollow sections are hardly mentioned in the books on structural engineering of this time period.

2. Due to the closed form of hollow sections, the calculation and design of hollow section structures require knowledge, which goes significantly beyond that for general structural engineering. In addition to this, comes the fact that the theoretical determination of the load bearing resistances of hollow section joints, particularly for predominantly applied welded lattice joints, in a form of general validity has not been successful until today on account of the non-uniformity of the stress distribution in them (Finite element analysis, which can be applied for this purpose from case to case, is very expensive!). Therefore in the last thirty years, numerous practical tests were performed, fortunately coordinated on an international basis, which led to the semi-empirical design formulae for the static and fatigue resistances of hollow section joints. However, the research works were carried out not only to determine the load bearing capacity, but also in other fields already mentioned above, where the application of hollow sections differs considerably from that of the conventional open profiles. This scenario changed gradually at the beginning of the eighties with the aid of these efforts, as many important research projects all over the world came to a close. Adequate design rules and recommendations could be worked out, which have now been incorporated into the national standards of the U.S.A. (partly), Japan, Canada and Australia as well as the European standard EN (Eurocode) and the international standard ISO.

Unfortunately, the informations regarding the mentioned activities are not yet available in a compiled form. Often the publications are not even accessible to the interested public. The author as one of the international co-ordinators of the world-wide research works on the applications of hollow sections has made it his objective to present these results as a digest in this book.

Special thanks go to the following persons and friends, whose intensive efforts and numerous research works have contributed to attain the recent high level of knowledge concerning constructions with steel hollow sections:

Prof. Dr.-Ing. Reinhard Bergmann
University of Bochum, Germany

Prof. Dr.-Ing. Ömer Bucak
Technical High School Munich, Germany

Dr.-Ing. Dietmar Grotmann,
Salzgitter AG, Germany

Prof. Dr. Yoshiaki Kurobane
Kumamoto University, Japan

Prof. Dr.-Ing. habil. Friedrich Mang (retired)
University of Karlsruhe, Germany

Prof. Dr. Jeff Packer
University of Toronto, Canada

Prof. Dr.-Ing. Ram Puthli
University of Karlsruhe, Germany

Prof. Dr. Jacques Rondal
University of Liege, Belgium

Prof. Dr. Jaap Wardenier
Delft University of Technology,
The Netherlands

Dipl.-Ing. Karl-Gerd Wuerker (retired)
Mannesmannroehren-Werke AG, Germany

Dr. Xiao-Ling Zhao
Monash University, Australia

The author apologises to all those, whose names have not been mentioned.

Further, grateful acknowledgement is made to the firm V & M Tubes and the organisation CIDECT for making the photographs available.

March 2002   Dipak Dutta

Contents

1	**Introduction**	1
2	**Manufacture of steel hollow sections**	21
2.1	General	21
2.2	The manufacture of seamless tubes	21
2.2.1	The skew roll piercing and pilger rolling process	23
2.2.2	The push bench process	24
2.2.3	The plug rolling process	24
2.2.4	The extrusion process	25
2.2.5	The continuous rolling process	25
2.3	The manufacture of welded tubes	25
2.3.1	The submerged-arc welding process	26
2.3.2	The gas-shielded arc welding process	27
2.3.3	The continuous pressure welding process (Fretz-Moon)	28
2.3.4	The electric resistance pressure welding process	28
2.4	The stretch-reducing or reducing process	28
2.5	The manufacture of rectangular and square hollow sections	29
2.6	The manufacture of hollow sections with further shapes	29
2.7	The testing of steel hollow sections	29
3	**Structural hollow section steel grades**	33
3.1	General	33
3.2	Selection of steel grades for structural design	41
4	**Dimensions, tolerances and sectional properties of circular, square and rectangular hollow sections**	45
4.1	General	45
4.2	Geometrical and sectional properties of hollow sections	47
4.2.1	Circular hollow sections (CHS)	48
4.2.2	Rectangular, including square, hollow sections (RHS)	48
4.2.3	Larger dimensions of square and rectangular hollow sections (RHS)	49
5	**Design bases for hollow sections in steel structures**	51
5.1	General	51
5.2	Partial safety factors	52
5.2.1	Actions and their partial safety factors	53
5.2.2	Resistances and their partial safety factors	54
5.3	Design resistances of cross sections	56
5.3.1	Hollow sections in axial force	56
5.3.2	Hollow sections in bending and shear	57
5.3.2.1	Design for pure uni-axial bending	57

5.3.2.2	Combined uni-axial bending and axial force (without considering shear)	58
5.3.2.3	Combined bi-axial bending and axial force (without considering shear)	58
5.3.2.4	Stress design of hollow sections in shear	59
5.3.2.5	Design for bending considering shear	59
5.3.2.6	Equivalent stress for combined axial force and shear	61
5.3.3	Stability cases – flexural (overall) and local buckling of hollow sections	61
5.3.3.1	Design for flexural buckling	61
5.3.3.2	Local buckling of hollow sections	66
5.3.3.3	Hollow sections in combined compression and bending	77
5.3.3.4	Buckling length	82
5.3.3.5	Design examples	86
5.3.4	Hollow sections in torsion	88
5.4	List of symbols	90
6	**Structural elements and constructions with hollow sections under predominantly static loading**	**93**
6.1	Single section beams	94
6.2	Columns	94
6.2.1	Column-to-truss connections	96
6.2.2	Beam-to-column connections	98
6.3	Arched girders	102
6.4	Knee joints	103
6.4.1	Knee joints with circular hollow sections	107
6.5	End-to-end connections	108
6.5.1	Indirectly joined bolted connections	108
6.5.2	Indirectly joined welded connections	121
6.5.3	Directly joined welded connections	122
6.6	Lattice girders (trusses)	126
6.6.1	Single plane lattice girders with hollow sections	128
6.6.1.1	Economical aspects regarding the selection of lattice girder types with hollow sections	130
6.6.2	Design arrangement of truss joints with hollow sections consisting of directly welded chord and bracing members	130
6.6.3	Weld seams in trusses with hollow sections consisting of directly welded bracing and chord members	135
6.6.3.1	Design of welds in truss joints	138
6.6.3.2	Weld sequences for truss joints	141
6.6.3.3	Weld lengths in truss joints	142
6.6.4	Analysis of the strength (resistance) behaviour of a truss with hollow sections	144
6.6.4.1	Design strengths of truss joints with directly welded bracing and chord members	147
6.6.4.2	Design strengths of reinforced truss joints under predominantly static loading	171
6.6.4.3	Design of truss joints with cranked chord	175
6.6.4.4	Design strengths of uni-planar welded truss joints with I or H section as chord and hollow sections as bracings	175
6.6.4.5	Design strengths of uni-planar welded truss joints with ⊏-profile (channel) as chord and hollow sections as bracings	179
6.6.5	Design of uni-planar trusses with directly welded hollow sections	183

6.6.6	Design strengths of special uni-planar welded truss joints of circular hollow sections (CHS)	186
6.6.7	Design strengths of multi-planar welded truss joints	186
6.6.8	Joints with bracings of circular hollow sections with flattened ends	193
6.6.9	Joints with rectangular or square hollow sections (RHS) in a double-chord truss	204
6.7	Hollow section joints in Vierendeel trusses loaded by bending moment	207
6.7.1	RHS T joints under in-plane bending moment M_{ip}	208
6.7.2	RHS X joints under in-plane bending moment M_{ip}	212
6.7.3	Interaction between axial force N_i and in-plane bending moment M_{ip} in RHS T and X joints	212
6.7.4	Design of welded RHS T joints under axial force N_i and in-plane bending moment M_{ip} with the aid of the design diagrams	212
6.7.5	CHS T, Y and X joints under M_{ip}	214
6.7.6	RHS T joints in Vierendeel girders	216
6.7.7	Design resistances of CHS and RHS joints under out-of-plane bending moment M_{op}	219
6.7.7.1	Design resistances of RHS joints under M_{op}	219
6.7.7.2	Design resistances of CHS joints under M_{op}	219
6.8	Design resistances of T and X connections of plates, I-profiles or RHS sections to CHS or RHS	220
6.8.1	T and X connections with CHS chord members	220
6.8.2	T connections with RHS chord members	223
6.9	Design of beam-to-column connections	223
6.9.1	RHS column with I-beam	223
6.9.2	I section column with RHS beam	225
6.10	Special joint of RHS chord and RHS bracing with "bill-shaped" or "bird-mouth" end	226
6.11	List of symbols	227
7	**Design of welded hollow section joints subjected to fatigue loading**	**231**
7.1	General	231
7.2	Fatigue resistance for cumulative load (load spectrum)	235
7.3	Effect of residual stresses on the fatigue resistance of hollow section joints	236
7.4	Effect of corrosive environment on the fatigue of hollow section joints	238
7.5	Stress or strain distribution in hollow section joints	238
7.5.1	Stress concentration factor SCF and strain concentration factor SNCF	243
7.5.1.1	Experimental measurement of strains with strain gauges	243
7.5.1.2	Theoretical determination of stresses or strains using "finite element" analysis	245
7.5.1.3	Determination of stress or strain concentration factors (SCF or SNCF)	247
7.5.1.4	Determination of total hot spot stress range $S_{r,hs,total}$ using stress concentration factors	249
7.5.1.5	Parametric formulae for the determination of stress concentration factors for welded hollow section joints	251
7.6	Effect of secondary bending moments on the fatigue strength of RHS and CHS truss joints of K and N type	283
7.7	Basic "$S_{r,hs}$ vs. N_F" lines for welded uni-planar CHS and RHS (square) joints (T, X, K, N and KT)	284

7.7.1	Correction factors accounting for the wall thickness of the applicable member (chord and bracing) being checked for fatigue cracking and the fatigue lines "$S_{r,hs}$ vs. N_F"	286
7.7.2	Application of high strength steels	288
7.8	Basic "$S_{r,hs}$ vs. N_F" lines for welded multi-planar CHS and square hollow section joints (TT, XX and KK)	290
7.9	Design procedure for welded, uni-planar or multi-planar CHS or RHS (square) truss joints subjected to fatigue loading	291
7.10	Fatigue strength design procedure for hollow section connections and truss joints using the "classification" method	293
7.10.1	CHS connections with splice plates	296
7.10.2	Recommendation for the modification of the detail categories for the hollow section joints according to EC 3	300
7.11	Effect of the local reinforcement by plates on the fatigue strength of RHS joints	303
7.12	Repair and reconstruction of hollow section joints in the critical zones susceptible to ruptures by cracks	303
7.12.1	Application of a crack stopper in the form of a drilled hole	306
7.12.2	Grinding or gouging of the crack and rewelding	306
7.12.3	Mounting a plate or a shell on the chord flange (crack in the chord flange)	307
7.12.4	Fitting RHS angle pieces	308
7.13	Improvement of fatigue strength applying mechanical and thermal processes	308
7.13.1	Post-weld heat treatment	309
7.13.2	Shot peening and hammer peening	309
7.13.3	Overloading of a structural element	309
7.13.4	Stress relieving vibration	310
7.13.5	Grinding of weld transition zones	310
7.13.6	TIG or plasma post-treatment	311
7.14	Fatigue of bolted flange connections	311
7.15	List of symbols	314
8	**Fabrication, assembly and transport of hollow section structures**	**317**
8.1	General	317
8.2	Cutting	318
8.2.1	Flame cutting	318
8.2.1.1	Manual flame cutting	319
8.2.1.2	Automatic flame cutting by machines	320
8.2.2	Sawing	322
8.2.3	Plasma cutting	325
8.2.4	Laser cutting	325
8.3	Slotting	326
8.4	Flattening of hollow section ends	327
8.5	Bending (arching) of hollow sections	327
8.5.1	Cold bending of CHS	329
8.5.1.1	Cold bending by pressing	329
8.5.1.2	Cold bending using a "former" box	329
8.5.1.3	Cold bending with a three-roller bender	329
8.5.1.4	Arches by means of mitre cuts	330
8.5.2	Cold bending of RHS	330

8.5.2.1	Cold bending by pressing	330
8.5.2.2	Arches by means of a) mitre cuts or b) "V" cutouts	330
8.5.2.3	Cold bending with a three-roller bender	330
8.5.3	Hot bending of hollow sections	331
8.5.3.1	Hot bending of hollow sections filled up with sand	331
8.5.3.2	"Hamburger arch" (CHS only)	331
8.5.3.3	Bending with inductive heating	331
8.5.3.4	Hot bending with a three-roller bender	331
8.5.3.5	Cambering	332
8.6	Bolting	332
8.6.1	Blind bolting	333
8.6.1.1	Flow drill	333
8.6.1.2	Lindapter Hollo-Fast	334
8.6.1.3	Huck Ultra-Twist	334
8.7	Welding	335
8.7.1	Hollow Section steel grades and their weldability	335
8.7.2	Methods for welding hollow section joints	336
8.7.2.1	Manual shielded metal arc welding with stick electrodes coated with shielding flux material	337
8.7.2.2	Gas metal arc welding GMAW	338
8.7.2.3	Flux cored arc welding FCAW	339
8.7.2.4	Submerged arc welding SAW	339
8.7.3	Preparation of the welds in hollow section structures	339
8.7.4	Welding positions and sequences	339
8.7.5	Tack welding	340
8.7.5.1	Plug and slot welding	341
8.7.6	Post-heat treatment of welded constructions of hollow sections	342
8.7.7	Residual stress and deformation due to welding and their reduction measures	342
8.7.8	Weld defects and their repairs	343
8.7.9	Inspection of welds	343
8.7.9.1	Visual inspection	345
8.7.9.2	Magnetic particle test	345
8.7.9.3	Dye penetration test	345
8.7.9.4	Ultrasonic inspection	346
8.7.9.5	Radiographic inspection by X or γ rays	346
8.7.10	Performance qualification tests for welders and welding workshops	347
8.7.11	Welding of cold formed hollow sections	348
8.7.12	Stud welding	348
8.7.13	Laser welding	349
8.7.14	General recommendations for welding	349
8.8	Nailing	351
8.9	Application of cast steel elements in hollow section structures	352
8.10	Assembly	361
8.11	Transport of hollow sections and their structures	363
8.12	List of symbols	364
9	**Space structures**	**367**
9.1	General	367
9.2	Constructional elements of space structures	368

9.3	Assembly and erection of space frames	370
9.4	Calculation of space structures	372
9.5	Economic optimization of space structures	373
9.6	Further remarks to the design of space structures	373
10	**End fixity of rectangular hollow section columns in a concrete fundament**	**377**
11	**Hollow sections in composite construction**	**381**
11.1	Hollow section composite columns	381
11.1.1	Design strength (resistance) of hollow section composite columns	382
11.1.1.1	General	382
11.1.1.2	Methodically concentric axial compression	383
11.1.1.3	Effect of long-term behaviour of concrete on the design strength (resistance) of slender columns	384
11.1.1.4	Increased design strength (resistance) for compact CHS columns filled up with concrete	384
11.1.1.5	Design strength (resistance) of composite hollow section columns under compression and uni-axial bending	385
11.1.1.6	Ultimate strength (resistance) of a hollow section cross section under compression and bending	387
11.1.1.7	Design strength (resistance) of composite hollow section columns under compression and bi-axial bending	388
11.1.1.8	Approximate calculation for M-N-interaction in composite hollow section columns	393
11.1.1.9	Influence of shear forces	395
11.1.1.10	Load introduction	395
11.2	Truss joints in RHS with concrete-filled chord members	397
11.3	Fabrication of concrete-filled hollow section columns	399
11.3.1	Structural components	399
11.3.1.1	Hollow sections	399
11.3.1.2	Concrete	399
11.3.1.3	Reinforcements	400
11.3.2	Concrete filling operation methods	401
11.3.2.1	Connections of concrete-filled hollow section columns from floor to floor in a building	403
11.4	List of symbols	404
12	**Corrosive behaviour and protection against corrosion of steel hollow sections and their structures**	**405**
12.1	General	405
12.2	Internal corrosion of hollow sections and their structures	405
12.3	External corrosion of hollow sections and their structures	407
12.4	Protection measures against corrosion	408
12.4.1	Coatings for protection against corrosion	408
12.4.1.1	Corrosion protection with a "shop primer"	409
12.4.2	Metallic coating by spraying	410
12.4.2.1	Hot dip galvanization	410
12.4.2.2	Zinc coating in galvanic bath	413

12.4.3	Electrochemical polarisation	413
12.4.4	Application of hollow sections of weathering steels	413
13	**Structural hollow sections exposed to fire**	**415**
13.1	General	415
13.2	Design resistance of unprotected and unfilled steel hollow sections	418
13.2.1	Design resistance of hollow section members under tension	419
13.2.2	Design resistance of hollow section members with a uniform temperature θ_a under combined axial tension and bending moment	419
13.2.3	Design buckling resistance of hollow section members with class 1, class 2 or class 3 cross sections with a uniform temperature θ_a under axial compression	420
13.2.4	Design buckling resistance of hollow section members with class 1 und class 2 cross sections with a uniform temperature θ_a under axial compression and bending moment	420
13.2.5	Simplified design of a hollow section column exposed to fire under concentric and eccentric axial compressive load	420
13.3	External protection against fire	421
13.3.1	External fire insulations	421
13.3.2	Fire protection paints	422
13.3.3	Design of external insulation	422
13.4	Fire protection of steel hollow sections by water cooling	423
13.4.1	Fundamentals to water cooling systems	423
13.4.2	Design of water cooling installations	427
13.5	Fire protection of steel hollow section columns by filling with concrete	428
13.5.1	Fundamentals	428
13.5.2	Fire design of concrete filled hollow section columns without external insulation	431
13.5.2.1	Level 1 design: tabulated data	431
13.5.2.2	Level 2 design: simple design diagrams	431
13.5.2.3	Level 3 design: general calculation method	462
13.5.3	Fire protection of hollow section columns filled by concrete with steel fibre reinforcement without external insulation	462
13.6	Fire protection of hollow section column to beam connections	463
13.6.1	Unfilled hollow section columns with or without external insulation	463
13.6.2	Concrete filled hollow section columns	463
13.6.3	Water cooled hollow section columns	465
13.7	List of symbols	465
14	**Wind loads on circular and rectangular hollow sections and their lattice structures**	**467**
14.1	General	467
14.2	Wind loads on single circular cylinders	467
14.3	Wind loads on single square cylinders	469
14.4	Wind loads on lattice structures in circular hollow sections	470
14.4.1	Approximate calculation of wind loads on lattice masts	471
14.5	Wind force coefficients for circular hollow section cylinders and their lattice structures according to DIN 1055-4, Issue 8/86 and EC 1	476
14.6	Wind loads of uni-planar and multi-planar lattice structures with square hollow sections	477

14.6.1	Uni-planar lattice structures	480
14.6.2	Multi-planar (space) lattice structures with square or rectangular plan	480
14.6.3	Design example	481
14.7	Wind loads according to Eurocode 1	483
14.7.1	Wind loads	483
14.7.1.1	Wind force coefficients C_f for rectangular hollow sections with rounded-off corners	483
14.7.1.2	Wind force coefficients C_f for CHS members with $\lambda = l/d$	483
14.7.1.3	Wind force coefficients C_f for lattice structures and scaffolding	483
14.8	List of symbols	486

Appendix I	Nominal sizes and geometric properties of cold formed circular hollow sections according to EN 10219-2	489
	Nominal sizes and geometric properties of hot finished circular hollow sections according to EN 10219-2	496
Appendix II	Nominal sizes and geometric properties of cold formed square hollow sections according to EN 10219-2	504
	Nominal sizes and geometric properties of hot finished square hollow sections according to EN 10219-2	509
Appendix III	Nominal sizes and geometric properties of cold formed rectangular hollow sections according to EN 10219-2	513
	Nominal sizes and geometric properties of hot finished rectangular hollow sections according to EN 10219-2	522
Appendix IV	New ultimate joint resistance formulae for uni-planar and multi-planar CHS T, X and XX joints	530
Appendix V	New ultimate resistance formulae for welded, uni-planar and multi-planar I beam to RHS column connections	534
Appendix VI	New ultimate joint resistance formulae for uni-planar and multi-planar RHS X, XX and TX joints	537

References	541
Subject index	565

1 Introduction

The history of the development of the manufacturing technology of steel hollow sections (SHS) is old and young at the same time. As the summarized development data for the manufacture of steel hollow sections shown in Table 1-1 manifest, the first steel tubes got off the ground in 1825 in England as forge-welded tubes nearly together with the rail sections and angles. On the other hand, the industrial production of the square and rectangular hollow sections (RHS) started first in the second half of the twentieth century.

Table 1-1 Historical overview of the development of the manufacturing processes of rolled steel profiles

1800–1820	Rail sections and angles in England
1825	Forge-welded circular tubes (CHS) from strip steel or skelp by C. Waterhouse in England
1831	Angle profiles in Germany
1835	Railway profiles in Germany
1845	Welded CHS (continuously butt welded by "bell" drawing process) (Albert Poensgen) in Germany
1849	First I-profiles in Zores, France
1857	I-profiles in Germany
1886–1889	Seamless CHS in Germany (Mannesmann brothers Max and Reinhard)
1930–1940	Electric welded CHS
1959	Welded rolled RHS in England
1962	Seamless rolled RHS in Germany

Already the prehistoric man understood the advantages of the hollow section as transmission pipe as well as structural element. He learnt the application of closed circular hollow sections from the nature. These cross-sections serve as feed channels for water or nutrient liquid in animal organisms or plants as well as skeletons of vertebrate animals, which are light and robust at the same time. Also the stems of grasses, cereals, reeds and in particular bamboos conform to this natural law (see Fig. 1-1).

Thousands of years ago, the inhabitants of India and Indo-China started to build the load-bearing components of their huts, such as studs, columns, cross-beams and roof-trees with bamboos. Till today, the application of bamboos for the scaffold constructions of even multi-storey buildings continues in these countries at the stage of economic take-off.

Until 1784 James Watt used cast iron tubes for the very first time in his prototype steam engine, the application of the tubular elements in iron had been restricted for a long time to cannons and firearms only.

Fig. 1-1 A bird sitting on a rye stem

During the industrial revolution starting at the end of the eighteenth century began the triumphant progress of iron and steel. The development of steelmaking processes and manufacturing technology led to the industrial production of the classical rolled steel profiles of I, L and U forms and further to the welded and seamless circular hollow sections.

William Murdoch produced the screw thread in 1815 to join water and gas pipes from the barrels of the muskets left behind by the Napoleonic wars. Using this process of Murdoch the small town of Fredonia in the U.S.A. built in 1821 a natural gas distribution installation with pipes for street lighting. However, the industrial manufacture of steel tubes really succeeded first when the Englishman Cornelius Whitehouse produced them by the so-called "bell" method in 1825. By this method a steel strip is brought up to white heat and then formed to an open tube, the edges of which are pressed firmly together thus affecting the metallurgical welding of the tube.

Albert Poensgen took up the manufacture of welded tubes in 1845 on the European continent and even built a factory of considerable size in Duesseldorf, Germany.

With the perfection of the industrial steel manufacturing process by Bessemer in 1854, the steel won the superiority over all other materials regarding the production of tubes especially because of its exceptionally favourable properties for this purpose.

A plant for manufacturing welded tubes was founded by August Thyssen in Muelheim on the Ruhr in 1871. A further decisive impulse came from the brothers Max and Reinhard Mannesmann, who developed the manufacturing process for seamless tubes by "skew roller piercer" in 1886 and later complemented this with the invention of pilger rolling.

It was however necessary to wait right up to the second half of the twentieth century to see the steel manufacturers produce the square and rectangular hollow sections, when the English firm Stewarts & Lloyds started their industrial production by hot forming in 1959 and the German firm Mannesmann manufactured them in 1962 from seamless tubes. It was about the same time that the square and rectangular hollow sections were produced by cold forming.

Although other hollow sections, such as of triangular, hexagonal or octagonal forms are manufactured at present, they could not gain much importance except for some special cases.

The manufacture of welded and seamless circular and rectangular hollow sections has a significant placing in the world economy today. Their manufacturing processes as well as their applications are numerous. The intention of this book is to describe to the reader the important manufacturing processes of this product and inform them about its increasing applications.

The best possible precision in calculation and design of structural elements of a construction exploiting all their favourable statical and dynamical properties, the application of appropriate fabrication and jointing techniques and the protection against corrosion as well as the rationalisation of manufacture, assembly, transport and errection are the most important criteria besides the material costs to analyse the economical aspect and to judge the quality and efficiency of an applied profile for a particular project. Although valid in general, they stand in particular for hollow sections.

Compared to open profiles (such as L (angle), [(channel) and I (beam) forms), the hollow section appears to be uneconomic due to its higher price per ton. Considering the above mentioned aspects however, the structures made with hollow sections can often be very effective and less expensive.

Due to the specific sectional properties of the hollow section with respect to compression, multi-axial bending and torsion, the hollow section is highly superior to the conventional open profiles, when subjected to these loadings. These sectional properties shown in Fig. 1-2 demonstrate this clearly, where these values for various cross sections with approximately the same weight per meter are confronted with one another.

1 Introduction

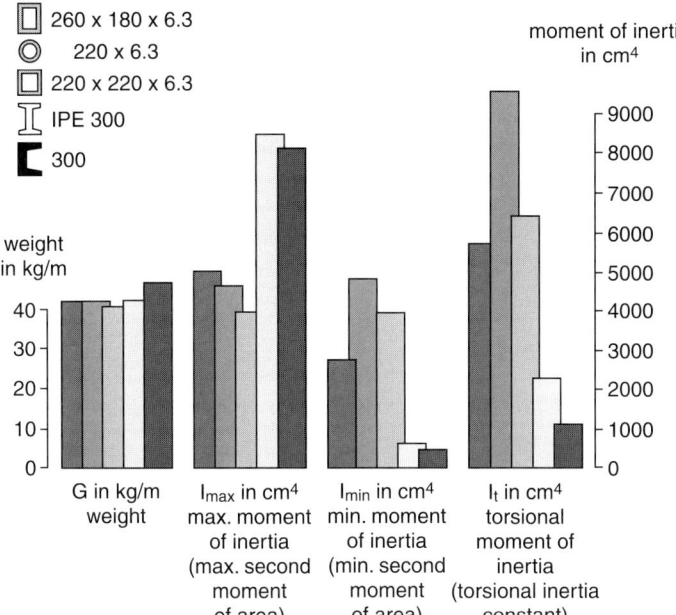

Fig. 1-2 Comparison of the sectional properties of a number of cross sections

They are in the following sequence from left to right:

RHS   260 × 180 × 6.3 mm
CHS   220 ⌀ × 6.3 mm
RHS (square) 220 × 220 × 6.3 mm
I beam   IPE 300
[steel (channel) [300

Larger buckling strength of the hollow sections compared to that of open profiles is indicated by its higher minimum moment of inertia I_{min} about the weak axis. In circular and square hollow sections, the principal axes are symmetrical. Consequently, no so-called "weak" axis exists for them, which has to be taken into account while designing a member under compression. Although IPE beams or [steels (channel) possess higher maximum moment of inertia I_{max} about the "strong" axis, this drops down substantially for the "weak" axis.

Further, the torsional moment of inertia I_t of the hollow section is 3 to 9 times larger than that of open profiles (see Fig. 1-2). For the same moment, the torsional angle of the hollow section amounts to a small fraction of that of open profiles (Fig. 1-3).

Further, in many projects the reason for the choice of hollow sections in structural engineering and vehicle construction (Fig. 1-4) as well as in light steel structures is based on the above mentioned facts, as the constructions

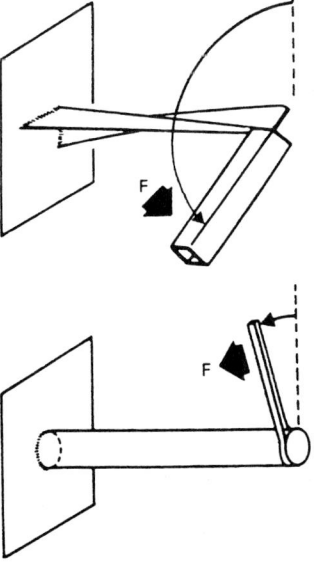

Fig. 1-3 Torsional loading of hollow section and open profile

Fig. 1-4 Examples for the application of hollow sections. a) Automobile transporter, b) rotary tower crane

like window or door frames, which are not considered as load bearing in general, have often projections requiring high torsional rigidity in order to avoid deformations.

Under uni-axial bending, I beam shows higher moment of inertia I_{max} about the "strong" axis than hollow section. The application of hollow section is however more advantageous for bi-axial or multi-axial bending. Under universal bending, CHS is best suitable.

A further special advantage of the hollow section is its low drag to wind or water current. Due to this property, the loads due to wind and water current can be lowered substantially. In this respect, CHS is highly superior to open rolled profiles (U, L, I and [ ) (Fig. 1-5). They are therefore often used in wind loaded structures in open air (such as towers, masts, transmission bridges etc.) as well as dolphins and offshore constructions under loadings by water current (Fig. 1-6).

Also the square and rectangular hollow sections with rounded corners yield smaller drag coefficients C_W than sharpedged rectangular elements and open rolled profiles.

As has been already mentioned, for structures subjected to forces by wind and water current, hollow section yields smaller load, which in its turn leads to saving in material. This subject will be further discussed in detail in Chapter 14.

Regarding the fabrication, a fundamental difference exists between hollow sections and conventional open profiles. The external surface of a hollow section is solely accessible in general, while there is only restricted access to the inside. This characteristic has to be especially accounted for, while designing with hollow sections. Naturally, the appropriate selection of the jointing technique is one of the most important design criteria in this respect.

Considering the usual jointing techniques, for steel structures, namely bolting, welding and gluing, the application of welding is most frequent in the case of constructions with hollow

1 Introduction

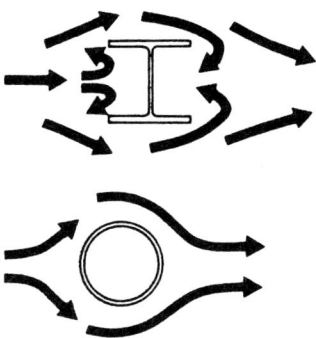

Fig. 1-5 Aerodynamically efficient CHS compared to I beam

Fig. 1-6 Dolphins on a ship berthing terminal

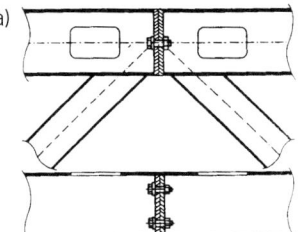

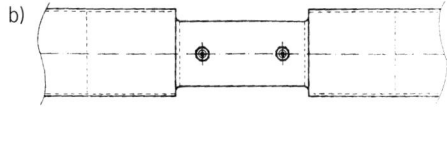

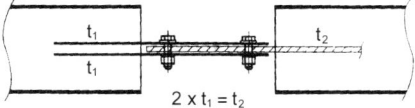

Fig. 1-7 Indirectly bolted connections with hollow sections. a) Hidden, bolted butt joints with head plates (bolting can be done through hand access hole), b) splice joint, c) CHS truss (chords and bracings are indirectly bolted together via gusset plates, also flange joints are used for jointing two chords end-to-end on site)

sections. Due to restricted access to the inside, a directly bolted joint of two hollow sections or between a hollow section and an open profile using standard bolts for steel structures is not possible, as they have to be tightened from inside. This can however be achieved by special measures such as hand access holes cut in the members or indirect connections with welded connection plates, stiffeners, head plates etc. (Fig. 1-7). The indirectly bolted connections are preferably used for assembly on site.

Welding can be applied both to direct and indirect connections of hollow sections without any restriction. In end-to-end connections hollow sections can be directly welded with each other from one side. It is necessary to prepare the weld for this purpose, which depends on whether the weld is to be carried out with or without a backing ring.

Figure 1-8 shows the trusses with hollow sections, where the chord and bracing members are welded directly or indirectly to one another. The welding is performed externally using fillet and partial or full penetration groove welds over the total circumference of the hollow section.

In the last forty years the ratio of the labour to the material costs has gone up substantially in all industrially developed countries. This has compelled the producers of steel structures to pay special attention to the simple manufacture of joints in a structure. For their economical manufacture, the application of connection plates and stiffeners has to be avoided as far as possible, thereby lowering the fabrication and welding costs. This can be attained by welded hollow section constructions more easily, as it is simpler to weld them directly to one another than in the case of open profiles. The expense of welding in a directly welded connection of hollow sections is significantly less than that for the connections with open sections.

In the case of RHS joints, where plane cut by sawing is sufficient for the end preparation of the members (see Fig. 1-8 c) and the connection is made by a fillet weld on the circumference, the manufacture is particularly economical.

Application of gluing in steel hollow section structures has earned only small significance up to now. Also very little research has been done for hollow section joints in this field.

Chapter 8 has been devoted extensively to the fabrication, assembly and transport of hollow section structures.

Principally, the load bearing properties of a construction depend not only on the profiles used but also on their connections. While designing a hollow section structure, it is therefore not only necessary that sufficient knowledge about the material used is available, furthermore intensive attention has to be paid to the geometry of the structural elements as well as to the configurations of their connections.

a)

b)

c)

Fig. 1-8 Trusses consisting of a) directly welded CHS to one another, b) indirectly welded CHS to one another via gusset plate, c) directly welded RHS to one another

The calculation and design of joints of hollow sections demand the knowledge, which goes markedly beyond that of conventional structural engineering. In order to design the welded joints, not only the load bearing strength of welds but also the design strength, particularly depending on the geometry of the members and joint configurations, have to be considered. The joint strength is expressed and governed mainly by the local deformations (Fig. 1-9).

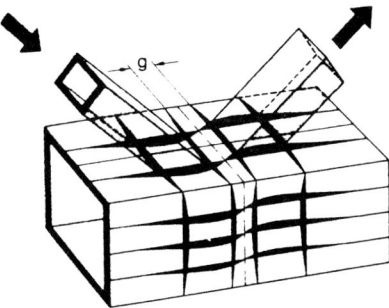

Fig. 1-9 Deformations of a RHS joint in a plane lattice girder

Due to the non-uniformity of the stress distribution, it is very difficult to determine theoretically the design strength (resistance) of hollow section joints, in particular of K and N forms. Therefore in the last thirty years, analytical models were developed and numerous series of practical tests were carried out in the testing centres all over the world to determine the hollow section joint strengths under static and fatigue loading. Fig. 1-10 shows two test rigs as examples. Chapter 6 of this book deals extensively with the design procedures of various hollow section connections (as for example joints in trusses, kneejoints, joints in Vierendeel girders etc.) under predominantly static load, which were determined basing on these test results.

The internal space of a hollow section can be used by various means:
- Housing of waste water pipes or sundry cables of wiring within columns
- Air or water circulation for air conditioning or fire protection

The use of concrete filled steel hollow sections as composite columns (Fig. 1-11) in buildings and also in bridges makes a relatively new field of application, which has obtained significant impulse in the last years through the research of their strength behaviour and the design guide lines emerging from it. Chapter 11 is devoted to this topic in particular.

By filling with concrete together with reinforcement in general, not only an increase of the load bearing strength compared to unfilled hollow section columns can be achieved, but also with appropriate reinforcement, a fire resistance time up to 120 minutes can be reached even without external fire insulation for hollow sections. This subject including external fire protection and water cooling has been dealt with in Chapter 13.

As for all steel structures, the protection against corrosion is of considerable importance also for constructions with hollow sections. Hollow sections yield in this respect considerable advantages technically and economically. The surface area of hollow section structures is as a rule significantly smaller than that of conventional open profiles. In the case of structures with wide spans, this can result up to 50% saving in painting surface. Saving of coating material and labour is the consequence. The maintenance of hollow section structures, where coating against corrosion has to be repeated from time to time, is significantly more economical than that of structures with open rolled sections.

Hollow sections have no sharp edges (Fig. 1-12). This makes it simple to paint them uniformly, which favours the long-time behaviour.

Directly joined, welded hollow section connections without gusset plates offer less possibilities of attack e.g. hollow spaces, water or snow traps and in-going angles, to the aggressive mediums for corrosion. In a joint with gusset plate, the plate must have an out-let hole for the water to flow out (Fig. 1-13).

Together with the harmonization of the national standards of the member countries of the European Union, a number of European product standards and design codes (European Standard EN, Eurocode) has been worked out. This job has only been partly accomplished and will take further time to be completed. As long as the European standards already published are not finally released by the CEN ("Comité Européen de Normalisation", European Committee for Standardisation), both national and European standards will hold good parallel to one another. However, a mixing up of the standards of both types is not allowed.

In this book, also the international standards ISO (International Standards Organisation) and IIW (International Institute of Welding) and other important standards such as of United States, Australia, Canada and Japan, have been mentioned in brief, as far as it has been found necessary.

As already described, the following factors play a major role in order to decide over the choice of hollow sections for a structure:

– Optimum weight to strength ratio
– Smaller surface area to be protected and maintained
– Easy and simple cleaning
– Clean, aesthetic appearance

In order to demonstrate this, the manifold possibilities to use hollow sections are shown and a number of special areas of application is described at the end of this chapter.

a)

b)

Fig. 1-10 a) Strength test on a single joint, b) strength test on joints in a plane lattice girder

1 Introduction

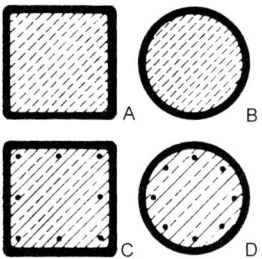

Fig. 1-11 Cross sections of concrete filled hollow sections. A and B: without reinforcement bars, C and D: with reinforcement bars

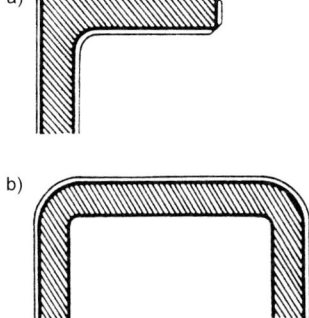

Fig. 1-12 a) Broken or weakened coating due to sharp edges of open profiles, b) uniform coating layer on hollow sections

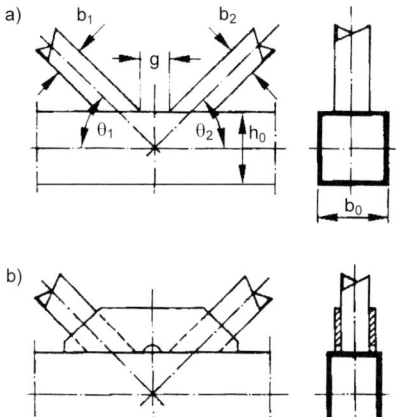

Fig. 1-13 Hollow section joints with paint coating against corrosion. a) Joint without gusset plate, b) joint with gusset plate (unfavourable for corrosion protection, remedial measure by out-let hole in the gusset plate)

Typical fields of hollow section application are:

Structural engineering: columns, beams, frames, roof trusses and lattice girders for industrial installations and halls, bridges, scaffoldings, racking systems, tribunes, sporting centres and exhibition halls as well as kiosks, greenhouses and mine equipment.

Architectural applications: visible steel columns (water cooled or concrete filled for fire protection), window arches, church buildings, frames for wall elements, staircases and platforms, guardrails, grid constructions, gates and door frames.

Vehicles for road and rail: frames and superstructures of lorries, busses, trailers, caravans, bicycles, locomotive trains, tank wagons, tram cars, mobile platforms, washing installations, chassis for diverse vehicles, snow ploughs.

Offshore technique: fixed and mobile offshore platforms, helicopter decks, legs for hydraulic platforms, substructures of offshore platforms.

Mechanical engineering: fundaments and supporting structures for machines, mechanical equipments, containers and vessels, traverses, rigs.

Applications in agriculture and forestry: plough machines, drag cultivators, crop or beet harvesters, harrowing and crumbling machines, wheel barrows, frames for silos, dredgers.

Conveying and handling technique: mobile cranes, portal cranes, rotary tower cranes, dock side cranes, diggers, substructures for conveyor belts, elevator shaft constructions, supporting structures for escalators, container lifters.

Miscellaneous uses: scaffoldings, containers, ships' docks, racks, road sign portals, highway safety fences, cable support gantries, radiotelescopes, telescope masts, installations in amusement parks.

Fig. 1-14 Reception hall of a hotel

Fig. 1-15 Entrance hall of the airport in Frankfurt

1 Introduction

Fig. 1-16 View of an exhibition hall in Leipzig

Fig. 1-17 Swimming hall with a roof structure of uni-planar RHS lattice girders

Fig. 1-18 Triangular arched lattice girders as a load bearing structure

Fig. 1-19 Cycle path footbridge over railway and carpark

Fig. 1-20 Triangular girder bridge made of CHS (Lully, Switzerland)

Fig. 1-21 Church hall with a multi-planar lattice roof structure

1 Introduction

Fig. 1-22 Retractable arched roof structure of Toronto Skydrome in construction

Fig. 1-23 Filigree façade construction

Fig. 1-24 Glass roof construction in the form of a half-opened umbrella

Fig. 1-25 Staircase construction with railings

Fig. 1-27 Chassis of a coach

Fig. 1-26 An architecturally interesting construction of a railway station with hollow sections

Fig. 1-28 Semi-submersible platform with helicopter deck

Fig. 1-29 Substructure of an offshore platform

Fig. 1-31 Access platform for loading an aeroplane

Fig. 1-30 Wind energy installation

Fig. 1-32 Harrowing and crumbling machine for sowing

Fig. 1-33 Mobile crane

Fig. 1-34 The roof construction of the International Airport Düsseldorf, Germany: View from the street to the framework trusses

Fig. 1-35 Subconstruction of escalator

Fig. 1-36 Warehouse construction with high level racking system

Fig. 1-37 Radar mast

Fig. 1-38 Roller coaster

Fig. 1-39 Tubular substructure of a bridge (Photo ZIS Industrietechnik)

Fig. 1-40 Erection of a lattice arch bridge (photo: ZIS)

Fig. 1-41 Footbridge in Weimar (photo: ZIS)

Correct placing of bearings

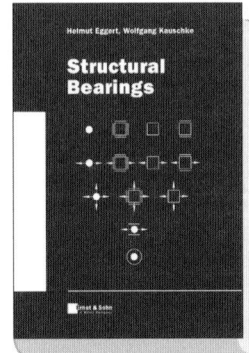

Helmut Eggert,
Wolfgang Kauschke
Structural Bearings
2002. Approx.
430 pages with
numerous illustrations
Hardcover.
€ 149,–* / sFr 250,–
ISBN 3-433-01238-5
May 2002

Bearings are used in construction mainly in the building of bridges for the discribution of loads between different elements and for compensating stresses. This volume describes their construct, function, calculation and applications, supplemented by normative regulations and research results.

Ernst & Sohn
Verlag für Architektur und
technische Wissenschaften GmbH

For orders and customer service:
Verlag Wiley-VCH
Boschstraße 12
69469 Weinheim
Telefon: (06201) 606-152
Telefax: (06201) 606-184
Email: service@wiley-vch.de

Ernst & Sohn
A Wiley Company

www.ernst-und-sohn.de

* € price is valid in Germany only.

2 Manufacture of steel hollow sections

2.1 General

The manufacturing processes for a fluid transmitting pipe and a circular hollow section intended for structures do not differ fundamentally from each other. The basis is a cylinder that is produced over a wide range of diameters and wall thicknesses with steels of various grades manufactured using various processes. The starting material is either a solid ingot for seamless tubes or a flat plate or strip for welded tubes. Square and rectangular hollow sections (also hexagonal and octagonal forms) are in general produced in the same way passing through the circular blank phase and applying the forming operation in hot or cold condition. This is independent of whether they are seamless or welded.

In the following Sections, the important methods for the manufacture of seamless and welded tubes as well as square, rectangular and other hollow section forms have been described.

2.2 The manufacture of seamless tubes

Seamless tubes are produced mainly by hot rolling processes, which basically consist of two phases:

- In the first phase, an ingot of round, square or polygonal shape is pierced by means of a piercing mandrel to create a hollow shell, a so-called "bloom"
- In the second phase, the hollow bloom is elongated into a finished steel tube

Most steel grades can be used to form seamless tubes provided the material is homogenous and has sufficient elongation capacity in hot condition.

Various manufacturing processes for seamless tubes have their specific technical and economical quality zones, which depend on diameter, wall thickness and required tolerances. They cover in accord with the rolling capacity a diameter range up to 660 mm. Beyond this diameter, seamless tubes can be expanded to a diameter up to 800 mm in hot condition.

The most important manufacturing processes in the sequence of their developments in date and their production ranges are given below:

1. The skew roll piercing and pilger rolling process (Mannesmann brothers)
   (Fig. 2-1 a and b)
   Outside diameter ca. 60 to 660 mm
   Wall thickness   ca. 3.2 to above 100 mm

2. The push bench process (Heinrich Erhardt)
   (Fig. 2-1 c and d)
   Outside diameter ca. 17 to 159 mm
   Wall thickness   ca. 2 to 10 mm

3. The plug rolling process (Fig. 2-1 e and f)
   Outside diameter ca. 100 to 355.6 mm
   Wall thickness   ca. 3.6 to 25 mm

4. The extrusion process (Fig. 2-1 g)
   Outside diameter ca. 60 to 260 mm
   Wall thickness   ca. 5 to 40 mm

5. The continuous rolling process (Fig. 2-2)
   Outside diameter ca. 21.3 to 177.8 mm
   Wall thickness   ca. 2 to 30 mm

A further process called the pierce and draw process (Fig. 2-3) is similar to the push bench process, but has a significantly larger production range:

   Outside diameter ca. 220 to 1450 mm
   Wall thickness   ca. 20 to 270 mm
                     (according to diameter)

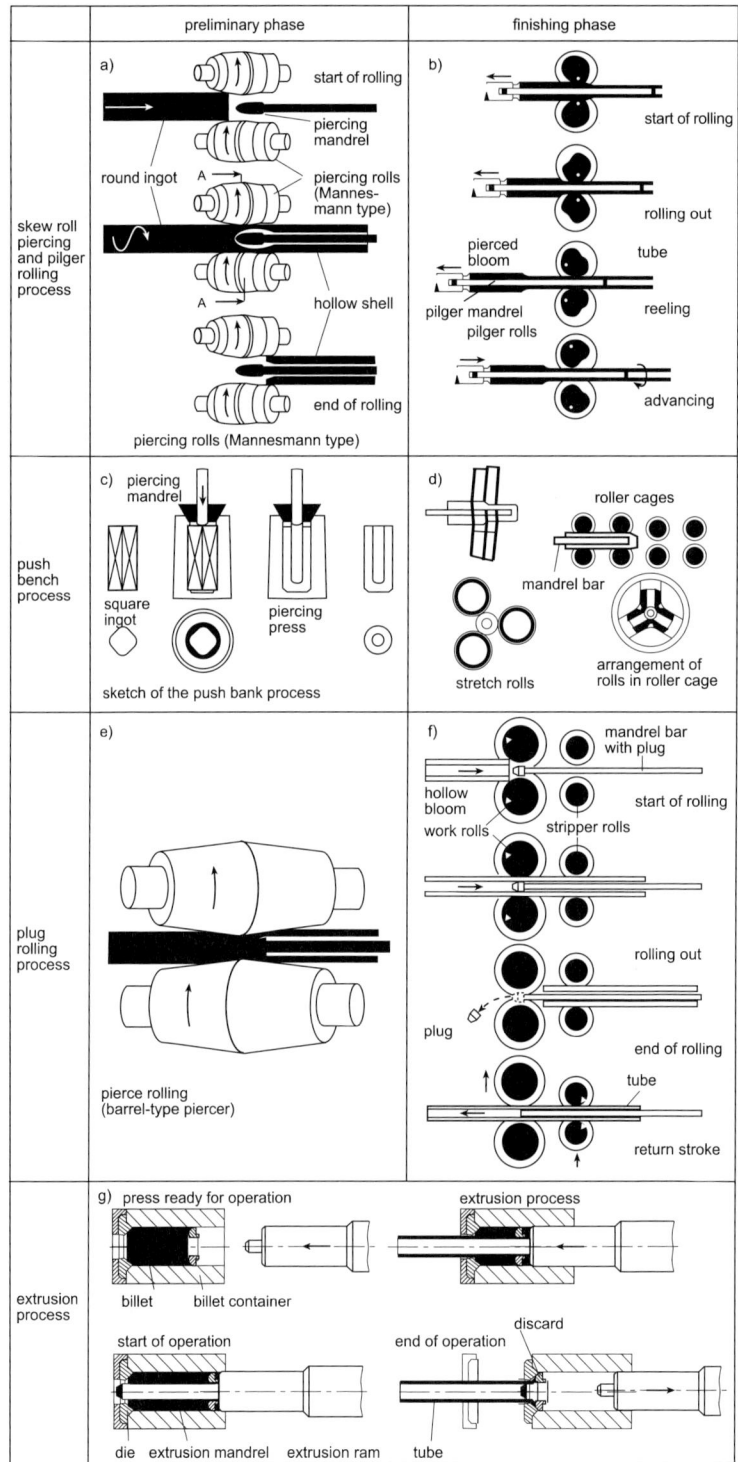

Fig. 2-1 Manufacturing processes of seamless tubes

2.2 The manufacture of seamless tubes

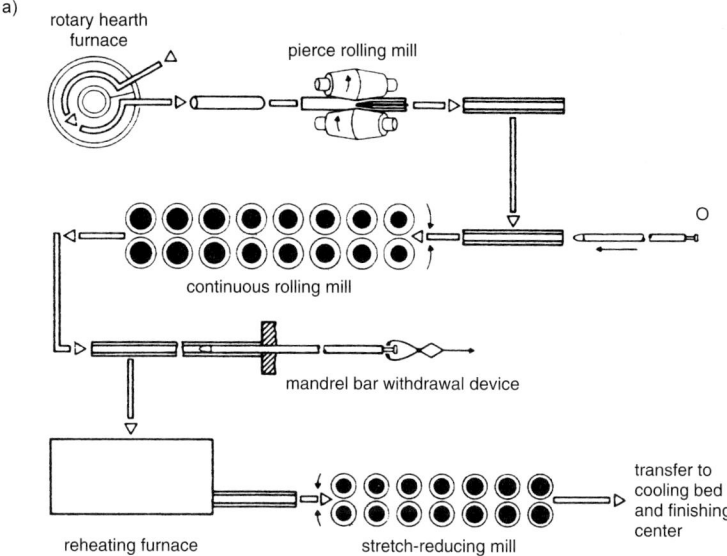

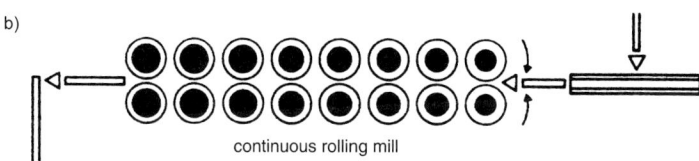

Fig. 2-2 Continuous rolling mill. a) Design, b) continuous rolling mill

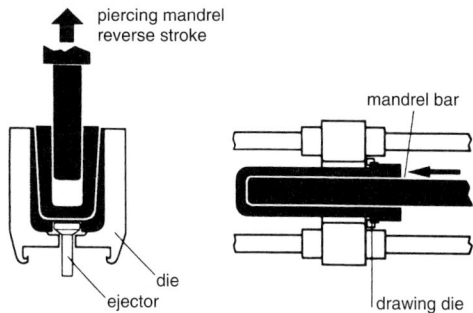

Fig. 2-3 Pierce and draw process

2.2.1 The skew roll piercing and pilger rolling process

The most important parts of the plant for this combined manufacturing process to produce seamless tubes, which is widely known as Mannesmann process, named after its inventors, are (see Fig. 2-1 a and b):

– Rotary hearth or pit furnaces for heating the starting material (round or polygonal ingots) to hot forming temperature between 1200 and 1400 °C depending on the steel grade used
– Piercing press for piercing the solid ingot to create a hollow shell with integral bottom (for large sizes with diameter up to 660 mm)
– Piercing mill for piercing the steel rounds (for small sizes) or for the reduction of wall thickness by drawing, and penetrating the bottom of the pierced ingot (for large sizes)
– Pilger rolling stand for drawing the hollow shell or bloom to a finished tube
– Size rolling mill to obtain the desired tube dimensions or the size reduction mill

A sketch of the skew roll piercing procedure shown in Fig. 2-1 a is described in the following:

The ingots are removed, one by one, from the rotary hearth furnace in time with the rolling mill cycle, and descaled. In the piercing mill the solid material is pierced, thus adopting the shape of a thick-walled hollow shell. As a further result of this process, this hollow body undergoes a longitudinal elongation, which corresponds to a reduction in cross section.

For the production of tubes of large sizes, round or polygonal cast ingots are heated to rolling temperature in rotary hearth or pit furnaces and subsequently pre-pierced in a press. The resulting hollow shell with integral bottom is elongated in the piercing mill, whereby the wall thickness is reduced and the bottom is penetrated.

In the pilger rolling mill (Fig. 2-1 b), two specially shaped rolls, whose axes are inclined at 3 to 6° towards the horizontal workpiece axis, are rotated in the same direction. A pilger mandrel, which functions as an internal tool, is positioned in the roll gap center, being supported via a rod by an external thrust block.

Following pierce rolling, the thick-walled hollow body, while still at rolling temperature, is rolled to its final shape in the pilger rolling stand. The pierced bloom is pushed over a lubricated cylindrical pilger mandrel, the diameter of which is roughly equal to the desired internal diameter of tube. An advancer then transfers the bloom to the pilger rolls. As soon as the bloom is gripped by the tapered portion of the work pass, a "wave" of material is pressed off the bloom outside circumference. This "wave" is then stretched to the desired wall thickness on the mandrel by the smoothing portion of the work pass with the mandrel plus pierced bloom moving backwards, that is opposite to the rolling direction, until it reaches the idler pass and is released. This process with its recurring backwards and forwards stepwise motion was given its name because of its similarity to the Echternach (Germany) dancing procession, in which the pilgrims take three steps forward and two back.

When the pilgering process is completed, the finished tube is pulled off the mandrel and the remaining unworked part of the bloom is cut off. Afterwards, if necessary, the tube is reheated before passing through either a sizing or a reducing mill.

In the sizing mill, the tube is rolled to an exactly defined outside diameter and a perfectly round cross section. Generally, the sizing mill consists of three stands with two or three-roll arrangement. The paired rolls, which form a closed pass, are offset to one another.

In the reducing mill, the tube outside diameter is reduced to a greater extent than in the sizing mill, to permit the production of intermediate sizes.

2.2.2 The push bench process

In this process, also known by the name of its inventor, the Erhardt process, a square, circular or polygonal billet is heated to rolling temperature in a rotary furnace and fed into the cylindrical mould of a piercing press. A piercing mandrel is then pushed into the billet (Fig. 2-1 c), which emerges as a thick walled bloom still closed at one end. This travels then over rollers to the rolling mill comprising of three skew rolls or stepped rolls. The resulting bloom is then fed to the push bench where it is threaded over a mandrel and pushed with the mandrel through a series of three roll stands set in line, each of them corresponding to a size, which gradually reduces from one stand to the next. Thereby, a thin walled tube is produced from a thick walled bloom. Finally, the mandrel is extracted from the tube and its both ends are cut off with a hot saw.

2.2.3 The plug rolling process

The plug rolling process plant consists of the following main parts:

- Rotary hearth furnace for preheating the starting material blocks to hot forming temperature of about 1280 °C
- Pierce rolling mill (with two biconical rollers) for piercing the solid starting material to create a hollow shell
- Plug rolling stand for elongating the hollow shell to tube length
- Size rolling mill, where the tube receives its final dimensions or a stretch-reducing mill (Fig. 2-10)

After preheating the ingots in the furnace, their surfaces are descaled and they pass through the pierce rolling mill yielding thin walled hollow shells (Fig. 2-1 e). The forming of the hollow shell into tube takes place in the subsequent two-high plug rolling stand.

The operating sequence of the plug rolling process is schematically illustrated in Fig. 2-1 f. The hollow bloom is thrust into the mill by a pneumatic pusher, seized by the rolls and rolled over the plug. Thereby the outside diameter and wall thickness are reduced. After the completion of rolling, the tube rests on the rod and the plug falls through a gap into a cooling and changing device. In order to return the tube to the pass entry side, the upper work roll is raised and at the same time, the stripper rolls are set in motion. Subsequently, the tube passes through a sizing mill, where the outside diameter is rolled to a defined dimension, followed by cooling on the cooling bed.

2.2.4 The extrusion process

In this process, the starting materials as forged or rolled round bar are cut to processing length after the elimination of surface imperfections by peeling, turning or grinding. The resulting billets are through-bored with a small diameter tool. These prepared billets are heated in induction furnaces. They are then lubricated and by means of a press, the internal diameters are expanded. When they have passed through a reheating furnace, the billets are internally and externally provided with a special lubricant, which also services as a separating agent, and laid in the billet container. The extrusion mandrel and die are so positioned that an annular gap is created between them, which determines the tube outside diameter and wall thickness. Four stages of the extrusion process are schematically illustrated in Fig. 2-1 g.

2.2.5 The continuous rolling process

The main components of the continuous rolling mill are (Fig. 2-2):
– Rotary hearth furnace for heating the starting material, especially continuously cast rounds, up to 1280 °C

– Piercing mill for piercing the round material, resulting in a 2 to 4-fold elongation
– Continuous rolling mill, in which the bloom undergoes a further 4-fold elongation
– Reheating furnace, where the now up to 30 m long continuously rolled blooms are reheated to approximately 980 °C
– Stretch-reducing mill with 28 stands for a further max. 10-fold elongation of the bloom
– Cooling bed for tube lengths of up to 160 m

The piercing mill employed by modern continuous tube mills has the barrel-shaped work rolls, which are arranged vertically one above the other and inclined at 12° towards the work piece axis.

In the continuous rolling mill, eight two-high stands are arranged in line and very close behind one another. Each stand is offset 90° to its predecessor and inclined 45° towards the horizontal. The pass diameter decreases from stand to stand. A long bar of uniform diameter is introduced into the hollow bloom. This bar serves as an internal tool during the rolling process. Bloom plus bar are transferred to the continuous rolling mill, where without being rotated the bloom is elongated as it travels from stand to stand, which are equipped with oval rolls. The last stand is designed as a round pass.

After the completion of the rolling process, the mandrel bar is removed and prepared for the next rolling pass, when the tube is outside the production flow.

The tube ends are then cropped, and the tube is reheated in a walking beam furnace as a preparatory measure for the subsequent stretch-reducing process (see Section 2.4).

2.3 The manufacture of welded tubes

In all manufacturing processes for welded steel tubes, a flat product is used as starting material. Depending on the desired tube diameter, this may be plate, wide strip or strip steel (so-called "coils" when rolled together). The manufacturing diameter range of welded tubes is significantly larger than that of seamless tubes. Principally, this is only restricted by the transportability of the finished tubes. The achievable wall thicknesses however, are smaller than those for seamless tubes.

The manufacture of welded tubes is carried out in two subsequently following phases:

- In the first phase, the flat product is shaped into a cylinder with a slit
- In the second phase, the edges of the slit are brought together and welded longitudinally, except in the case of spirally welded tubes, which are produced in the form of a helix

Welding processes are to be categorized into pressure welding and fusion welding. The most important welding methods are:

1. Submerged arc welding (fusion welding)
2. Gas-shielded arc welding (fusion welding)
3. Forge welding (Fretz-Moon; continuous pressure welding)
4. Electric resistance welding (pressure welding)

2.3.1 The submerged-arc welding process

This is an electric fusion welding method with covered arc, which is applied to manufacture longitudinally welded and spiral weld tubes. In this process, a bare wire is used as the filler metal, which melts under a layer of welding flux. The heat generated by the electric arc causes the filler metal and plate edges as well as a portion of the flux to melt. The molten flux forms a viscous slag which seals the weld and prevents pick-up of oxygen and nitrogen from the surrounding air. Submerged-arc weld is applied to tubes of large diameter, generally above 500 mm, as longitudinal as well as spiral weld.

Three methods of forming a tube with slit are usually applied today:

a) Three-roller bending procedure for the shaping of a plate into a cylinder with slit in hot or cold condition according to the tube diameter, wall thickness and steel grades (Fig. 2-4)

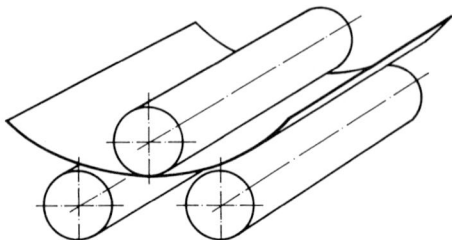

Fig. 2-4 Shaping using three-roller bending system. Max. diameter: depends on the transportability only. Max. wall thickness: about 50 to 100 mm or more depending on the facilities of the plant

b) UOE method (U = U-form, O = O-form, E = Expanding) (Fig. 2-5) to shape a plate in cold condition
Diameter range: 610 to 1626 mm
Wall thickness range: 7 to 40 mm

The plate is firmly clamped and both longitudinal edges are machined resulting highly accurate, plane parallel double-Y bevel along each longitudinal edge. Then in the plate edge crimping press, follows the bending of an area 200 to 400 mm wide of both plate edges, including run-in and run-off tabs, simultaneously to the final tube radius. The actual shaping of the plate into a cylinder is performed in two matching steps: shaping into a "U" cross section in the U shaping press, followed by shaping into an open tube in the O shaping press.

The open tubes are now transferred to a continuous tack welding machine, where the welding edges are exactly aligned and pressed together. The tubes are then tack-welded externally over entire length using a twin-head MAG-welding machine. The tack-welded tubes are first submerged-arc welded internally and the outside weld is then produced in the same manner.

c) Shaping of hot rolled wide strip (coils) to spiral weld tube in cold condition (Fig. 2-6)
Diameter range: 508 to 2032 mm
Wall thickness range: 6.3 to 14.2 mm

After reeling off, the leading end of the incoming strip is welded to the trailing end of the already inserted strip and then passed through a straightening machine. Shears trim the strip edges, which are then bevelled by an

2.3 The manufacture of welded tubes

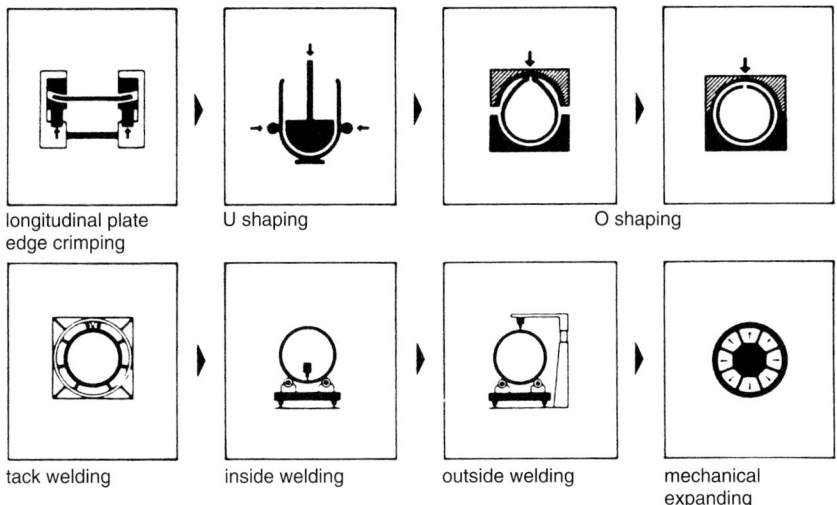

Fig. 2-5 Manufacture of welded tubes using UOE method

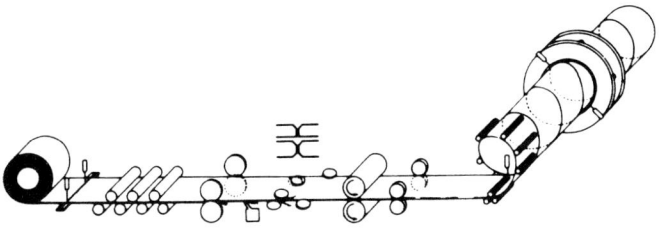

Fig. 2-6 Sketch of a spiral weld plant

edge miller. Finally, the strip reaches the forming section, where, in a three-roller bending device followed by a star-shaped inner sizing tool, it is bent helically into a cylindrical tube. The longitudinal edges of the incoming and already shaped strip are brought into low point contact, and the resulting butt joint is continuously welded using the submerged-arc process, internally at first and externally after a half turn of the tube. In this way, an endless tube string is created.

2.3.2 The gas-shielded arc welding process

This is a fusion welding process, where arc welding is carried out in an inert gas (Argon or Helium) or an active gas (CO_2 or gas mixture 80% Ar + 15% CO_2 + 5% O_2) atmosphere. The gas shield protects both arc and weld against the surrounding air.

Gas-shielded welding processes are preferably used for stainless austenitic and ferritic steels, nickel-base alloys and titanium. For longitudinally welded tubes made from these materials, the TIG (tungsten inert gas) process has proved specially advantageous, where an arc in a straight argon atmosphere is struck between a non-melting tungsten electrode and the open tube.

Tube production is carried out continuously. The strip cut to processing width is uncoiled,

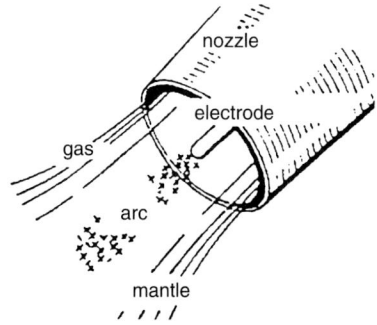

Fig. 2-7 Principles of the gas-shielded arc welding

formed to an open tube by rotating roller pairs and then subjected to arc welding.

2.3.3 The continuous pressure welding process (Fretz-Moon)

Diameter range: 13.5 to 88.9 mm
Wall thickness range: 1.8 to 7.1 mm

This fully continuous manufacturing process was developed in the mid 1920's applying the world's oldest tube welding method – the bell drawing process. The heating of the total strip takes place in a through type furnace after having been deburred on both sides. It is then shaped into an open tube in the first stand with forming rolls. The edges of the strip are heated to welding temperature with blowers and pressed firmly together in the next stand with welding rolls, affecting the welding connection (Fig. 2-8).

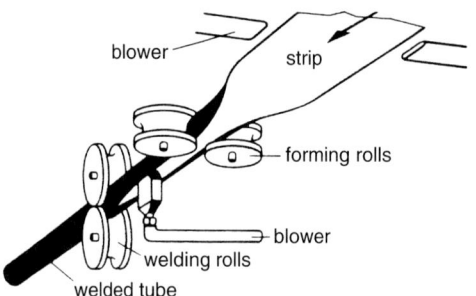

Fig. 2-8 Schematic illustration of the Fretz-Moon welding process

2.3.4 The electric resistance pressure welding process

Diameter range: about 20 to 508 mm
Wall thickness range: 1 to 12.5 mm
(according to diameter)

In this process using inductive or conductive electric current, the high-frequency induction welding has prevailed upon the other methods. The starting material for longitudinally welded tubes is hot or cold rolled strip. The coils are unwound at the start of the tube plant and the trailing and leading ends of the coils are welded together forming an endless skelp. In the edge trimming machine, the skelp is machined to the exact width required

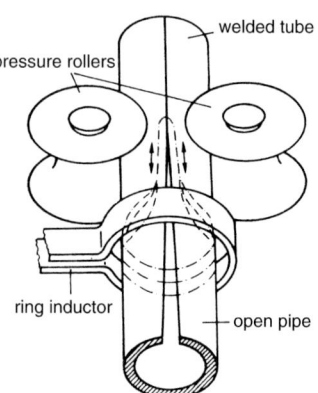

Fig. 2-9 Schematic illustration of the induction welding

for the tube diameter. It is then fed to a multiple-stand forming section, where at first both edges are slightly crimped and then, as the skelp passes through a variety of forming rolls, it is continuously bent to an open tube. In the welding section of the plant, the open tube passes through an inductor by which it is totally enclosed (Fig. 2-9). The skelp edges are heated to welding temperature generated by the high frequency induction current, which are then firmly pressed by means of pressure rollers yielding the weld seam from the tube material without using any filler metal.

2.4 The stretch-reducing or reducing process

Diameter range: 21 to 219 mm
Wall thickness range: 2 to 25 mm

In today's modern high-capacity plants to manufacture seamless and welded tubes, only a limited number of large basic dimensions up to about 219 mm is produced. The smaller sizes are manufactured in stretch-reducing or reducing mills, without using internal tools. A number of rolling stands (up to 28) are arranged in line. Each stand carries three rolls, with progressively decreasing groove sizes. The circumferential speed increases from stand to stand, corresponding to the decreasing tube diameter and the increasing length (Fig. 2-10). By varying the longitudinal traction between the stands, the wall thickness can be influenced.

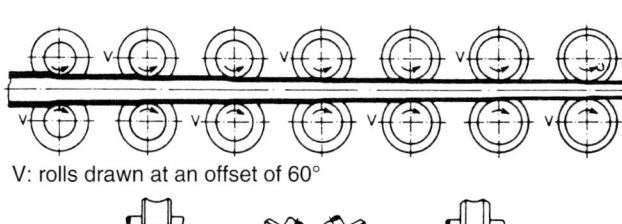

V: rolls drawn at an offset of 60°

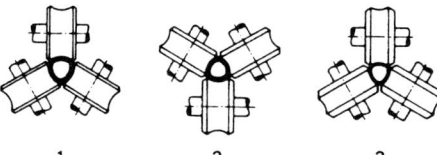

stand: 1   2   3

Fig. 2-10 Stretch-reducing process (sketch)

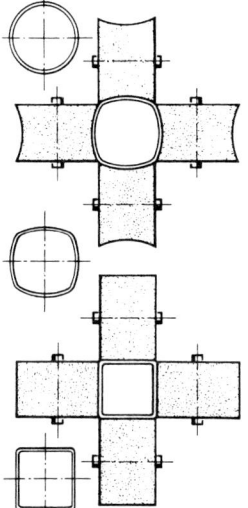

Fig. 2-11 Shaping of CHS to RHS (schematic illustration)

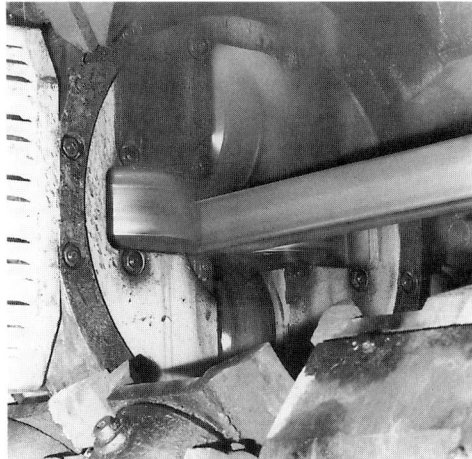

Fig. 2-12 Square hollow section in a roller cage

2.5 The manufacture of rectangular and square hollow sections

Rectangular and square hollow sections (RHS) are produced by transforming seamless and welded circular hollow sections (CHS) in hot (annealing temperature about 890–950 °C) as well as cold condition (Figs. 2-11 and 2-12).

A circular hollow section is introduced into roller cages (two or three) arranged in line behind one another and obtains there the desired shape of rectangular or square hollow section. Cold formed profiles are, if necessary, stress relief annealed at 530–580 °C to reduce residual stresses in the sections.

2.6 The manufacture of hollow sections with further shapes

Cold formed hollow sections with various cross sectional forms, which may partly be quite complicated, are produced by passing through rollers, also by multiple cold drawing through a die (sometimes with drawing on the internal bar or plug). In order to remove the strain hardening effect of cold working and to restore the elongation capacity of the steel, the cross sections are annealed. If several drawing passes are required to achieve a defined cross section, intermediate annealing may be necessary in some cases.

2.7 The testing of steel hollow sections

In the technical delivery requirements for steel hollow sections described in national and international standards and specifications, the scope of testing and the methods to

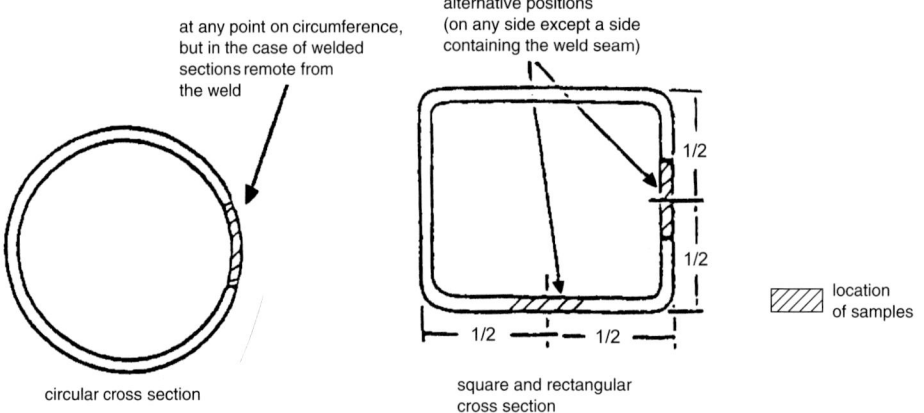

Fig. 2-13 Positions of test samples for tensile and impact test

be applied are laid down, which principally differ from one another depending on the later application of the hollow sections.

The control of the starting material takes place with the determination of the chemical composition by means of ladle analysis and macrography.

For further checks, destructive and non-destructive tests are to be differentiated. The former comprises mechanical and technological tests, carried out on specimens taken from the relevant product. Fig. 2-13 shows an example illustrating the location and orientation of test samples. The following tests are normally required for structural hollow sections:

a) Tensile tests to determine the strength and strain values: yield strength, tensile strength and elongation:
Tensile tests are generally carried out at room temperature. The type of the test sample, which often covers the total wall thickness, and the location on the hollow section (longitudinal or transversal) are laid down in the technical delivery requirements (see Tables 3-9 and 3-10, test samples according to EN 10002-1 [1])

b) Impact bend test to determine the impact energy absorbed by the test specimen as a reference value for the toughness of a steel:

Elongation and impact energy values classify the toughness behaviour of a steel. The impact energy is measured in Joules (J). During the impact test, the behaviour of steel, exposed to high rate of strain and multi-axial stress (constraint), is determined. For low-temperature applications, the testing temperature is decreased accordingly (see Tables 3-9 and 3-10, test samples Charpy-V according to EN 10045-1 [2])

Non-destructive tests belong to the second type of tests, the great advantage of which lies in the possibility of testing also the internal part of the test piece for material flaws without having to destroy it. They are for structural hollow sections as follows:

a) Ultrasonic test (see Fig. 2-14)

An acoustic transmitter probe generates a sound beam – generally at frequencies between 2 and 4 MHz – which is applied to the test

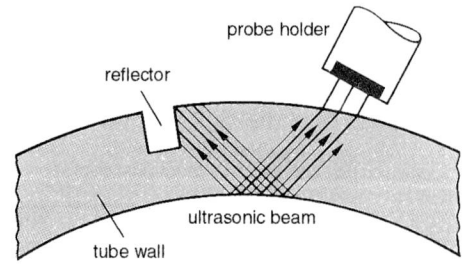

Fig. 2-14 Ultrasonic test

2.7 The testing of steel hollow sections

piece, using, in most cases, water as a couplant. Imperfections in the test piece, when being struck by the beam, reflect part of the sound energy. The reflected sound is partly received by a receiver probe and causes an indication.

In a further developed method, the ultrasonic beam is electro-magnetically generated, and the otherwise necessary couplant is not required.

b) Magnetic stray flux test (magnetic particle test) (see Fig. 2-15)

This involves magnetizing the test piece, after which the surface is powdered by magnetic particles. Defects in the surface are indicated by local magnetic stray flux.

Non-destructive tests are particularly applied to test the welds in welded constructions. Be-

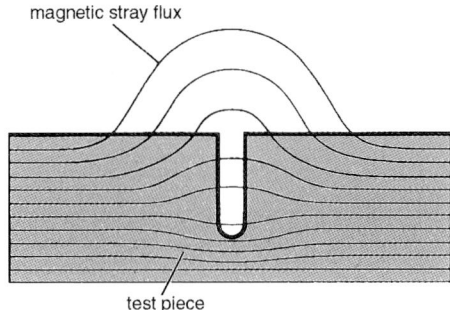

Fig. 2-15 Magnetic stray flux test

sides the ultrasonic and the magnetic stray flux tests, further methods such as X-ray test and dye penetration test are also used for this purpose, which have been dealt with in the Sections 8.7.9.3 and 8.7.9.5.

The interdependence of design and construction

Bernhard Göhler
Incrementally launched bridges
Design and construction
2000. 208 pages, 126 illustrations
Softcover. € 69,-* / sFr 120,-
ISBN 3-433-01793-X

The incremental launching method is characterised by design as a result of the construction process and its specific requirements.

This text deals in great detail with the calculation of forces on substructure components during launching, illustrated by practical examples.

One complete chapter analyses the requirements for precision construction, while special sections address such key topics as coupling tendons and anchoring bearings – also of interest in connection with the de-sign of prestressed concrete bridges.

Finally, the advantages and disadvantages of various launching techniques are critically assessed, with the durability of incrementally launched bridges considered from the perspective of design and experience.

Civil engineers will find here a wealth of detailed and upto date information on the principles and suppositions for both design and construction, as well as on the latest launching equipment, with a thorough discussion of the as-pects of super- and substructure.

Ernst & Sohn
Verlag für Architektur und
technische Wissenschaften GmbH

For orders and customer service:
Verlag Wiley-VCH
Boschstraße 12
69469 Weinheim
Telefon: (06201) 606-152
Telefax: (06201) 606-184
Email: service@wiley-vch.de

A Wiley Company
www.ernst-und-sohn.de

* € price is valid in Germany only.

3 Structural hollow section steel grades

3.1 General

Tables 3-1 to 3-3 contain the chemical compositions and the mechanical properties of the general purpose structural steels according to the international standard ISO 630 [1].

The following ISO standards show these values for the high strength steels and the steels with improved corrosion resistance:

ISO 4951 High yield strength steel bars and sections
ISO 4952 Structural steels with improved atmospheric corrosion resistance

Nearly in all industrially developed countries, corresponding national material standards are used, which may partly deviate from the ISO standards. However these standards show more or less comparable values. Whether the interchangeability of the steels is possible, has to be decided by the buyer from case to case. The steels described in a number of national standards are confronted with those according to ISO 630 in Table 3-4.

Besides the general purpose non-alloy structural steels, fine grain steels are also used, the strengths of which go partly beyond those of the former. They obtain a special fine grain microstructure by the addition of Aluminium and other alloys e.g. Niobium, Titanium or

Table 3-1 Chemical compositions (ladle analysis) of the structural steels according to ISO 630[1] [1]

Steel grade	Quality	Thickness mm	C % max.	P % max.	S % max.	N_2 %[2] max.	Method of deoxidation[3]
Fe 360	B	≤ 16 > 16	0.18 0.20	0.050	0.050	0.009	
	C D		0.17 0.17	0.045 0.040	0.045 0.040	0.009	NE GF
Fe 430	B	≤ 40 > 40	0.21 0.22	0.050	0.050	0.009	NE
	C D		0.20 0.20	0.045 0.040	0.045 0.040	0.009	NE GF
Fe 510[4]	B C	≤ 16 > 16 ≤ 35 > 35	0.22 0.20 0.22 0.20 0.22	0.050 0.045 0.040	0.050 0.050 0.040		NE NE GF
	D						

[1] It applies to steel plates with thicknesses of 3 mm and over, wide strip in coils wider than or equal to 600 mm wide and greater than 6 mm in thickness, wide flats and bars and "hot rolled" (synonymous with "hot finished") sections including hollow sections generally used in the as-delivered condition and normally intended for bolted, riveted or welded structures.
[2] For steel treated with aluminium, the maximum nitrogen content may be increased to 0.015%. The nitrogen contents are specified but they will be verified only if this is stated on the order. For steel made in an electric furnace, the maximum value can be 0.012%.
[3] NE = non rimming
GF = these steels shall have a content of elements sufficiently high to produce a fine-grain structure, for example total aluminium greater than 0.02%.
[4] The contents for Mn and Si shall not exceed 1.60% and 0.55% respectively.

Table 3-2 Permissible deviation for the product analysis in relation to the specified ladle analysis [1]

Element	Specified limits	Permissible deviation	
		Rimming steel	Non-rimming steel
C	≤ 0.24	+0.05	+0.03
P	≤ 0.060	+0.015	+0.005
S	≤ 0.050	+0.015	+0.005
Mn	≤ 1.60		+0.10
Si	≤ 0.55		+0.05
N_2	≤ 0.009	+0.002	+0.002

Table 3-3 Mechanical properties of the structural steels according to ISO 630 [1]

Steel grade	Quality	Min. yield strength N/mm²			Tensile strength[1] N/mm²	Min. elongation % $L_0 = 5.65\sqrt{S_0}$ [2]	Impact test (V = notch)	
		t ≤ 16 mm	t > 16 mm t ≤ 40 mm	t > 40 mm t ≤ 63 mm			Test temp. °C	Energy min. Joule
Fe 360	B	235	225	215	360–460	25	+20	27
	C	235	225	215		25	0	27
	D	235	225	215		25	−20	27
Fe 430	B	275	265	255	430–530	22	+20	27
	C	275	265	255		22	0	27
	D	275	265	255		22	−20	27
		t ≤ 16 mm	16 < t ≤ 35 mm	35 < t ≤ 50 mm				
Fe 510	B	355	345	335	490–630	21	+20	27
	C	355	345	335		21	0	27
	D	355	345	335		21	−20	27

[1] For the tensile strength of wide strip (coils), only the minimum value of the range is applicable.
[2] Normally the test piece used shall have a proportional prismatic and cylindrical shape and have an original gauge length given by the formula $L_0 = 5.65\sqrt{S_0}$, where S_0 is the original cross-sectional area.

Table 3-4 Equivalent grades of steel as used in a number of selected countries compared with those according to ISO 630 [1]

DIN 17100 [2] DIN 17119 [3] DIN 17120 [4] DIN 17121 [5]	BS 4360 [6]	NF 49–501 [7] NF 49–541 [8]	UNI 7806 [9] UNI 7810 [10]	ASTM	JISG 3444 [20] JISG 3466 [21]	ISO 630 [1]
Germany	United Kingdom	France	Italy	U.S.A	Japan	
St 37–2 USt 37–2 RSt 37–2	40 B	E 24–2	Fe 360 B	A 36 [11] A 500 [13] A 501 [14]	STK 41	Fe 360 B
St 37–3	40 C+D	E 24–3+4	Fe 360 C+D	A 572 [16]		Fe 360 C+D
St 44–2	43 B	E 28–2	Fe 430 B	A 500 [11] A 529 [15]		Fe 430 B
St 44–3	43 C+D	E 28–3+4	Fe 430 C+D	A 633 [19]		Fe 430 C+D
St 52–3	50 C+D	E 36–3+4	Fe 510 C+D	A 441 [12] A 588 [17] A 618 [18] A 633 [19]	STK 50	Fe 510 C+D

Vanadium or a combination of such elements together with less susceptibility to brittle fracture. Further to the basic steels, low temperature steels with recommended minimum impact energy at −50 °C are also available.

In the course of the product harmonization in the European Union, two product standards for steel structural hollow sections have been published – EN 10210-1 [22] for the hot finished and EN 10219-1 [23] for the cold formed sections. Cold formed hollow sections with subsequent heat treatment to obtain equivalent metallurgical conditions to those by normalizing rolling are deemed to meet the requirements of the standard for hot finished hollow section. These European standards, the part 1 [22, 23] of which describes the technical delivery requirements for circular, square and rectangular hollow sections, cover non-alloy general purpose structural steels as well as fine grain structural steels.

It has been agreed that the corresponding national standards of the member countries of the European Union are to be fully substituted by these European standards. Meanwhile after the publication of these standards by CEN (Comité Européen de Normalisation, European Committee for Standardization), this agreement has already been fulfilled.

Tables 3-5 to 3-10 show the chemical compositions and the mechanical properties of the structural steels according to EN 10210-1 (hot finished) [22] and EN 10219-1 (cold formed). The steel designations in these tables are explained as follows:

Non-alloy structural steels

– The capital letter S for structural steel
– The indication of the minimum specified yield strength for wall thickness ≤ 16 mm expressed in N/mm^2
– The capital letters JR for the qualities with specified impact properties at room temperature
– The capital letter J and a number 0 or 2 for the qualities with specified impact properties at 0 °C and −20 °C respectively
– The capital letter H to indicate hollow section

Table 3-5 Chemical composition (ladle analysis) of hot finished (HF) and cold formed (CF) structural hollow sections of non-alloy steels according to EN 10210-1 [22] and EN 10219-1 [23]

Designation	Type of deoxidation[c]	Quality[d]	C % max.		Si % max.	Mn % max.	P % max.	S % max.	N$_2$ % max.
			Nominal wallthickness						
			t ≤ 40 mm	40 < t ≤ 65 mm					
S 235 JRH+HF[a]	FN	BS	0.17	0.20	–	1.40	0.045	0.045	0.009
S 275 JOH+HF	FN	QS	0.20	0.22	–	1.50	0.040	0.040	0.009
S 275 J2H+HF	FF	QS	0.20	0.22	–	1.50	0.035	0.035	–
S 355 JOH+HF	FN	QS	0.22	0.22	0.55	1.60	0.040	0.040	0.009
S 355 J2H+HF	FF	QS	0.22	0.22	0.55	1.60	0.035	0.035	–
			Nominal wallthickness						
			t ≤ 40 mm						
S 235 JRH+CF[b]	FF		0.17		All these values are identical to those according to EN 10210-1 [22]				
S 275 JOH+CF	FF		0.20						
S 275 J2H+CF	FF		0.20						
S 355 JOH+CF	FF		0.22						
S 355 J2H+CF	FF		0.22						

[a] HF = hot finished.
[b] CF = cold formed (only circular hollow sections are available in thicknesses over 24 mm).
[c] FN = rimming steel not permitted.
  FF = fully killed steel containing nitrogen binding elements in amounts sufficient to bind available nitrogen (e.g. 0.020% total Al or 0.015% soluble Al).
[d] BS = base steel, QS = quality steel.

Table 3-6 Permissible deviations of the product analysis from the specified limits of the ladle analysis [23] for non-alloy steels

Element	Permissible maximum content in the ladle analysis %	Permissible deviation of the product analysis from the specified limits for the ladle analysis %
C [a]	≤ 0.20 > 0.20	+0.02 +0.03
Si	≤ 0.60	+0.05
Mn	≤ 1.60	+0.10
P	≤ 0.045	+0.010
S	≤ 0.045	+0.010
N	≤ 0.025	+0.002

[a] For S 235 JRH for thicknesses $t \leq 16$ mm, the permissible deviation = 0.04 % C and for thicknesses $16 < t \leq 40$ mm the permissible deviation = 0.05 % C.

Table 3-7 Chemical composition (ladle analysis) of hot finished (HF) and cold formed (CF) structural hollow sections of fine grain steels according to EN 10210-1 [22] and EN 10219-1 [23]

Steel designation	Type of deoxidation [1]	Quality [2]	C % max.	Si % max.	Mn %	P % max.	S % max.	Nb % max.	V % max.	Al$_{total}$ % min. [3]	Ti % max.	Cr % max.	Ni % max.	Mo % max.	Cu % max. [4]	N % max.
S 275 NH+HF [a] S 275 NLH+HF [a]	GF	QS	0.20	0.40	0.50–1.40	0.035 0.030	0.030 0.025	0.05	0.05	0.02	0.03	0.30	0.30	0.10	0.35	0.015
S 355 NH+HF [a] S 355 NLH+HF [a]	GF	QS	0.20 0.18	0.50	0.90–1.65	0.035 0.030	0.030 0.025	0.05	0.12	0.02	0.03	0.30	0.50	0.10	0.35	0.015
S 460 NH+HF [a] S 460 NLH+HF [a]	GF	SS	0.20	0.60	1.00–1.70	0.035 0.030	0.030 0.025	0.05	0.20	0.02	0.03	0.30	0.80	0.10	0.70	0.025
S 275 NH+CF [b] S 275 NLH+CF [b]								All these values are identical to those according to EN 10210-1 [22]								
S 355 NH+CF [b] S 355 NLH+CF [b]																
S 460 NH+CF [b] S 460 NLH+CF [b]																

[a] Wall thickness $t \leq 65$ mm.
[b] Wall thickness $t \leq 40$ mm, only circular hollow sections are available over 24 mm.
[1] GF = fully killed steel containing nitrogen binding elements in amounts sufficient to bind the available nitrogen and having a fine grain microstructure.
[2] QS = quality steel, SS = special steel.
[3] If sufficient N-binding elements are present, the minimum total Al content does not apply.
[4] If the copper content is greater than 0.30 %, then the nickel content shall be at least half of the copper content.

Table 3-8 Permissible deviations of the product analysis from the specified limits of the ladle analysis [22, 23] for fine grain steels

Element	Permissible maximum content in the ladle analysis %	Permissible deviation of the product analysis from the specified limits of the ladle analysis %
C	≤ 0.20 > 0.20	+0.02 +0.03
Si	≤ 0.60	+0.05
Mn	≤ 1.70	−0.05 +0.10
P	≤ 0.035	+0.005
S	≤ 0.030	+0.005
Nb	≤ 0.060	+0.010
V	≤ 0.20	+0.02
Ti	≤ 0.03	+0.01
Cr	≤ 0.30	+0.05
Ni	≤ 0.80	+0.05
Mo	≤ 0.10	+0.03
Cu	≤ 0.35 0.35 < Cu ≤ 0.70	+0.04 +0.07
N	≤ 0.025	+0.002
Al_{total}	≥ 0.020	−0.005

Fine grain structural steels

- The capital letter S for structural steel
- The indication of the minimum specified yield strength for wall thicknesses ≤ 16 mm expressed in N/mm^2
- The capital letter N to indicate normalized or normalized rolled*
- The capital letter L for the qualities with specified minimum values of impact energy at a temperature of −50 °C
- The capital letter H to indicate hollow section

For a comparison, the variety of steel types and grades in conformance with ASTM A 500 [13] with their chemical compositions and mechanical properties is given in Table 3-10 a.

---

* Normalized rolling is a process, in which the final deformation is carried out in a certain temperature range leading to a material condition equivalent to that obtained after normalizing, so that the specified values of the mechanical properties are retained even after normalizing.

Structural hollow sections can also be produced in special steels with yield strengths higher than 460 N/mm^2. A high strength fine grain structural steel StE 690 in quenched and tempered condition is available for seamless circular, rectangular and square hollow sections according to DASt-Ri 011 (Guide lines by German committee for steel structures [24]). Tables 3-11 and 3-12 show the chemical composition and the mechanical properties of this steel.

Further development has been achieved to manufacture seamless hollow sections in high strength weldable fine grain steels in quenched and tempered condition with yield strengths higher than that of St E 690. St E 770 and St E 790 belong to this category, mechanical properties of which are given in Table 3-13. They are mostly applied to construct mobile cranes, where the reduction of the dead weight is of high importance.

Table 3-9 Mechanical properties of hot finished (HF) and cold formed (CF) structural hollow sections of non-alloy steels according to EN 10210-1 [22] and EN 10219-1 [23]

Steel designation	Min. yield strength N/mm²			Tensile strength N/mm²			Min. elongation % ($L_o = 5.65\sqrt{S_o}$)					Impact properties	
							longitudinal		transversal			Test temperature °C	Min. average absorbed energy for standard test pieces in Joule
	t ≤ 16 mm	t > 16 mm t ≤ 40 mm	t > 40 mm t ≤ 65 mm	t < 3 mm	t ≥ 3 mm t ≤ 65 mm		t ≤ 40 mm	t > 40 mm t ≤ 65 mm	t ≤ 40 mm	t > 40 mm t ≤ 65 mm			
S 235 JRH+HF	235	225	215	360–510	340–470		26	25	24	23		+20	27
S 275 JOH+HF	275	265	255	430–580	410–560		22	21	20	19		0	27
S 275 J2H+HF												−20	27
S 355 JOH+HF	355	345	335	510–680	490–630		22	21	20	19		0	27
S 355 J2H+HF												−20	27
							Nominal wallthickness t ≤ 40 mm						
S 235 JRH+CF	235	225[1]	–	These values are identical to those according to EN 10210-1 [22]			24					These values are identical to those according to EN 10210-1 [22]	
S 275 JOH+CF	275	265[1]	–				20						
S 275 J2H+CF													
S 355 JOH+CF	355	345[1]	–				20						
S 355 J2H+CF													

[1] Only circular hollow sections are available in thickness over 24 mm.

3.1 General

Table 3-10 Mechanical properties of hot finished (HF) and cold formed (CF) structural hollow sections of fine grain steels according to EN 10210-1 [22] and EN 10219-1 [23]

Steel designation	Min. yield strength N/mm^2			Tensile strength N/mm^2	Min. elongation % $L_o = 5{,}65\sqrt{S_o}$		Impact property	
	t ≤ 16 mm	t > 16 mm t ≤ 40 mm	t > 40 mm t ≤ 65 mm	t ≤ 65 mm	t ≤ 65 mm long.	t ≤ 65 mm trans.	Test temp. °C	Energy Joule
S 275 NH+HF S 275 NLH+HF	275	265	255	370–510	24	22	−20 −50	40 27
S 355 NH+HF S 355 NLH+HF	355	345	335	470–630	22	20	−20 −50	40 27
S 460 NH+HF S 460 NLH+HF	460	440	430	550–720	17	15	−20 −50	40 27
				t ≤ 40 mm[1)]	t ≤ 40 mm[1)]			
S 275 NH+CF S 275 NLH+CF	275	265	–	These values are identical to those according to EN 10210-1 [22]	24		These values are identical to those according to EN 10210-1 [22]	
S 355 NH+CF S 355 NLH+CF	355	345	–			22		
S 460 NH+CF S 460 NLH+CF	460	440				17		

[1)] Only circular hollow sections are available in thickness over 24 mm.

Table 3-10a Chemical compositions and mechanical properties of cold formed structural hollow sections of steel according to ASTM A 500 [13]

	Chemical requirement			
Element	Composition %			
	Grade A and B		Grade C	
	Heat analysis	Product analysis	Heat analysis	Product analysis
C, max	0.26	0.30	0.23	0.27
Mn, max	–	–	1.35	1.40
P, max	0.04	0.05	0.04	0.04
S, max	0.05	0.063	0.05	0.063
Cu, when copper steel is specified, min	0.20	0.18	0.20	0.18

	Tensile requirement					
	CHS			RHS		
	Grade A	Grade B	Grade C	Grade A	Grade B	Grade C
Tensile strength, min, psi (N/mm^2)	45000 (310)	58000 (400)	62000 (427)	45000 (310)	58000 (400)	62000 (427)
Yield strength, min, psi (N/mm^2)	33000 (228)	42000 (290)	46000 (317)	39000 (269)	46000 (317)	50000 (345)
Elongation in 2 inches (50.8 mm), min, %	25	23	21	25	23	21

Table 3-11 Chemical composition of the steel St E 690 (ladle analysis) of circular and rectangular hollow sections [24]

Steel designation	Type[1]	C %	Si %	Mn %	P %	S %	Cr %	Cu %	Nb %	Ni %	V %	Mo %
T St E 690 V	V	≤ 0.20	0.15/0.80	≤ 1.7	≤ 0,025	≤ 0.025	≤ 1.00	≤ 0,50	≤ 0,05	≤ 1.50	≤ 0.12	≤ 0.60
E St E 690 V	V	≤ 0.18	0.15/0.80			≤ 0.015						

[1] These steels contain sufficient alloy elements to obtain fine grain microstructure. N-content of max. 0.020% forms nitrides. V = quenched and tempered.

Table 3-12 Mechanical properties of the steel St E 690 for circular and rectangular hollow sections [24]

Steel designation	Type[1]	Upper yield strength[2] N/mm² min.				Tensile strength N/mm²				Min. elongation %		Min. Impact energy (Joule) at test temperature −40 °C −20 °C for wallthicknesses			
		≤12 mm	>12 ≤20 mm	>20 ≤40 mm	>40 ≤50 mm	t ≤ 20 mm	>20 ≤40 mm	>40 ≤50 mm		long.	trans.	≥10 ≤20 mm	>20 ≤50 mm	≥10 ≤20 mm	>20 ≤50 mm
T St E 690 V	V	690	690	650	615	770–960	720–900	670–850		16	14	long. 40 trans. 27	30 25	50 30	40 27
E St E 690 V	V	690	690	690	650	770–960	770–960	700–880		16	14	long. 45 trans. 30	40 27	55 35	45 30

[1] V = quenched and tempered.
[2] When no distinct yield point is available, the value for 0.2% elongation limit shall be used.

Table 3-13 Mechanical properties of the high strength, weldable, quenched and tempered fine grain steel St E 770 and St E 790 for seamless hollow sections

Steel designation	Tensile strength for wallthickness ≤ 20 mm	Upper yield strength for wallthickness in mm				Min. elongation ($L_o = 5\, d_o$)	Min. impact energy (ISO-V long.) at $-20°$C and for wallthickness ≤ 20 mm
		≤12	>12 ≤20	>20 ≤40	>40 ≤50		
	N/mm²	min. N/mm²				%	Joule
St E 770	820–1000	770	750	700	670	long. 15 trans. 13	long. 50 trans. 30
St E 790	850–1030	790	790	750	710	long. 15 trans. 13	long. 55 trans. 35

There are also the weather resistant structural steels with equivalent mechanical properties as the general purpose structural steels, which possess certain resistance against atmospheric corrosion under particular conditions. DASt-Ri 007 [25] as well as EN 10155 [27] inform about the alloy elements, such as chromium, copper, Nickel, partly a higher percentage of phosphorus in order to attain weather resistance. Further informations regarding the chemical compositions of these steels, their mechanical properties as well as their fabrication and handling methods can be obtained from the tube manufacturers.

3.2 Selection of steel grades for structural design

Fig. 3-1 shows a diagram "stress f vs. strain ε" determined by tensile test. This characterizes the mechanical properties of a steelgrade – yield strength f_y, ultimate tensile strength f_u and elongation to failure ε_u, which act as the basis for designing structural elements.

The yield strength f_y is fundamentally the basis for designing, as this controls the deformation. While designing, adequate attention has to be given, so that the deformations are not too large. On the other hand, sufficient deformation or rotation capacity must be available, when yielding of structural members or that at particular locations in a structure has to provide the redistribution of loads (as for example, in statically indeterminate structures).

Under tensile load, a structural member made of a steel with adequate ductility may act as brittle if a particular cross section of it is weakened e.g. by holes. This is due to the fact that the ultimate tensile strength f_u of the steel is surpassed in the weakened cross section before the whole member yields. It is therefore necessary to pay special attention to the ratio f_u/f_y in particular applications to ensure that the structure acts in a ductile manner. Eurocode 3 [27] prescribes the following minimum value for the ratio:

$$\frac{f_u}{f_y} \geq 1.2$$

A constructor working in the field of structural or mechanical engineering is often confronted with dangerous situations caused by brittle fracture. This cleavage brittleness is

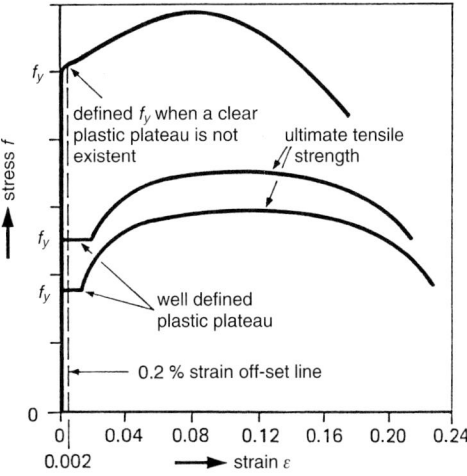

Fig. 3-1 "Stress vs. strain" curves

more dangerous, because the failure occurs without any prior notice. Impact energy measured by the Charpy-V tests represents an important indication for the inclination to brittle fracture (see Section 2.7). This shows that a structural element has to be sufficiently ductile (high minimum failure energy) when subjected to a shock load. With regard to ductility, the selection of steelgrades in steel structures can be made in accordance with the Appendix C "Design against brittle fracture" of Eurocode 3 [27] or DASt-Ri 009 (Guide lines by German committee for steel structures) [28]. State of stress, severity of damage risk (importance of a structural element for the safety of the total structure) and temperature lead to the classification, which determines the steel grade taking also the thickness of the material into account.

Besides, the metallurgical purity plays a big role. In case the impurities like sulphur, phosphorus and free nitrogen are low and the steel is fully killed by the addition of aluminium, the inclination to brittle fracture decreases [28, 29].

The brittle fracture behaviour depends also on the constructional details in a structure. Large material concentrations and rigidity jumps (abrupt change of cross section) should be avoided.

Adequate impact energy values are of high importance for the applications at low temperature ranges, as they become worse with decreasing temperature (Fig. 3-2).

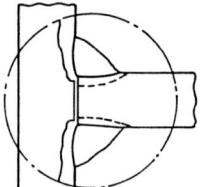

Fig. 3-3 Lamellar tearing

In the European standards [22, 23], a required minimum energy value of 27 Joule has been prescribed for low temperatures. Thick-walled sections loaded in the thickness direction must possess adequate strength and ductility to withstand lamellar tearing (Fig. 3-3). The source of this crack is the inclusion of non-metallic manganese sulphide and manganese silicate in the microstructure of a steel grade.

In principle, the crack initiation of this type does not take place, if the sulphur content is very low or the sulphur is combined with other elements such as calcium. The so-called "Z" quality can be checked by tensile tests, where the requirement for a certain reduction R_{AZ} of the cross sectional area of the test piece is to be fulfilled.

For the application of the cold formed hollow sections, the following points have to be considered:

- In the course of cold forming of hollow sections, the yield strength f_y and to a less extent the ultimate tensile strength f_u are increased, which consequently leads to a decrease of the ratio f_u/f_y. Further, the elongation to failure ε_u is somewhat decreased.

The increase of yield strength through hardening by cold forming can be taken into account for design using the calculation method in Table 3-14. This method of determining the average yield strength f_{ya} of cold formed hollow sections is given in Eurocode 3 [27].

- In order to ensure sufficient ductility, cold formed hollow sections shall possess the limiting values for the ratio "corner radius r/wall thickness t" shown in Table 3-15. As cold forming affects brittle fracture significantly, the values in the table prescribe when the cold worked zone can be welded at all. Otherwise, the welding has to be per-

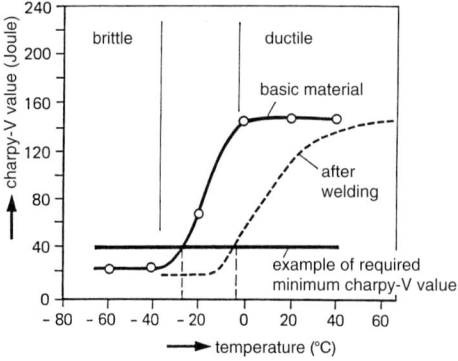

Fig. 3-2 Impact energy in relation with temperature

3.2 Selection of steel grades for structural design

formed at a minimum distance of $5t$ from the corner.

As regards the fatigue behaviour of different steels, designers should be aware that the use of non-alloy, mild steels (S 355) or thermo-mechanically (TM) milled steels is more economical than that of high strength steels. This is because of the higher material costs of high strength steels in general.

The notch effect is the governing factor with relation to the fatigue behaviour of the mild and the high strength steels (S 460, StE 690). It has been found that the fatigue quality of hollow section structures with (not machined) weld defects or other notches cannot be substantially improved by higher mechanical properties of high strength steels [30]. The more severe is the notch or defect, the less is that influence. However, if the notches are not sharp, a better fatigue behaviour is indicated by high strength steels.

For test specimens, which are fully unnotched and polished e.g. plates without weld or cut with smooth edges or plates having only drilled holes or machined specimens (that means, no defects exist, which are the starting points of the crack growth), a correlation exists between the ultimate tensile strength f_u and the high-cycle fatigue strength f_F (= fatigue limit at 10^6 to 10^7 loading cycles, see Fig. 7-8):

f_F = approximately $0.5 f_u$

In these cases, the fatigue strength increases with the increasing ultimate tensile strength, which indicates the beneficial effect of the application of the high strength steels.

Further, high strength steels are also more suitable than mild steels, when welded structures with defects are treated with some improvement methods (see Section 7.13).

For mild steel, peak loads reach the yield limit faster than high strength steels, so that local plastification occurs earlier.

Thermo-mechanical rolling is a process, in which the final deformation is carried out in a certain temperature range leading to a mate-

Table 3-14 Increase in yield strength due to cold forming of RHS

Average yield strength:
The average yield strength f_{ya} may be determined from full size section tests or as follows:

$$f_{ya} = f_{yb} + (k \cdot n \cdot t^2/A_g) \cdot (f_u - f_{yb})$$

where:
f_{yb}, f_u = specified tensile yield strength and ultimate tensile strength of the basic material in N/mm²
t = material thickness in mm
A_g = gross cross-sectional area in mm²
k = coefficient depending on the type of forming ($k = 7$ for cold forming)
n = number of 90° bends in the section with an internal radius $< 5t$ (fractions of 90° bends should be counted as fractions of n)
f_{ya} should not exceed f_u or $\leq 1.2 f_{yb}$

The increase in yield strength due to cold working should not be utilised for members which are annealed[1] or subjected to heating over a length with a high heat input after forming, which may produce softening.

Basic material:
Basic material is the flat hot rolled sheet material, out of which sections are made by cold forming.

[1] Stress relief annealing at more than 580 °C or for over one hour may lead to deterioration of the mechanical properties.

Table 3-15 Conditions for welding of cold formed RHS in the corner zone

Steel designation [23]	Wall thickness range t (mm)	min. (r/t)
S 235	$12 < t \leq 16$	3.0
S 275	$8 < t \leq 12$	2.0
S 355	$6 < t \leq 8$	1.5
	$t \leq 6$	1.0

rial condition with certain properties, which cannot be achieved by heat treatment alone. Heating above 580 °C lowers the strength values remarkably. The superiority of the thermo-mechanically milled steels is based on excellent toughness properties with values higher than 50 J at –20 °C (or lower). That causes significant lower crack growth under service conditions during lifetime. Therefore, a less remaining cross section and greater crack length can be tolerated, which allows longer inspection intervals.

4 Dimensions, tolerances and sectional properties of circular, square and rectangular hollow sections

4.1 General

The sizes and masses of the seamless and welded hollow sections are standardized. Similar to the material standards (see chapter 3), the national standards for the masses, sizes, tolerances and sectional properties of the hollow sections prescribed by the corresponding authorities of the member countries of the European Union have already been replaced by the corresponding harmonized European standards – EN 10210-2 [1] and EN 10219-2 [2] published by CEN (Comité Européen de Normalisation, European Committee for Standardization). This book deals mainly with these two standards.

The dimensions, tolerances and sectional properties of structural hollow sections have also been standardized on an international basis in ISO 657-14 [3] and ISO 4019 [4] for hot finished and cold formed sections respec-

Table 4-1 Comparison of the dimensional tolerances for circular hollow sections according to EN 10210-2 [1] and EN 10219-2 [2]

	EN 10210-2 [1] hot finished	EN 10219-2 [2] cold formed
Outside diameter D	$\pm 1\%$ of diameter with a minimum of ± 0.5 mm and a maximum of ± 10 mm	same as [1]
Wall thickness T	-10% [1] [2]	for $D \leq 406.4$ mm, $T \leq 5$ mm: $\pm 10\%$ $T > 5$ mm: $\pm 0,5$ mm for $D > 406.4$ mm, $\pm 10\%$ with a maximum of ± 2 mm
Mass M	$\pm 6\%$ on individual lengths [3]	same as [1]
Straightness $e/l \times 100\%$	0.2% of total length	same as [1]
Length (exact)	for ≥ 2000 mm to 6000 mm $+10$ mm -0 for >6000 mm $+15$ mm -0	for ≤ 6000 mm $+5$ mm -0 for >6000 to 10000 mm $+15$ mm -0 for >10000 mm $+5$ mm $+ 1$ mm/m -0

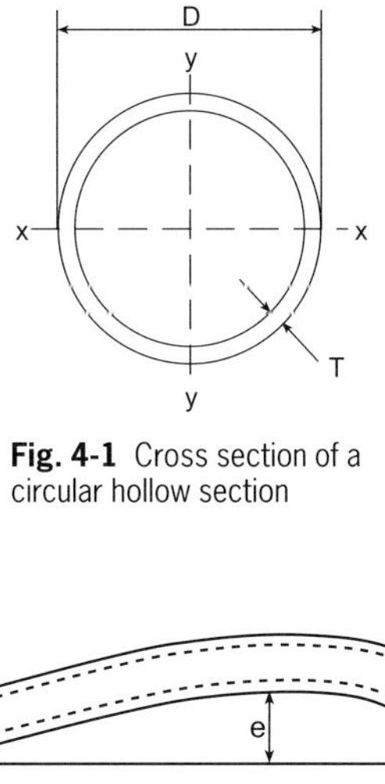

Fig. 4-1 Cross section of a circular hollow section

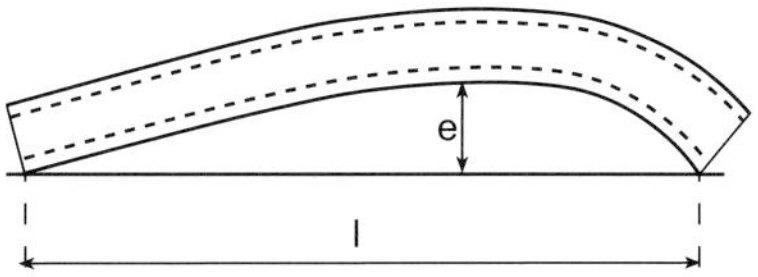

Fig. 4-2 Value for straightness $e/l \cdot 100\%$

[1] The positive deviation is limited by the tolerance on mass.
[2] For seamless sections, thicknesses of less than 10% but not less than 12.5% of the nominal thickness may occur in smooth transitional areas over not more than 25% of the circumference.
[3] For seamless sections, mass tolerance can be +8%.

Table 4-2 Comparison of the dimensional tolerances for square and rectangular hollow sections according to EN 10210-2 [1] and EN 10219-2 [2]

	EN 10210-2 [1] hot finished	EN 10219-2 [2] cold formed
Outside dimensions B, H	±1% with a minimum of ±0.5 mm	H, B < 100 mm, ±1% with a minimum of ±0.5 mm 100 ≤ H, B ≤ 200 mm, ±0.8% H, B > 200 mm, ±0.6% 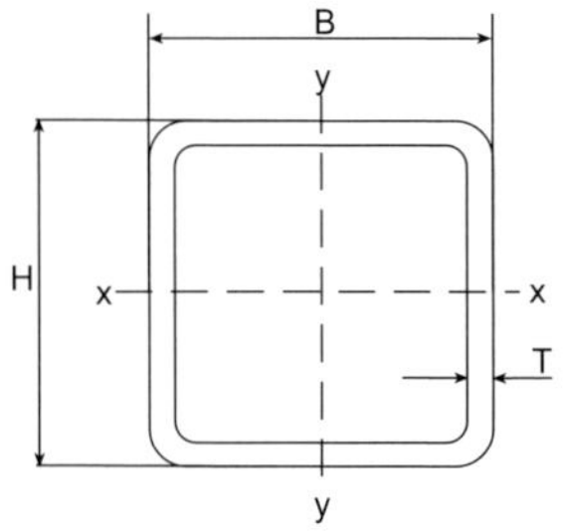
Thickness T	−10% [1) 2)]	T ≤ 5 mm: ±10% T > 5 mm: ±0,5%
Mass M	±6% [3)] on individual lengths	same as [1]
Straightness	0.2% of total length	0.15% of total length
Length (exact)	for ≥ 2000 to 6000 mm +10 mm −0 for > 6000 mm +15 mm −0	for ≤ 6000 mm +5 mm −0 for > 6000 to 10000 mm +15 mm −0 for > 10000 mm +5 mm + 1 mm/m −0
Squareness of sides 90°−Θ	90°±1°	90°±1° 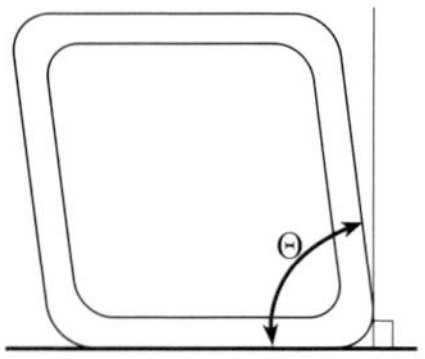
Outside corner radius r_0	3 T maximum at each corner	for T ≤ 6 mm, 1.5 T to 2.4 T for 6 < T ≤ 10 mm, 2.0 T to 3.0 T for T > 10 mm, 2.4 T to 3.6 T

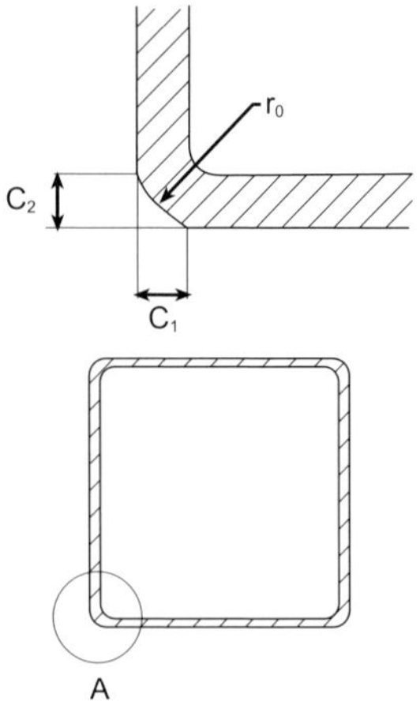

Fig. 4-3 Cross section of a square or rectangular hollow section

Fig. 4-4 Deviation of squareness

Fig. 4-5 Deviation of outside corner radius of square or rectangular hollow section

[1, 2, 3)] See footnotes for Table 4-1.

4.2 Geometrical and sectional properties of hollow sections

Table 4-2 (continued)

	EN 10210-2 [1] hot finished	EN 10219-2 [2] cold formed	
Concavity/ Convexity %	1%	max. 0.8% with a minimum of 0.5 mm	
Twist	2 mm plus 0.5 mm/m length	same as [1]	

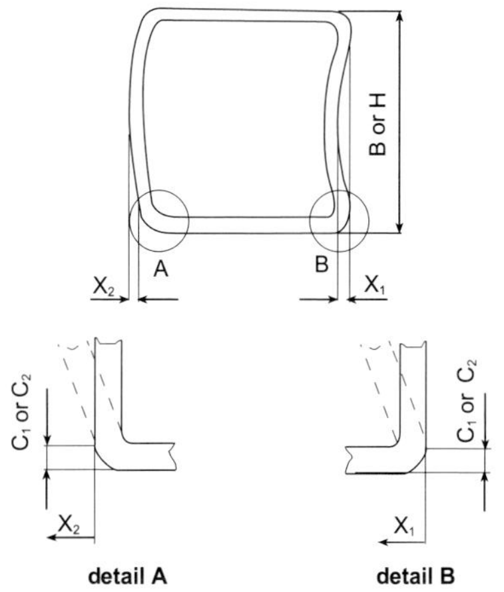

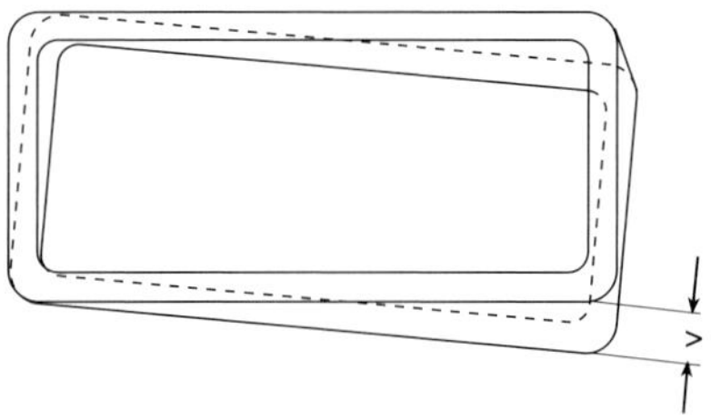

Fig. 4-6 Concavity and convexity of a square or rectangular hollow section X_1/B, X_2/B; X_1/H, X_2/H in %

Fig. 4-7 Twist of a square and rectangular hollow section

tively. The drafts for the renewed versions of these standards lean strongly on the above mentioned European standards.

The sizes given in the standards in Canada [5] and U.S.A. [6–8] do not exactly comply with those given in the European standards, because their origin lies in the imperial unit system (inch, pound).

In Tables 4-1 and 4-2, the tolerances for the hot finished and cold formed circular and rectangular hollow sections in accord with EN 10210-2 [1] and EN 10219-2 [2] respectively are compared to each other. They show very small differences. However, various national standards may deviate considerably from one another regarding the geometrical tolerances, because they depend fundamentally on the different manufacturing facilities in the mills of different countries. Compared to open sections, the dimensional tolerances of hollow sections are in general lower.

4.2 Geometrical and sectional properties of hollow sections

Although the ISO [3, 4] and EN [1, 2] standards regarding the sizes, tolerances and sectional properties are harmonized meanwhile, the sizes produced by some manufacturers go beyond those given by the above mentioned standards. They manufacture either intermediate dimensions, not included in the standards or leave some dimensions given in the standards. This depends mainly on the manufacturing facilities of individual mills. Details can be obtained from the information materials of the manufacturers.

The appendices I through III show the sizes and the sectional properties of the circular, square and rectangular hollow sections listed in EN 10210-2 [1] and EN 10219-2 [2]. In the following sections, the recommended formulae to calculate the sectional properties are given. This will make a quick calculation of the sectional properties for the extra cross sections possible.

4.2.1 Circular hollow sections (CHS)
(see Fig. 4-1)

Nominal outside diameter D (mm)
Nominal wall thickness $\quad T$ (mm)
Nominal inside diameter $\quad D_i = (D - 2\,T)$ (mm)

- Cross-sectional area

$$A_g = \frac{\pi(D^2 - D_i^2)}{4 \times 10^2}\ (\text{cm}^2)$$

- Superficial area per metre length

$$A_s = \frac{\pi D}{10^3}\ (\text{m}^2/\text{m})$$

- Mass per unit length

$$M = 0.785 \times A_g\ (\text{kg/m})$$

- Second moment of area

$$I = \frac{\pi(D^4 - D_i^4)}{64 \times 10^4}\ (\text{cm}^4)$$

- Radius of gyration

$$i = \sqrt{\frac{I}{A_g}}\ (\text{cm})$$

- Elastic section modulus

$$W_{el} = \frac{2I \times 10}{D}\ (\text{cm}^3)$$

- Plastic section modulus

$$W_{pl} = \frac{D^3 - D_i^3}{6 \times 10^3}\ (\text{cm}^3)$$

- Torsional inertia constant
  (Polar moment of inertia)

$$I_t = 2\,I\ (\text{cm}^4)$$

- Torsional modulus constant

$$C_t = 2\,W_{el}\ (\text{cm}^3)$$

4.2.2 Rectangular, including square, hollow sections (RHS)

Nominal length of shorter side
of RHS $\qquad B$ (mm)
Nominal length of longer side
of RHS $\qquad H$ (mm)
Nominal wall thickness $\qquad T$ (mm)
Nominal external corner radius $\quad r_o$ (mm)
Nominal internal corner radius $\quad r_i$ (mm)

- Cross-sectional area

$$A_g = \frac{2T}{10^2}(B + H - 2T) - (4 - \pi)(r_o^2 - r_i^2)\ (\text{cm}^2)$$

- Superficial area per metre length

$$A_s = \frac{2}{10^3}(H + B - 4r_o + \pi r_o)\ (\text{m}^2/\text{m})$$

- Mass per unit length

$$M = 0.785 \times A_g\ (\text{kg/m})$$

- Second moment of area:
Major axis

$$I_{xx} = \frac{1}{10^4}\left[\frac{BH^3}{12} - \frac{(B-2T)(H-2T)^3}{12}\right.$$
$$\left. -4\left(I_{zz} + A_z h_z^2\right) + 4\left(I_{\xi\xi} + A_\xi h_\xi^2\right)\right]\ (\text{cm}^4)$$

Minor axis

$$I_{yy} = \frac{1}{10^4}\left[\frac{HB^3}{12} - \frac{(H-2T)(B-2T)^3}{12}\right.$$
$$\left. -4\left(I_{zz} + A_z h_z^2\right) + 4\left(I_{\xi\xi} + A_\xi h_\xi^2\right)\right]\ (\text{cm}^4)$$

- Radius of gyration

$$i_{xx} = \sqrt{\frac{I_{xx}}{A_g}}\ (\text{cm})$$

$$i_{yy} = \sqrt{\frac{I_{yy}}{A_g}} \text{ (cm)}$$

- Elastic section modulus

$$W_{el_{xx}} = \frac{2 I_{xx}}{H} \times 10 \text{ (cm}^3\text{)}$$

$$W_{el_{yy}} = \frac{2 I_{yy}}{B} \times 10 \text{ (cm}^3\text{)}$$

- Plastic section modulus

$$W_{pl_{xx}} = \frac{1}{10^3} \left[\frac{BH^2}{4} - \frac{(B-2T)(H-2T)^2}{4} \right.$$
$$\left. -4(A_z h_z) + 4(A_\xi h_\xi) \right] \text{ (cm}^3\text{)}$$

$$W_{pl_{yy}} = \frac{1}{10^3} \left[\frac{HB^2}{4} - \frac{(H-2T)(B-2T)^2}{4} \right.$$
$$\left. -4(A_z h_z) + 4(A_\xi h_\xi) \right] \text{ (cm}^3\text{)}$$

- Torsional inertia constant

$$I_t = \frac{1}{10^4} \left[\frac{T^3 h}{3} + 2 K A_h \right] \text{ (cm}^4\text{)}$$

- Torsional modulus constant

$$C_t = 10 \left[\frac{I_t}{T + \frac{K}{T}} \right] \text{ (cm}^3\text{)}$$

where:

$$A_z = \left[1 - \frac{\pi}{4} \right] r_o^2 \text{ (mm}^2\text{)}$$

$$A_\xi = \left[1 - \frac{\pi}{4} \right] r_i^2 \text{ (mm}^2\text{)}$$

$$h_z = \frac{H}{2} - \left(\frac{10 - 3\pi}{12 - 3\pi} \right) r_o \text{ (mm)}$$

Major axis (for minor axis substitute B for H)

$$h_\xi = \frac{H - 2T}{2} - \left(\frac{10 - 3\pi}{12 - 3\pi} \right) r_i \text{ (mm)}$$

Major axis (for minor axis substitute B for H)

$$I_{zz} = \left[\frac{1}{3} - \frac{\pi}{16} - \frac{1}{3(12 - 3\pi)} \right] r_o^4 \text{ (mm}^4\text{)}$$

$$I_{\xi\xi} = \left[\frac{1}{3} - \frac{\pi}{16} - \frac{1}{3(12 - 3\pi)} \right] r_i^4 \text{ (mm}^4\text{)}$$

$$h = 2[(B-T) + (H-T)] - 2 R_c (4-\pi) \text{ (mm)}$$

$$A_h = (B-T)(H-T) - R_c^2 (4-\pi) \text{ (mm}^2\text{)}$$

$$K = \frac{2 A_h T}{h} \text{ (mm}^2\text{)}$$

$$R_c = \frac{r_o + r_i}{2} \text{ (mm)}$$

As can be observed in the Appendices II and III, the values for the sectional properties for RHS differ, when they are calculated according to EN 10210-2 [1] and EN 10219-2 [2] respectively. This is due to the fact that the corner radii, on which the calculations are based, differ from one another.

EN 10210-2:
$r_0 = 1.5\ T$ (mm)
$r_i = 1.0\ T$ (mm)

EN 10219-2:
for $T \leq 6$ mm, $r_0 = 2.0\ T$ (mm) and
$r_i = 1.0\ T$ (mm)
for $6 < T \leq 10$ mm, $r_0 = 2.5\ T$ (mm) and
$r_i = 1.5\ T$ (mm)
for $T > 10$ mm, $r_0 = 3.0\ T$ (mm) and
$r_i = 2.0\ T$ (mm)

Depending on the wall thickness, the values of the sectional properties for the hot finished RHS can be up to about 20% larger than those for the cold formed RHS.

4.2.3 Larger dimensions of square and rectangular hollow sections (RHS)

In a number of countries, additional RHS sizes are produced, which are significantly larger than those given in the European standards [1, 2]. They are particularly suited for building multi-storey structures and offshore installations.

Japan:

Cold rolled, welded square hollow sections, electric resistance welding

300 × 300 × 6–19 mm
350 × 350 × 9–22 mm
400 × 400 × 9–22 mm
450 × 450 × 9–22 mm
500 × 500 × 9–22 mm
550 × 550 × 12–22 mm

Cold pressed, welded square hollow sections, submerged arc welding

300 × 300 × 9–22 mm
350 × 350 × 9–25 mm
400 × 400 × 9–22 mm
450 × 450 × 9–36 mm
500 × 500 × 9–40 mm
550 × 550 × 9–40 mm
600 × 600 × 9–40 mm
700 × 700 × 12–40 mm
750 × 750 × 16–40 mm
800 × 800 × 16–40 mm
850 × 850 × 16–40 mm
900 × 900 × 16–40 mm
950 × 950 × 19–40 mm
1000 × 1000 × 19–40 mm

United Kingdom:

Hot finished, welded square hollow sections

350 × 350 × 19–25 mm
400 × 400 × 22–25 mm
450 × 450 × 12–32 mm
500 × 500 × 12–32 mm
550 × 550 × 16–40 mm
600 × 600 × 25–40 mm
650 × 650 × 25–40 mm
700 × 700 × 25–40 mm

5 Design bases for hollow sections in steel structures

5.1 General

Fundamental considerations for designing steel structures are based on the limiting states, where the design values represent the direct or indirect actions F (self-weight, wind load, snow load, movable imposed load, effect of temperature or settlement) and the resistances R (strength of materials, structural members or connections, rigidity obtained from the geometrical data etc.). Limit states are states beyond which a structure or a structural element does not fulfil the design performance requirements.

Limit state conditions are classified into two categories, which prescribe the rules for the structural design corresponding to the different design situations:

1. Ultimate limit states consider the loss of static equilibrium of a structure or a part of it (buckling, shearing) and the rupture or excessive stress or deformation of a section, member or connection (fatigue included).

2. Serviceability limit states correspond to states beyond which specified service criteria are no longer met. They include deformations or deflections which adversely affect the appearance or effective use of the structure including the proper functioning of machines or services or cause damage to finishes or non-structural elements. They consider also the vibration, which causes discomfort to people and damage to the building or its contents or limits its its functional effectiveness. To avoid them, it is necessary to limit deformations, deflections and vibrations.

Regarding the fulfilment of the ultimate limit states condition the following verifications have to be made:

1. For static equilibrium or gross displacements or deformations of the structure,

$$E_{d,dst} \leq E_{d,stb}$$

where $E_{d,dst}$ is the design effect of the destabilizing actions and $E_{d,stb}$ is the design effect of the stabilizing actions

When considering a limit state of stability, it has to be verified that instability does not occur unless actions F exceeds their design values F_d.

2. For rupture or excessive deformation of a section, member or connection (fatigue included),

$$S_d \leq R_d \qquad (5\text{-}1)$$

where S_d is the design loading value given by an internal force or moment (or of a respective vector of several internal forces or moments)

and R_d is the corresponding design resistance (load bearing strength) determined generally directly from material and sectional properties

When written as indices, S_d and R_d are given by Sd and Rd.

For serviceability limit states condition, the following condition is to be fulfilled:

$$E_d \leq R_d$$

where E_d is the design effect of action and R_d is the corresponding design resistance

The calculation of the design load is carried out using elastic or yield hinge or yield zone theory while that of design resistance basing on elastic or plastic load bearing strength.

Table 5-1 Bending moments and shear forces in hollow sections for full plastification; approximation $d_m, h_m, b_m \gg t$, corner radii have not been taken into account. Exact calculation of M_{pl} and α can be done using W_{pl} and W_{el} formulae in Sections 4.2.1 and 4.2.2.

Profile	Plastic bending moment $M_{pl} = W_{pl} \cdot f_y$	Plastic shear	Formfactor $\alpha = \dfrac{W_{pl}}{W_{el}}$
(circular, d_m, t)	$M_{pl} = t \cdot d_m^2 \cdot f_y$	$V_{pl} = Q_{pl} = 2t \cdot d_m \cdot \dfrac{f_y}{\sqrt{3}}$	$\dfrac{4}{\pi} = 1.273$
(square, b_m, t)	$M_{pl} = \dfrac{3}{2} \cdot t \cdot b_m^2 \cdot f_y$	$V_{pl} = Q_{pl} = 2t \cdot b_m \cdot \dfrac{f_y}{\sqrt{3}}$	$\dfrac{9}{8} = 1.125$
(rectangular, b_m, h_m, t)	$M_{pl,x} = \left(t \cdot b_m \cdot h_m + \dfrac{1}{2} t \cdot h_m^2\right) f_y$	$V_{pl} = Q_{pl} = 2t \cdot h_m \cdot \dfrac{f_y}{\sqrt{3}}$	$\dfrac{1 + \dfrac{1}{2}\dfrac{h_m}{b_m}}{1 + \dfrac{1}{3}\dfrac{h_m}{b_m}}$

Under bending load the utilization of the plastic strength demands from the wall thicknesses of hollow sections more stringent requirements than the strength determined by using elastic theory. In this case, the ductility of steel is to be taken into consideration. The plastic calculation of a cross section is in general more economic than that using simple elastic limit load, where the first reaching of the cross section edge by the yield limit is decisive.

Fig. 5-1 shows the stress distributions in hollow cross sections for full plastification, when they are loaded separately by bending moment and transverse (shear) force.

The bending moments and shear forces of hollow sections for full plastification can be determined according to Table 5-1. The form factor α is the ratio $M_{pl}/M_{el} = W_{pl}/W_{el}$ and indicates the plastic reserve of a cross section (W_{pl} and W_{el} are plastic and elastic section modulus respectively).

5.2 Partial safety factors

Considerations regarding the safety of a structure [1–5], start basing on a semi-probabilistic method using split safety and combination factors. Characteristic values for the design values of actions F_k and of resistances R_k are defined in the design standards. Material properties for steel structures are generally represented by nominal values used as characteristic values. These values are supported by the corresponding partial safety factors in accordance with the existing risk. This risk may consist of the magnitude and frequency of loadings or the simultaneity of their appearances. It may also depend on the accuracy of modelling of the actions and on the precision of the transmission of the results of the model tests performed to predict the real structural behaviour. Further, favourable and unfavourable design effects of actions as well as per-

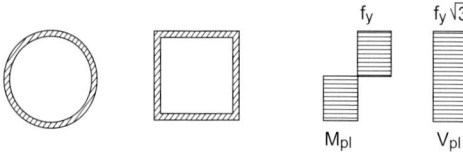

Fig. 5-1 Stress distribution for full plastification of hollow sections loaded by bending moment and shear force separately

5.2 Partial safety factors

manent, transient, variable and accidental design situations can be accounted for by means of partial safety factors.

The design values of internal force or moment S_d (F_d is the design value of an action in general) and the design resistance R_d are calculated considering the partial safety factors γ_F and γ_M respectively. These factors take also the scatter of the characteristic values of loads and resistances into consideration.

Load bearing strength of a structure with relation to applied load is to be verified as follows:

$$S_d \triangleq F_d = \gamma_F \cdot F_k \leq R_d = R_k/\gamma_M \quad (5\text{-}2)$$

5.2.1 Actions and their partial safety factors

In general, γ_F (partial safety factor for actions) and F_k (characteristic value for actions) are laid down in relevant design standards [3]. Further, F_k may also be agreed between the client and the designer, where mostly the minimum requirements are based on the existing standards or the rules prescribed by the competent national or international authorities.

The actions are differentiated as follows:

- By their variation in time
  - Permanent actions (G, characteristic value G_k), e.g. self-weight of structures, fittings, ancillaries and fixed equipment
  - Variable actions (Q, characteristic value Q_k), e.g. imposed loads, traffic loads, wind loads and snow loads
  - Accidental actions (A, characteristic value A_k), e.g. explosion and impact from vehicles
- By their spatial variation
  - Fixed actions, e.g. self-weight
  - Free actions, e.g. movable imposed loads, wind loads and snow loads

Various safety models have been described in the literature. As an example, the formulae for the determination of the design values of the actions F_d related to their combinations and the frequency of their appearances in accordance with Eurocode 3 [2] are given in the following:

1. Persistent and transient design situations (fatigue is not included)

$$F_d = \sum_j \gamma_{F,G,j} \cdot G_{k,j} + \gamma_{F,Q,1} \cdot Q_{k,1} + \sum_{i>1} \gamma_{F,Q,i} \cdot \psi_{o,i} \cdot Q_{k,i} \quad (5\text{-}3)$$

2. Accidental design situations

$$F_{da} = \sum_j \gamma_{F,GA,j} \cdot G_{k,j} + \gamma_{F,A} \cdot A_k + \psi_{1,1} \cdot Q_{k,1} + \sum_{i>1} \psi_{2,i} \cdot Q_{k,i} \quad (5\text{-}4)$$

where:
- $G_{k,j}$ the characteristic values of the permanent actions
- $Q_{k,1}$ the characteristic value of the most unfavourable variable action
- $Q_{k,i}$ the characteristic values of the other variable actions
- A_k the characteristic value of the accidental action
- $\gamma_{F,A}$ the partial safety factor for the accidental action
- $\gamma_{F,G,j}$ the partial safety factor for the permanent action $G_{k,j}$
- $\gamma_{F,GA,j}$ the partial safety factor for the permanent action in accidental design situation
- $\gamma_{F,Q,1}$ the partial safety factor for the most unfavourable one of the variable actions $Q_{k,1}$
- $\gamma_{F,Q,i}$ the partial safety factor for the other variable actions $Q_{k,i}$
- $\psi_{0,i}$ the combination factor ⎫
- ψ_1 the frequency factor ⎬ given in relative standards e.g. [2]
- ψ_2 the factor for quasi-permanence ⎭

For building structures, Eq. (5-3) may be replaced by whichever of the following combinations gives the larger value:

- Considering only the most unfavourable variable action

$$F_d = \sum_j \gamma_{F,G,j} \cdot G_{k,j} + \gamma_{F,Q,1} \cdot Q_{k,1} \quad (5\text{-}5)$$

- Considering all unfavourable variable actions

$$F_d = \sum_j \gamma_{F,G,j} \cdot G_{k,j} + 0.9 \sum_{i \geq 1} \gamma_{F,Q,i} \cdot Q_{k,i} \quad (5\text{-}6)$$

Table 5-2 Partial safety factors γ_F for the ultimate limit state condition [1, 2]

Permanent action $\gamma_{F,G}$	Most unfavourable variable action $\gamma_{F,Q,1}$	Accompanying variable action $\gamma_{F,Q,i}$
1.35	1.5	1.5

Table 5-3 Combinations of actions according to [1, 2]

1. $\gamma_{F,G,j} \cdot \sum G_{k,j} + \gamma_{F,Q,1} \cdot Q_{k,1}$ $= 1.35 \sum G_{k,j} + 1.5 \cdot Q_{k,1(max)}$ 2. $\gamma_{F,G,j} \cdot \sum G_{k,j} + 0.9 \sum \gamma_{F,Q,i} \cdot Q_{k,i}$ $= 1.35 \sum G_{k,j} + 0.9 \sum 1.5 \, Q_{k,i}$ $= 1.35 \, G_{k,j} + 1.35 \sum Q_{k,i}$	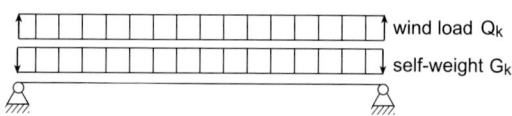 wind load Q_k self-weight G_k In case G_k and Q_k work in directions opposite to each other, then $\gamma_{F,G} = 1.0$ In case Q_k works in the opposite direction to the dominant load, then $\gamma_{F,Q} = 0$

The partial safety factors for the actions according to Eqs. (5-5) and (5-6) are listed in Table 5-2.

Table 5-3 shows the combinations of actions using the partial safety factors in Table 5-2.

The following simple formulae are used to calculate the combinations of actions for the serviceability limit condition according to Eurocode 3 [2]. The most unfavourable combination among them shown by Eqs. (5-7) and (5-8) is to be applied.

$$F_d = \sum_j G_{k,j} + Q_{k,1 \, (max)} \quad (5\text{-}7)$$

$$F_d = \sum_j G_{k,j} + 0.9 \sum_{i \geq 1} Q_{k,i} \quad (5\text{-}8)$$

However, there are standards, such as DIN 18800, part 1 [1], which recommend that the partial safety factors and the combination factors as well as the loading combinations under serviceability limit conditions are to be agreed between the client, the designer and the competent authority as long as they are not prescribed by the specific industrial standards.

5.2.2 Resistances and their partial safety factors

The design resistances R_d are determined by dividing their characteristic values R_k with the corresponding partial safety factors γ_M.

$$R_d = \frac{R_k}{\gamma_M} \quad (5\text{-}9)$$

The characteristic values R_k represent the material mechanical properties e.g. yield strength f_y and ultimate tensile strength f_u (see Tables 3-3, 3-9 and 3-10) as well as the rigidity values (bending stiffness EI_{max}, buckling stiffness EI_{min}, torsional stiffness GI_t), which are calculated basing on the nominal values for the sectional properties (see Appendices I through III) and the characteristic values for the modulus of elasticity E and the shear modulus G (see Table 5-4).

Table 5-4 Physical properties of structural steels

Modulus of elasticity	$E = 210\,000 \text{ N/mm}^2$
Shear modulus	$G = \dfrac{E}{2(1+\nu)} = 81\,000 \text{ N/mm}^2$
Poisson coefficient	$\nu = 0.3$
Coefficient of linear expansion	$\alpha = 12 \cdot 10^{-6}/°C$
Density	$\varrho = 7850 \text{ kg/m}^3$

In order to design according to limit state method, the structural cross sections are categorized in four classes which show the interrelationships between the loads and the load bearing strengths. Corresponding models are available, which are necessary for the verification (see Table 5-5).

Class 1 cross section

Verification procedure: plastic-plastic
The cross section can form a plastic hinge with the rotation capacity required for plastic analysis. The ultimate limit state is reached

5.2 Partial safety factors

Table 5-5 Cross section classification and design models

Cross section class	Class 1	Class 2	Class 3	Class 4
Design resistance (load bearing strength)	Full plasticity in the cross section, full rotation capacity	Full plasticity in the cross section, restricted rotation capacity	Elastic cross section, yield stress in the extreme fibre	Elastic cross section, necessary to make explicit allowances for the effects of local buckling
Stress distribution and rotation capacity				
Procedure for determination of design load	plastic	elastic	elastic	elastic
Procedure for determination of design resistance (load bearing strength)	plastic	plastic	elastic	elastic

when the number of plastic hinges is sufficient to produce a mechanism. The system must remain in static equilibrium.

Class 2 cross section (compact)

Verification procedure: elastic-plastic
The cross section can develop its plastic resistance, but have limited rotation capacity. The load resultants are determined based on elastic analysis and are compared to plastic resistances. The formation of the first plastic hinge indicates that the ultimate limit state is reached.

Class 3 cross section (semi-compact)

Verification procedure: elastic-elastic
The design resistance of the cross section is indicated, when the calculated stress in its extreme compression fibre reaches its yield strength (pure elastic calculation of design load and design resistance). As local buckling is however liable to prevent the development of plastic moment resistance, the wall thickness of the section should not be too small.

Class 4 cross section

Verification procedure: elastic-elastic
The determination of the ultimate moment or compression resistance of the cross section allows explicitly to consider the effect of local buckling. That means, the wall thicknesses must be thinner than those of class 3. The calculation is to be made taking effective width b_1 or effective depth h_1 ($b_1 = b - 3\,t$, see Table 5-16; $h_1 = h - 3\,t$, see Table 5-15) into consideration.

In order to avoid the local buckling of the cross sections or their parts before achieving the ultimate limit state, the b/t- and h/t-ratio for rectangular and d/t-ratio for circular hollow section must not exceed certain maximum values (see Section 5.3.3.2), which are different for the classes 1 through 3 (see Tables 5-17 and 5-24 in accordance with Eurocode 3 [2]). Countries other than those belonging to European Union e.g. U.S.A., Japan, Canada, Australia may have design codes with slightly different values [9].

The design load and the design resistance of a cross section must be calculated basing on the least favourable (highest) class of the cross sectional elements under compression and/or bending.

The partial safety factors γ_M of the design resistances of the structural cross sections of different classes as well as of the connections

Table 5-6 Partial safety factors γ_M according to Eurocode 3 [2] and DIN 18800 [1]

	Eurocode 3	DIN 18800
Resistance of class 1, 2 or 3 cross section	$\gamma_{M0} = 1.1$	$\gamma_{M0} = 1.1$
Resistance of class 4 cross section	$\gamma_{M1} = 1.1$	$\gamma_{M1} = 1.1$
Resistance of member to buckling	$\gamma_{M1} = 1.1$	$\gamma_{M1} = 1.1$
Resistance of welded connections	$\gamma_{Mw} = 1.25$	$\gamma_{Mw} = 1.1$
Resistance of bolted connections (of net section at bolt holes)	$\gamma_{Mb} = 1.25$	$\gamma_{Mb} = 1.1$
Resistance of rivetted connections	$\gamma_{Mr} = 1.25$	$\gamma_{Mr} = 1.1$
Resistance of pin connections	$\gamma_{Mp} = 1.25$	$\gamma_{Mp} = 1.1$
Resistance of joints in hollow section lattice girders	$\gamma_{Mj} = 1.1$	$\gamma_{Mj} = 1.1$

according to Eurocode 3 [2] are put together in Table 5-6. Compared to the national standards, such as DIN 18800 [1], differences in some cases may exist.

5.3 Design resistances of cross sections

This section deals with the corresponding verification procedures according to Eurocode 3 [2], which however do not differ fundamentally from those given by the national standards of the countries not belonging to European Union e.g. Canada, U.S.A., Japan, Australia. Additional background informations to some rules have also been added.

5.3.1 Hollow sections in axial force

Eq. 5-10 describes the limiting stresses produced by the axial forces:

$$f_{Rd} = f_y / \gamma_M \quad (5\text{-}10)$$

Axial stresses are to be verified by:

$$\frac{f_{Sd}}{f_{Rd}} \leq 1 \quad (5\text{-}11)$$

For hollow section in axial tension, the design value of the tensile force $N_{t,Sd}$ shall be related to the design tension resistance of the cross section $N_{t,Rd}$ as follows:

$$N_{t,Sd} \leq N_{t,Rd} \quad (5\text{-}12)$$

$N_{t,Rd}$ is to be taken as the smaller of the following values:

- The design plastic resistance of the gross cross section

$$N_{t,Rd} = N_{pl,Rd} = A_g \cdot f_y / \gamma_{M0} \quad (5\text{-}13)$$

Gross cross-sectional areas A_g are based on specified dimensions of hollow sections without deducting holes for fasteners, but making allowances for larger openings.

- The design ultimate resistance of the net cross section at holes for fasteners

$$N_{t,Rd} = N_{u,Rd} = 0.9 \, A_{net} \cdot f_u / \gamma_{Mb} \quad (5\text{-}14)$$

Net cross-sectional areas A_{net} are determined by making appropriate deductions for all holes and other openings from gross cross sectional area A_g.

For connections designed to be slip-resistant at the ultimate limit state (preloaded high strength bolts), the plastic design resistance of the net cross section at holes for fasteners $N_{net,Rd}$ shall not be larger than $A_{net} \cdot f_y / \gamma_{M0}$.

Where ductile behaviour is required, the design ultimate resistance of the net section at fastener holes $N_{u,Rd}$ is not to be less than the design plastic resistance $N_{pl,Rd}$

$$N_{u,Rd} \geq N_{pl,Rd}$$

$$\frac{0.9 \cdot A_{net} \cdot f_u}{\gamma_{Mb}} \geq \frac{A_g \cdot f_y}{\gamma_{M0}} \quad (5\text{-}15)$$

$$\frac{0.9 \cdot A_{net}}{A_g} \geq \frac{f_y}{f_u} \cdot \frac{\gamma_{Mb}}{\gamma_{M0}} \quad (5\text{-}16)$$

5.3 Design resistances of cross sections

For hollow section in axial compression, the design value of the compressive force $N_{c,Sd}$ at each cross section has to satisfy:

$$N_{c,Sd} \leq N_{c,Rd} \quad (5\text{-}17)$$

where $N_{c,Rd}$ is the design compression resistance of the cross section. The smaller of the following values is to be taken for $N_{c,Rd}$:

- The design plastic resistance of the gross cross section

$$N_{pl,Rd} = A_g \cdot f_y / \gamma_{M0} \quad (5\text{-}18)$$

(cross section class 1, 2 and 3)

- The design local buckling resistance of the gross cross section

$$N_{c,Rd} = A_{eff} \cdot f_y / \gamma_{M1} \quad (5\text{-}19)$$

(cross section class 4)

where A_{eff} is the effective area of the cross section based on the effective widths of the compression elements (see Section 5.3.3.2).

Section 5.3.3 deals in detail with the verification of stability of hollow sections in compression against global and local buckling.

Further, it is to be mentioned that no sectional reduction for hollow sections with holes shall be made except when the holes are oversized or they are slotted.

5.3.2 Hollow sections in bending and shear

For simple bending load with or without shear force, there is no fundamental difference between the calculations for the structural hollow sections and the conventional open profiles.

Local buckling of the walls of very thinwalled hollow sections has to be checked in particular (see Section 5.3.3.2).

In general, lateral-torsional buckling resistance need not be checked for circular hollow sections and rectangular hollow sections normally used in practice ($b/h \geq 0{,}5$) due to their very large polar moment of inertia I_t.

5.3.2.1 Design for pure uni-axial bending

In this case, the design value of the bending moment M_{Sd} at each cross section shall satisfy:

$$M_{Sd} \leq M_{Rd}$$

For cross section class 1 and 2 (plastic calculation):

$$M_{Rd} = W_{pl} \cdot f_y / \gamma_{M0} \quad (5\text{-}20)$$

For cross section class 3 (elastic calculation):

$$M_{Rd} = W_{el} \cdot f_y / \gamma_{M0} \quad (5\text{-}21)$$

For cross section class 4 (elastic calculation taking local buckling into account):

$$M_{Rd} = W_{eff} \cdot f_y / \gamma_{M1} \quad (5\text{-}22)$$

where W_{eff} is the effective section modulus basing on the effective widths of the compression elements (see Section 5.3.3.2).

In order to determine the ultimate moment resistance of the net section at bolt holes $M_{u,Rd}$, similar calculation as with Eq. (5-14) is to be made. For hollow sections, only oversized or slotted holes are to be considered.

Verification of lateral-torsional buckling

As has already been stated, it is not in general necessary to verify the lateral-torsional buckling resistance of a beam made of hollow section. The elastic critical moment for lateral-torsional buckling decreases with the increasing beam length l. Table 5.7 shows the beam length to average depth ratio of rectangular hollow sections made of various steel grades, below which no lateral-torsional buckling check is necessary [9].

The following equation has been established using the non-dimensional slenderness for lateral-torsional buckling $\bar{\lambda}_{LT} = \sqrt{\frac{f_y}{f_{cr,LT}}} = 0{,}4$ according to Eurocode 3 [2]:

$$\frac{l}{h-t} \leq \frac{113400}{f_y} \cdot \frac{\alpha_x^2}{1+3\alpha_x} \sqrt{\frac{3+\alpha_x}{1+\alpha_x}} \quad (5\text{-}23)$$

where:
f_y = yield strength of material in N/mm^2
$f_{cr,LT}$ = critical elastic stress for lateral-torsional buckling in N/mm^2

$$\alpha_x = \frac{b-t}{h-t} = \frac{b_m}{h_m}$$

Table 5-7 Limiting $l/(h-t)$ ratio for a rectangular hollow section, below which lateral-torsional buckling need not be checked

α_x	$l/(h-t)$			
	$f_y = 235$ N/mm²	$f_y = 275$ N/mm²	$f_y = 355$ N/mm²	$f_y = 460$ N/mm²
0.5	73.7	63.0	48.8	37.7
0.6	93.1	79.5	61.6	47.5
0.7	112.5	96.2	74.5	57.5
0.8	132.0	112.8	87.4	67.4
0.9	151.3	129.3	100.2	77.3
1.0	170.6	145.8	112.9	87.2

$$\alpha_x = \frac{b-t}{h-t} = \frac{b_m}{h_m}$$

Eq. (5-23) is based on pure bending of a beam (most unfavourable loading case, on the safe side for design) for elastic stress distribution (cross section class 3). However, it is also valid for plastic stress distribution (cross section classes 1 and 2).

5.3.2.2 Combined uni-axial bending and axial force (without considering shear) [2]

In this case, the following criterion is to be satisfied (cross section class 1 and 2):

$$M_{Sd} \leq M_{N,Rd} \qquad (5\text{-}24)$$

where $M_{N,Rd}$ is the reduced design plastic moment resistance allowing for the axial force.

The following approximation equations may be applied:

- For square hollow section:

$$M_{N,Rd} = 1.26\, M_{pl,Rd}(1-n) \leq M_{pl,Rd} \qquad (5\text{-}25)$$

- For rectangular hollow section:

$$M_{N,x,Rd} = 1.33\, M_{pl,x,Rd}(1-n) \leq M_{pl,x,Rd} \qquad (5\text{-}26)$$

$$M_{N,y,Rd} = M_{pl,y,Rd}(1-n)/\left(0.5 + \frac{ht}{A_g}\right) \leq M_{pl,y,Rd} \qquad (5\text{-}27)$$

- For circular hollow sections:

$$M_{N,Rd} = 1.04\, M_{pl,Rd}(1-n^{1.7}) \leq M_{pl,Rd} \qquad (5\text{-}28)$$

If $\frac{V_{Sd}}{V_{pl,Rd}} \leq 0.25$, the following exact equation can be used:

$$\frac{M_{Sd}}{M_{pl,Rd}} \leq \cos\left(\frac{N_{Sd}}{N_{pl,Rd}} \cdot \frac{\pi}{2}\right)$$

where $\sqrt{M_{x,Sd}^2 + M_{y,Sd}^2}$

For Eqs. (5-25) to (5-28),

$$n = \frac{N_{Sd}}{N_{pl,Rd}} = \frac{N_{Sd} \cdot \gamma_{M0}}{A_g \cdot f_y} \qquad (5\text{-}29)$$

For circular hollow sections, the following exact and simple equation [8] is valid instead of Eq. (5-28):

$$\frac{M_{Sd}}{M_{pl,Rd}} \leq \cos\left(\frac{N_{Sd}}{N_{pl,Rd}} \cdot \frac{\pi}{2}\right) \qquad (5\text{-}30)$$

where $M_{Sd} = \sqrt{M_{x,Sd}^2 + M_{y,Sd}^2}$

In this case, the shear force V_{Sd} must be limited to:

$$V_{Sd} \leq V_{pl,Rd} \cdot 0.25 \text{ (for } V_{pl,Rd} \text{ see Eq. (5-41))}$$

5.3.2.3 Combined bi-axial bending and axial force (without considering shear) [2]

For the plastic design in this case (cross section classes 1 and 2), the following conservative approximation may be used:

$$\frac{N_{Sd} \cdot \gamma_{M0}}{A_g \cdot f_y} + \frac{M_{x,Sd} \cdot \gamma_{M0}}{W_{pl,x} \cdot f_y} + \frac{M_{y,Sd} \cdot \gamma_{M0}}{W_{pl,y} \cdot f_y} \leq 1 \qquad (5\text{-}31)$$

5.3 Design resistances of cross sections

or

$$\frac{N_{Sd}}{N_{pl,Rd}} + \frac{M_{x,Sd}}{M_{pl,x,Rd}} + \frac{M_{y,Sd}}{M_{pl,y,Rd}} \le 1 \quad (5\text{-}32)$$

The following approximate criterion may also be applied:

$$\left[\frac{M_{x,Sd}}{M_{N,x,Rd}}\right]^\alpha + \left[\frac{M_{y,Sd}}{M_{N,y,Rd}}\right]^\beta \le 1 \quad (5\text{-}33)$$

where α, β are exponents

For circular hollow sections, $\alpha = \beta = 2$ (5-34)

For square and rectangular hollow sections:

$$\alpha = \beta = \frac{1.66}{1 - 1.13\,n^2} \le 6 \quad (5\text{-}35)$$

where n is according to Eq. (5-29)

For elastic design (cross section class 3), the following simple, linear equation may be applied:

$$\frac{N_{Sd}}{A_g \cdot f_{yd}} + \frac{M_{x,Sd}}{W_{el,x} \cdot f_{yd}} + \frac{M_{y,Sd}}{W_{el,y} \cdot f_{yd}} \le 1 \quad (5\text{-}36)$$

where $f_{yd} = \dfrac{f_y}{\gamma_{M0}}$

Eq. (5-36) may also be used, as a lower bound but more simple to use, for plastic design of cross section classes 1 and 2 instead of Eq. (5-31) or (5-33).

5.3.2.4 Stress design of hollow sections in shear

Fig. 5-2 shows the distribution of shear stress in a circular and a rectangular hollow section loaded by shear force.

In general, the following verification of shear stress τ (elastic design) has to be made:

$$\tau_{max} = \frac{Q(\text{or } V_{Sd}) \cdot \max S}{I \cdot t} \le \tau_{Rd} = \frac{f_y}{\sqrt{3} \cdot \gamma_{M0}}$$

(5-37)

where:

Q or V_{Sd} = shear force
S = static moment in relation to the neutral axis for the area of the section between the free edge and the fibre being considered
I = second moment of inertia for the hollow section
t = wall thickness of the hollow section

τ_{max} can be determined with sufficient degree of precision by using the following equations:

- For circular hollow sections:

$$\tau_{max} = \frac{2Q}{A_g} = \frac{2Q}{d_m \cdot \pi \cdot t} \quad (5\text{-}38)$$

where $d_m = d - t$

- For rectangular hollow sections:

$$\tau_{max} = \frac{Q}{2\,h_m \cdot t} \quad (5\text{-}39)$$

where $h_m = h - t$ (b_m instead of h_m, when shear force is parallel to width b)

5.3.2.5 Design for bending considering shear

Design shear force V_{Sd} in a structural element shall satisfy the following criterion:

$$V_{Sd} \le V_{pl,Rd} \quad (5\text{-}40)$$

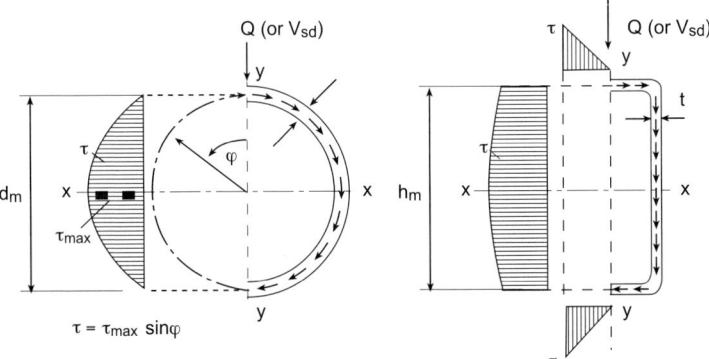

Fig. 5-2 Shear stress distribution in a circular and a rectangular hollow section

$V_{pl,Rd}$ is the design plastic resistance of a cross section to be calculated as follows:

$$V_{pl,Rd} = \frac{A_v \cdot f_y}{\gamma_{M0} \cdot \sqrt{3}} \quad (5\text{-}41)$$

The shear area A_v of hollow sections may be determined using the following equations:

- For rectangular hollow section:

a) $A_v = \dfrac{A_g \cdot h}{b+h}$ (when shear force V_{Sd} acts parallel to depth h)

b) $A_v = \dfrac{A_g \cdot b}{b+h}$ (when shear force V_{Sd} acts parallel to width b)

- For circular hollow section:

$$A_v = \frac{2A_g}{\pi}$$

The design shear force V_{Sd} may be neglected, when the condition $V_{Sd} \leq 0.5\, V_{pl,Rd}$ is fulfilled. This is satisfied in nearly all practical cases.

However, it is to be mentioned that the limiting value for $V_{Sd}/V_{pl,Rd}$, below which the design shear force may be neglected, may be significantly less according to other design standards (see, for example [1]).

For $V_{Sd} > 0.5\, V_{pl,Rd}$, the design resistance of the cross section to combinations of bending moment, axial force and shear force shall be calculated using a reduced yield strength for the shear area.

$$\text{red. } f_y = (1-\varrho)\, f_y \quad (5\text{-}42)$$

$$\text{where } \varrho = \left(2 \cdot \frac{V_{Sd}}{V_{pl,Rd}} - 1\right)^2 \quad (5\text{-}43)$$

Table 5-8 Reduced design plastic moment resistance of hollow sections $M_{NV,Rd}$ in combination with axial and shear forces

Hollow section			$V_{Sd} \leq V_{pl,Rd}$	$V_{Sd} > 0.5\, V_{pl,Rd}$
○		$N_{Sd} \leq 0.25\, N_{pl,Rd}$	$\dfrac{M_{N,V,Rd}}{M_{pl,Rd}} = 1$	$\dfrac{M_{N,V,Rd}}{M_{pl,Rd}} = 1-\varrho$
		$N_{Sd} > 0.25\, N_{pl,Rd}$	$\dfrac{M_{N,V,Rd}}{M_{pl,Rd}} = 1.04\,(1-n^{1.7})$	$\dfrac{M_{N,V,Rd}}{M_{pl,Rd}} = 1.04\left[1-\varrho-\dfrac{n^{1.7}}{(1-\varrho)^{1.7}}\right]$
□		low	$\dfrac{M_{N,V,Rd}}{M_{pl,Rd}} = 1$	$\dfrac{M_{N,V,Rd}}{M_{pl,Rd}} = 1-\varrho$
		high	$\dfrac{M_{N,V,Rd}}{M_{pl,Rd}} = 1.26\,(1-n)$	$\dfrac{M_{N,V,Rd}}{M_{pl,Rd}} = 1.26\,(1-n-\varrho)$
x---x		low	$\dfrac{M_{N,V,Rd}}{M_{pl,x,Rd}} = 1$	$\dfrac{M_{N,V,Rd}}{M_{pl,x,Rd}} = 1-\varrho$
		high	$\dfrac{M_{N,V,Rd}}{M_{pl,x,Rd}} = 1.33\,(1-n)$	$\dfrac{M_{N,V,Rd}}{M_{pl,x,Rd}} = 1.33\,(1-n-\varrho)$
y---y		low	$\dfrac{M_{N,V,Rd}}{M_{pl,y,Rd}} = 1$	$\dfrac{M_{N,V,Rd}}{M_{pl,y,Rd}} = 1-\varrho$
		high	$\dfrac{M_{N,V,Rd}}{M_{pl,y,Rd}} = \dfrac{1-n}{0.5+\dfrac{ht}{A_g}}$	$\dfrac{M_{N,V,Rd}}{M_{pl,y,Rd}} = \dfrac{1-n-\varrho}{0.5+\dfrac{ht}{A_g}}$
		low level } $N_{Sd} \leq 0.25\, N_{pl,Rd}$		high level } $N_{Sd} > 0.25\, N_{pl,Rd}$

$$n = \frac{N_{Sd}}{N_{pl,Rd}} = \frac{N_{Sd}}{\left(\dfrac{A \cdot f_y}{\gamma_{M0}}\right)} \qquad \varrho = \left(2\,\frac{V_{Sd}}{V_{pl,Rd}} - 1\right)^2$$

5.3 Design resistances of cross sections

Table 5-8 shows the reduced design plastic moment resistance $M_{NV,Rd}$ of circular, square and rectangular hollow sections taking axial and shear force into consideration.

For circular hollow sections, the following exact but simple equation may be used without making any reduction of f_y [8] (compare with Eq. 5-42):

$$\frac{M_{Sd}}{M_{pl,Rd}} \leq \eta \cdot \cos\left(\frac{N_{Sd}}{\eta \cdot N_{pl,Rd}} \cdot \frac{\pi}{2}\right) \quad (5\text{-}44)$$

where:

$$\eta = \sqrt{1 - \left(\frac{V_{Sd}}{V_{pl,Rd}}\right)^2} \quad (5\text{-}45)$$

$$V_{Sd} = \sqrt{V_{x,Sd}^2 + V_{y,Sd}^2} \quad (5\text{-}46)$$

$$M_{Sd} = \sqrt{M_{x,Sd}^2 + M_{y,Sd}^2} \quad (5\text{-}47)$$

$V_{pl,Rd}$ according to Eq. (5-41)

5.3.2.6 Equivalent stress for combined axial force and shear

Equivalent stress f_v produced by the combination of axial stresses f_x, f_y and f_z and shear stresses τ_{xy}, τ_{xz} and τ_{yz} shall be calculated by using the following equation:

$$f_v = \sqrt{f_x^2 + f_y^2 + f_z^2 - f_x \cdot f_y - f_x \cdot f_z - f_y \cdot f_z + 3\tau_{xy}^2 + 3\tau_{xz}^2 + 3\tau_{yz}^2} \quad (5\text{-}48)$$

The following verification is to be made:

$$\frac{f_v}{f_{Rd}} \leq 1 \quad (5\text{-}49)$$

where $f_{Rd} = f_y/\gamma_{M0}$

5.3.3 Stability cases – flexural (overall) and local buckling of hollow sections

The phenomenon of buckling of a column under concentric compressive force represents the oldest problem of structural stability, which was already investigated by Euler in 1750 and has been further dealt with to work out various design standards [1, 2] even in the recent time. During failure of a column due to buckling, displacement and torque occur

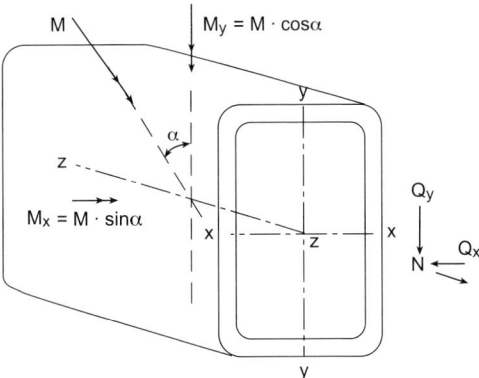

Fig. 5-3 Directions of displacement and torque about the axes of a rectangular hollow section

about the column axis (see Fig. 5-3). They can also take place simultaneously.

It is to be differentiated between the flexural buckling and the lateral-torsional buckling. During flexural buckling, displacement has to be taken into account while torque shall be neglected. Displacement as well as torque have to be considered in the case of lateral-torsional buckling.

As has been already described in Section 5.3.2, lateral-torsional buckling need not be verified for hollow sections in general, since very large torsional rigidity of hollow sections in comparison with that of open sections prevents any torsional buckling. It is sufficient to verify the resistance to flexural buckling.

5.3.3.1 Design for flexural buckling

At the present time, the design resistance to flexural buckling of a column, which is systematically centrally compressed with constant force and consists of straight parts of constant equal cross sections, is determined by using the so called "European buckling curves" in most European countries, also in Canada and Australia [2]. Other standards adopt a single buckling curve, presumably due to the fact that emphasis is given on simplicity. In order to set up these buckling curves, more than one thousand buckling tests on various profile forms (including circular hollow sections) were carried out in Belgium, Germany, France, United Kingdom, the Netherlands and former Yugoslavia [10] with the help of ECCS (**E**uropean **C**onvention for **C**onstructional **S**teelwork). Further tests on circular, square and rectangular hollow sections, three hundred in total, were performed with the support of CIDECT (**C**omité **I**nternational pour le **D**eveloppement et l'**E**tude de la **C**onstruction **T**ubulaire, International Committee for the Development and Study of Tubular Structures) [11, 12]. Later a theoretical study of the buckling phenomenon based on numerical simulation, initiated by ECCS and carried out by Schulz and Beer [13, 14] led to numerical solutions of the equations for buckling taking the structural imperfections (residual stress, non-uniform distribution of yield strength) and the geometrical imperfections (eccentric loading, non-linear member axis) into consideration. The result of these and further investigations was the establishment of five dimensionless buckling curves a_0 and a–d. Each single profile form is assigned to one of these curves in accordance with the distribution of residual stress in it (Fig. 5-4). Table 5-9 shows the selection of the bucking curves a–c for hollow sections, which have residual stresses of different magnitudes depending on their different manufacturing processes.

For hot finished hollow sections in compression made of steel grade S 460, the higher buckling curve a_0 can be applied. This has been confirmed by new tests [16] and numerical investigations and is based on the fact that

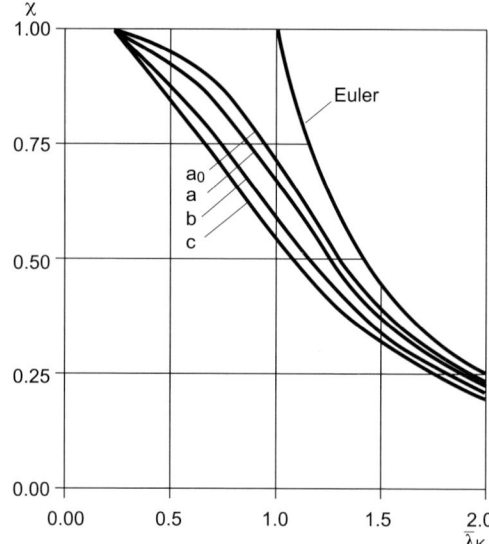

Fig. 5-4 European buckling curves

the buckling behaviour of sections of high strength steel is less influenced by the structural and geometrical imperfections than those of conventional structural steels.

The verification of the design buckling load of a hollow section column in axial compression (cross section classes 1, 2 or 3) shall be made using the following relationship:

$$N_{c,Sd} \leq N_{K,Rd} \qquad (5\text{-}50)$$

where:
$N_{c,Sd}$ = Design compressive force on the column (γ_F times service load)
$N_{K,Rd}$ = Design buckling resistance of the column

Table 5-9 European buckling curves for hollow sections according to manufacturing process [1, 2]

Cross section	Manufacturing process	Buckling curve
	hot forming	a
	cold forming (f_{yb} used)	b
	cold forming (f_{ya} used)	c

f_{yb} = yield strength of the basic (not cold formed) material.
f_{ya} = yield strength of the material after cold forming (see Table 3-14).

5.3 Design resistances of cross sections

Table 5-10a Buckling curve a_o – reduction factor χ

$\bar{\lambda}_K$	0	1	2	3	4	5	6	7	8	9
0.00	1.0000	1.0000	1.0000	1.0000	1.0000	1.0000	1.0000	1.0000	1.0000	1.0000
0.10	1.0000	1.0000	1.0000	1.0000	1.0000	1.0000	1.0000	1.0000	1.0000	1.0000
0.20	1.0000	0.9986	0.9973	0.9959	0.9945	0.9931	0.9917	0.9903	0.9889	0.9874
0.30	0.9859	0.9845	0.9829	0.9814	0.9799	0.9783	0.9767	0.9751	0.9735	0.9718
0.40	0.9701	0.9684	0.9667	0.9649	0.9631	0.9612	0.9593	0.9574	0.9554	0.9534
0.50	0.9513	0.9492	0.9470	0.9448	0.9425	0.9402	0.9378	0.9354	0.9328	0.9302
0.60	0.9276	0.9248	0.9220	0.9191	0.9161	0.9130	0.9099	0.9066	0.9032	0.8997
0.70	0.8961	0.8924	0.8886	0.8847	0.8806	0.8764	0.8721	0.8676	0.8630	0.8582
0.80	0.8533	0.8483	0.8431	0.8377	0.8322	0.8266	0.8208	0.8148	0.8087	0.8025
0.90	0.7961	0.7895	0.7828	0.7760	0.7691	0.7620	0.7549	0.7476	0.7403	0.7329
1.00	0.7253	0.7178	0.7101	0.7025	0.6948	0.6870	0.6793	0.6715	0.6637	0.6560
1.10	0.6482	0.6405	0.6329	0.6252	0.6176	0.6101	0.6026	0.5951	0.5877	0.5804
1.20	0.5732	0.5660	0.5590	0.5520	0.5450	0.5382	0.5314	0.5248	0.5182	0.5117
1.30	0.5053	0.4990	0.4927	0.4866	0.4806	0.4746	0.4687	0.4629	0.4572	0.4516
1.40	0.4461	0.4407	0.4353	0.4300	0.4248	0.4197	0.4147	0.4097	0.4049	0.4001
1.50	0.3953	0.3907	0.3861	0.3816	0.3772	0.3728	0.3685	0.3643	0.3601	0.3560
1.60	0.3520	0.3480	0.3441	0.3403	0.3365	0.3328	0.3291	0.3255	0.3219	0.3184
1.70	0.3150	0.3116	0.3083	0.3050	0.3017	0.2985	0.2954	0.2923	0.2892	0.2862
1.80	0.2833	0.2804	0.2775	0.2746	0.2719	0.2691	0.2664	0.2637	0.2611	0.2585
1.90	0.2559	0.2534	0.2509	0.2485	0.2461	0.2437	0.2414	0.2390	0.2368	0.2345
2.00	0.2323	0.2301	0.2280	0.2258	0.2237	0.2217	0.2196	0.2176	0.2156	0.2136
2.10	0.2117	0.2098	0.2079	0.2061	0.2042	0.2024	0.2006	0.1989	0.1971	0.1954
2.20	0.1937	0.1920	0.1904	0.1887	0.1871	0.1855	0.1840	0.1824	0.1809	0.1794
2.30	0.1779	0.1764	0.1749	0.1735	0.1721	0.1707	0.1693	0.1679	0.1665	0.1652
2.40	0.1639	0.1626	0.1613	0.1600	0.1587	0.1575	0.1563	0.1550	0.1538	0.1526
2.50	0.1515	0.1503	0.1491	0.1480	0.1469	0.1458	0.1447	0.1436	0.1425	0.1414
2.60	0.1404	0.1394	0.1383	0.1373	0.1363	0.1353	0.1343	0.1333	0.1324	0.1314
2.70	0.1305	0.1296	0.1286	0.1277	0.1268	0.1259	0.1250	0.1242	0.1233	0.1224
2.80	0.1216	0.1207	0.1199	0.1191	0.1183	0.1175	0.1167	0.1159	0.1151	0.1143
2.90	0.1136	0.1128	0.1120	0.1113	0.1106	0.1098	0.1091	0.1084	0.1077	0.1070
3.00	0.1063	0.1056	0.1049	0.1043	0.1036	0.1029	0.1023	0.1016	0.1010	0.1003
3.10	0.0997	0.0991	0.0985	0.0979	0.0972	0.0966	0.0960	0.0955	0.0949	0.0943
3.20	0.0937	0.0931	0.0926	0.0920	0.0915	0.0909	0.0904	0.0898	0.0893	0.0888
3.30	0.0882	0.0877	0.0872	0.0867	0.0862	0.0857	0.0852	0.0847	0.0842	0.0837
3.40	0.0832	0.0828	0.0823	0.0818	0.0814	0.0809	0.0804	0.0800	0.0795	0.0791
3.50	0.0786	0.0782	0.0778	0.0773	0.0769	0.0765	0.0761	0.0756	0.0752	0.0748
3.60	0.0744	0.0740	0.0736	0.0732	0.0728	0.0724	0.0720	0.0717	0.0713	0.0709

Table 5-10b Buckling curve a – reduction factor χ

$\bar{\lambda}_K$	0	1	2	3	4	5	6	7	8	9
0.00	1.0000	1.0000	1.0000	1.0000	1.0000	1.0000	1.0000	1.0000	1.0000	1.0000
0.10	1.0000	1.0000	1.0000	1.0000	1.0000	1.0000	1.0000	1.0000	1.0000	1.0000
0.20	1.0000	0.9978	0.9956	0.9934	0.9912	0.9889	0.9867	0.9844	0.9821	0.9798
0.30	0.9775	0.9751	0.9728	0.9704	0.9680	0.9655	0.9630	0.9605	0.9580	0.9554
0.40	0.9528	0.9501	0.9474	0.9447	0.9419	0.9391	0.9363	0.9333	0.9304	0.9273
0.50	0.9243	0.9211	0.9179	0.9147	0.9114	0.9080	0.9045	0.9010	0.8974	0.8937
0.60	0.8900	0.8862	0.8823	0.8783	0.8742	0.8700	0.8657	0.8614	0.8569	0.8524
0.70	0.8477	0.8430	0.8382	0.8332	0.8282	0.8230	0.8178	0.8124	0.8069	0.8014
0.80	0.7957	0.7899	0.7841	0.7781	0.7721	0.7659	0.7597	0.7534	0.7470	0.7405
0.90	0.7339	0.7273	0.7206	0.7139	0.7071	0.7003	0.6934	0.6865	0.6796	0.6726
1.00	0.6656	0.6586	0.6516	0.6446	0.6376	0.6306	0.6236	0.6167	0.6098	0.6029
1.10	0.5960	0.5892	0.5824	0.5757	0.5690	0.5623	0.5557	0.5492	0.5427	0.5363
1.20	0.5300	0.5237	0.5175	0.5114	0.5053	0.4993	0.4934	0.4875	0.4817	0.4760
1.30	0.4703	0.4648	0.4593	0.4538	0.4485	0.4432	0.4380	0.4329	0.4278	0.4228
1.40	0.4179	0.4130	0.4083	0.4036	0.3989	0.3943	0.3898	0.3854	0.3810	0.3767
1.50	0.3724	0.3682	0.3641	0.3601	0.3561	0.3521	0.3482	0.3444	0.3406	0.3369
1.60	0.3332	0.3296	0.3261	0.3226	0.3191	0.3157	0.3124	0.3091	0.3058	0.3026
1.70	0.2994	0.2963	0.2933	0.2902	0.2872	0.2843	0.2814	0.2786	0.2757	0.2730
1.80	0.2702	0.2675	0.2649	0.2623	0.2597	0.2571	0.2546	0.2522	0.2497	0.2473

Table 5-10b (continued)

$\bar{\lambda}_K$	0	1	2	3	4	5	6	7	8	9
1.90	0.2449	0.2426	0.2403	0.2380	0.2358	0.2335	0.2314	0.2292	0.2271	0.2250
2.00	0.2229	0.2209	0.2188	0.2168	0.2149	0.2129	0.2110	0.2091	0.2073	0.2054
2.10	0.2036	0.2018	0.2001	0.1983	0.1966	0.1949	0.1932	0.1915	0.1899	0.1883
2.20	0.1867	0.1851	0.1836	0.1820	0.1805	0.1790	0.1775	0.1760	0.1746	0.1732
2.30	0.1717	0.1704	0.1690	0.1676	0.1663	0.1649	0.1636	0.1623	0.1610	0.1598
2.40	0.1585	0.1573	0.1560	0.1548	0.1536	0.1524	0.1513	0.1501	0.1490	0.1478
2.50	0.1467	0.1456	0.1445	0.1434	0.1424	0.1413	0.1403	0.1392	0.1382	0.1372
2.60	0.1362	0.1352	0.1342	0.1332	0.1323	0.1313	0.1304	0.1295	0.1285	0.1276
2.70	0.1267	0.1258	0.1250	0.1241	0.1232	0.1224	0.1215	0.1207	0.1198	0.1190
2.80	0.1182	0.1174	0.1166	0.1158	0.1150	0.1143	0.1135	0.1128	0.1120	0.1113
2.90	0.1105	0.1098	0.1091	0.1084	0.1077	0.1070	0.1063	0.1056	0.1049	0.1042
3.00	0.1036	0.1029	0.1022	0.1016	0.1010	0.1003	0.0997	0.0991	0.0985	0.0978
3.10	0.0972	0.0966	0.0960	0.0954	0.0949	0.0943	0.0937	0.0931	0.0926	0.0920
3.20	0.0915	0.0909	0.0904	0.0898	0.0893	0.0888	0.0882	0.0877	0.0872	0.0867
3.30	0.0862	0.0857	0.0852	0.0847	0.0842	0.0837	0.0832	0.0828	0.0823	0.0818
3.40	0.0814	0.0809	0.0804	0.0800	0.0795	0.0791	0.0786	0.0782	0.0778	0.0773
3.50	0.0769	0.0765	0.0761	0.0757	0.0752	0.0748	0.0744	0.0740	0.0736	0.0732
3.60	0.0728	0.0724	0.0721	0.0717	0.0713	0.0709	0.0705	0.0702	0.0698	0.0694

Table 5-10c Buckling curve b – reduction factor χ

$\bar{\lambda}_K$	0	1	2	3	4	5	6	7	8	9
0.00	1.0000	1.0000	1.0000	1.0000	1.0000	1.0000	1.0000	1.0000	1.0000	1.0000
0.10	1.0000	1.0000	1.0000	1.0000	1.0000	1.0000	1.0000	1.0000	1.0000	1.0000
0.20	1.0000	0.9965	0.9929	0.9894	0.9858	0.9822	0.9786	0.9750	0.9714	0.9678
0.30	0.9641	0.9604	0.9567	0.9530	0.9492	0.9455	0.9417	0.9378	0.9339	0.9300
0.40	0.9261	0.9221	0.9181	0.9140	0.9099	0.9057	0.9015	0.8973	0.8930	0.8886
0.50	0.8842	0.8798	0.8752	0.8707	0.8661	0.8614	0.8566	0.8518	0.8470	0.8420
0.60	0.8371	0.8320	0.8269	0.8217	0.8165	0.8112	0.8058	0.8004	0.7949	0.7893
0.70	0.7837	0.7780	0.7723	0.7665	0.7606	0.7547	0.7488	0.7428	0.7367	0.7306
0.80	0.7245	0.7183	0.7120	0.7058	0.6995	0.6931	0.6868	0.6804	0.6740	0.6676
0.90	0.6612	0.6547	0.6483	0.6419	0.6354	0.6290	0.6226	0.6162	0.6098	0.6034
1.00	0.5970	0.5907	0.5844	0.5781	0.5719	0.5657	0.5595	0.5534	0.5473	0.5412
1.10	0.5352	0.5293	0.5234	0.5175	0.5117	0.5060	0.5003	0.4947	0.4891	0.4836
1.20	0.4781	0.4727	0.4674	0.4621	0.4569	0.4517	0.4466	0.4416	0.4366	0.4317
1.30	0.4269	0.4221	0.4174	0.4127	0.4081	0.4035	0.3991	0.3946	0.3903	0.3860
1.40	0.3817	0.3775	0.3734	0.3693	0.3653	0.3613	0.3574	0.3535	0.3497	0.3459
1.50	0.3422	0.3386	0.3350	0.3314	0.3279	0.3245	0.3211	0.3177	0.3144	0.3111
1.60	0.3079	0.3047	0.3016	0.2985	0.2955	0.2925	0.2895	0.2866	0.2837	0.2809
1.70	0.2781	0.2753	0.2726	0.2699	0.2672	0.2646	0.2620	0.2595	0.2570	0.2545
1.80	0.2521	0.2496	0.2473	0.2449	0.2426	0.2403	0.2381	0.2359	0.2337	0.2315
1.90	0.2294	0.2272	0.2252	0.2231	0.2211	0.2191	0.2171	0.2152	0.2132	0.2113
2.00	0.2095	0.2076	0.2058	0.2040	0.2022	0.2004	0.1987	0.1970	0.1953	0.1936
2.10	0.1920	0.1903	0.1887	0.1871	0.1855	0.1840	0.1825	0.1809	0.1794	0.1780
2.20	0.1765	0.1751	0.1736	0.1722	0.1708	0.1694	0.1681	0.1667	0.1654	0.1641
2.30	0.1628	0.1615	0.1602	0.1590	0.1577	0.1565	0.1553	0.1541	0.1529	0.1517
2.40	0.1506	0.1494	0.1483	0.1472	0.1461	0.1450	0.1439	0.1428	0.1418	0.1407
2.50	0.1397	0.1387	0.1376	0.1366	0.1356	0.1347	0.1337	0.1327	0.1318	0.1308
2.60	0.1299	0.1290	0.1281	0.1272	0.1263	0.1254	0.1245	0.1237	0.1228	0.1219
2.70	0.1211	0.1203	0.1195	0.1186	0.1178	0.1170	0.1162	0.1155	0.1147	0.1139
2.80	0.1132	0.1124	0.1117	0.1109	0.1102	0.1095	0.1088	0.1081	0.1074	0.1067
2.90	0.1060	0.1053	0.1046	0.1039	0.1033	0.1026	0.1020	0.1013	0.1007	0.1001
3.00	0.0994	0.0988	0.0982	0.0976	0.0970	0.0964	0.0958	0.0952	0.0946	0.0940
3.10	0.0935	0.0929	0.0924	0.0918	0.0912	0.0907	0.0902	0.0896	0.0891	0.0886
3.20	0.0880	0.0875	0.0870	0.0865	0.0860	0.0855	0.0850	0.0845	0.0840	0.0835
3.30	0.0831	0.0826	0.0821	0.0816	0.0812	0.0807	0.0803	0.0798	0.0794	0.0789
3.40	0.0785	0.0781	0.0776	0.0772	0.0768	0.0763	0.0759	0.0755	0.0751	0.0747
3.50	0.0743	0.0739	0.0735	0.0731	0.0727	0.0723	0.0719	0.0715	0.0712	0.0708
3.60	0.0704	0.0700	0.0697	0.0693	0.0689	0.0686	0.0682	0.0679	0.0675	0.0672

5.3 Design resistances of cross sections

Table 5-10d Buckling curve c – reduction factor χ

$\bar{\lambda}_K$	0	1	2	3	4	5	6	7	8	9
0.00	1.0000	1.0000	1.0000	1.0000	1.0000	1.0000	1.0000	1.0000	1.0000	1.0000
0.10	1.0000	1.0000	1.0000	1.0000	1.0000	1.0000	1.0000	1.0000	1.0000	1.0000
0.20	1.0000	0.9949	0.9898	0.9847	0.9797	0.9746	0.9695	0.9644	0.9593	0.9542
0.30	0.9491	0.9440	0.9389	0.9338	0.9286	0.9235	0.9183	0.9131	0.9078	0.9026
0.40	0.8973	0.8920	0.8867	0.8813	0.8760	0.8705	0.8651	0.8596	0.8541	0.8486
0.50	0.8430	0.8374	0.8317	0.8261	0.8204	0.8146	0.8088	0.8030	0.7972	0.7913
0.60	0.7854	0.7794	0.7735	0.7675	0.7614	0.7554	0.7493	0.7432	0.7370	0.7309
0.70	0.7247	0.7185	0.7123	0.7060	0.6998	0.6935	0.6873	0.6810	0.6747	0.6684
0.80	0.6622	0.6559	0.6496	0.6433	0.6371	0.6308	0.6246	0.6184	0.6122	0.6060
0.90	0.5998	0.5937	0.5876	0.5815	0.5755	0.5695	0.5635	0.5575	0.5516	0.5458
1.00	0.5399	0.5342	0.5284	0.5227	0.5171	0.5115	0.5059	0.5004	0.4950	0.4896
1.10	0.4842	0.4790	0.4737	0.4685	0.4634	0.4583	0.4533	0.4483	0.4434	0.4386
1.20	0.4338	0.4290	0.4243	0.4197	0.4151	0.4106	0.4061	0.4017	0.3974	0.3931
1.30	0.3888	0.3846	0.3805	0.3764	0.3724	0.3684	0.3644	0.3606	0.3567	0.3529
1.40	0.3492	0.3455	0.3419	0.3383	0.3348	0.3313	0.3279	0.3245	0.3211	0.3178
1.50	0.3145	0.3113	0.3081	0.3050	0.3019	0.2989	0.2959	0.2929	0.2900	0.2871
1.60	0.2842	0.2814	0.2786	0.2759	0.2732	0.2705	0.2679	0.2653	0.2627	0.2602
1.70	0.2577	0.2553	0.2528	0.2504	0.2481	0.2457	0.2434	0.2412	0.2389	0.2367
1.80	0.2345	0.2324	0.2302	0.2281	0.2260	0.2240	0.2220	0.2200	0.2180	0.2161
1.90	0.2141	0.2122	0.2104	0.2085	0.2067	0.2049	0.2031	0.2013	0.1996	0.1979
2.00	0.1962	0.1945	0.1929	0.1912	0.1896	0.1880	0.1864	0.1849	0.1833	0.1818
2.10	0.1803	0.1788	0.1774	0.1759	0.1745	0.1731	0.1717	0.1703	0.1689	0.1676
2.20	0.1662	0.1649	0.1636	0.1623	0.1611	0.1598	0.1585	0.1573	0.1561	0.1549
2.30	0.1537	0.1525	0.1514	0.1502	0.1491	0.1480	0.1468	0.1457	0.1446	0.1436
2.40	0.1425	0.1415	0.1404	0.1394	0.1384	0.1374	0.1364	0.1354	0.1344	0.1334
2.50	0.1325	0.1315	0.1306	0.1297	0.1287	0.1278	0.1269	0.1260	0.1252	0.1243
2.60	0.1234	0.1226	0.1217	0.1209	0.1201	0.1193	0.1184	0.1176	0.1168	0.1161
2.70	0.1153	0.1145	0.1137	0.1130	0.1122	0.1115	0.1108	0.1100	0.1093	0.1086
2.80	0.1079	0.1072	0.1065	0.1058	0.1051	0.1045	0.1038	0.1031	0.1025	0.1018
2.90	0.1012	0.1006	0.0999	0.0993	0.0987	0.0981	0.0975	0.0969	0.0963	0.0957
3.00	0.0951	0.0945	0.0939	0.0934	0.0928	0.0922	0.0917	0.0911	0.0906	0.0901
3.10	0.0895	0.0890	0.0885	0.0879	0.0874	0.0869	0.0864	0.0859	0.0854	0.0849
3.20	0.0844	0.0839	0.0835	0.0830	0.0825	0.0820	0.0816	0.0811	0.0806	0.0802
3.30	0.0797	0.0793	0.0789	0.0784	0.0780	0.0775	0.0771	0.0767	0.0763	0.0759
3.40	0.0754	0.0750	0.0746	0.0742	0.0738	0.0734	0.0730	0.0726	0.0722	0.0719
3.50	0.0715	0.0711	0.0707	0.0703	0.0700	0.0696	0.0692	0.0689	0.0685	0.0682
3.60	0.0678	0.0675	0.0671	0.0668	0.0664	0.0661	0.0657	0.0654	0.0651	0.0647

Further,

$$N_{K,Rd} = \chi \cdot N_{pl,Rd} = \chi \cdot \frac{A_g \cdot f_y}{\gamma_{M1}} \quad (5\text{-}51)$$

where:

χ = reduction factor (to be taken from Fig. 5-4 or Table 5-10 or to be derived from Eq. (5-54)), expresses the relationship between the design buckling resistance $N_{K,Rd}$ and the design plastic resistance $N_{pl,Rd}$
A_g = total area of the cross section
f_y = yield strength of the material used
γ_{M1} = partial safety factor for buckling resistance

Further derivation leads to:

$$\chi = \frac{N_{K,Rd}}{N_{pl,Rd}} = \frac{f_{K,Rd}}{f_y/\gamma_{M1}} \quad (5\text{-}52)$$

where:

$f_{K,Rd}$ = design cross-sectional buckling stress
$= \dfrac{N_{K,Rd}}{A_g}$

χ is dependent on the dimensionless slenderness $\bar{\lambda}$ and the buckling curves (see Fig. 5-4) as well as on decisive stability case (χ_{min} according to I_{min}).

The dimensionless slenderness of a column shall be determined as follows:

$$\bar{\lambda}_K = \frac{\lambda_K}{\lambda_E} = \sqrt{\frac{N_{pl,Rd}}{N_{KE}}} = \sqrt{\frac{f_y/\gamma_{M1}}{f_{KE}}} \qquad (5\text{-}53)$$

where:

λ_K = slenderness = $\dfrac{l_K}{i}$

$\lambda_E = \pi \cdot \sqrt{\dfrac{E}{f_y}}$ = Eulerian slenderness, dependent on steel grade used (see Table 5-11)

l_K = effective buckling length of column (see Section 5.3.3.4)

i = radius of gyration of column cross section = $\sqrt{\dfrac{I_{min}}{A_g}}$

I = second moment of inertia of column cross section

E = modulus of elasticity of structural steel = 210 000 N/mm²

A_g = cross sectional area of column

$N_{KE} = \dfrac{\pi^2 \cdot E \cdot I}{l_K^2}$ = Eulerian buckling load

$f_{KE} = \dfrac{\pi^2 \cdot E}{\lambda_K^2}$ = Eulerian buckling stress

As the European buckling curves are based on tests on columns and simulation calculations, there are no exact analytical formulations for them, but only tabled values. The following formula has however prevailed:

$$\chi = \frac{1}{\phi + \sqrt{\phi^2 - \bar{\lambda}_K^2}}, \text{ however } \chi \leq 1 \qquad (5\text{-}54)^*$$

where:

$\phi = 0.5\,[1 + \alpha\,(\bar{\lambda}_K - 0.2) + \bar{\lambda}_K^2]$

α = Imperfection factor for the buckling curves to be taken from Table 5-12

According to the methods used in U.S.A. and Japan, the results differ only slightly from those obtained from the European buckling curves. The differences between the buckling curves used in various countries of the world are described in [15]. Similar to [1, 2], several

---

* According to DIN 18800 [1], part 2:

For $\bar{\lambda}_K \leq 0{,}2$, $\chi = 1$

for $\bar{\lambda}_K > 0{,}2$, $\chi = \dfrac{1}{\phi + \sqrt{\phi^2 - \bar{\lambda}_K^2}}$

for $\bar{\lambda}_K > 3{,}0$, $\chi = 1/\bar{\lambda}_K\,(\bar{\lambda}_K + \alpha)$

Table 5-11 Eulerian slenderness λ_E in relation with structural steel grade

Steel grade	S 235	S 275	S 355	S 460
f_y (N/mm²)	235	275	355	460
λ_E	93.9	86.8	76.4	67.1

Table 5-12 Imperfection factor α for various buckling curves for hollow sections

Buckling factor	a_0	a	b	c
Imperfection factor α	0.13	0.21	0.34	0.49

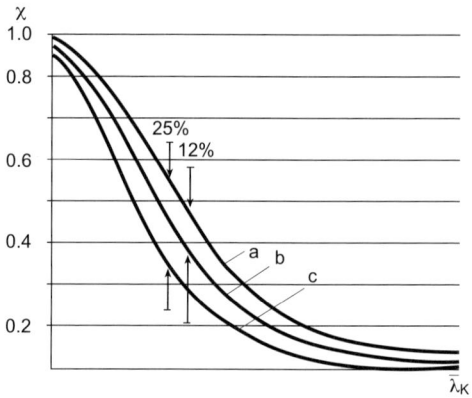

Fig. 5-5 The curve a shows up to 25% higher χ value than the curve c and up to 12% higher χ value than the curve b

buckling curves are used in Australia and Canada. A number of design standards applies only one buckling curve for the sake of simplicity. The differences in the middle zone of the dimensionless slenderness $\bar{\lambda}_K$ for the curves a, b and c are illustrated in Fig. 5-5.

5.3.3.2 Local buckling of hollow sections

Circular, square or rectangular hollow sections with much smaller wall thicknesses than the corresponding diameter d, width b or depth h $(d \gg t, b \gg t, h \gg t)$ have to be considered as thin-walled structural elements. In addition to the verification of the design elastic or plastic resistances and the resistances for flexural buckling, the local stability of the walls against buckling has to be investigated in this case. Local buckling causes significant

5.3 Design resistances of cross sections

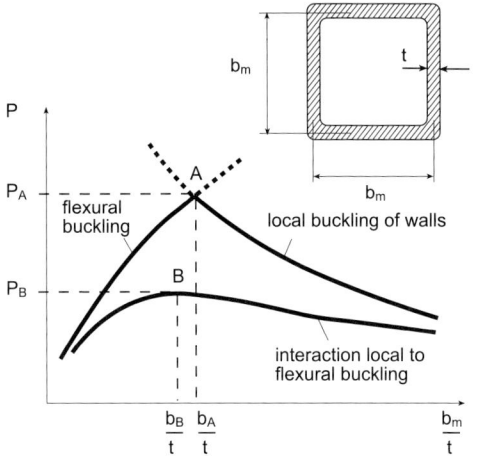

Fig. 5-6 Comparison of the failure loads P of a square hollow section column with constant cross sectional area $A_g = 4 \cdot b_m \cdot t$ with varying b_m/t ratio

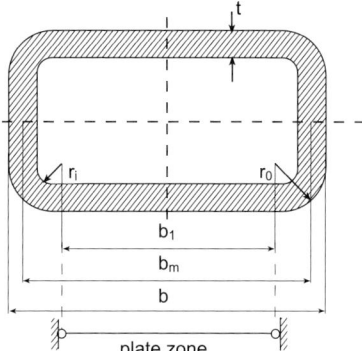

Fig. 5-7 Plate buckling of a rectangular hollow section. Definitions: $b_1 = b - 2(t + r_i)$ where r_i = internal corner radius

loss of design resistance and may lead to premature failure. The unavoidable imperfections of the profile involve an interaction between local buckling in the cross section and flexural buckling of the column. This decreases the resistances to both types of buckling. Due to this interaction, a significantly lower design buckling resistance may result in certain ranges of dimensionless local buckling slenderness $\bar{\lambda}_B$ and of dimensionless flexural buckling slenderness $\bar{\lambda}_K$ than that obtained by separate calculation of local and flexural buckling resistances. This is illustrated in Fig. 5-6 taking a square hollow section as an example.

Local buckling of square and rectangular hollow sections [2, 8, 9]

Square and rectangular hollow sections in compression, bending or shear may buckle locally in their flat areas. The magnitude of this buckling load depends on the ratio width b or depth h/wall thickness t. This is a case of the occurrence of plate buckling (see Fig. 5-7).

For sole compression or shear on a plate zone, the ideal plate buckling stress is given by:

$$f_{B,P\text{ideal}} \text{ (or } \tau_{B,P\text{ideal}}) = k_P \text{ (or } k_\tau) \cdot \frac{\pi^2 \cdot E}{12(1-v^2)} \cdot \left(\frac{t}{b_1}\right)^2 \quad (5\text{-}55)$$

Table 5-13 Plate buckling factor k_P and k_τ

RHS	Stress distribution (compression positive)	Plate buckling factor k_P or k_τ
	compression rectangle with b_1	4.0
	triangular with + only	7.81
	bending with $+$ and $-f$	23.9
	shear τ	5.34

$f_{B,P\text{ideal}}$ or $\tau_{B,P\text{ideal}}$ = Eulerian stress for pin-ended plate strip, bending rigidity is to be substituted by plate rigidity

where k_p or k_τ is plate buckling factor depending on:
- stress type and stress distribution (see Table 5-13)
- side length ratio α of the buckling zone (for long, unstiffened hollow section, $\alpha = \infty$)
- support condition: pin-ended on both sides according to Fig. 5-7, edge restraint has not been considered in the calculation

E = modulus of elasticity (210 000 N/mm² for structural steels)
v = transversal contraction coefficient (Poisson's ratio = 0.3 for steel)
t = wall thickness

b_1 = width of the plate zone (in some design standards, the average width b_m or the full width b is used instead of b_1)

Analogous to flexural buckling, the plate buckling slenderness $\lambda_{B,P}$ may be defined as:

$$\lambda_{B,P} = \pi \cdot \sqrt{\frac{E}{f_{B,P}}} \quad (5\text{-}56)$$

Referring to the Eulerian slenderness $\lambda_E = \pi \cdot \sqrt{\frac{E}{f_y}}$ (f_y = yield strength of material), the dimensionless plate buckling slenderness $\bar{\lambda}_{B,P}$ can be obtained as follows:

$$\bar{\lambda}_{B,P} = \frac{\lambda_{B,P}}{\lambda_E} = \sqrt{\frac{f_y}{f_{B,P}}} = \sqrt{\frac{1}{\bar{f}_{B,P}}} \quad (5\text{-}57)$$

where:

$$\bar{f}_{B,P} = \frac{f_{B,P}}{f_y} = \text{dimensionless plate buckling stress}$$

Basing on Eqs. (5-55) and (5-57), the dimensionless plate buckling slenderness can be derived:

$$\bar{\lambda}_{B,P} = \frac{1}{0.95} \cdot \frac{1}{\sqrt{k_P}} \cdot \sqrt{\frac{f_y}{E}} \cdot \frac{b_1}{t} \quad (5\text{-}58)$$

The ideal plate buckling stress $f_{B,P,\text{ideal}}$ is valid for the supercritical, elastic zone. A reduced buckling stress $f_{B,P}$ is to be determined at the elastic-plastic transition (see Fig. 5-8). While the German guideline 012 (DASt-Ri 012 [19]) uses the Euler-curve according to Eq. (5-57) in the supercritical (elastic) range, which joins $\bar{f}_{B,P} = \frac{f_{B,P}}{f_y} = 0.6$ and $\bar{f}_{B,P} = 1.0$ with a straight line, other rules recommend through curves up to $\bar{f}_{B,P} = 1.0$, which lie clearly above the Euler-curve in the slender zone (that means large b/t-ratio). This is also due to the fact that the thin plates, especially with large side length ratio α for the buckling zone, have large supercritical reserves.

The dimensionless limiting plate buckling slenderness $\bar{\lambda}_{B,P,\text{limit}}$ is of special importance, below which yielding occurs before buckling, that means $\bar{f}_{B,P} = 1.0$. This limits the higher b_1/t range, where the buckling need not be checked.

Max. b_1/t is to be determined using Eq. (5-58):

$$\max. \frac{b_1}{t} = 0.95 \cdot \bar{\lambda}_{B,P,\text{limit}} \cdot \sqrt{k_P} \cdot \sqrt{\frac{E}{f_y}} \quad (5\text{-}59)$$

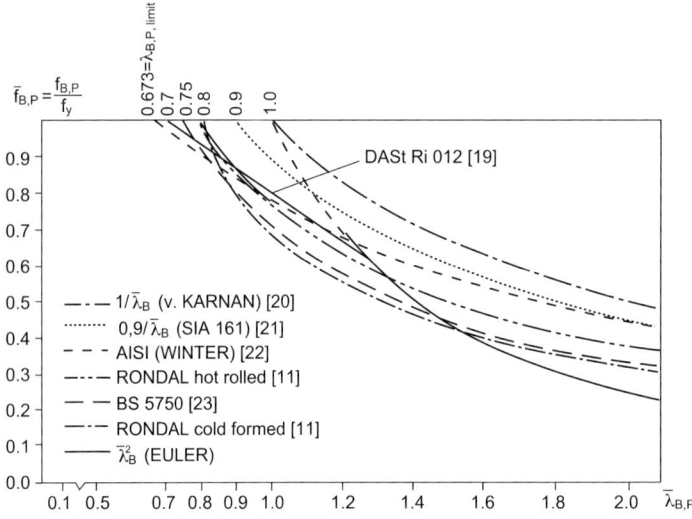

Fig. 5-8 Dimensionless plate buckling stress $\bar{f}_{B,P}$ in relation with dimensionless plate buckling slenderness $\bar{\lambda}_{B,P}$ for RHS, $\bar{\lambda}_{B,P} = \frac{1}{1.9}\sqrt{\frac{f_y}{E}} \cdot \frac{b_1}{t}$

5.3 Design resistances of cross sections

For $k_p = 4$ (concentric compression or compressed flange in bending, see Table 5-13),

$$\max. \frac{b_1}{t} = 1.9 \cdot \bar{\lambda}_{B,P,\text{limit}} \cdot \sqrt{\frac{E}{f_y}} \qquad (5\text{-}60)$$

A number of guidelines defines the limiting plate buckling slenderness using full or average width (b or b_m) of RHS. Fig. 5-8 shows the buckling curves for the dimensionless plate buckling stress $\bar{f}_{B,P}$ according to various guidelines [20–23].

Table 5-14 contains the maximum width/wall thickness-ratios for the transition $\bar{\lambda}_{B,P,\text{limit}}$ from plastic to elastic-plastic range, recommended by several design standards (for constant compressive stress, $k_p = 4$).

Table 5-14 Maximum width/wall thickness – ratios for RHS without plastification, when local buckling need not be checked ($k_P = 4$)

	$\bar{\lambda}_{B,P,\text{limit}}$	Maximum $\frac{\text{width}}{\text{wall thickness}}$-ratio when buckling check is not necessary	
		Value	Reference
ECCS 1978 [24]	0.8	$1.52\sqrt{\frac{E}{f_y}}$	b_m/t
SIA 161, 1979 [21]	0.9	$1.71\sqrt{\frac{E}{f_y}}$	b_m/t
DASt-Ri 012, 1980 [19]	0.7	$1.33\sqrt{\frac{E}{f_y}}$	b_1/t

Eurocode 3 [2] gives the b/t or h/t ratios for RHS of the cross section classes 1, 2 and 3, where the cross sections or their parts do not buckle locally before reaching the limiting design resistance (see Tables 5-15 to 5-17).

Compression resistance of rectangular hollow section column belonging to cross sectional class 4

In case the wall thickness of a hollow section is thinner than that of hollow sections belonging to the cross section class 1, 2 or 3 (see Table 5-17), it falls in the cross section class 4. In order to determine its design resistance, local buckling has to be taken into account.

The design resistance of a hollow section of cross section class 4 is to be calculated using the following equation:

$$N_{B,Rd} = \chi \cdot \beta_A \cdot A_g \cdot f_y / \gamma_{M1} \qquad (5\text{-}61)$$

where:

$$\beta_A = A_{\text{eff}} / A_g \qquad (5\text{-}62)$$

The effective sectional properties (A_{eff}, i_{eff}, W_{eff}) of hollow sections belonging to cross section class 4 under compression, bending and flexural buckling are determined using the effective (reduced) width b_{eff} (see Table 5-18). The effective width arises by the increasing disengagement of the fibers in the middle of plate zone (buckling field) from the load application. A simplified calculation model is used assuming that the uneven distribution of stresses, which occurs when a wall area is deformed by local buckling, is taken into account.

This is done by basing the calculation on the stresses present at the edges (corners) uniformly applied over a width ($b_{e1} + b_{e2}$), which is lower than the width of the plate zone b_1 (see Fig. 5-7). Using the properties of this reduced cross section based on the reduced width b_{eff}, the design resistance and the stability of the unreduced section shall be determined.

These effective widths of the compressed plate elements according to Eurocode 3 [2] are listed in Table 5-18. In compliance with EC 3, the plate buckling reduction factor ϱ has been described in Table 5-19 [9].

The effective width b_{eff} of a flange may be determined by means of the stress ratio $\psi = f_2/f_1$ (see Table 5-18) with the full (not reduced) cross section of a hollow section. In order to calculate the effective depth of the web (h_{eff}), the effective cross section of the flange ($b_{\text{eff}} \cdot t$) and the full cross section of the web ($h \cdot t$) are to be taken. This simplification makes a direct calculation of the effective width b_{eff} possible. A precise calculation can be made by iteration.

Table 5-15 Upper limits for h_1/t ratios in RHS webs

Webs: (internal element perpendicular to the bending axis)

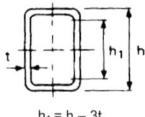

$h_1 = h - 3t$

		Web subjected to bending	Web subjected to compression	Web subjected to bending and compression
Plastic stress distribution in web (compression positive)				
Cross section class				
1		$\dfrac{h_1}{t} \leq 72\,\varepsilon$	$\dfrac{h_1}{t} \leq 33\,\varepsilon$	when $\alpha > 0.5$, $\dfrac{h_1}{t} \leq 396\,\varepsilon/(13\alpha - 1)$ when $\alpha < 0.5$, $\dfrac{h_1}{t} \leq 36\,\varepsilon/\alpha$
2		$\dfrac{h_1}{t} \leq 83\,\varepsilon$	$\dfrac{h_1}{t} \leq 38\,\varepsilon$	when $\alpha > 0.5$, $\dfrac{h_1}{t} \leq 456\,\varepsilon/(13\alpha - 1)$ when $\alpha < 0.5$, $\dfrac{h_1}{t} \leq 41.5\,\varepsilon/\alpha$
Elastic stress distribution in web (compression positive)				
Cross section class				
3		$\dfrac{h_1}{t} \leq 124\,\varepsilon$	$\dfrac{h_1}{t} \leq 42\,\varepsilon$	when $\psi > -1$, $\dfrac{h_1}{t} \leq 42\,\varepsilon/(0.67 + 0.33\,\psi)$ when $\psi < -1$, $\dfrac{h_1}{t} \leq 62\,\varepsilon(1-\psi)\sqrt{(-\psi)}$

$\varepsilon = \sqrt{\dfrac{235}{f_y}}$	f_y (N/mm^2)	235	275	355	460
	ε	1	0.92	0.81	0.72

It may result under bending load that only one flange becomes active for an effective (reduced) width. A mono-symmetrical cross section is initiated with a corresponding shift δ of the neutral axis. It is therefore necessary to calculate the effective section modulus W_{eff} by accounting for the new neutral axis (see Table 5-20).

Another method to calculate the design resistance of RHS in compression taking local buckling into consideration is based on the use of local buckling curves shown in Fig. 5-8 $\left(\bar{\lambda}_{B,P} = 1/1.9 \cdot \sqrt{f_y/E} \cdot b_1/t\right)$. The curves proposed by Rondal [11] may be recommended, which have been developed with the help of the results of the tests supported by ECSC

5.3 Design resistances of cross sections

Table 5-16 Upper limits for b_1/t ratios in RHS flanges

Flanges: (internal element parallel to bending axis)

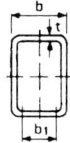

$b_1 = b - 3t$

Cross section class	Plastic stress distribution in flange and cross section (compression positive)	Section in bending	Section in compression
1		$b_1/t \leq 33\,\varepsilon$	$b_1/t \leq 42\,\varepsilon$
2		$b_1/t \leq 38\,\varepsilon$	$b_1/t \leq 42\,\varepsilon$
	Elastic stress distribution in flange and cross section (compression positive)		
3		$b_1/t \leq 42\,\varepsilon$	$b_1/t \leq 42\,\varepsilon$

$\varepsilon = \sqrt{\dfrac{235}{f_y}}$	f_y (N/mm^2)	235	275	355	460
	ε	1	0.92	0.81	0.72

Table 5-17 Upper limits of $\dfrac{b}{t}$ and $\dfrac{h}{t}$ ratios in RHS for the cross section classes 1, 2 and 3, where $\dfrac{b}{t} = \left(\dfrac{b_1}{t}+3\right)$ and $\dfrac{h}{t} = \left(\dfrac{h_1}{t}+3\right)$

Hollow section	Cross section	Loading Element	Class f_y (N/mm^2)	1				2				3			
				235	275	355	460	235	275	355	460	235	275	355	460
RHS	com-pression*	com-pression		45	41.6	36.6	32.2	45	41.6	36.6	32.2	45	41.6	36.6	32.2
RHS	bending	com-pression		36	33.3	29.3	25.7	41	37.9	33.4	29.3	45	41.6	36.6	32.2
RHS	bending	bending		75	69.3	61.1	53.6	86.0	79.5	70.0	61.5	127	117.3	103.3	90.8

* For the total cross section under compression only, there is no difference between the $\dfrac{b}{t}$ and $\dfrac{h}{t}$ ratios for the cross section classes 1, 2 and 3

Table 5-18 Effective widths b_{eff} and plate buckling factors k_P for thin-walled RHS

Stress distribution (compression positive) $b_1 = (b-3t)$ oder $(h-3t)$	Effective width b_{eff}
[diagram: uniform compression with f_1, f_2, b_{e1}, b_{e2}, b_1]	$b_{eff} = \varrho \cdot b_1$ $b_{e1} = 0.5 \cdot b_{eff}$ $b_{e2} = 0.5 \cdot b_{eff}$
[diagram: trapezoidal stress with f_1, f_2, b_{e1}, b_{e2}, b_1]	$b_{eff} = \varrho \cdot b_1$ $b_{e1} = \dfrac{2 b_{eff}}{5 - \psi}$ $b_{e2} = b_{eff} - b_{e1}$ $\psi = \dfrac{f_2}{f_1}$
[diagram: stress reversal with f_1, f_2, b_c, b_t, b_{e1}, b_{e2}, b_1]	$b_{eff} = \varrho \cdot b_c$ $b_{e1} = 0.4 \cdot b_{eff}$ $b_{e2} = 0.6 \cdot b_{eff}$

$\psi = f_2/f_1$	+1	$+1 > \psi > 0$	0	$0 > \psi > -1$	-1	$-1 > \psi > -2$
Plate buckling factor $k_P{}^*$	4.0	$\dfrac{8.2}{1.05 - \psi}$	7.81	$7.81 - 6.29\psi + 9.78\psi^2$	23.9	$5.98(1-\psi)^2$

$$* \quad k_P = \frac{16}{\sqrt{(1+\psi)^2 + 0.112(1-\psi)^2} + (1+\psi)} \quad \text{where } 1 \geq \psi \geq -1 \tag{5-63}$$

Table 5-19 Plate buckling reduction factor ϱ

For $\bar{\lambda}_{B,P} \leq 0.673$, $\quad \varrho = 1.0$

For $\bar{\lambda}_{B,P} > 0.673$, $\quad \varrho = \dfrac{\bar{\lambda}_{B,P} - 0.22}{\bar{\lambda}_{B,P}} \leq 1.0$ $\hspace{2cm}$ (5-64)

where $\bar{\lambda}_{B,P}$ is the dimensionless buckling slenderness of the flat element in compression

$$\bar{\lambda}_{B,P} = \sqrt{\frac{f_y}{f_{B,P}}} \quad \text{(see Eq. (5-57))}$$

$$\bar{\lambda}_{B,P} = \frac{b_1/t}{28.4\,\varepsilon\sqrt{k_P}}$$

where $f_{B,P}$ = critical plate buckling stress
$\quad\quad\;\, f_y$ = yield strength of the steel used
$\quad\quad\;\, k_P$ = plate buckling factor

According to Eurocode 3 [2], the influence of the internal corner radius need not be considered provided that:

$\quad r_i \leq 5\,t$ (see Fig. 5-7)
$\quad r_i/b_1 \leq 0.15$

These conditions are fulfilled by nearly all actually produced square and rectangular hollow sections

5.3 Design resistances of cross sections

Table 5-20 Effective sectional properties of thin-walled RHS

Effective cross-sectional area A_{eff} and effective radius of gyration i_{eff}:

$A_{eff} = 2 \cdot t \cdot (b_{eff} + h_{eff} + 4 \cdot t)$

$i_{eff,x} = 0.289 \cdot h_m \sqrt{3 - \left(\dfrac{h_{eff} + 2t}{h_m}\right)^2 \left(\dfrac{3h_m - h_{eff} - 2t}{b_{eff} + h_{eff} + 4t}\right)}$

$i_{eff,y} = 0.289 \cdot b_m \sqrt{3 - \left(\dfrac{b_{eff} + 2t}{b_m}\right)^2 \left(\dfrac{3b_m - b_{eff} - 2t}{b_{eff} + h_{eff} + 4t}\right)}$

Shift of the neutral axis δ and effective section modulus W_{eff}:

$\delta_x = \left(\dfrac{h_m}{2}\right)\left(\dfrac{b_m - b_{eff} - 2t}{2h_m + b_m + b_{eff} + 2t}\right)$

$\delta_y = \left(\dfrac{b_m}{2}\right)\left(\dfrac{h_m - h_{eff} - 2t}{2b_m + h_m + h_{eff} + 2t}\right)$

$W_{eff,x} = t\left[(b_{eff} + 2t)\left(\dfrac{3h_m}{2} - \delta_x\right) - 2\left(\dfrac{h_m}{2} - \delta_x\right)(h_m + b_{eff} + 2t) + \dfrac{b_m\left(\dfrac{h_m}{2} - \delta_x\right)^2 + \dfrac{2}{3}h_m^3}{\dfrac{h_m}{2} + \delta_x}\right]$

$W_{eff,y} = t\left[(h_{eff} + 2t)\left(\dfrac{3b_m}{2} - \delta_y\right) - 2\left(\dfrac{b_m}{2} - \delta_y\right)(b_m + h_{eff} + 2t) + \dfrac{h_m\left(\dfrac{b_m}{2} - \delta_y\right)^2 + \dfrac{2}{3}b_m^3}{\dfrac{b_m}{2} + \delta_y}\right]$

When $t \ll b$
$t \ll h$

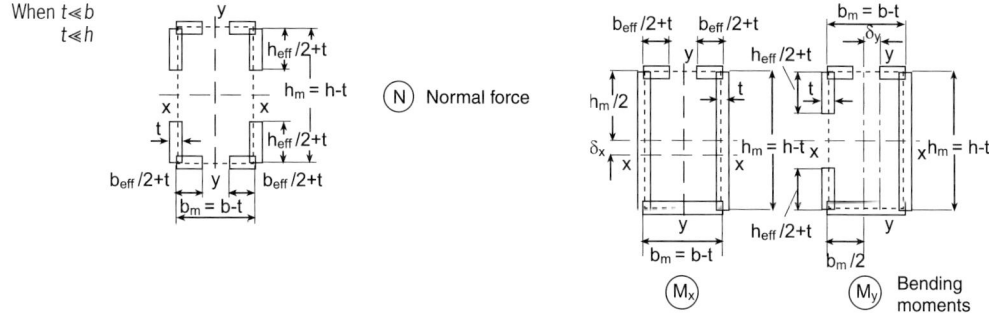

Bending moments

and CIDECT [11, 12]. This method can also be applied using any other buckling curve (see Fig. 5-8).

Comparing the test results [11, 12] a calculation method is proposed, which pursues the European buckling curves (Fig. 5-4) in an analogous way. This method is based on the idea that the design buckling resistance $N_{K,Rd}$ (or the design cross-sectional buckling stress $f_{K,Rd}$) is smaller than or at the largest equal to the failure load due to the local buckling of the wall (or the buckling stress $f_{B,P}$). This is finally smaller or equal to the design plastic resistance $N_{pl,Rd}$ (or yield strength f_y/γ_{M1}).

For RHS, one starts with European buckling curve "a" (Fig. 5-4) modifying however the parameter on the abscissa and ordinate in a way that the yield strength f_y is substituted by buckling stress $f_{B,P}$ (compare with Eqs. (5-56) and (5-57), see also Fig. 5-9).

The dimensionless slenderness $\left(\lambda_E = \pi \cdot \sqrt{\dfrac{E}{f_y}}\right)$ is modified as follows:

$$\lambda_{BK} = \pi \cdot \sqrt{\frac{E}{f_{B,P}}} = \pi \cdot \sqrt{\frac{E}{\bar{f}_{B,P} \cdot f_y}} \quad (5\text{-}65)$$

(The index "BK" means local and flexural buckling interaction.)

The dimensionless slenderness is expressed by the following equation:

$$\bar{\lambda}_{BK} = \frac{\lambda_K}{\lambda_{BK}} = \sqrt{\frac{N_{B,P}}{N_{KE}}} = \sqrt{\frac{\bar{f}_{B,P} \cdot f_y}{f_{KE}}} \quad (5\text{-}66)$$

where:

$$\lambda_K = \frac{l_K}{i}$$

$$f_{KE} = \frac{\pi \cdot E}{\lambda_K^2}$$

From the ordinate of the buckling curve "a", the dimensionless buckling stress considering local and flexural buckling interaction $\bar{f}_{BK}$ is to be read instead of $f_K/f_y = \bar{f}_K$.

$$\bar{f}_{BK} = \frac{f_{BK}}{f_{B,P}} = \frac{f_{BK}}{\bar{f}_{B,P} \cdot f_y}$$

For $\bar{f}_{B,P} = 1$ (that means according to [11] for $\bar{\lambda}_{B,P} \leq \bar{\lambda}_{BK} = 0.8$), the local and flexural bucklings transform into simple flexural buckling.

The buckling curve "a" with modified coordinates $\bar{\lambda}_{BK}$ and $\bar{f}_{BK}$ describes the test results from [11] with good precision (see Fig. 5-9).

Basing on the test results [11], different buckling curves have been established for hot finished and cold formed RHS. They are classified in Table 5-9 in accordance with the differences in buckling behaviour.

Local buckling of circular hollow sections

It is more complex to assess the buckling behaviour of thin-walled CHS than that of plates. This can be explained by the behaviour of a cylindrical shell related to buckling stability. The important characteristics of the cylindrical shells are high susceptibility to imperfections and sudden reduction of design resistance without reserve.

The first approaches to the solution of the buckling problems of axially compressed, thin-walled circular cylindrical shell came from R. Lorenz (1908) [25] and S. Timoshenko (1910) [26]. Assuming an exact circular cylinder (that means, without geometrical imperfection), they developed theoretically the well known classical formula for the ideal buckling stress $f_{B,S,ideal}$ in thin-walled CHS under pure elastic load.

$$f_{B,S,ideal} = \frac{E}{\sqrt{3(1-v^2)}} \cdot \frac{t}{r_m} = 0.605 \cdot E \cdot \frac{t}{r_m} \quad (5\text{-}67)$$

where:
E = elastic modulus
t = wall thickness of CHS
v = transversal contraction coefficient, 0.3 for structural steel
r_m = average radius of CHS

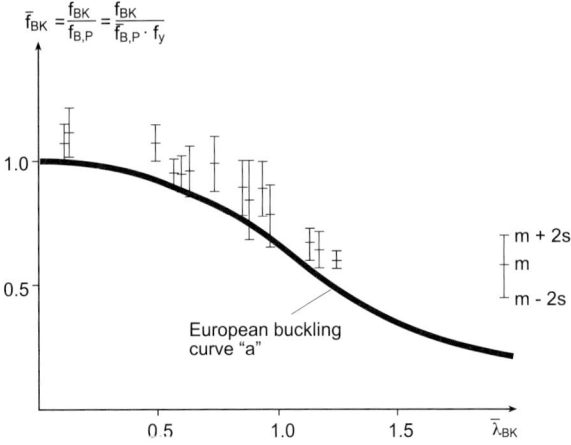

Fig. 5-9 Comparison of the modified buckling curve "a" with the results of the six test series on thin-walled RHS [11]

5.3 Design resistances of cross sections

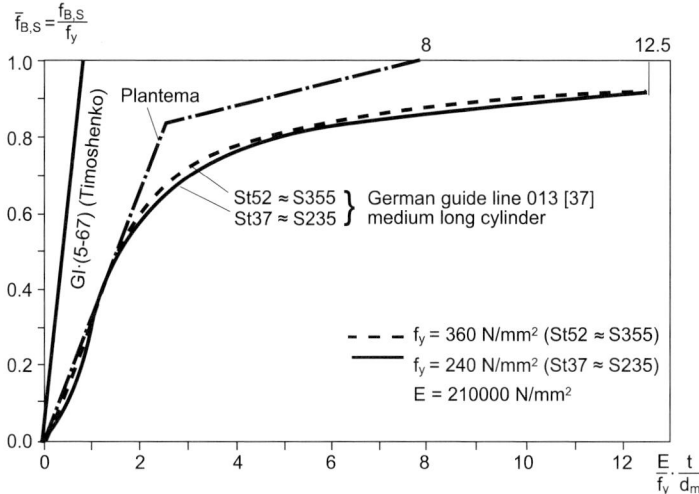

Fig. 5-10 Dimensionless buckling stress $f_{B,S}/f_y$ for CHS

The later tests [20, 27–34] demonstrated the significant dependence of the buckling stress on the magnitude of local pre-buckling (geometrical imperfection), which lowers $f_{B,S,\text{real}}$ to 1/2 to 1/6 of $f_{B,S,\text{ideal}}$. J. Plantema assessed the then available test results in a diagram (see Fig. 5-10) and came to the following equation for the elastic range:

$$f_{B,S,\text{real}} = 0.166 \cdot E \cdot \frac{t}{d_m} \quad (5\text{-}68)$$

According to Plantema, for $f_{B,S}/f_y = 1.0$,

$$\frac{E}{f_y} \cdot \frac{t}{d_m} = 8 \quad \text{(see Fig. 5-10)}$$

This limit can be found, often shifted to the safe side, in the recommendations for calculations in the design standards of various countries. Beyond these values of $(E/f_y \cdot t/d_m)$, the local buckling of the CHS walls need not be checked (see Table 5-21). The stated limiting values are valid for compression and also for bending, when the cross section resistance is to be determined elastically.

The German guide line 013 [31] as well as the guide lines proposed by the European Convention for Constructional Steelwork ECCS [28] take the recent evaluations of the available test results into consideration. Formulae and diagrams for limiting buckling stresses $f_{B,S}$ with reference to f_y in the elastic and elastic-plastic zones are given (see Fig. 5-10) mentioning also the reliable geometrical imperfections. For $E/f_y \cdot t/d_m \geq 12.5$, local buckling need not be considered.

The maximum d/t-ratios recommended by various design standards partly deviate from one another significantly. This is manifested in Table 5-22, which contains the max. d/t-ratios calculated with the yield strengths of the German steels St 37 ($f_{y,k} = 240$ N/mm^2) and St 52 ($f_{y,k} = 360$ N/mm^2).

Table 5-21 Limiting values of $\left(\dfrac{E}{f_y} \cdot \dfrac{t}{d_m}\right)$ for CHS, beyond which local buckling need not be verified

	Plantema [35]	AISI [22]	SIA 161 [21]	DNV [36]	DASt-Ri 013 [37]
$\left(\dfrac{E}{f_y} \cdot \dfrac{t}{d_m}\right) \geq$	8.0	9.21	8.0	9.0	12.5

Table 5-22 Max. $\frac{d}{t}$-ratios for CHS recommended by various design standards

	AISI [22]	SIA [21]	DNV [36]	DASt-Ri 013 [37]
$\frac{d}{t} \leq$ $f_y = 240$ N/mm²	96	110	98	71
$\frac{d}{t} \leq$ $f_y = 360$ N/mm²	64	74	66	48

Table 5-23 Upper limits of $\frac{d}{t}$-ratios for CHS according to EC 3 [2], when local buckling need not be checked

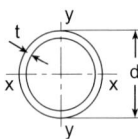

Cross section class	Axial compression and/or bending
1	$\frac{d}{t} \leq 50\, \varepsilon^2$
2	$\frac{d}{t} \leq 70\, \varepsilon^2$
3	$\frac{d}{t} \leq 90\, \varepsilon^2$

$\varepsilon = \sqrt{\dfrac{235}{f_y}}$

According to EC 3 [2], the maximum d/t-ratios of the cross section classes 1, 2 and 3 are then obtained, when the cross section or part of it does not buckle locally before the limiting design resistance is reached (see Tables 5-23 and 5-24).

Circular hollow sections of cross section class 4 in compression

When the d/t-limits go beyond those for the cross section class 3, the design resistance of CHS has to be verified taking local buckling into account. This means that the cross section belongs to the class 4. In general, this occurs seldom in steel structures as CHS with $d/t > 50$ is not or seldom applied for this purpose.

In the case of the rare application of thin-walled CHS with $d/t >$ limiting values according to Table 5-24, the procedure on the safe side can be that f_y (yield strength) is substituted by real buckling stress $f_{B,S,\text{real}}$.

The real buckling stresses $f_{B,S,\text{real}}$ (short or medium long cylinder) can be determined according to [1, part 4]. The following equations are applied:

$$f_{B,S,\text{real}} = 0.605 \cdot C \cdot E \cdot \frac{t}{r_m} \tag{5-69}$$

$$C = 1 + 1{,}5 \left(\frac{r_m}{l}\right)^2 \cdot \frac{t}{r_m} \tag{5-70}$$

for $\dfrac{l}{r_m} \leq 0.5 \sqrt{\dfrac{r_m}{t}}$

Table 5-24 Maximum $\frac{d}{t}$-limits for the CHS cross section classes 1, 2 and 3 in steel grades S235, S275, S355 and S460 (local buckling check is not necessary)

	Class	1				2				3			
Cross section loading	f_y N/mm²	235	275	355	460	235	275	355	460	235	275	355	460
Axial compression and/or bending		50	42.7	33.1	25.5	70.0	59.8	46.3	35.8	90.0	76.9	59.6	46.0

5.3 Design resistances of cross sections

For simplification, C may be taken to be equal to 1.

l = length of cylinder

Dimensionless shell slenderness $\bar{\lambda}_{B,S}$:

$$\bar{\lambda}_{B,S} = \sqrt{\frac{f_y}{f_{B,S,\text{ideal}}}} \qquad (5\text{-}71)$$

For f_y, the value for $f_{y,k}$ in [1, part 1] is to be used.

Real buckling stress $f_{B,S,\text{real}}$:

$$f_{B,S,\text{real}} = \kappa \cdot f_y \qquad (5\text{-}72)$$

where:
κ = reduction factor = $f(\bar{\lambda}_{B,S})$

Reduction factor κ:

In cases of shell buckling normally susceptible to imperfection,

for $\bar{\lambda}_{B,S} \leq 0.4$, $\kappa = 1.0$ (5-73 a)

for $0.4 < \bar{\lambda}_{B,S} < 1.2$, $\kappa = 1.274 - 0.686\,\bar{\lambda}_{B,S}$ (5-73 b)

for $1.2 \leq \bar{\lambda}_{B,S}$, $\kappa = 0.65/\bar{\lambda}_{B,S}^2$ (5-73 c)

In cases of shell buckling highly susceptible to imperfection,

for $\bar{\lambda}_{B,S} \leq 0.25$, $\kappa = 1$ (5-74 a)

for $0.25 < \bar{\lambda}_{B,S} \leq 1.0$, $\kappa = 1.233 - 0.933\,\bar{\lambda}_{B,S}$ (5-74 b)

for $1.0 < \bar{\lambda}_{B,S} \leq 1.5$, $\kappa = 0.3/\bar{\lambda}_{B,S}^2$ (5-74 c)

for $1.5 < \bar{\lambda}_{B,S}$, $\kappa = 0.2/\bar{\lambda}_{B,S}^2$ (5-74 d)

Partial safety factor γ_{M1} [1, part 4]:

For Eq. (5-73), $\gamma_{M1} = 1.1$

For Eq. (5-74 a), $\gamma_{M1} = 1.1$

For Eq. (5-74 b–d),

$$\gamma_{M1} = 1.1\left(1 + 0.318\,\frac{\bar{\lambda}_{B,S} - 0.25}{1.75}\right)$$

Verification for shell buckling:

$$\frac{f_{\text{axial}} \cdot \gamma_{M1}}{f_{B,S,\text{real}}} \leq 1 \qquad (5\text{-}75)$$

where:
f_{axial} = axial stress due to the design value of the internal stresses calculated using elastic theory

5.3.3.3 Hollow sections in combined compression and bending

Besides systematically concentric compression, this is the most frequent loading case for steel structures. The design procedures for columns in axial force together with uni-axial as well as bi-axial bending moment are described in Eurocode 3 [2]. They are valid for the cross section classes 1, 2 and 3. For the sake of comparison, the design calculation according to DIN 18800 [1] is also shown here.

Hollow sections under combined load of compression and uni-axial bending

Verification according to DIN 18800, part 2 [1]

The verification using the elastic theory of the second order is based on the recommendations by ECCS [24], which integrate besides edge moments and moments due to shear forces also the moment due to the axial force with the substituted imperfection e. This imperfection represents a combination of all imperfections such as non-linearity, residual stresses, inhomogeneity of yield strength, hardness etc.

The cross-sectional stress f'' due to the moment about x-axis shall be calculated as follows:

$$f'' = \frac{N_{c,Sd}}{A_g} + \frac{\beta_m \cdot M_{x,Sd} + N_{c,Sd} \cdot e}{W_{x,el,Rd}}$$

$$\cdot \frac{1}{1 - \dfrac{N_{c,Sd}}{N_{KE}}} \leq f_{y,k} \qquad (5\text{-}76)$$

where:
$M_{x,Sd}$ = bending moment of the first order
$W_{x,el,Rd}$ = elastic section modulus

For other terms, see Eq. 5-82.

The substituted imperfection e is determined in a way, so that the conditions for Eq. (5-51) are fulfilled, in case $M_x = 0$. This means that

Table 5-25 Equivalent uniform moment factors β_m [1]

Moment diagram	β_m for flexural buckling	β_M for lateral-torsional buckling or bi-axial bending
End moments M_1 ... $\psi \cdot M_1$ $-1 \leq \psi \leq 1$	$\beta_{m,\psi} = 0.66 + 0.44\psi$ however $\beta_{m,\psi} \geq 1 - \dfrac{1}{\eta_{KE}*}$ and $\beta_{m,\psi} \geq 0.44$	$\beta_{M,\psi} = 1.8 - 0.7\psi$
Moments due to in-plane lateral loads M_Q (upward) M_Q (downward)	$\beta_{m,Q} = 1.0$	$\beta_{M,Q} = 1.3$ $\beta_{M,Q} = 1.4$
Moments due to in-plane lateral loads plus end moments M_1, M_Q, ΔM (three diagrams)	$\psi \leq 0.77$: $\beta_m = 1.0$ $\psi > 0.77$: $\beta_m = \dfrac{M_Q + M_1 \cdot \beta_{m,\psi}}{M_Q + M_1}$	$\beta_M = \beta_{M,\psi} + \dfrac{M_Q}{\Delta M}(\beta_{M,Q} - \beta_{M,\psi})$ $M_Q = \|\max M\|$ due to lateral load only $\Delta M = \begin{cases} \|\max M\| & \text{for moment diagram without change of sign} \\ \|\max M\| + \|\min M\| & \text{for moment diagram with change of sign} \end{cases}$

* $\eta_{KE} = \dfrac{N_{KE}}{N_{c,Sd}}$, where N_{KE} is the Eulerian buckling load

this fully conforms with the imperfection integrated in the European buckling curves (see Fig. 5-4). Further, the buckling resistance about the y-axis (see Fig. 5-3) (without moments) shall be checked.

A procedure called "substituted column method", which was proposed later, is an approximation based on the yield hinge theory of the second order. This starts from the Eq. (5-76) (however with plastic forces or moments) leading to the following derivation for the substituted imperfection e [39, 40]:

$M_x = 0$, $N_{pl,Rd} = A_g \cdot f_{y,k}/\gamma_{M1}$,
$M_{pl,Rd} = W_{pl} \cdot f_{y,k}/\gamma_{M1}$, $N_{KE} = N_{pl,Rd}/\bar{\lambda}_K^2$,
$N_{K,Rd} = \chi \cdot N_{pl,Rd}$,

$$e = \frac{(1-\chi)(1-\chi \cdot \bar{\lambda}_K^2)}{\chi} \cdot \frac{M_{pl,Rd}}{N_{pl,Rd}} \quad (5\text{-}77)$$

The dimensionless χ-$\bar{\lambda}_K$ relationship can be described according to [4] with good precision as follows:

$$(1-\chi)(1-\chi \cdot \bar{\lambda}_K^2) = \eta \cdot \chi \quad (5\text{-}78)$$

5.3 Design resistances of cross sections

where $\eta = \alpha(\bar{\lambda}_K - 0.2)$ (5-79)

$\alpha =$ to be obtained from Table 5-12

Using Eqs. (5-78) and (5-79), the Eq. (5-77) can be transformed to:

$$e = \eta \cdot \frac{M_{pl,Rd}}{N_{pl,Rd}} = \alpha(\bar{\lambda}_K - 0.2) \cdot \frac{M_{pl,Rd}}{N_{pl,Rd}} \quad (5-80)$$

Setting Eq. (5-77) in Eq. (5-76) and using $N_{KE} = \frac{N_{pl,Rd}}{\bar{\lambda}_K^2}$, this yields finally:

$$\frac{N_{c,Sd}}{\chi_x \cdot N_{pl,Rd}} + \frac{1}{1 - \frac{N_{c,Sd}}{\chi_x \cdot N_{pl,Rd}} \cdot \chi_x^2 \cdot \bar{\lambda}_K^2} \cdot$$

$$\cdot \frac{\beta_m \cdot M_{x,Sd}}{M_{pl,x,Rd}} \leq 1 \quad (5-81)$$

Transforming further:

$$\frac{N_{c,Sd}}{\chi_x \cdot N_{pl,Rd}} + \frac{\beta_m \cdot M_{x,Sd}}{M_{pl,x,Rd}} \leq 1 - \Delta n \quad (5-82)$$

where:

$$\Delta n = \chi_x^2 \cdot \bar{\lambda}_K^2 \cdot \frac{N_{c,Sd}}{\chi_x \cdot N_{pl,Rd}} \left(1 - \frac{N_{c,Sd}}{\chi_x \cdot N_{pl,Rd}}\right)$$

(5-83)

Notations:

$N_{c,Sd}$ = axial compression (with γ_F times load)

$M_{x,Sd}$ = maximum bending moment about x-axis ($M_{y,Sd}$ is maximum bending moment about y-axis)

β_m = equivalent uniform moment factor according to moment diagram in Table 5-25

χ_x = reduction factor, to be taken from Fig. 5-4 or Table 5-10 or to be derived from Eq. 5-54 (for moment about y-axis: χ_y)

$f_{y,k}$ = characteristic yield strength (to be taken from [1, part 1])

γ_{M1} = partial safety factor (see Table 5-6) = 1.1

Δn = additional part (this can be simplified by approximation on the safe side)

$\bar{\lambda}_K$ = dimensionless slenderness according to Eq. (5-53)

As Fig. 5-11a shows, the expression $\frac{N_{c,Sd}}{\chi_x \cdot N_{pl,Rd}} \left(1 - \frac{N_{c,Sd}}{\chi_x \cdot N_{pl,Rd}}\right)$ yields 0.25 at the maximum. Therefore, the following approximation can be made:

$$\Delta n = 0.25 \chi_x^2 \cdot \bar{\lambda}_K^2 \quad (5-84)$$

For further simplification, the part Δn can be taken to be equal to ~ 0.1 on the safe side (see Fig. 5-11b).

For double symmetrical cross sections, where the web area has a minimum share of 18%, $M_{pl,x,Rd}$ in Eq. (5-82) may be substituted by 1.1 $M_{pl,x,Rd}$, when $N_{c,Sd}/N_{pl,Rd} > 0.2$.

Further for small axial compression $N_{c,Sd}$, which fulfils the conditions given by Eq. (5-85), $N_{c,Sd}$ may be neglected.

$$\frac{N_{c,Sd}}{\chi \cdot N_{pl,Rd}} < 0.1 \quad (5-85)$$

a)
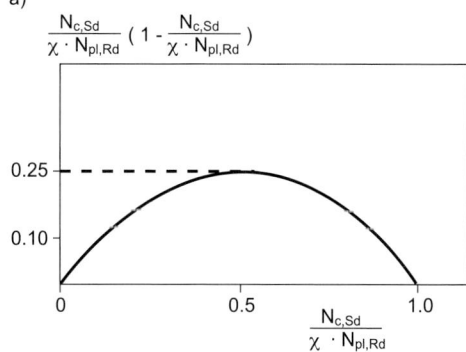

b)
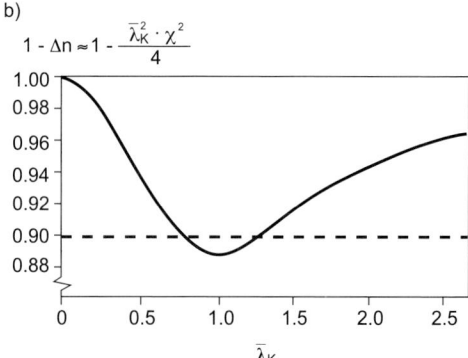

Fig. 5-11 Diagram for the right side of the Eqs. (5-82) and (5-83) with χ following the European buckling curve "a"

Verification according to Eurocode 3 [2]

The following condition (cross section classes 1 and 2) applies in this case, where the European buckling curves (see Fig. 5-4 or Eq. (5-54) or Table 5-10) are used:

$$\frac{N_{c,Sd}}{\chi_x \cdot N_{pl,Rd}} + \frac{\kappa_x \cdot M_{x,Sd}}{M_{pl,x,Rd}} \leq 1 \qquad (5\text{-}86)$$

where:

$M_{x,Sd}$ = largest absolute value of bending moment according to the theory of the first order without considering imperfections

$$\frac{N_{c,Sd} \cdot \gamma_{M1}}{\chi_x \cdot A_g \cdot f_y} + \frac{\kappa_x \cdot M_{x,Sd} \cdot \gamma_{M1}}{W_{pl,x} \cdot f_y} \leq 1 \qquad (5\text{-}87)$$

where:

$$\kappa_x = 1 - \frac{\mu_x \cdot N_{c,Sd}}{\chi_x \cdot A_g \cdot f_y}, \text{ however } \kappa_x \leq 1.5 \quad (5\text{-}88)$$

$$\mu_x = \bar{\lambda}_{K,x}(2\beta_{M,x}^* - 4) + \left\{\frac{W_{pl,x} - W_{el,x}}{W_{el,x}}\right\},$$

however $\mu_x \leq 0.9$ \hfill (5-89)

$\beta_{M,x}$ = see Table 5-26
γ_{M1} = 1.1
f_y = minimum yield strength from [22, 23] of Chapter 3

For elastic calculation (cross section class 3), $W_{pl,x}$ in Eq. (5-89) is to be substituted by $W_{el,x}$.

Further, the following additional requirement has to be fulfilled:

$$\frac{N_{c,Sd}}{\chi_y \cdot N_{pl,Rd}} = \frac{N_{c,Sd} \cdot \gamma_{M1}}{\chi_y \cdot A_g \cdot f_y} \leq 1 \curvearrowright$$

$$N_{c,Sd} \leq \chi_y \cdot \frac{A_g \cdot f_y}{\gamma_{M1}} \qquad (5\text{-}90)$$

Hollow sections under combined load of compression and bi-axial bending

Verification according to DIN 18800, part 2 [1]

The design load has to be verified basing on the following condition:

$$\frac{N_{c,Sd}}{\chi_{min} \cdot N_{pl,Rd}} + \frac{\kappa_x \cdot M_{x,Sd}}{M_{pl,x,Rd}} + \frac{\kappa_y \cdot M_{y,Sd}}{M_{pl,y,Rd}} \leq 1 \qquad (5\text{-}91)$$

where:

χ_{min} = min (χ_x, χ_y) = reduction factor for the buckling curves (see Fig. 5-4 or Table 5-10)

$M_{x,Sd}, M_{y,Sd}$ = largest absolute value of bending moments according to the theory of the first order without considering imperfections

$$\kappa_x = 1 - \frac{\mu_x \cdot N_{c,Sd}}{\chi_x \cdot A \cdot f_{y,k}}, \text{ however } \kappa_x \leq 1.5$$

(compare with Eq. (5-88))

$$\mu_x = \bar{\lambda}_{K,x}(2\beta_{M,x}^* - 4) + \left\{\frac{W_{pl,x} - W_{el,x}}{W_{el,x}}\right\},$$

however $\mu_x \leq 0.8$ (compare with Eq. (5-89))

$$\kappa_y = 1 - \frac{\mu_y \cdot N_{c,Sd}}{\chi_y \cdot A_g \cdot f_{y,k}}, \text{ however } \kappa_y \leq 1.5 \quad (5\text{-}92)$$

$$\mu_y = \bar{\lambda}_{K,y}(2\beta_{M,y}^* - 4) + \left\{\frac{W_{pl,y} - W_{el,y}}{W_{el,y}}\right\},$$

however $\mu_y \leq 0.8$ \hfill (5-93)

$\beta_{M,x}^*, \beta_{M,y}^*$ = equivalent uniform moment factors according to Table 5-25, column 3 or Table 5-26

$M_{pl,x,Rd}, M_{pl,y,Rd}$ = elastic design bending moment resistance

$M_{pl,x,Rd} = W_{pl,x} \cdot f_{y,k}/\gamma_{M1}$

$M_{pl,y,Rd} = W_{pl,y} \cdot f_{y,k}/\gamma_{M1}$

For elastic calculation (cross section class 3), $W_{pl,x} = W_{el,x}$ and $W_{pl,y} = W_{el,y}$ are to be used in Eqs. (5-87), (5-89) and (5-93).

Verification according to Eurocode 3 [2]

This method is almost identical to that proposed in DIN 18800, part 2 [1]. Following differences have to be considered:

1. μ_x, μ_y according to Eqs. (5-89) and (5-93): ≤ 0.9

---

* The equivalent uniform moment factors $\beta_{M,x}$ and $\beta_{M,x}$ are to be taken in accordance with the moment diagrams between the lateral supports from Table 5-25, 3. column:

Moment diagram	Moment about the axis	Lateral support in the direction
$\beta_{M,x}$	x–x	y–y
$\beta_{M,y}$	y–y	x–x

5.3 Design resistances of cross sections

Table 5-26 Equivalent uniform moment factors β_M [2]

Moment diagram	β_M
End moments M_1 ⟋⟋⟋ $\psi \cdot M_1$ $-1 \leq \psi \leq 1$	$\beta_{M,\psi} = 1.8 - 0.7\,\psi$
Moments due to in-plaine lateral loads ↓↓↓ M_Q (upward) ↓↓↓ M_Q (upward)	$\beta_{M,Q} = 1.3$ $\beta_{M,Q} = 1.4$
Moments due to in-plane lateral loads plus end moments M_1, M_Q, ΔM M_1, M_Q, ΔM M_1, M_Q, ΔM M_1, M_Q, ΔM	$\beta_M = \beta_{M,\psi} + (\beta_{M,Q} - \beta_{M,\psi}) \cdot M_Q / \Delta M$ $M_Q = |\max M|$ due to lateral load only $\Delta M = \begin{cases} |\max M| & \text{for moment diagram without change of sign} \\ |\max M| + |\min M| & \text{for moment diagram with change of sign} \end{cases}$

2. The characteristic yield strength $f_{y,k}$ according to [1, part 1] is to be substituted by the minimum yield strength f_y according to ([22, 23] of Chapter 3).

3. The equivalent uniform moment factors $\beta_{M,x}$, $\beta_{M,y}$ are to be taken from the Table 5-26.

Substituted column method according to DIN 18800, part 2 [2]

This second verification method has been proposed by Roik [42] for the profiles, which are not affected by lateral-torsional buckling:

$$\frac{N_{c,Sd}}{\chi_{min} \cdot N_{pl,Rd}} + \frac{\kappa_x \cdot \beta_{m,x} \cdot M_{x,Sd}}{M_{pl,x,Rd}} +$$

$$+ \frac{\kappa_y \cdot \beta_{m,y} \cdot M_{y,Sd}}{M_{pl,y,Rd}} + \Delta n \leq 1 \qquad (5\text{-}94)$$

where:

$\Delta n = \chi_{min}^2 \cdot \bar{\lambda}_K^2 \cdot \dfrac{N_{c,Sd}}{\chi_{min} \cdot N_{pl,Rd}} \cdot$

$\cdot \left(1 - \dfrac{N_{c,Sd}}{\chi_{min} \cdot N_{pl,Rd}}\right)$ (compare with Eq. 5-83)

$\chi = \min(\chi_x, \chi_y)$; $\bar{\lambda}_K = \max(\bar{\lambda}_{K,x}, \bar{\lambda}_{K,y})$

$M_{x,Sd}$, $M_{y,Sd}$ = largest absolute value of bending moment according to the theory of the first order without considering imperfections

$M_{pl,x,Rd}$, $M_{pl,y,Rd}$ = design bending moment resistance in full plastic condition

$\beta_{m,x}$, $\beta_{m,y}$ = equivalent uniform moment factors for flexural buckling according to Table 5-25, column 2

For $\chi_x < \chi_y$, $\kappa_x = 1$, $\kappa_y = c_y$

For $\chi_x = \chi_y$, $\kappa_x = 1$, $\kappa_y = 1$

For $\chi_y < \chi_x$, $\kappa_x = c_y$, $\kappa_y = 1$

$c_y = \dfrac{1}{c_x} = \dfrac{1 - \dfrac{N_{c,Sd}}{N_{pl,Rd}} \cdot \bar{\lambda}_{K,x}^2}{1 - \dfrac{N_{c,Sd}}{N_{pl,Rd}} \cdot \bar{\lambda}_{K,y}^2}$

5.3.3.4 Buckling length

According to the definition, the effective buckling length l_K of a structural member in compression is equal to the system length l_0 of a member with hinged pin supports at both

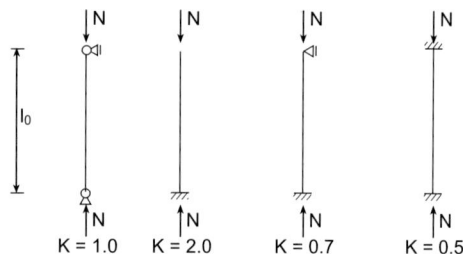

Fig. 5-12 Buckling length coefficient K

ends, where it possesses the same second moment of area and shows the same critical buckling load as the real compression member. Other possible restraining conditions at the member ends (see Fig. 5-12) lead to a reduction of the system length resulting in the effective buckling length.

Effective buckling length $l_K = K \cdot l_0$ (5-95)

where:
K = buckling length coefficient

In general, the effective buckling length is calculated using the following equation:

$$l_K = \pi \cdot \sqrt{\frac{EI}{N_{KE}}} \qquad (5\text{-}96)$$

where:
N_{KE} = see Eq. (5-53)

Buckling length of members in welded lattice girders made of hollow sections

In usual static calculations of lattice girders, the forces in the chord and bracing members are determined assuming the joints to be hinged. Actually however, the members are partially fixed at the joint. As a consequence, the effective buckling lengths of the members are reduced.

The buckling length coefficients $K(= l_K/l_0)$ are evaluated differently in different countries. For welded lattice girders made of hollow sections with laterally supported ends, K is taken on the safe side to be equal to 1.0 in Germany according to DIN 18808 [43]. In United Kingdom and Canada, K is proposed to be 0.7. API [44] recommends $K = 0.8$, while DNV [45] proposes for circular hollow sections to

5.3 Design resistances of cross sections

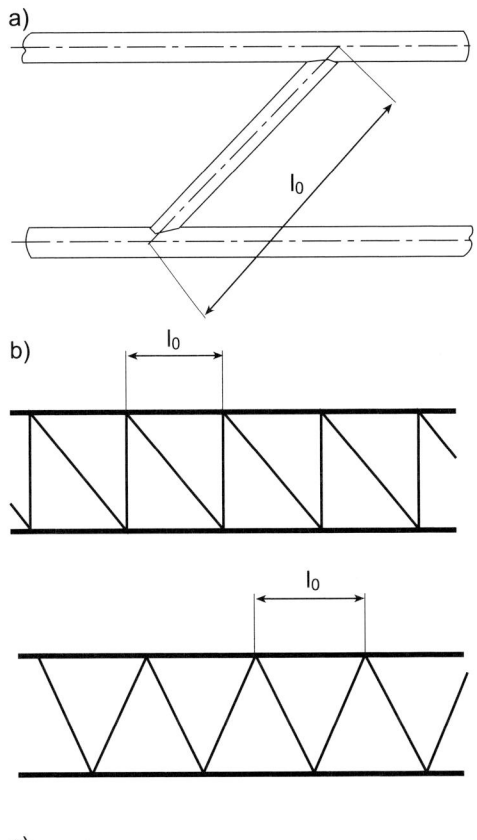

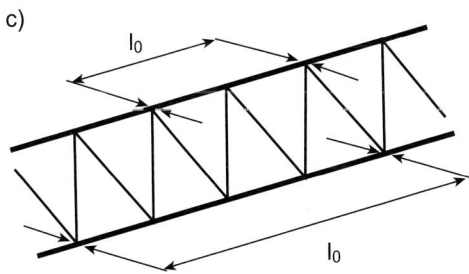

Fig. 5-13 System length l_0 of the members of a lattice girder. a) Distance between the intersections of neutral axes of the bracing and chord members, b) distance between the neighbouring joints, c) distance between two lateral supports

take K-values between 0.7 and 1.0 depending on the ratio wall thickness/diameter.

In accordance with EC 3 [2], the effective buckling length l_K for chord members in general and bracing members for buckling out-of-plane of the lattice girders has to be fixed to be equal to the system length l_0, when no precise calculation is carried out. For in-plane buckling check of bracing members, the effective buckling length l_K can be assumed to be less than the system length l_0, when the twist of the bracing ends can be obstructed by the chord members and the joint construction (welded or bolted by two bolts at least). Provided these conditions are fulfilled, the effective buckling length l_K of the bracing members can be taken to be equal to 0.9 l_0 for in-plane buckling.

Basically the effective buckling length l_K of the members of a lattice girder can be determined theoretically, when the joint rigidities for the corresponding loads are known. However, only a few data regarding the joint rigidities for a combination of normal forces and bending moments are available. Later, the theoretical and experimental research works [46] have met this need partially. They deliver the basis for the solution of the problem of stability of the members of a lattice girder with regard to their dimensions considering the joint rigidities.

As the excellent torsional rigidity of hollow sections (see Fig. 1-2) was taken in the design standards already mentioned into account only insufficiently, CIDECT* initiated the investigations to develop more precise calculation methods to determine the effective buckling length l_K of the bracing members in a lattice girder of hollow sections, where they are welded to the chord members [47, 48].

In this case, a simple theoretical method to determine the effective buckling length of the bracing members is not possible, as the chord wall experiences a small deformation under the compression applied through the bracing member. This results in a change of the elastic restraint coefficient (method for the calculation of the elastic hinges [49]) with relation to the magnitude of the applied force. Therefore, destructive tests on girders of hollow sections could not be dispensed with in this case. The test specimens and arrangements are shown in Fig. 5-14.

---

* Comité International pour le Developpement et l'Etude de la Construction Tubulaire, Geneva.

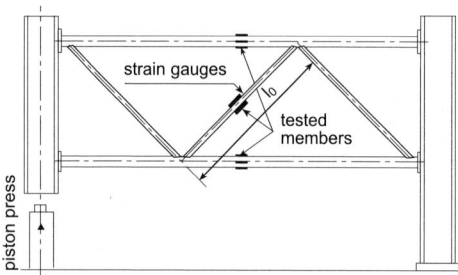

Fig. 5-14 Test specimen and test arrangement [47, 48] (short girder with three fields and long girder with twelve fields)

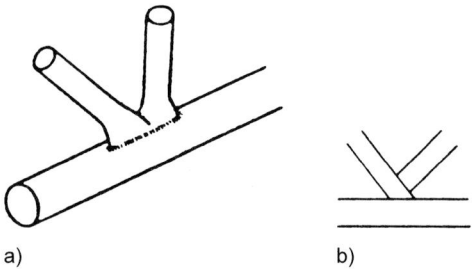

Fig. 5-16 a) CHS joint with flattened bracing ends, b) hollow section joint with fully overlapped bracing member

An analysis of the test results to determine the European buckling curve "a" (see Fig. 5-4) led to the empirical relationship between the effective buckling length l_K of the bracing member and the typical parameters defined in Fig. 5-15.

Rondal [50] modified the equations given in [47, 48] by means of a statistical evaluation.

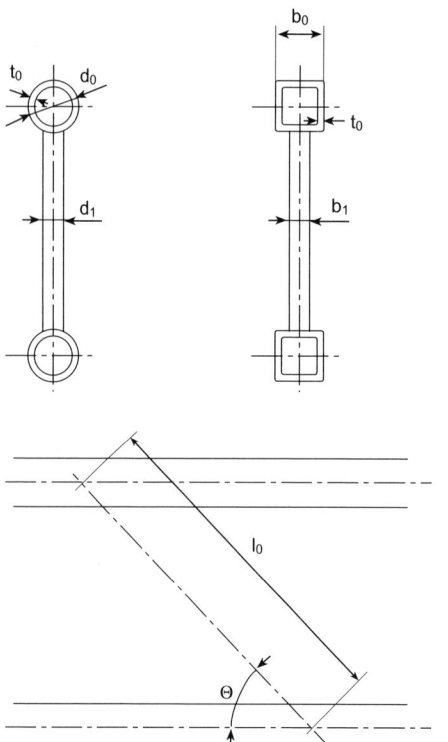

Fig. 5-15 Test parameters [47, 48]

CIDECT recommended them for application (see Table 5-27 [9]).

1. The equations given in Table 5-27 are only valid for the bracing members, which are welded to the chords along the full perimeter length. For the bracing members with overlap in a joint, an effective buckling length l_K equal to the system length l_0 has to be used.

2. For bracing members with cropped or flattened ends (Fig. 5-16a) or when they are fully overlapped (Fig. 5-16b), $l_K = l_0$ (buckling in-plane and out-of-plane)

3. For bracing members with flattened ends, which are bolted to a plate and the plate is welded to the chord member in its turn (Fig. 5-17), the effective buckling length l_K in-plane or out-of-plane can be taken to be equal to l_0.

4. For bracing members of hollow sections, which are welded to chord members made of I or H profiles, l_K can be used to be equal to l_0 in all planes. For the effective buckling length l_K of the chord members of I or H profiles, $l_K = 0.9\, l_0$ (in-plane) and $l_K = l_0$ (out-of-plane) may be applied, when smaller values are not justified.

Effective buckling length of chords of lattice girders, whose joints are not supported laterally

The both known calculation methods to solve these problems are difficult and lengthy, as they are based on an iterative procedure and require the use of a computer. Therefore, only reference literatures are mentioned here [47,

5.3 Design resistances of cross sections

Table 5-27 Recommendations by CIDECT [9] and Eurocode 3 [2] for the calculation of the effective buckling length of the chord and bracing members in a lattice structure of hollow sections (see items 1–4 in the text) (conditions: a) upper and lower chords are parallel or almost parallel to each other, b) upper and lower chords have identical sizes[a])

Chord members:

Buckling in-plane: $l_K = 0.9 \times l_0$ (system length between joints)

Buckling out-of-plane: $l_K = 0.9 \times l_0$ (system length between the lateral supports)

Bracing members: for all β-ratios

Buckling in-plane and out-of-plane: $l_K = 0.75 \times l_0$ (system length between joints)

In case $\beta < 0.6$ (in general $0.5 \leq l_K/l_0 \leq 0.75$), a more precise calculation can be made using the following formulae:

l_K/l_0	Chord member	Bracing member
$2.20 \, [d_1^2/(l_0 \cdot d_0)]^{0.25}$	circular, d_0	circular, d_1
$2.35 \, [d_1^2/(l_0 \cdot b_0)]^{0.25}$	square, $b_0 \times h_0$	circular, d_1
$2.30 \, [b_1^2/(l_0 \cdot b_0)]^{0.25}$	square, $b_0 \times h_0$	square, $b_1 \times h_1$

l_K – effective buckling length
l_0 – system length
d_1 – outer diameter of bracing member
d_0 – outer diameter of chord member
b_1 – external width of square or rectangular bracing member
b_0 – external width of square or rectangular chord member

$\beta = \dfrac{d_1}{d_0}$ or $\dfrac{d_1}{b_0}$ or $\dfrac{b_1}{b_0}$

[a] When the upper and lower chord members are of different sizes, it is recommended to determine the buckling length coefficient K for the joints at each point of intersection of the bracing member. The larger value has to be taken as decisive [47]. Lit. [2] recommends to use the average value of them. A further proposal on the conservative side is as follows: 1) For RHS chord members: b_0 is to be substituted by h_0 when $h_0 < b_0$. 2) For RHS bracing members: b_1 is to be substituted by h_1, when $h_1 > b_1$. Lit. [2] recommends to use b_1 and b_0 for the buckling in-plane and h_1 and h_0 for the buckling out-of-plane.

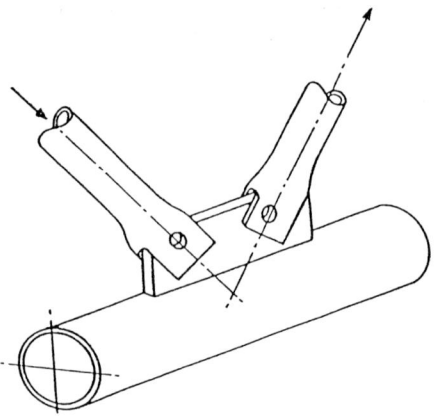

Fig. 5-17 Flattened bracing ends bolted to a plate

51, 52], by means of which the calculation can be carried out. The results of the calculations show that the effective buckling length of chords of lattice girders without lateral support can be considerably smaller than the real (full) span without support.

In order to simplify the application in the mostly occurring cases (laterally supported girders), 64 design diagrams are given in [47].

5.3.3.5 Design examples
RHS column in planned concentric compression

Given data:

RHS 140 × 80 × 4 mm (hot finished)

$A_g = 16.8 \text{ cm}^2$
$i_x = 5.12 \text{ cm}$
$i_y = 3.31 \text{ cm}$

Material: steel grade S 235, $f_y = 235 \text{ N/mm}^2$

Effective buckling length: $l_{K,x} = 6.0 \text{ m}$
$l_{K,y} = 3.0 \text{ m}$

Applied load: permanent force $N_G = 80 \text{ kN}$
variable force $N_Q = 40 \text{ kN}$

Design load:

$N_{c,Sd} = 1.35 \, N_G + 1.5 \, N_Q = 1.35 \cdot 80 + 1.50 \cdot 40$
$= 168 \text{ kN}$

Full plastic design resistance for the cross section:

$$N_{pl,Rd} = \frac{A_g \cdot f_y}{\gamma_{M1}} = \frac{16.8 \cdot 23.5}{1.1} = 358.91 \text{ kN}$$

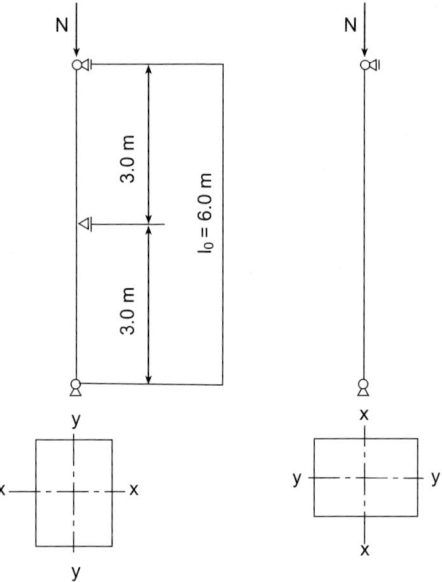

Fig. 5-18 Column in concentric compression

$$\max \cdot \frac{b_1}{t} = \frac{140 - 3 \cdot 4}{4} = 32 < 42$$

(compare with Tables 5-15 and 5-16)

Buckling check is not necessary.

Axis $x-x$:

$$\lambda_{K,x} = \frac{l_{K,x}}{i_x} = \frac{600}{5.12} = 117.19 \qquad \lambda_E = 93.9$$

(see Table 5-11)

$$\bar{\lambda}_{K,x} = \frac{\lambda_{K,x}}{\lambda_E} = \frac{117.19}{93.9} = 1.25 \rightarrow \chi_x = 0.4993$$

(see Table 5-10b: buckling curve "a")

Axis $y-y$:

$$\lambda_{K,y} = \frac{l_{K,y}}{i_y} = \frac{300}{3.31} = 90.63 \qquad \lambda_E = 93{,}9$$

$$\bar{\lambda}_{K,y} = \frac{\lambda_{K,y}}{\lambda_E} = \frac{90.63}{93.9} = 0.97 \rightarrow \chi_y = 0.6865$$

According to Eq. (5-51):

$N_{K,Rd} = \chi_{min} \cdot N_{pl,Rd} = 0.4993 \cdot 358.91$
$= 179.2 \text{ kN} > 168 \text{ kN}$

5.3 Design resistances of cross sections

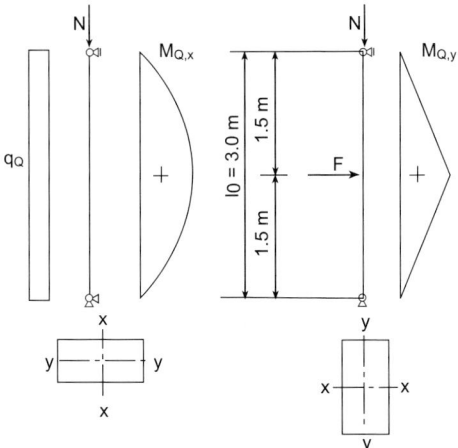

Fig. 5-19 Column in axial compression and bi-axial bending

RHS column under combined axial compression and bi-axial bending moment

Given data:

RHS 200 × 100 × 12 (cold formed)

A_g = 60.1 cm²
i_x = 6.59 cm
i_y = 3.82 cm
$W_{pl,x}$ = 350 cm³ $W_{el,x}$ = 261 cm³
$W_{pl,y}$ = 215 cm³ $W_{el,x}$ = 175 cm³

Material: steel grade S 355, f_y = 355 N/mm²

Average yield strength after cold forming f_{ya} according to Table 3-14 (k = 7, n = 4, A_g = $2t \cdot (b+h-2t) \approx (b+h) \cdot 2t$):

$$f_{ya} = 355 + \frac{14 \cdot 12}{200 + 100}(490 - 355)$$
$$= 430.6 \text{ N/mm}^2 \approx 1.2 \cdot 355 = 426 \text{ N/mm}^2$$

Effective buckling length: $l_{K,x} = l_{K,y}$ = 3.0 m

Applied load: permanent force N_G = 130 kN
variable force N_Q = 45 kN
F_Q = 30 kN
q_Q = 10 kN/m

Design loads (theory of the first order): The force F_Q has the largest loading influence ($\gamma_{F,Q}$ = 1.5).

$$N_{c,Sd} = \gamma_{F,G} \cdot N_G + 0.9 \gamma_{F,Q} \cdot N_Q$$
$$= 1.35 \cdot N_G + 1.35 \cdot N_Q$$
$$= 1.35(130 + 45) = 236 \text{ kN}$$

$$M_{Q,x} = 0.9 \gamma_{F,Q} \cdot \frac{q_Q \cdot l_0^2}{8} = 1.35 \cdot \frac{10 \cdot 3^2}{8}$$
$$= 15.2 \text{ kNm}$$

$$M_{Q,y} = \gamma_{F,Q} \cdot \frac{F_Q \cdot l_0}{4} = 1.5 \cdot \frac{30 \cdot 3}{4}$$
$$= 33.7 \text{ kNm}$$

$$\left.\begin{array}{l}\max \cdot \dfrac{b_1}{t} = \dfrac{200 - 3 \cdot 12}{12} = 13.7 \\[2mm] \max \cdot \dfrac{h_1}{t} = \dfrac{100 - 3 \cdot 12}{12} = 5.3\end{array}\right\} < 33 \cdot \varepsilon =$$

$$= 33 \cdot \sqrt{\frac{235}{430.6}} = 24.3$$

The cross section fulfils the conditions for the class 1.

For f_{ya} = 430.6 N/mm² (f_y = 355 N/mm²),

$$\lambda_E = 93.9 \cdot \sqrt{235/430.6} = 69.4$$

Axis x–x:

$$\lambda_{K,x} = \frac{l_{K,x}}{i_x} = \frac{300}{6.59} = 45.52$$

$$\bar{\lambda}_{K,x} = \frac{\lambda_{K,x}}{\lambda_E} = \frac{45.52}{69.4} = 0.66 \rightarrow \chi_x = 0.7493$$

(see Table 5-10d, buckling curve "c")

Axis y–y:

$$\lambda_{K,y} = \frac{l_{K,y}}{i_y} = \frac{300}{3.82} = 78.53$$

$$\bar{\lambda}_{K,y} = \frac{\lambda_{K,y}}{\lambda_E} = \frac{78.53}{69.4} = 1.13 \rightarrow \chi_y = 0.4685$$

(see Table 5-10d, buckling curve "c")

Equivalent uniform moment factors (according to Table 5-26):

$\beta_{M,x}$ = 1.3
$\beta_{M,y}$ = 1.4

The verification of the design resistance is to be performed following the Eq. (5-91).

$$\mu_x = \bar{\lambda}_{K,x}(2\beta_{M,x}-4) + \left\{\frac{W_{pl,x} - W_{el,x}}{W_{el,x}}\right\}$$

$$= 0.66(2 \cdot 1.3 - 4) + \left\{\frac{350 - 261}{261}\right\}$$

$$= -0.583$$

$$\mu_y = \bar{\lambda}_{K,y}(2\beta_{M,y}-4) + \left\{\frac{W_{pl,y} - W_{el,y}}{W_{el,y}}\right\}$$

$$= 1.13(2 \cdot 1.4 - 4) + \left\{\frac{215 - 175}{175}\right\}$$

$$= -1.128$$

$$\kappa_x = 1 - \frac{\mu_x \cdot N_{c,Sd}}{\chi_x \cdot A_g \cdot f_{ya}}$$

$$= 1 - \frac{(-0.583) \cdot 236 \cdot 10}{0.7493 \cdot 60.1 \cdot 430.6}$$

$$= 1.071$$

$$\kappa_y = 1 - \frac{\mu_y \cdot N_{c,Sd}}{\chi_y \cdot A_g \cdot f_{ya}}$$

$$= 1 - \frac{(-1.128) \cdot 236 \cdot 10}{0.4685 \cdot 60.1 \cdot 430.6}$$

$$= 1.220$$

Check according to Eq. (5-91):

$$\frac{N_{c,Sd} \cdot \gamma_{M1}}{\chi_{min} \cdot A_g \cdot f_{ya}} + \frac{\kappa_x \cdot M_{Q,x} \cdot \gamma_{M1}}{W_{pl,x} \cdot f_{ya}} +$$

$$+ \frac{\kappa_y \cdot M_{Q,y} \cdot \gamma_{M1}}{W_{pl,y} \cdot f_{ya}} = \frac{236 \cdot 10 \cdot 11}{0.4685 \cdot 60.1 \cdot 430.6} +$$

$$+ \frac{1.071 \cdot 15.2 \cdot 1.1 \cdot 1000}{350 \cdot 430.6} +$$

$$+ \frac{1.220 \cdot 33.7 \cdot 1.1 \cdot 1000}{215 \cdot 430.6} =$$

$$= 0.214 + 0.119 + 0.489 = 0.822 < 1.0$$

The verification of the adequate design resistance of the cross section can be carried out by means of the "elastic" Eq. (5-36) on the safe side.

$$\frac{N_{Sd} \cdot \gamma_{M1}}{A_g \cdot f_{ya}} + \frac{M_{Q,x} \cdot \gamma_{M1}}{W_{el,x} \cdot f_{ya}} + \frac{M_{Q,y} \cdot \gamma_{M1}}{W_{el,y} \cdot f_{ya}}$$

$$= \frac{236 \cdot 1.1}{6010 \cdot 0.4306} + \frac{15.2 \cdot 1000 \cdot 1.1}{261 \cdot 430.6}$$

$$+ \frac{33.7 \cdot 1000 \cdot 1.1}{175 \cdot 430.6} = 0.100 + 0.149 + 0.492$$

$$= 0.741 < 1.0$$

5.3.4 Hollow sections in torsion

As Fig. 1-2 demonstrates, hollow sections represent the most efficient profile form with regard to torsional resistance. Due to the uniform distribution of the material around the axis of the member, CHS has the best design resistance against torsion among all closed sections. Therefore they are preferably used to transmit torsion in the constructions of machines, vehicles, airplanes, ships etc. However in numerous cases of application, in particular for agricultural machinery, cranes and buildings, square and rectangular hollow sections are preferred on account of fundamental design reasons (as for example, more economical fabrication), although they show slightly less favourable torsional behaviour than that of CHS.

A precise analysis of the load carrying behaviour of profiles in torsion can be done basing on the advanced theory of the strength of materials. When the applied torsional moment is not uniform or when the free warping of the flat cross sections is prevented, the torsional load is divided into a uniform "Saint-Venant" torsional moment and a warping torsional moment. A complex stress calculation for the sections is required to determine the distribution of the warping torsional stresses. Therefore often in practice, one makes an approximate estimation of the torsional shear stresses and neglects the resultant additional stresses in longitudinal and transversal directions of a profile arising from the prevention of warping and the profile deformation. A special advantage of hollow sections in relation to open sections represents that the warping torsional moment can be neglected in this case.

Fig. 5-20 shows, for various sections, the portion of the torsional moment, which is ab-

5.3 Design resistances of cross sections

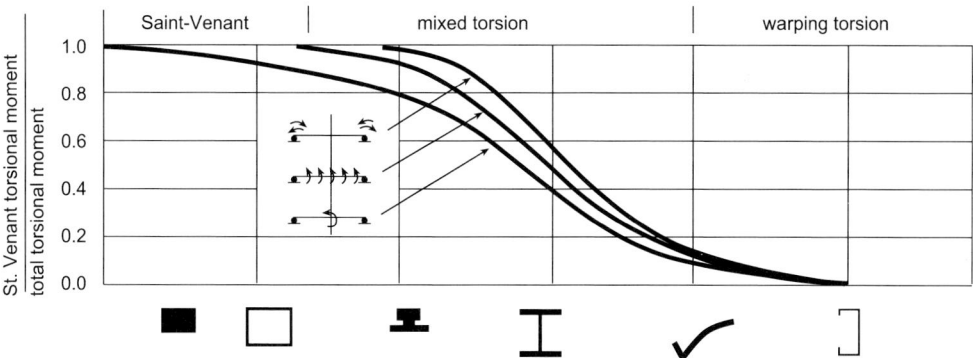

Fig. 5-20 Ratio of the "Saint-Venant" torsional moment to the total torsional moment for various profile types

sorbed as uniform or "Saint-Venant" torsion [53]. It is observed that the hollow sections in all cases can be checked for the "Saint-Venant" torsion with sufficient accuracy.

The design torsion of CHS and RHS, which are free of warping, can be calculated using the theory of Saint-Venant under the following conditions:

- The cross-sectional form is maintained by means of stiffeners or bulkheads at the place of load transmission. If stiffeners are not applied, the load bearing strength of a hollow section may be overestimated in particular cases.
- The applied torsional moment is constant.
- A pure shear stress condition exists. Only shear stresses and glidings occur. Axial stresses and elongations do not emerge.

Starting from the theory of elasticity, a simple calculation of the torsional shear stresses can be made on the basis of the "Saint-Venant" theory of torsion.

A twisting moment M_T acts on a member to twist two cross sections at a distance l from each other. The torsional angle θ is proportional to the length l of the member and can be calculated using the following equation:

$$\theta = \frac{M_T}{I_T \cdot G} \cdot l \qquad (5\text{-}97)$$

In degrees (°), $\theta = \dfrac{M_T \cdot l}{I_T \cdot G} \cdot \dfrac{180}{\pi}$

where:
M_T = torsional or twisting moment in Ncm
l = member length in cm
I_T = torsional inertia constant in cm^4
G = gliding modulus
$= \dfrac{E}{2(1+v)} = \dfrac{21 \cdot 10^6}{2(1+0.3)} = 8.1 \cdot 10^6$ N/cm^2
E = modulus of elasticity (Young's modulus)
$= 21 \cdot 10^6$ N/cm^2 (for steel)
v = transversal contraction coefficient
 (Poisson ratio) = 0.3 for steel

For CHS, the torsional inertia constant I_T is equal to the polar moment of inertia. Torsional inertia constant for other profile types is always smaller than polar moment of inertia.

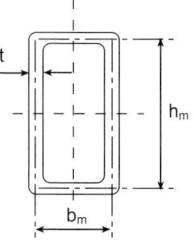

Fig. 5-21 RHS cross section, definition of b_m and h_m

For thinwalled RHS, the following is valid (Fig. 5-21):

$$I_T = \frac{4 \cdot A_m^2 \cdot t}{U} \qquad (5\text{-}98)$$

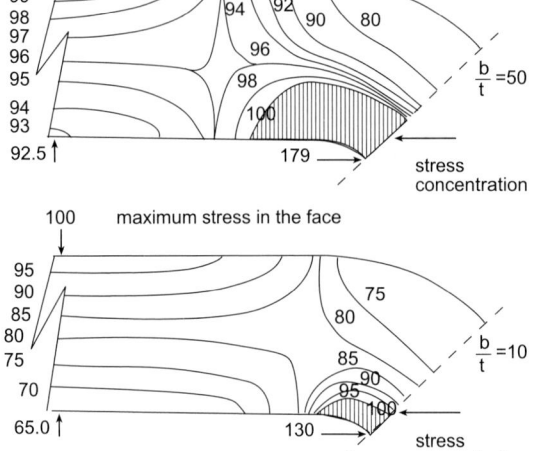

Fig. 5-22 Stress (τ_T) concentrations in the inside corner area of RHS

where:
$A_m \approx h_m \cdot b_m$
$U \approx 2\,(h_m + b_m)$

The St. Venant torsional shear stress can be determined very simply:

$$\tau_{T,Rd} = \frac{M_{T,Sd}}{W_T \cdot \gamma_{M0}}$$

W_t is the torsional section modulus.

For CHS, W_T is equal to polar section modulus.

CHS: $W_T = \dfrac{\pi}{16} \dfrac{d^4 - d_i^4}{d}$

The Bredt's formula, that is valid for thin-walled RHS, is described as follows:

$W_t = 2\,A_m \cdot t$

where:
$A_m \approx h_m \cdot b_m$

For RHS, the torsional shear stress is strongly affected by the wall thickness. This is due to the distribution of torsional shear stress over the wall thickness, which is not constant; τ_T decreases towards inside.

A stress concentration takes place on the inside of the rounded corners of RHS. In order to calculate the stress concentration factor, Timoshenko recommended a formula [54], which may give a value higher than 2. Fig. 5-22 shows the torsional stress diagrams for two RHS with $b/t = 50$ and $b/t = 10$ respectively, which have been determined with the help of the finite element analysis. As the corner region with stress concentrations has a share of only 1.13% to 1.15% of the total cross section, in practice the stress concentrations for b/t ratios lying between 10 to 50 need not be taken into consideration in calculation.

5.4 List of symbols

CHS	circular hollow section
RHS	rectangular (also square) hollow section
A_g	(total or gross) cross-sectional area
A_{net}	net cross-sectional area
A_v	shear area
E	modulus of elasticity (Young's modulus)
G	gliding (shear) modulus
G	permanent action (load)
I	second moment of area
K	buckling length coefficient $\left(I = \dfrac{l_K}{l_0}\right)$
M	mass
M	moment
M_x, M_y	bending moment about x and y axis
N, P	normal or axial force
Q, V	transversal force, shear force
Q	variable action (load)

5.4 List of symbols

W	section modulus
$b, B; h, H$	external side lengths of RHS
b_1	width of a flat element (see Table 5-16)
h_1	depth of a flat element (see Table 5-15)
b_m	average width of RHS $= (b-t)$
b_c	
b_{e1}, b_{e2}	
b_{eff}	see Table 5-18
b_1	
d, D	outer diameter of CHS
d_m	average diameter of CHS $= (d-t)$
d_i, D_i	inner diameter of CHS
f	axial (normal) stress
$f_{cr,LT}$	critical elastic stress for lateral-torsional buckling
f_u	ultimate tensile strength of the basic material
f_v	equivalent stress
f_y	tensile yield strength
f_{ya}	average increased tensile yield strength of a cold formed hollow section
f_{yb}	tensile yield strength of the basic material of a hollow section
h_m	average depth of RHS $= (h-t)$
i	radius of gyration
k_p or k_τ	plate buckling factor
l	length
l_0	system length
l_K	effective buckling length
r	outer radius of CHS
r_0	outer corner radius of RHS
r_i	inner corner radius of RHS
r_m	average diameter of CHS $= (d-t)$
t, T	wall thickness
$x-x$	major axis of the cross section
$y-y$	minor axis of the cross section
$z-z$	member axis
α	coefficient of linear expansion
α	imperfection factor for the European buckling curves (see Table 5-12) or width ratio of the buckling zone (see Fig. 5-7)
β_M, β_m	equivalent uniform moment factor (see Tables 5-25 and 5-26)
γ_F	partial safety factor for action (load)
γ_M	partial safety factor for resistance (see Table 5-6)
δ_x, δ_y	shift of the neutral axis of a thin-walled hollow section (see Table 5-20)
ϱ	density
ϱ	plate buckling reduction factor
ε_u	ultimate strain
ε_y	yield strain
λ	slenderness of a column
λ_E	Eulerian slenderness
$\bar{\lambda}$	dimensionless slenderness of a column
$\bar{\lambda}_{LT}$	dimensionless slenderness of a member for lateral-torsional buckling
κ_x, κ_y	coefficients (see Eqs. (5-88) and (5-92))
μ_x, μ_y	factors (see Eqs. (5-89) and (5-93))
ν	transversal contraction coefficient (Poisson ratio)
χ	reduction factor according to the European buckling curves
ψ	ratio of the stresses (see Figs. 5-15 and 5-18)
τ	shear or torsional stress

Subscripts:

b	bolts
B	local buckling
BK	interaction local and flexural buckling
c	compression
eff	effective
el	elastic
k	characteristic
K	flexural buckling
max	maximum
min	minimum
net	net
pl	plastic
p	plate
Rd	design resistance
r	rivet
s	shell
Sd	design load (action)
t	tension
T	torsion
LT	lateral-torsion
x	$x-x$ axis
y	$y-y$ axis

6 Structural elements and constructions with hollow sections under predominantly static loading

As Fig. 1-2 demonstrates, the most important preferences for hollow sections are based on their characteristic properties as load bearing elements for compression, torsion and bi-axial as well as multi-axial bending. Therefore in all structural applications of hollow sections, attention has to be paid that these properties are adequately taken into account.

Further, as has been already stated in the chapter for introduction, the welding in hollow section structures has a prime position over other jointing techniques. The significant reason for this lies in the fact that two hollow sections can be directly connected to each other only by welding, as one-sided access is sufficient for this purpose. In case the bolting has to be used as jointing technique the two-sided access, which is necessary to lock the bolt from inside, may be a problem; the connection can only be made indirectly through welded plates, angles or similar, where a combined application of welding and bolting takes place (see Fig. 1-7). In some cases, long bolts are also used for bolting through the hollow section (see Fig. 6-27). However, a few blind bolts have been developed, which make directly bolted connection of two hollow sections possible. This subject will be further dealt with in Section 8.5.

In principle, hollow section joints can be classified in two main types (Fig. 6-1):

- Direct connections, where the structural elements are joined to one another directly.

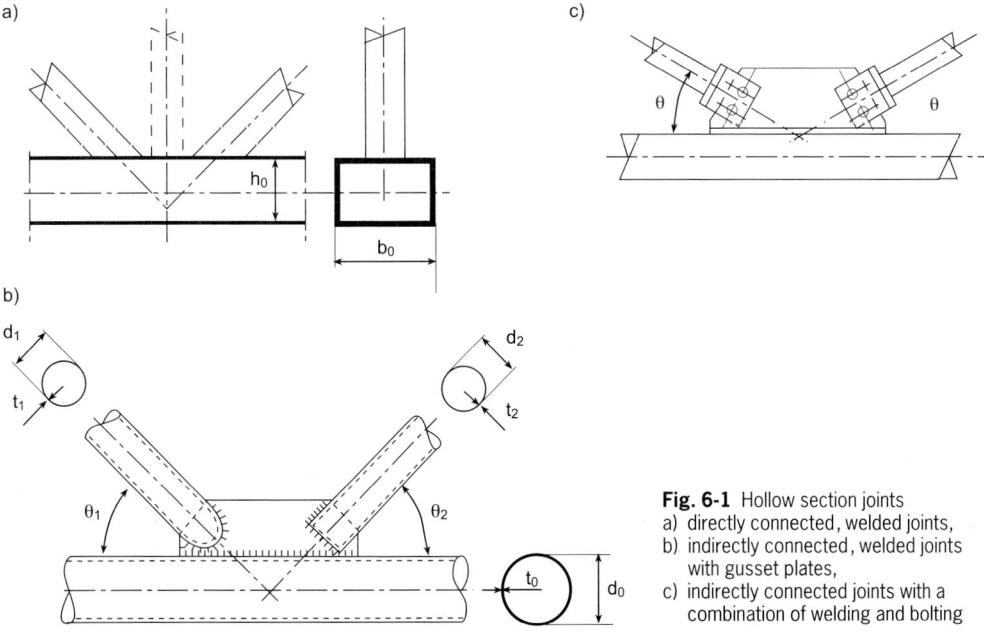

Fig. 6-1 Hollow section joints
a) directly connected, welded joints,
b) indirectly connected, welded joints with gusset plates,
c) indirectly connected joints with a combination of welding and bolting

- Indirect connections, where the structural elements are joined to one another through head plates, gusset plates, angles etc.

The structural integrity and technical security of a construction with directly connected members is clearly more favourable than that, where the members are joined indirectly. This is due to the fact that the transmission of force or moment in a direct joint takes place directly from one hollow section to the other, while this occurs twice in an indirect joint – first from one hollow section to the gusset plate and then from the plate to the other hollow section. Therefore statistically, the sources of error in design and fabrication are double as many for indirect joints as those for direct ones (Fig. 6-1).

This fact speaks in favour of the direct connections in hollow section structures, when all aspects of a project are taken into consideration. They are e.g. calculation and design, manufacture, assembly, erection, maintenance and repair as well as technical security, economy and aesthetics. It is therefore recommended to use direct connections as far as possible.

Naturally in many cases, e.g. field assembly or restricted transport sizes (when a structure has to be transported in smaller sections) or avoiding field welding, detachable bolted joints are preferably used. Prefabricated, welded units of a structure are simply indirectly bolted to one another on site. Special attention has to be given to the observance of the fabrication tolerances.

6.1 Single section beams

Various solutions are offered by rolled I or H sections, RHS, castellated beams etc., which are suitably applied as girders under bending moment in structures depending on the span length and load magnitude (Fig. 6-2) (see Section 5.3.2.1). RHS girders with their long sides set in the plane of bending are usually applied for uni-axial bending, as they possess maximum bending rigidity EI_{max} in this position. It is evident that a rolled I beam presents a more economical solution for this type of load than hollow sections.

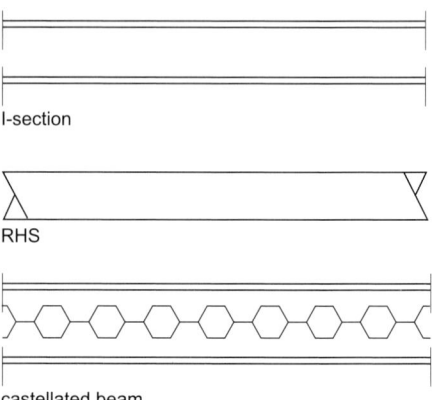

I-section

RHS

castellated beam

Fig. 6-2 Solid web girder

Square and circular hollow sections are more suitable for bi-axial bending. Due to its identical load bearing strengths in all planes, CHS can be applied at the best for multi-axial bending.

For small spans, hollow sections show good shear properties even without any stiffening. For longer spans, RHS demonstrates excellent lateral-torsional stability (see verification of lateral-torsional resistance in Section 5.3.2.1).

6.2 Columns

Due to the superior stability behaviour (flexural and local buckling, lateral-torsion) of hollow sections to other structural profiles, they are used most frequently as columns. This is done in buildings as well as in various other fields. In general, hollow sections are preferred as structural elements in compression.

Fig. 6-3 shows an example, which demonstrates higher saving of material in kg/m, when CHS/RHS is used for a 3 m long column instead of open profiles. The structural arrangement at the base of a column is to be made according to the moment to be transmitted. The most usual case represents a single, fairly thick flat plate, which is simply welded to the end of the column (Fig. 6-4). The column end is either plane-cut or milled to a plan surface to bear evenly on the base plate.

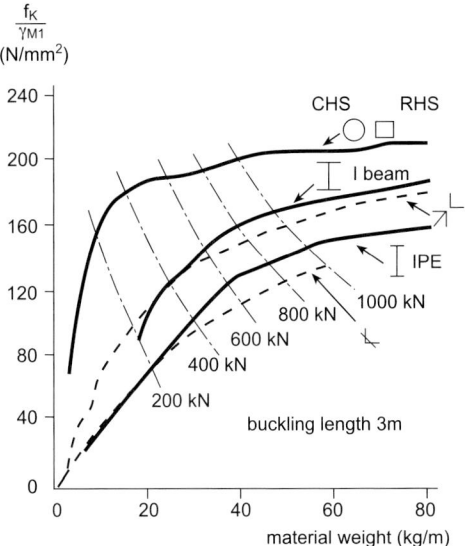

Fig. 6-3 Comparison of the masses of hollow and open sections in compression in relation with loading

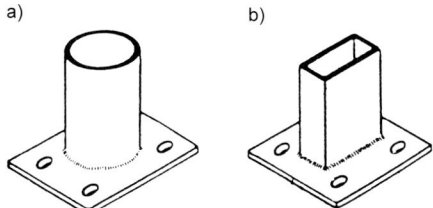

Fig. 6-4 Single leg column welded to base plate. a) CHS, b) RHS

The columns can also comprise of a lattice construction instead of a single hollow section (Fig. 6-5).

The arrangements shown in Fig. 6-4 are normally suitable for transmitting small bending moments. When larger moments have to be withstood, stiffeners consisting of flat plates and angles can be applied (Fig. 6-6). The stif-

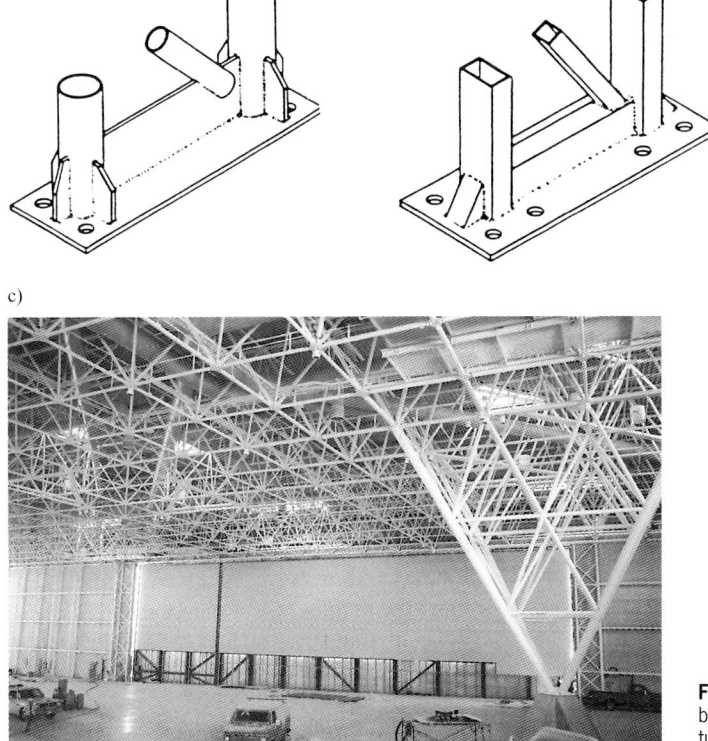

Fig. 6-5 Lattice column welded to base plate. a) CHS, b) RHS, c) braced tubular column support for a roof construction

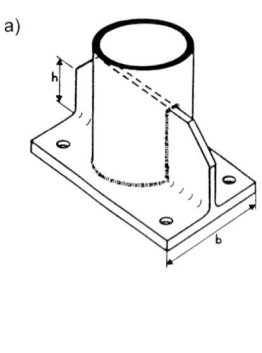
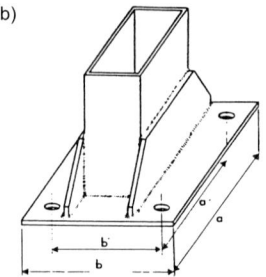
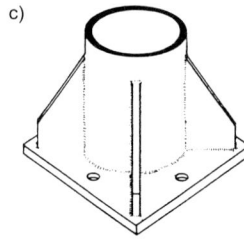
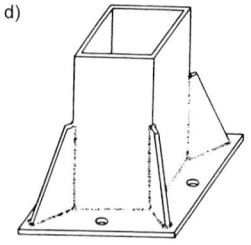
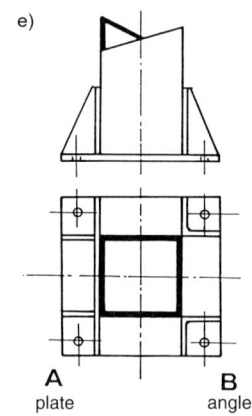

A
plate

B
angle

Fig. 6-6 Single leg columns with various arrangement of stiffeners.
a) and b) Uni-axial bending,
c)–e) multi-axial bending

feners are located depending on the directions of the bending moments. In principle, because of economical fabrication, stiffeners should be avoided, if possible, by substituting the stiffeners by a thicker base plate.

There may also be cases, where the design assumptions specify an effective hinge in a particular plane, which makes a rotation about an axis possible. Fig. 6-7 shows a number of these column bases with hinged supports.

Column bases can also be made adjustable allowing for height as well as out of plumb adjustments (Fig. 6-8).

Fig. 6-9 illustrates an arrangement for connecting an internal rain water down pipe at the foot of a hollow section column. An elbow, in plastic or cement, is bedded into the concrete foundation and connected to the rain water down pipe. A locating template is desirable, as the possibilities of adjustment are somewhat restricted here. However, measures are to be taken for the protection of the inside of the column against corrosion. This can be done either by galvanizing the hollow section or applying a seal between the pipe and the hollow section at the head and the foot of the column.

6.2.1 Column-to-truss connections

The truss rests upon the top of the column. This type of construction is mostly found at the supporting locations of trusses on hollow section columns. Usually a bolting construction is applied in order to attain economy by avoiding site welding for assembly and erection. Truss units are welded in workshops, transported to building locations and bolted to the head of the column with a head plate on site. Fig. 6-10a and b illustrate applications, where the joint is made over a continuous upper or lower chord of a truss.

6.2 Columns

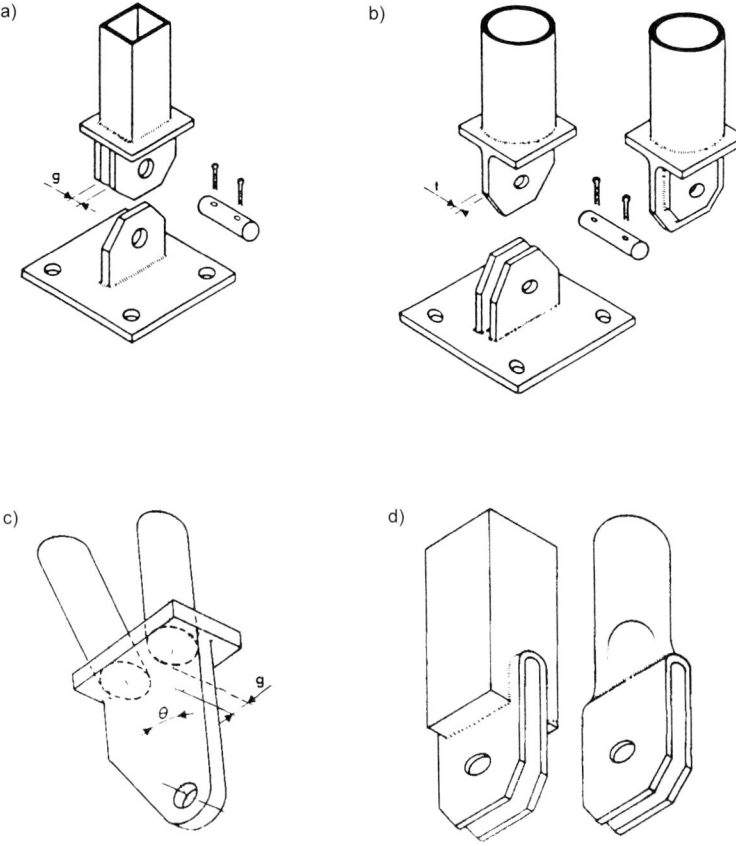

Fig. 6-7 Column base with hinged support

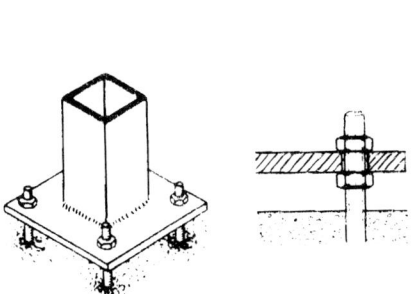

Fig. 6-8 Column base with adjustable support

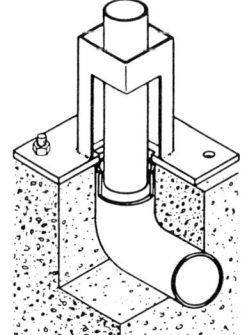

Fig. 6-9 Hollow section column base with a rain water down pipe

a)

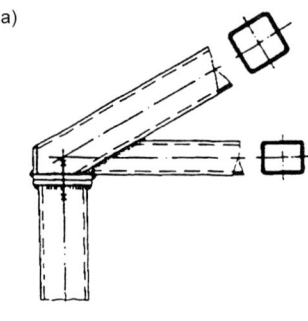

b)

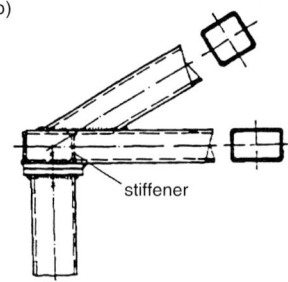
stiffener

Fig. 6-10 Column-to-truss connections.
a) Continuous upper chord, b) continuous lower chord

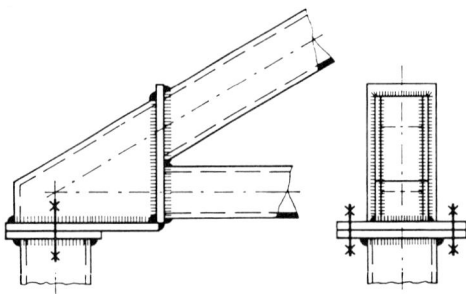

Fig. 6-11 Preset shoe bolted to the column head plate and welded to the truss chords

a)

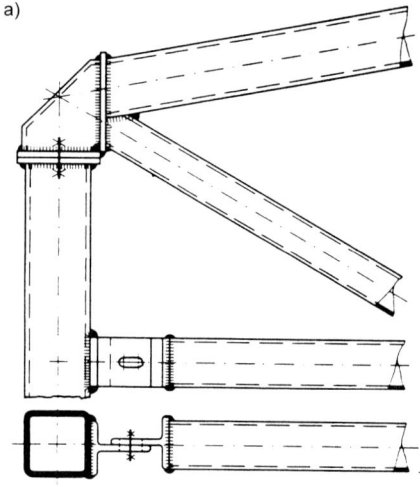

b)

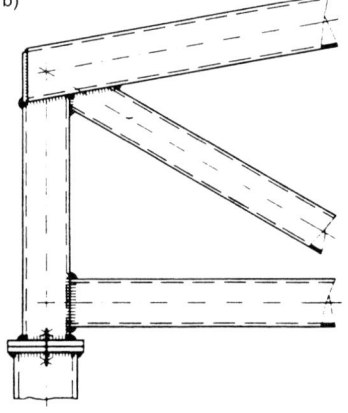

Fig. 6-12 Upper and lower chord of a truss welded or bolted to a hollow section column at separate locations. a) Console welded to the upper chord and the lower chord welded to a tee section is bolted to another tee section, which is welded to the hollow section column, b) complete truss bolted to the column through a head plate

For transmitting high load through lower chord, an additional stiffening plate can be welded (see Fig. 6-10b).

As Fig. 6-11 shows, it is also possible to weld a shoe to the converging upper and lower chords and use it as a load bearing foot.

Upper and lower chords can also be welded or bolted at separate locations on hollow section columns using tee sections or flat plates (Fig. 6-12).

A further alternative is shown in Fig. 6-13, where the truss is set against the side of the

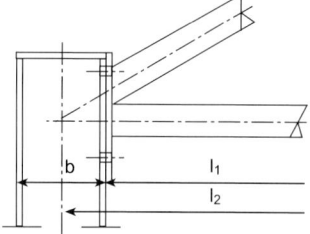

Fig. 6-13 Truss bolted laterally to the hollow section column

column. In this case, the difference between l_1 and l_2 is accommodated by inserting a distance piece.

6.2.2 Beam-to-column connections

In order to realise a simple design of a beam-to-column connection, it is necessary that this is provided with sufficient flexibility and rotation capacity to accommodate beam end rotation with the deflection of the beam. These indirect connections are almost always made with a combination of welding and bolting. Often in skeleton structures, beams of rolled I sections are joined with hollow section columns.

The given examples in Fig. 6-14 can be used in a wide variety of constructions: Single and multiple storeyed buildings, light and heavy structures. The selection of the type of the connection depends fundamentally on the size of the beam and the magnitude of the force to be transmitted. The connections can be assumed to be hinged or very nearly hinged. Fig. 6-15 shows the basic arrangement of a conventional type of connection and indicates the various kinds of loading:

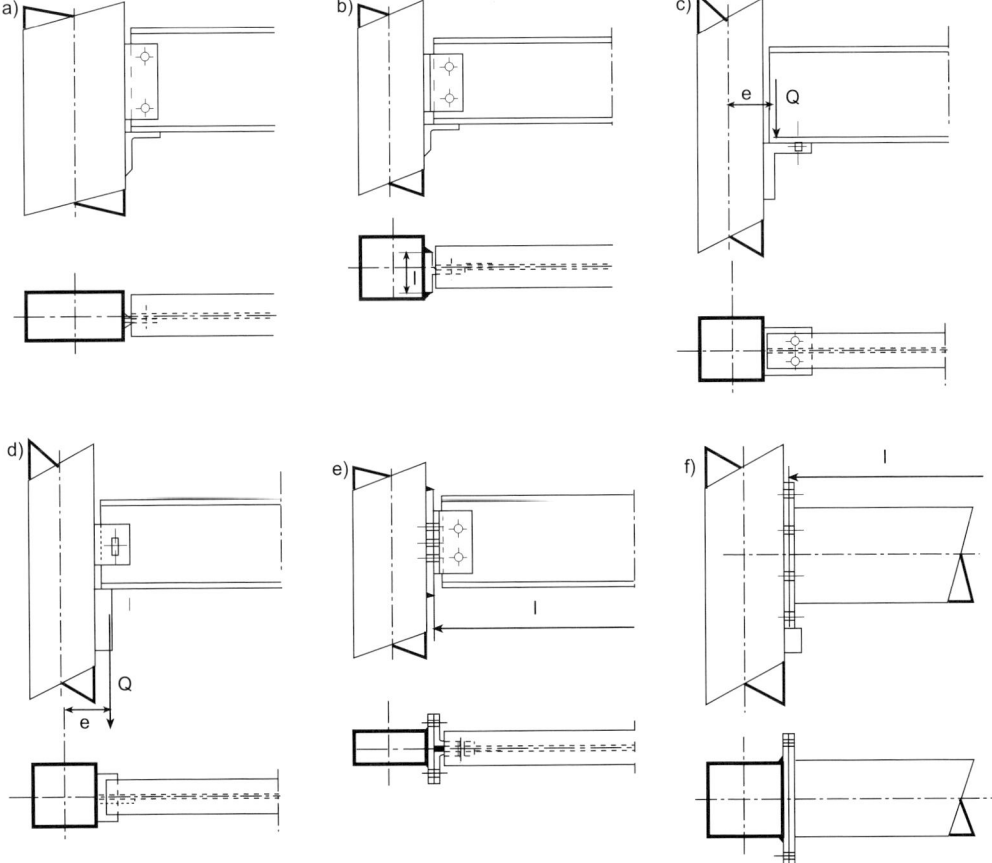

Fig. 6-14 Various beam-to-column connections: a) with console and plate welded to column, b) with console and tee section welded to column, c) with angle welded to column face and bolted to bottom flange of beam, d) plates welded to column face horizontally and vertically, where the vertical plate is joined by bolting to the beam web (long hole) while the bottom flange of beam is set on the horizontal plate, e) a pair of angles bolted to plate, which is welded to column face; angles are further joined to the web of beam by bolting, f) plate on column face bolted to beam end plate

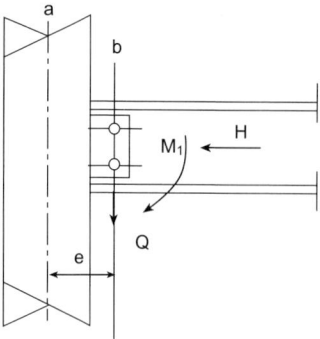

Fig. 6-15 Basic arrangements of a conventional beam-to-column connection

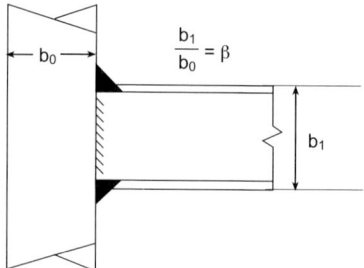

Fig. 6-16 I beam directly welded to RHS column face

Q: Vertical reaction (action)
H: Horizontal reaction (action)
M_1: Bending moment transmitted by the beam
M_2: Bending moment due to the eccentricity of the attachment $= Q \cdot e$

According to the arrangement (one or more bolts), the action of the bending moment takes place differently:

– The hinge is located on the axis "a" and an arrangement with two bolts gives a slight fixity of the joint upon the beam. The resulting bending moment M_1 acts on the beam.
– The hinge is located on the axis "b" and an arrangement with one bolt results in the bending moment $M_2 = Q \cdot e$ acting on the column.

The connections shown in Fig. 6-14a and b are very common and are used for medium loads. Fig. 6-14c and d suggest connections to transmit high shear loads, where the bending moment $M = Q \cdot e$ is reacted by the column. Fig. 6-14e and f show further connections of different rigidities, where Fig. 6-14f illustrates a connection with both beam and column consisting of RHS. This is however a very rare case in practice.

If an I-section beam with a relatively wide flange is directly welded to RHS column face (Fig. 6-16), the in-plane rigidity of the column flange is very small for $\beta \ll 1.0$ against the concentrated load from the beam flange. The column flange may collapse as a consequence and the column web may buckle due to the bending moment in the beam flange.

This deformation can be prevented by a arranging stiffeners at the level of the beam flanges, through which the flange stresses are transmitted to the opposite side of the column and the crippling of the column web also does not take place. Three types of diaphragms are shown in Fig. 6-17, which act as stiffeners as well as serve as a mounting platform for beams.

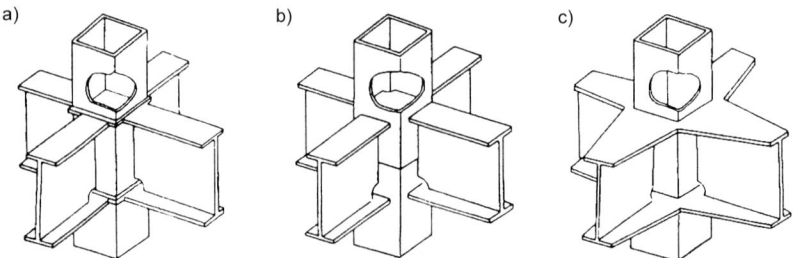

Fig. 6-17 Types of diaphragms to increase the rigidity of the beam-to-column connections.
a) Through diaphragm, b) interior diaphragm, c) exterior diaphragm

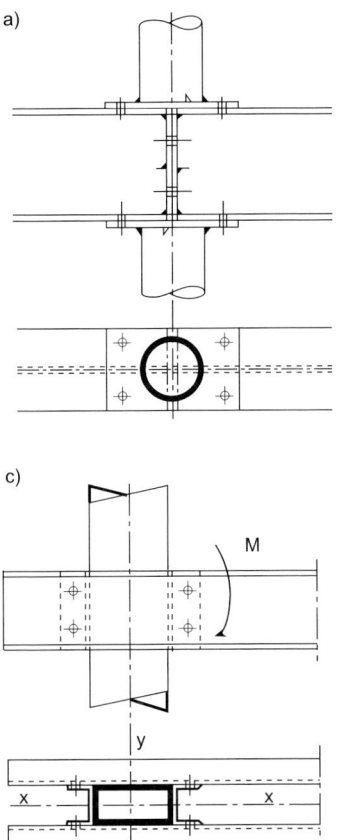
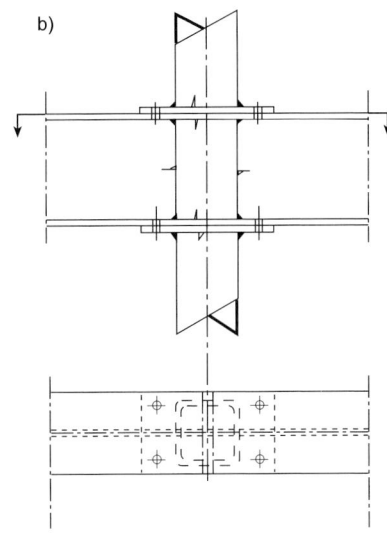

Fig. 6-18 Beam-to-column connections with continuous beams

Fig. 6-18 illustrates three further proposals for connections, which are, although complex, often applied. In these cases, the beams are continuous and extend on both sides of the column. They can be designed to transmit large forces and bending moments.

It is here worth mentioning about a Swedish development for a roof construction with a continuous RHS column and a "welded top hat" section as a simply supported beam (Fig. 6-19). The beam support comprises an end plate, welded on to the beam, which is carried on a cleat on the column by bolting. The bolts are located in the extended end plate outside the hollow section column. The floor slab can be laid on the bottom flanges of the beam (Fig. 6-20). In this position, there is no beam protruding below the floor. As a consequence, the fire protection measures can be partly or wholly omitted.

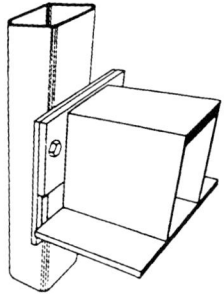

Fig. 6-19 "Welded top hat" beam support to RHS column

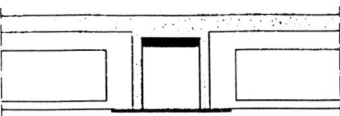

Fig. 6-20 "Welded top hat" beam is built into concrete floor. This construction saves space and requires no or only partly fire protection

Fig. 6-21 Roof construction with arches of single RHS

6.3 Arched girders

With the progressive development of the bending technique, the architects use more and more bent steel hollow sections as girders while designing attractive structures such as domes and vaults. Arches are made of single sections (Fig. 6-21) as well as with lattice type girders (Fig. 6-22). Due to the simple fabrication, single hollow sections are often bent in the form of an arch. The lattice type may well be circular, parabolic or elliptical.

According to the type of construction, the ends of the arches spring directly from their abutments or are supported at the heads of vertical columns.

Three basic configurations of arches are usually applied: 3 pinned arch, 2 pinned arch and fixed end arch (Fig. 6-23).

The design of the foot of the arch has to be made basing on the soil properties and the possible setting behaviour of the fundament.

In the case of raised arches on columns, a tie or even a tie-strut is commonly added in order

Fig. 6-22 Arched triangular girder with the foot fixed in a concrete foundation

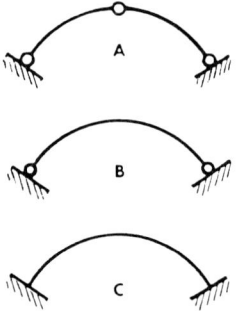

Fig. 6-23 Static systems for the arched girders:
A) 3 pinned arch, B) 2 pinned arch, C) fixed end arch

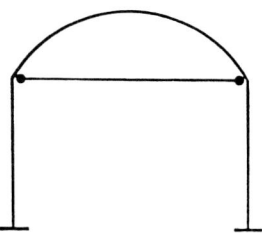

Fig. 6-24 Tie-strut balancing the horizontal spring force in the arch

to balance the thrusts at the springing points (Fig. 6-24).

In order to ensure the stability of a vault, hollow sections are preferably applied because of their excellent torsional rigidity and the resulting high lateral-torsional stability. Fig. 6-25 shows a vault construction consisting of arched girders and bracing members. It is possible to achieve very wide spans even without any interconnecting bracing between the arched girders. According to Fig. 6-25, the arched girders can be made of single hollow sections as well as in the form of lattice girders (triangular in particular).

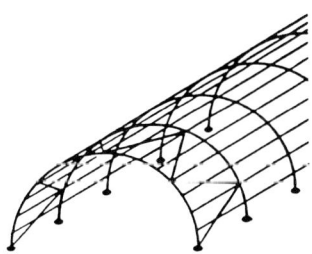

Fig. 6-25 Vault construction with arched girders and interconnecting bracings between them

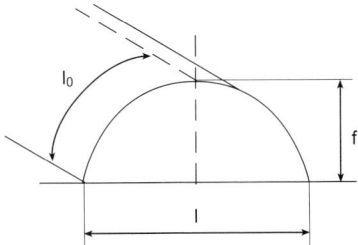

Fig. 6-26 Definition of the effective buckling length of an arch

The calculation of arched girders can be made using the method for the straight segments. The verification of the in-plane buckling resistance of a circular or parabolic arch can be carried out as follows [2]:

Effective buckling length $l_K = B \cdot l_0$ \hfill (6-1)

The values of the coefficient B can be taken from Table 6-1.

Table 6-1 Values of the coefficient B (see Eq. 6-1 and Fig. 6-26)

Ratio f/l	0.05	0.20	0.30	0.40	0.50
3 pinned arch	1.20	1.16	1.13	1.19	1.25
2 pinned arch	1.00	1.06	1.13	1.19	1.25
Fixed end arch	0.70	0.72	0.74	0.75	0.76

6.4 Knee joints

As has been already described in Section 6.2.2 for beam-to-column connections, the knee joints of hollow sections can be made as welded (flexurally stiff) or bolted (detachable) connections.

Fig. 6-27 illustrates a number of examples of detachable, bolted knee joints.

Fig. 6-27a and b illustrate two possible joints. The bored holes in the hollow section tie beams can be sealed inside by means of welded tubular jackets, so that the rusting of the inside of the tie beam is prevented (Fig. 6-27a). The holes in the tie beam is not required, if gusset plates are welded to the tie beam, which are then bolted to the counterplates welded to the post (Fig. 2-27b).

Fig. 6-27c shows a simple and economical knee joint, where slant head plates welded to the members are bolted together.

Welded, flexurally rigid knee joints with RHS are used very often. These joints consisting of legs of identical sizes and lintel cross sections are directly welded connections (Fig. 6-28a). Knee joints can also be fabricated with intermediary stiffening plates to adapt different member sizes (Fig. 6-28b). Although RHS are applied more frequently to these joints,

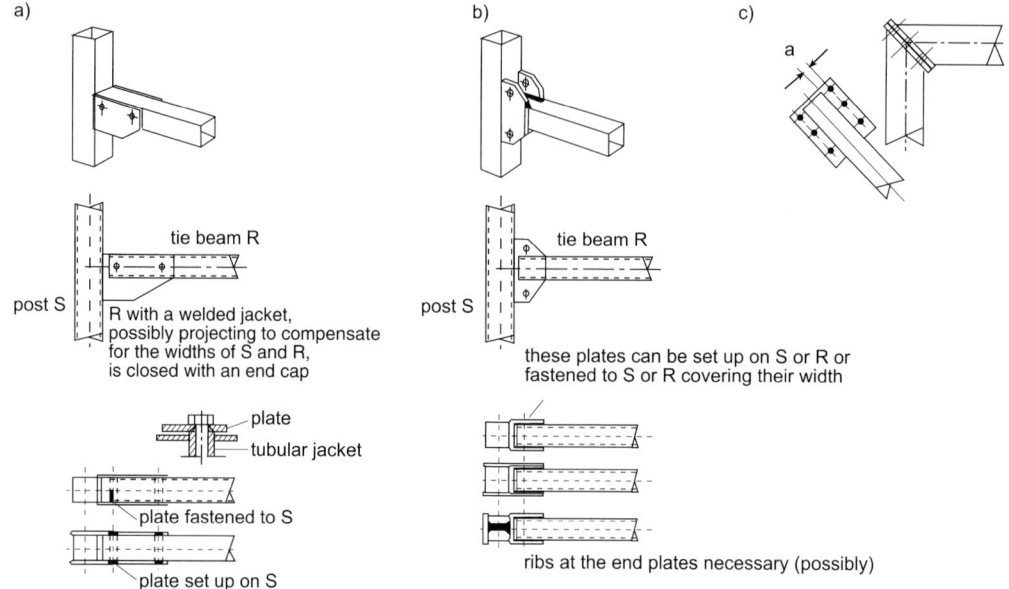

Fig. 6-27 Detachable, bolted knee joints: a) with continuous, detachable post and fish plate, b) with continuous, detachable post and double fish plate, c) with discontinued post

the use of CHS is also possible for this purpose.

The directly welded knee joints without stiffening plates are meant for low loads. They tend to fail under compression due to excessive deformation of the transverse face of the hollow section.

The knee joints with intermediary plates are usually used when high joint strength is required. Excessive deformation can only take place for very thinwalled hollow sections.

The following conditions apply while determining the thickness of the intermediary plate of a stiffened knee joint [3]:

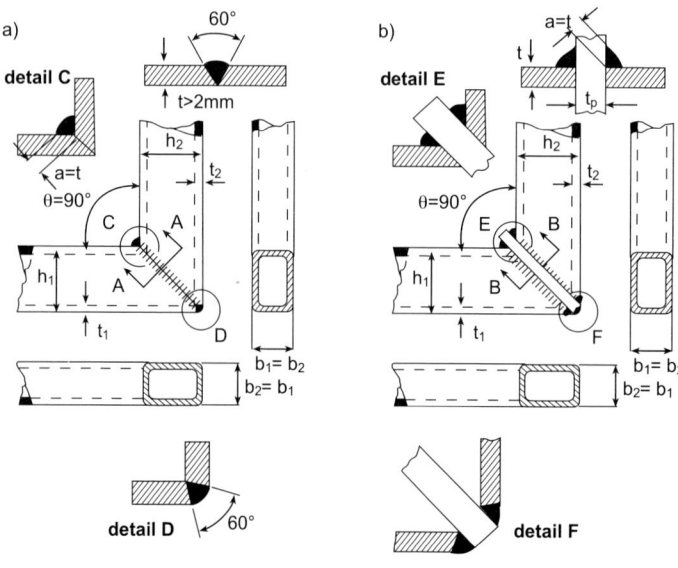

Fig. 6-28 Bending resistant, welded knee joints:
a) without stiffening plate,
b) with stiffening plate

6.4 Knee joints

$t_p \geq 1.5\, t_i\ (i=1\ \text{or}\ 2)$ and $t_p \geq 10\ \text{mm}$ \hfill (6-2)

Research works to work out design rules for RHS knee joints with or without stiffening plates were carried out at the Testing Centre for Steel, Timber and Stone of the University of Karlsruhe [4]. The proposed design recommendations have been adopted by [3, 5, 6] (for further literature, see [7–9]).

RHS joints with stiffening plates are used in structures requiring large rotation capacity, which meet the design requirements for stiffened structures [7]. In knee joints without stiffeners, this can be attained by applying thickwalled RHS of the cross section class 1 (Fig. 6-29) [8].

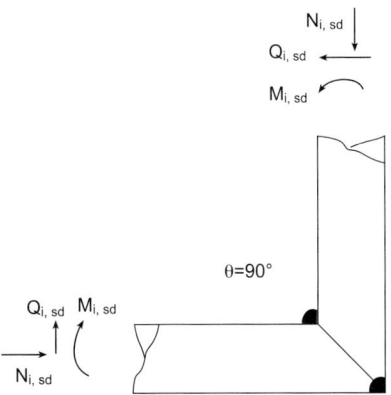

Fig. 6-29 Definition of the design loading on knee joints without stiffeners

The test results show that an adequate design resistance of an unstiffened knee joint with mitre cut of 90° under bending moment and axial force can be calculated by means of multiplication of the material yield strength f_y with a reduction factor α. The following relationship is valid according to [4]:

$$\frac{M_{i,Sd}}{W_{i,pl}} + \frac{N_{i,Sd}}{A_{g,i}} \leq \alpha \cdot \frac{f_{y,i}}{\gamma_{MW}} \quad (6\text{-}3)$$

where:
$M_{i,Sd}$ = design bending moment on the hollow section i (1 or 2) at the system point of knee joint

$N_{i,Sd}$ = design axial force on the hollow section i
$W_{i,pl}$ = plastic section modulus of the hollow section i
$A_{g,i}$ = cross-sectional area of the hollow section i
f_{yi} = yield strength of the material of the hollow section i
α = reduction factor
γ_{MW} = partial safety factor for welded joint resistance

The reduction factor α can be read from the diagrams in Fig. 6-30. Table 6-2 contains the validity ranges of these diagrams. Further following conditions have to be fulfilled:

– Weld thickness a = minimum t
– Weld preparation according to Fig. 6-28
– Shear in the connection zone should meet the following relationship:

$$Q_{i,Sd} \leq \frac{1}{3} \cdot \frac{f_{y,i}}{\sqrt{3}} \cdot A_v \cdot \frac{1}{\gamma_{MW}} \quad (6\text{-}4)$$

A_v = cross-sectional area of the RHS web $(2 \cdot h_i \cdot t_i)$

– Weld verification according to DIN 18800, part 1, section 8.4 [10], weld verification is not necessary when $\alpha \leq 0.84$ for S 235 and $\alpha \leq 0.71$ for S 355

$\alpha = 1.0$ is to be assumed for the welded knee joints with intermediary plates. Use Eq. (6-2) for determining the thickness of stiffening plate.

The angle of inclination between the axes of the hollow section members may exceed 90° (obtuse angle $90° < \theta < 180°$) (Fig. 6-31). These joints can in any case be designed similar to 90°-joints, since their strength is higher than that of 90°-joints.

The increase of the design strength of knee joints by using obtuse angles between the axes of the members has been exploited to advantage by Eurocode 3 [3] as follows:

$$\frac{N_{i,Sd}}{N_{pl,i,Rd}} + \frac{M_{i,Sd}}{M_{pl,i,Rd}} \leq \alpha \quad (6\text{-}5)$$

where for $\theta \leq 90°$,

$$\alpha = \frac{\sqrt{b_i/h_i}}{(b_i/t_i)^{0.8}} + \frac{1}{\left(1+\dfrac{2b_i}{h_i}\right)} \quad (6\text{-}6)$$

and for $90° < \theta \leq 180°$,

$$\alpha = 1 - \sqrt{2}\cos\left(\frac{\theta}{2}\right)(1 - \alpha_{90°}) \quad (6\text{-}7)$$

$\alpha_{90°}$ = value of α for $\theta = 90°$ according to Eq. (6-6).

Applying Eqs. (6-5), (6-6) and (6-7), the following conditions are to be met:

- For simple bending, RHS has to belong to the cross section class 1.

- $N_{i,Sd} \leq 0.2\, N_{pl,i,Rd} = 0.2\, A_g \cdot f_{yi}/\gamma_{M0}$ (6-8)

- Execution of weld according to Fig. 6-28

- $V_{i,Sd} \leq 0.5\, V_{pl,i,Rd}$

$$= 0.5 \cdot \frac{f_{yi}}{\sqrt{3}}(2 \cdot h_i \cdot t_i)\frac{1}{\gamma_{M0}} \quad (6\text{-}9)$$

In case the condition according to Eq. (6-9) is not fulfilled, the joint strength can still then be accepted as adequate, when the interaction of axial force, bending moment and shear can be verified as positive (see Table 5-8).

An alternative design to increase the joint strength is to reinforce the joint by welding a haunch, which can be made of an offcut of RHS (Fig. 6-32a). It is also possible to prevent local buckling by welding two lateral plates (Fig. 6-32b). A single plate however, cannot substitute a haunch (Fig. 6-32c).

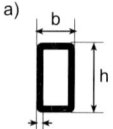

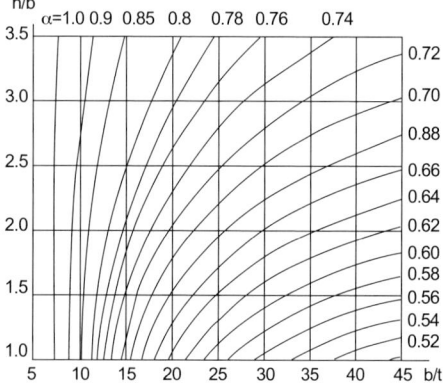

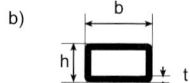

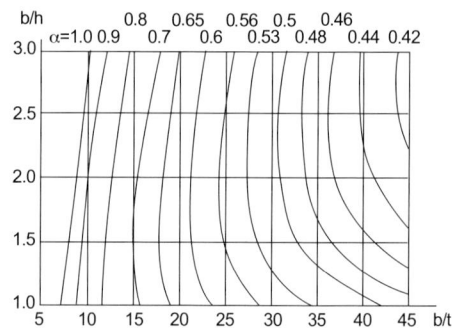

Fig. 6-30 Reduction factor α for bending resistant, unstiffened knee joints of RHS

Table 6-2 Ranges and rules for member sizes of flexurally rigid RHS knee joints [5]

Validity ranges for knee joints with mitre cut member ends	
with stiffening plate	without stiffening plate
$b_i \leq 400$ mm	$b_i \leq 300$ mm
$h_i \leq 400$ mm	$h_i \leq 300$ mm
$0.33 \leq h_i/b_i \leq 3.5$	$033 \leq h_i/b_i \leq 3.5$
$t_i \geq 2.5$ mm	$t_i \geq 2.5$ mm
For S235: $t_i \leq 30$ mm	For S235: $t_i \leq 30$ mm
For S355: $t_i \leq 25$ mm	For S355: $t_i \leq 25$ mm
For S235: $b_i/t_i \leq 43$	For S235: $b_i/t_i \leq 43$
$h_i/t_i \leq 43$	$h_i/t_i \leq 43$
For S355: $b_i/t_i \leq 36$	For S355: $b_i/t_i \leq 36$
$h_i/t_i \leq 36$	$h_i/t_i \leq 36$

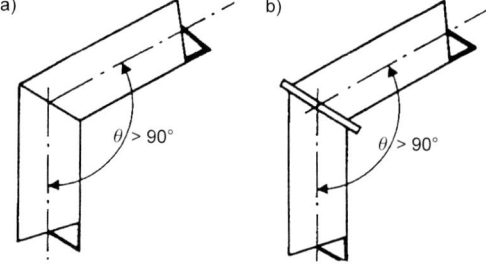

Fig. 6-31 Knee joints with obtuse angle between RHS axes

6.4 Knee joints

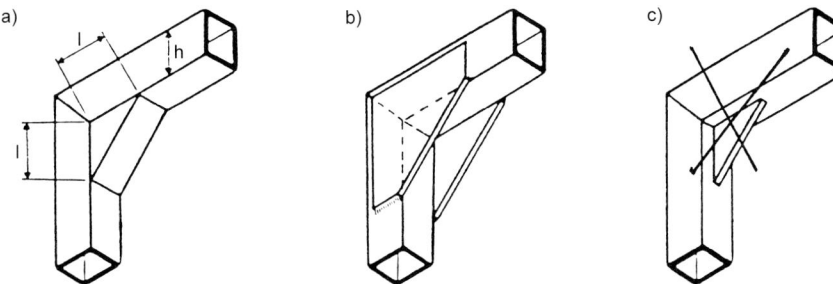

Fig. 6-32 Reinforcement measures for unstiffened knee joints

Taking the width of the haunch equal to that of the RHS members, it is simple to fabricate a haunch as an offcut of RHS of the same width as the members. If the length of the haunch is adequate so that the bending moment is not exceeded by the yielding moment of the members $W_i \cdot f_{yi}$, it is not then further necessary to check the joint strength.

With regard to the welding of the haunch, the weld gap g must be at the most equal to 3 mm (Fig. 6-33a). As a simple alternative, the ratio $b_1/b_0 \leq 0.85$ can be taken (Fig. 6-33b).

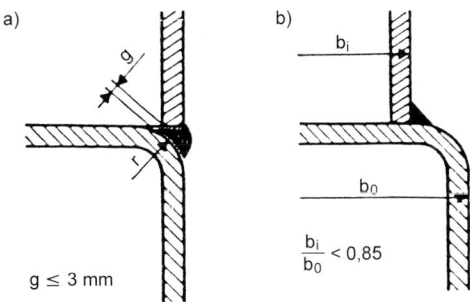

Fig. 6-33 Weld design of the haunch in a knee joint

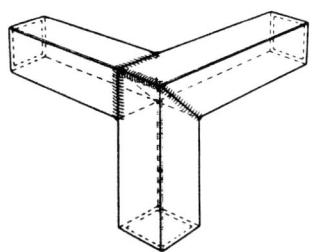

Fig. 6-34 Welded, three-dimensional RHS joint without stiffener

Fig. 6-34 shows further a welded, three-dimensional (each 90°) unstiffened joint with RHS members of identical size.

6.4.1 Knee joints with circular hollow sections

The research results, reported by the Testing Centre for Steel, Timber and Stone of the University of Karlsruhe [11], offer the method to verify the design strength of welded, bending resistant and unstiffened knee joints with CHS.

The Eqs. (6-3) and (6-4) can be applied also to CHS joints, but the value for the reduction factor α has to be read in Fig. 6-35.

In accordance with the investigations carried out [11], the following ranges of validity are decisive:

$d_i \leq 300$ mm
$d_i/t_i \leq 67$
$\theta_i \leq 90°$

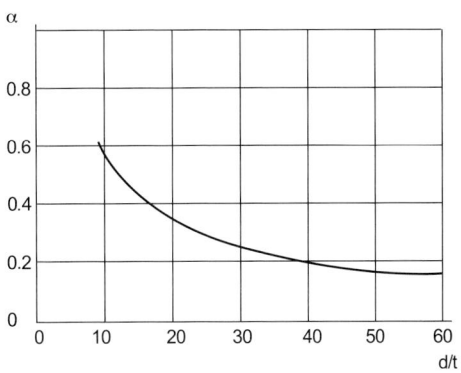

Fig. 6-35 Reduction factor α for welded, bending resistant CHS knee joint without stiffening plate

"A_v" value in Eq. 6-4 has to be calculated for CHS joints as follows:

$$A_v \approx 2\, d_i \cdot t_i \qquad (6\text{-}10)$$

6.5 End-to-end connections

Two fundamental types are described in this Section – indirectly joined, detachable connections made by welding or by a combination of welding and bolting and directly joined connections with welded members.

6.5.1 Indirectly joined bolted connections

The calculation of the bolted connections with hollow sections is made normally following the procedure used in conventional structural engineering. However, there are distinctive features in the load carrying behaviour of hollow section connections amongst themselves or in combination with other profiles or plates, which affect their strength significantly and had therefore to be investigated in the last decades.

Fig. 6-36 illustrates a variety of simple detachable bolted end-to-end connections, where the intermediaries e.g. plates, angles, tees, channels etc. are welded to the hollow sections, thereby ensuring a natural corrosion protection of the inside volume of the hollow sections (Fig. 6-36 a–g) by closing their ends. While determining the dimension of the tee section in Fig. 6-36c (tee section can be fabricated also by welding two flat plates together), it has to be attended to that the flange of the tee section is sufficiently thick, so that the load can be transmitted effectively to the hollow section [12].

A simple calculation of a CHS connection with a tee section welded to the hollow section end (assumption: joint strength ≥ full design resistance of the CHS cross section) can be carried out as follows:

$$\pi \cdot d_1 \cdot a \cdot f_{a,\text{weld}} \geq A_1 \cdot f_{y1} \qquad (6\text{-}11)$$

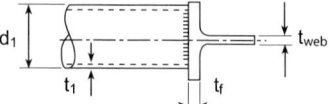

a) with angle welded to hollow section end — connection is suitable for the transmission of load in the shown direction

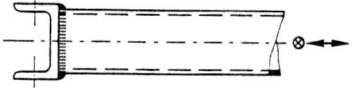

b) with channel welded to hollow section end

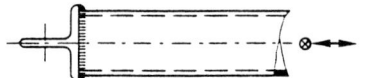

c) with tee section welded to hollow section end

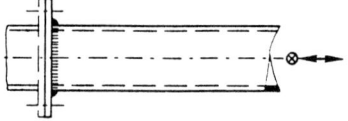

d) with flange plate welded to hollow section end

e) with flat plate inserted into the hollow section slit and then welded

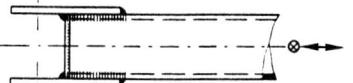

f) with flat plates welded to the two sides of RHS

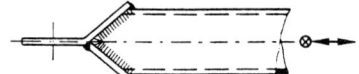

g) with a "Y" piece made of flat plates welded to RHS

Fig. 6-36 Bolted end-to-end connections of hollow sections with open profiles or plates welded to them

$$A_{T,\text{webplate}} \cdot f_{y,\text{web}} \geq A_1 \cdot f_{y1} \qquad (6\text{-}12)$$

$$t_f \geq \frac{d_1 - t_{\text{web}}}{5} \qquad (6\text{-}13)$$

(valid also for Fig. 6-37a) [230]

For larger sizes, double-tongued forks can be fabricated from plates and then welded to the hollow section ends (Fig. 6-37a). Another alternative is to manufacture the fork ends by steel forgings or by machining solid bars for

6.5 End-to-end connections

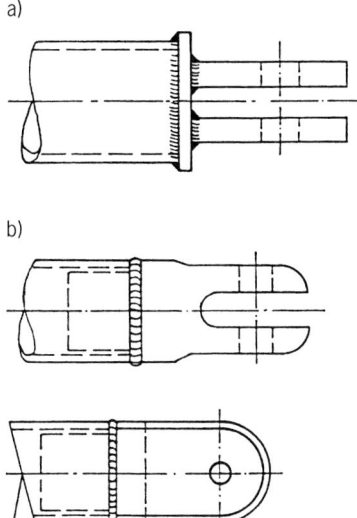

Fig. 6-37 End-to-end connections of hollow sections: a) with double-tongued forks, b) with forged or machined fork ends

smaller dimensions (Fig. 37b). According to [230], Eq. (6-13) can also be applied to RHS connections:

$$t_f \geq \frac{b_1 - t_{web}}{5} \quad \text{or} \quad t_f \geq \frac{h_1 - t_{web}}{5}$$

Fig. 6-38 shows head plate or flange plate connections, the loadings of which have to be lower than the allowable ranges and the rigidity of which must be so large that a deformation of the head or flange plate does not take place. The possible modes of failure are:

a) Plastification of the hollow section
b) Rupture of the weld between the hollow section and the head or flange plate
c) Plastification or rupture of the high strength bolts
d) Plastification or rupture of the head or flange plate

Flange plates can be made of a ring (X) or solid (Y) plate (Fig. 6-38). Corresponding to the circular, rectangular or square hollow sections, the flange plates can have various forms (Z). For a head or flange plate connection with CHS in axial tension, a circular plate with equal spacing of bolts is most suitable.

The failure mode a) can be easily avoided by proper selection of the dimensions and the materials of hollow sections.

The failure mode b) can be prevented by proper configuration and design of the weld between the hollow section and the head or flange plate. Fillet welds as well as full or partial penetration groove welds can be applied. The welds can be calculated in accord with various standards and recommendations e.g. DIN 18800, part 1 [10], Eurocode 3 [3] and AWS [13].

Under compressive load, also in the compressive zone under bending, the thickness of the fillet weld must not be lower than 0.4 × wall-thickness of the hollow section. In head or flange plate connections under tensile load, the weld thickness "a" has to be equal to the wall thickness "t" of the hollow section.

The failure modes c) and d) depend on the load distribution in the bolts and the plate, which is fundamentally characterized by the following geometrical parameters:

– Dimensions of the head or flange plate (diameter or width, depth, thickness)

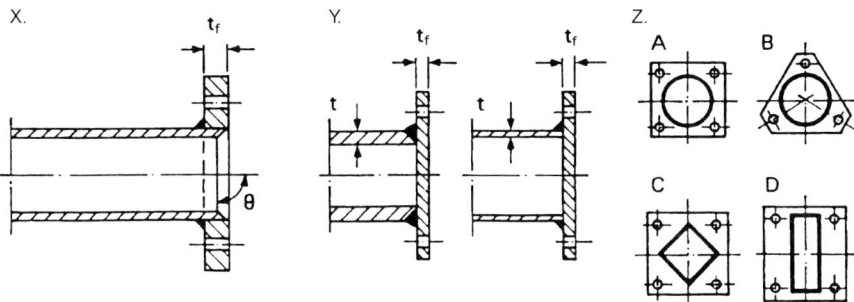

Fig. 6-38 Various flange plate connections with hollow sections

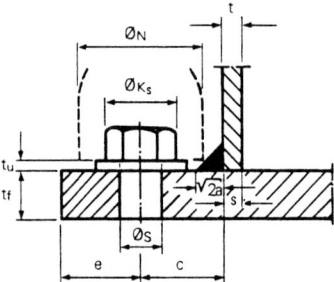

Fig. 6-39 Definitions for a head or flange plate connection

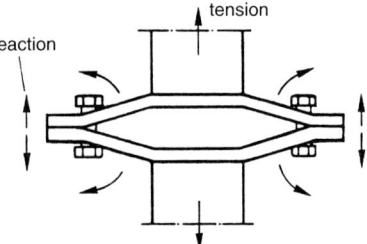

Fig. 6-40 Enlargement of space between head plates by tension

- Diameter and number of the bolts
- Arrangement of the bolts

The distance between the side of the hollow section and the bolt axis "c" and that between the head plate edge and the bolt axis "e" (see Fig. 6-39) must be kept as small as possible so that the bending in the headplates and the enlargement of space between them (see Fig. 6-40) do not become large. There must however be enough space for the socket wrench to tighten the bolts.

The theoretically minimum distance "c" can be determined as follows:

$$\text{required } c = \sqrt{2} \cdot a + \frac{\emptyset N}{2} - t_u$$

where:
a = weld thickness
$\emptyset N$ = outer diameter of the socket wrench
t_u = thickness of the supporting disc

Due to practical reasons, c has to be increased by 5 mm. The distances of the plate edge to the axes of the high tensile, controlled torque bolts and the spacings between the bolts are recommended in various standards e.g. DASt-Ri 010 (German guidelines) [20], CAN CSA-S16.1-M89 [21]. The strength behaviour of large plate connections with CHS in axial tension was investigated extensively by Rockey and Griffith [14–18] in the seventies. Effort was made to avoid the reaction on the plate edge fully by enlarging the plate thickness. By nature, this procedure was not economical.

In general, head plates under tensile load are calculated using the elastic theory for circular plates [19]. Assuming continuously applied forces in the bolts, the required minimum plate thickness can be determined by means of Eq. 6-14. This equation is valid for the number of bolts higher than or equal to 6 and for a spacing between bolts of at least 5 d_s, where d_s is the bolt diameter.

$$t_f \geq \sqrt{\frac{k \cdot N_{i,Sd}}{f_{y,P}}} \qquad (6\text{-}14)$$

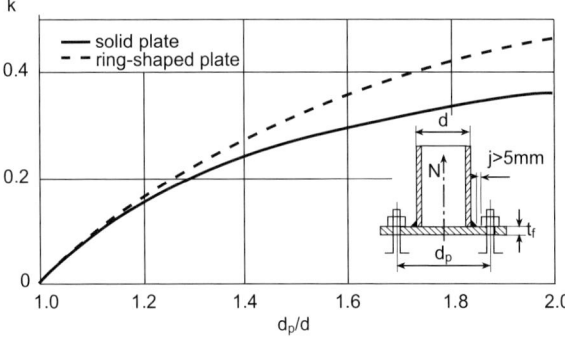

Fig. 6-41 Coefficient k to calculate head or flange connections with bolts arranged in the form of a ring

where:
t_f = thickness of the head or flange plate
k = coefficient (to be read in Fig. 6-41)
$N_{i,Sd}$ = tensile force (γ_F times load)
$f_{y,P}$ = yield strength of the plate material

In the eighties, a number of research works involving ring-shaped flange plate connections with CHS were carried out in Japan [22, 23], which resulted in the relatively simple design proposals [24]. The most recent recommendation [23] is based on the assumption that the CHS-flange plate connection fails by large plastic deformation of the ring-shaped flange plate and not by the rupture of the high strength bolts.

According to [23], the thickness of the flange plate t_f can be determined as follows:

$$t_f \geq \frac{2 N_{i,Sd} \cdot \gamma_{M0}}{f_{y,P} \cdot \pi \cdot f_3} \qquad (6-15)$$

where:
$N_{i,Sd}$ = axial tensile force
$f_{y,P}$ = yield strength of the flange plate material
γ_{M0} = partial safety factor

$$f_3 = \frac{1}{2 k_1} \left[k_3 + \sqrt{k_3^2 - 4 k_1 (1 + k_1 k_2)} \right]$$

$\left. \begin{array}{l} k_1 = \ln(r_2/r_3) \\ k_2 = r_4/r_3 \\ k_3 = 2 + k_1 (k_2 + 1) - k_2 \end{array} \right\}$ see Fig. 6-42

According to the yield line analysis, the plastic strength of the CHS flange plate connection has two circular yield curves with the radii r_1 and r_2.

The number of bolts n can be determined by:

$$n \geq \frac{N_{i,Sd} \left[1 - \dfrac{1}{f_3} + \dfrac{1}{f_3 \cdot \ln(r_1/r_2)} \right] \gamma_{M0}}{T_u} \qquad (6-16)$$

where:
$r_1 = (d_i/2 + e_1 + e_2)$
$r_2 = (d_i/2 + e_1)$
T_u = ultimate design tensile resistance of a high strength bolt
f_3 = to be read from Fig. 6-43

Further, it is assumed that the yield strength of CHS is reached by the application of load.

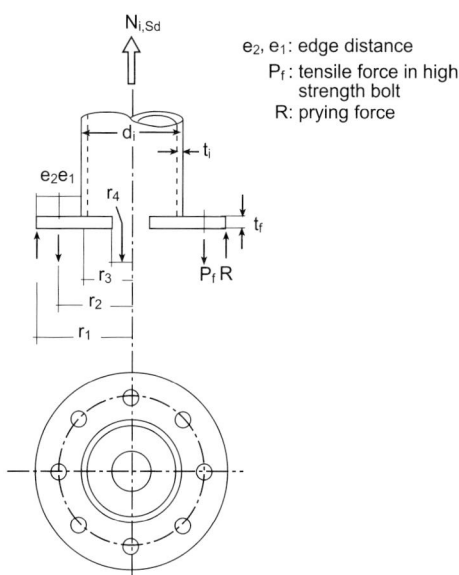

Fig. 6-42 Definitions for determining f_3

The design resistance of the total flange plate connection is then equal to $A_g \cdot f_y$ of the hollow section.

In Eq. 6-16, T_u has to be reduced by 1/3 of its value in order to take the additional prying force from the flange moment into account.

Fig. 6-44 shows a detail to reduce the thickness of the flange plate by means of stiffening ribs [23]. Because of the negative effect by the local buckling of the CHS wall through the ribs as well as the additional costs for the

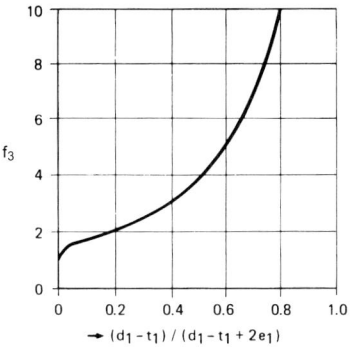

Fig. 6-43 Parameter f_3 for the calculation of the CHS flange plate connection (see Eq. 6-16)

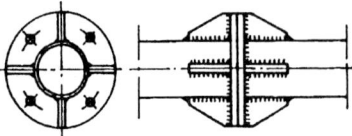

Fig. 6-44 CHS-flange plate connection with stiffening ribs

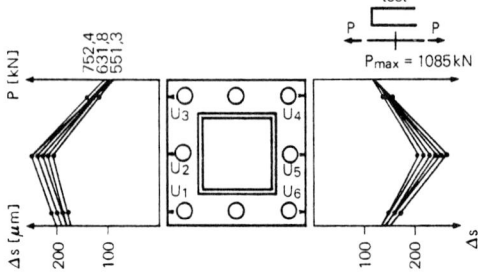

Fig. 6-45 Space enlargement between the plates in RHS (square)-head plate connections under tensile load

fabrication, protection and maintenance, this construction is however not recommended.

It is more favourable to use a thicker plate without stiffeners. Special attention is however to be paid to the material quality. In very thick plates, a low sulphur component in the material composition is necessary, so that no lamellar tearing can occur (load in the direction of thickness).

Only relatively less or incomplete investigations on RHS-flange plate connections have been performed. It is therefore necessary to carry out further research works in this sector in order to establish secured design methods.

In [25], it has been reported on tests with RHS (square) head plate connections under tensile load. Fig. 6-45 shows the distribution of forces and deformations.

It is apparent that the bolts at the corners receive lower loads and therefore an arrangement of bolts shown in Fig. 6-46 takes the load carrying behaviour better into account.

The tests on RHS-flange plate connections under bending load has also been reported in [25]. The test device is shown in Fig. 6-47.

Fig. 6-48 illustrates the bolt arrangements, which have been investigated.

The form III with bolts preferably in the tensile part of the flange plate delivered the best results. It is interesting that the bolts in the

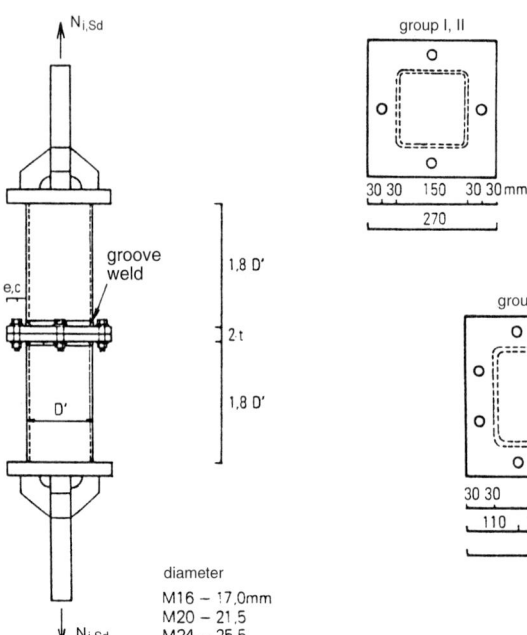

Fig. 6-46 Bolted RHS (square)-head plate connections, tensile tests [26]

6.5 End-to-end connections

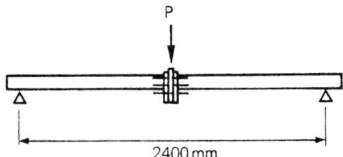

Fig. 6-47 Test device for RHS-head plate connection under bending load

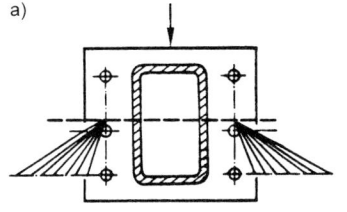

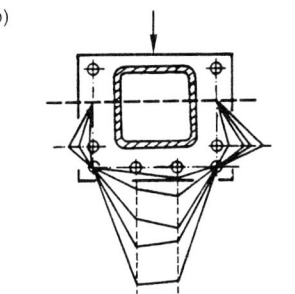

Fig. 6-49 Measured elongations on contact bolts in bending tests [25]

upper corners do not or seldom bear any load (compare with Fig. 6-49b). In Fig. 6-49a, the measured elongations shows, as expected, that the bolts in the level of the tensile zone receive the highest loads.

The connections according to the form I require thickest plates, as the plate is loaded as a uni-axially bent beam.

The evaluation of the test results given in [25] is fundamentally of qualitative nature:

– An increased number of evenly spaced bolts in the tensile zone (corresponding to the form III, see Fig. 6-48) is recommended. The bolts in the corners can eventually be dispensed with.
– In principle, small bolt diameters and large number of bolts are to be preferred (better distribution of bolts, bolt position close to the hollow section, lower loading of the plate)
– It is recommended for members in tension as well as in bending that the thickness of the head plate comes to 1.5 times bolt diameter at the minimum, as far as no precise verification is carried out.

In [26], the investigation of the connections of square hollow sections with bolts arranged along all four sides, as shown in Fig. 6-46, has been reported. Based on the yield line theory, a calculation model has been established estimating the prying force. A later research work [8] has proved that the strength of the

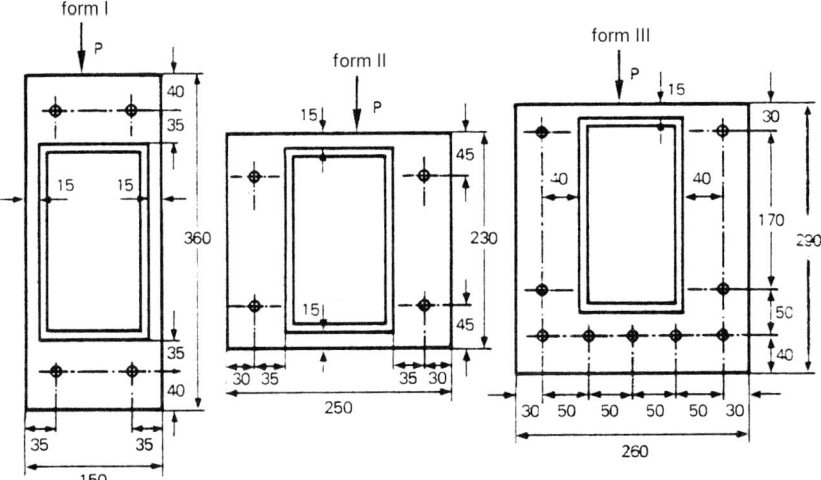

Fig. 6-48 Investigated RHS-head plate connections under bending load

connection has been overestimated by about 25% and the result is therefore unreliable.

Lit. [26] mentions for practical application a simple method of verification, which was derived from the results of the tests on flange or head plate connections with square hollow sections of 150 and 200 mm width (Fig. 6-46):

- For $n = 4$, the flange plate thickness $t_f \geq$ bolt diameter $\emptyset_s$
- For $n = 8$, $t_f \geq$ bolt diameter $\emptyset_s + 3$ mm

The tests with bolts M16, M20 and M24 resulted that the yield load for the head plate connection P_y came to 0.8 times the sum of the prestressing forces (tension) in the bolts and that the force in each bolt P_f due to external load was not more than 75% of the tensile resistance of each bolt T_u.

The following can be derived:

Design load (γ_F-times) $\leq P_y$

where:

$P_y \leq 0.8 \cdot \Sigma$ (prestressing forces in all bolts) \hfill (6-17)

$P_f \leq 0.75 \, T_u$ \hfill (6-18)

As it is somewhat insecure to apply the above mentioned design proposal at present (yield strength of the material of the head plate is not considered in the calculation), it is recommended to take $t_f \geq 1.5 \times$ bolt diameter for the connections (Fig. 6-46) to be on the safe side.

It has been reported in [27, 28] on extensive experimental investigations regarding RHS-flange plate connections with bolts along two sides opposite to each other (see Fig. 6-50) and their design. It has been shown that the ultimate tensile strength of the members can be attained also by this bolt arrangement through suitable selection of connection parameters.

A calculation model was developed by modifying the calculation model for T-stub in use [29]. Meanwhile, the calculation model for T-stub has been adopted by Eurocode 3 [3]. The modified procedure involves a redefinition of various parameters in the T-stub design

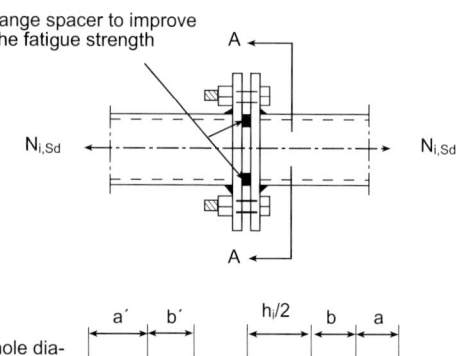

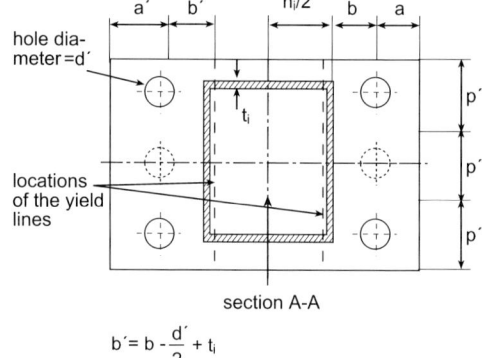

Fig. 6-50 RHS-flange plate connection with a bolt arrangement along two sides opposite to each other

method. In order to represent the real strength and deformation behaviour by the more complex analytical model of the connection, the distance b has been adjusted to b' (see Fig. 6-50). This reflects the observed location of the inner plastic hinge line close to the hollow section walls.

The design procedure has been put together in the following [6] (definitions can be taken from Fig. 6-50):

- The number, grade and size of the required bolts are to be estimated for the given tensile force N_i taking also prying forces into account. In general, the applied external load per bolt amounts to only 60 to 80% of the bolt tensile resistance anticipating bolt load amplification due to prying. In general the bolt pitch p'' has to be taken to be about 4 to 5 times bolt diameter (smaller bolt pitch is possible) and the edge distance a is to be equal to about $1.25 \, b$. The prying force R decreases as "a" increases up to $1.25 \, b$, beyond which there is no advantage.

6.5 End-to-end connections

- Calculation of δ (ratio of the net cross-sectional area along the bolts to the gross cross-sectional area at the side of the hollow section)

$$\delta = 1 - \frac{d'}{p''} \quad (6\text{-}19)$$

where:
d' = bolt hole diameter
p'' = bolt pitch

- Determination (first estimation) of flange plate thickness t_f:

$$[KP_f/(1+\delta)]^{0.5} \leq t_f \leq (K \cdot P_f)^{0.5} \quad (6\text{-}20)$$

where:

$$P_f = \frac{N_{i,Sd}}{n} = \text{external tensile design force per bolt}$$

n = number of bolts

$$K = 4b'/(0.9 f_{y,P} \cdot p'') \quad (6\text{-}21)$$

- Calculation of α (ratio of the bending moment per unit plate width at the bolt row to the bending moment per unit plate width at the inner yield line) for the equilibrium assuming that the bolts are loaded to their ultimate design tensile resistance:

Preselected: number, grade and sizes of the bolts as well as the first estimation of t_f,

$$\alpha = \left[\left(KT_u/t_f^2\right) - 1\right]\left[\left(a+\frac{d'}{2}\right)/\delta(a+b+t_i)\right]$$
$$(6\text{-}22)$$

where:
$a \leq 1.25\, b$ and T_u = ultimate tensile resistance per bolt

- Calculation of the design tensile resistance of the RHS-flange plate connection $N^*_{i,Rd}$ using α according to Eq. (6-22) (set $\alpha = 0$ if $\alpha < 0$):

$$N^*_{i,Rd} = t_f^2 (1+\delta\alpha) n/K \quad (6\text{-}23)$$

$N^*_{i,Rd}$ must be $\geq N_{i,Rd}$

- If required, the total tensile load of the bolts T_f including the effect of the prying force has to be checked:

$$T_f = P_f \left[1 + \left(\frac{b'}{a'}\right)\left\{\frac{\delta\alpha}{1+\delta\alpha}\right\}\right] \quad (6\text{-}24)$$

where:
$a' = a$ (however $\leq 1.25\, b$) + $\dfrac{d'}{2}$

$$\alpha = \left[\left(\frac{KP_f}{t_f^2}\right) - 1\right]\left(\frac{1}{\delta}\right) \quad (6\text{-}25)$$

This value of α is not necessarily the same as that from Eq. (6-22), which is based on the assumption that the bolts are loaded to their full tensile resistance. Eq. (6-25) corresponds to Eq. (6-20) for $\alpha = 0$ or $\alpha = 1.0$.

This design method should be restricted to flange plate thickness between 12 and 26 mm, as it has been investigated experimentally and analytically within this range.

For repeated loading, the flange plate must be made thick and stiff enough, so that no deformation of the flange plate takes place ($\alpha \leq 0$). In most structural standards, it is required that the bolts under tensile load are prestressed; this requirement is essential for fatigue loading. If range spacers are set between the plates parallel to the bolt row (see Fig. 6-50), prying action can be precluded and thereby the fatigue strength is improved [30].

Fig. 6-36e illustrates an end-to-end bolted connection with a plate with drilled holes welded into longitudinal slots at the end of CHS. Lit. [6] gives the following simple steps for the calculation: It is to be assumed that the strength of the connection is equal to or higher than the full tensile strength of CHS.

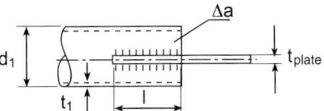

Assumption: $t_{plate} \geq t_1$

$$4l \cdot a \cdot f_{s,weld} + 2 \cdot t_{plate} \cdot a \cdot f_{a,weld} \geq f_{y1} \cdot A_1 \quad (6\text{-}26)$$

$$4l \cdot t_1 \cdot \frac{f_{y1}}{\sqrt{3}} + 2 \cdot t_{plate} \cdot t_1 \cdot f_{y1} \geq f_{y1} \cdot A_1 \quad (6\text{-}27)$$

$$A_{plate} \cdot f_{y,plate} \geq A_1 \cdot f_{y1} \quad (6\text{-}28)$$

Table 6-3 Design recommendations for CHS-plate (through) connections [24]

Connection type	Load type	Required weld length	Design resistance	Validity range
A	tension	$l \geq 1.2\, d_1$	$0.85\, A_1 \cdot f_{y1}$	$18 \leq \dfrac{d_1}{t_1} \leq 50$
A	compression		$0.85\, A_1 \cdot f_{y1}$	
B	tension	$l \geq 0.6\, d_1$	$0.85\, A_1 \cdot f_{y1}$	$18 \leq \dfrac{d_1}{t_1} \leq 50$
B	compression		$0.85\, A_1 \cdot f_{y1}$	
C	tension	$l \geq 1.2\, d_1$ $l \geq 1.5\, d_1$	$0.85\, A_1 \cdot f_{y1}$	$20 \leq \dfrac{d_1}{t_1} \leq 34$ $34 \leq \dfrac{d_1}{t_1} \leq 42$
C	compression	$l \geq 1.2\, d_1$	$0.85\, A_1 \cdot f_{y1}$	$20 \leq \dfrac{d_1}{t_1} \leq 42$

A_1 = cross-sectional area of CHS; d_1 = CHS outer diameter; t_1 = CHS wall thickness; f_{y1} = minimum yield strength of the CHS material

In order to avoid a premature crack of the weld at the tip of the plate, the factor 0.85 ($A_{\text{plate}} \cdot f_{y,\text{plate}}$) is recommended. The width of the plate has to be larger than the outer diameter of CHS (or the width of RHS) by $4 \cdot t_{\text{plate}}$, so that a flawless weld is guaranteed.

Various types of CHS-plate (through) connections were investigated experimentally and analytically in Japan [31]. Based on the results of this research work, a table with the design recommendations has been published by AIJ (see Table 6-3). They involve the following failure modes:

- Connection under tensile load fails by cracks in CHS wall, which are initiated at the tip of the plate and propagate along the weld.
- Connection under compression fails by local buckling in CHS wall in the neighbouring zone of the tip of the plate.

If a square hollow section is used instead of a CHS, the length of the weld seam l has to be increased from $1.2\, d_1$ to $1.5\, b_1$ and from $1.5\, d_1$ to $1.9\, b_1$.

See Lit. [32] for further references about other research works.

Fig. 6-51 shows a mode of failure of a hollow section-fork plate connection, where the single flat plate is substituted by the fork plate. Two separate CHS segments are here subjected to eccentric load. The figure demonstrates that the CHS walls fail by combined tension and bending, as it produces a significant reduction of the load bearing strength of the connection.

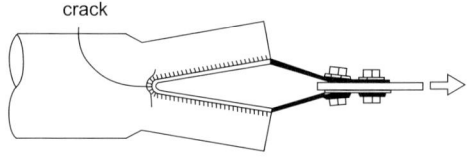

Fig. 6-51 Failure mode of a CHS-fork plate connection

6.5 End-to-end connections

A simple calculation of a CHS-fork plate connection is given in the following:

Presumption: connection resistance ≥ full resistance of CHS

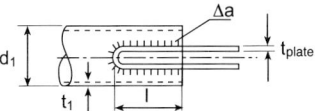

Assumption: $t_{plate} \geq t_1$

$$4l \cdot a \cdot f_{s,weld} + 4 t_{plate} \cdot a \cdot f_{a,weld} \geq A_1 \cdot f_{y1} \quad (6\text{-}29)$$

$$4l \cdot t_1 \cdot \frac{f_{y1}}{\sqrt{3}} + 4 t_{plate} \cdot t_1 \cdot f_{y1} \geq A_1 \cdot f_{y1} \quad (6\text{-}30)$$

$$A_{plate} \cdot f_{y,plate} \geq A_1 \cdot f_{y1} \quad (6\text{-}31)$$

Since the two halves of CHS are eccentrically loaded, the bolted connection with the fork plate has to be able to withstand the moment due to eccentricity.

Instead of inserting a plate through the slot in the hollow section, two plates can be welded to the flanks of RHS (see Fig. 6-36 f).

A somewhat uniform load distribution on all four side walls of RHS can be achieved by cutting the hollow section ends at a 60° angle and welding them to a "Y" piece made of two plates (see Fig. 6-36 g).

Fig. 6-52 illustrates a further alternative of in-line joints with bolted splice plates (inserted), which can be either exposed or covered with cover plates. The cover plate adds to a smooth external appearance of the total construction.

A number of typical examples of bolted RHS-fish plate connections is shown in Fig. 6-53, which are simple to handle on site. The occurring failure modes – bolt shear, rupture of net cross section, inadequate bearing pressure resistance of bolt hole – are known in general. General rules for the calculation of these connections are to be found in the standards [5, 10, 33]. A fundamental difference to the bolted connections with hollow sections already mentioned lies in the bolts inserted through the hollow section and tightened on the opposite side by means of nuts.

Contrary to the usual fish plate connections, the local support of the hole zone in the inside

a)

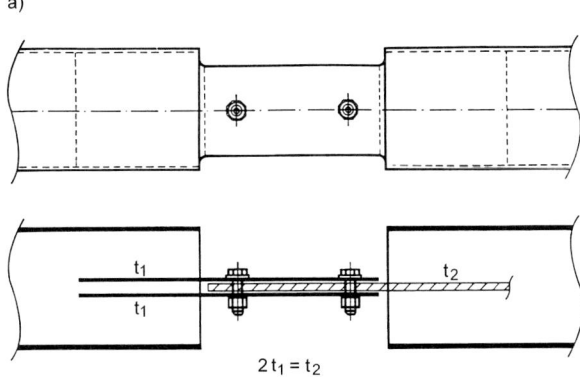

b)

Fig. 6-52 Hollow section splice (inserted) joint. a) Exposed, b) with cover plate

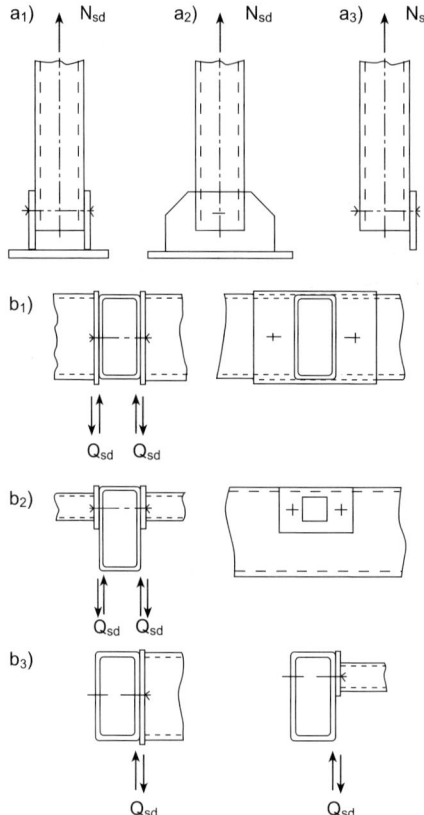

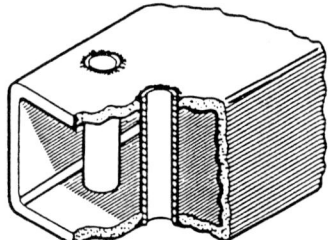

Fig. 6-54 Drilling the hollow section and welding a spacer tube in position

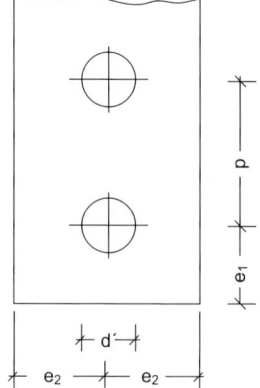

Fig. 6-53 RHS-fish plate connections with through bolts. a) Axial load, b) shear load

Fig. 6-55 Definitions of edge and bolt hole spacings

of RHS by plate as well as by bolt head and nut lacks in hollow section connections. Welding a spacer tube in position to permit bolting through the hollow section is possible in principle, but the required expense is high and not practicable. In addition, large bending loads occur in the bolts, especially in the unsymmetrical connections.

Lit. [33, 34] contain the results of the investigations on RHS-fish plate connections under axial force or shear.

For hollow sections in axial force, which is being dealt with in this Section, the bolts are loaded perpendicular to their axes. The main objective of this investigation is to determine the strength of the connection with regard to the inadequate bearing pressure resistance of bolt hole and the rupture of net cross section, but not the bolt shear.

The bearing pressure resistance of bolt hole is mainly dependent on the edge and bolt hole spacings (Fig. 6-55) [35, 36].

For the failure by the rupture of the net cross section of hollow section, the following strength of the symmetrical connection is recommended [33, 34]:

$$N_{t,Rd} = 1.0 \cdot A_{net} \cdot f_u/\gamma_{Mb} \qquad (6\text{-}32)$$

instead of Eq. 5–14 according to Eurocode 3 [3]

For the failure due to inadequate bearing pressure resistance of bolt hole, the strength of the connection can be calculated as follows:

For symmetrical connections with one bolt,

$$N_{b,Rd} = 2.5 f_u \cdot d_{shaft} \cdot t_{RHS} \cdot \alpha/\gamma_{Mb} \qquad (6\text{-}33)$$

6.5 End-to-end connections

where:
α = smaller value between $e_1/3\,d'$ and 1.0 (see Fig. 6-55)
f_u = ultimate tensile strength of RHS material
d_{shaft} = shaft diameter of bolt
t_{RHS} = wall thickness of RHS
γ_{Mb} = partial safety factor = 1.25 for bolts according to Eurocode 3 [3]
$N_{b,Rd}$ = design bearing pressure resistance of bolt hole

For symmetrical connections with two bolts,

$$N_{b,Rd} = 2.5\,f_u \cdot d_{shaft} \cdot t_{RHS}\,[\alpha_1 + (n-1)\,\alpha_2] \quad (6\text{-}34)$$

where:
α_1 = smaller value between $e_1/3\,d'$ and 1.0
α_2 = smaller value between $(p/3\,d' - 0.25)$ and $(p/3\,d' - 1.0)$

The number of tests carried out in the range $e_1 > 3\,d'$ as well as on connections with two bolts is according to Lindner [34] not sufficient to make a statistical evaluation precisely. He proposes to apply Eqs. (6-32) through (6-34) for symmetrical connections within the following restricted range (see Fig. 6-55):

t_{RHS} = 2.7 to 4.6 mm
edge distance e_1 = 1 to 6 d'
bolt spacing p = 2 to 4 d'
material grade S 235
bolt diameter M 16

ratio $\dfrac{\text{RHS wall thickness}}{\text{bolt diameter}} > 0{,}18$

The unsymmetrical connections show markedly lower strength than equivalent symmetrical connections. All test specimens of this type with $e_1/d' > 2$ failed due to bolt shear. Other failure modes occurred for very small e_1 [34].

Simple in-line connections under tensile load with one, two or three bolt arrangements as shown in Fig. 6-56 were tested by Mang [37], where the decisive parameters according to EC 3 [3] were so oriented that the limiting shear in the bolt was always larger than the limiting bearing pressure resistance of the bolt hole (see Table 6-4 and Fig. 6-57).

Following conclusions can be derived from the tests:

- The thickness of the fish plate must be larger than that of the hollow section so that the bearing pressure on the bolt hole can take place in the hollow section.
- The bolts should not be prestressed; this can initiate large deformations in the hollow section.
- Limiting bolt shear and bearing pressure resistance of bolt hole as well as limiting tensile force can be determined based on EC 3 [3] (Table 6-4). Limiting bolt shear has to be larger than the limiting bearing pressure on bolt hole.
- The edge distance e_1 of the bolts affects the strength of the connection significantly. The influence of the bolt hole spacing p_1 on the connection strength is not so strong; however a minimum value has not to be fallen short.
- Within the range $e_1 \geq 3\,d'$ and $p_1 \geq 3.5\,d'$, the maximum value of α according to EC 3 [3] has to be reduced by 10%.

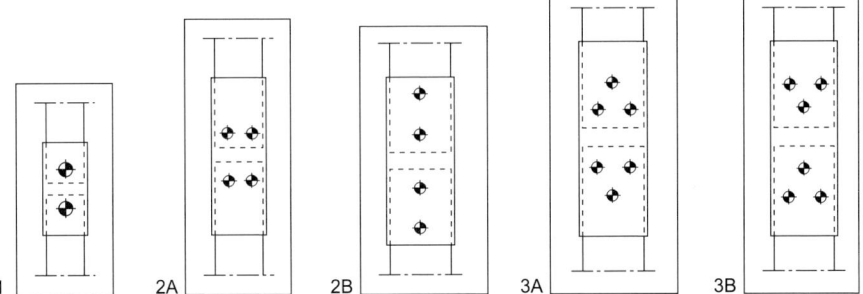

Fig. 6-56 In-line connections in tension with various bolt arrangements

Table 6-4 Design resistances for bolts [3]

Shear resistance per shear plane $N_{V,Rd}$

If the shear plane passes through the threaded portion of the bolt:

- for bolt strength grades 4.6, 5.6 und 8.8:

$$N_{V,Rd} = \frac{0.6 \cdot f_{ub} \cdot A_s}{\gamma_{Mb}}$$

- for bolt strength grades 4.8, 5.8, 6.8 and 10.9:

$$N_{V,Rd} = \frac{0.5 \cdot f_{ub} \cdot A_s}{\gamma_{Mb}}$$

If the shear plane passes through the unthreaded portion of the bolt:

$$N_{V,Rd} = \frac{0.6 \cdot f_{ub} \cdot A_g}{\gamma_{Mb}}$$

Bearing pressure resistance of bolt hole $N_{b,Rd}$

$$N_{b,Rd} = \frac{2.5\,\alpha \cdot f_u \cdot d \cdot t}{\gamma_{Mb}}$$

where α is the smallest of:

$$\frac{e_1}{3d'};\ \frac{p_1}{3d'} - \frac{1}{4};\ \frac{f_{ub}}{f_u}\ \text{or}\ 1.0$$

Tensile resistance $N_{t,Rd}$

$$N_{t,Rd} = \frac{0.9\,f_{ub} \cdot A_s}{\gamma_{Mb}}$$

A_g = gross cross-sectional area of bolt
A_s = tensile stress area of bolt
d = bolt shaft diameter
d' = bolt hole diameter
f_u = ultimate tensile strength of hollow section material
f_{ub} = ultimate tensile strength of bolt material
γ_{Mb} = partial safety factor of net section at bolts

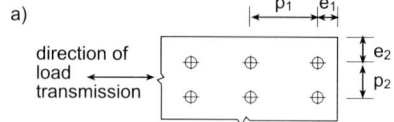
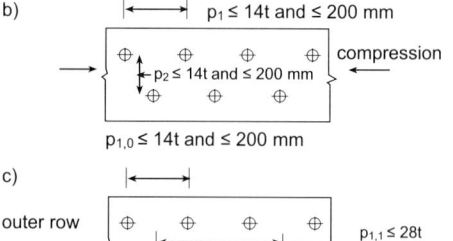
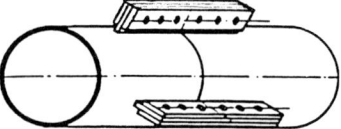

Fig. 6-57 Edge and bolt spacings according to EC 3 [3].
a) Designations for bolt hole spacings, b) staggered hole arrangement of the structural elements in compression, c) bolt hole spacings in the structural elements in tension

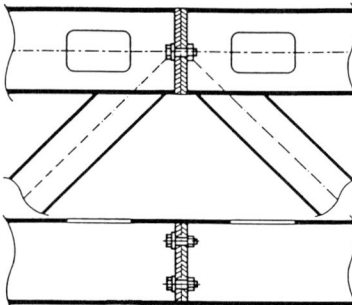

Fig. 6-58 Fish plate connection with strips

- Connections with one bolt arrangement can be designed according to EC 3 [3]. Multiple bolt connections can attain higher strength by varying bolt arrangements. For two bolt connections, the connections 2A have higher connection strength than 2B (see Fig. 6-56); for three bolt connections, 3A is stronger than 3B.

Fig. 6-58 illustrates a fish plate connection of another type. In this connection, four, six or eight strips are welded longitudinally on the periphery of the hollow section and connected by sets of double lap plates, one on each side. Heavy loads can be transmitted by this connection. Special attention has to be paid to the prevention of corrosion in this construction.

Fig. 6-59 Hidden, bolted butt-joint with head plates in the chord of a lattice girder

Two hollow sections with welded head plates are to be bolted to each other in a truss chord. In order to lend the connection a smooth appearance, the head plates must not extend beyond the width or the diameter of the hollow section. In this case, the accessibility to the inside of the members is required to fix the bolts inside. Fig. 6-59 presents an application,

6.5 End-to-end connections

where hand access holes are cut in the members to reach the bolts.

Fig. 6-60 illustrates a screwed tensioner. For smaller CHS diameter, a bolt may be welded to one end of the section and a nut to the end of the other.

Instead of a bolt, a screwed pin can be welded to the end of a CHS (Fig. 6-61) and a screwed coupling can be used instead of a nut. This type of connection is not suitable for load bearing structures. In general, they are applied in hand railings, guard railings, cattle pens etc.

Another type of connection using a length of a solid bar is shown in Fig. 6-62, which is particularly suitable for hand railings or crash barriers. The bar is plug welded in the inside of the hollow section and fixed in its position. At the free end of the bar, a further plug weld or a drilled hole for bolting on site is required.

Fig. 6-63 illustrates a further type of plug welded RHS connection with a sleeve incorporating two shaped steel strips.

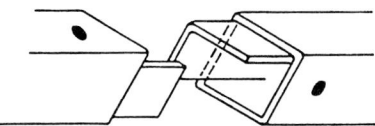

Fig. 6-63 End-to-end RHS connection with plug welded strips

6.5.2 Indirectly joined welded connections

In certain cases, also indirectly joined in-line connections with hollow sections are fabricated only by welding. Fig. 6-64 shows an example with an internal backing sleeve, where it provides a supporting face for the liquid weld. Outside diameter of the sleeve profile corresponds to the inside diameter of the CHS to be joined. This connection is then recommended, when the gap between the ends of the hollow sections is large and no other constructive solution is possible. Reversedly, an external sleeve may also be set as a butt strap

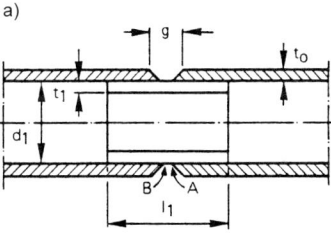

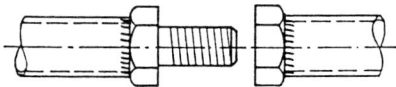

Fig. 6-60 A bolt and a nut welded to CHS ends

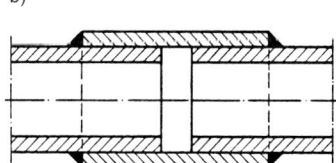

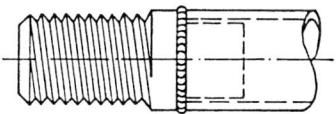

Fig. 6-61 A screwed pin welded to CHS ends

Fig. 6-64 Welded in-line connections with hollow sections as sleeves: a) with inner sleeve, b) with external sleeve

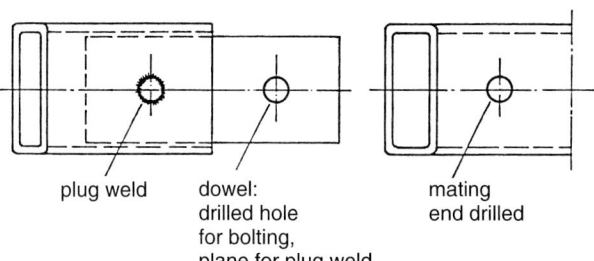

Fig. 6-62 End-to-end connection with a plug welded solid bar

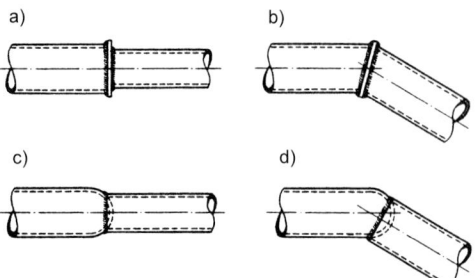

Fig. 6-65 End-to-end connections with fillet welds

(Fig. 6-64b). Fillet welds are applied instead of butt welds in this case.

Fig. 6-65a–d shows further means to construct end-to-end connections with hollow sections. While in Fig. 6-65a the joint is made by means of a plate welded in-between, which is applicable to both CHS and RHS, Fig. 6-65c illustrates a CHS connection, where one CHS is dished semicircularly at the end and another CHS, possibly of smaller size, is welded to the dished end of the other CHS by means of fillet weld. Both types of connections can also be made with deviated ends (Fig. 6-65b and d).

6.5.3 Directly joined welded connections

Directly joined end-to-end connections with hollow sections are made principally by groove welding. Three following cases can occur to apply this simple joining method (see Fig. 6-66).

Case 1. For thinwalled members, no weld preparation at the member ends is necessary.

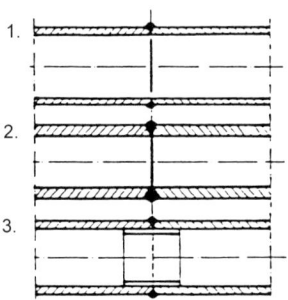

Fig. 6-66 Groove welded end-to-end connections of hollow sections

Case 2. For thickwalled members, member ends are bevelled in I, V and Y form for welding without back gauge weld.

Case 3. Bevels at the hollow section ends can also be backed by an internal backing ring, which is set below the weld root to support the liquid weld during through welding. The width of the backing ring ranges from about 30 to 40 mm, the thickness from 3 to 5 mm. Corresponding to the required adjustment, the backing ring can be fabricated in one piece as well as in two pieces (Fig. 6-67).

For the installation of a backing ring within a RHS, special care has to be taken while fitting in the corner part. Bending of the corner part of the backing ring must be carried out very precisely and during installation any mismatch has to be corrected by local heating and hammering. The gap between two backing plates is to be filled up by weld metal in order to prevent any initiation of defects during welding.

Technically, the groove weld without internal backing ring is more advantageous, as this can be checked more easily. However, qualified welders are required to execute this job. Further, groove welds are more suitable for the connections under fatigue loading.

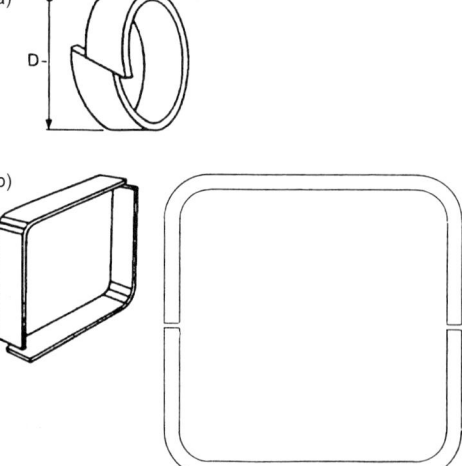

Fig. 6-67 Detail of a backing ring: a) in one piece, b) in two pieces

6.5 End-to-end connections

Usual joint forms for groove welded end-to-end connections according to ENV 1090-4 [38] are listed in Table 6-5. The national standards of the countries of the European Union e.g. DIN 2559 [39] and DIN 8551 [40] deviate from [38] only slightly.

The strength of the end-to-end connections with complete penetration groove welds is given by the strength of the weaker member.

Appropriate selection of the welding electrodes suited to member material is essential here. A complete penetration and full fusion of weld and basic material over the total thickness of the connecting hollow sections is required for through welding.

The weld thickness of a not fully penetrated groove weld is to be taken to be equal to the depth of the penetration.

Table 6-5 Groove weld preparation for end-to-end connections with hollow sections [38]

Joint form	Weld thickness in mm	Gap g in mm		Leg depth R in mm		Backing plate thickness t_P in mm	
		min.	max.	min.	max.	min.	max.
I seam without backing ring	up to 3	0	3				
I seam with backing ring	3	3	5			3	3
	5	5	6			3	5
	6	6	8			3	6
V seam without backing ring	up to 20	2	3	1	2.5		
V seam with backing ring	up to 20	5	8	1	2.5	3	6
V seam with backing ring (flame cut)	20–30	8	10	1	3	3	10

The following aspects of the demonstrable requirements for the external configuration of the butt welds by tests have to be checked:

The thickness of the weld must be known when the cross section of the weld is not completely welded. Excessive convexity, concavity and root depth of weld can be measured by gauges. Such a self-fabricated gauge is illustrated schematically below.

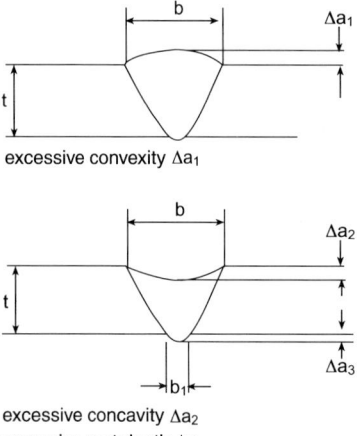

excessive convexity Δa_1

excessive concavity Δa_2
excessive root depth Δa_3

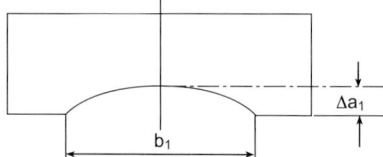

Fig. 6-68 illustrates the stresses in groove and fillet weld seams, where the equivalent stress $f_{v,\text{weld}}$ is to be calculated as follows:

$$f_{v,\text{weld}} = \sqrt{f_\perp^2 + \tau_\perp^2 + \tau_\parallel^2} \qquad (6\text{-}35)$$

The weld stress $f_\parallel$ in the longitudinal direction need not be taken into account.

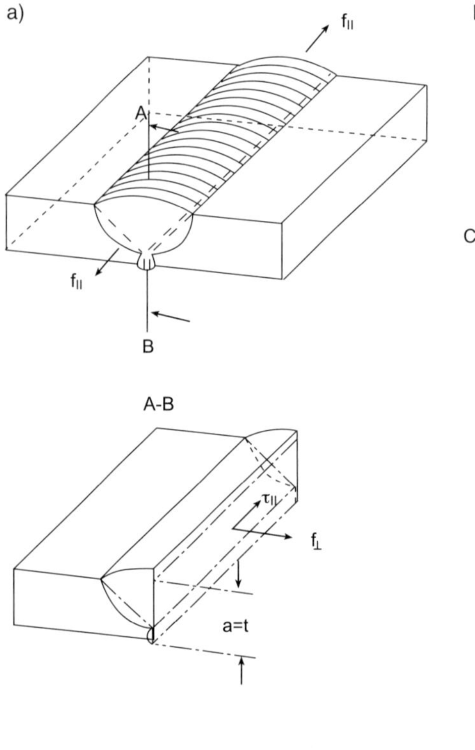

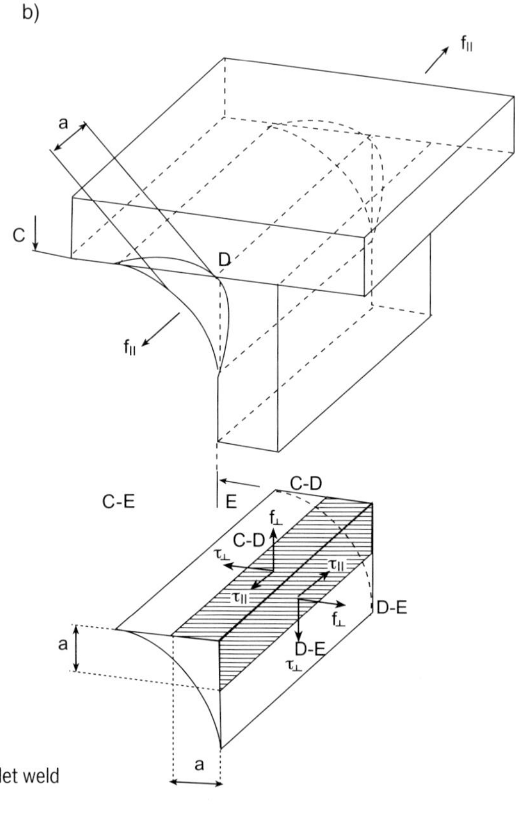

Fig. 6-68 Stresses in weld seam. a) Groove weld, b) fillet weld

6.5 End-to-end connections

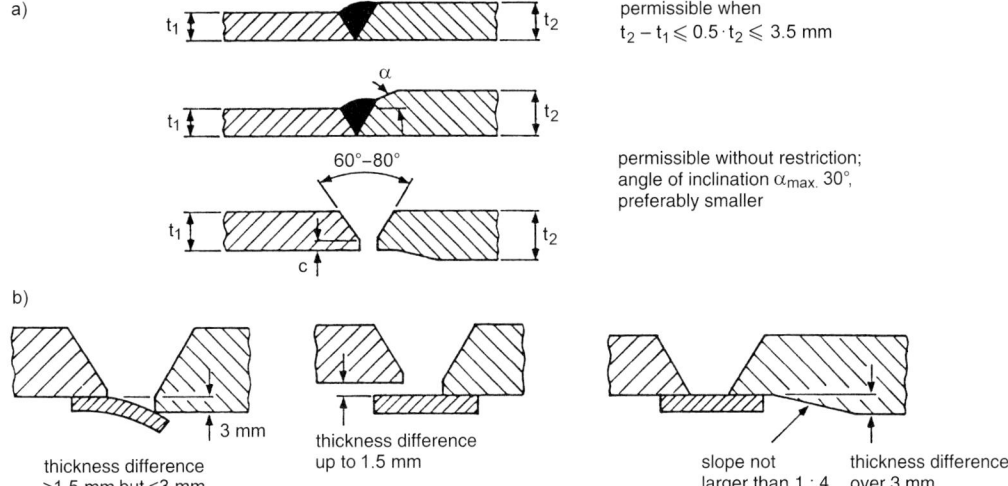

Fig. 6-69 Weld preparation for groove welded end-to-end connections of hollow section members:
a) without backing ring, b) with backing ring

Fig. 6-69 shows the end preparation of the members for the groove welded connections, when the wall thicknesses of the members differ from each other. In these cases, the transition has to be made as smooth as possible, especially for constructions under fatigue loading. It is also possible to join hollow sections of different wall thicknesses by groove welds with internal backing ring (see Fig. 6-69 b).

In some cases, end-to-end connections have to be vertically groove welded on site. It is recommended to support the liquid weld by means of backing ring. The angle of inclination of the weld in the upper member shall be larger than that in the lower member (see Fig. 6 70).

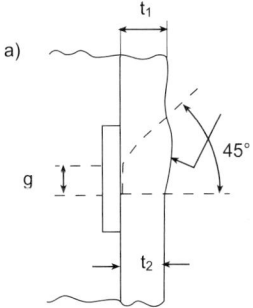

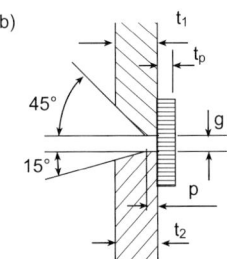

Fig. 6-70 Vertically groove welded end-to-end connections of hollow sections with backing ring.
a) $t_1 = t_2$, b) $t_a > t_2$

Verification of design resistance of groove weld in end-to-end connection

As has been already mentioned, the design resistance of a groove weld with full penetration is based on that of the weaker of the members to be joined:

$$N_{W,Rd} = f_{y,W} \cdot (\sum L \cdot a)/\gamma_{MW} \geq N_{Rd} = A \cdot f_y/\gamma_M$$

(6-36)

where:

$f_{y,W}$ = characteristic yield strength of weld material

γ_{MW} = partial safety factor for the design resistance of weld (see Table 5-6)

$N_{W,Rd}$ = design resistance of weld

N_{Rd} = design resistance of structural member

γ_M = partial safety factor for the design resistance of structural member

L = weld length (for the verification, the weld lengths are to be taken, which, based on their positions, bear the existing forces and moments)

a = weld thickness (see Fig. 6-68) ≈ thickness of the weaker structural member

A = cross sectional area of the weaker structural member

f_y = characteristic yield strength of the material of the structural member

The design resistance of a partially penetrated groove weld in an end-to-end connection is to be determined based on the depth of the penetration, as is also the case for the fillet weld.

Table 6-6 Combinations of profile forms in lattice girders

Bracing members	Chord members
○	○
○	□ □ □
□ □ □	□ □ □
○	I I
□ □ □	I I
○	⌐
□ □ □	⌐

6.6 Lattice girders (trusses)

Lattice girders, uni-planar as well as multi-planar (triangular and quadrangular), are for hollow sections typical and most frequently used structural constructions. Fig. 6-71 and Table 6-6 show the combinations of the profile forms for the manufacture of the hollow section trusses.

Further, based on various types of connections or joints, the lattice girders are to be differentiated fundamentally:

- Direct connections or joints, where the chord and the bracing members are directly welded to one another (Figs. 1-8 a and b, 6-72 a and b)
- Indirect connections or joints, where the chord and the bracing members are connected via gusset or side plates to one another by means of bolting and/or welding (Figs. 1-8 c, 6-72 c and d)

The priority of the hollow sections for the application in lattice girders can be attributed to the following essential aspects:

- **Static properties** (see Fig. 1-2)

With respect to the axial compressive resistance, CHS as well as RHS are superior to open profiles (I, U, L). This property contributes to higher critical buckling load of columns or members of lattice girders under compression, when they are made of hollow sections. The European buckling curves take this fact into account (see Fig. 5-4).

Hollow sections and open profiles are of the equal status with regard to axial tension. As a

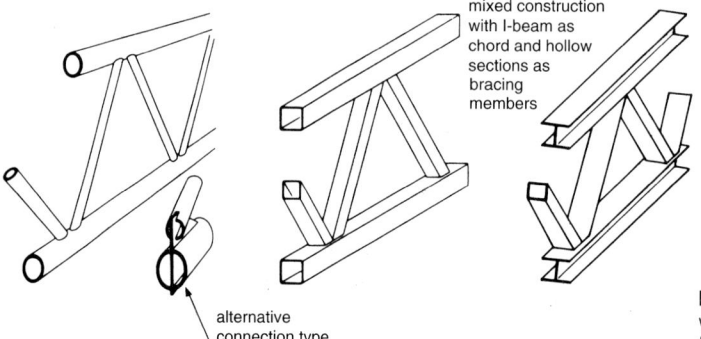

Fig. 6-71 Uni-planar lattice girders with various combinations of profile forms

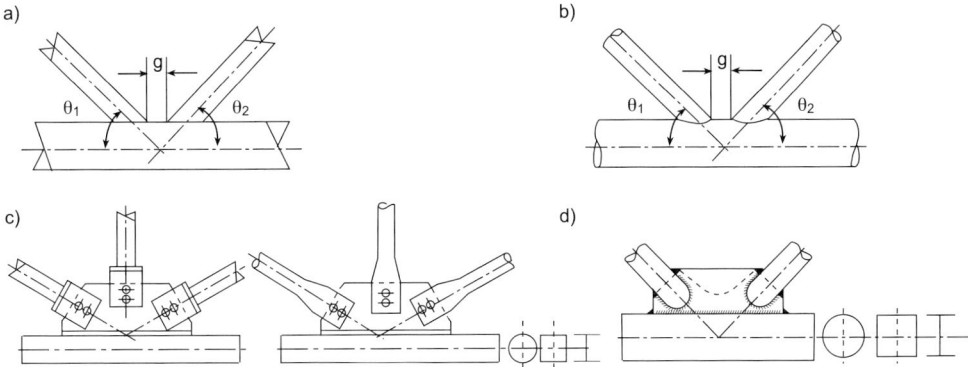

Fig. 6-72 Truss connections in hollow sections: a) and b) direct connections, c) and d) indirect connections. In mixed constructions, the chord members can consist of open profiles (I or ⊏ profiles). a) RHS joint (plane cut in the bracing end), b) CHS joint (profile cut in the bracing end), c) hollow section joint (the bracing end is bolted to a gusset plate), d) hollow section joint (the bracing end is welded to a gusset plate)

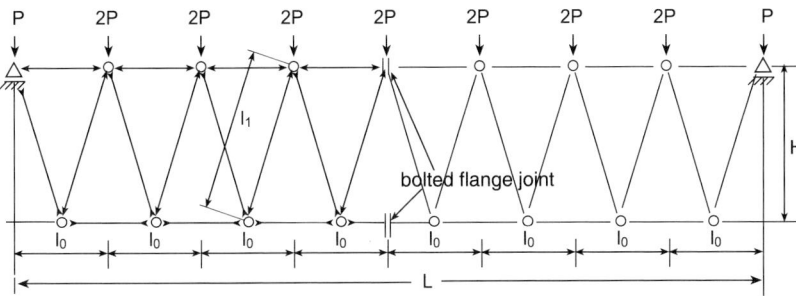

Fig. 6-73 Plane lattice girder with pin joints

beam under uni-axial bending, I-profile however carries significantly more load than a single hollow section.

The elastic calculation of lattice girders is done assuming that all members are pin jointed and they are loaded axially only (see Fig. 6-73). This method of calculation affects the application of hollow sections favourably as hereby the saving of material can be attained for the chord and bracing members under axial compression. In average lattice girders, about 50% of the total material weight is contributed by the chord members under compression, roughly 30% by the chord members under tension and about 20% by the bracing members. Considering this aspect, the compressive members (particularly chords under compression) should be optimised by using thinwalled hollow sections (larger radius of gyration → larger buckling resistance). It will be reported later about the influence of the dimensions of the hollow sections on the design joint resistance, which demands inversely thickwalled compressive chord members (see Section 6.6.4.1). The final compressive chord dimension ($\triangleq d_0/t_0$ or b_0/t_0) will be a compromise between the joint strength and the buckling strength of the member and relatively stocky sections will be chosen.

It is to be especially mentioned that the calculation of a lattice girder assuming flexurally stiff joints (increased moment in bracing members) in plane and triangular girders with directly welded chord and bracing members is not recommended, as the axial forces are also the same in this case as when pin joints are assumed.

Due to the above mentioned reasons, a lattice girder in hollow sections is a competitive construction to an I-beam (single profile) under bending. One must consider in this case that the sizes of the I-profiles in the production programmes of the profile manufacturers are significantly larger than those of the hollow sections. Single hollow sections, which are usually available, will not therefore be able to substitute the larger I-beams (Chapter 4). Using one of the largest available size 260 × 260 × 17.5 mm for the chord of a plane lattice girder, a large span of about 50 m can be achieved.

The characteristics of a lattice girder are mainly given by the span L, the depth H, the lattice geometry and the distance l_0 between two neighbour joints. The depth H is determined considering the span L, the loads in chord and the maximal allowed deflection of the truss. Truss forces may be lowered by increasing the truss depth H, increasing also the lengths l_1 of the bracing members at the same time. The ideal ratio span L/depth H lies between 10 and 15. As regards increase of length of the bracing members, large torsional stiffness and the associated high lateral-torsional stability of the hollow sections show to their best advantage. Large effective buckling lengths of compression members can be achieved without problem using hollow sections (see Section 5.3.2.1).

As structural element loaded by bi-axial or multi-axial bending, the hollow sections, square as well as circular, are more suitable than conventional open profiles. As they also yield low drag coefficients when subjected to wind-, water- and wave loading, constructions under multi-axial bending, such as masts, towers, offshore structures etc. are often manufactured as trusses in hollow sections.

- **Fabrication**

A welded truss with joints made of directly welded chord and bracing members of hollow sections requires less welding than that with open sections, which are often joined via gusset or connection plates. For the end preparation of the members for welding, RHS has a significant advantage over CHS as they require only plane cuts. For CHS as chord and bracing members, a multi-planar profile cut is necessary, which requires more labour and expenditure.

However automatic flame cutting machines can be used to conduct this job, which is time- and cost saving. The subject "fabrication" of structures with hollow sections will be extensively dealt with in Chapter 8.

- **Protection against corrosion**

The protection by painting and spraying and later the maintenance of trusses in hollow sections cause lower costs for structures in hollow sections. This is firstly due to the smaller surface area and further because of the simple welded connections without gusset plates, accessible corners, lack of water or snow traps etc. (see Fig. 1-13). In general, no protective measure of hollow sections against internal corrosion is necessary, provided they are airtight sealed at all openings.

- **Transport and assembly**

Trusses in hollow sections have a low material weight and lead thereby to a better economy in transport and assembly.

- **Appearance**

On account of the external aesthetic appearance, the architects often prefer the trusses in hollow sections to those in other profiles.

6.6.1 Single plane lattice girders with hollow sections

In the following, a number of typical uni-planar lattice girders in hollow sections and their merits and demerits will be described. The configuration of joints and their constructive details play for designing trusses an important role in order to obtain optimum uniform stress distribution in the construction (particularly in connections under repeated loads). This knowledge to work out the guide lines for the design strength of trusses in hollow sections is of fundamental importance. In this context, one has to consider that the stress distribution in a hollow section joint is difficult to determine theoretically. Therefore, tests and measurements such as brittle lacquer method or strain measurements with strain gauges have to be applied.

6.6 Lattice girders (trusses)

• **Warren type girder** (Fig. 6-74)
These girders with upper and lower chord members, parallel or non-parallel to one another, provide not only an architecturally aesthetic but also an economical solution. The advantage lies in the potential that the bracing members under compression can be designed with larger effective buckling lengths. A lower value of the angle of inclination θ between the chord and the bracing members decreases the number of joints in the lattice girder. With the reduction of the number of joints, the associated fabrication costs are also minimised. Optimisation studies regarding the lattice webs have indicated that the most favourable angle of inclination θ lies in the range of 40° to 50°. A further reduction of θ may lead to more economy in fabrication. However, a minimum θ of 30° is recommended, so that problems to weld at the heel of the web member can be avoided (the theoretical weld root cannot be fully penetrated when $\theta < 30°$ (see Table 6-7 and Figs. 6-94 to 6-97).

In case it is required to support all loading points in a chord e.g. to reduce chord bending moments, additional vertical members can be added (KT joint, member M in Fig. 6-7a).

A Warren type girder is a relatively light construction, which also offers sufficient space for the arrangement of service pipes, cables and other facilities in the lattice.

• **Pratt type girder** (Fig. 6-75)
The joints of a Pratt type girder are of N form and consist of a vertical and an inclined bracing member connected to the chord. As for a Warren type girder, the upper and lower chords of a Pratt type girder can be parallel or non-parallel to one another depending on the roof inclination. The arrangement of the members in a Pratt type girder results in a higher number of bracing members than that in an Warren type girder and thereby in more joints. It is therefore a less economical solution due to the increased costs for fabrication and protection against corrosion. It is however important to note that the Pratt type girder can be economical within its application range, when the compressive loads are mainly transmitted through the shortest members i.e. the verticals.

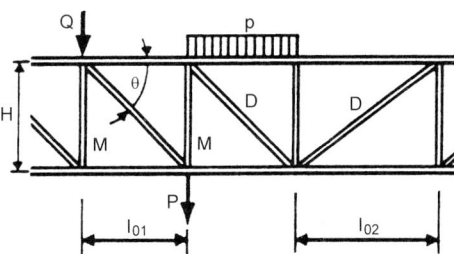

Fig. 6-75 Pratt type girder with N joints

• **Howe type girder** (Figs. 6-76 and 6-77)
The girder of this type consists of "X" lattices with or without vertical members. A further form of this girder type is made of two half depth girders, which are welded complete in the workshop, and offers special advantage re-

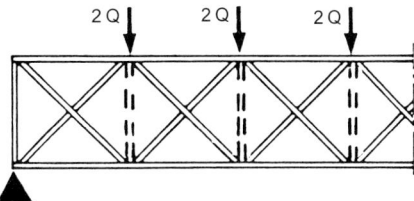

Fig. 6-76 Howe type girder with X lattices (in case the loads in the verticals are small, they can be dispensed with)

a)

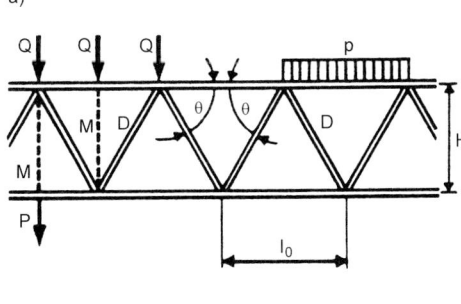

b)

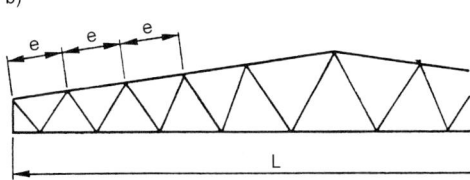

Fig. 6-74 Warren type girder with K joints (proposal for modification with vertical members: KT joints, see Fig. 6-74a)

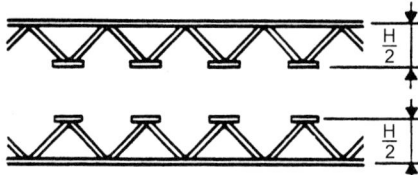

Fig. 6-77 Howe type girder with two half depth girders for easy transport to the site and bolting on site to X joints

garding the transport to the site due to its low material weight and smaller dimension. The assembly on site is made by bolting the half depth girders at the intersections of the bracing members (see Fig. 6-77).

- **K-form lattice girder** (Fig. 6-78)

Due to the relatively large number of structural members and joints, the fabrication time and labour costs for this type of girder are high. However it may be preferred for particularly deep girders, since the effective buckling length of the bracing members can be reduced by means of this joint arrangement. The manufacturing principle for Howe type girder with two half depth girders can also be applied here.

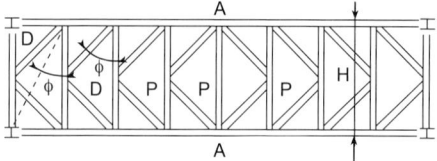

Fig. 6-78 K-lattice girder with K joints on the verticals

- **Arched lattice girder** (Fig. 6-79)

The arched chords represent the main feature of this girder type, where different lattice types are made by the bracing members. Fig. 6-79 shows two examples.

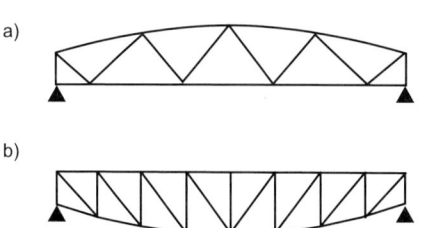

Fig. 6-79 Arched lattice girder. a) Warren type, b) Pratt type

6.6.1.1 Economical aspects regarding the selection of lattice girder types with hollow sections

In a lattice girder construction, the major share of the fabrication costs belongs to the bracing members. While selecting a lattice girder form, it is therefore important to take care that the number of the bracing members is as low as possible. Fig. 6-80 demonstrates an example comparing three lattice girders with N, KT and K joints [41]. It is evident that the Warren type girder with K joints results in the lowest number of bracing members and joints.

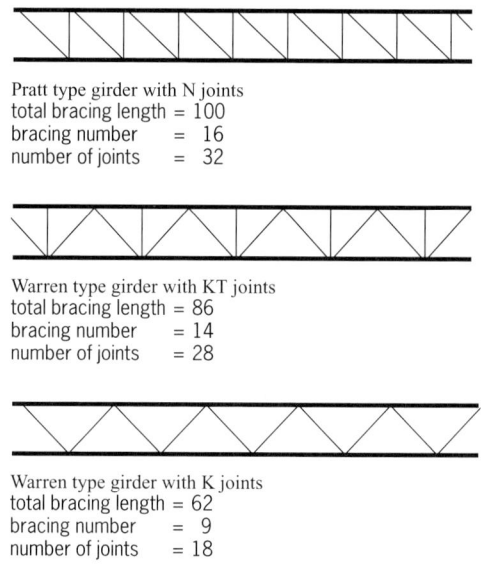

Pratt type girder with N joints
total bracing length = 100
bracing number   = 16
number of joints = 32

Warren type girder with KT joints
total bracing length = 86
bracing number   = 14
number of joints = 28

Warren type girder with K joints
total bracing length = 62
bracing number   = 9
number of joints = 18

Fig. 6-80 Comparison of various lattice girder types

6.6.2 Design arrangement of truss joints with hollow sections consisting of directly welded chord and bracing members

Among the joint types with directly welded members shown in Fig. 6-81 (Table 6-6 lists the combinations of profile types as chords and bracings), K and N joints are used most frequently. Gap and overlap are two features, which affect the joint strength significantly. It is therefore essential to define these characteristics precisely (Fig. 6-82).

The gap g is defined as the distance, measured along the length of the connecting face of the chord, between the toes of the adjacent

6.6 Lattice girders (trusses)

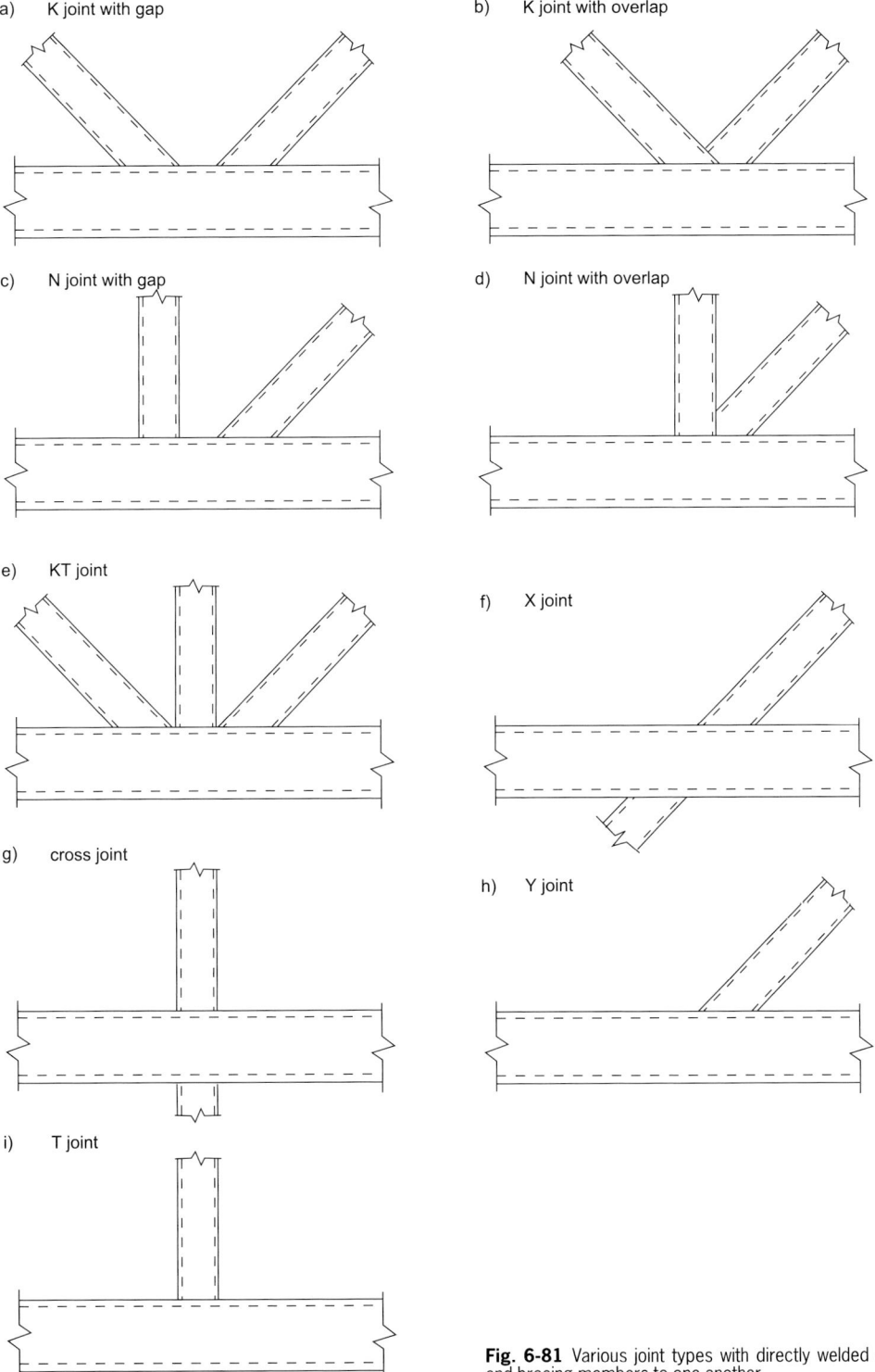

a) K joint with gap
b) K joint with overlap
c) N joint with gap
d) N joint with overlap
e) KT joint
f) X joint
g) cross joint
h) Y joint
i) T joint

Fig. 6-81 Various joint types with directly welded chord and bracing members to one another

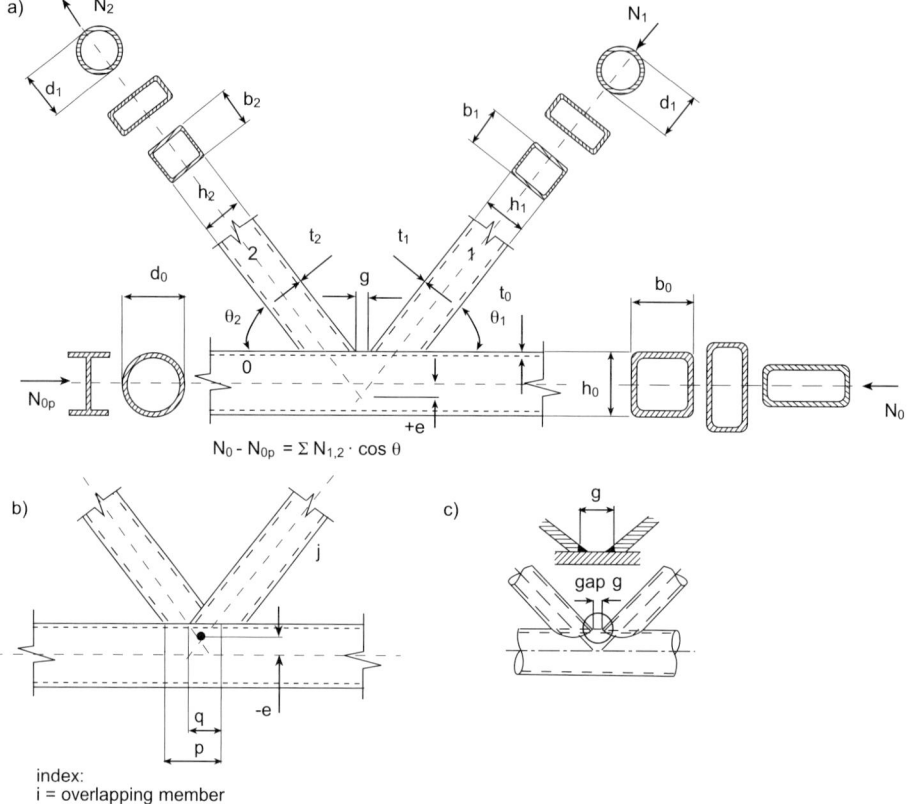

index:
i = overlapping member
j = overlapped member

overlap grade $\lambda_{ov} = \dfrac{q}{p} \cdot 100\%$

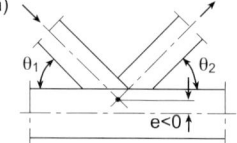

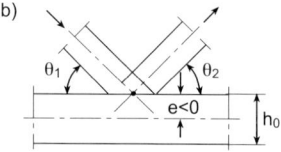

Fig. 6-82 K joint with gap or overlap (N joint is a special form of K joint)

bracing members without considering the weld (see Fig. 6-82c). In general, there must be a minimum gap $g \geq (t_1 + t_2)$ (sum of wall thicknesses of the adjacent bracing members) in order to avoid the overlap of the welds.

The overlap of the bracing members is differentiated in partial or full overlap (Fig. 6-83). Overlap is defined in Fig. 6-82b, which is given in percentage.

The fabrication costs for gap connections are lower than those with overlap, where the fully overlapped joints may be preferred to those with partial overlap from the viewpoint of the economy of fabrication. RHS with gap requires a simple plane cut of the ends of the both bracings with the same cutting angle, while for joints with partial overlap, one bracing member has double plane cuts at its end

Fig. 6-83 Joints with overlap. a) Partial overlap, b) full overlap

and the other requires a single plane cut. Gap connections are easiest to prepare, fit and weld. The joints with 100% overlap require plane cuts at the ends of the both bracings, but the cutting angles may be different.

6.6 Lattice girders (trusses)

The fabrication costs for joints depend on the joint configurations and profile combinations (CHS, RHS, I and ⊏ profiles) as well as the applied cutting methods (plane or profile cut, see Fig. 6-72 a and b). Also the availability of the fabricating machines e.g. cutting and welding machines, plays an important role.

The direct joining of CHS chord and bracing members necessitates nearly always a "profile" cut. This can however be made also by a plane cut in case the diameter of the bracing member is less than one third of the chord diameter (see Fig. 6-84 c).

The selected joint configuration affects the fabricating preparation of the structural members significantly and hence the manufacturing costs of the total girder also with regard to the fitting tolerances.

Due to the high fitting tolerance, the joint with gap is most suitable for the manufacture of a girder. Small adjustments can be made in each joint in order to secure the positions of the joints.

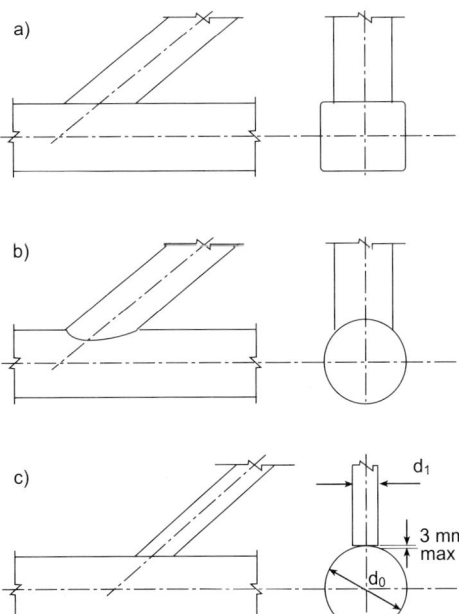

Fig. 6-84 Cross sections of joints with gap. a) Simple plane cut with RHS as chord and RHS or CHS as bracings (I or ⊏ profile can also be used as chord), b) profile cut with CHS as chord and bracing, c) simple plane cut with CHS as chord and bracing (in case $d_1 \leq 1/3\, d_0$)

The manufacture of a girder with fully overlapped joints is more difficult than that with gap joints. The fitting possibility is smaller; if the lengths of the bracing members are not exactly cut, the accumulated inaccuracy may lead to a dislocation of the position of the joints. The fitting of a joint with partial overlap is more problematic than that of a fully overlapped joint and consequently the manufacture of girders with partially overlapped joints may be more complex.

In the following, the relative costs are listed in rising order:

Cheapest:
1. Gap joint with RHS as chord and RHS or CHS as bracing
2. Fully overlapped joint with RHS as chord and bracing
3. Gap joint with CHS as chord and bracing
4. Partially overlapped joint with RHS as chord and bracing
5. Fully overlapped joint with CHS as chord and bracing

Dearest: Partially overlapped joint with CHS as chord and bracing

Partially overlapped joints are mostly fabricated in the workshop and it is particularly important to select suitable joint details in order to reduce the manufacturing costs as far as possible.

Fig. 6-85 illustrates three solutions for K joints with partial overlap consisting of CHS. In Fig. 6-85 a, the connection between two diagonals is made via a vertically welded plate. At the location of the joint, the bracing members are plane cut. In Fig. 6-85 b, the external diameters and the wall thicknesses of the both diagonals are about equal and the angle of inclination $\theta_1 \neq \theta_2$. In this case, the tensile bracing member has firstly to be welded to the chord. The partial overlap of the tensile diagonal by the compressive diagonal can be made by welding (double profile cuts in the compressive diagonal end).

If the difference between the diameters of the bracings is large (Fig. 6-85 c), the larger bracing has firstly to be welded to the chord. The smaller member overlaps subsequently the lar-

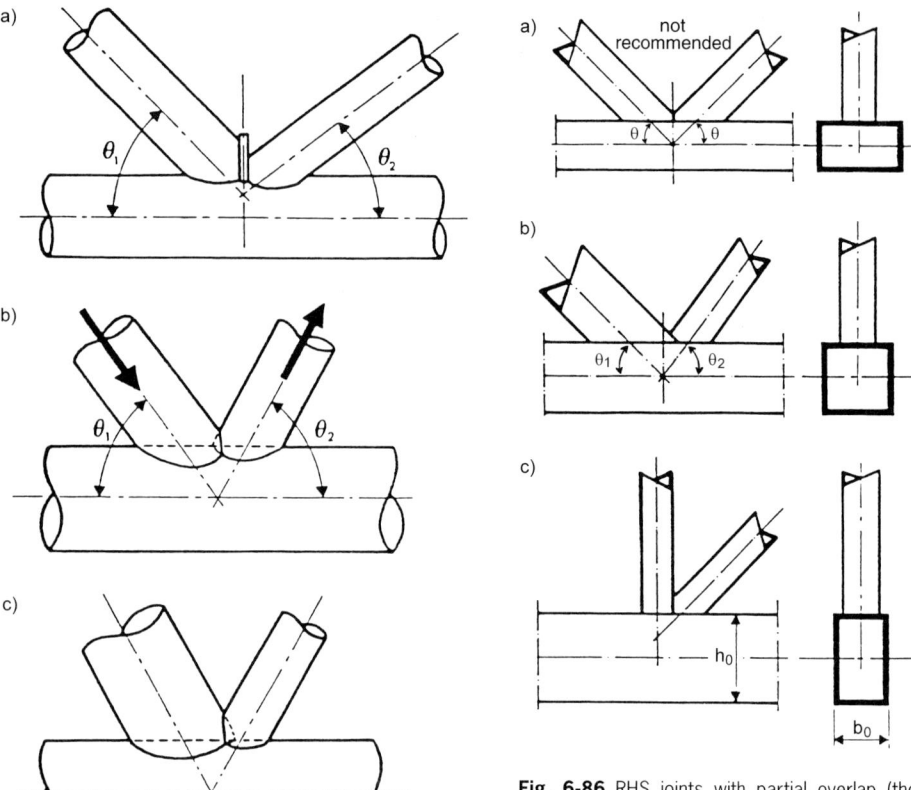

Fig. 6-85 Various CHS joints with partial overlap

Fig. 6-86 RHS joints with partial overlap (the bracing members may be made with RHS or CHS and the chord members may consist of RHS or I or ⊏ profiles)

ger one by welding. In this case, the smaller member will necessitate double profile cuts.

K joints with partial overlap made of RHS are shown in Fig. 6-86. In Fig. 6-86a, the joint is configured in a manner, so that both bracing members have double plane cuts at the ends. This joint is not recommended, because the joint has a strength lower than that for a joint with bracings fully welded to the chord.

Fig. 6-86b shows an asymmetrical K joint with the bracings of different widths and with different angles of inclination. The wider bracing has to be welded to the chord at first. The end of the more slim bracing member is then double plane cut and overlaps the wider bracing partially by welding.

For N joint with partial overlap (Fig. 6-86c), the following fabrication steps have to be observed:

- The vertical member has to be welded to the chord at first.
- The diagonal member end is then double plane cut and overlaps the vertical member partially by welding. The execution of this job with this sequence is however not always possible.

In a 90°/45° joint, the load in the diagonal member is 40% higher than that in the vertical, which will probably mean that the diagonal member has larger dimension and/or thicker wall than the vertical one. In this case, the diagonal has to be welded first to the chord.

Fig. 6-87 shows KT joints with gap and partial overlap, a number of them in mixed combinations. In general, the load in the vertical is much less than in the diagonals.

In joints with mixed combinations, the chord members made of hollow sections in Fig. 6-86

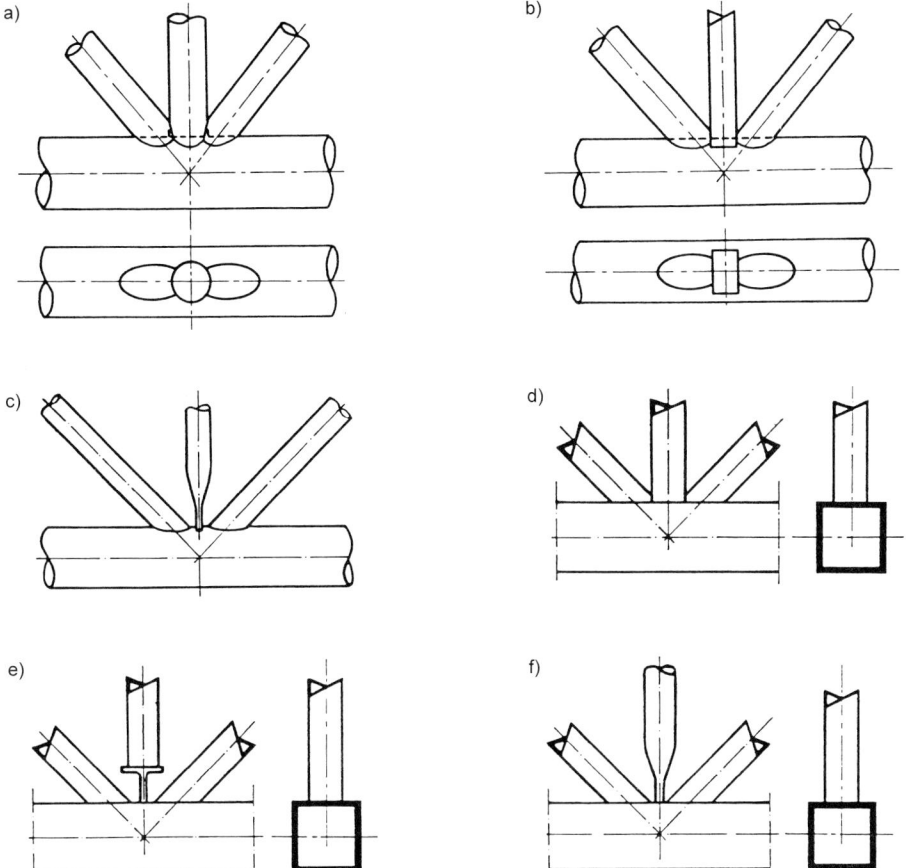

Fig. 6-87 KT joint with partial overlap and gap. a) KT joint of CHS with partial overlap, b) KT joint with partial overlap (RHS as vertical, all other members of CHS), c) KT joint with gap (vertical of CHS with cropped end, all other members of CHS), d) KT joint with partial overlap in RHS, the bracing members may also consist of CHS, e) KT joint with gap (vertical consist of T profile welded to CHS or RHS, all other members of RHS, bracing members may also be of CHS), f) KT joint with gap (vertical of CHS with cropped end, all other members of RHS, bracing members may also be of CHS)

and 6-87 can be substituted by I sections or channel profiles.

6.6.3 Weld seams in trusses with hollow sections consisting of directly welded bracing and chord members

Hollow sections in truss joints are in general welded to one another by fillet and partial or complete penetration groove welds or a combination of both of them. Selection of the weld type depends mainly on the angle of inclination θ of the bracing to the chord as well as the wall thickness of the bracing, which is according to the design rules [3, 5, 13, 24, 38, 42, 43] usually smaller than the chord wall thickness.

Fillet weld is the most frequently used weld type. The weld thickness a in a fillet weld can be configured arbitrarily. It can be shaped as convex, flat or concave form. The weld thickness a is defined as the depth of the seam described by the isosceles triangle shown in Fig. 6-88.

For fillet welds, the depth of the isosceles triangle is not always measurable. The weld seam measuring gauges have however been developed, by means of which the weld thick-

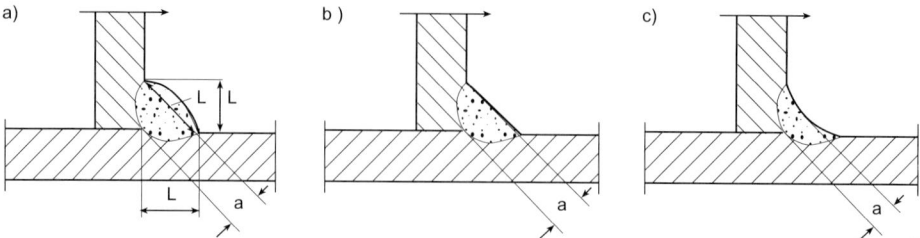

Fig. 6-88 Fillet weld types a) convex, b) flat, c) concave

ness a of a flat fillet weld and a concave fillet weld (with some problems) can be precisely determined [44]. For the convex fillet weld, the camber is estimated and then deducted from the measured value. Contrary to this, the weld thickness a and the unequal sides "Δz" can be exactly determined with gauges, which measure the seam legs "z_1" and "z_2" (Fig. 6-89).

For calculating the weld thicknesses of only flat and convex fillet welds, the weld legs z_1 and z_2 are first measured and the shorter of the both is used to describe the following equation:

$$a = 0.5 \cdot \sqrt{2} \cdot z = 0.71 \cdot z \qquad (6\text{-}37)$$

The selfconstruction of the measuring gauges for more precision is necessary, when the welds cannot be measured with gauges in standard business practice. If the fillet weld seam is not right-angled, this will be required in general. As an example for the selfconstruction of a measuring gauge (only to measure weld thickness and the given throat angle), the following steps are described:

A plate is cut with the same angle as the inclined workpieces colliding with each other; the required weld thickness a is measured from the corner. In order to be able to measure the concave tapered weld, two half arches are prepared. Fig. 6-90 illustrates two such gauges.

This gauge construction method can also be applied for right-angled parts colliding with each other and fillet welds with unequal legs.

A large number of weld measuring gauges is available on the market, which are more or

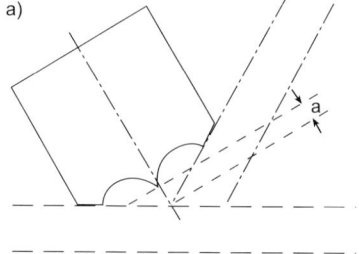

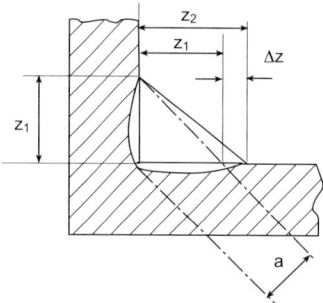

Fig. 6-89 Characteristic results regarding the fillet weld. a = actual weld thickness, z = weld leg, Δz = measure of the inequality of the weld legs

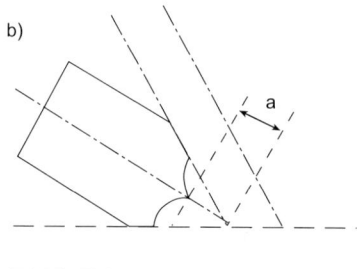

Fig. 6-90 Selfconstructed gauges to determine the fillet weld thickness a: a) for throat angle $> 90°$, b) for throat angle $< 90°$

6.6 Lattice girders (trusses)

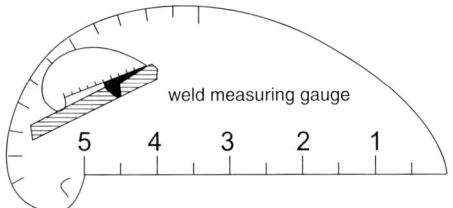

Fig. 6-91 Template weld gauge, suitable to measure fillet welds of the thickness between 3 to 15 mm; measuring range for excessive weld convexity: 0 to 5 mm; reading range: from about 0.2 to 0.5 mm; throat angle deviation: allowed to a small extent; weld thickness a for convex weld cannot be measured

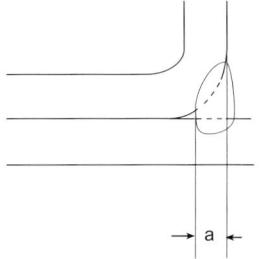

Fig. 6-92 Concave fillet weld between RHS and flat plate

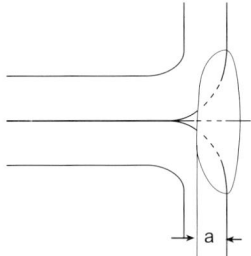

Fig. 6-93 Concave fillet weld between two RHS

less suitable to perform all or some particular jobs. Fig. 6-91 shows an example. This gauge is placed on the throat with its curved part in a way that this touches the workpiece and the fillet weld at three points. The convexity of the butt welds can be measured with the straight part.

Table 6-7 lists the recommendations [3, 5, 10, 38] to calculate the fillet weld thicknesses of hollow section joints in lattice girders taking undercuts into consideration.

IIW [49] recommends as follows:

– $a \geq t_1$ for S 235/275
– $a \geq 1.1\, t_1$ for S 355

It is to be mentioned here that the concave fillet welds (Fig. 6-88c) show a better load carrying behaviour under fatigue load due to the smooth and continuierlich transition from the weld to the basic material.

Figs. 6-92 and 6-93 illustrate two special applications of the concave fillet welds with RHS. In these cases, the effective weld thickness is determined using the average values of welds measured by means of tests on structural members for every set of procedures. The welds on structural members are intersected and measured, so that the measured, achieved weld thicknesses can be examined during the fabrication.

Table 6-7 Weld thickness of the fillet weld in a truss of hollow sections

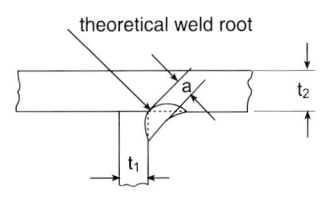	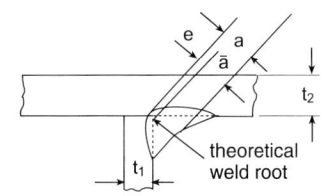
According to Eurocode 3 [3] and ENV 1090-4 [38]	
a: see Table 6-8 (determined according to material grade) a = minimum 3 mm	$a = \bar{a} + e$ For deep undercut, the enlarged weld thickness may be considered in the calculation, provided the undercut beyond the theoretical weld root is found by tests, $a = t_1$

Table 6-8 Fillet weld thickness according to Eurocode 3 [3] depending on the steel grade

For steel grades according to EN 10025 [45]:
- For Fe 360 (S 235*), $a/t \geq 0.84\,\alpha$
- For Fe 430 (S 275*), $a/t \geq 0.87\,\alpha$
- For Fe 510 (S 355*), $a/t \geq 1.01\,\alpha$

For steel grades according to pr EN 10113 [46]:
- For FeE 275 (S 275**), $a/t \geq 0.91\,\alpha$
- For FeE 355 (S 355**), $a/t \geq 1.05\,\alpha$

* according to [47, 48] non-alloy steels
** according to [47, 48] fine grain steels

α is to be determined as follows:

$$\alpha = \frac{1.1}{\gamma_{Mj}} \times \frac{\gamma_{MW}}{1.25}$$

where
γ_{Mj} = partial safety factor for truss joints
γ_{MW} = partial safety factor for weld

According to [3], $\gamma_{Mj} = 1.1$ and $\gamma_{MW} = 1.25$; this leads to $\alpha = 1.0$.

The equivalent stress for the existing weld stresses in the fillet weld (see Fig. 6-68b) can be calculated using the Eq. (6-35).

6.6.3.1 Design of welds in truss joints

As already mentioned at the outset of the Section 6.6.3, the truss joints with directly welded chord and bracing members are manufactured by fillet or partial or complete penetration groove welds or combinations of them. The selection of the weld seams depends on the angle between the axes of the chord and the bracing as well as on the wall thickness of the bracing, which is usually smaller than the chord wall thickness (see Fig. 6-82).

Fig. 6-94 or 6-95 shows the fillet weld forms in truss joints of CHS or RHS for the bracing wall thickness $t_{1,2} < 8$ mm.

Groove welds in truss joints of CHS or RHS for the bracing wall thickness $t_1 \geq 8$ mm are illustrated in Figs. 6-96 and 6-97.

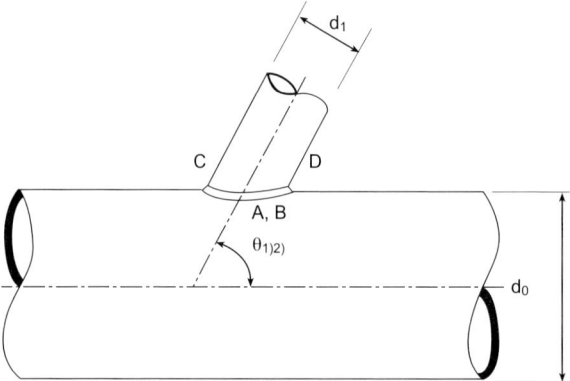

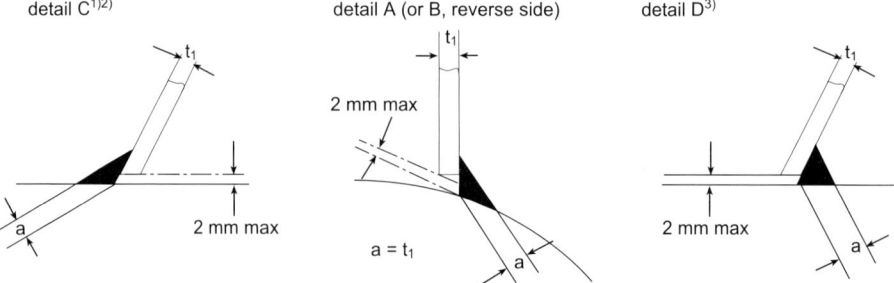

Fig. 6-94 Fillet weld types for truss joints of CHS ($t_1 < 8$ mm). 1) θ shall not be smaller than 30°. 2) In case $\theta < 60°$, the groove weld detail C in Fig. 6-96 is to be used. 3) For smaller angle of inclination, the full penetration to the weld root is not required, if sufficient weld thickness is available

6.6 Lattice girders (trusses)

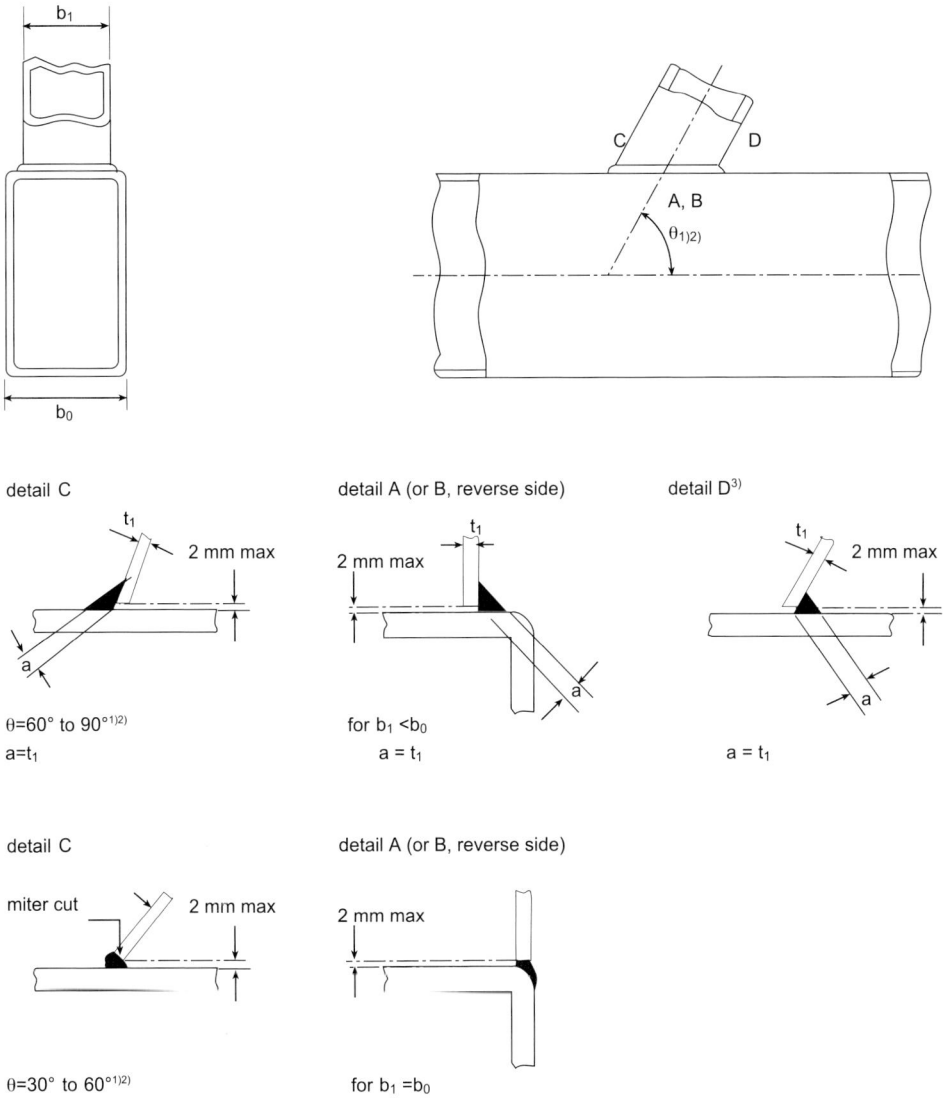

Fig. 6-95 Fillet weld types in truss joints of RHS ($t_1 < 8$ mm). 1) θ shall not be smaller than 30°. 2) In case $\theta < 60°$, the groove weld detail C in Fig. 6-97 is to be used. 3) For smaller angle of inclination, the full penetration to the weld root is not required, if sufficient weld thickness is available

For mixed combinations (I- or ⊏ profile as chord and RHS or CHS as bracing), the execution of the welds are similar to those shown in Figs. 6-95 and 6-97.

It has to be especially mentioned here that prequalified joint details for partial and complete joint penetration groove welds as well as simple fillet welds in T, Y and K joints have been elaborately illustrated in [13].

The transition of the weld, which is executed as a combination of fillet and groove weld (partial and complete penetration), must be smooth and uniform. As the details demonstrate (see Figs. 6-94 to 6-97), the angle of the cut surface between the continuous member and the hollow section set on it changes gradually.

For RHS joints with $b_1 = b_0$ (Fig. 6-98), the cross sectional area of the weld is dependent

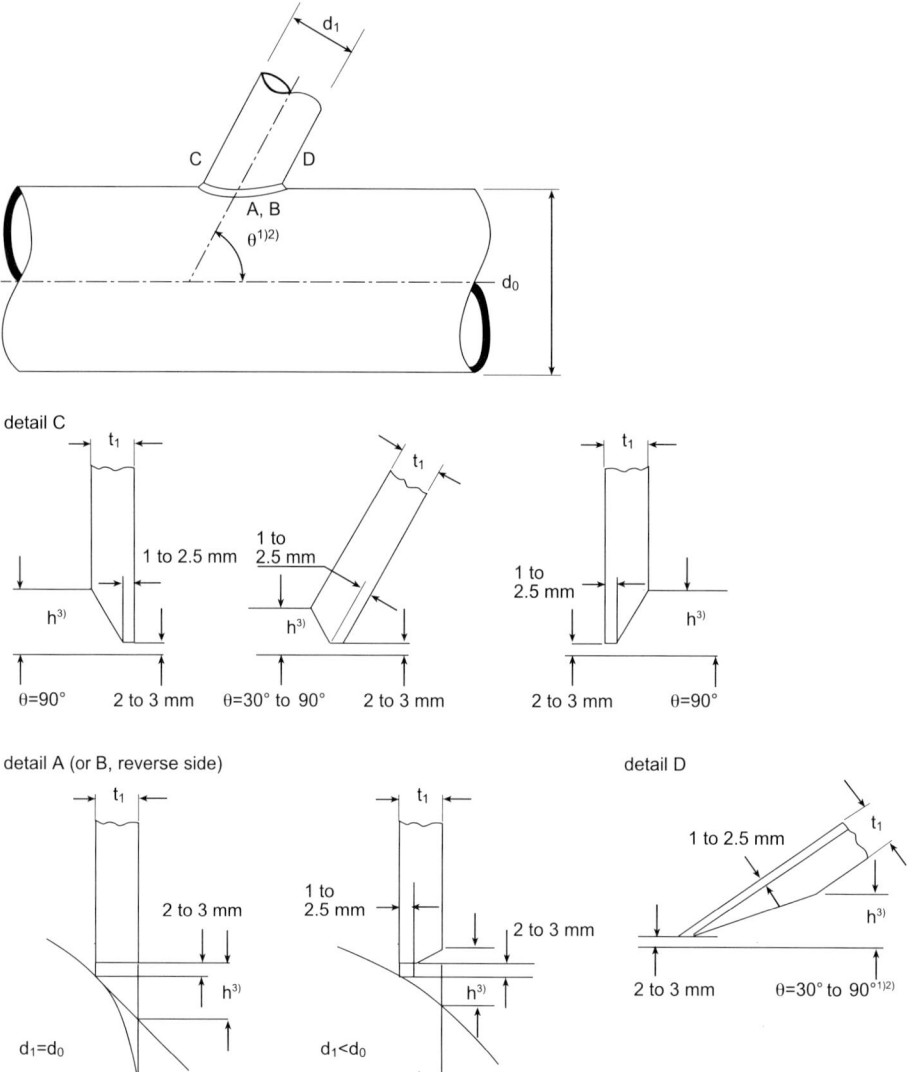

Fig. 6-96 Partial or complete joint penetration groove welds in truss joints of CHS ($t_1 \geq 8$ mm). 1) θ shall not be smaller than 30°. 2) In case $\theta \geq 60°$, the fillet weld detail D in Fig. 6-94 is to be used. 3) $h_{min} = t_1$. 4) For smaller angle of inclination, the complete penetration to the weld root is not required, if sufficient weld thickness is available

on the corner radius of the chord hollow section. If possible, the gap $g \leq 3$ mm is to be achieved by reducing the bracing width b_1. In unavoidable cases, the gap has to be filled up by weld metal, which is however expensive.

One specific feature for the joints with partial overlap has to be considered in its application (see Fig. 6-99). In the manufacturing workshop, it is usual to assemble the structural members of a truss on a jig by means of tack welding. The final continuous welding takes place in a separate working process.

This sequence makes it impossible to weld the seam in the covered part "A". However, tests have shown that the strength of a joint is not generally affected, when the weld at "A" is left out. On the other hand, this weld must be made, if the vertical components of the loads

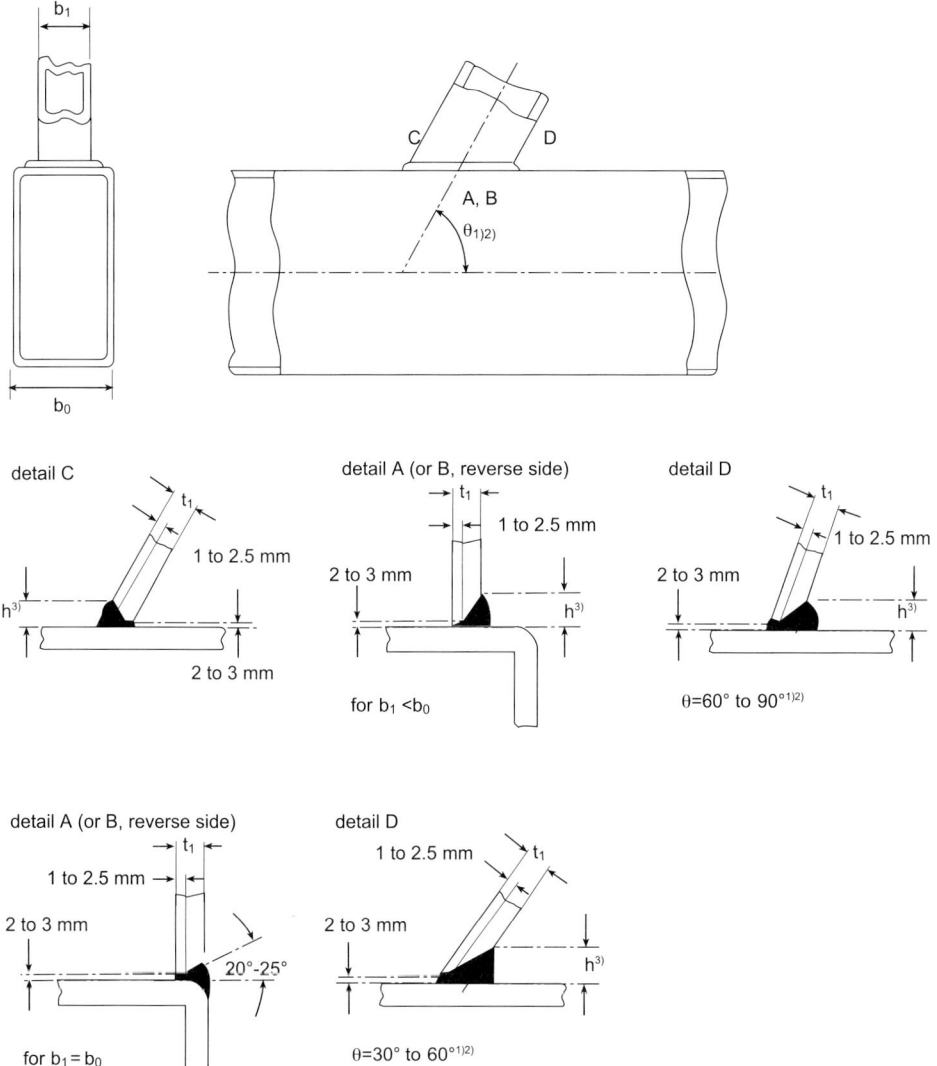

Fig. 6-97 Partial or complete joint penetration groove welds in truss joints of RHS ($t_1 \geq 8$ mm). 1) θ shall not be smaller than 30°. 2) In case $\theta \geq 60°$, the fillet weld detail D in Fig. 6-95 is to be used. 3) $h_{min} = t$. 4) For smaller angle of inclination, the complete penetration to the weld root is not required, if sufficient weld thickness is available

in the two bracings deviate from each other by more than 20%. That means, the tack welding of the hidden part "A" is to be substituted by the full welding in this case.

6.6.3.2 Weld sequences in truss joints

While welding hollow sections in a truss joint, special care has to be taken that high, weld-initiated residual stresses and unallowable deformations do not occur. In uni-planar and multi-planar hollow section joints, non-uniform heating of the chord member can lead to undesirable deformations and residual stresses. They can be primarily reduced by selecting appropriate welding sequences. In principle, the weld sequence has to be selected in a manner that the individual structural members can shrink freely as long as possible. In lattice girders, for example, welding of the

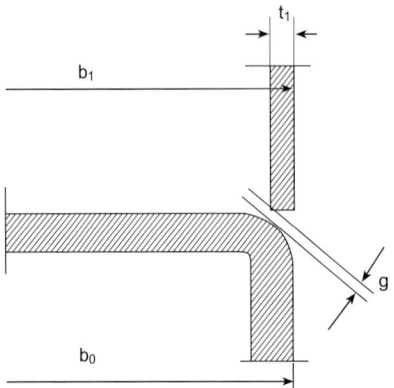

Fig. 6-98 RHS joint with $b_1 = b_0$

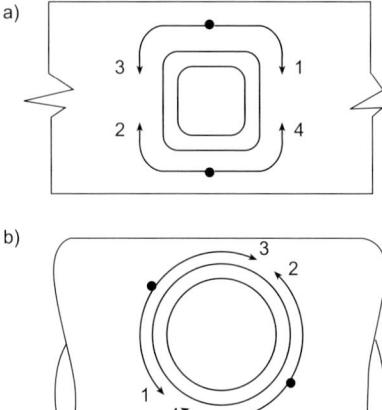

Fig. 6-100 Recommended welding sequences for truss joints consisting of a) RHS (square), b) CHS

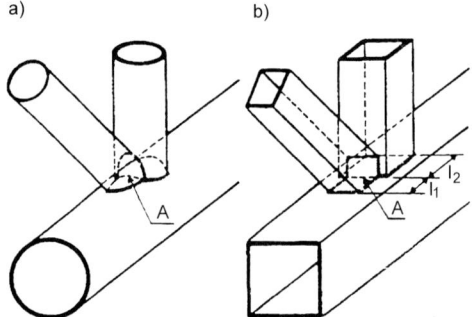

Fig. 6-99 Execution of weld in a truss joint with partial overlap consisting of: a) CHS, b) RHS A – Welding of the hidden toe may be dispensed with if the vertical load components in the bracings differ from each other by not more than 20%

bracings should start at the centre and work outwards towards the end. This results in favourable compressive residual stresses being initiated in the weld metal rather than undesirable tensile residual stresses.

Fig. 6-100 shows the sequences of the execution of welds in truss joints as recommended by [38].

For RHS joints, stop/start positions of the welding operation shall not be located at or close to the corner of the hollow section i.e. one starts at the middle of the plane of the hollow section and then weld outwards alternately along both sides.

For CHS joints, stop/start positions shall not be located at or close to the toe (crown) or lateral flank (saddle) position.

Overlap of welds has to be avoided, which may occur while welding two structural members subsequently.

A complete welding of the total cross section of a hollow section is desired, even if this total length of weld is not always required for strength reasons. Fig. 6-99 illustrates an exception.

6.6.3.3 Weld lengths in truss joints

The theoretical length l_W of the fillet and groove weld is given by the geometrical length of the root line and can be calculated as follows:

6.6 Lattice girders (trusses)   143

1. Connection of a CHS to a flat plate or a RHS

   For $\theta = 90°$,
   weld length $l_W = \pi \cdot d_1$

   For other θ-values,
   weld length $l_W = a + b + \sqrt{a^2 + b^2} \cdot 3$

   where $a = \dfrac{d_1}{2\cos\theta}$ and $b = \dfrac{d_1}{2}$

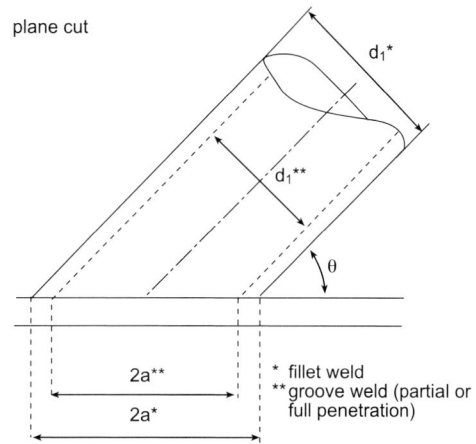

* fillet weld
** groove weld (partial or full penetration)

2. Connection of a RHS to a flat plate or another RHS

   For $\theta = 90°$,
   weld length $l_W = 2h_1 + 2b_1$

   For other θ-values,

   weld length $l_W = \dfrac{2h_1}{\cos\theta} + 2b_1$

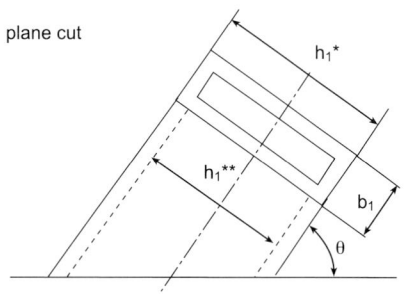

* fillet weld
** groove weld (partial or full penetration)

3. Connection of a CHS to another CHS

   Weld length $l_W = a + b + \sqrt{a^2 + b^2} \cdot 3$

   where

   $a = \dfrac{d_1}{2\cos\theta}$ and $b = \dfrac{d_0 \cdot \Phi}{4}$ (Φ in radian)

   $\sin\dfrac{\Phi}{2} = \dfrac{d_1}{d_0}$

   The theoretical weld area amounts to:

   $A_W = \sum a \cdot l_W$

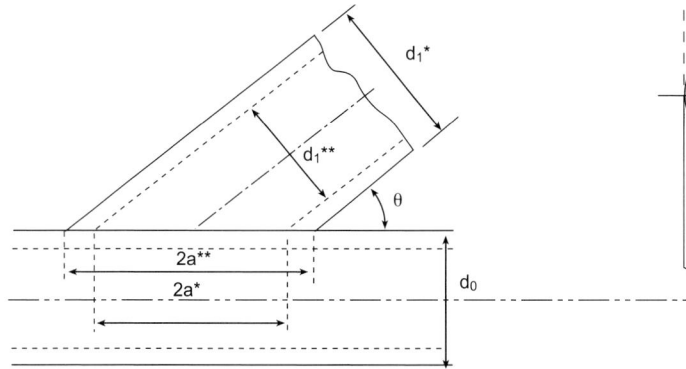

* fillet weld
** groove weld (partial or full penetration)

6.6.4 Analysis of the strength (resistance) behaviour of a truss with hollow sections

The determination of bending moments and axial forces in a truss is carried out by means of elastic analysis under the assumption of pin joints between the connecting members (see Fig. 6-73), where the members are subjected to axial forces. The difference between the trusses of hollow sections and those made with open profiles lies in the joint eccentricities (Fig. 6-101), which is often produced by the intersecting points of the centrelines of the members. These eccentricities create primary bending moments, which need to be taken into account for the equilibrium of loads. Joint eccentricities are initiated by the gap between the toes of the neighbouring bracings or their overlaps, which is required either to increase the joint strength or to improve the fabrication.

The joint configurations (see Fig. 6-82) depend mutually on the gap (+g), the overlap (−g) and the eccentricity e. The eccentricity (±e) and the gap (+g) or the overlap (−g) can be calculated based on the sizes of the truss members and the angle of inclination θ using the following equations:

Assumption: $\theta_1 \geq 0°, \theta_2 \leq 180°$

$$(+g) \text{ or } (-q) = \frac{\pm e + D}{C} - (A + B) \qquad (6\text{-}38)$$

$$\pm e = C[A + B(+g) \text{ or } (-q)] - D \qquad (6\text{-}39)$$

where:

$$A = \frac{h_1 \text{ or } d_1}{2 \sin \theta_1} \qquad (6\text{-}40)$$

$$B = \frac{h_2 \text{ or } d_2}{2 \sin \theta_2} \qquad (6\text{-}41)$$

$$C = \frac{\sin \theta_1 \cdot \sin \theta_2}{\sin (\theta_1 + \theta_2)} \qquad (6\text{-}42)$$

$$D = \frac{h_0 \text{ or } d_0}{2} \qquad (6\text{-}43)$$

Special case: $e = 0$

$$(+g) \text{ or } (-q) =$$
$$= \frac{h_0 \text{ (or } d_0) \cdot \cos \theta_1 - h_1 \text{ (or } d_1)}{2 \sin \theta_1} +$$
$$+ \frac{h_0 \text{ (or } d_0) \cdot \cos \theta_2 - h_2 \text{ (or } d_2)}{2 \sin \theta_2} \qquad (6\text{-}44)$$

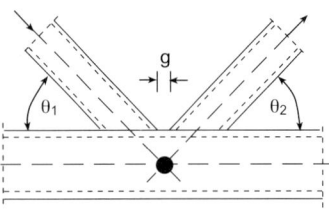

gap joint with e=0

gap joint with e>0

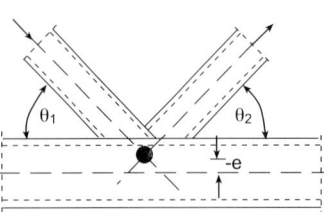

partial overlap joint with e<0

full overlap joint witht e<0

Fig. 6-101 Examples of joint eccentricities

6.6 Lattice girders (trusses)

Primary bending moments due to the joint eccentricity can be neglected, when they lie within the following ranges:

$$-0{,}55 \leq \frac{e}{h_0} \text{ or } \frac{e}{d_0} \leq +0{,}25 \quad \text{(according to [3])}$$

When the eccentricities are within the above mentioned limits, the influence of the primary bending moments is already integrated in the given formulae for the joint design resistance (see Tables 6-9 to 6-14).

If these eccentricity limits are exceeded, the chord members under compression have to be designed also for the eccentricity moments. Due to the mostly used low stiffness of the bracings, these moments in the bracings are not taken into consideration and the total moment is distributed in the chord according to Fig. 6-102.

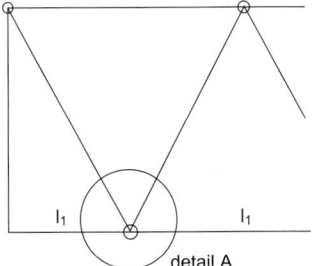

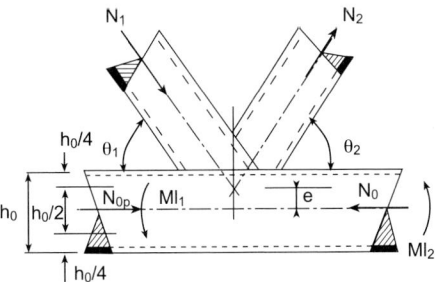

$M_{Total} = e(N_1 \cdot \cos\theta_1 + N_2 \cdot \cos\theta_2)$

$Ml_1 = M_{Total} \left(\dfrac{I_1/l_1}{I_1/l_1 + I_2/l_2} \right)$

$Ml_2 = M_{Total} \left(\dfrac{I_2/l_2}{I_1/l_1 + I_2/l_2} \right)$

I = second moment of area

Fig. 6-102 Distribution of the primary bending moment due to joint eccentricity e in the chord member

For the verification of the compression chord taking account of the eccentricity moment (M_{l1} or M_{l2}), the interaction between the axial compression N_0 or the preload N_{0p} in the chord and the eccentricity moment have to be considered. Eq. (5-86) can be applied for this purpose.

The eccentricity moments can be ignored for the design of tension chord and bracings.

Further, a primary bending moment can also occur when transverse loads are applied to either chord away from the panel points. While designing chords in a truss, this moment must be taken into account.

Fig. 6-103 shows the model of a truss, which is used as a rule for the computer plane frame programs to calculate the bending moments and axial forces. The continuous chords are pin connected with the bracing members at distances of $+e$ or $-e$ from the system lines of chords. It has to be assumed that the links to the pins are extremely stiff. This modelling leads to a distribution of bending moments automatically generated throughout the truss, for the cases in which bending moments must be considered in the chord design.

Although the design of hollow section trusses is, as already described, carried out assuming that all members are pin connected, secondary bending moments are actually initiated, as the bracings are welded to the continuous chords producing semi-rigid joints. This is caused by the real deformations in the structure (Fig. 6-104a) or the joint stiffnesses (Fig. 6-104b). In principle, the secondary bending moments are not required for the equilibrium of loads. They can be ignored while designing structural members and joints, provided that there is adequate deformation or rotation capacity, which allows the redistribution of stresses by local plastification at the joints.

In order to assume pin joints in trusses, one of the following conditions has to be fulfilled [7]:

– The joints have higher strength than the connected members and the geometrical parameters of structural members allow a redistribution of stresses.

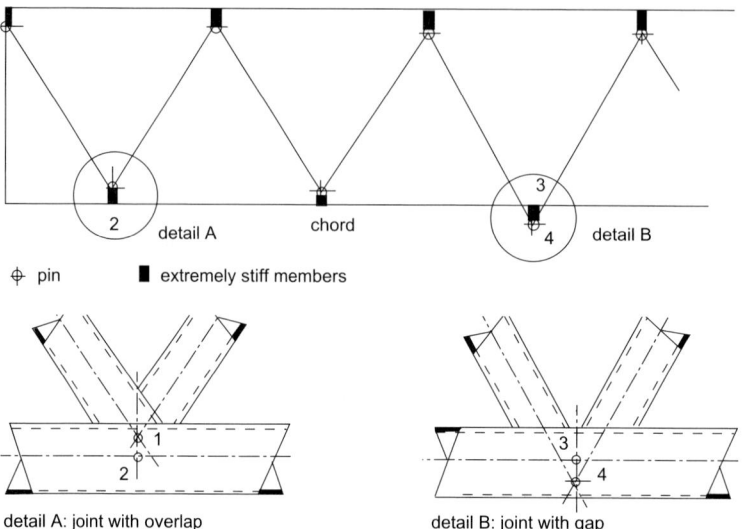

Fig. 6-103 Plane frame connection modelling assumptions to obtain realistic forces for member design

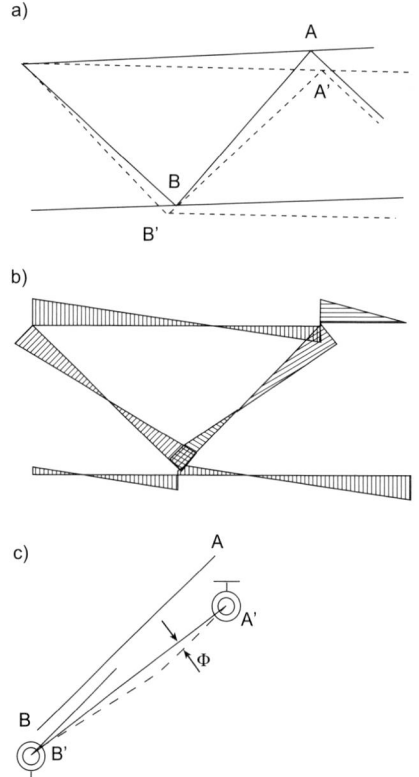

Fig. 6-104 Actual moments in a truss. a) Actual deformations, b) actual moment distribution due to joint stiffness, c) required rotation capacity

– The joints have lower strength than the connected members, but they possess in the limit state adequate deformation and rotation capacity, which makes a redistribution of stresses possible.

Further, the welds in a joint must have a resistance capacity, which facilitates adequate stress redistribution before premature failure. The recommended fillet weld thicknesses for the bracing members given in Section 6.6.3 fulfil this condition [3, 49]. These recommendations have been made on the safe side and permit a certain degree of weld defects. The welded truss joints corresponding to Lit. [3, 49] are designed for any load in the structural members.

The following has to be considered while designing a weld for a certain load in the bracing:

The total length of a weld (the periphery of the bracing) cannot be effective in all cases and this has to be taken into consideration when the strength and the deformation of a weld are to be determined.

Plastic design can also be used in order to determine forces and moments in a truss. It is permitted to calculate the sizes of the chord members by means of modelling a truss consisting of chords as continuous beams with pin supports for the web members. For this, the following prerequisites have to be met:

- The structural members fulfil the requirements of class 1 sections (see Tables 5-5, 5-15 to 5-17, 5-23, 5-24).
- The minimum design resistance of the weld must be equal to the cross-sectional design resistance of the connected bracings.

6.6.4.1 Design strengths of truss joints with directly welded bracing and chord members

Fig. 6-105 illustrates two examples of non-uniform stress distribution in CHS or RHS truss joints. Complex force and moment transmission as well as non-linear rigidity distribution in truss joint make it very difficult to determine the joint strength theoretically with adequate precision. The joint configurations (Fig. 6-81) and the geometrical ratios of the dimensions of the structural members on the one hand and the local effective loads, axial and transversal to the profile axis, given by the joint geometry on the other hand, have to be considered in the calculation. This strength is primarily described by the local deformations of the connection surface of the chord, which is relatively yielding, as it is unstiffened in most cases. Fig. 1-9 presents the local deformation behaviour in a K joint of RHS graphically. The local peak stress can be reduced by plastic deformation and stress redistribution. The effect of this process is significant for static loads.

Due to the reasons described above, extensive research works (establishment of theoretical models and practical tests) had to be carried out to work out design rules and develop the formulae for design resistances of truss joints of hollow sections under predominantly static load. The publications of these research reports are listed in the chapter "Literature", as far as they are known to the author. Based on the research results, the design recommendations for hollow section joints were made, which were taken to formulate various national and international standards or guidelines into consideration [3, 5, 6, 13, 24, 49, 148, 226].

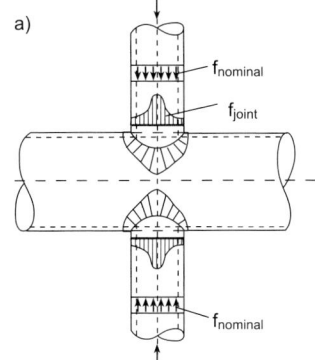

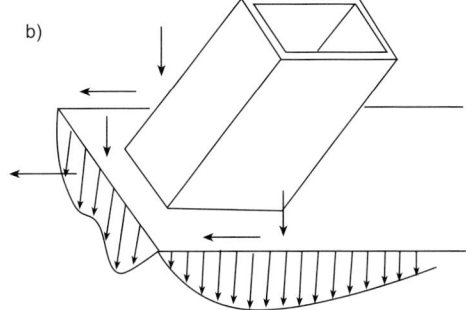

Fig. 6-105 Stress distribution in hollow section joints. a) CHS (X joint), b) RHS (Y joint)

In this context, it has to be mentioned that the majority of these tests was performed on single joints. The transferability of the results of the tests on single joints and the conclusions derived from them on the joints in complete trusses has been additionally checked by the investigations with complete trusses (see Fig. 1-10 b). These tests confirm the relevant transferability.

In order to design the structural members of a hollow section truss basing on the statics for the lattice girders, it is required to verify the joint design resistance additionally. The following parameters are decisive for this purpose (see Fig. 6-82):

- Rigidity of chord cross section depending on $\gamma = \dfrac{d_0}{2\,t_0}$ or $\dfrac{b_0}{2\,t_0}$
- Diameter or width ratio
  $\beta = \dfrac{d_1}{d_0}$ or $\dfrac{d_0}{b_0}$ or $\dfrac{b_1}{b_0}$ or $\dfrac{d_1 + d_2}{2\,d_0}$ or

$$\frac{b_1+b_2}{2b_0} \text{ or } \frac{b_1+b_2+h_1+h_2}{4b_0} \text{ etc.}$$

- Wall thickness ratio bracing/chord $\tau = \dfrac{t_1}{t_0}$
- Gap g or overlap $\lambda_{ov}\%$
- Ratio depth/width $\eta = \dfrac{h_i}{b_i}$
- Angle of inclination of the bracing to the chord θ_i
- Function to take prestress in the chord into consideration $f(n') = \dfrac{N_{op}}{A_0 \cdot f_{y0}} + \dfrac{M_0}{W_0 \cdot f_{y0}}$

Regarding the effect of the geometrical ratios on the design resistances, the following comments are made:

Deformations in hollow section joints appear mainly on the connection surface of the chord. They can preferably be reduced by using stiffer chord members (γ ratio).

Smaller γ ratios increase the load transmissibility by the frame effect of hollow cross sections. This statement is of significant importance: Chord members have to be made of larger wall thickness and smaller width as far as possible in order to increase the design resistance capacity.

A further facility to attain more rigid joints lies in enlarging the β ratios and in reduction of the gap g.

Besides γ, the ratio τ determines the load distribution in the connection zone of the bracing and the chord members decisively.

Characteristic features to assess the joint strength

In general, the static strength of a hollow section joint is characterized by the following criteria:

- Ultimate load (5)
- Load based on deformation limit (2)
- Visual sign of crack (4)

Fig. 6-107 shows the forces in a K joint mounted on a test jig (Fig. 6-108). The load measuring devices are shown in Fig. 6-109.

The ultimate compressive load (point 5) is clearly defined and therefore is to be taken as the basis for the design of joints.

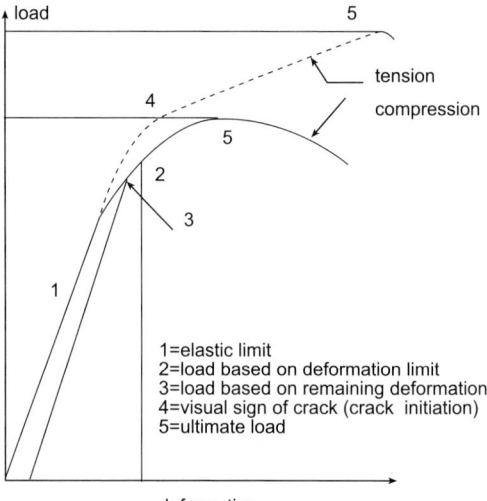

Fig. 6-106 Load vs. deformation curves for a compression and a tension member in a truss joint of hollow sections

Because of the non-linear load-deformation behaviour of hollow section joints, there is no international agreement regarding the deformation limit (point 2) or the determination of yield strength of a hollow section joint (1).

As Fig. 6-106 demonstrates, the ultimate tensile load in a member is considerably higher than the ultimate load in a member under compression.

Further, the deformation in T, Y and X joints under tensile load is large, if the diameter or width ratio β is small. On the other hand, the deformation limit decreases with the increase of β. This is the reason why, in general, the ultimate tensile load is not used for joint design.

The typical load vs. deformation curves for a hollow section joint (Fig. 6-106) show kinks at the point for ultimate compressive load (point 5) as well as at the point for the initia-

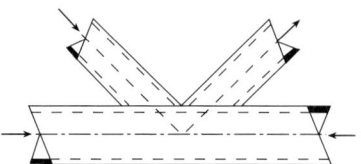

Fig. 6-107 Load directions in a K-type test joint

Fig. 6-108 Test jig for hollow section joints

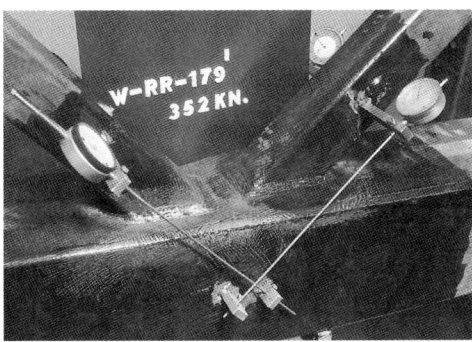

Fig. 6-109 Measuring devices for testing strengths of hollow section joints (gauges for measuring elongation)

tion of the first crack under tensile load at about the same load level. In order to avoid too large deformation and to achieve additional safety for joints with small deformation capacity, it is recommended to apply compressive load as the basis for the joint design for the members under compression as well as tension.

The tests to determine joint strengths on various types of welded hollow section joints have demonstrated that the load vs. deformation curve is smooth and runs uniformly with increasing load till the joint fails (Fig. 6-110).

This behaviour indicates that a sudden joint instability does not occur for usual CHS and RHS sizes applied in practice. A joint failure, as described by the dashed curve in Fig. 6-110, can take place in the following cases:

– In joints with extremely thin hollow sections, failure can occur by local instability, even when the elastic limit is not exceeded.
– In joints consisting of materials, which possess a lower ductility

The case mentioned at first has to be considered while designing a joint e.g. by selecting adequately large $\gamma \left(= \dfrac{h_0}{2\,t_0} \text{ or } \dfrac{d_0}{2\,t_0} \right)$ ratios. The second case can be avoided by approving only appropriate materials for the steel structures. In some cases, the lacking ductility of the materials, brought forth by i.e. large ratio yield limit/ultimate strength, is compensated by the higher demand for safety.

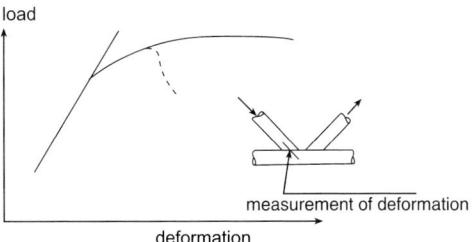

Fig. 6-110 Load vs. deformation diagram for hollow section joints

Theoretical models to analyse the load carrying behaviour of hollow section joints

Besides the strength tests on various joint types described in Section 6.6.4.1, also the approach was made to determine the design resistances theoretically [76–79]. These calculations are in general restricted to CHS T joints. The calculated design strengths agree fairly well with the experimental results for small d_i/d_0 values, but not in other ranges.

Lit. [7] contains simple models to determine the decisive parameters for the design strengths of CHS and RHS joints. It is to be noted here that the analytical models considering all influencing parameters are in general very complicated. Therefore they are dispensed with in most cases.

A theoretical treatment of the frequently used K and N joint is very difficult due to the complicated load transmission in them. This is also not possible for CHS joint, as no usable method is available for this purpose. However, the methods to calculate the strength of a K joint made of RHS theoretically have been developed [179].

The ring model and the punching shear model are mainly applied to calculate CHS joint resistances. The resistance behaviour of RHS joints is generally described by the yield line model and that with the effective width.

- **Ring model**

The ring model as illustrated in Fig. 6-111 was first used in [52]. In this case, a X joint is simply schematized to a ring with an effective length B_e.

The bracing stresses are assumed to be concentrated at the "saddle" locations, where the load in the bracing is schematized to two line loads at a distance of $c_1 \cdot d_1$ from each other. c_1 has a constant value, which is smaller than 1.0.

If the effect of the axial and shear stresses are neglected, the plastic failure load of a X joint can be determined as follows:

$$2m_p = \frac{N_1 \cdot \sin\theta_1}{2}(1 - \sin\varphi)\left(\frac{d_0 - t_0}{2}\right) \quad (6\text{-}45)$$

where:

$$m_p = \frac{B_e \cdot t_0^2}{4} \cdot f_{y0} \quad (6\text{-}46)$$

$$\sin\varphi \approx c_1 \cdot \beta \quad (6\text{-}47)$$

$$\frac{d_0 - t_0}{2} \approx \frac{d_0}{2} \quad (6\text{-}48)$$

$$N_1 = \frac{2B_e}{d_0} \cdot \frac{1}{1 - c_1 \cdot \beta} \cdot \frac{f_{y0} \cdot t_0^2}{\sin\theta_1} \quad (6\text{-}49)$$

The effective length is to be determined experimentally:

$$B_e \approx 3\,d_0 \quad (6\text{-}50)$$

The following equation leads to the resistance of X joint:

$$N_1 = \frac{c_2}{1 - c_1 \cdot \beta} \cdot \frac{f_{y0} \cdot t_0^2}{\sin\theta_1} \cdot f(n') \quad (6\text{-}51)$$

In Eq. (6-51), c_1 and c_2 are two coefficients, which are to be determined by tests. $f(n')$ represents the effect of the additional load in the chord member.

$f(n') = 1.0$ for tensile load and
$f(n') \leq 1.0$ for compressive load

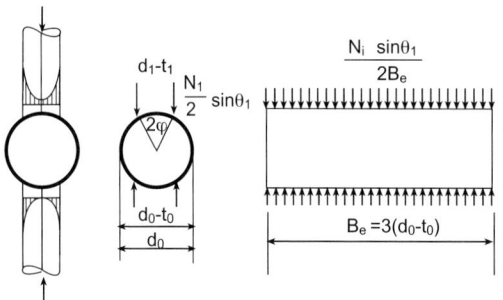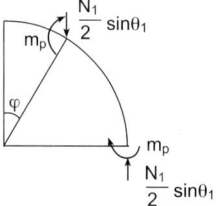

Fig. 6-111 Ring model

6.6 Lattice girders (trusses)

The load transmission is significantly complex for T joints, more particularly for K and N joints. Following functions are used in general as the basis for the determination of the joint resistance:

$$N_1 = f(\beta) \cdot f(\gamma) \cdot f(g) \cdot f(n') \cdot \frac{f_{y0} \cdot t_0^2}{\sin \theta_1} \quad (6\text{-}52)$$

The functions for β, γ, g and n' are determined mainly by means of tests.

- **Punching shear model**

Fig. 6-112 shows the punching shear model for a CHS joint in a schematized form. It is assumed that the punching shear stress v_p is uniformly distributed over the punching shear area. This results in

$$N_1 \cdot \sin \theta_1 = k_a \cdot \pi d_1 t_0 \cdot v_p \quad (6\text{-}53)$$

where:

$$v_p = 0.58 f_{y0} \text{ and } k_a \approx \frac{1 + \sin \theta_1}{2 \sin \theta_1}$$

$$N_1 = 0.58 f_{y0} \cdot \pi d_1 t_0 \cdot \frac{1 + \sin \theta_1}{2 \sin^2 \theta_1} \quad (6\text{-}54)$$

In general, this failure criterion can only then be critical, when β is small.

In RHS joints (Fig. 6-113), the rigidity of the bracing sides is significantly larger than that in the middle of the chord flange.

For $\beta \approx 1.0$, the total perimeter of the bracing can be assumed to be effective. The punching shear strength $\hat{N}_1$ for T, Y and X joints can be calculated as follows:

$$\hat{N}_1 = v_p \cdot t_0 \cdot \left(\frac{2 h_1}{\sin \theta_1} + 2 b_{e,p} \right) \cdot \frac{1}{\sin \theta_1} \quad (6\text{-}55)$$

The full perimeter of the bracing can be effective only when b_0/t_0 as well as β are small.

The effective width $b_{e,p}$ for the punching shear has to be determined experimentally. In a joint with more than one bracing member, the rigidity distribution at the location of the intersection of the bracing with the chord determines the punching shear area. Compared to the width ratio β, the gap g has to be restricted, so that the effective width does not become too small.

- **Model with the effective bracing width for a RHS joint**

The punching shear model is principally used for RHS joints with relatively thickwalled bracing members. The effective width b_e of the bracing can however be critical for joints with thinwalled bracings. The strength in this case can be expressed in a similar way as for the punching shear model, but is now related to the bracing dimensions and the bracing material properties. Accordingly, the strength of T, Y and X joints is given by the following equation:

$$\hat{N}_1 = f_{y1} \cdot t_1 (2 h_1 - 4 t_1 + 2 b_e) \quad (6\text{-}56)$$

The effective bracing width becomes larger with decreasing b_0/t_0 and t_1/t_0 ratios.

- **Yield line model**

The yield line model is mainly applied to joints made of square and rectangular hollow sections. Fig. 6-114 shows a simplified yield line model to determine the resistances of T, Y and X joints with $\beta \leq 0.8$, when yield limit of the material is reached. This gives an upper bound solution for the yield load. In the simplified model, the effects of the membrane action and the strain hardening are ignored. Thereby, a cautious estimation of the actual resistances is made. This is in particular valid

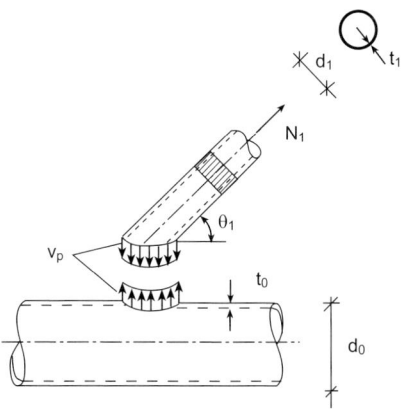

Fig. 6-112 Punching shear model for a CHS joint

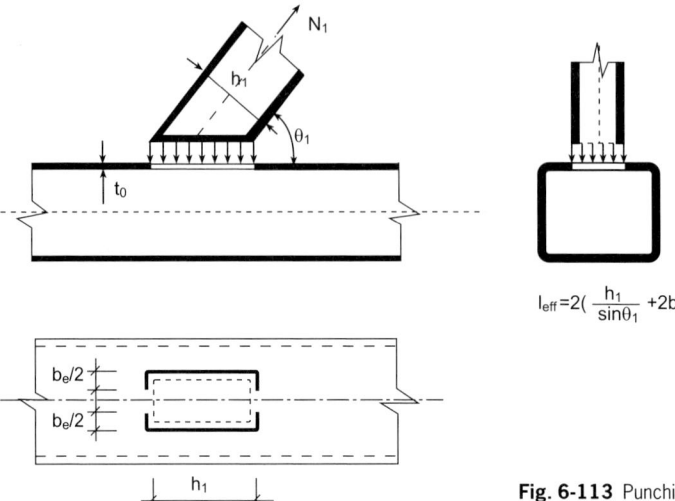

Fig. 6-113 Punching shear model for a RHS joint

for joints with small β, which undergo a very large deformation at the ultimate load. For T, Y and X joints, the load at the yield limit is taken as the basis, so that a large deformation is avoided in the actual design.

Membrane actions are taken into account in the yield line models for K and N joints [227]. The procedure consists in equating the works of the external forces with those by the plastic hinge systems. An example with a Y joint:

$$N_1 (\sin \theta_1 \cdot \delta) = \sum l_i \cdot \varphi_i \cdot m_{pi} \qquad (6\text{-}57)$$

where:

δ is the deformation due to the load N_1, l_i is the length of the yield line i and φ_i is the rotation for the yield line i.

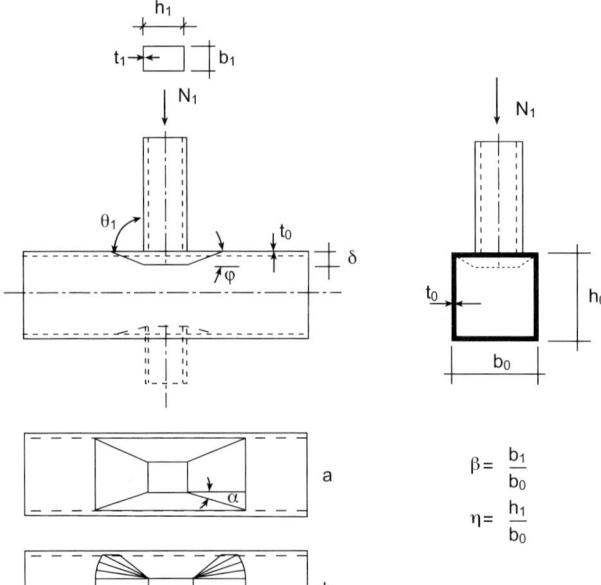

Fig. 6-114 Yield line model for a RHS joint of T type

6.6 Lattice girders (trusses)

Eq. (6-57) describes a function of the angles between the yield lines. The minimum for the load N_1 can be achieved by differentiation:

$$N_1 = \frac{f_{y0} \cdot t_0^2}{1-\beta} \left(\frac{2\eta}{\sin\theta_1} + 4\sqrt{1-\beta} \right) \frac{1}{\sin\theta_1} \quad (6\text{-}58)$$

- **Chord web bearing model for a RHS joint**

As Fig. 6-115 demonstrates, T, Y and X joints of RHS can fail by yielding or local buckling of the side walls of the chord.

For full width joint ($\beta = 1.0$), the joint strength under tensile load can be calculated using the following equation:

$$N_1 = 2 f_{y0} \cdot t_0 \left(\frac{h_1}{\sin\theta_1} + 5 t_0 \right) \cdot \frac{1}{\sin\theta_1} \quad (6\text{-}59)$$

In order to determine the strength of a joint loaded by compression, the yield strength f_{y0} of the chord in the Eq. (6-59) has to be substituted by the critical buckling stress f_K:

$$N_1 = 2 f_K \cdot t_0 \left(\frac{h_1}{\sin\theta_1} + 5 t_0 \right) \cdot \frac{1}{\sin\theta_1} \quad (6\text{-}60)$$

Lit. [228] states that the result for the joints with $h_0 < b_0$ using Eq. (6-60) is very conservative on the safe side, in case f_K is determined according to the European buckling curves (Fig. 5-4). The slenderness λ_K for this purpose is calculated by:

$$\lambda_K = 3.46 \left(\frac{h_0}{t_0} - 2 \right) \sqrt{\frac{1}{\sin\theta_i}}$$

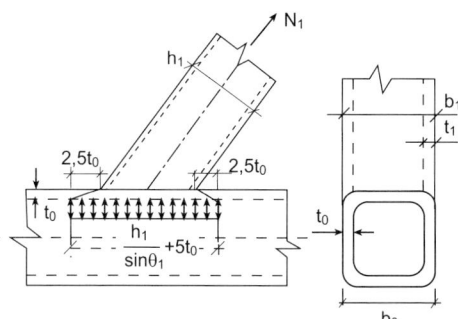

Fig. 6-115 Chord web bearing model for a RHS joint

- **Chord shear yield model for a RHS joint** (Fig. 6-116)

The joints with large β and/or small h_0/b_0 ratio can also fail by chord side wall shear. This failure depends on the strength of the chord cross section between the neighbouring bracing toes. The following equation can be used to calculate the maximum shear strength of a RHS joint:

$$\hat{N}_1 = \frac{f_{y0}}{\sqrt{3}} (2 h_0 \cdot t_0 + \alpha \cdot b_0 \cdot t_0) \cdot \frac{1}{\sin\theta_1} \quad (6\text{-}61)$$

The effect of the upper chord flange dependent on g/t_0 ratio is represented by α. The effective shear are A_v of the chord flange is equal to $\alpha \cdot b_0 \cdot t_0$, where

$$\alpha = \sqrt{\frac{1}{1 + \frac{4 g^2}{3 t_0^2}}} \quad (6\text{-}62)$$

The axial joint resistance can be determined according to "Huber Hencky-von Mises" criterion:

$$\hat{N}_1 = (A_0 - A_v) f_{y0} + A_v \cdot f_{y0} \cdot \sqrt{1 - \left\{ \frac{V}{\frac{f_{y0}}{\sqrt{3}} \cdot A_v} \right\}^2} \quad (6\text{-}63)$$

Modes of failure of hollow section joints

Depending on the various combinations of the profile forms (see Table 6-6), the joint types (Fig. 6-81), the geometrical parameters (β, γ, τ, g, λ_{ov}, θ) and the loading conditions as well as the preload in the chord of a hollow section joint, different modes of failure take place (see Fig. 6-117).

Besides, the combinations of the above mentioned failure modes can also lead to failure of hollow section joints in many cases. For example, the failure modes c and f_2 are illustrated as the failure based on the effective width of the bracing and treated accordingly. In both cases, the ultimate resistance is determined basing on the effective cross section of the critical bracing member. Another example is given by the meeting of the failure modes a and b for $\beta = 0.6 - 0.8$ with a thinwalled bra-

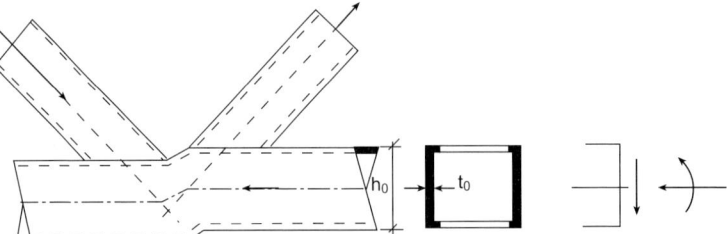

Fig. 6-116 Chord shear yield model

cing. The failure mode c can also occur together with a, in case the bracing is relatively thinwalled. Further it is to be noted that the failure mode f_2 for joints with overlap takes place most frequently.

In order to define the ultimate load capacity of hollow section joints with CHS or RHS chords, where load vs. deformation or moment vs. rotation diagrams do not show a pronounced peak load, a failure criterion based on the deformation limits has been established and is often used in various research works. As has been already described, the behaviour of hollow section joints is mainly dependent on the local plastification of the chord, if the connection is designed in a manner that the failure of the bracing does not occur before the connection failure. Therefore, the deformation limit criterion has been based on the local deformation of the chord face at the intersection between the bracing and the chord. The design of a joint should ensure that the conditions regarding ultimate resistance, deformation capacity and serviceability are satisfied. [267] proposes the deformation limit of 3% b_0 (or d_0) based on the local deformation of the chord or column face, which corresponds to the ultimate resistance of a joint. If this ultimate deformation limit is not used, an additional serviceability deformation limit criterion has to be adopted for the design. For axially loaded hollow section joints, an indentation of 1% of the chord width b_0 (diameter d_0) at the chord face is generally used as the serviceability deformation limit as given by IIW [268]. This is consistent with the deformations, which have been found for hollow section joints by tests or FE analysis.

Further, the load in the thickness direction in a thickwalled hollow section can lead to "lamellar tearing". This subject has already been discussed in Section 3.1.

Determination of the design resistance (strength) of hollow section joints

The design of hollow section joints is carried out semi-empirically by means of a combination of theoretical approaches (analytical models) and practical test results. This means that the influencing parameters are to be determined applying simplified theoretical models. The final design formulae are established by a statistical evaluation of the test results through the modification of the theoretical results. It must be attended to that various influencing parameters – all of them as far as possible – are taken into consideration. The characteristic resistances N_k are derived from the design resistances N_{Rd} considering all variables e.g. the scatter of the test results regarding joint strengths as well as different mechanical properties, dimensions and manufacturing tolerances. The characteristic resistance is described by the following correlation (see Fig. 6-118):

$$N_k = N_{um} - K \cdot s \qquad (6\text{-}64)$$

where N_{um} is the arithmetical average value of the test results and s is the standard deviation. According to [143], K is taken to be equal to 1.64 (95% fractile).

The characteristic resistance is to be divided by the partial safety factor γ_M in order to compare with the applied load (characteristic load

6.6 Lattice girders (trusses)

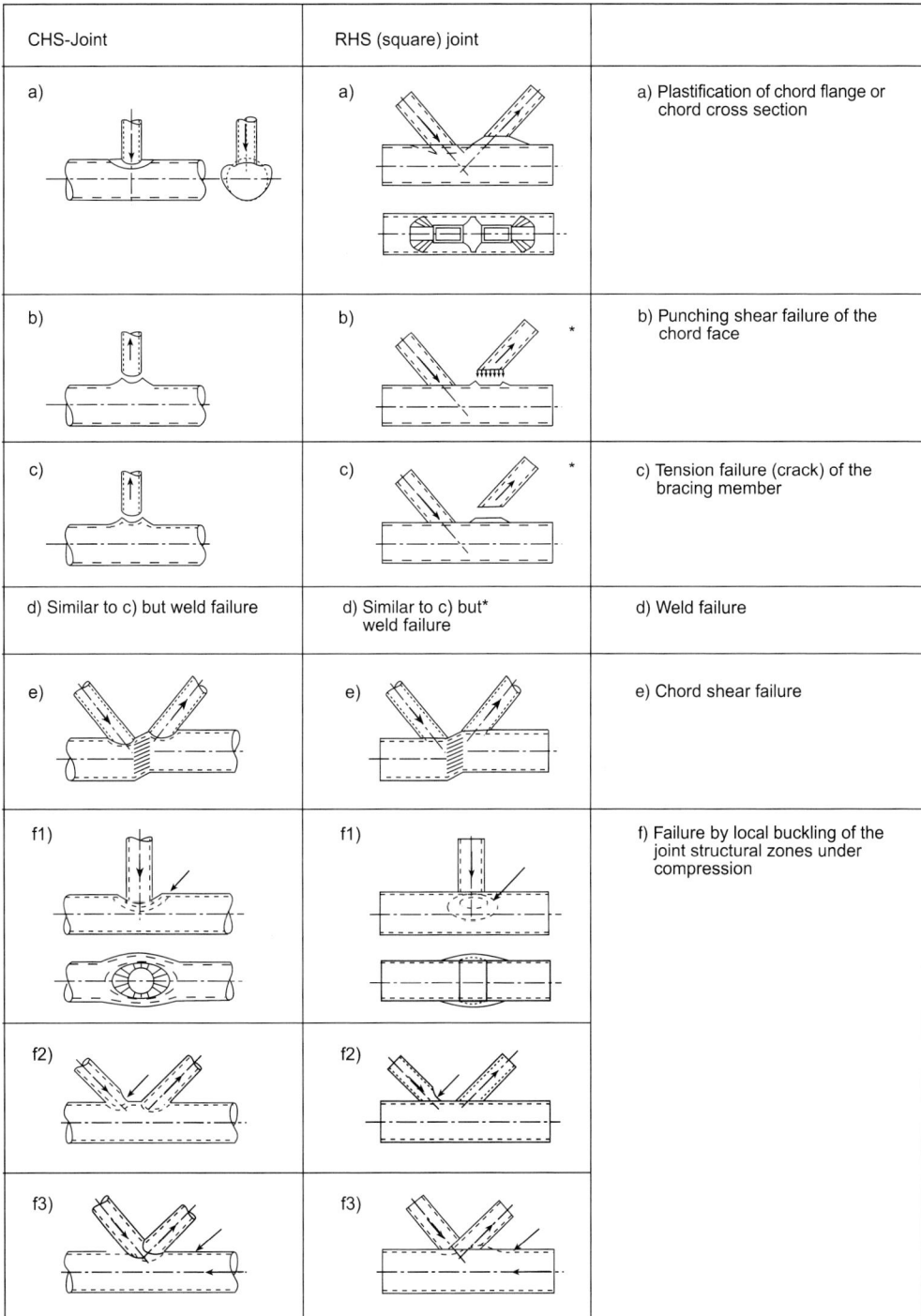

Fig. 6-117 Failure modes for hollow section joints

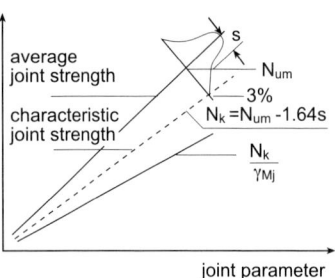

Fig. 6-118 Principles to determine the joint resistance (strength) by tests [143]

(action) Q_k multiplied by appropriate partial safety factor for the load γ_F, $Q_k \cdot \gamma_F$):

$$N_{Rd} = \frac{N_k}{\gamma_{Mj}} \qquad (6\text{-}65)$$

$$Q_k \leq \frac{N_{Rd}}{\gamma_F} = \frac{N_k}{\gamma_{Mj} \cdot \gamma_F} \qquad (6\text{-}66)$$

Following Lit. [3], the partial safety factor for the hollow section joint resistance γ_{Mj} is set to be 1.1. In this case, it is recommended to take $\gamma_F = 1.5$.

When the number of the available test results is insufficient, so that a satisfactory statistical analysis is not possible, the characteristic resistance is then estimated by the lowest bound line describing the test results.

While designing with respect to the crack initiation in a joint (see Fig. 6-106, point 4), the involved parameter ranges have to be restricted in a way, so that the crack initiation does not represent a design criterion.

Design resistances of the hollow section joints in plane trusses

Most tests worldwide have been performed on isolated uni-planar joints (see Literature to Chapter 6). Extensive evaluations by IIW (International Institute of Welding) [49] and CIDECT (Comité International pour le Développement et l'Etude de la Construction Tubulaire) [6, 150, 226] led to the latest calculation rules as well as the formulae for the design resistances for the plane T, Y, X, K and N joints, which have gained acceptance through out the world by now. These formulae have already been included in the structural standards of the European Union, Canada and Japan and partly by the U.S.A.

The design resistances of the plane truss joints in CHS and RHS can be calculated using the formulae in the following tables:

- Table 6-9 Design resistances of uni-planar welded joints between circular chord and bracing members.
- Table 6-10 Validity ranges for the application of Table 6-9.
- Table 6-11 Design resistances of uni-planar welded joints between square or circular bracings and a square hollow section as chord.
- Table 6-12 Validity ranges for the application of Table 6-11.
- Table 6-13 Design resistances of uni-planar welded joints between square or rectangular or circular bracings and a rectangular hollow section as chord.
- Table 6-14 Validity ranges for the application of Table 6-13.

The validity ranges according to the Tables 6-10, 6-12 and 6-14 are set up in a way, so that a deformation limit is not exceeded when serviceability is concerned.

T, Y, X, N and K joints between circular chord and bracing members (Table 6-9)

T, Y and X joints with axially loaded members bear about 1.5 times load under tension compared to compression. The tensile load is however associated with large local deformations. Enlarged tensile resistance is therefore not taken into consideration, which means that the design resistances under tension and compression for the given equations are taken to be identical. A verification of deformation need not be made.

Fundamentally, the design resistances of circular hollow section joints are derived from the ring model. In order to take the effect of the membrane actions (γ), gap ($g' = g/t_0$) and preload in the chord (n') into account, modifications are also made by means of the function of β and other experimental functions. A reduc-

6.6 Lattice girders (trusses)

Table 6-9 Design resistances of uni-planar welded joints between circular chord and bracing members

Type of joint	Design resistance (i = 1, 2, 3, j = overlapped bracing)
T and Y joint	Chord plastification
	$N_{1,Rd} = \dfrac{f_{y0} \cdot t_0^2}{\sin \theta_1} \cdot (2.8 + 14.2 \beta^2) \cdot \gamma^{0.2} \cdot f(n') \left(\dfrac{1.1}{\gamma_{Mj}} \right)$ (6-67)
X joint	Chord plastification
	$N_{1,Rd} = \dfrac{f_{y0} \cdot t_0^2}{\sin \theta_1} \cdot \left[\dfrac{5.2}{1 - 0.81 \beta} \right] \cdot f(n') \cdot \left(\dfrac{1.1}{\gamma_{Mj}} \right)$ (6-68)
K and N joint with gap or overlap	Chord plastification
	$N_{1,Rd} = \dfrac{f_{y0} \cdot t_0^2}{\sin \theta_1} \cdot \left(1.8 + 10.2 \dfrac{d_1}{d_0} \right) \cdot f(\gamma, g') \cdot f(n') \cdot \left(\dfrac{1.1}{\gamma_{Mj}} \right)$ (6-69)
	$N_{2,Rd} = N_{1,Rd} \dfrac{\sin \theta_1}{\sin \theta_2}$ (6-70)
General	Punching shear
Punching shear check for T, Y and X joints as well as K, N and KT joints with gap, if $d_i \leq d_0 - 2 t_0$	$N_{i,Rd} = \dfrac{f_{y0}}{\sqrt{3}} \cdot t_0 \pi d_i \cdot \dfrac{1 + \sin \theta_i}{2 \sin^2 \theta_i} \cdot \left(\dfrac{1.1}{\gamma_{Mj}} \right)$ (6-71)
Functions	
$f(n') = 1.0$ für $n' \leq 0$ (tension) $n' = \dfrac{f_{op}}{f_{y0}}$ $f(n') = 1 - 0.3 n' - 0.3 n'^2$ for $n' > 0$ (compression) however ≤ 1.0	$f(\gamma, g') = \gamma^{0.2} \cdot \left[1 + \dfrac{0.024 \gamma^{1.2}}{\exp(0.5 g' - 1.33) + 1} \right]$ (6-72) (see Fig. 6-119)

tion of the design resistance of the joint (factor $f(n')$) takes place due to an additional axial compressive load N_{op} (see Fig. 6-82). Only the prestress in the chord need to be considered here. Horizontal components of the forces transmitted by the bracings are ignored.

Regarding the effect of the prestress in the chord on the design resistance of a joint in a lattice girder, a number of general comments will be made here. In a lattice girder on two supports, the prestresses at the girder ends are small. The joints at these locations bear largest loads in the bracings and have therefore to be particularly checked during the design. Only small bracing loads and large chord preloads occur in the middle of the girder. Often the verification of a joint at this location is not required.

Large bracing loads as well as high preloads in the chords occur at the inner supports of

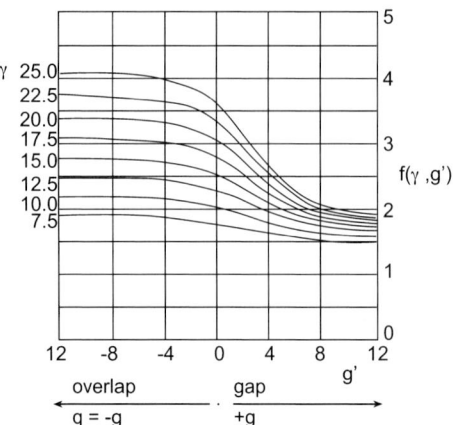

Fig. 6-119 $f(\gamma, g')$ depending on γ and $(g' = g/t_0)$ (see Eq. (6-72) and Table 6-9)

Table 6-10 Validity ranges for the application of Table 6-9

$0.2 \leq \dfrac{d_i}{d_0} \leq 1.0$	
$5 \leq \dfrac{d_i}{2\,t_i} \leq 25$	
$5 \leq \dfrac{d_0}{2\,t_0} \leq 25$	
$5 \leq \dfrac{d_0}{2\,t_0} \leq 20$ for X joints	
$-0.55 \leq \dfrac{e}{d_0} \leq 0.25$	
$\dfrac{t_i}{t_j} \leq 1.0$	
$\lambda_{ov} \geq 25\%$	
$g \geq t_1 + t_2$	
$f_y \leq 355\ N/mm^2\ (S\,355)$	

Upper limit of the ratio $\dfrac{N_{1,Rd}}{A_1 \cdot f_{y1}}$ for compression bracings

For f_{y1} up to	For $\dfrac{d_1}{t_1}$				
	30	35	40	45	50
235 N/mm²	1.00	1.00	1.00	0.98	0.93
275 N/mm²	1.00	1.00	0.96	0.88	0.86
355 N/mm²	0.98	0.88	0.85	0.78	0.76

Eq. (6-69) describes the effect of gap or overlap in the K and N joints of CHS, so that a continuous function $f(\gamma, g')$ covers the whole range of gap and overlap.

Eq. (6-71) results in the punching shear for T, Y, X, K and N joints with gap (Fig. 6-117b). For usual ratios, the punching shear is not decisive for CHS chords.

T, Y and X joints with chord and bracing members of rectangular or square hollow sections (Tables 6-11 and 6-13)

The load bearing behaviour has been investigated regarding the width ratio $\beta = b_i/b_0$ separately in three zones:

$\beta \leq 0.85$
$0.85 < \beta < 1.0$
$\beta = 1.0$

The difference between the design resistances in Table 6-11 (square hollow section as chord) and those in Table 6-13 (rectangular hollow section as chord) is based mainly on the fact that the formulae given in Table 6-11 have been simplified by restricting the validity ranges (Table 6-12), which however cover the zone for practical application. Simplifying by assuming $h_i = b_i$ and restricting the range by $\beta \leq 0.85$, it is achieved that various failure modes do not become critical at all. The failure is restricted to chord flange plastification for T, Y and X joints. This limitation is based on the yield line model with the deformation limit for the serviceability.

Table 6-13 contains the joint design resistances taking further failure modes into consideration e.g. chord web failure, punching

continuous lattice girders and the joints must be made strong accordingly.

The following table gives a further guideline, which can serve as a design aid. This shows the upper limits of the ratio $\dfrac{N_{1,Rd}}{A_1 \cdot f_{y1}}$ for the compression bracings in relation with $\dfrac{d_1}{t_1}$ ratio, so that they do not undergo premature local failure.

Table 6-11 Design resistances of uni-planar welded joints between square or circular bracings and a square hollow section as chord

Type of joint	Design resistance ($i = 1$ or 2, j = overlapped bracing)	
T, Y and X joint	Chord plastification	$\beta \leq 0.85$
	$N_{1,Rd} = \dfrac{f_{y0}\, t_0^2}{(1-\beta)\sin\theta_1} \cdot \left[\dfrac{2\beta}{\sin\theta_1} + 4(1-\beta)^{0.5}\right] \cdot f(n) \cdot \left(\dfrac{1.1}{\gamma_{Mj}}\right)$	(6-73)
K and N joint with gap	Chord plastification	$\beta \leq 1.0$
	$N_{i,Rd} = \dfrac{8.9\, f_{y0}\, t_0^2}{\sin\theta_i} \cdot \left[\dfrac{b_1 + b_2}{2\, b_0}\right]\left[\dfrac{b_0}{2\, t_0}\right]^{0.5} f(n) \cdot \left(\dfrac{1.1}{\gamma_{Mj}}\right)$	(6-74)
K and N joint with overlap [a]	Effective width	$25\% \leq \lambda_{ov} < 50\%$
	$N_{i,Rd} = f_{yi}\, t_i \left[\dfrac{\lambda_{ov}}{50}(2 h_i - 4 t_i) + b_e + b_{e,ov}\right] \cdot \left(\dfrac{1.1}{\gamma_{Mj}}\right)$	(6-75)
	Effective width	$50\% \leq \lambda_{ov} < 80\%$
	$N_{i,Rd} = f_{yi}\, t_i \left[2 h_i - 4 t_i + b_e + b_{e,ov}\right] \cdot \left(\dfrac{1.1}{\gamma_{Mj}}\right)$	(6-76)
	Effective width	$\lambda_{ov} \geq 80\%$
	$N_{i,Rd} = f_{yi}\, t_i \left[2 h_i - 4 t_i + b_i + b_{e,ov}\right] \cdot \left(\dfrac{1.1}{\gamma_{Mj}}\right)$	(6-77)
Circular bracings	Multiply the formulae by $\pi/4$ and replace b_1 and h_1 by d_1 and b_2 and h_2 by d_2.	
	Functions	

$n = f_0 / f_{y0}$

$f(n) = 1.3 - \dfrac{0.4}{\beta} \cdot n \quad \text{für } n > 0 \text{ (compression)}$

$f(n) = 1.0 \quad \text{für } n \leq 0 \text{ (tension)}$

however $f(n) \leq 1.0$

$b_e = \dfrac{10}{b_0/t_0} \cdot \dfrac{f_{y0} \cdot t_0}{f_{yi} \cdot t_i} \cdot b_i$

however $b_e \leq b_i$

$b_{e,ov} = \dfrac{10}{b_j/t_j} \cdot \dfrac{f_{yj} \cdot t_j}{f_{yi} \cdot t_i} \cdot b_i$

however $b_{e,ov} \leq b_i$

[a] Only overlapping bracing member has to be checked. However the efficiency (the joint design resistance divided by the full yield resistance of the bracing member) of the overlapped bracing member is not to be taken higher than that of the overlapping bracing member.

Table 6-12 Validity ranges for the application of Table 6-11

Type of joint	Joint parameters ($i = 1$ or 2, j = overlapped bracing)						
	b_i/b_0 d_i/b_0	b_i/t_i d_i/t_i		b_0/t_0	$(b_1 + b_2)/2\,b_i$ b_i/b_j t_i/t_j	Gap/overlap g/q	Material yield strength f_y	Joint eccentricity e
		Compression	Tension					
T, Y and X joint	$b_i/b_0 \geq 0.25$ [a] however ≤ 0.85	$b_i/t_i \leq 1.25\sqrt{\dfrac{E}{f_{yi}}}$	$b_i/t_i \leq 35$	$b_0/t_0 \geq 10$ [a] however ≤ 35			$\leq 355\ \text{N/mm}^2$ [b]	$-0.55 \leq \dfrac{e}{h_0} \leq 0.25$
K and N joint with gap	$b_i/b_0 \geq 0.35$ and $\geq 0.1 + 0.01\,b_0/t_0$	$b_i/t_i \leq 35$		$b_0/t_0 \geq 15$ [a] however ≤ 35	$(b_1 + b_2)/2\,b_i \geq 0.6$ but ≤ 1.3	$g/b_0 \geq 0.5(1-\beta)$ [c] however $\leq 1.5(1-\beta)$ $g \geq t_1 + t_2$		
K and N joint with overlap	$b_i/b_0 \geq 0.25$	$b_i/t_i \leq 1.1\sqrt{\dfrac{E}{f_{yi}}}$		$b_0/t_0 \leq 40$	$t_i/t_j \leq 1.0$ $b_i/b_j \geq 0.75$	$\lambda_{ov} \geq 25\%$ however $\leq 100\%$		
Circular bracings	$d_i/b_0 \geq 0.40$ however ≤ 0.8	$d_i/t_i \leq 1.5\sqrt{\dfrac{E}{f_{yi}}}$	$d_i/t_i \leq 50$		Limitations as above for $d_i = b_i$			

[a] Outside the range of validity, further failure criteria can be governing, e. g. punching shear, effective width, chord shear or local buckling. If these particular limits of validity are violated, the joints have still to be checked as if they have rectangular chords using Table 6-13, provided the limits of validity given by Table 6-14 are still met.
[b] f_{yi}, $f_{yj} \leq 355\ \text{N/mm}^2$, f_{yi} (or f_{yj})/f_{ui} (or f_{uj}) ≤ 0.8.
[c] If $g/b_0 > 1.5(1-\beta)$, check as for T and Y joints.

Table 6-13 Design resistances of uni-planar welded joints between square or rectangular or circular bracings and a rectangular hollow section as chord

Type of joint	Design resistance ($i = 1, 2$)		
T, Y and X joint	Chord flange plastification		$\beta \leq 0.85$
	$N_{1,Rd} = \dfrac{f_{y0} t_0^2}{(1-\beta) \sin \theta_1} \left[\dfrac{2\eta}{\sin \theta_1} + 4\sqrt{(1-\beta)} \right] \cdot f(n) \left(\dfrac{1.1}{\gamma_{Mj}} \right)$		(6-78)
	Chord web failure [a]	$\beta = 1.0$	$0.85 \leq \beta \leq 1.0$
	$N_{1,Rd} = \dfrac{f_k t_0}{\sin \theta_1} \left[\dfrac{2 h_1}{\sin \theta_1} + 10 t_0 \right] \dfrac{1.1}{\gamma_{Mj}}$ (6-79)		use linear interpolation of chord flange plastification and chord web failure criteria
	Effective width		$\beta > 0.85$
	$N_{1,Rd} = f_{y1} t_1 [2 h_1 - 4 t_1 + 2 b_e] \left(\dfrac{1.1}{\gamma_{Mj}} \right)$		(6-80)
	Punching shear		$0.85 \leq \beta \leq 1 - 1/\gamma$
	$N_{1,Rd} = \dfrac{f_{y0} t_0}{\sqrt{3} \sin \theta_1} \left[2 \dfrac{h_1}{\sin \theta_1} + 2 b_{e,p} \right] \cdot \left(\dfrac{1.1}{\gamma_{Mj}} \right)$		(6-81)
K and N joint with gap	Chord flange plastification		
	$N_{i,Rd} = 8.9 \dfrac{f_{y0} t_0^2}{\sin \theta_i} \left[\dfrac{b_1 + b_2 + h_1 + h_2}{4 b_0} \right] \left[\dfrac{b_0}{2 t_0} \right]^{0.5} f(n) \cdot \left(\dfrac{1.1}{\gamma_{Mj}} \right)$		(6-82)
	Chord shear		
	$N_{i,Rd} = \dfrac{f_{y0} A_v}{\sqrt{3} \sin \theta_i} \left(\dfrac{1.1}{\gamma_{Mj}} \right)$		
	also $N_{0,Rd \text{ (in gap)}} \leq (A_0 - A_v) f_{y0} + A_v \cdot f_{y0} \left[1 - \left(\dfrac{V}{V_p} \right)^2 \right]^{0.5} \cdot \left(\dfrac{1.1}{\gamma_{Mj}} \right)$		(6-83)
	Effective width		
	$N_{i,Rd} = f_{yi} t_i [2 h_i - 4 t_i + b_i + b_e] \cdot \left(\dfrac{1.1}{\gamma_{Mj}} \right)$		(6-84)
	Punching shear		$\beta \leq 1 - 1/\gamma$
	$N_{i,Rd} = \dfrac{f_{y0} t_0}{\sqrt{3} \sin \theta_i} \left[\dfrac{2 h_i}{\sin \theta_i} + b_i + b_{e,p} \right] \cdot \left(\dfrac{1.1}{\gamma_{Mj}} \right)$		(6-85)
K and N joint with overlap	Similar to joints of square hollow sections (Table 6-11). Additional check for chord shear failure, when $h_0/b_0 < 1.0$		
Circular bracings	Multiply the formulae by $\pi/4$ and replace b_i and h_i by d_i.		

Functions

Tension: $f_k = f_{y0}$
Compression: $f_k = f_K$
(T and Y joint)
$f_k = 0.8 \cdot f_K \sin \theta_i$
(X joint)
f_K buckling stress according to the structural standard, use of slenderness:

$\lambda_K = 3.46 \left(\dfrac{h_0}{t_0} - 2 \right) \sqrt{\dfrac{1}{\sin \theta_i}}$ (6-86)

$f(n) = 1.0$ for $n \leq 0$ (tension) $n = f_0/f_{y0}$

$f(n) = 1.3 - \dfrac{0.4 n}{\beta}$ for $n > 0$ (compression)

however $f(n) \leq 1.0$

$V_p = \dfrac{f_{y0} \cdot A_v}{\sqrt{3}}$ $\alpha = \sqrt{\dfrac{1}{1 + \dfrac{4 g^2}{3 t_0^2}}}$

For square and rectangular bracings: $A_v = (2 h_0 + \alpha b_0) t_0$
For circular bracing: $\alpha = 0$

$b_e = \dfrac{10}{b_0/t_0} \cdot \dfrac{f_{y0} t_0}{f_{yi} t_i} \cdot b_i$
however $b_e \leq b_i$

$b_{e,p} = \dfrac{10}{b_0/t_0} b_i$
however $b_{e,p} \leq b_i$

[a] For X joints with inclination angle $\theta < 90°$, the chord web must be additionally checked for shear.

shear and failure involving the effective width of bracing. Table 6-14 lists further validity ranges.

In the range $\beta \leq 0.85$, the joint fails in general due to the yielding of chord flange. Eq. (6-78) has been established based on material yielding shown in Fig. 6-114.

The calculated values using Eq. (6-78) are at the lower limit in comparison with the test results. In this case, the safety interval increases with decreasing $\beta = b_i/b_0$. As however the local deformation of a joint enlarges with the reduction of β, this additional safety can be considered as desirable.

An additional compression N_{op} in the chord diminishes the joint design resistance. This effect rises with the reduction of β. As the test results for the joint resistance with additional N_{op} are higher than the calculated values by the Eq. (6-78), it is not required to consider N_{op} while applying this equation. Only the compression N_0 in the chord member is taken into account.

Fundamentally, this can be made sure that the deformation at the serviceability limit remains acceptable, when the joint design resistance $N_{1,Rd}$ is restricted to the joint resistance at the yield limit.

In the zone $\beta = 1.0$, the following failure criteria are decisive:

- Due to the yielding of the chord webs under tensile load, the yield strength of the chord is used to determine the joint design strength.
- The chord webs buckle under compression. The joint design strength is determined using Eq. (6-86) by means of the standardized buckling curve (European buckling curves, see Fig. 5-4) for the slenderness λ_K.

The Eq. (6-79) is in principle valid for the joint design resistance under tension. Substituting however the yield strength of the chord material by the critical buckling stress f_K, this equation can also be used for compression in the bracing. For X joints with $\beta = 1.0$, the deformations of the chord webs are larger than those for T and Y joints. The characteristic value for X joints is reduced accordingly:

$$f_k = f_K \cdot 0.8 \cdot \sin\theta_1$$

A comparison of the test results with the calculated joint design strengths according to Eq. (6-79) has demonstrated a satisfactory agreement.

Additional chord compression exercises hardly any effect for $\beta = 1.0$. No reduction of the joint resistance could be found for additional tension in chord.

The test results to determine the joint design resistance show large scatters [158]. The insecurity due to scatter and the possibility of an abrupt failure lead to the recommendation of a safety factor $\gamma_{Mj} = 1.25$ for the joints under compression [143]. However Eurocode 3 [3] prescribes $\gamma_{Mj} = 1.1$ for the sake of harmonized simplification.

In the range $0.85 < \beta \leq 1.0$, a gradual transition from the chord flange plastification ($\beta = 0.85$) to the chord web buckling or yielding ($\beta = 1.0$) takes place. It is recommended to make a linear interpolation between the joint design resistances for $\beta = 0.85$ and $\beta = 1.0$ in this range.

Besides, strength verifications have to be carried out using Eqs. (6-80) and (6-81).

Eq. (6-80) is applied to check the bracing wall thickness for $\beta > 0.85$, taking the effective length of the intersection of the bracing member with the chord flange into consideration. This intersection length is made up with the both side lengths $2(h_i - 2 t_i)$ and the effective widths $2 b_e$.

For $0.85 \leq \beta \leq \dfrac{b_0 - 2t_0}{b_0}\left(=1 - \dfrac{1}{\gamma}\right)$, Eq. (6-81) takes account of the possible punching shear in the chord flange (see Fig. 6-117b), where $\beta = \left(1 - \dfrac{1}{\gamma}\right)$ represents the upper limit of the shear area.

T, Y and X joints between circular bracing and square or rectangular chord

The local stress concentration for this profile combination and for $\beta = 1.0$ is very high due to the small length of the CHS periphery on

6.6 Lattice girders (trusses)

Table 6-14 Validity ranges for the application of Table 6-13

Type of joint	Joint parameters ($i = 1$ or 2, j = overlapped bracing)							
	b_i/b_0, h_i/b_0, d_i/b_0	b_i/t_i, h_i/t_i, d_i/t_i		h_0/b_0, h_i/b_i	b_0/t_0, h_0/t_0	Gap/overlap b_i/b_j, t_i/t_j	Joint eccentricity	Material yield strength f_y
		Compression	Tension					
T, Y, X	≥ 0.25	$\leq 1.25\sqrt{\dfrac{E}{f_{yi}}}$			≤ 35		$-0.55 \leq \dfrac{e}{h_0} \leq 0.25$	≤ 355 N/mm² [a]
K and N joint with gap	$\beta \geq 0.35$ and $\geq 0.1 + 0.01\, b_0/t_0$	≤ 35	≤ 35	$h_i/b_i \geq 0.5$ however ≤ 2.0	≤ 35	$g/b_0 \geq 0.5(1-\beta)$ however $\leq 1.5(1-\beta)$ [b] $g \geq t_1 + t_2$		
K and N joint with overlap	$b_i/b_0 \geq 0.25$	$\leq 1.1\sqrt{\dfrac{E}{f_{yi}}}$			≤ 40	$\lambda_{ov} \geq 25\%$ however $\leq 100\%$ $t_i/t_j \leq 1.0$ $b_i/b_j \geq 0.75$		
Circular bracing	$d_i/b_0 \geq 0.40$ however ≤ 0.8	$\leq 1.5\sqrt{\dfrac{E}{f_{yi}}}$ ≤ 50				limitations as above for $d_i = b_i$		

[a] f_{yi}, $f_{yj} \leq 355$ N/mm², f_{yi} (or f_{yj})/f_{ui} (or f_{uj}) ≤ 0.8.
[b] If $g/b_0 > 1.5(1-\beta)$, check as for T and Y joints.

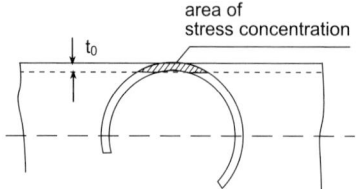

Fig. 6-120 Stress concentration on the chord side wall in a joint with CHS bracing for $\beta = d_i/b_0 = 1{,}0$

the chord side wall (see Fig. 6-120). Such a design detail has therefore to be avoided. It is recommended to select $\beta = d_i/b_0 \leq 0.8$ while designing the joint (see Tables 6-12 and 6-14).

Below this range, a modification of the Eq. (6-73) or (6-78) (multiplication with $\pi/4 \triangleq$ periphery CHS/perimeter RHS and substitution of b_i and h_i by d_i) has to be carried out.

Due to the limitation $\beta \leq 0.8$, the punching shear and the effective bracing periphery cannot be critical.

K and N joints with gap between circular or square or rectangular bracings and square or rectangular chord (Tables 6-11 and 6-13)

In general, the bracing width is smaller than the width of chord ($\beta < 1.0$) in K and N joints. Due to this, the failure takes place normally with large deformation and yielding of the upper chord face according to Eq. (6-74) or (6-82) (see Fig. 6-117, failure mode a). After the deformation, a crack appears mostly at the foot of the tensile bracing on the connecting face of the chord. The quality of the weld connecting the bracing with the chord plays a significant role, as a wrong execution of the weld can lead to premature weld rupture, before the local peakstresses are reduced by plastic stress redistribution. Section 6.6.3 deals with the execution of welds in truss joints.

The given ranges for the gap lengths in Table 6-12 and 6-14 ensure a relatively uniform stress distribution, which also takes acceptable manufacturing tolerances into consideration.

Fig. 6-121 shows the qualitative distribution of the forces at the transition from the bracing to the chord flange depending on the gap length. From Fig. 6-121, this can be clearly identified that a uniform stress distribution on all sides can be attained by means of a certain gap length. This gap corresponds to about $g = b_0 - b_i$.

For $\beta = 1.0$, a uniform stress distribution in a joint can be achieved with $g = 0$. However, a minimum gap length is required in order to execute a correct welding between tension and compression bracings and chord flange. The influence of the possible defects in the closely neighbouring connecting welds of the tension and the compression bracings can be reduced, when the toes of the welds lying opposite to each other is bridged by a further weld seam.

It can be derived from Eqs. (6-74) and (6-82) to (6-85) that the design resistances of K or N joints of square and rectangular hollow sections are practically independent of the gap g.

As for T, Y and X joints, the test results show that the additional compression N_{op} in the chord need not be accounted also for K and N joints. Only N_0 is to be taken into consideration ($f(n)$), which can be determined with the force components in the bracings $\Sigma N_{i,Rd} \cdot \cos \theta_i$ (see also Section 6.6.4.1).

Eq. (6-74) is only valid for K and N joints with gap consisting of square hollow sections, where the functions of β and γ are determined experimentally. As has been already mentioned, Eq. (6-74) is mainly based on the failure mode "chord flange plastification", which is reflected by the plastic moment in the chord flange per unit length ($f_{y0} \cdot t_0^2/4$). In this case, it is not necessary to check further failure modes.

The gap K and N joints with RHS as chord has to be additionally examined for the failure modes b, c, e and f_2 (see Fig. 6-117). Eq. (6-84) has been derived based on the relationship of the effective width of bracing intersection to the chord, while Eq. (6-85) represents the punching shear.

For the failure mode e "chord shear" in the gap (Eq. 6-83), the chord cross section is di-

6.6 Lattice girders (trusses)

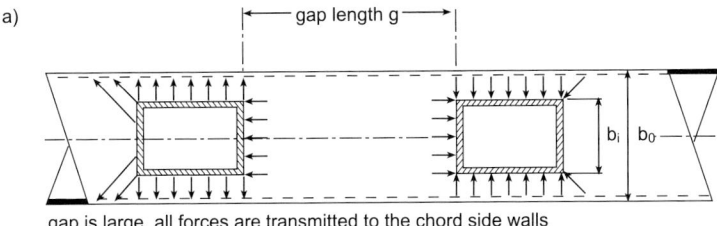

a) gap is large, all forces are transmitted to the chord side walls

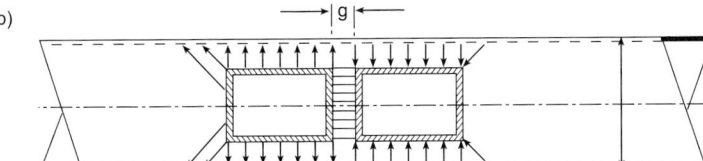

b) gap is small, the major part of the forces is transmitted through the zone between the diagonals

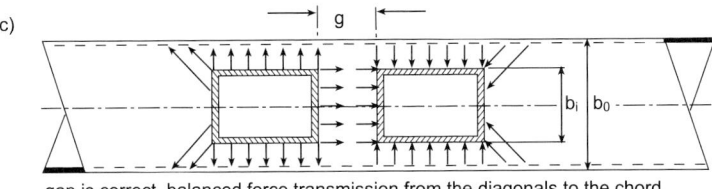

c) gap is correct, balanced force transmission from the diagonals to the chord

Fig. 6-121 Effect of gap length g on the distribution of the forces at the transition from bracing to chord

vided in the shear area A_v (comprising the area of the chord side walls plus a part of the area of the top flange, see Fig. 6-122) and the remaining area $A_0 - A_v$. The first carries both shear and axial forces interactively, while the second is effective in carrying axial forces but not shear.

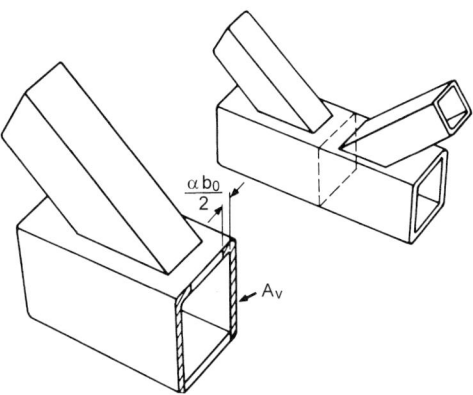

Fig. 6-122 Shear area (A_v) of the chord web in the gap region of a RHS K joint

In order to calculate the design resistances of CHS joints, Eq. (6-74) in Table 6-11 and Eqs. (6-82) to (6-85) in Table 6-13 are to be multiplied with $\pi/4$ and $b_{1,2}$ and $h_{1,2}$ are to be replaced by $d_{1,2}$. The application is restricted to $\beta \leq 0.8$.

K and N joints with overlap between circular or square or rectangular bracings and square or rectangular chord (Tables 6-11 and 6-13)

The overlap $\lambda_{ov}\%$ in K or N joint is defined by Fig. 6-82 b.

Often the design resistance of the joint with overlap used in practice is as high as that for the bracing itself. The required verification of this joint type is based on the effective width of the overlapped bracing. A minimum overlap of 25% is necessary, where the relative width of the bracings b_2/b_1 should be at least equal to 0.75. The effective widths b_e, $b_{e,p}$ and $b_{e,ov}$ in the Table 6-11 and 6-13 are defined in Fig. 6-123.

The design resistance increases linearly with overlap λ_{ov} from 25% to 50% (Eq. 6-75) and

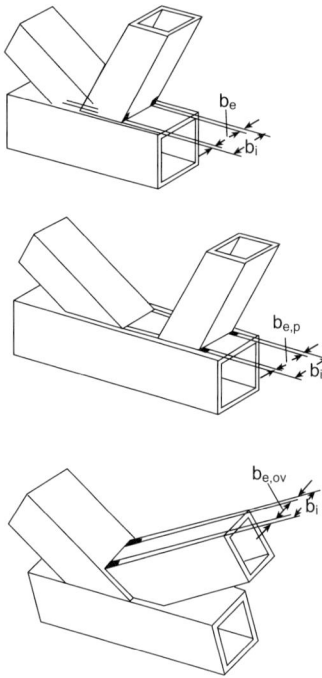

Fig. 6-123 Physical interpretation of the effective widths b_e, $b_{e,p}$ and $b_{e,ov}$ in Tables 6-11 and 6-13

is constant from 50% to <80% (Eq. 6-76). From 80% on, this remains constant at a higher level (Eq. 6-77).

When circular bracings are used, the procedure is to be applied, which has already been mentioned (Tables 6-11 and 6-13, K and N joint with gap or overlap between the bracings).

Design resistances of KT joints of hollow sections

Various applications of KT joints of CHS or RHS with gap or overlap are illustrated in Fig. 6-87.

The calculation of the design resistances of KT joints with gap is done based on the corresponding calculations for K and N joints according to Tables 6-9, 6-11 or 6-13, where the gap between the two bracings (compression and tension bracing) with highest loads is taken as the basis for the calculation. The vertical load components $N_{i,Sd} \cdot \sin \theta_i$ of the two bracing members acting in the same sense are

added together. The joint design component, normal to the chord $N_{i,Rd} \cdot \sin \theta_i$ of the remaining diagonal is required to exceed the added loads initiated by the two other bracings.

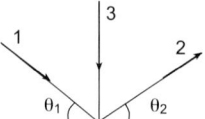

$$N_{1,Sd} \cdot \sin \theta_1 + N_{3,Sd} \cdot \sin \theta_3 \leq N_{2,Rd} \cdot \sin \theta_2$$

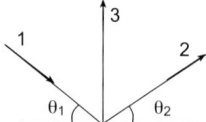

$$N_{2,Sd} \cdot \sin \theta_2 + N_{3,Sd} \cdot \sin \theta_3 \leq N_{1,Rd} \cdot \sin \theta_1$$

$N_{i,Rd}$ values are to be determined with the formulae for the design resistances in Tables 6-9, 6-11 or 6-13. The β values are calculated as follows:

- for Table 6-9: $\dfrac{d_1 + d_2 + d_3}{3 \, d_0}$

- for Table 6-11: $\dfrac{b_1 + b_2 + b_3}{3 \, b_0}$ or $\dfrac{d_1 + d_2 + d_3}{3 \, b_0}$

- for Table 6-13: $\dfrac{b_1 + b_2 + b_3 + h_1 + h_2 + h_3}{6 \, b_0}$ or $\dfrac{b_1 + b_2 + b_3}{3 \, b_0}$ or $\dfrac{d_1 + d_2 + d_3}{3 \, b_0}$

For KT joints with overlap, the design resistance $N_{i,Rd}$ of each overlapping bracing must be larger than or equal to the design load $N_{i,Sd}$ (see also the footnote for Table 6-11). While checking the bracing effective width, care should be taken to ensure that the member sequence of overlapping is properly accounted for.

Example: KT joint with gap on compression chord
Material: S 355 ($f_y = 355$ N/mm^2)

6.6 Lattice girders (trusses)

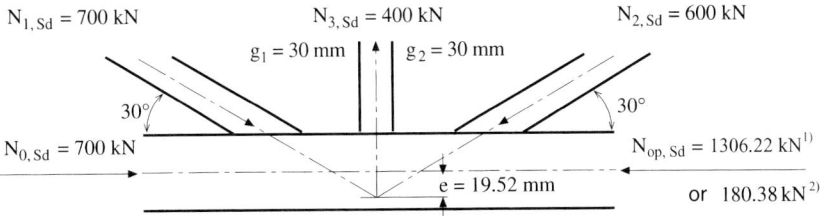

[1] For a joint consisting of the chord and the bracings 1 and 3.
[2] For a joint consisting of the chord and the bracings 2 and 3.

Compression chord:
CHS selected: 219.1 Ø × 10 mm
Cross sectional area $A_0 = 65.7$ cm²

Compression bracing (i = 1):
CHS selected: 139.7 Ø × 7.1 mm
Axial load in compression bracing (γ_F times load): $N_{1,Sd} = 700$ kN
Included angle between the compression bracing (i = 1) and the chord: $\theta_1 = 30°$

Compression bracing (i = 2):
CHS selected: 139.7 Ø × 6.3 mm
Axial load in compression bracing (γ_F times load): $N_{2,Sd} = 600$ kN
Included angle between the compression bracing (i = 2) and the chord: $\theta = 30°$

Tension vertical (i = 3):
CHS selected: 108 Ø × 6.3 mm
Axial load in tension vertical (γ_F times load): $N_{3,Sd} = 400$ kN
Included angle between the tension vertical (i = 3) and the chord: $\theta_3 = 90°$

Design resistance of N joint consisting of the chord and the bracings 1 and 3

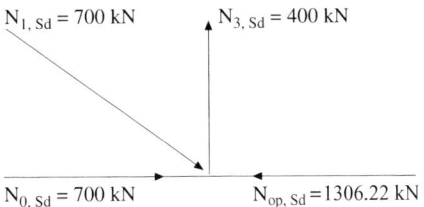

Verification of the application ranges (see Table 6-10)

Diameter ratio β (KT joint)

$$= \frac{d_1 + d_2 + d_3}{3\,d_0} = \frac{139.7 + 139.7 + 108.0}{3 \cdot 219.1}$$

$= 0.59 > 0.2$

$\dfrac{d_1}{2\,t_1} = \dfrac{139.7}{2 \cdot 7.1} = 9.84 \to 5 < 9.84 < 25$

$\dfrac{d_3}{2\,t_3} = \dfrac{108}{2 \cdot 6.3} = 8,57 \to 5 < 8.57 < 25$

$\dfrac{d_0}{2\,t_0} = \dfrac{219.1}{2 \cdot 10} = 11 \to 5 < 11 < 25$

$g \geq t_1 + t_3 \to 30 > (7.1 + 6.3) = 13.4$ mm

$\dfrac{g}{t_0} = \dfrac{30}{10} = 3$

$e = 19.52$ mm $\to \dfrac{e}{d_0} = \dfrac{19.52}{219.1}$

$= 0.089 < 0,25$

Eccentricity need not be taken into consideration.

$N_{op,Sd} = N_{0,Sd} + N_{1,Sd} \cdot \cos\theta_1$
$= 700 + 700 \cdot \cos 30° = 1306.22$ kN

$f_{op} = \dfrac{N_{op,Sd}}{A_0} = \dfrac{1306220}{6570} = 198.81$ N/mm²

$n' = \dfrac{f_{op}}{f_{y0}} = \dfrac{198.81}{355} = 0.56$

$f(n') = 1 - 0.3\,n'(1 + n') = 0.738$

$f(\gamma,g') = \gamma^{0.2} \cdot \left(1 + \dfrac{0.024 \cdot \gamma^{1.2}}{1 + e^{(0.5g/t_0 - 1.33)}}\right)$

$= 11^{0.2} \cdot \left(1 + \dfrac{0.024 \cdot 11^{1.2}}{1 + e^{(0.5 \cdot 30/10 - 1.33)}}\right) = 1,93$

Joint design resistance (see Table 6-9).

Failure criterion:

Chord flange plastification

$N_{1,Rd} = \dfrac{f_{y0} \cdot t_0^2}{\sin\theta_1} \cdot \left(1.8 + 10.2\,\dfrac{d_1}{d_0}\right)$

$f(\gamma,g') \cdot f(n') \cdot \dfrac{1.1}{\gamma_{Mj}}$

$$N_{1,Rd} = \frac{355 \cdot 10^2}{\sin 30°}$$

$$\cdot \left(1.8 + 10.2 \cdot \frac{139.7}{219.1}\right) \cdot 0.738 \cdot 1.93$$

$$\triangleq 839.72 \text{ kN} > 700 \text{ kN}$$

$$N_{3,Rd} = N_{1,Rd} \frac{\sin \theta_1}{\sin \theta_3} = 839.72 \cdot \frac{\sin 30°}{\sin 90°}$$

$$= 419.86 \text{ kN} > 400 \text{ kN}$$

Failure criterion:
Punching shear in chord

$$N_{1,Rd} = \frac{f_{y0}}{\sqrt{3}} \cdot t_0 \cdot \pi \cdot d_1 \cdot \frac{1 + \sin \theta_1}{2\sin^2 \theta_1} \cdot \frac{1.1}{\gamma_{Mj}}$$

$$= \frac{355}{\sqrt{3}} \cdot 10\pi \cdot 139.7 \cdot \frac{1 + \sin 30°}{2\sin^2 30°} \cdot \frac{1.1}{1.1}$$

$$= 2698.57 \text{ kN} > 700 \text{ kN}$$

$$N_{3,Rd} = \frac{355}{\sqrt{3}} \cdot 10\pi \cdot 108 \cdot \frac{1 + \sin 90°}{2\sin^2 90°} \cdot \frac{1.1}{1.1}$$

$$\triangleq 695.41 \text{ kN} > 400 \text{ kN}$$

The result of the joint verification (bracings 1 and 3) is positive.

Joint design resistance of N joints consisting of the chord and the bracings 2 and 3:

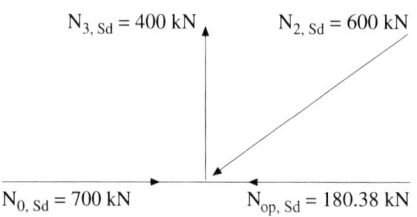

$N_{3, Sd} = 400$ kN, $N_{2, Sd} = 600$ kN
$N_{0, Sd} = 700$ kN, $N_{op, Sd} = 180.38$ kN

Verification of the application ranges (see Table 6-10)
Diameter ratio β (KT joint)
= 0.52 > 0.2

$$\frac{d_2}{2 t_2} = \frac{139.7}{2 \cdot 6.3} = 11.1 \to 5 < 11.1 < 25$$

$$\frac{d_3}{2 t_3} = \frac{108}{2 \cdot 6.3} = \to 5 < 8.57 < 25$$

$$\frac{d_0}{2 t_0} = \frac{219.1}{2 \cdot 10} = 11 \to 5 < 11 < 25$$

$$g \geq t_2 + t_3 \to 30 > (6.3 + 6.3) = 12.6 \text{ mm}$$

$$\frac{g}{t_0} = \frac{30}{10} = 3$$

$$e = 19.52 \text{ mm} \to \frac{e}{d_0} = \frac{19.52}{219.1}$$

$$= 0.089 < 0.25$$

Eccentricity need not be taken into consideration.

$$N_{op,Sd} = N_{0,Sd} - N_{2,Sd} \cdot \cos 30°$$
$$= 700 - 600 \cdot \cos 30° = 180.38 \text{ kN}$$

$$f_{op} = \frac{N_{op,Sd}}{A_0} = \frac{180380}{6570} = 27.46 \text{ N/mm}^2$$

$$n' = \frac{27.46}{355} = 0.08$$

$$f(n') = 1 - 0.3 n'(1 + n') = 0.97$$

$$f(\gamma,g) = \gamma^{0.2} \cdot \left(1 + \frac{0.024 \cdot \gamma^{1.2}}{1 + e^{(0.5g/t_0 - 1.33)}}\right)$$

$$= 11^{0.2} \cdot \left(1 + \frac{0.024 \cdot 11^{1.2}}{1 + 1.185}\right) = 1.93$$

Joint design resistance $N_{2,Rd}$ (see Table 6-9)
Failure criterion:
Chord flange plastification

$$N_{2,Rd} = \frac{f_{y0} \cdot t_0^2}{\sin \theta_2} \cdot \left(1.8 + 10.2 \cdot \frac{d_2}{d_0}\right)$$

$$f(n') \cdot f(\gamma,g') \cdot \frac{1.1}{\gamma_{Mj}}$$

$$N_{2,Rd} = \frac{355 \cdot 10^2}{\sin 30°}$$

$$\cdot \left(1.8 + 10.2 \cdot \frac{139.7}{219.1}\right) \cdot 0.97 \cdot 1.93 \cdot \frac{1.1}{1.1}$$

$$\triangleq 1103.70 \text{ kN} > 600 \text{ kN}$$

$$N_{3,Rd} = N_{2,Rd} \frac{\sin 30°}{\sin 90°} = 1103.70 \cdot 0.5$$

$$= 551.85 > 400 \text{ kN}$$

Failure criterion:
Punching shear in chord

$$N_{2,Rd} = \frac{f_{y0}}{\sqrt{3}} \cdot t_0 \cdot \pi \cdot d_2 \cdot \frac{1 + \sin \theta_2}{2\sin^2 \theta_2} \cdot \frac{1,1}{\gamma_{Mj}}$$

$$= \frac{355}{\sqrt{3}} \cdot 10\pi \cdot 139.7 \cdot \frac{1 + \sin 30°}{2\sin^2 30°} \cdot \frac{1.1}{1.1}$$

$$\triangleq 2698.57 \text{ kN} > 600 \text{ kN}$$

$$N_{3,Rd} = 695.41 \text{ kN} > 400 \text{ kN}$$

6.6 Lattice girders (trusses)

The result of the joint verification (bracings 2 and 3) is positive.

A further condition for the verification of the KT joint is as follows:

$N_{3,Rd} \sin 90°$
$\geq N_{1,Sd} \cdot \sin 30° + N_{2,Sd} \cdot \sin 30°$
$695.41 > (700 \cdot 0.5 + 600 \cdot 0.5) = (350 + 300)$
$= 650$ kN

This condition is also fulfilled.

Design resistances for truss joints consisting of hollow sections and purlin attachments

Fig. 6-124a and b shows two fundamentally different types of purlins for trusses of hollow sections.

Single sections are simple, light applications, while trusses are then used when lengths of the span are large.

A number of purlin attachments of single sections e. g. I beam, RHS and cold rolled section is illustrated in Fig. 6-125. A frequently applied system is shown in Fig. 6-125a, where an I beam is attached to the upper chord of a truss by means of a folded corner plate.

The use of square or rectangular hollow section as purlin (see Fig. 6-125b) is favourable, as its lateral-torsional stability is excellent.

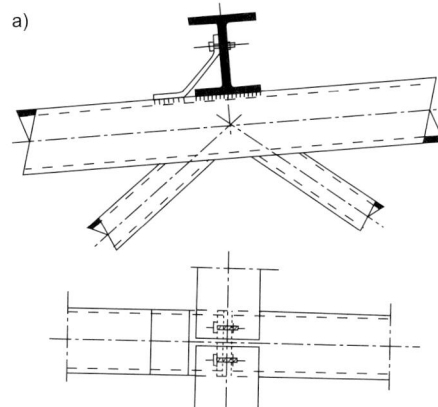

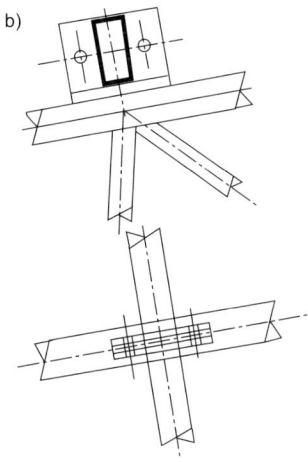

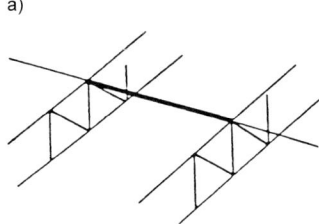

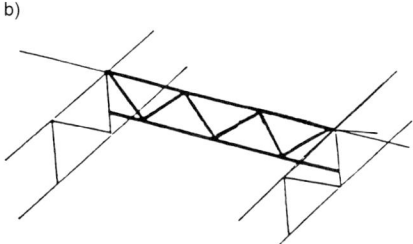

Fig. 6-124 Purlin types for trusses. a) Single section, b) lattice girder

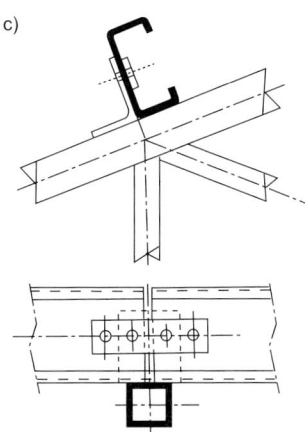

Fig. 6-125 Attachments of single section purlins with hollow section trusses. a) I section, b) RHS, c) profile made of folded strip

This means that intermediate ties can often be omitted.

The purlins of cold rolled sections (folded strips) are attached with splices (see Fig. 6-125 c), which however are not always essential. But they assist in lining up the holes on erection. These splices can be quite large and become fish plates ensuring the continuity of the purlins. This design system is mostly favoured for trusses with square or rectangular chords. The variant as shown in Fig. 6-125 b is also suitable for trusses with circular chords.

Fig. 6-126 shows examples of lattice purlins, where the lower chords of the lattice purlins are connected to the truss; thus it is possible to make continuous purlins to brace the tie of the truss.

Two alternatives for the detail of the connection at the location 1 are illustrated in Fig. 6-127, where the detail "a" allows for the length adjustment by means of distance pieces. The arrangement shown in the detail "b" does not have more length adjustment than the bolt hole tolerance.

Depending on the angle of inclination θ, one of the both alternatives shown in Fig. 6-128

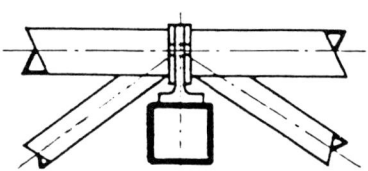
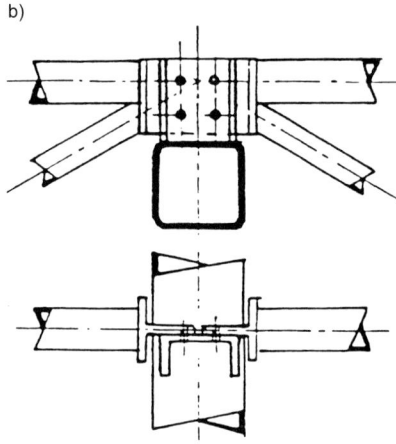

Fig. 6-127 Connection alternatives, detail 1 in Fig. 6-126

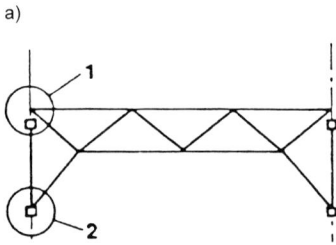
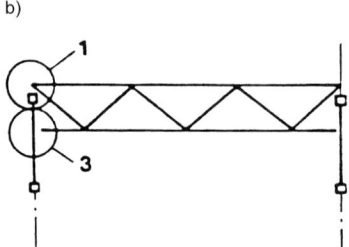

Fig. 6-126 Examples of the attachments of lattice purlins with lattice girders

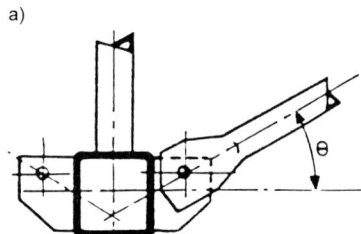
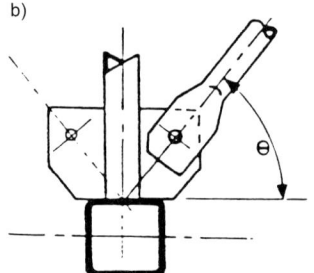

Fig. 6-128 Connection alternatives, detail 2 in Fig. 6-126 a

6.6 Lattice girders (trusses)

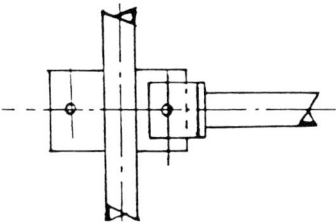

Fig. 6-129 Connection detail 3 in Fig. 6-126b

can be selected as the attachment detail at the location 2 in Fig. 6-126a. The offset of the intersection of the member axes is generally acceptable in the case of bracing members. It can, however, be avoided by an appropriate layout of the main members (in continuous purlins for example).

Further, the detail 3 of Fig. 6-126b with the attachment of the lower chord of the lattice purlin with the vertical of the main truss is illustrated in Fig. 6-129.

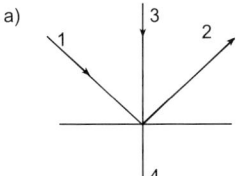

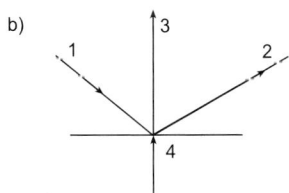

When there is a cross-chord load (for instance from a purlin or a hanger), which acts in the same direction as the combined load components e.g.

a) $N_{1,Sd} \cdot \sin\theta_1 + N_{3,Sd} \cdot \sin\theta_3 + N_{4,Sd}$
b) $N_{2,Sd} \cdot \sin\theta_2 + N_{3,Sd} \cdot \sin\theta_3 + N_{4,Sd}$

the joint design resistance of the remaining bracing members needs to be verified directly:

a) $N_{2,Rd} \cdot \sin\theta_2 \geq N_{2,Sd} \cdot \sin\theta_2$
b) $N_{1,Rd} \cdot \sin\theta_1 \geq N_{1,Sd} \cdot \sin\theta_1$

$N_{1,Rd}$ and $N_{2,Rd}$ are calculated using Table 6-9 or 6-11 or 6-13.

Besides, the loading $N_{4,Sd}$ on the purlin or hanger connection to the chord may require an additional check. A further verification must be made as for X joint.

Tests have led to the following conclusions:

- The design resistances of K and N joints in the presence of cross-chord loadings (i.e. as for X joint loading) are identical to those given by the resistance formulae for K and N joints without cross loads. The same is valid for KT joints.
- If all the bracing forces on one side of a joint act in the same sense or if only one bracing member is loaded, the joint has to be checked as an X joint using an equivalent bracing member size.

6.6.4.2 Design strengths of reinforced truss joints under predominantly static loading

In principle, it is to recommend hollow section joints to be used without reinforcements, as the manufacturing costs increase significantly by the reinforcing measures. It is however sometimes difficult or not possible to do without reinforcement construction, such as for a repair or when the joint is made undersized due to a mistake while designing.

Depending on the decisive failure modes e.g. chord flange plastification, and chord shear for K and N joints as well as chord flange plastification and chord side wall failure for T, Y and X joints, the hollow section joints are stiffened by the chord flange plate, chord side wall plate or vertical plate welded to bracing members. Design and calculation procedures are recommended for this purpose by ENV 1993-1-1, A1 [230]. They are valid for reinforced RHS joints. Recommendation is made to apply them [8] also to CHS joints (see Fig. 6-130).

Formulae for design resistances of the reinforced RHS joints according to ENV 1993-1-1, A1 [230]

These formulae for design resistances given by [230] have mainly been developed basing on the design resistances for the unreinforced

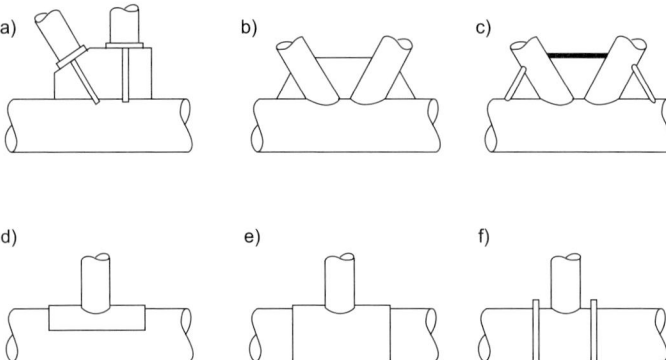

Fig. 6-130 Types of reinforcement for CHS joints

RHS joints according to the Table 6-11 and 6-13, as the failure criteria for RHS joints with or without stiffeners are similar in principle.

The joint checks are to be carried out following the joint resistances $N_{i,Rd}$ of the Table 6-15 (T, Y and X joint) and Table 6-16 (K and N joint).

When the chord face yielding controls the design resistances of T, Y and X joints under compression bracing loading, the joint design resistance can be increased by using chord flange plate reinforcement (Eq. 6-89). This will usually occur when $\beta_P \leq 0.85$. When the chord side wall failure governs, the joint resistance can be enlarged by reinforcing with a pair of chord side plates (Eq. 6-90). This failure mode will usually govern when $\beta_P = 1.0$.

For T, Y and X joints stiffened with a chord flange plate, a difference can be observed in the behaviour of the stiffening plate depending on the sense of the load in the bracing member. With a tension load in the bracing, the plate tends to lift off the chord member and behave as a plate clamped (welded) along its four edges. The joint design resistance thereby depends only on the plate geometry and the material properties of the plate. According to the yield line theory [7] (see also Section 6.6.4.1, yield line model), Eq. (6-88) gives an estimation of the design resistance of the tensile bracing member of T, Y and X joints.

For K and N joints with gap, usually the chord flange reinforcement is applied against the plastification of the chord face (Eq. 6-91), when $\beta_P \leq 1.0$ and when a square hollow section is used as a structural member. The design resistances can be increased by welding a pair of side wall plates against chord shear (Eq. 6-92), when mostly the conditions $\beta = 1.0$ and $h_0 < b_0$ are valid.

Lit. [231] contains the design recommendations for K joints stiffened by chord flange plates. They are based on an elastic deformation requirement of the connection plate under service load.

Lit. [7] recommends an approximation to calculate the required thicknesses of the stiffening plates for K joints with gap. This is done by substituting the chord wall thickness t_0 with the stiffening plate thickness t_P in the equations for the joint design resistances according to the Table 6-11 and 6-13. Further, the yield strength of the chord material f_{y0} has to be replaced by the yield strength $f_{y,P}$ of the material of the stiffening plate.

The dimensioning of stiffening plates is related to the cross-sectional resistances of the bracing members $(A_i \cdot f_{yi})$. This is achieved provided $t_P \geq 2\,t_1$ and $2\,t_2$.

The welds joining the stiffening plate and the chord face should have a weld throat size at least equal to the wall thickness of the adjacent bracing member.

A minimum gap g between the bracings is required in order to prevent the overlap of the adjacent welds: $g \geq t_1 + t_2$.

In general, all-round welding is necessary to connect the stiffening plate to the chord face

6.6 Lattice girders (trusses)

Table 6-15 Design resistances of reinforced, welded T, Y and X joints between rectangular, square or circular bracing members and rectangular or square chords

Type of joint	Joint design resistance (i = 1, 2)
T, Y and X joints reinforced with chord flange plates *Tension loading* $\beta_P = b_i/b_P$ $\eta = h_i/b_P$ *Compression loading*	Chord flange plastification, $\quad \beta_P \leq 0.85$ brace failure (effective width) or punching shear $$l_P \geq \frac{h_i}{\sin \theta_i} + \sqrt{b_P(b_P - b_i)} \quad (6\text{-}87)$$ and $\geq 1.5 h_i/\sin \theta_i$ $b_P \geq b_0 - 2 t_0$ $$N_{i,Rd} = \frac{f_{y,P} \cdot t_P^2}{\left(1 - \frac{b_i}{b_P}\right) \sin \theta_i}$$ $$\cdot \left(\frac{2 h_i/b_P}{\sin \theta_i} + 4\sqrt{1 - b_i/b_P}\right)\left(\frac{1.1}{\gamma_{Mj}}\right)$$ (6-88) $\beta_P \leq 0.85$ $$l_P \geq \frac{h_i}{\sin \theta_i} + \sqrt{b_P(b_P - h_i)} \quad (6\text{-}89)$$ and $\geq 1.5 h_i/\sin \theta_i$ $b_P \geq b_0 - 2 t_0$ $N_{i,Rd}$ for T, Y and X joints is to be taken from Table 6-13 but with t_0 replaced by t_P for chord flange plastification, brace failure (effective width) and punching shear only and set $f(n) = 1.0$
T, Y and X joints reinforced with a pair of chord side wall plates *Tension or compression loading*	Chord side wall buckling or chord side wall shear $$l_P \geq 1.5 h_i/\sin \theta_i \quad (6\text{-}90)$$ $N_{i,Rd}$ for T, Y and X joints is to be taken from Table 6-13 but with t_0 replaced by $(t_0 + t_P)$

Note: For circular bracing, replace b_i and h_i by d_i.
Validity range: $f_y \leq 355$ N/mm^2.

Table 6-16 Design resistances of reinforced welded K and N joints between rectangular, square or circular bracing members and rectangular or square chords

Type of joint	Joint design resistance ($i = 1, 2$)
K and N joints reinforced with chord flange plate 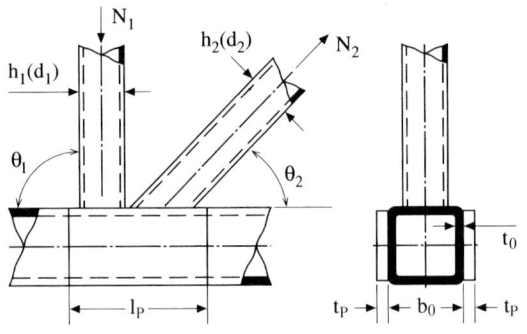	Chord flange plastification, brace failure (effective width) or punching shear $$l_P \geq 1.5\left(\frac{h_1}{\sin\theta_1} + g + \frac{h_2}{\sin\theta_2}\right) \quad (6\text{-}91)$$ $b_P \geq b_0 - 2t_0$ $t_P \geq 2t_1$ und $2t_2$ $N_{i,Rd}$ for K and N joints is to be taken from Table 6-13 substituting t_0 by t_P, b_0 by b_P and f_{y0} by $f_{y,P}$
K and N joints reinforced with a pair of chord side wall plates 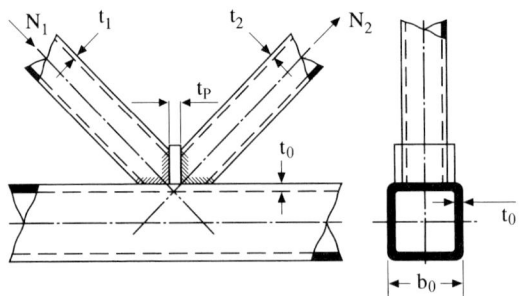	Chord shear failure $$l_P \geq 1.5\left(\frac{h_1}{\sin\theta_1} + g + \frac{h_2}{\sin\theta_2}\right) \quad (6\text{-}92)$$ $N_{i,Rd}$ for K and N joints is to be taken from Table 6-13 substituting t_0 by $(t_0 + t_P)$
K and N joint with a vertical plate between the bracing members	Insufficient overlap of the bracing members $t_P \geq 2t_1$ and $2t_2$ $\quad (6\text{-}93)$ $N_{i,Rd}$ for K and N joint with $\lambda_{ov} < 80\%$ overlap is to be taken from Table 6-13 and $b_{e,ov}$ from Table 6-11 substituting b_j, t_j and f_{yj} by b_P, t_P und $f_{y,P}$

Note: For circular bracing, replace b_i and h_i by d_i.
Validity range: $f_y \leq 355$ N/mm^2.

to prevent the access of air and water to the inner surfaces, which consequently leads to the protection against corrosion.

In order to avoid partial overlapping of one bracing member on to another in a K joint, each bracing member may be welded to a vertical stiffener. It is recommended as follows:

$t_P \geq 2\ t_1$ and $2\ t_2$

6.6.4.3 Design of truss joints with cranked chord

Fig. 6-131 illustrates a cranked-chord joint in a Pratt truss, which is characterized by a crank or bend in the chord member at the joint noding point. Two common sections are welded together by groove or concave fillet weld at the appropriate angle. The intersection of the centre lines of the three hollow section members is usually made coincident, which means zero eccentricity. The cranked chord part takes up the function of an "equal width bracing member". Although this joint has an appearance similar to T or Y joint, its strength behaviour is quite different. An experimental research programme with square and rectangular members [232] has revealed that cranked-chord RHS joints behave as K or N joints with overlap and their design resistances can be determined using Table 6-13 correspondingly.

As shown in Fig. 6-131 b, one part (mb) of the chord member can be given an imaginary extension and the cranked-chord part (ma) is to be treated as an overlapped bracing member (Eqs. (6-75) to (6-77)).

Eqs. (6-69) and (6-70) in Table 6-9 can be applied to determine the joint design resistances for CHS cranked-chord joints.

6.6.4.4 Design strengths of uni-planar welded truss joints with I or H section as chord and hollow sections as bracings

Joints of this type are often executed with stiffening ribs between the chord flanges (Fig. 6-132). They can be checked using verifying procedures according to the conventional structural engineering. In general, it is sufficient to make a check of the weld stresses.

The design resistance formulae given in Table 6-17 [230] are related to the welded joints, where I section chords are not provided with stiffening ribs. Table 6-18 contains the ranges of validity for the application parameters.

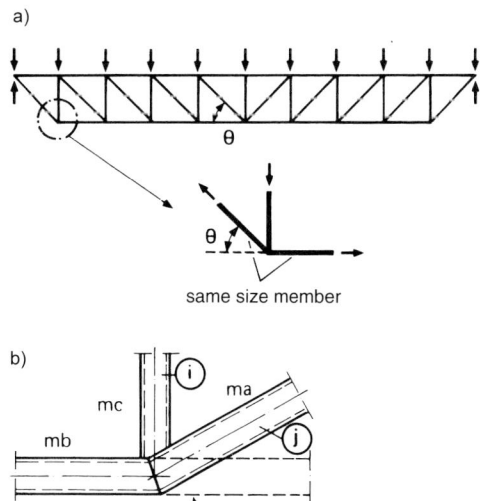

Fig. 6-131 Cranked-chord joint represented as an overlapped K or N joint

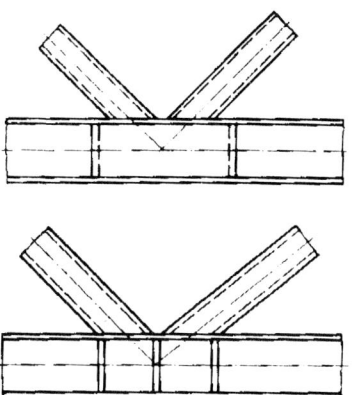

Fig. 6-132 Joints with I or H section as chord with stiffening ribs

Table 6-17 Design resistances of uni-planar, welded truss joints between square or rectangular or circular bracing members and I or H section as chord

Type of joint	Design resistance ($i = 1, 2, j$ = overlapped bracing)	
T, Y and X joint	Chord web yielding	
	$N_{1,Rd} = \dfrac{f_{y0} t_w b_w}{\sin \theta_1} [1.1/\gamma_{Mj}]$	(6-94)
	Effective width (brace failure)	
	$N_{1,Rd} = 2 f_{y1} t_1 b_e [1.1/\gamma_{Mj}]$	(6-95)
K and N joint with gap [i = 1 or 2]	Chord web stability	Brace failure need not be checked, if:
	$N_{i,Rd} = \dfrac{f_{y0} t_w b_w}{\sin \theta_i} [1.1/\gamma_{Mj}]$ (6-96)	$g/t_f \leq 20 - 28\beta$ $\beta \leq 1.0 - 0.03\gamma$
	Effective width[b]	and for CHS:
	$N_{i,Rd} = 2 f_{yi} t_i b_e [1.1/\gamma_{Mj}]$ (6-97)	$0.75 \leq d_1/d_2 \leq 1.33$
	Chord shear	or for RHS:
	$N_{i,Rd} = \dfrac{f_{y0} A_v}{\sqrt{3} \sin \theta_1} [1.1/\gamma_{Mj}]$ (6-98)	$0.75 \leq b_1/b_2 \leq 1.33$
K and N joint with overlap[a]	Effective width[b]	$25\% \leq \lambda_{ov} < 50\%$
	$N_{i,Rd} = f_{yi} t_i \left(b_e + b_{e.ov} + \dfrac{\lambda_{ov}}{50}(2 h_i - 4 t_i)\right)[1.1/\gamma_{Mj}]$ (6-99)	
	Effective width[b]	$50\% \leq \lambda_{ov} < 80\%$
	$N_{i,Rd} = f_{yi} t_i (b_e + b_{e.ov} + 2 h_i - 4 t_i)[1.1/\gamma_{Mj}]$ (6-100)	
	Effective width[b]	$\lambda_{ov} \geq 80\%$
	$N_{i,Rd} = f_{yi} t_i (b_i + b_{e.ov} + 2 h_i - 4 t_i)[1.1/\gamma_{Mj}]$ (6-101)	
$A_v = A_0 - (2 - \alpha) b_0 t_f + (t_w + 2 r_0) t_f$ For RHS bracing: $\alpha = \left[\dfrac{1}{1 + \dfrac{4 g^2}{3 t_f^2}}\right]^{0.5}$ For CHS bracing: $\alpha = 0$	$b_e = t_w + 2 r_0 + 7 \left(\dfrac{f_{y0}}{f_{yi}}\right) \cdot t_f$ but $b_e \leq b_i$	$b_w = \dfrac{h_i}{\sin \theta_i} + 5(t_f + r_0)$ but $b_w \leq 2 t_i + 10(t_f + r_0)$
	$b_{e.ov} = \dfrac{10}{b_j/t_j} \dfrac{f_{yj} t_j}{f_{yi} t_i} b_i$ but $b_{e.ov} \leq b_i$	
CHS bracings	Multiply the expressions with $\pi/4$. Replace b_1 and h_1 by d_1 and b_2 and h_2 by d_2.	

[a] Only the overlapping bracing member need to be checked. The bracing efficiency (i. e. the design resistance of the joint divided by the design plastic resistance of the bracing) of the overlapped bracing should not exceed the efficiency of the overlapping bracing.

[b] Brace failure $\beta = \dfrac{b_i}{b_0}$; $\gamma = \dfrac{b_0}{2 t_f}$

Table 6-18 Validity ranges for the application of Table 6-17

Type of joint	Joint parameter (i = 1 or 2, j = overlapped bracing)							
	h_0/t_w	b_i/t_i and h_i/t_i or d_i/t_i		h_i/b_i	b_0/t_f	b_i/b_j	b_i/b_0	Eccentricity	Material yield strength f_y
		Compression	Tension					
X joint		$\dfrac{h_i}{t_i} \leq 1.1 \sqrt{\dfrac{E}{f_{yi}}}$	$\dfrac{h_i}{t_i} \leq 35$	≥ 0.5 but ≤ 2.0		–		
T or Y joint	$\dfrac{h_0}{t_w} \leq 1.2 \sqrt{\dfrac{E}{f_{y0}}}$ and $h_0 \leq 400$ mm	$\dfrac{b_i}{t_i} \leq 1.1 \sqrt{\dfrac{E}{f_{yi}}}$	$\dfrac{b_i}{t_i} \leq 35$	1.0	$\dfrac{b_0}{t_f} \leq 0.75 \sqrt{\dfrac{E}{f_{y0}}}$	–	≤ 1.0	$-0.55 \leq \dfrac{e}{h_0} \leq 0.25$	≤ 355 N/mm²
K joint with gap N joint with gap	$\dfrac{h_0}{t_w} \leq 1.5 \sqrt{\dfrac{E}{f_{y0}}}$ and $h_0 \leq 400$ mm	$\dfrac{d_i}{t_i} \leq 1.5 \sqrt{\dfrac{E}{f_{yi}}}$	$\dfrac{d_i}{t_i} \leq 50$	≥ 0.5 but ≤ 2.0		≥ 0.75		
K joint with overlap N joint with overlap								

All formulae are based on the test results and their evaluations, which have been described in [150, 155].

Depending on the type of joints, joint parameters and loading conditions, several modes of failure may occur:

- Chord web failure by buckling (crippling) (under compression, Fig. 6-133 a) or by material yielding (Fig. 6-133 b)
- Plastification of the chord cross section by shear and axial load (Fig. 6-134)
- Buckling and crack of the bracing (Fig. 6-135)

Eq. (6-94) in Table 6-17 describes the yielding of the chord web. The effective area of the web is calculated using the effective length b_w (see Fig. 6-136); Table 6-17 gives the formula to calculate b_w.

In the case of failure by crack in the bracing under tension, which can occur close to the

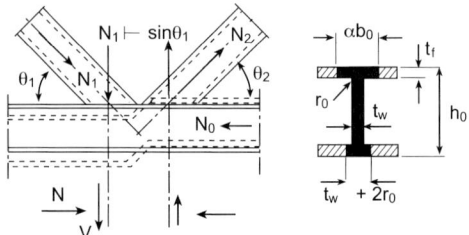

☒ reduced cross-sectional area for axial load
■ cross-sectional area A_V for shear load

Fig. 6-134 Chord plastification by shear and axial load

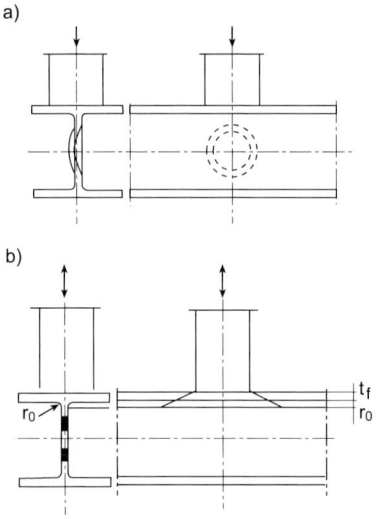

Fig. 6-133 Chord web failure: a) buckling, b) yielding

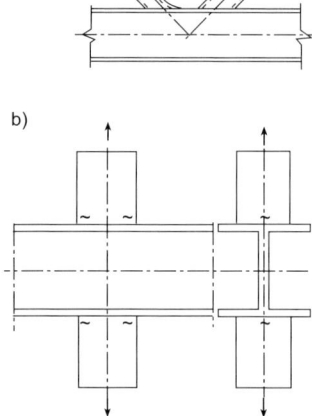

Fig. 6-135 Bracing failure: a) buckling, b) crack

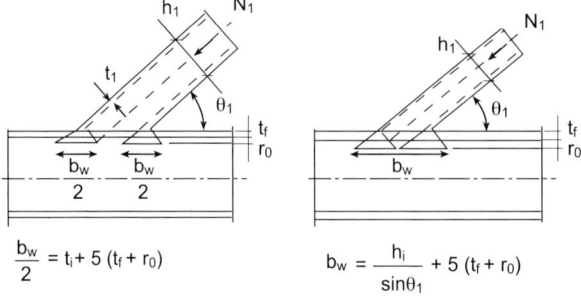

$$\frac{b_w}{2} = t_i + 5(t_f + r_0) \qquad b_w = \frac{h_i}{\sin\theta_1} + 5(t_f + r_0)$$

Fig. 6-136 Effective bearing length b_w of the joint with I section as chord

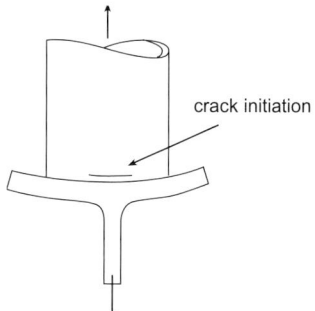

Fig. 6-137 Crack in the bracing under tension of a joint with I section as chord

connecting weld, a concentration of the force distribution in the direction of web can be expected due to low stiffness of the chord web (Fig. 6-137).

This concentration is considered assuming the effective periphery b_e (Fig. 6-138) of the connected bracing, which is determined experimentally [150]:

$$b_e = t_w + 2r_o + 7(f_{y0}/f_{yi})t_f \qquad (6\text{-}102)$$

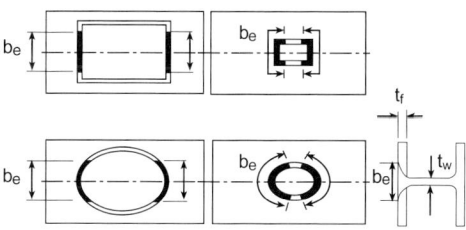

Fig. 6-138 Definition of the effective periphery b_e

Further, the interaction between bending moment and axial force has also to be checked:

$$\frac{N_{i,Sd}}{N_{i,Rd}} + \frac{M_{ip,i,Sd}}{M_{ip,i,Rd}} \le 1{,}0 \; (N_{i,Rd} \text{ and } M_{ip,i,Rd} \text{ can}$$

be calculated according to Tables 6-17 and 6-33).

For joints with gap, the axial design resistance $N_{op(\text{in gap}),Rd}$ or $N_{0(\text{in gap}),Rd}$ in the gap zone has to be checked taking the transversal forces $V_{i,Sd} = N_{i,Sd} \cdot \sin\theta_i$ transmitted through the bracing members into consideration:

In case $\dfrac{V_{i,Sd}}{V_{pl,i,Rd}} \le 0{,}5$,

$$N_{op(\text{in gap}),Rd} \text{ or } N_{0(\text{in gap}),Rd} \le \frac{A_0 \cdot f_{y0}}{\gamma_{M0}}$$

In case $\dfrac{V_{i,Sd}}{V_{pl,i,Rd}} > 0{,}5$,

$N_{op(\text{in gap}),Rd}$ bzw. $N_{0(\text{in gap}),Rd} \le$

$$\left[A_0 - A_v \left(\frac{2V_{i,Sd}}{V_{pl,i,Rd}} - 1 \right)^2 \right] \frac{f_{y0}}{\gamma_{M0}}$$

6.6.4.5 Design strengths of uni-planar welded truss joints with ⊏-profile (channel) as chord and hollow sections as bracings

The only research project involving this joint type was carried out in the Testing Centres of the Delft University of Technology and the Institutes TNO [155, 164, 233] and the formulae for the design resistances of the K and N joints (Fig. 6-139) are recommended [7, 150] basing on theoretical analysis and test results.

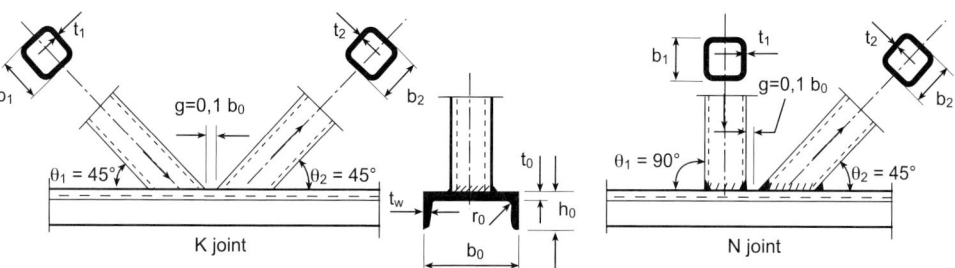

Fig. 6-139 K and N joint with ⊏-profile as chord

Depending on the joint parameters

$$\beta = \frac{b_1 + b_2}{2\,b_0}$$

$$\beta^* = \frac{b_1 + b_2}{2\,b_0^*} \quad \text{where } b_0^* = b_0 - 2\,(t_W + r_o)$$

$$\frac{b_i}{t_i},\ \frac{h_i}{t_i},\ \frac{d_i}{t_i},\ \frac{h_i}{b_i},\ g\,(\text{gap}),\ \lambda_{ov}\,(\text{overlap}),\ \theta_i$$

the following failure modes are possible (Fig. 6-140):

a) Plastification of the chord cross section by shear load, bending moment and axial load
b) Yielding of chord flange
c) Crack in the bracing member or in the weld
d) Chord flange shear

A further mode of failure by local buckling of the thinwalled bracing or the chord flange in compression can also taken place.

Table 6-19 lists the design resistance formulae and their validity ranges for K and N joints with gap. Table 6-20 contains the same for K and N joints with overlap.

The comparison between the analytical models and the test results leads to the following conclusions:

• The joint eccentricity in joints with gap is usually large, which initiates large bending moments in the joint components.

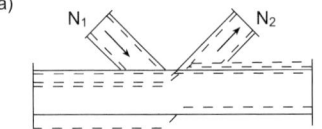
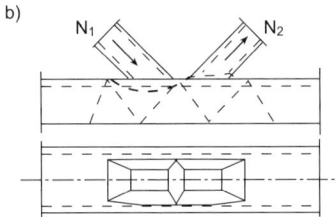
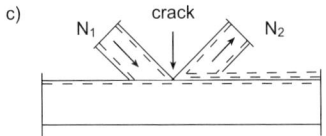
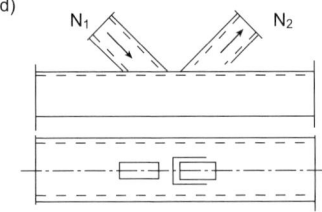

Fig. 6-140 Failure modes for joints with ⊏-profile (channel) as chord and hollow sections as bracings. a) Plastification of chord cross section by shear load, bending moment and axial load, b) yielding of chord flange, c) crack in the bracing member or in the weld, d) chord flange shear

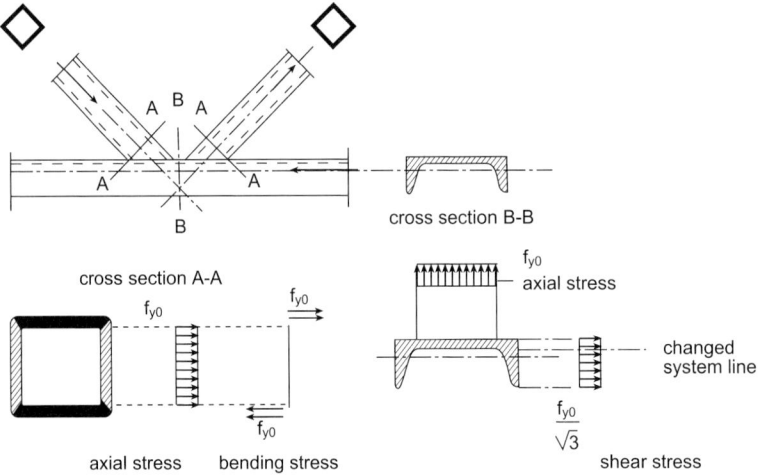

Fig. 6-141 Moment of eccentricity fully taken over by the bracings

6.6 Lattice girders (trusses)

Table 6-19 Design resistances of uni-planar, welded K and N joints with gap between rectangular, square and circular bracings and ⊏-profile as chord [150]

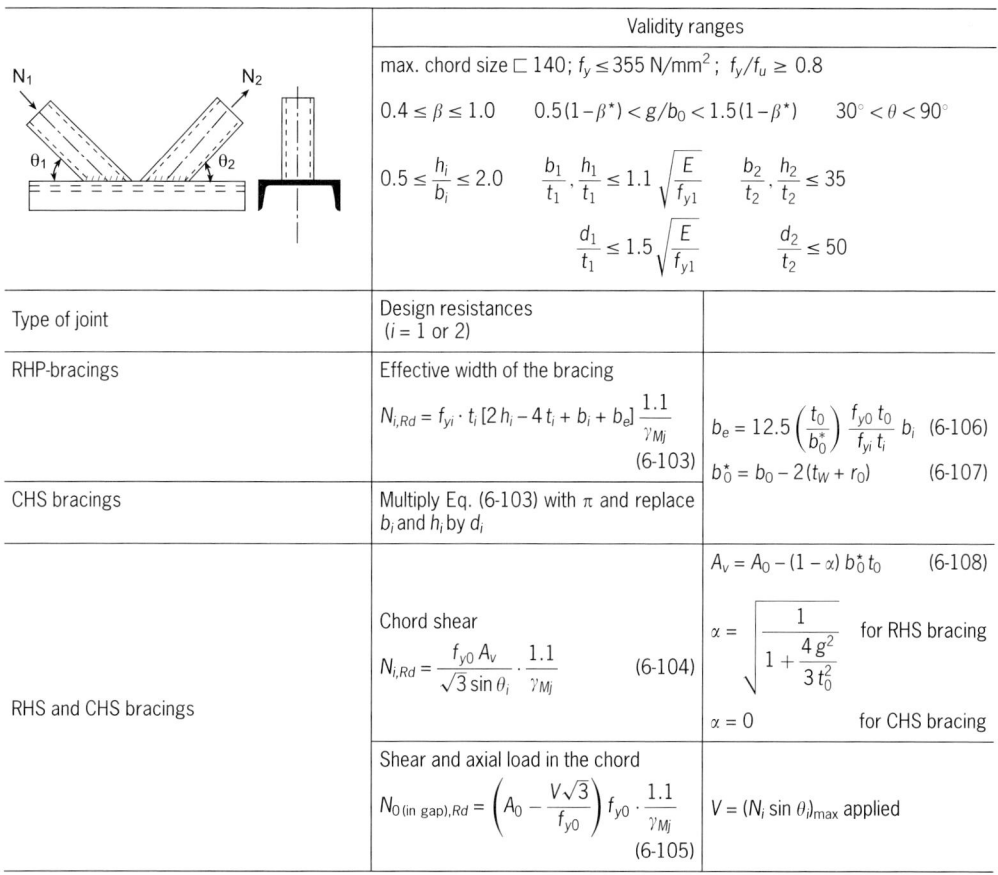

Type of joint	Design resistances ($i = 1$ or 2)	
RHP-bracings	Effective width of the bracing $$N_{i,Rd} = f_{yi} \cdot t_i [2 h_i - 4 t_i + b_i + b_e] \frac{1.1}{\gamma_{Mj}} \quad (6\text{-}103)$$	$b_e = 12.5 \left(\dfrac{t_0}{b_0^*}\right) \dfrac{f_{y0}\, t_0}{f_{yi}\, t_i} b_i \quad (6\text{-}106)$ $b_0^* = b_0 - 2(t_w + r_0) \quad (6\text{-}107)$
CHS bracings	Multiply Eq. (6-103) with π and replace b_i and h_i by d_i	
RHS and CHS bracings	Chord shear $$N_{i,Rd} = \frac{f_{y0} A_v}{\sqrt{3} \sin \theta_i} \cdot \frac{1.1}{\gamma_{Mj}} \quad (6\text{-}104)$$ Shear and axial load in the chord $$N_{0\,(in\,gap),Rd} = \left(A_0 - \frac{V\sqrt{3}}{f_{y0}}\right) f_{y0} \cdot \frac{1.1}{\gamma_{Mj}} \quad (6\text{-}105)$$	$A_v = A_0 - (1 - \alpha) b_0^* t_0 \quad (6\text{-}108)$ $\alpha = \sqrt{\dfrac{1}{1 + \dfrac{4 g^2}{3 t_0^2}}}$ for RHS bracing $\alpha = 0$ for CHS bracing $V = (N_i \sin \theta_i)_{max}$ applied

Validity ranges:
max. chord size ⊏ 140; $f_y \le 355$ N/mm²; $f_y/f_u \ge 0.8$
$0.4 \le \beta \le 1.0$; $0.5(1-\beta^*) < g/b_0 < 1.5(1-\beta^*)$; $30° < \theta < 90°$
$0.5 \le \dfrac{h_i}{b_i} \le 2.0$; $\dfrac{b_1}{t_1}, \dfrac{h_1}{t_1} \le 1.1 \sqrt{\dfrac{E}{f_{y1}}}$; $\dfrac{b_2}{t_2}, \dfrac{h_2}{t_2} \le 35$
$\dfrac{d_1}{t_1} \le 1.5 \sqrt{\dfrac{E}{f_{y1}}}$; $\dfrac{d_2}{t_2} \le 50$

In a plastic calculation, the moment due to eccentricity is taken over only by the bracings, when they have adequate load bearing capacity and the chord member is loaded by shear and axial load or solely by shear load up to the yield limit (Fig. 6-141). Otherwise, the moment due to eccentricity has to be distributed among all structural members as favourably as possible (Fig. 6-142).

The structural members are to be checked using the following formula for the interaction of axial load and bending moment [234]:

$$\left(\frac{N_i}{N_{pl,i}}\right)^\chi + \left(\frac{M_i}{M_{pl,i}}\right) \le 1.0 \quad (6\text{-}111)$$

where:
$i = 1, 2$

$\chi = 1.5$ for $0.5 \le \dfrac{h_i}{b_i} \le 2$

$\chi = 1.2$ for $\dfrac{h_i}{b_i} \le 0.5$

$\chi = 1.0$ for ⊏-profile (assumption, as no interaction formula for ⊏-profile is available)

The effect of shear stress induced by the moment due to eccentricity is small and can therefore be neglected.

- The design for shear resistance of the chord cross section in the gap region can be deter-

Table 6-20 Design resistances of uni-planar, welded K and N joints with overlap between rectangular, square or circular bracings and ⊏-profile as chord [7]

Type of joint	Design resistances ($i = 1, 2, j$ = overlapped bracing)	
K and N joint with overlap (RHS bracings)	$\lambda_{ov} = 100\%$	
	$N_{i,Rd} = f_{yi} \cdot t_i [2 h_i - 4 t_i + b_i + b_{e,ov}] \cdot \dfrac{1.1}{\gamma_{Mj}}$	(6-109)
	$30\% \leq \lambda_{ov} < 100\%$	
	$N_{i,Rd} = f_{yi} \cdot t_i [2 h_i - 4 t_i + b_e + b_{e,ov}] \cdot \dfrac{1.1}{\gamma_{Mj}}$	(6-110)
CHS bracings	Multiply Eq. (6-109) or (6-110) with π and replace b_i and h_i by d_i	
Definitions	Validity ranges	
$b_e = 12.5 \left(\dfrac{t_0}{b_0^*}\right) \cdot \dfrac{f_{y0} \cdot t_0}{f_{yi} \cdot t_i} \cdot b_i$ with $1 \leq \dfrac{f_{y0} \cdot t_0}{f_{yi} \cdot t_i} \leq 2$ $b_0^* = b_0 - 2(t_w + r_0)$ $b_{e,ov} = \dfrac{12.5}{(b_j/t_j)} \cdot \dfrac{f_{yj} \cdot t_j}{f_{yi} \cdot t_i} \cdot b_i$ with $1 \leq \dfrac{f_{yj} \cdot t_j}{f_{yi} \cdot t_i} \leq 2$	max. chord size ⊏ 400; $f_y \leq 355$ N/mm²; $\dfrac{f_y}{f_u} \leq 0.8$ $\beta \geq 0.25$ $\dfrac{b_i}{b_j} \geq 0.75$ $30\% \leq \lambda_{ov} \leq 100\%$ $0.5 \leq \dfrac{h_i}{b_i} \leq 2$ $\dfrac{b_1}{t_1}, \dfrac{h_1}{t_1} \leq 1.1 \sqrt{\dfrac{E}{f_{y1}}}$ $\dfrac{b_2}{t_2}, \dfrac{h_2}{t_2} \leq 35$ $\dfrac{d_1}{t_1} \leq 1.5 \sqrt{\dfrac{E}{f_{y1}}}$ $\dfrac{d_2}{t_2} \leq 50$	

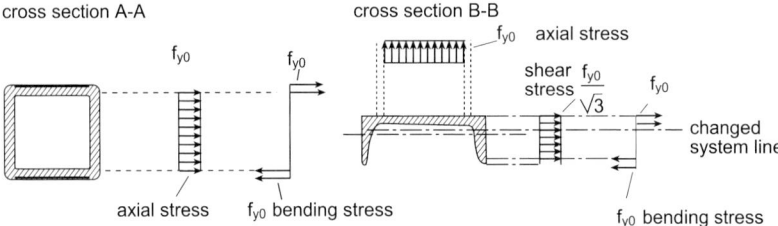

Fig. 6-142 Moment of eccentricity distributed over all structural components (see also Fig. 6-141)

mined using Eq. (6-104) in Table 6-19. Eq. (6-105) describes the design shear and axial resistance of the chord cross section in gap (Fig. 6-143).

- For $\beta = \dfrac{b_1 + b_2}{2 b_0} < 0{,}4$, the failure modes 1) plastification of chord flange and 2) chord flange shear are critical. In this range, no check was done by tests.

- For larger chord sizes, the perimeter for effective width and punching shear will not be fully effective. Further, the moment due to eccentricity becomes larger. Due to this, the axial design resistance of the members reduces considerably. In this case, it is recommended to apply the joint with overlap instead of that with gap.

- Although no test has been carried out on joints with overlap, it is proposed to calculate the design resistances of these joints treating them as joints with rectangular or square hollow section as chord (see Table 6-20) [7]. In

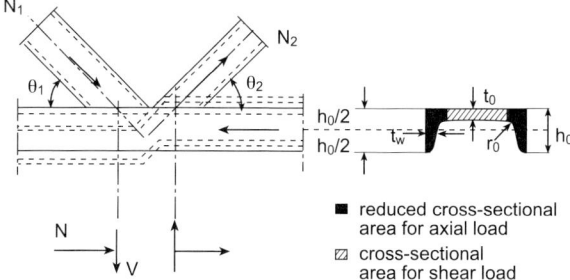

- reduced cross-sectional area for axial load
- cross-sectional area for shear load

Fig. 6-143 Shear and axial load in channel chord cross section in the gap

this case, the moment due to eccentricity can be ignored while designing the joint. However, it has to be taken into consideration, when the sizes of the structural members are to be determined.

6.6.5 Design of uni-planar trusses with directly welded hollow sections

The fundamental difference between the design procedures of a truss consisting of the welded and bolted open profiles as well as the hollow sections and open profiles joined via gusset plates by welding or bolting (see Fig. 6-72c and d) and a truss consisting of the directly welded hollow sections (see Fig. 6-82) lies in the aspect that the planner in the first case can fix the chord and bracing sizes without paying attention to the joint details, which in general belongs to the job of the stress analyst and constructor. In the second case, verifications of the structural members and the joints (see Section 6.6.4.1) are closely associated with each other, as the joint design resistances are dependent on the geometrical parameters and the resistances of the truss members (see Tables 6-9, 6-11 and 6-13). It is therefore necessary to take joint resistances into account while sizing the truss members. If the designer does not consider the joint behaviour right from the beginning, this may lead to undesirable stiffening of joints later due to restricted deformation and rotation capacity, which must be eliminated.

However, this does not mean that the joints have to be designed in detail right at the conceptual phase. It is required that the chord and the bracing members have to be chosen in such a way that the governing joint parameters e.g. γ, τ, β, g, λ_{ov}, θ provide an adequate joint resistance, an economical fabrication and a low cost maintenance (corrosion protection).

The following fundamental guidelines have to be followed to arrive at optimum design: Although lattice structures are designed assuming pin-jointed members, secondary bending moments may occur due to the actual end fixities of the members. With respect to design both members and joints, they can be ignored, provided there is adequate deformation and rotation capacity (see Fig. 6-104). This can be achieved by restricting the slenderness $\left(\text{dependent on } \dfrac{b_i \text{ or } d_i}{t_i}\right)$ of certain structural members, particularly the compression bracing.

If the conditions for the validity ranges given in Tables 6-10, 6-12 and 6-14 are fulfilled, the secondary bending moment need not be taken into consideration while applying the design resistance formulae in Tables 6-9, 6-11 and 6-13.

The following steps have to be undertaken in order to obtain an economical and technically efficient truss construction in hollow sections:

1. Determination of the truss layout (see Fig. 6-73) e.g. span L, depth H, panel length l_0 (depends on the truss form; see also Section 6.6.1.1) and purlin interval (mostly equal to the panel length).

Hereby the construction of the total structure e.g. truss spacings and lateral supports of the truss has also to be considered. Effort has to be made to keep the nodal joints to a minimum, which often results in an Warren-type truss. A low number of joints saves the manufacturing costs significantly.

2. Determination of loads at the joints and on the members of the truss. They have to be simplified to equivalent loads at the joints.

3. Determination of the axial forces in all members assuming that the joints are pinned and that the centre lines of all joint members are noding (eccentricity $e = 0$).

4. First determination of the chord member sizes by considering axial loading and member slenderness (for CHS, usual $d_0/t_0 = 20$ to 30, for RHS, usual $d_0/t_0 = 15$ to 25).

As the major part of the material weight of a truss is used for the chords in compression (about 50%), it is possible to save the total material weight by selecting thin walled sections for them (large d_0/t_0 or b_0/t_0 ratio). However, the material weight cannot be saved for the chords in tension (30% share) as well as for the bracing members under tensile load. The material saving for the bracings in compression is only small.

At the same time, attention has to be paid that the outer surface area of the truss members is small, so that the costs for the paints or sprays against corrosion are minimized.

The design resistance of a truss joint increases with decreasing d_0/t_0 or b_0/t_0 as well as increasing t_0/t_i ratio. As a result, a compromise between adequate joint resistance and member stability has to be found out for the chord in compression. Mostly, a relatively compact section is selected.

For the chord in tension, the d_0/t_0 or b_0/t_0 ratio shall be as small as possible.

Further, the selection of only a single outer diameter or a single width is preferred for the chord due to the manufacturing as well as aesthetic reasons.

The number of the end-to-end connections in a truss depends on the single profile lengths supplied by the profile manufacturing mills. Special profile lengths can however be ordered and supplied for larger projects.

Effective buckling lengths for the chords in compression are determined as follows:

According to the recommendations by CIDECT [6, 226] and Eurocode 3 [230], $l_K = 0.9\ l_0$ (see also Table 5-27 and Fig. 6-73).

The application of the steel grade S 355 instead of S 235 is recommended due to the following reasons:

- The yield strength of S 355 is 50% higher than that of S 235. This can be fully exploited by the structural members in tension.

- Also in members in compression, the application of S 355 can lead to low slenderness, which affects the function of a truss favourably.

- The joint design resistance in a truss is increased due to the higher values of the material properties of S 355, when it is applied to the chord (not to the bracings). However, the application of S 235 only to the bracings can also bring economical advantage (e.g. S 235 for bracings, S 355 for chords).

5. First determination of the bracing member sizes based on axial loading taking following points into consideration:

- The wall thicknesses of the bracings should be smaller than those of the chords, as larger t_0/t_i ratio increases the joint design resistance.

- For overlapped bracings (see Fig. 6-83), the width or the outer diameter of the overlapped bracing should be larger than that of the overlapping bracing. This makes the execution of welds easier.

- Effective buckling length of the bracing in compression is determined as follows: According to the recommendations by CIDECT [6, 226] and Eurocode 3 [230], $l_K = 0.75\ l_1$ (see Table 5-27). For bracings of joints with overlap [6, 226], $l_K = l_1$.

- Minimizing the number of bracing sizes and steel grades to a few selected dimensions (perhaps even two only). It is important that the availability of the selected sections in the market is considered. Due to the aesthetic reasons, a single outside width of the members is preferably selected (with

6.6 Lattice girders (trusses)

varying wall thicknesses, if necessary). In this case, special quality control is required in order to prevent their mixing up in the workshop.

6. Checking the joint geometry and configurations considering the following points:

• Gap g (Fig. 6-82), partial and full overlap (Fig. 6-83). From the fabrication point of view, gap joints are most economical. A full overlap (with an eccentricity $e = -0.55 \, d_0$ or b_0 or h_0) provides a more straight forward fabrication and results in higher joint resistance than that for a gap as well as for a partial overlap.

Verification of joint geometry and parameters (member sizes and ratios $\gamma, \tau, \beta, g, \lambda_{ov}, \theta, f(n)$ or $f(n')$) to determine the joint design resistances according to Eurocode 3 [3] (see Tables 6-9, 6-11 and 6-13).

Their values must be within the validity ranges given in Tables 6-10, 6-12 and 6-14, so that secondary bending moments can be ignored.

• If the joint resistances are not adequate, the joint layout (e.g. overlap instead of gap) and/or the bracing or chord sizes are to be modified. In general, only a few joints need checking.

• Checking the joint geometry with regard to eccentricity (see Fig. 6-101).

The primary bending moment due to joint eccentricity can be neglected, if the following condition is fulfilled:

$-0.55 \leq e/h_0$ or e/b_0 or $e/d_0 \leq 0.25$ (according to [3, 230])

Beyond this range, the moment can be ignored for the chord in tension.

However the moment due to joint eccentricity beyond the above mentioned range has to be considered for the chord in compression. According to Fig. 6-102 (Section 6.6.4) the moment due to eccentricity is to be distributed in the chord. The sizing of the compression chord has to be done considering the interaction between the axial compression and the eccentricity moment (see Fig. 6-144).

The chord in compression is to be checked according to Eurocode 3 [3] taking the moment due to eccentricity into consideration:

$$\frac{N_0 \text{ or } N_{op} \cdot \gamma_M}{\chi \cdot A_0 \cdot f_{y0}} + \frac{M_0 \cdot \gamma_M \cdot \kappa}{W_{pl,0} \cdot f_{y0}} \leq 1$$

(compare with Eqn. (5-86))

κ = factor depending on slenderness, cross section class and moment distribution according to the theory of the second order (see Section 5.3.3.3) [3]

Further, the following condition for the interaction of the combined bending (in-plane and out-of-plane) and axial force in a joint with the CHS chord member has to be met [226]:

$$\frac{N_{i,Sd}}{N_{i,Rd}} + \left[\frac{M_{ip,i,Sd}}{M_{ip,i,Rd}}\right]^2 + \left[\frac{M_{op,i,Sd}}{M_{op,i,Rd}}\right] \leq 1.0$$

For a joint with the RHS chord, the condition for the interaction is as follows:

$$\frac{N_{i,Sd}}{N_{i,Rd}} + \frac{M_{ip,i,Sd}}{M_{ip,i,Rd}} + \frac{M_{op,i,Sd}}{M_{op,i,Rd}} \leq 1.0$$

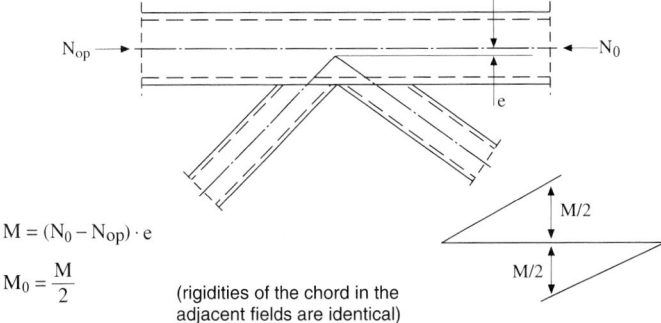

$M = (N_0 - N_{op}) \cdot e$

$M_0 = \dfrac{M}{2}$ (rigidities of the chord in the adjacent fields are identical)

Fig. 6-144 Interaction "axial compression – eccentricity moment" in the chord of a hollow section joint

$M_{ip,i,Rd}$ and $M_{op,i,Rd}$ values can be calculated using the formulae in Tables 6-25 and 6-29.

For a gap joint with RHS chord, the axial design resistance in the gap region of the chord has to be checked taking the shear force transmitted by the bracings into consideration:

If $\dfrac{V_{i,Sd}}{V_{pl,i,Rd}} \leq 0.5$,

$$N_{0,(in\ gap),Rd} \leq \dfrac{A_0 \cdot f_{y0}}{\gamma_{Mo}}$$

If $\dfrac{V_{i,Sd}}{V_{pl,i,Rd}} > 0.5$,

$$N_{0,(in\ gap),Rd} \leq \left[A_0 - A_v \left(2\dfrac{V_{i,Sd}}{V_{pl,i,Rd}} - 1\right)^2\right] \dfrac{f_{y0}}{\gamma_{M0}}$$

(compare with Eq. (6-83))

7. If required, the serviceability condition of overall truss deflection under specified (unfactored) loads is to be checked using correct loading positions. Trusses with gap joints are analysed assuming all members to be pin-jointed. If the joints are overlapped throughout, the truss deflection is checked by assuming continuous chords and pin-ended bracing members taking account of the moments due to eccentricity.

In the first case (RHS gap joints), the truss deflection is underestimated by around 12 to 15% [6, 235–238] because of the flexibility of the joints. Thus, the maximum deflection of a RHS truss with gap joints can be estimated conservatively by multiplying the truss deflection calculated by a pin-jointed analysis with 1.15 [6].

8. Checking the welds:

Figs. 6-94 to 6-97 describe the recommended weld executions for truss joints of hollow sections (see also Section 6.6.3). The prescribed design rules for the welds in the countries belonging to the European Union are available [38], the observance of which leads to the prequalification of welds.

The welds are designed in accordance with the material yield strengths of the connecting hollow sections. They are therefore automatically considered to be prequalified for any member load.

Special consideration has to be given to the following points:

- In gap joints, a gap $g \geq t_1 + t_2$ is recommended, so that the adjacent welds do not overlap each other.
- For overlap joints, the overlap percentage $\lambda_{ov}\%$ must not be less than 25%.
- The weld at the toe of a bracing, particularly of an overlapped one, has to be executed with special care. In case $\theta < 60°$, special weld preparations have to be undertaken. A partial or full penetration groove weld is usually applied here.
- $\theta \geq 30°$ is recommended to assure appropriate penetration of the weld in the heel region of a bracing member.
- As the weld volume depends on $t_{1,2}^2$, the welding of a thinwalled bracing is more economical than that of a bracing with thick walls.

6.6.6 Design strengths of special uni-planar welded truss joints of circular hollow sections (CHS)

In tubular structures, various other joint configurations with special loading constellations exist, which are used from time to time. The design resistance formulae for some of these joints of CHS or RHS are listed in Table 6-21 or 6-22. Their design rules are based on those for the joints described in Tables 6-9, 6-11 and 6-13.

6.6.7 Design strengths of multi-planar welded truss joints

Fig. 6-145a and b show two triangular girders made with CHS and RHS, which offer easy handling during fabrication and are frequently used as exposed structures. They have the special advantages of large horizontal resistance, high torsional rigidity and increased lateral stability. Due to the symmetrical form of the CHS chords from all sides, the bracing members can be joined in each plane. The same is also valid for quadrangular girders (Fig. 6-145c and d), where the bracings are welded to the chords in two planes. Due to aesthetical reasons, the triangular girders of hollow sections are preferably applied to build

6.6 Lattice girders (trusses)

Table 6-21 Design resistances of special types of welded uni-planar joints between CHS bracings and CHS chord

Type of joint	Design criterion
YY joint	$N_{1,Sd} \leq N_{1,Rd}$ $N_{1,Rd}$ is the value of $N_{1,Rd}$ for an X joint from Table 6-9
KT joint	$N_{1,Sd} \sin\theta_1 + N_{3,Sd} \sin\theta_3 \leq N_{1,Rd} \sin\theta_1$ $N_{2,Sd} \sin\theta_2 \leq N_{1,Rd} \sin\theta_1$ $N_{1,Rd}$ is the value of $N_{1,Rd}$ for a K joint from Table 6-9, but with d_1/d_0 replaced by: $$\frac{d_1 + d_2 + d_3}{3\,d_0}$$
XX joint	$N_{1,Sd} \sin\theta_1 + N_{2,Sd} \sin\theta_2 \leq N_{x,Rd} \sin\theta_x$ $N_{x,Rd}$ is the value of $N_{x,Rd}$ for an X joint from Table 6-9, where $N_{x,Rd} \cdot \sin\theta_x$ is the larger of $\lvert N_{1,Rd} \sin\theta_1 \rvert$ and $\lvert N_{2,Rd} \sin\theta_2 \rvert$
KK joint	$N_{i,Sd} \leq N_{i,Rd}$ $N_{i,Rd}$ is the value of $N_{i,Rd}$ for a K joint from Table 6-9, provided that, in a gap joint, at the section 1-1 the chord satisfies: $$\left[\frac{N_{0,Sd}}{N_{0,pl,Rd}}\right]^2 + \left[\frac{V_{0,Sd}}{V_{0,pl,Rd}}\right]^2 \leq 1.0$$

Table 6-22 Design resistances of special types of welded uni-planar joints between RHS bracings and RHS chord

Type of joint	Design criterion
YY joint	$N_{1,Sd} \leq N_{1,Rd}$ $N_{1,Rd}$ is the value of $N_{1,Rd}$ for X joint from Table 6-13
KT joint	$N_{1,Sd} \sin\theta_1 + N_{3,Sd} \sin\theta_3 \leq N_{1,Rd} \sin\theta_1$ $N_{2,Sd} \sin\theta_2 \leq N_{1,Rd} \sin\theta_1$ $N_{1,Rd}$ is the value of $N_{1,Rd}$ for K joint from Table 6-13 with $\dfrac{b_1 + b_2 + h_1 + h_2}{4 b_0}$ substituted by $\dfrac{b_1 + b_2 + b_3 + h_1 + h_2 + h_3}{6 b_0}$
XX joint	$N_{1,Sd} \sin\theta_1 + N_{2,Sd} \sin\theta_2 \leq N_{x,Rd} \sin\theta_x$ $N_{x,Rd}$ is the value of $N_{x,Rd}$ for X joint from Table 6-13, where $N_{x,Rd}$ is the larger value of $\lvert N_{1,Rd} \sin\theta_1 \rvert$ and $\lvert N_{2,Rd} \sin\theta_2 \rvert$
KK joint	$N_{i,Sd} \leq N_{i,Rd}$ $N_{i,Rd}$ is the value of $N_{i,Rd}$ for K joint from Table 6-13 assuming that in the cross section 1-1 the following condition is fulfilled: $\left[\dfrac{N_{0,Sd}}{N_{0,pl,Rd}}\right]^2 + \left[\dfrac{V_{0,Sd}}{V_{0,pl,Rd}}\right]^2 \leq 1.0$

6.6 Lattice girders (trusses)

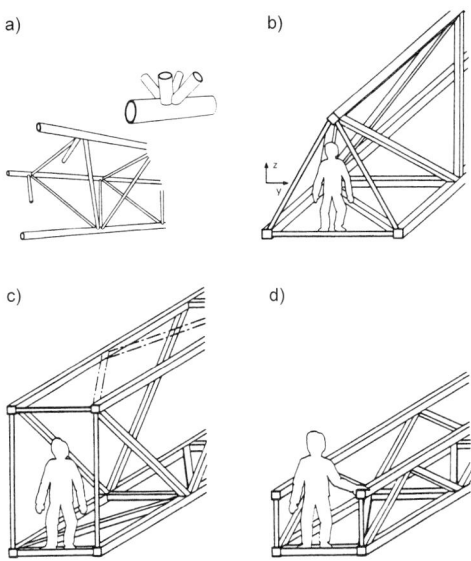

Fig. 6-145 Triangular girder made with a) CHS, b) RHS. Quadrangular girder made with RHS: c) closed form, d) open form, three sided

Fig. 6-146 End piece of a curved triangular girder in the air terminal, Hamburg

the modern representative halls (see Figs. 6-146 and 6-147). They are often used for structures with large spans instead of expensive space structures (see Chapter 9).

Until about ten years, test results on multi-planar joints of hollow sections were not available. Although a large number of test series on TT [98, 134, 135, 241, 243], XX [116] and KK joints [90, 93, 94, 99, 127, 191] as well as numerical investigations on TT [116, 243], TX [116, 240], XX [109, 116] and KK joints [124, 219, 239] were carried out in the last years [see Fig. 6-148], only the following two structural standards have been worked out, which contain the design recommendations for multi-planar joints of hollow sections:

– AWS 1992, D1.1–92 [13]
– Eurocode 3 [230]

The AWS rules are based only on the elastic analysis without taking the test results into account.

Eurocode 3 [230] recommends Table 6-23 (CHS joints) and Table 6-24 (RHS joints) with reduction factors, which have to be multiplied with the design resistances of the cor-

Fig. 6-147 Seven main girders in triangular form covering a hall area of 75 × 101 m in the air terminal, Hamburg

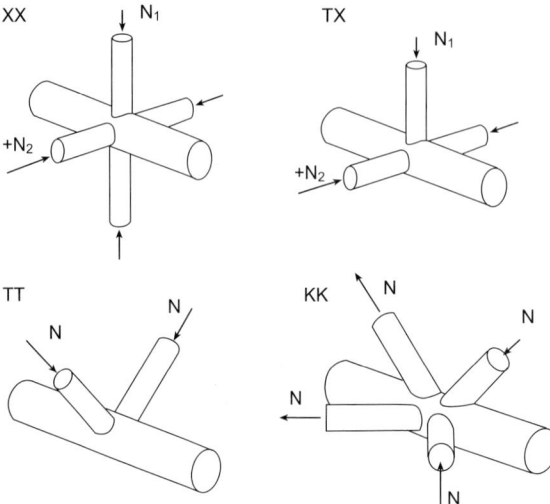

Fig. 6-148 Types of multi-planar joints

responding uni-planar joints in Tables 6-9 and 6-13 in order to determine the design resistances of the multi-planar joints. These factors have been proposed basing on the knowledge acquired by the tests for practical applications on the safe side. The tests and the "finite element" calculations carried out up to now lead to the following conclusions:

• In comparison with the uni-planar X joints, the multiplanar loadings affect the design resistances and the rigidities of the XX joints significantly.

The "finite element" calculations [109] with XX joints of CHS have shown that the design resistance of the XX joint will be reduced to one third of that for the uni-planar X joint, if the loads in the both planes are identical but applied in the opposite sense (tension against compression). However, there is a significant increase in joint resistance when the bracings are loaded in the same sense; the deformation is small at the same time (Fig. 6-149).

Little difference exists between the calculations of the design resistances of RHS XX [222] and CHS XX joint (see Eqs. 6-113 and 6-116).

• Experimental investigations on CHS TT joints ($\phi = 90°$) with compressive loads in both bracings [241] have shown that the design resistances of uni-planar and multi-planar joints do not differ from each other (see Eq. 6-112). For RHP TT joints and $60° \leq \phi \leq 90°$, a little difference was observed and a reduction of 10% is recommended (Eq. 6-115).

• For CHS and RHS KK joints, a reduction factor of 0.9 is recommended in order to simplify the calculation [242] and also to consider the out-of-plane bending moment M_{op}. However, Lit. [230] proposes that the interaction between the axial force and the bending moment in each relevant plane has to be checked:

$$\frac{N_{i,Sd}}{N_{i,Rd}} + \frac{M_{ip,i,Sd}}{M_{ip,i,Rd}} + \frac{M_{op,i,Sd}}{M_{op,i,Rd}} \leq 1.0 \quad (6\text{-}119)$$

For gap joints, the failure mode "chord shear" has to be checked, see Eqs. (6-114) and (6-118).

• Lit. [243] contains a comparison of the CHS joint resistances obtained by the tests on TT [135, 243] and KK [90, 243] joints and those given by Eurocode 3 [230], see also Table 6-23:

– The test results for joint resistances of TT joints are 1.13 to 2.04 times larger than those according to Eurocode 3.
– For KK joints, the joint resistances obtained by tests are 1.40 to 2.14 times larger than those given by Eurocode 3.

6.6 Lattice girders (trusses)

Table 6-23 Reduction factors for CHS multi-planar joints

Type of joint	Reduction factor μ related to uni-planar joints (see Table 6-9)
TT joint	$60° \leq \phi \leq 90°$ $\mu = 1.0$ (6-112)
XX joint	$\mu = 1 + 0.33\, N_{2,Sd}/N_{1,Sd}$ (6-113) Taking account of the sign of $N_{1,Sd}$ and $N_{2,Sd}$, where $\|N_{2,Sd}\| \leq \|N_{1,Sd}\|$
KK joint	$60° \leq \phi \leq 90°$ $\mu = 0.9$ (6-114) Provided that, in a gap joint, at section 1-1 the chord satisfies: $\left[\dfrac{N_{0,Sd}}{N_{pl,0,Rd}}\right]^2 + \left[\dfrac{V_{0,Sd}}{V_{pl,0,Rd}}\right]^2 \leq 1.0$ $V_{0,Sd}$ is the total shear in both planes

This demonstrates that the joint resistances according to Eurocode 3 have been established in a relatively conservative manner on the safe side.

Regarding the manufacturing geometry of the joints, the following comments are made:

- Overlap of the intersecting bracings in both planes of a CHS joint should be avoided as far as possible, as thereby the joint eccentricities can emerge (Fig. 6-150).

The joint eccentricity $e \leq 0.25\, d_0$ can be ignored. Beyond this limit, the moment due to eccentricity is to be distributed in the chord corresponding to its rigidity as is to be done for uni-planar joints.

- Further, the gap between the toes of the adjacent bracings has to be larger than or equal to the sum of the bracing wall thicknesses, so that the welds do not overlap each other.

Table 6-24 Reduction factors for RHS multi-planar joints

Type of joint	Reduction factor μ related to uni-planar joints (see Table 6-13)
TT joint	$60° \leq \phi \leq 90°$ $\mu = 0.9$ \hfill (6-115)
XX joint	$\mu = 0.9\,(1 + 0.33\, N_{2,Sd}/N_{1,Sd})$ (6-116) Taking account of the sign of $N_{1,Sd}$ and $N_{2,Sd}$, where $|N_{2,Sd}| \leq |N_{1,Sd}|$
KK joint	$60° \leq \phi \leq 90°$ $\mu = 0.9$ \hfill (6-117) Provided that, in a gap joint, at section 1-1 the chord satisfies: $\left[\dfrac{N_{0,Sd}}{N_{pl,0,Rd}}\right]^2 + \left[\dfrac{V_{0,Sd}}{V_{pl,0,Rd}}\right]^2 \leq 1.0$ (6-118) $V_{0,Sd}$ is the total shear in both planes

- In a RHS KK joint, the width ratio β is large, if the angle between the bracing member planes $\phi < 90°$ (see Fig. 6-151). If an eccentricity emerges in this case, the design resistance of the tensile chord flange of a multi-planar joint is larger than that of the chord flange of a uni-planar joint with identical sizes of the structural members.

Based on further tests and parameter studies, more precise design resistance formulae have been established for CHS multi-planar joints [116]. Appendix IV contains a number of these formulae. Analogously, further theoretical and experimental investigations on RHS multi-planar joints (also uni-planar joints) have been carried out [266], which have led to the more accurate design formulae for these joints. They are in good agreement with the test results. These design resistances are compiled in Appendix VI.

6.6 Lattice girders (trusses)

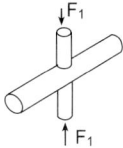
1. uni-planer X joint

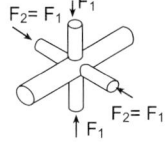
2. XX joints with loads in both planes applied in the same sense

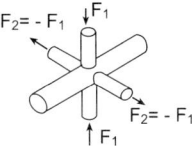
3. XX joints with loads in both planes applied in the opposite sense

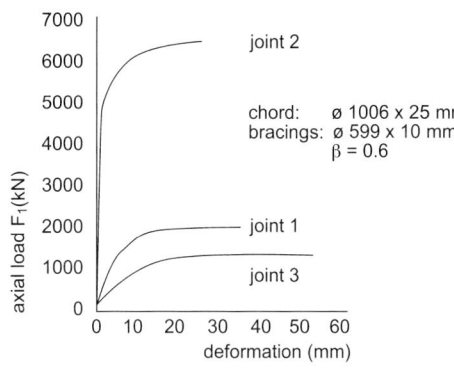

chord: ø 1006 × 25 mm
bracings: ø 599 × 10 mm
$\beta = 0.6$

Fig. 6-149 Effect of loading constellations on the design resistances of CHS XX joints [109]

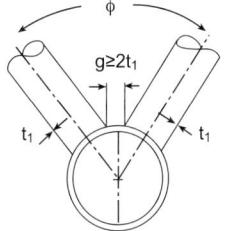

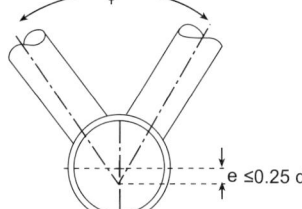

Fig. 6-150 Gap and eccentricity in a multi-planar joint between two planes

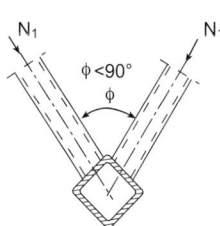

Fig. 6-151 Elevation view of a RHS KK joint with the chord in tension

6.6.8 Joints with bracings of circular hollow sections with flattened ends

In general, the joints of this type are used for small and temporary constructions.

As Fig. 6-152a illustrates, the bracing end preparation for the joints with RHS chord is done with plane cuts. The CHS bracings necessitate "profile" cuts with multi-planar intersection curves at the ends to fit CHS chords (Fig. 6-152b). In special cases, when $\beta = d_i/d_0$ values are not large and the weld gaps are small, the bracing ends can also be plane cut (Fig. 6-152c). In order to avoid "profile" cuts in CHS bracings, they are flattened at their ends fully or partially. This simplifies the fabrication of a truss and thereby reduces its production cost (Fig. 6-152d).

Based on the workshop practice and available equipments, various types of CHS end flattening are used (Fig. 6-153).

With regard to the application of the flattened RHS ends, their fabrication is very complex and they are not therefore applied usually.

When the chords (CHS or RHS) are directly joined to the bracings of CHS with flattened

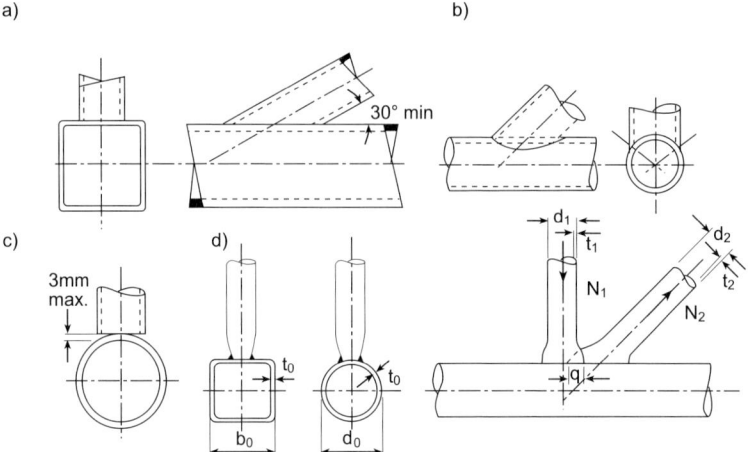

Fig. 6-152 Bracing end preparation for trusses in hollow sections

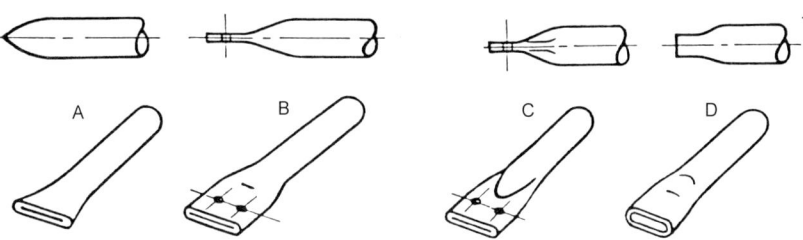

Fig. 6-153 Various forms of the flattened CHS ends. A) Cropped end, B) fully flattened end, C) fully flattened end in a press with a recessed die, D) partially flattened end reducing the stroke of the press and introducing a distance piece into the flattened portion

ends, only plane saw or flame cuts and simple fillet welds are required (Fig. 6-154).

However, detachable bolted joints with gusset plates are also possible. Fig. 6-155 shows such a construction, where the fully flattened bracing ends are reinforced with welded plates in order to compensate for the weakening of the cross section by drilled holes.

In general, the bolted joints of this type are used only as structural parts, which are permanently within a building or also in the open air for a limited time.

At present, there is no standard, which gives recommendations for cold or hot flattening of hollow sections. Cold flattening is used more frequently than flattening in hot condition because of its simplicity, quickness and low costs.

In the following, the hot flattening procedure for hollow sections has been described, which makes the reasons for the increase of fabricating costs evident:

The end area of a hollow section intended for flattening is to be heated up to a temperature range of 750 to 900 °C. Heating can be done by electricity, oxyacetylene torch or butane and propane burners.

Fig. 6-156 shows a hot flattening installation with a series of continuous heaters, which can be financially rewarding when a mass production is required. The danger of crack initiation is less for hot flattening.

During the cold flattening process, the material is plastically deformed and the deformations occur in both longitudinal and transversal directions. Cracks can be produced at the

Fig. 6-154 Welded joint with CHS chord and CHS bracings with flattened ends

Fig. 6-155 Indirectly bolted joint with CHS chord and CHS bracings with flattened ends

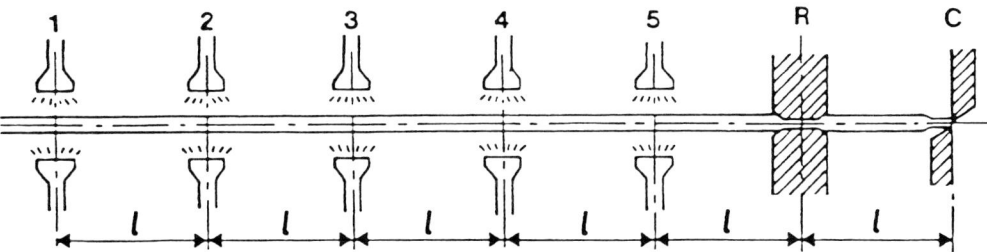
Fig. 6-156 Design arrangement of heaters for a hot flattening installation

flattened edges or at the tapering from the full section to the flattened location. A local peak strain may be over 200%. If cracks occur along the manufacturing weld seam of CHS, they can be avoided by offsetting the weld seam from the line of extreme deformation.

Further, a proper choice of d/t is necessary, as in general, the flattening process is easier with the increase of d/t. It has to be considered that the flattening reduces the compression capacity of a member when $d/t > 25$.

Various flattening equipments are used to manufacture the flattened forms illustrated in Fig. 6-153. The shapes of the dies determine the inclination and the length of the transition from the profile to the flat location for full or partial flattening, the adaptation of which is important to prevent cracks.

1. Cropped ends

Cropping is a very economical procedure, which is carried out by a shear or a guillotine or a notcher corresponding to the CHS sizes. The full flattening occurs at the end of the hollow sections in this process.

As shown in Fig. 6-157 at ①, the workpiece is placed horizontally. The fixed lower blade has to be adjusted so that it projects a distance nearly equal to the radius of the CHS, which prevents a non-symmetrical flattening.

At ②, the machine is a standard plate shear with a fixed blade at table level. The workpiece must be slightly inclined to avoid non-symmetrical flattening.

CHS are cut and fully flattened in one operation by the same tool. With the cut, the bevel for the weld is produced simultaneously (Fig. 6-158).

This process can be applied in a) cold as well as b) hot condition (Fig. 6-159).

For hot flattening with cropping, the heated length is determined based on CHS diameter (Fig. 6-160):

Heated length $l_w \approx 1.7\,d$

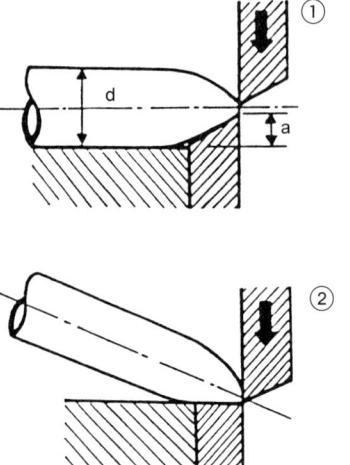

Fig. 6-157 Working principle of a cropping machine

2. Fully flattened ends

Fig. 6-161 illustrates the working principles indicating that a relatively long portion of the member is subjected to full flattening in this process. With a simple die, the length l of the transition zone is recommended to be within the range $1.2\,d \leq l \leq 1.5\,d$. During the flattening process the symmetrical forming takes place gradually, provided that the edges of the dies are rounded off, which prevents any transverse crack.

This process involves a greater risk of cracking and hence the need to keep an eye, specially in the case of cold flattening, on the mechanical properties of steel, particularly the elongation. However, normally the structural steels for the constructional hollow sections have minimum elongation between 22–26% (see Tables 3-3, 3-9 and 3-10).

The transition length l cannot be affected, when the shape of the die shown in Fig. 6-161 is used. The bottom dies shown in Fig. 6-162 determine however the form of the flattened member ends, especially the transition length l, often within the length range between $1.7\,d$ and $2.2\,d$ for the tapered part. The taper cone of CHS to the flattened location has to be lower than 25% (1 : 4).

In order to reduce the risk of crack initiation in general, an inner CHS section can be inserted into the ends of the CHS to be flattened and then the flattening process is undertaken. The reinforcement of the member end is specially favourable for bolted joints.

3. Fully flattened ends in a press with a recessed die

The flattening in this case is done in a press with two recessed dies embodying a gradual change of the tubular section (Fig. 6-163). The length of the transition area is about $2\,d$.

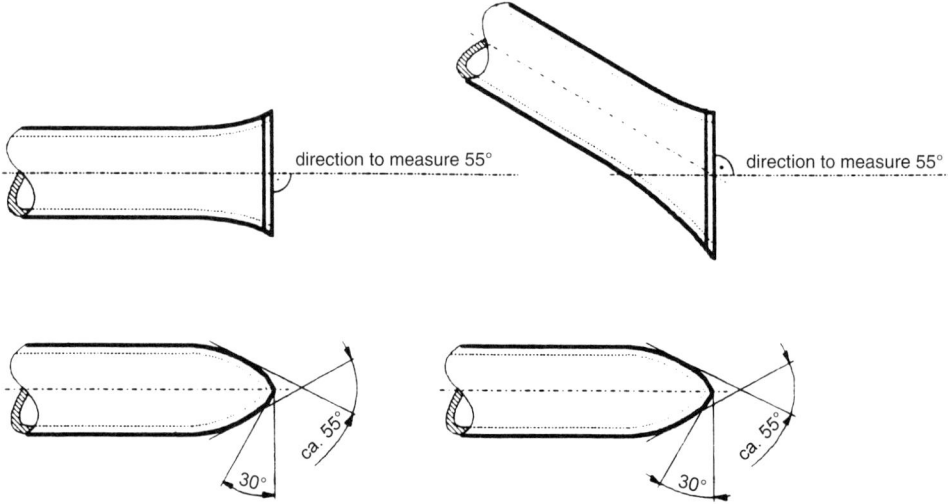

Fig. 6-158 Cropping with simultaneous separation cut

6.6 Lattice girders (trusses)

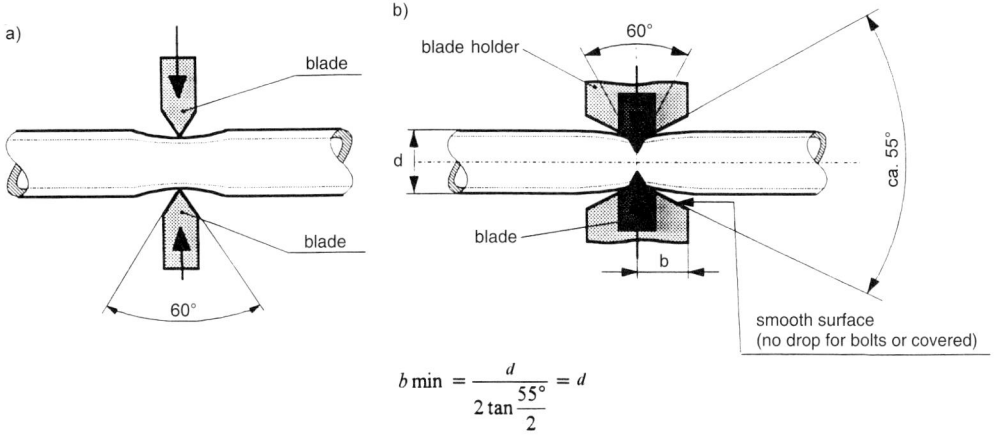

$$b\,\min = \frac{d}{2\tan\dfrac{55°}{2}} = d$$

Fig. 6-159 Cold and hot cropping. a) Cold (room temperature about 20 °C), CHS wall thickness ≤ 5 mm, b) hot (cutting temperature between 800–1200 °C), CHS wall thickness > 5 mm. Initiated grooves > 0.5 mm have to be grinded out or the blade holder has to be shaped, so that no groove is initiated

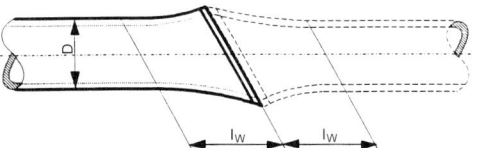

Fig. 6-160 Heated length for hot cropping

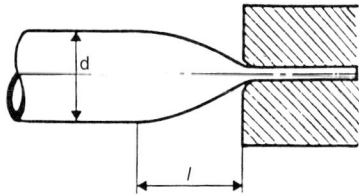

Fig. 6-161 Simple process for full flattening of CHS ends

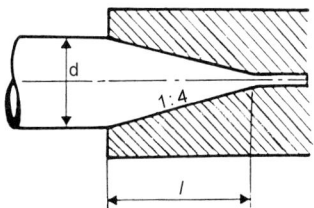

Fig. 6-162 Full flattening of CHS ends with a predetermined transition length l

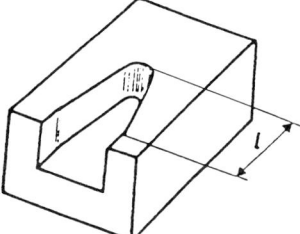

Fig. 6-163 A recessed die

This form is favourable for bolting and has a larger design resistance than the form manufactured applying the simple full flattening dies (Fig. 6-161).

The investment cost for the recessed dies is significantly higher than the plain flat ones. Inspite of that, their use can be economically justified for mass production on account of the case of flattening they impart and their lower rate of wear.

4. Flattened ends introducing a distance piece into the flattened portion (predetermined partial flattening)

This method is similar to the flattening according to Fig. 6-161 but with a limited stroke of the press. In order to obtain truly parallel faces (Fig. 6-164), a distance piece or tongue is introduced into the flattened portion during the flattening operation (Fig. 6-165). Thereby a predetermined partial flattening is achieved, which is often necessary to manufacture the joints shown in Fig. 6-164.

Finally, regarding the fabrication of the flattened CHS ends, the following investigations are described:

As no clearly defined standard concerning the flattening processes exists, it is recommended to carry out a number of pretests to select the appropriate process before starting larger structural projects. Eventually tests must be conducted, in case thickwalled CHS or CHS with large outer diameter is required.

In Canada, the resistances (under predominantly static loading) of the N type truss joints with cropped, flattened CHS bracing ends were investigated [203, 205–208]; the chord members consisted of CHS as well as square hollow sections (Fig. 6-152d). The CHS bracing ends were shaped in cold condition. The preliminary tests on joints shown in Fig. 6-166 led to the following conclusions:

1. The gap joints and the joints with overlap result about equal resistances.
2. The joints with overlap possess double rigidity at the least as the gap joints.
3. The resistances of the overlap joints consisting of bracings with cropped, flattened ends (Fig. 6-152d) are about equal to those of joints with bracings, which have "profile" (Fig. 6-152b) or plane (Fig. 6-152c) cuts at the ends.
4. The direction of flattening (Fig. 6-167)
   a) parallel to the chord axis (in the plane of the truss),
   b) perpendicular to the chord axis does not affect the joint resistances. The fabrication of the joints shown in Fig. 6-166a or c is simpler than that of the joints shown in Fig. 6-166b or d. Therefore, the joint configurations a or c are recommended.

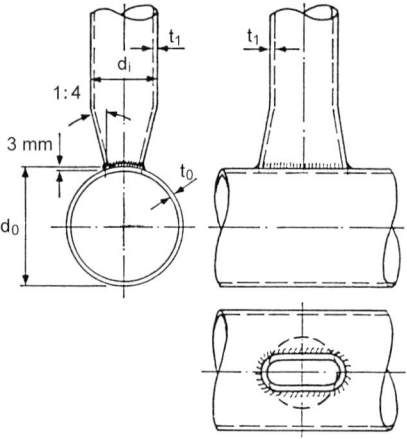

Fig. 6-164 Joint with predetermined, partly flattened CHS bracing end

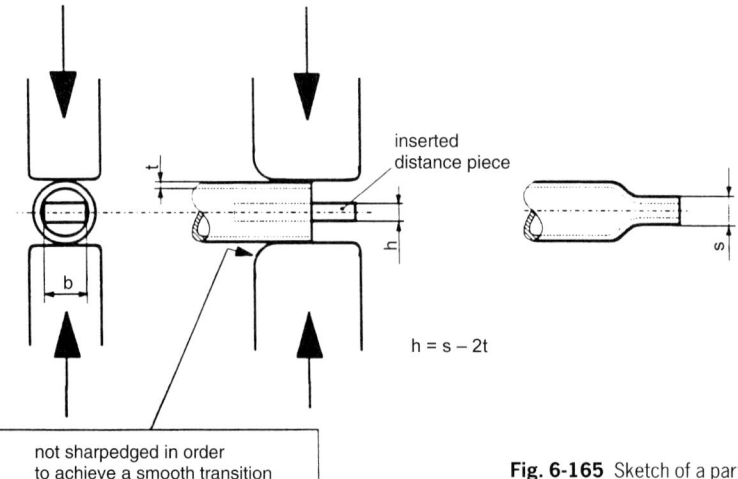

Fig. 6-165 Sketch of a partial flattening device with an inserted distance piece

6.6 Lattice girders (trusses)

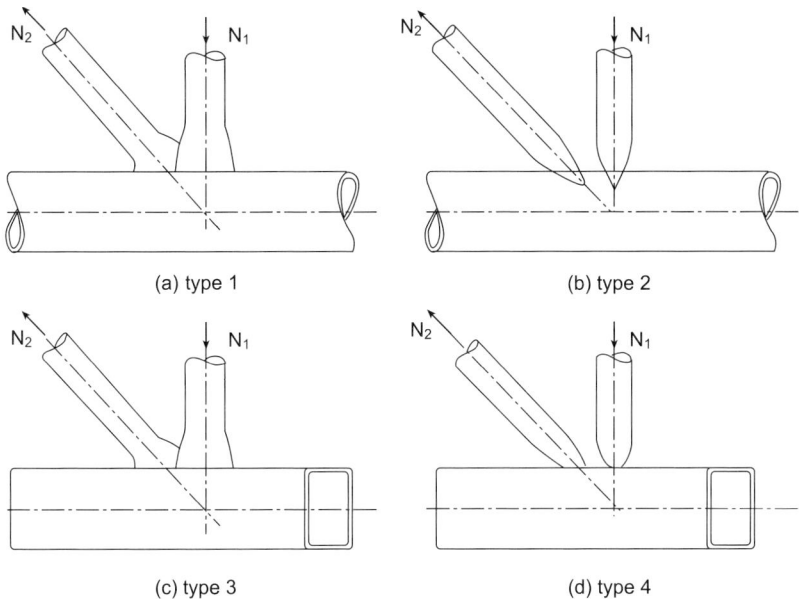

(a) type 1 (b) type 2 (c) type 3 (d) type 4

Fig. 6-166 Single joint test specimens for basic comparative tests: chord ⌀ 101.6 × 101.6 × 4.78 mm or ⌀ 114.3 × 4.78 or 6.53 mm, bracing (tensile) ⌀ 48.3 × 2.79 bzw. 3.81 mm, bracing (compressive) ⌀ 48.3 × 2.79 or 3.81 mm

In joints shown in Fig. 6-167b, the flattened bracing ends prevent the deformation of the chord flange, provided the width of the bracing ends are about equal to that of the chord.

5. The resistances of the joints with square chords are about 10% lower than those of the identical joints with CHS chords.
6. The rigidity of the joints with square chord is about equal to one third of the rigidity of the identical joints with CHS chords.

Based on the results of further tests on the joint type 3 shown in Fig. 6-166 and the following linear regression analysis, the joint resistance formula for N joint with square chord is recommended [203, 207]:

$$\frac{N_{1,Rd}}{t_0 \cdot b_0 \cdot f_{y0}} = 0.504 + 6.10 \left(\frac{d_1}{b_0}\right)^3 -$$
$$- 43.3 \left(\frac{d_1}{b_0}\right)^2 \cdot \left(\frac{t_0}{b_0}\right) \quad (6\text{-}120)$$

where:

$N_{1,Rd}$ = ultimate resistance of the joint (measured in compression bracing)

Fig. 6-167 Directions of the brace end flattening to the chord axis

f_{y0} = yield strength of the square chord material
b_0 = width of the square chord
t_0 = wall thickness of the square chord
d_1 = outer diameter of the CHS bracing in compression

The validity range for the application of this formula is as follows:

- width × depth of the chord member:
  102 × 102 – 152 × 152 mm
- wall thickness of the chord member:
  4.78 – 7.95 mm
- outer diameter of the bracing member:
  3.18 – 6.35 mm
- overlap: $\lambda_{ov} \leq 75\%$
- yield strength of the square chord material:
  ≤ 400 N/mm²
- $\theta_1 = 90°, \theta_2 = 45°$
- t_0/b_0: 0.031 – 0.078
- d_1/b_0: 0.279 – 0.716

The resistances can be read from the curves given in Fig. 6-168 without taking prestressing load N_{op} in the chord into account. The effect of the overlap $\lambda_{ov}\%$ on the resistance is not significant. Similarly, the tests on joint type 1 in Fig. 6-166 have led to the joint resistance formula for N joints with CHS chord [203, 208]:

$$\frac{N_{1,Rd} \cdot d_1}{t_1 \cdot t_0 \cdot d_0 \cdot f_{y0}} = 10.50 + 40.6 \left(\frac{d_1}{d_0}\right)^2 - 172.0 \left(\frac{t_0}{d_0}\right) \quad (6\text{-}121)$$

where:
$N_{1,Rd}$ = ultimate resistance of the joint (measured in compression bracing)
f_{y0} = yield strength of the CHS chord
d_0 = outer diameter of the CHS chord
t_0 = wall thickness of the CHS chord
d_{1or2} = outer diameter of the CHS bracing
t_{1or2} = wall thickness of the CHS bracing

This formula can be applied in the following parameter range:

- outer diameter of the CHS chord d_0:
  114.3 – 168.3 mm
- wall thickness of the CHS chord t_0:
  4.78 – 7.95 mm

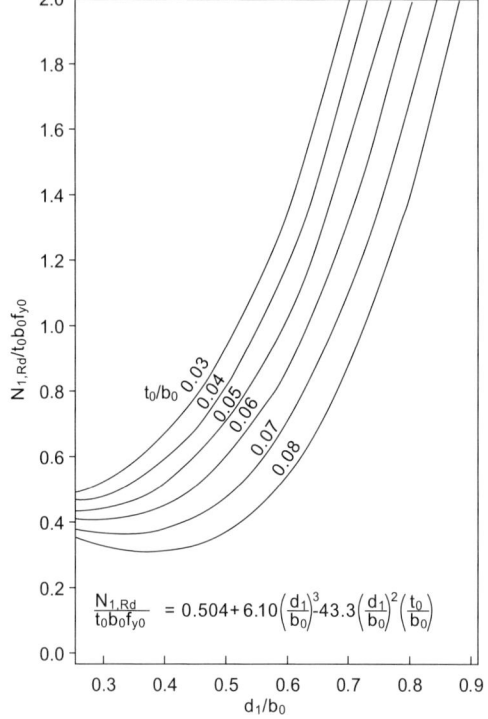

Fig. 6-168 Diagram for joint resistances of N joints with square chord and CHS bracings with cropped and flattened ends [203]

- outer diameter of the CHS bracing d_{1or2}:
  42.2 – 88.9 mm
- wall thickness of the CHS bracing t_{1or2}:
  3.18 – 4.78 mm
- $\theta_1 = 90°$ and $\theta_2 = 45°$
- yield strength of the CHS chord material:
  ≤ 400 N/mm²
- overlap λ_{ov}: $\leq 75\%$
- d_0/t_0: 15 – 36
- d_1/d_0: 0.251 – 0.778

Fig. 6-169 shows the design curves for the joint resistances, which have been set up using Eq. (6-121). Thereby the prestressing load in the chord N_{op} has been ignored.

As for the joints with RHS chord, the effect of overlap is also insignificant in this case.

Due to the safety reasons, the safety factor 1.25 to the Eqs. (6-120) and (6-121) has been proposed [226].

6.6 Lattice girders (trusses)

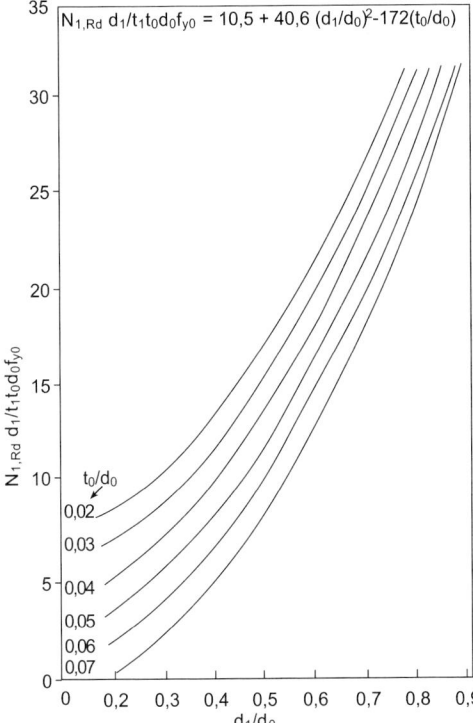

Fig. 6-169 Diagram for joint resistances of N joints with CHS chord and CHS bracings with cropped and flattened ends [203]

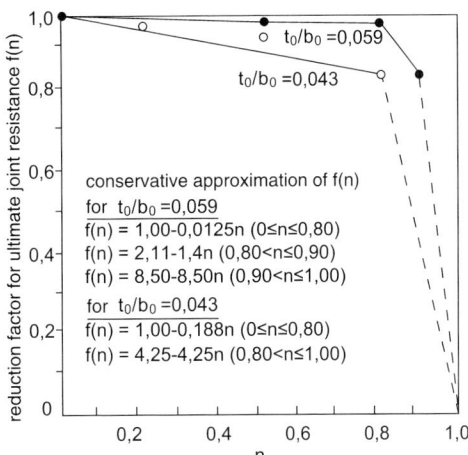

Fig. 6-170 Joint resistance reduction factor $f(n)$ for N joints with square chord taking the prestressing load in the chord in compression $\left(n = \dfrac{N_{op}}{A_0 \cdot f_{y0}} + \dfrac{M_0}{W_0 \cdot f_{y0}}\right)$

As has been already mentioned, the Eq. (6-120) has been established basing on the tests, without taking the prestressing load N_{op} in the chord into account. Further tests [244] to determine the effect of the precompression in the chord (pretension in the chord can be ignored) result in the reduction factor $f(n)$ dependent on n $\left(= \dfrac{N_{op}}{A_0 \cdot f_{y0}} + \dfrac{M_0}{W_0 \cdot f_{y0}}\right)$ for N joints with square chord (Fig. 6-170).

For joints with CHS chord, the effect of the prestressing load N_{op} on the joint resistance is considered as follows [226]: For tensile prestressing load in the chord,

$f(n') = 1.0$ for $n' \geq 0$

For the chords, which are precompressed up to 80% of the yield strength, the joint resistance has to be multiplied with $f(n') = 1 + 0.2\,n'$ ($0 > n' \geq -0.8$). Larger prestress cannot be used, as adequate test results are not available.

The Eqs. (6-120) and (6-121) are modified as follows:

For N joints with RHS chord and CHS bracings with cropped and flattened ends:

$$N_{1,Rd} = \dfrac{t_0\,b_0 \cdot f_{y0}}{1,25}$$
$$\left[0.504 + 6.10\left(\dfrac{d_1}{b_0}\right)^3 - 43.3\left(\dfrac{d_1}{b_0}\right)^2 \cdot \dfrac{t_0}{b_0}\right] f(n)$$
(6-122)

For N joints with CHS chord and CHS bracing with cropped and flattened ends:

$$N_{1,Rd} = \dfrac{t_1 \cdot t_0 \cdot d_0 \cdot f_{y0}}{1,25\,d_1}$$
$$\left[10.50 + 40.6\left(\dfrac{d_1}{d_0}\right)^2 - 172.0\left(\dfrac{t_0}{d_0}\right)\right] f(n')$$
(6-123)

where:
$f(n') = 1.0$ für $n' \geq 0$
$f(n') = 1 + 0.2\,n'$ for $0 \geq n' \geq -0,8$
$n' = \dfrac{N_{op}}{A_0 \cdot f_{y0}} + \dfrac{M_0}{W_0 \cdot f_{y0}}$

Besides N joints, also K joints without gap consisting of square chord and CHS bracings

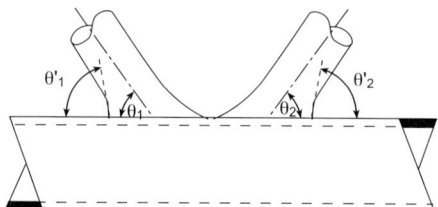
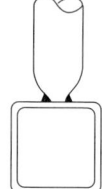

Fig. 6-171 Tested K joint without gap consisting of square chord and CHS bracings with cropped and flattened ends [223]

with cropped and flattened ends (Fig. 6-171) have been investigated experimentally [223].

The tested sizes and the geometrical parameters are given below:

b_0 = 102–152 mm
t_0 = 4–13 mm
d_1 = 42–102 mm
t_1 = 3–6 mm
$f_{y0}, f_{y1,2} \leq 400$ N/mm^2
b_0/t_0 = 12–26
d_1/b_0 = 0.41–0.67
d_1/d_2 = 1.0; $t_1/t_2 = 1.0$
$\theta_1 = \theta_2$ = 60°
gap $g = 0$; overlap $\lambda_{ov} = 0\%$

Analysing theoretically according to the yield line model, Lit. [224] proposes the following formula for the joint resistance of the K joints of this type:

$N_{2,Rd} = 0.4 N_{y1}$

$$\left[1 + 0.021 \frac{b_0}{t_0}\right]\left[1 + 1.71 \frac{d_1}{b_0}\right] \quad (6\text{-}124)$$

where:

$$N_{y1} = \frac{t_0^2 \cdot f_{y0}}{\sin \theta_1}$$

$$\left[\frac{\pi}{2} + \frac{b_1' + 2h_1'}{b_0' - b_1'} + \frac{1.32}{t_0}\sqrt{\frac{f_{y1}}{f_{y0}}} \cdot tg\theta' \cdot b_0' \cdot t_1\right] \cdot f(n)$$

(6-125)

$b_0' = b_0 - t_0$
$b_1' = 2t_1$ (width of flattened bracing cross section) or $2t_1 + 2a$ (a = fillet weld thickness)
$h_1' = \dfrac{\pi(d_1 - t_1) + t_1}{2\sin\theta_1}$

θ' = inclination of the bracing end (Fig. 6-171) related to the chord ($\theta_1' = \theta_1$) can be set on the safe side
$f(n)$ = as in Table 6-11

The design of T and X joints as well as of K joints consisting of CHS chord and CHS bracings with partially flattened ends (Fig. 6-153) can be carried out using the following procedure [245] (see Figs. 6-164 and 6-172):

The joint resistances $N_1.Rd$ of these joints can be determined according to the formulae in Table 6-9, which have to be modified as follows:

– T and X joints (see Eqs. 6-67 and 6-68)
d_1 is to be replaced by $d_{1,\min}$

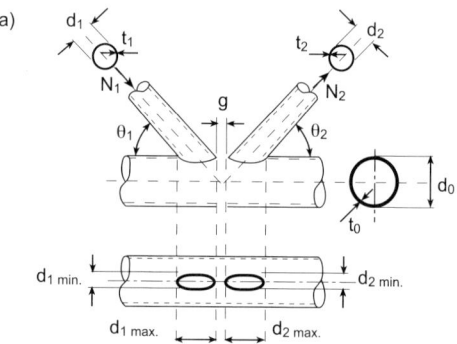

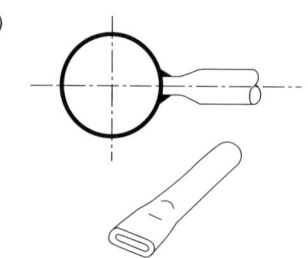

Fig. 6-172 K joint consisting of CHS chord and CHS bracing with partially flattened end

6.6 Lattice girders (trusses)

– K joints with gap (see Eqs. 6-69 to 6-71)

d_i is to be replaced by $\dfrac{d_i + d_{i,\min}}{2}$

Investigations concerning the effective buckling length of CHS compression bracing with cropped and flattened end (see Fig. 6-153A) [203, 246] have demonstrated that l_K is equal to l_0 in a truss with (supposed) pin joints (Fig. 6-173).

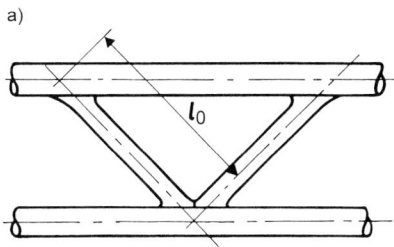

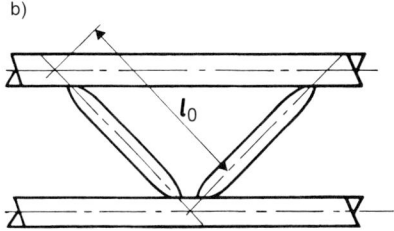

Fig. 6-173 Truss joints with cropped and flattened bracing ends. Direction of flattening. a) Parallel to chord axis, b) perpendicular to chord axis

For the test pieces with the uni-planar plate welded to the CHS members with cropped and flattened ends (high level of restraint), $l_K = 0.62\, l_0$ is recommended.

Due to the lack of adequate test results, it is however proposed to take $l_K = l_0$.

For truss bracings with partially flattened ends welded to CHS chord (Fig. 6-174), it is suggested in Lit. [247] to calculate the effective length of these bracings similarly as the bracings with "profile" cut ends (see Table 5-27), provided the dimension a shown in Fig. 6-174 fulfils the following condition:

$a \geq \dfrac{2}{3} d_{i(1,2)}$

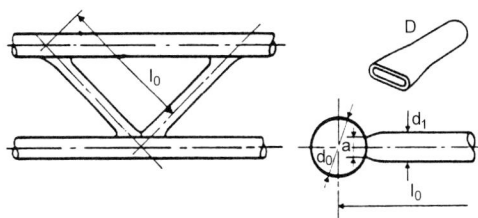

Fig. 6-174 Truss joint with CHS chord and CHS bracings with partially flattened ends

$a \geq \dfrac{1}{3} d_0$

The weld type applied to truss joints with CHS bracings directly welded to CHS or RHS chord depends on the included angle θ and the flattening forms (see Fig. 6-153). The welding can be conducted one-sided or from both sides. Fig. 6-175 shows the recommended weld types a) to d):

a) $\theta \geq 40°$
A – cropping and flattening
either partially or fully penetrated groove weld from one side or fillet weld on both sides

b) $40° > \theta \geq 30°$
A – cropping and flattening
partially or fully penetrated groove weld from the front

c) $\theta \geq 40°$
B and C – full flattening
partially or fully penetrated groove weld from the front or fillet weld on both sides

d) $40° > \theta \geq 20°$
B and C – full flattening
partially or fully penetrated groove weld from the front

In the case of partial flattening (Fig. 6-153 d), the application of fillet weld or of partially or fully penetrated groove weld around the periphery of the flattened end depends on the wall thickness of the CHS bracing (following Fig. 6-95 or 6-97).

Fig. 6-176 or 6-177 shows the application of CHS bracings with flattened ends in triangular or quadrangular girders.

A triangular girder can be manufactured with 3 CHS or 3 RHS chords. A special configura-

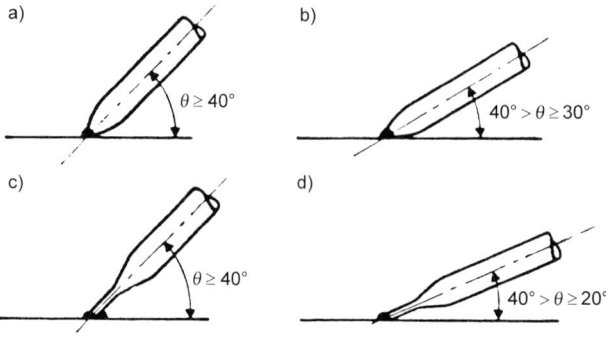

Fig. 6-175 Weld types applied to hollow section joint of CHS or RHS chord and CHS bracings with flattened ends

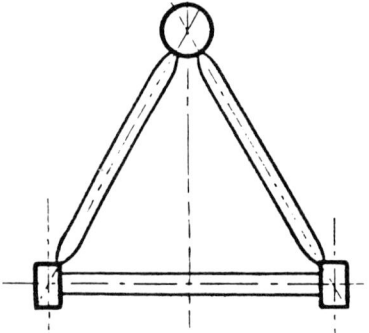

Fig. 6-176 Triangular girder with flattened bracing ends

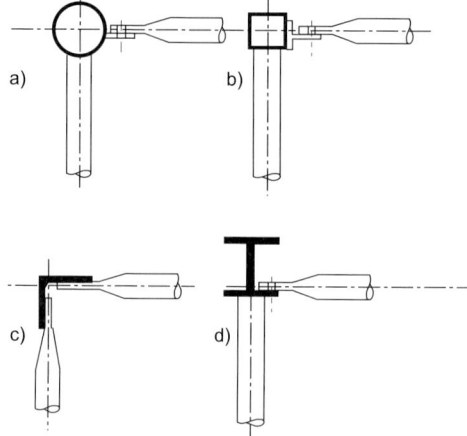

Fig. 6-178 Detachable bolted joints

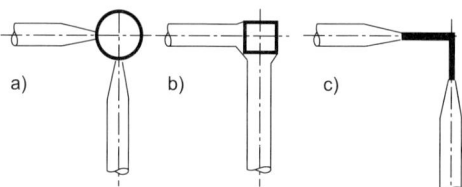

Fig. 6-177 Welded joints in a quadrangular girder

tion is given, when the RHS chords lie horizontally and the flattened bracing ends are welded to the corners of the RHS chord. This construction can be sometimes encountered.

In quadrangular girders, angle section chords are also used instead of CHS and RHS.

Fig. 6-178 illustrates detachable bolted joints, which are often applied in stiffened structures subjected to wind loading. In general these joints are designed according to the rules for the conventional structural engineering.

Simplified details of a joint at the end of a triangular girder before (Fig. 6-179) and after (Fig. 6-180) the bolting operation demonstrate an interesting application.

Fig. 6-181 shows an example of a joint in a space truss consisting of members with flattened ends. In this case, an economical transport of the detachable structural elements and an easy assembly on site are possible by avoiding welding.

6.6.9 Joints with rectangular or square hollow sections (RHS) in a double-chord truss

Fig. 6-182 illustrates three K type truss joints of RHS with double chords, which were investigated theoretically and experimentally in Canada [192, 214, 218]. Tests were conducted on isolated joints as well as on complete

6.6 Lattice girders (trusses)

Fig. 6-179 Plate connection end of a joint in a triangular girder before bolting

Fig. 6-180 Bolted end-to-end connection in a triangular girder

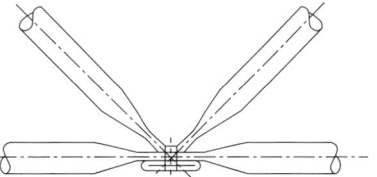

Fig. 6-181 Nodal joint in a space truss

trusses. Also an example of a practical application comes from Canada (Fig. 6-183).

The use of double RHS chord members enables longer clear spans, particularly contrary to the limited span that can be achieved by single chord trusses. This is due to the restricted RHS sizes available in the market. An usual single chord truss using the maximum size of the square hollow section 260 × 260 × 17.5 mm available in Germany can reach a span of about 50 meter. Beyond this span, which may come in question in the case of representative structures such as buildings for recreation centres, exhibition halls etc., the double chord trusses are applicable.

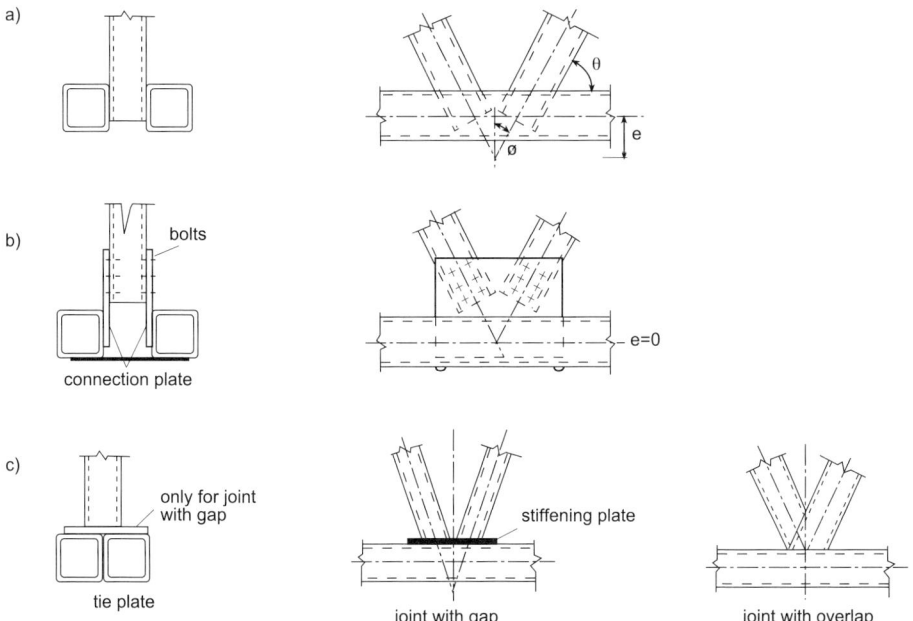

Fig. 6-182 K type joints in a double-chord truss of square hollow sections. a) Welded, separated chord joint, b) bolted, separated chord joint, c) welded, back-to-back chord joint

Fig. 6-183 Double chord girder with welded, separated chord joints (Hamilton Convention Centre, Canada)

In general, the double chord girders have more rigid joints than the girder with single chord. In a double chord girder with gap joints, the force in a bracing member is transmitted to the other one over the stiff inner webs of the chords. Contrary to this, in a gap joint with a single chord, the force from one bracing is transmitted to the other one over the chord flange, which has relatively low stiffness. Due to the high level of restraint of the bracings at the joints in a double chord girder, the effective buckling length l_K of the members is more favourable (smaller) than that in a single chord girder.

As adequate test results are not available, it is proposed to determine the buckling lengths of the members of a double chord girder as it is done for those in a girder with a single chord (see Table 5-27).

Besides, enhanced lateral stiffness of a double chord girder can reduce the number of the lateral stiffeners as well as their requirements.

The evaluation of the test results for all three joint types leads to the following conclusions:

- The welded, separated chord joints (a) failed by inner chord web shear or by chord shear in the gap. This can lead to the limitation of the transmission of the vertical shearing force in the joint to the shearing capacity of the inner chord webs.

- The bolted, separated chord joints (b) as well as the welded "back-to-back" chord joints (c) failed by plastic, local buckling of the chord, which demonstrated that the joints had higher design resistances than the structural members. In the joint type (c) with gap, a tie plate to stiffen the chord flange is required. The plate thickness shall be three times the chord wall thickness. A 100 % overlap of this type of joint is recommended, as it avoids the complexity of the determination of the effective width for partial overlap and also possesses the adequate rigidity.

- Although the joint types (b) and (c) are in no way less resistant than the joint type (a) under static loading, their fabrication is more expensive than that for the joint type (a).

The fabrication of the joint type (a) is particularly economical, as the bracings are set simply between the two chords. The lengths of the bracings and the angles of inclination of their end parts for fitting and welding operations during the manufacture of a girder can be easily checked, as it is not needed to fabricate them very precisely. The economy in the manufacture of these joints is as follows in the rising order:

Cheapest:
1. welded, separated chord joint
2. welded, back-to-back chord joint with full overlap
3. welded, back-to-back chord joint with gap

Dearest:
4. bolted, separated chord joint

It is therefore recommended to use the joint type (a) preferably. The forces and the moments in a double chord girder are calculated assuming pinned joints.

- The design of the joint type (a) is to be carried out as follows:

Failure mode: chord web shear

$$N_{i,Rd} = \left\{ \frac{f_{y0} \cdot A_v}{\sqrt{3} \cdot \sin \theta_i} \right\} \frac{1}{\gamma_{Mj}} \qquad (6\text{-}126)$$

$$A_v = 2.6 h_0 \cdot t_0 \quad \text{für} \quad \frac{h_0}{b_0} \geq 1 \qquad (6\text{-}127)$$

$$A_v = 2.0 h_0 \cdot t_0 \quad \text{für} \quad \frac{h_0}{b_0} < 1 \qquad (6\text{-}128)$$

Eqs. (6-127) and (6-128) take into account the reduced effectiveness of the chord outer side walls in resisting shear forces, at different chord aspect ratios (h_0/b_0).

Failure mode: interaction of axial and shear force in the gap region of the double chord

$$N_{0(in\ gap),Rd} \le \left\{ (2A_0 - A_v)f_{y0} + A_v \cdot f_{y0} \left[1 - \left(\frac{V_{Sd}}{V_{pl,Rd}} \right)^2 \right]^{0.5} \right\} \frac{1}{\gamma_{Mj}} \tag{6-129}$$

where:
A_0 = cross-sectional area of a chord member
A_v = effective shear area (see Eq. (6-127) or (6-128))
V_{Sd} = transversal force $N_{i,Sd} \cdot \sin\theta$ (no purlin load)
$V_{pl,Rd} = \dfrac{f_{y0} \cdot A_v}{\sqrt{3}}$

Following points have to be attended to:
– The joint eccentricity has little effect on joint resistance
– Secondary bending moments are ignored (assuming pinned joints)
– Axial and shear force interaction in the joint gap has to be checked:

$$\left(\frac{N_{Sd}}{N_{pl,Rd}} \right)^2 + \left(\frac{V_{Sd}}{V_{pl,Rd}} \right)^2 \le 1.0$$

6.7 Hollow section joints in Vierendeel trusses loaded by bending moment

Welded T joints appear mainly in frame constructions or frame trusses. Vierendeel trusses as frames were first proposed in 1896 by Arthur Vierendeel in Belgium. They are comprised of parallel or non-parallel chords, which are connected by vertical bracings with each other; they are nearly at 90° to the chords (Fig. 6-184).

Contrary to the trusses with diagonal and vertical bracings, which are designed assuming pinned joints and transmit only axial forces as decisive loads, the chord and vertical mem-

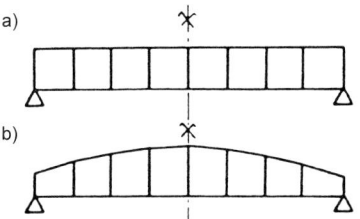

Fig. 6-184 Vierendeel girders with a) parallel, b) not parallel chords

bers in a Vierendeel girder are designed to withstand the bending moment in-plane M_{ip} (predominantly), the axial force N and the shear Q (Fig. 6-185). It has to be especially mentioned that the moment out-of-plane M_{op} does not occur in a Vierendeel girder. On the other hand, multi-planar constructions can be loaded with both M_{ip} and M_{op}.

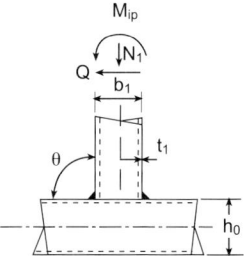

Fig. 6-185 Load types in a Vierendeel girder joint

Fig. 6-186 describes the joint behaviour for "non-linear moment M vs. rotation φ" under predominantly moment loading showing the moment resistance $M_{i,Rd}$, rotational stiffness $c_{i,\varphi}$ and rotational capacity φ_i. These curves can be established by means of tests. Fig. 6-187 shows the sketch of a test rig with measuring gauges [183].

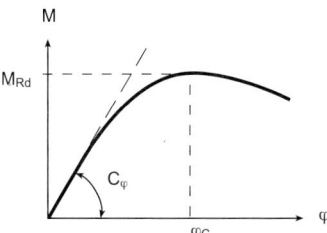

Fig. 6-186 M-φ curve

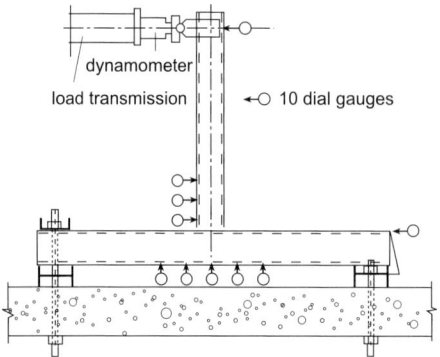

Fig. 6-187 Test rig for T joint under bending moment

6.7.1 RHS T joints under in-plane bending moment M_{ip}

Numerous investigations on square and rectangular single chord connections [7, 83, 86, 87, 89, 119, 121, 171, 174, 183, 190, 193, 194, 198, 202, 209–213, 249–254] loaded by in-plane bending moments have been carried out in various institutes of the world in the last twenty years. The most important objective is the determination of the moment resistance and the rigidity of T joints, which can be manufactured with or without reinforcement (Fig. 6-188).

As tests have shown that the maximum connection moment occurs at excessively large connection deformation, one has to consider the possible failure modes shown in Fig. 6-189, when the ultimate moment resistance is reached. The failure modes c), d) and f) are not further considered here for analytical solutions due to the following reasons:

c) Chord shear failure is strictly a member failure.
d) Bracing in the chord (chord punching shear) was not actually observed in any test, as this occurs only in the shear area between the bracing side weld and the chord wall.
f) Local buckling in the bracing can be avoided by limiting d_1/t_1 ratio.

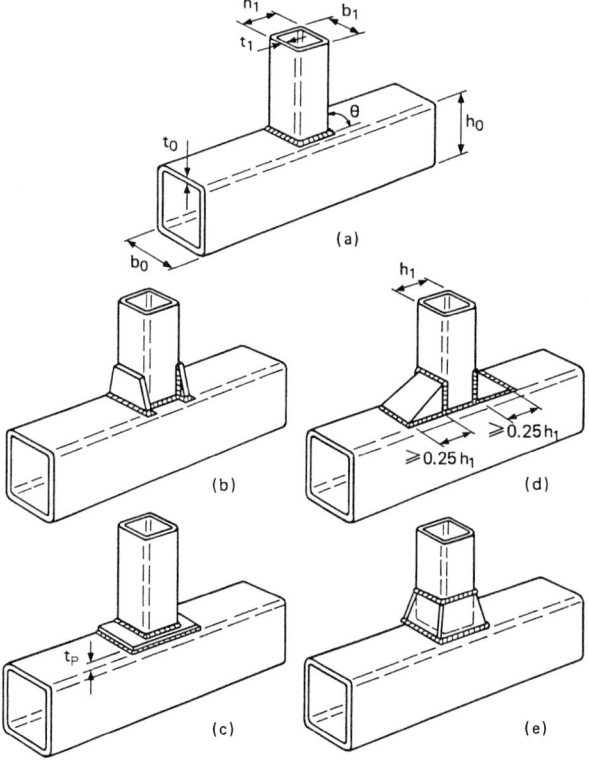

Fig. 6-188 Tested T joints under bending moment

6.7 Hollow section joints in Vierendeel trusses loaded by bending moment

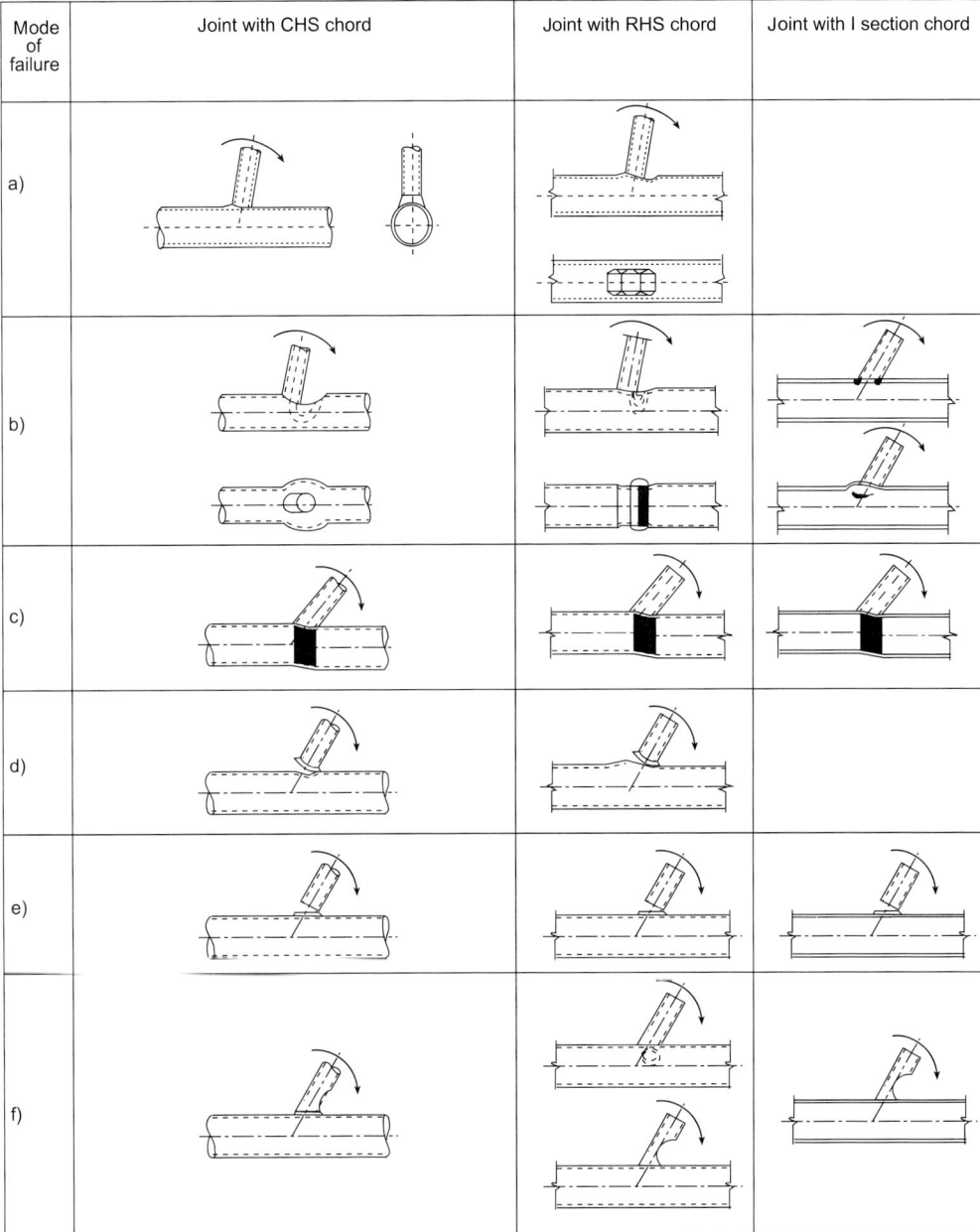

Fig. 6-189 Failure modes of the hollow section joints under moment loading. a) Plastification of chord flange or chord cross section, b) chord web failure by yielding or instability (buckling or crippling of the chord side walls), c) chord shear failure, d) crack in the chord (punching shear), tearing off of the bracing from the chord, e) bracing failure with reduced effective width (crack in weld or in bracing), f) local buckling in bracing

Regarding the dependence of the rigidity of an welded, unreinforced T joint (Fig. 6-190) on the geometrical parameters, the results of the tests on RHS joints show in general that the moment resistance and the flexural rigidity decrease as the chord slenderness ratio b_0/t_0 increases and as the bracing to chord width ratio b_1/b_0 decreases.

According to [210], a $b_1/b_0 \approx 1.0$ and a low b_0/t_0 may lead to almost full rigidity of T joints and the full moment in the bracing may be transmitted to the chord. For $\beta < 1.0$, a significant reduction of the joint rigidity takes place, which means that these joints are only partially stiff (semi-rigid).

However, the level of rigidity of the partially stiff joints can be raised by reducing the ratio b_0/t_0.

Referring to the failure modes a), b) and e) shown in Fig. 6-189, the analytical models [7] were developed and the formulae for the design moment resistances $M_{ip,i,Rd}$ and $M_{op,i,Rd}$ were derived. At present, they have gone into Eurocode 3 [230] and a number of other design recommendations [6, 8].

Failure mode a): plastification of chord flange

Fig. 6-191 illustrates the simple yield line model for RHS T joints loaded by moment in-plane M_{ip} (see also Section 6.6.4.1).

Neglecting the influence of membrane effects and strain hardening, the moment resistance is given by:

$$M_{ip,i,Rd} = 0.5 \cdot f_{y0} \cdot t_0^2 \cdot b_0$$
$$\left\{ 1 + \frac{4 h_1/b_0}{\sin \theta_1 \sqrt{1-\beta}} + \frac{2 (h_0/b_0)^2}{\sin^2 \theta_1 (1-\beta)} \right\} f(n)$$

(6-136)

For T and X joint, Eq. (6-130) in Table 6-25 is established applying $\theta_1 = 90°$.

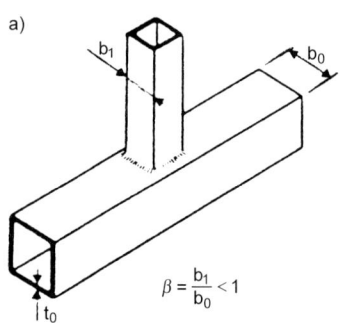

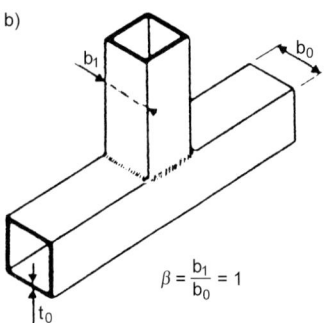

Fig. 6-190 Unreinforced RHS T joint.
a) $\beta < 1.0$, b) $\beta = 1.0$

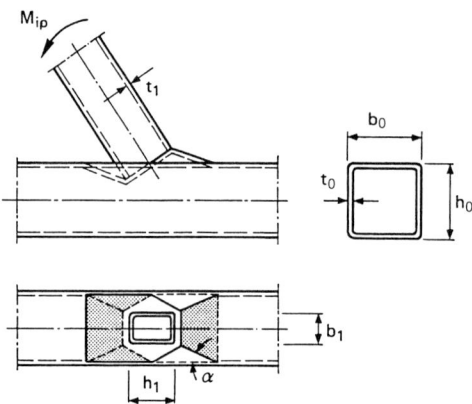

Fig. 6-191 Yield line model for the failure mode a) plastification of chord flange [7]

Failure mode b): chord web failure

Fig. 6-192 shows the model illustrating the strength and stability behaviour of the chord webs (compare with Section 6.6.4.1) assuming a plastic stress distribution. Similar to the joints with axial forces in the members, the calculation is based on the material yield stress f_k (compare with Table 6-13) resulting in Eq. (6-131) of Table 6-25.

6.7 Hollow section joints in Vierendeel trusses loaded by bending moment

Table 6-25 Design moment resistances for in-plane bending $M_{ip,i,Rd}$ and for out-of-plane bending $M_{op,i,Rd}$ in RHS T and X joints

T and X joint	Design moment resistance (i = 1 or 2)	
Moment in plane ($\theta \approx 90°$)	Chord flange plastification	$\beta \leq 0.85$
	$M_{ip,i,Rd} = f(n) \, f_{y0} \, t_0^2 \, h_i \left[\dfrac{1}{2 h_i/b_0} + \dfrac{2}{\sqrt{1-\beta}} + \dfrac{h_i/b_0}{1-\beta} \right] \left[\dfrac{1.1}{\gamma_{Mj}} \right]$	(6-130)
	Chord web failure (web crippling)	$0.85 < \beta \leq 1.0$
	$M_{ip,i,Rd} = 0.5 \, f_k \, t_0 \, (h_i + 5 t_0)^2 \, [1.1/\gamma_{Mj}]$	(6-131)
	$f_k = f_{y0}$ for T joint	
	$f_k = 0.8 \, f_{y0}$ for X joint	
	Bracing failure (effective width)	$0.85 < \beta \leq 1.0$
	$M_{ip,i,Rd} = f_{yi} \, [W_{pl,i} - (1 - b_e/b_i) \, b_i \, h_i \, t_i] \, [1.1/\gamma_{Mj}]$	(6-132)
Moment out-of plane ($\theta \approx 90°$)	Chord web failure	$0.85 < \beta \leq 1.0$
	$M_{op,i,Rd} = f_k \, t_0 \, (b_0 - t_0)(h_i + 5 t_0) \, [1.1/\gamma_{Mj}]$	(6-133)
	$f_k = f_{y0}$ for T joint	
	$f_k = 0.8 \, f_{y0}$ for X joint	
	Chord distortional failure (T joints only) [a]	$\beta \leq 0.85$
	$M_{op,i,Rd} = 2 \, f_{y0} \, t_0 \, [h_i \, t_0 + (b_0 \, h_0 \, t_0 \, (b_0 + h_0))^{0.5}] \, [1.1/\gamma_{Mj}]$	(6-134)
	Bracing failure (effective width)	$0.85 < \beta \leq 1.0$
	$M_{op,i,Rd} = f_{yi} \, [W_{pl,i} - 0.5(1 - b_e/b_i)^2 \, b_i^2 \, t_i] \, [1.1/\gamma_{Mj}]$	(6-135)
Parameter b_e and $f(n)$		Validity range
$b_e = \dfrac{10}{b_0/t_0} \cdot \dfrac{f_{y0} \, t_0}{f_{yi} \, t_i} \, b_i$ whereby $b_e \leq b_i$	For $n > 0$ (compression): $f(n) = 1.3 - \dfrac{0.4 \, n}{\beta}$ but $f(n) \leq 1.0$ For $n \leq 0$ (tension): $f(n) = 1.0$	$\dfrac{b_0}{t_0}$ and $\dfrac{h_0}{t_0} \leq 35$ $\theta = 90°$ $\dfrac{b_i}{t_i} \leq 1.1 \sqrt{\dfrac{E}{f_{yi}}}$

[a] This failure mode does not apply, when the chord deformation is prevented by other means.

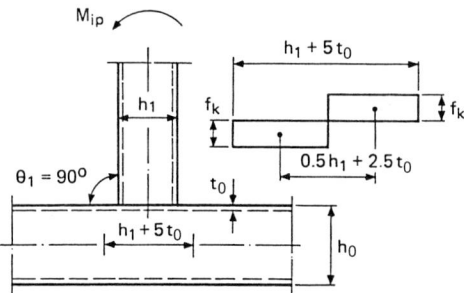

Fig. 6-192 Model for strength and stability behaviour of the chord web in RHS joint [7]

Failure mode e): bracing failure with reduced effective width

In this case, the effective width b_e is to be determined exactly as it is done for truss joints with axial forces in the members (see Table 6-13 and Section 6.6.4.1). The moment resistance can be calculated using Eq. (6-132).

The plastic section modulus W_{pl} has to be applied in the case of compact bracing members. In other cases, the elastic section modulus W_{el} is to be used.

As a concluding remark, it is to be mentioned here that CHS bracings should not be used to transmit moments in T joints, as hereby a low joint rigidity occurs and the load transmission behaviour is consequently unfavourable.

6.7.2 RHS X joints under in-plane bending moment M_{ip}

As Table 6-25 demonstrates, the moment resistance $M_{ip,i,Rd}$ of X joint is calculated exactly as for the T joints. The only difference lies in the Eq. (6-131) for the chord web failure, where $f_k = f_{y0}$ is substituted by $f_k = 0.8 f_{y0}$.

6.7.3 Interaction between axial force N_i and in-plane bending moment M_{ip} in RHS T and X joints

The effect of the axial force on the joint bending moment resistance depends on the critical failure mode. A set of interaction formulae can be proposed, which are partially very complex. Consequently, it is conservatively proposed that a linear interaction relationship be used to reduce the in-plane moment resistance on the safe side:

$$\frac{N_{i,Sd}}{N_{i,Rd}} + \frac{M_{ip,i,Sd}}{M_{ip,i,Rd}} \leq 1 \qquad (6\text{-}137)$$

6.7.4 Design of welded RHS T joints under axial force N_i and in-plane bending moment M_{ip} with the aid of the design diagrams

Leaning against DIN 18808 [5], this design procedure was developed in the framework of a research project by the Testing Centre of Steel, Timber and Stone of the University of Karlsruhe [174, 255]. This is based on the assumption that adequate joint resistance can be obtained, when existing $\left(\dfrac{t_u}{t_a}\right) \geq$ required $\left(\dfrac{t_u}{t_a}\right)$.

In this case,

t_u = wall thickness of the member set below (for T joint, $t_u \triangleq t_o$)

t_a = wall thickness of the member set above (for T joint, $t_a \triangleq t_i$)

In order to determine the resistances $M_{ip,Rd}$ and $N_{i,Rd}$, the design diagrams (Figs. 6-193 and 6-194) for the efficiencies μ_M (bending moment) and μ_N (axial force) against

$$\kappa = \frac{b_0 \cdot t_i}{t_0^2} \cdot \frac{f_{yi}}{f_{y0}}$$

have been established. The validity ranges for the application of these diagrams are shown in Table 6-26.

The efficiencies μ_M and μ_N are defined as follows:

$$\mu_M = \frac{M_{ip,a,Rd}}{M_{pl,a}} \qquad (6\text{-}138)$$

$$\mu_N = \frac{N_{a,Rd}}{N_{pl,a}} \qquad (6\text{-}139)$$

where:

$M_{ip,a,Rd}$ = in-plane bending moment resistance (yield load) in the connection zone of the member set above (determined by tests)

$N_{a,Rd}$ = axial force resistance (yield load) in the connection zone of the member set above (determined by tests)

$N_{pl,a} = f_{ya} \cdot A_a$
$M_{pl,a} = f_{ya} \cdot W_{pl,a}$

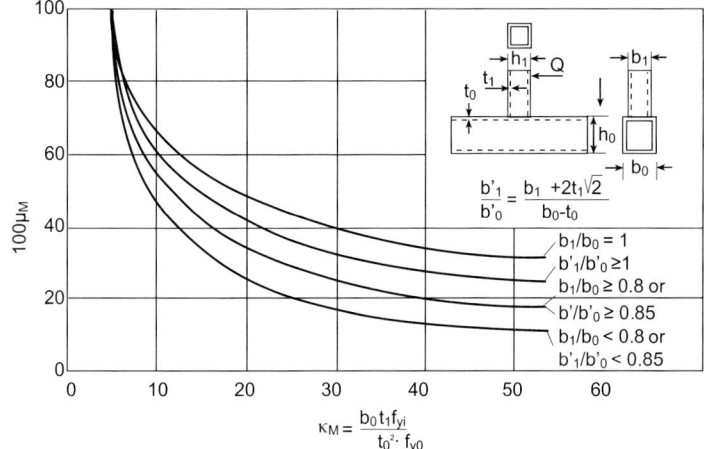

Fig. 6-193 Efficiency of T joints under in-plane bending moment [174]

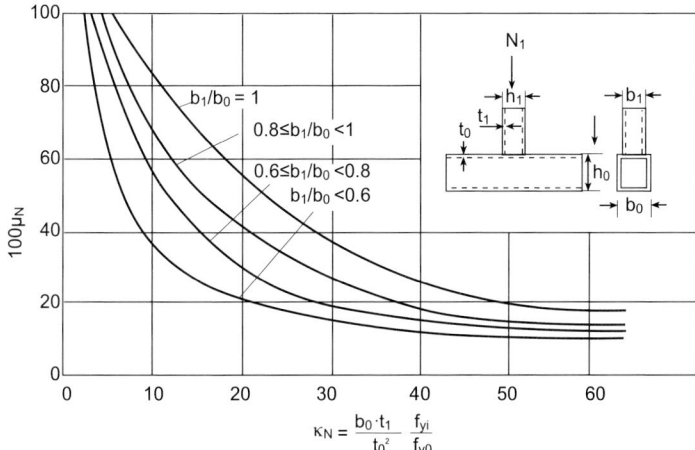

Fig. 6-194 Efficiency of T joints under axial force [174]

$W_{pl,a}$ = plastic section modulus of the member set above

A_a = cross-sectional area of the member set above

For the existing in-plane bending moment and the axial force, the corresponding μ_M and μ_N can be determined as follows:

$$\text{existing } \mu_M = \frac{\gamma_F \cdot \text{existing } M_{ip}}{M_{pl,i}} \qquad (6\text{-}140)$$

$$\text{existing } \mu_N = \frac{\gamma_F \cdot \text{existing } N_i}{N_{pl,i}} \qquad (6\text{-}141)$$

The values of κ_M and κ_N can be read from the diagrams (Figs. 6-193 and 6-194).

The required wall thickness ratio can be calculated using the following equations:

$$\text{required}\left(\frac{t_0}{t_i}\right) = \frac{1}{\kappa_M} \cdot \frac{b_0}{t_0} \cdot \frac{f_{yi}}{f_{y0}} \qquad (6\text{-}142)$$

and

$$\text{required}\left(\frac{t_0}{t_i}\right) = \frac{1}{\kappa_N} \cdot \frac{b_0}{t_0} \cdot \frac{f_{yi}}{f_{y0}} \qquad (6\text{-}143)$$

Besides, the interaction conditions given in Table 6-27 have to be fulfilled in order to meet the combined effect of the bending moment and the axial force.

Regarding the effect of additional axial compression in the chord, the tests led to the con-

Table 6-26 Validity ranges of the RHS sizes in T joints to apply the diagrams (Figs. 6-193 and 6-194)

Parameter	Validity range
b h	≤ 300 mm ≤ 300 mm
b/h	$0.4 \leq b/h \leq 2.0$
t	$t \geq 3.0$ mm S235: $t \leq 30$ mm S355: $t \leq 25$ mm
b_i/t_i for bracings in compression	S235: $b_i/t_i \leq 43$ S355: $b_i/t_i \leq 36$
b_0/t_0	S235: $b_0/t_0 \leq 35$ S355: $b_0/t_0 \leq 35$
b_i/b_0	S235: $b_i/b_0 \geq 0.35$ S355: $b_i/b_0 \geq 0.35$

clusion that the loss of the joint resistance in comparison with the T joints without additional axial force in the chord can be estimated with good approximation by multiplying with a reduction factor μ_0:

$$\mu_0 = 1.2 - \frac{0.5}{b_i/b_0}\left(f_0/f_{y0}\right); \; \mu_0 \leq 1.0 \quad (6\text{-}144)$$

where f_0 is the largest existing compressive stress in the connection zone of the chord.

Table 6-28 contains the formulae to determine the moment or axial force resistance of RHS T joints when each of the load types are applied alone (with or without additional compression in the chord). Simplified linear interaction relations for combined bending moment and axial force are given in Table 6-27.

Table 6-27 Interaction relations for combined loadings (axial load and bending moment)

	$\dfrac{\gamma_F \cdot V_{Sd}}{V_{pl}} \leq \dfrac{1}{3}$
$\dfrac{\gamma_F \cdot N_{Sd}}{N_{Rd}} \leq \dfrac{1}{11}$	$\dfrac{\gamma_F \cdot M_{Sd}}{M_{Rd}} \leq 1$
$\dfrac{1}{11} < \dfrac{\gamma_F \cdot N_{Sd}}{N_{Rd}} \leq 1$	$\dfrac{\gamma_F \cdot M_{Sd}}{1.1\, M_{Rd}} + \dfrac{\gamma_F \cdot N_{Sd}}{N_{Rd}} \leq 1$

Table 6-28 Bending moment and axial force resistances of RHS T joints

Joint and loading type		Joint design resistance $N_{i,Rd}$, $M_{ip,i,Rd}$
tension		$N_{i,Rd} = \mu_N \cdot N_{pl,i} \cdot 1/\sin\varphi$ (6-145)
compression		$N_{i,Rd} = \mu_N \cdot \mu_0 \cdot N_{pl,i} \cdot 1/\sin\varphi$ (6-146)
tension	M	$M_{ip,Rd} = \mu_M \cdot M_{pl,i}$ (6-147)
compression	M	$b_1/b_0 = 1$: $M_{ip,i,Rd} = \mu_M \cdot M_{pl,i}$ $b_1/b_0 < 1$: $M_{ip,i,Rd} = \mu_M \cdot \mu_0 \cdot M_{pl,i}$ (6-148)

6.7.5 CHS T, Y and X joints under M_{ip}

A number of tests on CHS joints under in-plane bending moment M_{ip} were carried out in the framework of various research works [131, 132, 145, 146, 264], although not so extensively as on the CHS truss joints under axial loads. Assuming that the structural members are not critical and the welds are sufficiently strong, only the failure modes a) plastification of chord cross section and e) crack in weld or in bracing among those shown in Fig. 6-189 have been investigated analytically in order to develop the formulae for the design moment resistances (see Table 6-29).

In order to simplify the design procedure, the diagram "efficiency c_{ip} vs. d_0/t_0" is made (Fig. 6-195):

$$C_{ip} = \frac{M_{ip,i,Rd}}{M_{pl,i}}$$

6.7 Hollow section joints in Vierendeel trusses loaded by bending moment

Table 6-29 Design resistances for in-plane bending moment $M_{ip,i,Rd}$ and out-of-plane bending moment $M_{op,i,Rd}$ in CHS joints

T, X and Y joint	Design resistances (i = 1 or 2)
	Chord plastification
(figure: T-joint with $M_{ip,1}$, d_1, θ_1, N_{op}, t_0, d_0)	$M_{ip,i,Rd} = 4.85 \dfrac{f_{y0} \, t_0^2 \, d_i}{\sin \theta_i} \sqrt{\gamma} \, \beta \, f(n') \left[\dfrac{1.1}{\gamma_{Mj}}\right]$ (6-149)
K, N, T, X and Y joint	**Chord plastification**
(figure: joint with $M_{op,1}$, d_1, θ_1, N_{op}, t_0, d_0)	$M_{op,i,Rd} = \dfrac{f_{y0} \, t_0^2 \, d_i}{\sin \theta_i} \dfrac{2.7}{1 - 0.81 \beta} f(n') \left[\dfrac{1.1}{\gamma_{Mj}}\right]$ (6-150)

K and N gap joint and T, X and Y joint   chord punching shear

For $d_i \leq d_0 - 2\,t_0$: $M_{ip,i,Rd} = \dfrac{f_{y0} \, t_0 \, d_i^2}{\sqrt{3}} \dfrac{1 + 3 \sin \theta_i}{4 \sin^2 \theta_i} \left[\dfrac{1.1}{\gamma_{Mj}}\right]$ (6-151)

$M_{op,i,Rd} = \dfrac{f_{y0} \, t_0 \, d_i^2}{\sqrt{3}} \dfrac{3 + \sin \theta_i}{4 \sin^2 \theta_i} \left[\dfrac{1.1}{\gamma_{Mj}}\right]$ (6-152)

Factor $f(n')$	Validity range
For $n' > 0$ (compression): $f(n') = 1 - 0.3\,n'(1 + n')$, but $f(n') \leq 1.0$ For $n' \leq 0$ (tension): $f(n') = 1.0$	see Table 6-10

The upper efficiency limit is given by the horizontal line based on the punching shear failure. The condition $d_1 \leq d_0 - 2\,t_0$ is to be fulfilled.

As in RHS joints, the rotational rigidity $c_{i,\varphi}$ in a statically undetermined system (e.g. frames or Vierendeelgirder) is significantly affected by the moment distribution. The joint rigidity under bending load can be calculated by means of an elastic "finite element" computer program. DNV [256] proposes the following equation to calculate the CHS joint rotational rigidity $c_{i,\varphi}$ for in-plane bending moment:

$$\frac{C_{i,\varphi}}{E \cdot d_0^3} = 0.054 \left(\frac{1}{\gamma} - 0.01\right)^{2.35 - 1.5\beta}$$

The joint rigidity of CHS T joints loaded by in-plane bending moment is graphically illu-

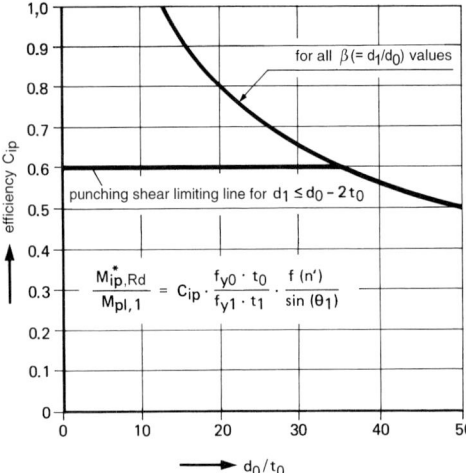

Fig. 6-195 Design diagram for CHS joints (T, Y, X) loaded by in-plane bending moment [226]

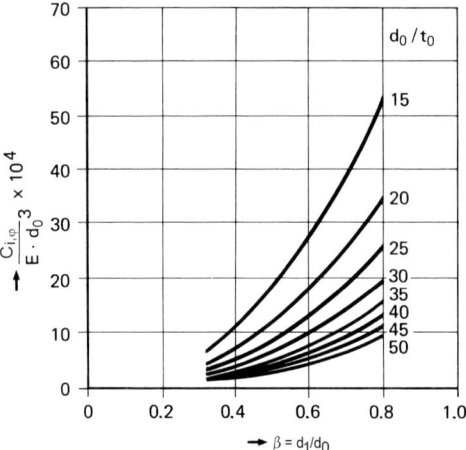

Fig. 6-196 Rotational rigidity of CHS T joints loaded by in-plane bending moment [226, 257]

strated in Fig. 6-196 [257]. The influence of the axial force on the joint rigidity is however considerable [112].

Eq. (6-137) describes the interaction of axial force and in-plane bending moment on CHS joints with the following modification:

$$\frac{N_{i,Sd}}{N_{i,Rd}} + \left(\frac{M_{ip,i,Sd}}{M_{ip,i,Rd}}\right)^2 \leq 1$$

6.7.6 RHS T joints in Vierendeel girders

The fundamentals regarding the properties of the Vierendeel girders and their joints have already been described in Sections 6.7 and 6.7.1 (see Figs. 6-184 to 6-186). The joint moment resistances can be calculated using Eqs. (6-130) to (6-132) in Table 6-25.

The T joints in these girders must be very rigid also without any additional stiffening device. Tests have shown that the transmission of the full bracing moment to the joint is allowed provided $\beta = b_1/b_0 = 1.0$ together with an adequately low b_0/t_0 ratio.

The decisive geometrical ratios b_0/t_0 and t_0/t_i (for $\beta = 1.0$ and identical steel grade for bracings and chord) can be determined according to [7], where the joint moment resistance reaches the full moment resistance of the bracing.

1. Moment resistance of T joints corresponding to the failure mode "chord web failure" with $h_0 = b_0 = h_1 = b_1$ (with identical sizes of the bracing and chord in square hollow sections) and $h_0/t_0 \leq 16$:

$$M_{ip,1,Rd} = 12 \cdot f_{y0} \cdot t_0^2 \cdot h_1$$

2. Plastic moment resistance of square hollow section bracing is given by (compare with Table 5-1):

$$M_{pl,1} = 1.5 b_1^2 \cdot t_1 \cdot f_{y1}$$

3. 1 and 2 yield:

$$\frac{M_{ip,1,Rd}}{M_{pl,1}} = \frac{8}{b_0/t_0} \cdot \frac{f_{y0}}{f_{y1}} \cdot \frac{t_0}{t_1} \quad (6\text{-}153)$$

4. Assuming $b_0/t_0 = 16$ and $t_0/t_1 = 2$, $M_{ip,1,Rd} = M_{pl,1}$

It is however well known that the axial forces occur in the bracings of a Vierendeel girder, which may be quite substantial.

Considering the interaction relation Eq. (6-137), the design moment resistance $M_{ip,i,Rd}$ has to be reduced due to the influence of the axial force in the bracing. In this case, the joint moment resistance $M_{ip,i,Rd}$ can exceed the plastic moment resistance $M_{pl,i}$ of the bracing. These joints can be taken to be fully rigid, when they are applied to Vierendeel gir-

6.7 Hollow section joints in Vierendeel trusses loaded by bending moment

ders. Otherwise, these joints can be considered as partially stiff or semi-rigid.

In order to determine the forces and moments in a construction with semi-rigid joints, the load vs. deformation behaviour has been investigated (see Fig. 6-186 or 6-187). This can also be done theoretically with the aid of "finite element" analysis.

Vierendeel girders with hollow sections (predominantly RHS) are frequently used due to architectural reasons, especially when the diagonal bracings are not desired. Because of the predominantly occurring bending moments, rigid frames are heavier than the corresponding lattice girders. As the Vierendeel girders are constructed with a statically indeterminate system, it is not simple to calculate them manually. They are mostly done with the aid of computer programs.

Usual approximation procedure for the calculation of Vierendeel girders is based on the assumption that the joints possess high rigidity. As a rule, this is not the case for hollow section joints. The deviation of the calculated values from the actual values is therefore large. Recently, the manual design procedure for a Vierendeel girder has been developed, which delivers relatively reliable values [6, 174, 226].

The moment distribution in the girder with fully rigid joints is relatively uniform (Fig. 6-197a). The moments in the verticals and the axial forces in the middle of the chord are maximum, while the deflection of the girder is small.

With partially rigid joints, the bending moment attains the maximum value in the middle of the girder, while the axial forces become less (Fig. 6-197b).

The total deflection of the Vierendeel girder becomes decisive for the calculation with the increasing joint deformation capacity. However the calculation procedure based on allowable girder deflection is more complex [174].

Various types of joint reinforcements shown in Fig. 6-188 can be applied to reduce the deflection of a Vierendeel girder. A number of the investigations on reinforced frame joints has been described in Lit. [183].

The reinforcements lead to an enlarged joint resistance as well as a higher rigidity related to unreinforced joints (Fig. 6-188a).

Lit. [183] evaluates the resistance and rigidity behaviour of the above mentioned reinforced frame joints as follows:

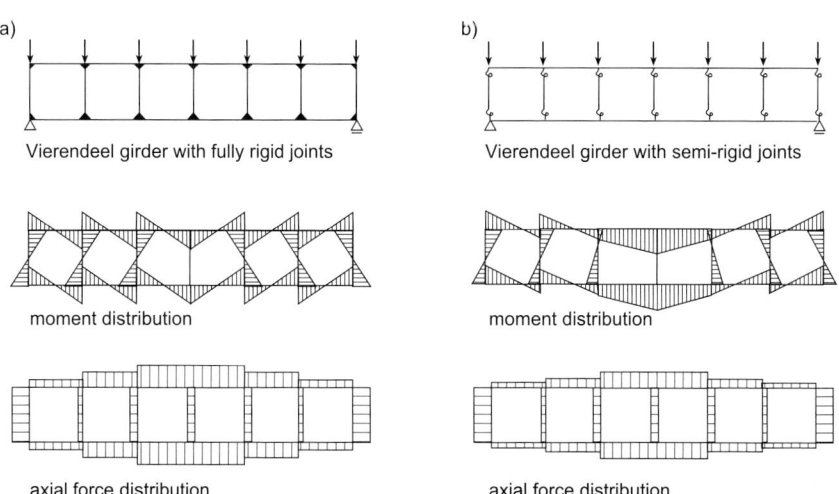

Fig. 6-197 Distribution of axial force and bending moment in a Vierendeel girder

1. Fig. 6-188 b – with bracing plate stiffeners:

For $b_1/b_0 \geq 0.75$, the joint resistance is disappointing; for $b_1/b_0 < 0.75$, the joint resistance is larger than those for other joint types, however at a low stress level.

The rigidity of this joint type is less than those of other joint types. Basically, this joint type is not recommended for Vierendeel girders.

2. Fig. 6-188 c – with chord flange stiffener

No buckling in the chord flange has been observed for $t_P = 2\ t_0$.

Transmission of fully plastic bracing moment to the joint is possible without any significant reduction of the joint rigidity.

For $b_1/b_0 \geq 0.75$, the joint resistance is sufficiently large; however, the joint rigidity is not as high as those for the joint types d) and e).

For $b_1/b_0 < 0.75$, the joint rigidity is about equal to that of the joint type d), although the joint has to be classified as partially rigid.

Considering the joint resistance and rigidity, this joint type can be recommended for application.

Fig. 6-198 illustrates for the joint type c) the failure moment derived from the tests.

Definitions are given below:

$$m_u = \frac{M_u \text{ moment at the failure of joint}}{M_{pl,1,Rd} \text{ plastic moment resistance of the bracing}}$$

$\rho = (b_0/b_1)^l \cdot (b_0/t_0)^m$

$M_{p.,1,Rd} = f_{y,1} \cdot W_{pl,1}$ = yield strength × plastic section modulus of the bracing

t_0 = wall thickness of the chord
l and m = exponents corresponding to joint type

In order to calculate the moment resistance $M_{ip,1,Rd}$ of the RHS joint in a Vierendeel girder, the following equation for the chord web failure can be used basing on Eq. (6-131):

$$M_{ip,i,Rd} = 0.5 f_k \cdot t_0 [h_i + 5(t_0 + t_P)]^2 \cdot \frac{1.1}{\gamma_{Mj}}$$

(6-154)

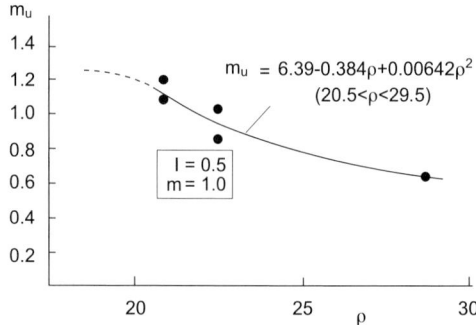

Fig. 6-198 "m_u-ρ" diagram for joint type c) (Fig. 6-188) with chord flange stiffener

Following conditions have been proposed in Lit. [8, 183] for the transmission of fully plastic bracing moment to the chord:

Width of the chord flange stiffening plate

$b_P \geq (b_0 - 4\ t_0)$

Length of the chord flange stiffening plate

$l_P = 2\ b_0$

Thickness of the chord flange stiffening plate

$t_P \geq 0.63 (b_1 \cdot t_1)^{0.5} - t_0$ (6-155)

3. Fig. 6-188 d – with haunch stiffeners

This type of joint possesses adequate moment resistance and rigidity and can be manufactured at an acceptable cost at the same time. The haunch sizes determine the joint resistance and the rigidity. However the haunches are classified as partially rigid.

Fig. 6-199 shows "m_u-ρ" diagram for the joint type d). They are recommended for the application in architecturally unpretentious structures.

Based on Eq. (6-130) (chord flange plastification) and Eq. (6-131) (chord web failure), the moment resistance $M_{ip,i,Rd}$ can also be calculated by using the following formulae [8, 183]:

- Modified Eq. (6-130) ($\beta \leq 0.85$)

$$M_{ip,i,Rd} = 3 f_{y0} \cdot t_0^2 \cdot h_i \left[\frac{1}{6 h_i/b_0} + \frac{2}{\sqrt{1-\beta}} + \frac{3 h_i/b_0}{1-\beta} \right] f(n) \cdot \frac{1.1}{\gamma_{Mj}}$$

(6-156)

6.7 Hollow section joints in Vierendeel trusses loaded by bending moment

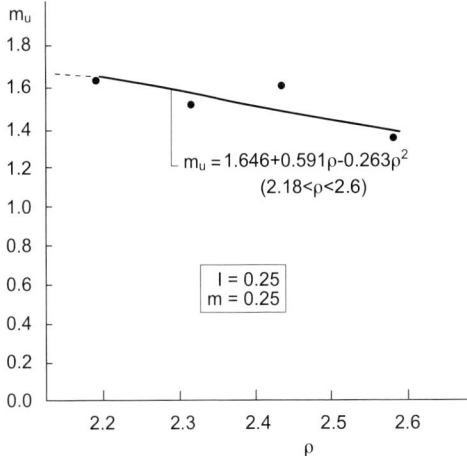

Fig. 6-199 "m_u-ρ" diagram for joint type d) (Fig. 6-192 d) with haunch stiffeners

where:

$$f(n) = 1.2 - \left(\frac{0.5}{\beta}\right) \cdot n \quad (n > 0, f(n) > 1, 0).$$

In this case, 45° haunches are used, which results in a total contact length of 3 h_i.

Modified Eq. (6-131) (0.85 < β ≤ 1.0)

$$M_{ip,i,Rd} = 0.5 f_k \cdot t_0 (3 h_i + 5 t_0)^2 \cdot \frac{1.1}{\gamma_{Mj}} \quad (6\text{-}157)$$

The assumption $f_k = f_{y0}$ (T joints) and $f_k = 0.8 f_{y0}$ (X joints) is only valid when $h_0/t_0 \leq 23$ [258].

For $h_0/t_0 > 23$, $f_k = f_K$ according to Table 6-13

4. Fig. 6-188 e – with truncated pyramid stiffeners

This joint type allows the fully plastic bracing moment to be transmitted to the chord and can be categorized as fully rigid.

The manufacturing cost is however very high and the architectural appearance is also not pretentious. These joints are therefore not selected in most cases.

6.7.7 Design resistances of CHS and RHS joints under out-of-plane bending moment M_{op}

Out-of-plane bending moment $M_{op,i,Rd}$ occurs often in the space structures.

6.7.7.1 Design resistances of RHS joints under M_{op}

RHS T and X joints under out-of-plane bending can be designed using Eqs. (6-133) to (6-135) (see Table 6-25) for corresponding β values and failure modes as can be done for in-plane bending.

The condition for the interaction of axial force, in-plane and out-of-plane bending moment has to be observed, which is recommended as follows on the safe side:

$$\frac{N_{i,Sd}}{N_{i,Rd}} + \frac{M_{ip,i,Sd}}{M_{ip,i,Rd}} + \frac{M_{op,i,Sd}}{M_{op,i,Rd}} \leq 1 \quad (6\text{-}158)$$

Following points have to be taken into consideration:

1. While applying Eq. (6-135), the plastic section modulus $W_{pl,i}$ of the bracing about the corresponding bending axis has to be used. The bracings must be assigned to the cross section class 2 at the least.
2. $M_{op,i,Rd}$ for X joints is to be determined analogous to T joints with the exception that the moment resistance for the chord web is calculated taking $f_k = 0.8 f_{y0}$.

6.7.7.2 Design resistances of CHS joints under M_{op}

Eqs. (6-150) and (6-152) in Table 6-29 are the design formulae for CHS joints (K and N joints with gap as well as T, X and Y joints) under out-of-plane bending moment, which have been derived based on the failure mode a) chord plastification or b) chord punching shear.

Plastic shear load initiated by the moment is given for the punching shear, while the function of the angle θ depends on the elastic basis.

Fig. 6-200 shows "efficiency C_{op} vs. β" graphically. Provided $d_1 \leq d_0 - 2 t_0$, the upper limit of the efficiency is given by the horizontal line (punching shear) in Fig. 6-200.

Further, the following relation of the interaction of axial force and in-plane and out-of-plane bending moment is valid:

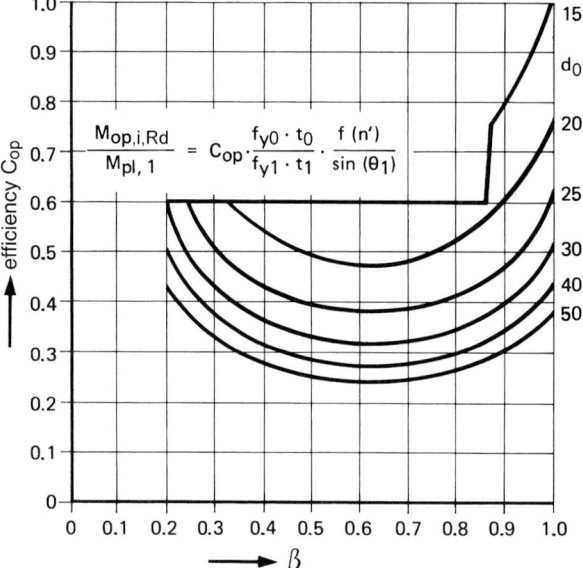

Fig. 6-200 Calculation diagram for CHS joints (K and N joints with gap as well as T, Y and X joints) loaded by out-of-plane bending moment [226]

$$\frac{N_{i,Sd}}{N_{i,Rd}} + \left(\frac{M_{ip,i,Sd}}{M_{ip,i,Rd}}\right)^2 + \frac{M_{op,i,Sd}}{M_{op,i,Rd}} \leq 1 \quad (6\text{-}159)$$

In this equation, it has been considered that the in-plane bending exercises a smaller influence than the out-of-plane bending.

6.8 Design resistances of T and X connections of plates, I profiles and RHS sections to CHS or RHS

Connections of this type are often used for the assembly and erection on site of the steel structures by means of bolting (indirect connection), as the welding on site is considerably more expensive than bolting. In particular, these constructions with bolted joints are often found in trusses, purlin connections and suspension attachments.

Two types of connection plates welded to the hollow sections are applied in the structures:

– Transversal or fin plates perpendicular to the hollow section axis
– Longitudinal or gusset plates parallel to the hollow section axis

Mainly, T and X joints between hollow section chords and plates have to be differentiated.

6.8.1 T and X connections with CHS chord members

Tables 6-30 and 6-31 contain the design resistances for the axial force $N_{i,Rd}$, the in-plane bending moment $M_{ip,i,Rd}$ and the out-of-plane bending moment $M_{op,i,Rd}$ [8, 226, 230], which have been developed by means of the tests and the theoretical model investigations done in Japan [24, 31, 61, 259]. Taking the closed ring model for T and X joints of CHS, the design formulae have been worked out, which agree relatively well with the practical test results for compression loading. The formulae for $N_{i,Rd}$ and $M_{ip,i,Rd}$ [8, 24, 226, 230] have to be further modified.

In the following, the design procedure for a CHS joint has been described in an example, where the bracings with flattened ends are bolted to the gusset plate, which in its turn is welded to the CHS chord (see Fig. 6-201).

Following conditions have to be fulfilled:

$$2l \cdot a \cdot f_{y,weld} \geq N_1 \cdot \cos\theta_1 + N_2 \cdot \cos\theta_2 \quad (6\text{-}187)$$

where:
a = weld thickness
$f_{y,weld}$ = yield strength of the weld material

6.8 Design resistances of T and X connections of plates, I profiles and RHS sections to CHS or RHS

Table 6-30 Design resistances of welded joints connecting fin and gusset plates with CHS members

Chord plastification	Design resistances
(diagram: plate with $M_{ip,i}$, N_i, t_1; $M_{op,i}$, b_i; d_0, t_0)	$N_{i,Rd} = f(n')\, f_{y0}\, t_0^2\, (4 + 20\beta^2)\, [1.1/\gamma_{Mj}]$ (6-160) $M_{ip,i,Rd} = 0$ (6-161) $M_{op,i,Rd} = 0.5\, b_i\, N_{i,Rd}$ (6-162)
(diagram: through plate with $M_{ip,i}$, N_i, t_1; $M_{op,i}$, b_i; d_0, t_0; lower N_i, $M_{ip,i}$)	$N_{i,Rd} = \dfrac{5\, f(n')\, f_{y0}\, t_0^2}{1 - 0.81\beta}\, [1.1/\gamma_{Mj}]$ (6-163) $M_{ip,i,Rd} = 0$ (6-164) $M_{op,i,Rd} = 0.5\, b_i\, N_{i,Rd}$ (6-165)
$t_i/d_0 \le 0.2$ (diagram: T-plate with $M_{ip,i}$, N_i, h_i; $M_{op,i}$, t_i; d_0, t_0)	$N_{i,Rd} = 5\, f(n')\, f_{y0}\, t_0^2\, (1 + 0.25\eta)\, [1.1/\gamma_{Mj}]$ (6-166) $M_{ip,i,Rd} = h_i\, N_{i,Rd}$ (6-167) $M_{op,i,Rd} = 0$ (6-168)
$t_i/d_0 \le 0.2$ (diagram: X-plate with $M_{ip,i}$, N_i, h_i; $M_{op,i}$, t_i; d_0, t_0; lower N_i, $M_{ip,i}$)	$N_{i,Rd} = 5\, f(n')\, f_{y0}\, t_0^2\, (1 + 0.25\eta)\, [1.1/\gamma_{Mj}]$ (6-169) $M_{ip,i,Rd} = h_i\, N_{i,Rd}$ (6-170) $M_{op,i,Rd} = 0$ (6-171)

Punching shear

$f_{max}\, t_i = (N_{i,Sd}/A_i + M_{i,Rd}/W_{i,el})\, t_i \le 2\, t_0 \left(f_{y0}/\sqrt{3} \right)\, [1.1/\gamma_{Mj}]$ (6-172)

Validity ranges	Factor $f(n')$
In addition to the validity ranges given in Table 6-9: $\beta \ge 0.4$ and $\eta \le 4$ where $\beta = b_i/d_0$ and $\eta = h_i/d_0$	for $n' > 0$ (compression): $f(n') = 1 - 0.3\, n'\, (1 + n')$ but $f(n') \le 1.0$ for $n' \le 0$ (tension): $f(n') = 1.0$

Table 6-31 Design resistances of welded joints connecting I, H or RHS sections to CHS members

Chord plastification	Design resistances
(I/H section, axial + in-plane moment on chord)	$N_{i,Rd} = f(n')\, f_{y0}\, t_0^2\, (4 + 20\beta^2)(1 + 0.25\eta)\, [1.1/\gamma_{Mj}]$ (6-173) $M_{ip,i,Rd} = h_i N_{i,Rd} / (1 + 0.25\eta)$ (6-174) $M_{op,i,Rd} = 0.5\, b_i\, N_{i,Rd}$ (6-175)
(I/H section, in-plane + out-of-plane moments)	$N_{i,Rd} = \dfrac{5\, f(n')\, f_{y0}\, t_0^2}{1 - 0.81\beta}\, (1 + 0.25\eta)\, [1.1/\gamma_{Mj}]$ (6-176) $M_{ip,i,Rd} = h_i N_{i,Rd} / (1 + 0.25\eta)$ (6-177) $M_{op,i,Rd} = 0.5\, b_i\, N_{i,Rd}$ (6-178)
(RHS section, axial + in-plane moment)	$N_{i,Rd} = f(n')\, f_{y0}\, t_0^2\, (4 + 20\beta^2)(1 + 0.25\eta)\, [1.1/\gamma_{Mj}]$ (6-179) $M_{ip,i,Rd} = h_i\, N_{i,Rd}$ (6-180) $M_{op,i,Rd} = 0.5\, b_i\, N_{i,Rd}$ (6-181)
(RHS section, in-plane + out-of-plane moments)	$N_{i,Rd} = \dfrac{5\, f(n')\, f_{y0}\, t_0^2}{1 - 0.81\beta}\, (1 + 0.25\eta)\, [1.1/\gamma_{Mj}]$ (6-182) $M_{ip,i,Rd} = h_i\, N_{i,Rd}$ (6-183) $M_{op,i,Rd} = 0.5\, b_i\, N_{i,Rd}$ (6-184)

Punching shear	
For I section: $f_{max}\, t_i = (N_{i,Sd}/A_i + M_{i,Rd}/W_{i,el})\, t_i \le 2 t_0 \left(f_{y0}/\sqrt{3}\right)\, [1.1/\gamma_{Mj}]$	(6-185)
For RHS section: $f_{max}\, t_i = (N_{Sd}/A + M_{Rd}/W_{el})\, t_i \le t_0 \left(f_{y0}/\sqrt{3}\right)\, [1.1/\gamma_{Mj}]$	(6-186)

Validity ranges	Factor $f(n')$
In addition to the validity ranges given in Table 6-9: $\beta \ge 0.4$ and $\eta \le 4$ where $\beta = b_i/d_0$ and $\eta = h_i/d_0$	for $n' > 0$ (compression): $f(n') = 1 - 0.3\, n'\,(1 + n')$ but $f(n') \le 1.0$ for $n' \le 0$ (tension): $f(n') = 1.0$

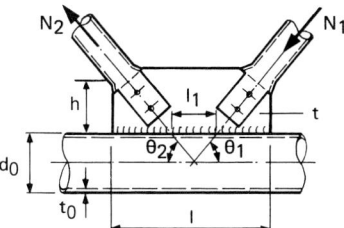

Fig. 6-201 K type bolted CHS joint with gusset plate

$$l \cdot t \cdot \frac{f_{y,P}}{\sqrt{3}} \geq N_1 \cdot \cos\theta_1 + N_2 \cdot \cos\theta_2 \quad (6\text{-}188)$$

$$2l \cdot t_0 \frac{f_{y0}}{\sqrt{3}} \geq N_1 \cdot \cos\theta_1 + N_2 \cdot \cos\theta_2 \quad (6\text{-}189)$$

$$\frac{t \cdot h^2}{6} \cdot f_{y,P} \geq N_1 \cdot \sin\theta_1 \cdot l_1 \quad (6\text{-}190)$$

The resistance of the CHS wall has to be checked using Eq. (6-166) in Table 6-30:

$$5.0 \left(1 + 0.25 \frac{l}{d_0}\right) \cdot f_{y0} \cdot t_0^2 \cdot f(n') \cdot l \geq$$

$$N_1 \sin\theta_1 \cdot l_1 \quad (6\text{-}191)$$

If the welds have a lower resistance than the plate, the welds have also to be checked for the combined effect of shear and bending moment.

6.8.2 T connections with RHS chord members

The design resistances for axial force $N_{i,Rd}$ and in-plane bending moment $M_{ip,i,Rd}$ in T type welded joints connecting gusset plates or I or H sections to RHS members (see Table 6-32) have been developed by means of tests and analytical investigations in the last years [7, 159, 193, 201, 260]. It has to be particularly mentioned that the joint with longitudinal connection plate is very yielding and therefore is not recommended. However, Eq. (6-195) in Table 6-32 serves the control of the deformation of these joints.

6.9 Design of I beam-to-RHS column connections

6.9.1 RHS column with I beam

Various connections between I beams and RHS columns have been illustrated in Figs. 6-14a–e and 6-15 (see Section 6.2.2). Although extensive test results regarding the load carrying behaviour of the directly welded RHS beam to the RHS column (see Section 6.7) are available and the corresponding design resistances have been developed, the joints with RHS column and I section beam had not been adequately investigated. However, based on the available test results, Eq. (6-197) in Table 6-32 has been proposed [230], which delivers the design moment resistance on the safe side for these joints. However, further research work had to be done in this field. This was accomplished by an extensive research work carried out in the Netherlands [265]. The ultimate resistance formulae obtained from [265] are listed in Appendix V.

The relative rigidity of the flange of an I section beam to that of the connecting face of a RHS column leads to different failure modes depending on the joint configurations. In order to prevent the collapse of the column flange or the failure of the column web (see Fig. 6-16), a stiffening plate is welded to the connecting flange of the RHS column.

Two connections of this type (Fig. 6-202) were investigated in Canada and their design guidelines were worked out [261, 262].

In these connections, the moment from the I beam is transmitted to the RHS column through the moment transmitting plates (tension or compression). These plates are welded to the beam flange and to the stiffening plate on the RHS column flange. The design is to be made following the guidelines given in [262]:

Table 6-32 Design resistances of welded T joints connecting fin or gusset plates or I or H sections to RHS members

	Design resistances	
Transversal plate	Bracing failure for all β	
	$N_{i,Rd} = f_{yi} \cdot t \cdot b_e [1.1/\gamma_{Mj}]$	(6-192)
	Chord web failure for $b_i \geq b_0 - 2 t_0$	
	$N_{i,Rd} = f_{y0} t_0 (2 t_i + 10 t_0)[1.1/\gamma_{Mj}]$	(6-193)
	Shear failure $b_i \leq b_0 - 2 t_0$	
$\beta = \dfrac{b_1}{b_0}$	$N_{i,Rd} = \dfrac{f_{y0} t_0}{\sqrt{3}}(2 t_i + 2 b_{e,p})[1.1/\gamma_{Mj}]$	(6-194)
Longitudinal plate (gusset plate)	Chord flange plastification for $t_i/b_0 \leq 0.2$	
	$N_{i,Rd} = \dfrac{f(n) f_{y0} t_0^2}{1 - t_i/b_0}(2 h_i/b_0 + 4\sqrt{1 - t_i/b_0})\left[\dfrac{1.1}{\gamma_{Mj}}\right]$	(6-195)
	$M_{ip,i,Rd} = 0.5\, N_{i,Rd} \cdot h_i$	(6-196)
$t_i/b_0 \leq 0.2$		
I section		
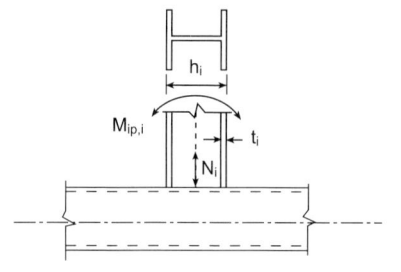	Conservatively, $N_{i,Rd}$ for an I or H section is to be based on the design resistance of two transverse plates similar to the flanges of the H section, determined as specified above.	
	$M_{ip,i,Rd} = N_{i,Rd} \cdot (h_i - t_i)$	(6-197)

Validity ranges

in addition to the validity ranges given in Table 6-14:

$0.5 \leq \beta \leq 1.0$

$\dfrac{b_0}{t_0} \leq 30$

Parameters b_e, $b_{e,p}$ and $f(n)$

$b_e = \dfrac{10}{b_0/t_0} \dfrac{f_{y0} \cdot t_0}{f_{yi} \cdot t_i} \cdot b_i$ but $b_e \leq b_i$	For $n > 0$ (compression): $f(n) = 1.3(1 - n)$ but $f(n) \leq 1.0$
$b_{e,p} = \dfrac{10}{b_0/t_0} \cdot b_i$ but $b_{e,p} \leq b_i$	For $n \leq 0$ (tension): $f(n) = 1.0$

6.9 Design of I beam-to-RHS column connections

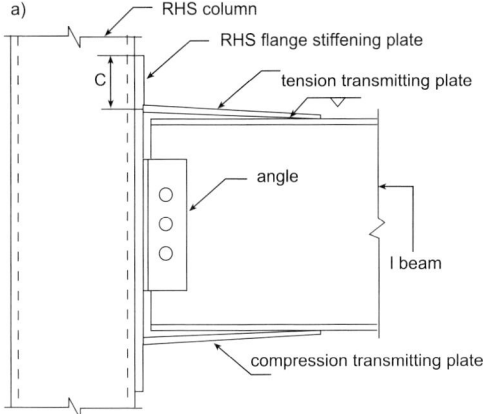

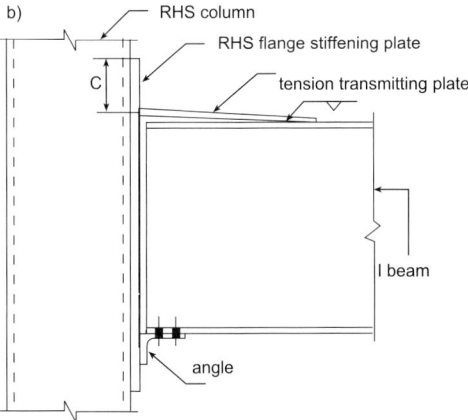

Fig. 6-202 Reinforced I beam-to-RHS column connection [261]

1. Transmission of tension

• The shear resistance V_P of the stiffening plate is to be calculated using the following equation derived from the test results:

$$V_P = \frac{2}{3} f_{y,P} \cdot t_P (b_1 + t_1) \qquad (6\text{-}198)$$

where:
$f_{y,P}$ = yield strength of the RHS stiffening plate
t_P = wall thickness of the RHS stiffening plate
b_1 = width of the tension transmitting plate
t_1 = thickness of the tension transmitting plate

Eq. (6-198) is valid for

$$\frac{\text{width of the tension transmitting plate } b_1}{\text{width of the RHS stiffening plate } b_P}$$

ratio lying between 0.4 and 0.8.

• It is recommended to weld the total periphery of the stiffening plate with fillet weld.

• The thickness of the tension transmitting plate within a realistic range has no effect on the shear resistance of the connection, provided the thickness of the RHS stiffening plate remains constant.

The resistance N_1 of the tensile transmitting plate can be calculated using the following equation, which is obtained by the modification of the design formulae in [7]:

$$N_1 = b_e \cdot t_1 \cdot f_{y1} \qquad (6\text{-}199)$$

where:

$b_e = \dfrac{10}{b_P/t_P} \cdot \dfrac{f_{yP} \cdot t_P}{f_{y1} \cdot t_1} \cdot b_1$ = effective width
b_P = width of the RHS stiffening plate
t_P = thickness of the RHS stiffening plate
b_1, t_1 = see Eq. (6-198)
$f_{y,P}$ = yield strength of the stiffening plate material
f_{y1} = yield strength of the tension plate material

• The minimum distance C (see Fig. 6-202) increases with the increase of the thickness of the stiffening plate. Tests have shown that minimum C has to be equal to 80 mm, in case $t_P = 12$ mm.

• A full penetration groove weld on the tension plate is adequate to avoid the weld failure.

2. Transmission of compression

• The resistance N_{web} of the RHS web can be determined by means of the following equation:

$$N_{\text{web}} = 2 f_{y0} \cdot t_0 [W_s + 4.4 (t_0 + t_P)] \qquad (6\text{-}200)$$

where:
f_{y0} = yield strength of RHS material
t_0 = wall thickness of RHS
t_P = thickness of the RHS stiffening plate
W_s = thickness of the compression transmitting plate including the weld

Table 6-33 Design in-plane bending moment resistances of welded joints between RHS bracing members and I or H section chords

T and Y joint	Design resistance ($i = 1$ or 2)
	Chord web yielding
(diagram)	$M_{ip,i,Rd} = 0.5 f_{y0} t_w b_w h_i [1,1/\gamma_{Mj}]$ (6-201)
	Bracing failure (effective width)
	$M_{ip,i,Rd} = f_{yi} t_i b_e (h_i - t_i)[1.1/\gamma_{Mj}]$ (6-202)

Parameter b_e and b_w		Validity range
$b_e = t_w + 2r + 7(f_{y0}/f_{yi}) t_f$	$b_w = \dfrac{h_i}{\sin \theta_i} + 5(t_f + r)$	see Table 6-18
but $b_e \leq b_i$	but $b_w \leq 2 t_i + 10 (t_f + r)$	

- In association with the angle, the load transmission to the column is more effective than when the compression flange of the I beam is directly welded to the column.

- The minimum C in the compression zone should amount to 100 mm; if the thickness of the stiffening plate lies between 8 and 12 mm.

6.9.2 I section column with RHS beam

Lit. [230] contains the design formulae for in-plane bending moment resistances $M_{ip,i,Rd}$ (Table 6-33), which have been worked out and modified basing on failure modes a) I section chord web yielding (see Figs. 6-133 and 6-136), b) RHS bracing failure (effective width) (see Figs. 6-135 and 6-138) [7].

6.10 Special joint of RHS chord and RHS bracing with "bill-shaped" or "bird-mouth" end

Fig. 6-203 illustrates a special structural joint arrangement in a RHP truss, where the bracings with bill-shaped "profile" ends are

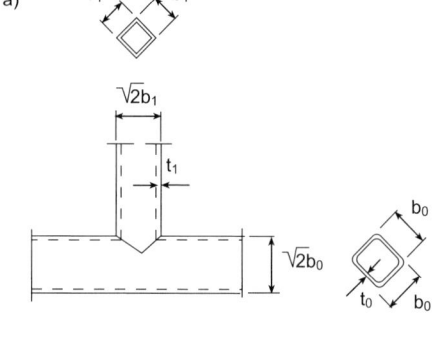

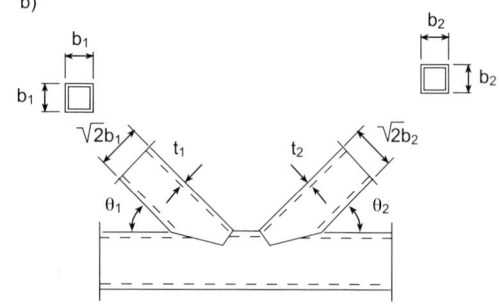

Fig. 6-203 RHS "bird-mouth" or "bill-shaped" T and K joints

welded to the RHS chord framing into the corners of the RHS. Herewith a larger joint resistance and stiffness can be achieved regardless of the bracing to chord width ratio.

Based on the test results and the analytical evaluations [263], the following formulae for the design joint resistance are recommended:

For T joint:

$$N_{1,Rd} = t_0^2 \cdot f_{y0} \left(\frac{1}{0.211 - 0.147(b_1/b_0)} + \frac{b_0/t_0}{1.794 - 0.942(b_1/b_0)} \right) \cdot f(n') \cdot \frac{1}{\gamma_{Mj}}$$

(6-203)

For K joint:

$$N_{i,Rd} = \frac{t_0^2 \cdot f_{y0}}{\sqrt{1 + 2\sin^2 \theta_i}} (4\alpha)(b_0/t_0) \cdot f(n') \cdot \frac{1}{\gamma_{Mj}}$$

(6-204)

where:
α   = to be read from Fig. 6-204 (for 45° K joints)
$f(n')$ = to be calculated according to Table 6-9

The validity ranges for the application of Eqs. (6-203) and (6-204) are as follows:

For T joints:

$$16 \leq \frac{b_0}{t_0} \leq 42$$

$$0.3 \leq \frac{b_1}{b_0} \leq 1.0$$

For K joints:

$$16 \leq \frac{b_0}{t_0} \leq 42$$

$$0.2 \leq \frac{b_i}{b_0} \leq 0.7$$

$$\theta_i = 45°$$

6.11 List of symbols

CHS	circular hollow section
RHS	rectangular hollow section (also square)
a	throat thickness of a fillet weld
A_g	total cross-sectional area
A_i	cross-sectional area of a structural member i ($i = 0, 1, 2, 3$)
A_v	shear area of the chord
b	width
b_i	external width of square or rectangular hollow section (RHS) member i ($i = 0, 1, 2, 3$)
b_e	effective width of a bracing member related to the chord
$b_{e,ov}$	effective width for overlapping bracing member connected to overlapped bracing member
$b_{e,p}$	effective width for punching shear
b_P	width of plate
b_w	effective width of chord web
$C_{i,\varphi}$	rotational stiffness of a joint
C_{ip}	joint efficiency for in-plane bending moment
C_{op}	joint efficiency for out-of-plane bending moment
d	diameter
d_i	external diameter of circular hollow section (CHS) i ($i = 0, 1, 2, 3$)
e	noding eccentricity for a joint (see Fig. 6-101)
E	modulus of elasticity
F_i, N_i	axial force in member i ($i = 0, 1, 2, 3$)
f	stress

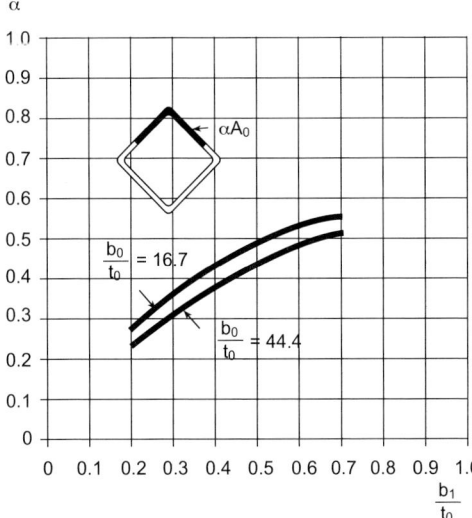

Fig. 6-204 Effective area coefficient α vs. b_1/b_0 for "bird-mouth" 45° K joints

f_{yi}	yield stress of the material of the member i ($i = 0, 1, 2, 3$)
$f_{y,P}$	yield stress of plate material
f_k	characteristic yield stress
f_K	critical buckling stress according to structural standard
f_i	axial stress in member i ($i = 0, 1, 2, 3$)
f_0	axial stress in chord
$f_{a,weld}$	design stress of the weld perpendicular to the weld
$f_{s,weld}$	design stress of the weld parallel to the weld
f_u	ultimate tensile stress
f_{op}	prestress in chord
$f(n)$	function in the joint resistance formulae which incorporate the influence of axial forces in the chord of RHS joints
$f(n')$	function in the joint resistance formulae which incorporate the influence of axial forces in the chord of CHS and "bird-mouth" joints
g	gap between the bracing member toes (ignoring welds) of a K, N or KT joints on the face of the chord (negative value of g represents the overlap of the bracings)
g'	gap divided by chord wall thickness, $g' = \dfrac{g}{t_0}$
h	depth
h_i	external depth of rectangular or square hollow section member i ($i = 0, 1, 2, 3$)
H	truss depth
I	second moment of area (moment of inertia)
l_1	system length (bracing)
l_K	effective buckling length
l_0	distance between the panel joints in chord of a truss
l_P	plate length
L	span
M	moment
M_i	bending moment in member i ($i = 0, 1, 2, 3$)
M_{ip}	in-plane bending moment
M_{op}	out-of-plane bending moment
M_{pl}	plastic bending moment
N_{op}	chord "preload" (additional axial force in the chord member at a joint which is not necessary for the equilibrium of the bracing horizontal load components)
$N_{0(in\ gap),Rd}$	reduced axial load resistance due to shear in the cross section of the chord at the gap
n	$\dfrac{f_0}{f_{y0}} = \dfrac{N_0}{A_0 \cdot f_{y0}} + \dfrac{M_0}{W_0 \cdot f_{y0}}$ or number of bolts
n'	$\dfrac{N_{0p}}{A_0 \cdot f_{y0}} + \dfrac{M_0}{W_0 \cdot f_{y0}}$
p	length of the projected contact area between the overlapping bracing and the chord of a K or N joint without the presence of the overlapped bracing (see Fig. 6-82 b)
p''	bolt pitch or length of flange plate attributed to each bolt (see Fig. 6-50)
P_f	external tensile force applied to a bolt
P_y	yield load of the head plate connection
Q	Transversal force
q	projected length of overlap between the bracings of a K or N joint at the chord face (see Fig. 6-50)
r_0	corner radius between flange and web of a I section or channel
t	thickness
t_i	wall thickness of member i ($i = 0, 1, 2, 3$)
t_f	flange plate thickness
t_P	plate thickness
t_w	web thickness of a I section
T_i	tensile force applied to a member i ($i = 0, 1, 2, 3$)
T_u	ultimate tensile resistance of a bolt
V_i	shear force applied to a member i ($i = 0, 1, 2, 3$)
V_p	shear resistance $\dfrac{f_{y0} \cdot A_v}{\sqrt{3}}$ (see Table 6-13)
$V_{pl,i}$	$\dfrac{A_i \cdot f_{yi}}{\sqrt{3}}$ (plastic shear force for a member i ($i = 0, 1, 2, 3$))
W_i	section modulus of a member ($i = 0, 1, 2, 3$)
W_{el}	elastic section modulus
W_{pl}	plastic section modulus
α	$\dfrac{2\,l_0}{d_0}$ or non-dimensional factor
β	width or diameter ratio between bracing member(s) and chord

6.11 List of symbols

$\beta = \dfrac{d_1}{d_0}, \dfrac{d_1}{b_0}, \dfrac{b_1}{b_0}$ (T, Y, X joints)

$\beta = \dfrac{d_1+d_2}{2d_0}, \dfrac{d_1+d_2}{2b_0}, \dfrac{b_1+b_2+h_1+h_2}{4b_0}$

(K and N joints)

$\beta = \dfrac{d_1+d_2+d_3}{3d_0}, \dfrac{d_1+d_2+d_3}{3b_0},$

$\dfrac{b_1+b_2+b_3+h_1+h_2+h_3}{6b_0}$

(KT joints)

or ratio of the plate width to the outer diameter or width of the chord

β_P width or diameter ratio between bracing member i and plate b_i/b_P

γ $\dfrac{b_0 \text{ or } d_0}{2t_0}$

γ_{Mj} partial safety factor for joint resistance

γ_{MW} partial safety factor for welded joint resistance

γ_F partial safety factor for loads (action)

τ shear stress

τ_i t_i/t_0

η h_i/b_0

η_P h_i/b_P

θ_i included angle between the bracing member i ($i = 1, 2, 3$) and the chord

λ_{ov} overlap grade $\dfrac{q}{p} \times 100\%$
(see Fig. 6-82 b)

ϕ angle between the bracing axes in a multiplanar joint

Subscripts:

0 designates chord

1 refers in general to the bracing for T, Y and X joints or it refers to the compression bracing member for K, N and KT joints

2 refers to the tension bracing member for K, N and KT joints

3 refers to the vertical member for KT joint

j refers to the overlapping bracing member for K and N overlap joint

k characteristic

K flexural buckling

a bracing member set above in a K or N joint

P plate

u ultimate

u bracing member set below in a K or N joint

Rd resistance

Sd applied load (action)

7 Design of welded hollow section joints subjected to fatigue loading

7.1 General

As nearly all structures are in practice more or less subjected to fatigue loading, they are categorized according to the nature of the load into two main groups:

- Structures subjected to predominantly static loading. The structural components are designed in this case on the basis of the statical strength data e.g. yield and ultimate strength, elongation etc.
- Structures subjected to repeated loading varying with time. The structural components are designed in this case on the basis of the fatigue strength data.

Variable load with time, which is in general defined as fatigue load, covers all regular and random loads illustrated in Figs. 7-1 and 7-2.

Random loading pattern represents arbitrarily varying loadings with regard to magnitude and frequency initiated in general by predominantly accidental forces and moments (Fig. 7-1).

Regular loading is of the periodically repeating type about a mean value, the course of which may be sinusoidal, trapezoidal or of thrust form (Fig. 7-2).

The fatigue process is characterized by three stages of damage: crack initiation, propagation of one dominant crack and final fracture, when a ductile or a brittle overload fracture exhausts the capacity of the cross section.

Although regular loadings are seldom applied in practice, particularly the sinusoidal type of load cycle illustrated by Fig. 7-2a is predominantly used for the tests in the research centres as well as the basis for the design of the structures. The results obtained from these tests are designated as the fatigue load or resistance.

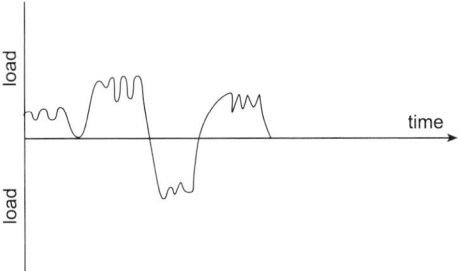

Fig. 7-1 Random "load vs. time" function

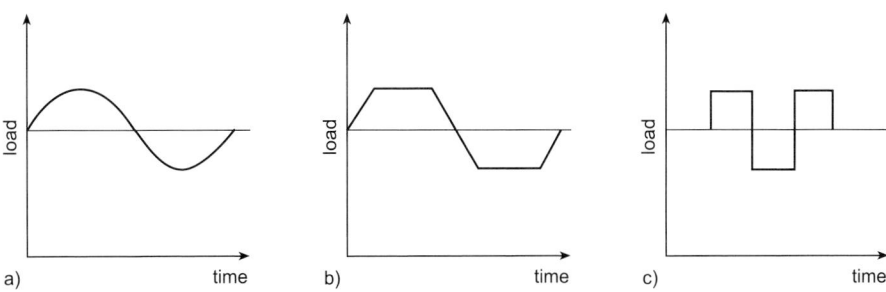

Fig. 7-2 Periodically repeating "load vs. time" function about a mean load: a) sine-curve loading, b) trapezoidal loading, c) thrust type loading

Fig. 7-3 Rotary tower cranes

Fig. 7-4 Overhang girder of a mobile crane

The fatigue load can be described by means of the following data (compare with Fig. 7-6):

f_{max}: maximum stress in the cycle
f_{min}: minimum stress in the cycle
f_m: mean stress in the cycle $\left(\dfrac{f_{max}+f_{min}}{2}\right)$
f_a: stress amplitude $\left(\dfrac{f_{max}-f_{min}}{2}\right)$

Fig. 7-5 Section of a giant wheel

R: stress ratio $\left(\dfrac{f_{min}}{f_{max}}\right)$
S_r: stress range $(f_{max}-f_{min}) = 2f_a$

In order to describe the periodically varying, sinusoidal loads, the following definitions are made:

Alternating load (Fig. 7-6a)
$R = -1$
$f_{min} = -f_{max}$
$f_m = 0$
$f_a = f_{max}$
$S_r = 2f_{max}$

Pure pulsating load (Fig. 7-6b)
$R = 0$
$f_{min} = 0$ and $f_{max} \neq 0$
$f_m = \dfrac{f_{max}}{2}$
$S_r = f_{max}$

Pulsating load (Fig. 7-6c)
$R > 0$

Static load
$R = +1$

Two fundamental terms are named to characterize the fatigue resistance in practice:

- Ultimate fatigue resistance and fatigue resistance for finite life for single-step

7.1 General

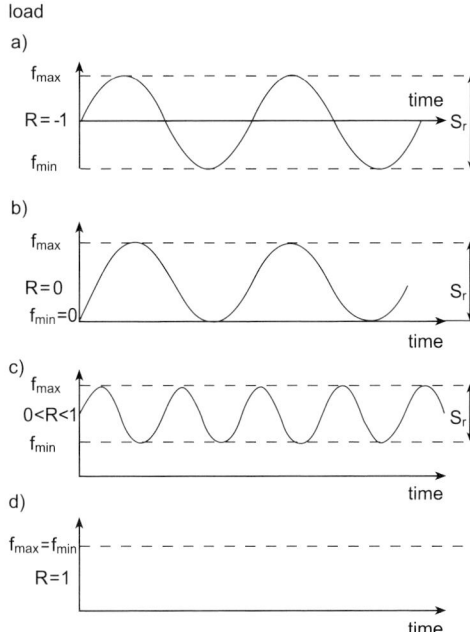

Fig. 7-6 Types of constant amplitude sinusoidal loading. a) Alternating loading, b) pure pulsating loading, c) pulsating loading, d) static loading

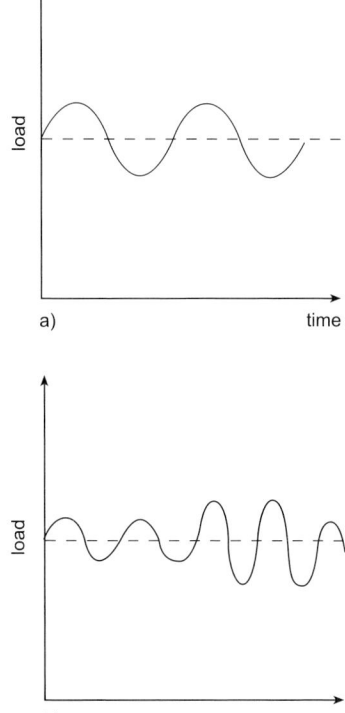

Fig. 7-7 a) Single-step periodic loading with constant amplitude, b) multistage periodic loading with variable amplitudes

(constant amplitude), periodically varying, sinusoidal loading (Fig. 7-7a).

Ultimate fatigue resistance (endurance limit) is defined as the maximum stress f_{max}, which a material or a structural element or a construction can withstand without rupture or excessively large deformation over an infinite number of loading cycles. This can also be defined as the corresponding stress range S_r for a certain stress ratio R.

Fatigue resistance for **finite life** is given by f_{max} or S_r, which a material or a structural part can just sustain over a certain number (not infinite) of loading cycles.

Fatigue life is generally defined by the number of cycles of stress or strain before failure of a specified character occurs. In most recommendations, a crack through the wall is considered as failure. Agreement can be made regarding other failure specifications e. g. size of the crack presumed to be sufficient to cause failure.

The most common of the cycle counting methods are listed below:

– Rainflow method program is convenient for long stress analysis and should be used with computer.

– Reservoir method is easy to use by hand for short stress histories.

• Fatigue resistance is determined following the method introduced by Woehler (1860), which consists of applying repeating loads (in stress or strain) to identical test pieces according to a constant amplitude sinusoidal curve pattern. This is the simplest form of load spectrum with a constant amplitude stress time history with constant mean stress f_m. Such load is used in the testing laboratories to form reproducible fatigue tests on specimens. Test specimens are subjected to the loading levels (f_{max}, f_{min}) or maximum stresses f_{max} or stress ranges S_r till failure occurs, where identical specimens have to undergo different loadings. Fatigue life is represented by the number N of loading cycles to failure (N_F). Fig. 7-8 shows a Woehler curve in principle, where the test results are set out in the form "f_{max} vs. N_F" or "S_r vs. N_F" curves. The horizontal asymptote of the Woehler curve, designated as f_F in Fig. 7-8, represents the ultimate fatigue resistance. The non-asymptotic range of the Woehler curve, which is arch-shaped in a linear scale (Fig. 7-8a) or is a straight line in a logarithmic scale (Fig. 7-8b), represents the fatigue resistance for **finite life**. Normally, a Woehler curve is drawn in logarithmic scale. In Fig. 7-8b, Woehler lines are shown for various stress ratios R.

When the Woehler curve manifests a distinct asymptotic horizontal line – this is the case for flat specimens in steel –, the ultimate fatigue resistance can be determined in a relatively precise manner. For welded constructions i.e. hollow section joints, no horizontal symptote representing the ultimate fatigue stress f_F or stress range $S_{r,F}$ exists. That is why it is accepted in practice that the number of cycles to failure corresponding to the ultimate fatigue stress or stress range is equal to a finite number of loading cycles estimated arbitrarily. When this is reached with a certain maximum stress f_{max} or stress range S_r just without any crack initiation, this f_{max} or S_r can be taken as the ultimate fatigue resistance f_F or $S_{r,F}$. For welded steel structures, the number of the loading cycles N to attain f_F or $S_{r,F}$ is accepted to be equal to $2 \cdot 10^6$ cycles. According to Eurocode 3 [42], the corresponding number is $5 \cdot 10^6$.

According to the practical fatigue tests carried out in the field of offshore technology as well as the relevant experiences from the projects in this field, it could be observed that there was no characteristic horizontal asymptote in the Woehler curve representing the ultimate fatigue resistance f_F or stress range $S_{r,F}$ for welded constructions. As a consequence, the number of cycles to failure N_F corresponding to f_F or $S_{r,F}$ has been arbitrarily chosen to be 10^7 cycles, sometimes also 10^8 cycles for this purpose.

The results of the fatigue tests scatter significantly, even if identical test specimens are prepared precisely with great care. Fig. 7-9 illustrates the scatter range of these test results. Investigations, made later, have proved that the prevalenting reason for the scatter lies in the imponderables of the material and the preparation of test specimens and not in the execution of tests.

Fatigue resistance cannot therefore be presumed to be a certain, precisely determined value. The Woehler curve has to be taken to represent a set of curves, each of which is associated with a certain failure probability P (Fig. 7-10). As a consequence, it is necessary to perform tests on a number of identical specimens at the same stress level in order obtain a Woehler curve. One restricts the stress levels to be investigated to an acceptable

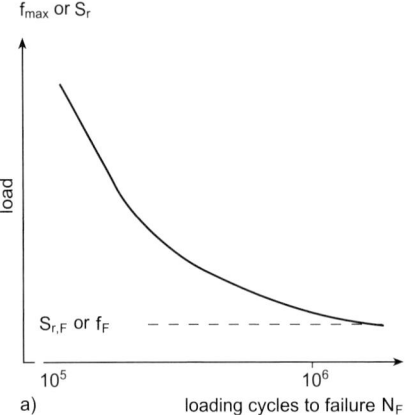

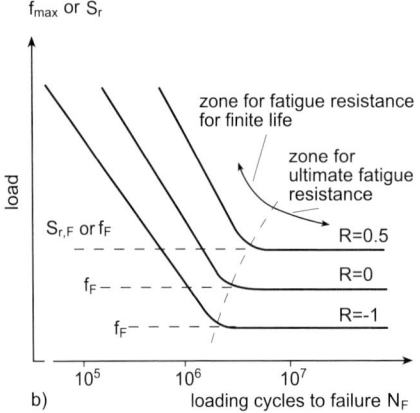

Fig. 7-8 Woehler curves a) in linear scale, b) in logarithmic scale

7.2 Fatigue resistance for cumulative load (load spectrum) 235

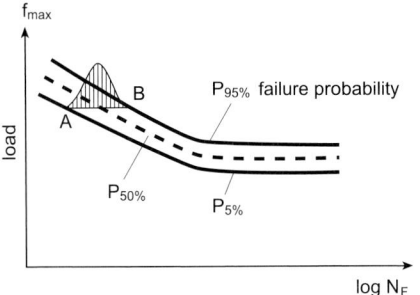

Fig. 7-9 Scatter of test results

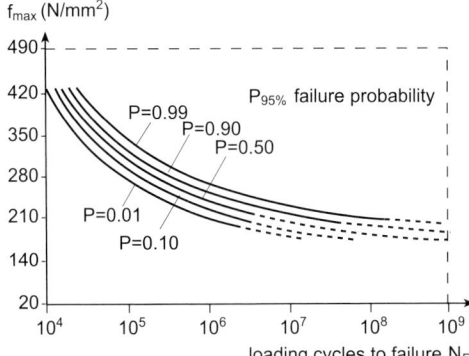

Fig. 7-10 Example of a diagram (f_{max} vs. N_F) with failure probability P

number, so that the number of tests do not become too large.

The experience demonstrates that the values for $\log_{10} N_F$ obey the Gauss-Laplace distribution law (Fig. 7-10).

7.2 Fatigue resistance for cumulative load (load spectrum)

The Woehler curve delivers the fatigue life N_F of an workpiece under constant amplitude fluctuating loading, where maximum stress f_{max} or stress range S_r and stress ratio R are the governing parameters. This has however no validity, when the loadings are of variable amplitudes. The simplest procedure to assess and solve this problem is based on the use of cumulative damage rule to describe the fatigue behaviour under spectrum loading: A structural element has a fatigue life to failure N_i at load level i when it is subjected to the constant amplitude load $S_{r,i}$ (see Fig. 7-11). When a number of loading cy-

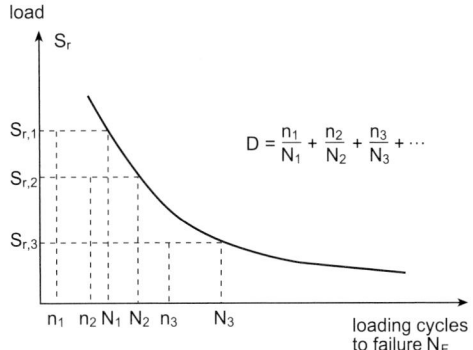

Fig. 7-11 Palmgren-Miner's rule for cumulative load (load spectrum)

cles n_i is absolved at this load level, the fatigue damage D_i amounts to:

$$D_i = \frac{n_i}{N_i}$$

This hypothesis is fully valid in the following two limiting cases: For $n_i = 0$, the damage D_i is equal to 0 and for $n_i = N_F$, the damage is equal to 1, which corresponds to the failure of the workpiece. When a workpiece subjected to a load spectrum (that means, variable amplitude loadings $S_{r,i}$ acting for cycles n_i), this can be assumed that the failure occurs when the following condition is met:

$$D = \sum \frac{n_i}{N_i} = 1$$

where N_i is the corresponding fatigue life for the load level i and D designates the total damage. The fatigue damage accumulates linearly with the number of cycles at a particular load level. This failure criterion was proposed by Palmgren (1924) and Miner (1945).

How far the Palmgren-Miner rule $D = 1.0$ is realistic in practice, has been investigated in the last years by many researchers with the aid of results of the fatigue tests under load spectrum [43, 44]. Different values for the total damage ($D \leq 1$) were determined assuming various parameters differently e.g. loading cycles, load type, material properties, environment etc. Therefore in particular cases, tests to determine the fatigue behaviour are performed applying a load spectrum. This is evidently very expensive. That is why the Palmgren-Miner rule has been taken up by some new structural standards e.g. Eurocode 3 [42] as fatigue verifying procedure due to its easy implementation.

The value of the stress range S_r is multiplied by the partial safety factor γ_F in order to assess the insecurity due to

– Load description
– Calculation method
– Application of Palmgren–Miner rule

This leads to the required reliability.

The value for the partial safety factor can be taken from the available literature on the subject. As for example, the accumulated damage $D = \sum \frac{n_i}{N_i} = 0.5$ (instead of 1) has been proposed by the German Lloyds for offshore constructions (which means that the partial safety factor γ_F is equal to 2.0).

In case the required load spectrum for a particular project is not available, the accumulated damage D can be determined through "load vs. time" curve by counting the loading cycle N_i and the corresponding stress range $S_{r,i}$. Thus the fatigue design calculation procedure for a load spectrum can be established.

In general, the following ranges will not be taken into consideration while establishing a load spectrum (see Fig. 7-12):

– Stress range $S_{r,i} \leq 0.55\, S_{r,F}$
– Loading cycle $N_F \leq 10^2$ (for $S_{r,\max}$)

The value $0.55\, S_{r,F}$ is originated from the test results. This corresponds to the assigned loading cycles of 10^8, if the real inclination of the Woehler curve above $5 \cdot 10^6$ loading cycles is considered.

The limiting 10^2 loading cycles appear to be justified for larger volumes of load spectrums. This is covered by the static tests. The magnitude of peak loads, which are vital for static design, but which occur only about 10 to 100 times in the life of the structure, does not influence the number of loading cycles, which are susceptible to fatigue. Investigations on the axially loaded RHS T joints [99] have however shown that the hot spot stress or strain fatigue design curves of the high cycle range can be

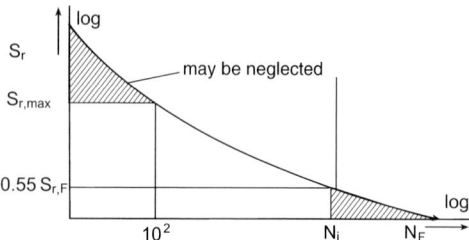

Fig. 7-12 Schematization of a stress spectrum with zones to be neglected

extrapolated to the low cycle range. No cut-off is required if the curves are based on hot spot stress or hot spot strain.

In the case of low cycle fatigue, fatigue fracture can be forced if maximum/minimum stress is at yield stress. In this case, some few cycles can cause fracture (example: earth quake, sometimes at 3 to 5 cycles). According to EC 3 [42], all nominal stresses shall be within the elastic limits of the material. The range of the design values of such stresses shall not exceed $1.5\, f_y$ for axial stress or $1.5\, f_y/\sqrt{3}$ for shear stresses.

The fatigue behaviour of hollow section structures dealt with in this book does not include the following:

1. Structures subjected to temperatures exceeding 150 °C

However, the following recommendation has been proposed regarding the effect of high temperature on fatigue strength [100], where a fatigue reduction factor $f(\text{temp})$ has to be multiplied with the fatigue strength:

$$f(\text{temp}) = 1.12 - 0.0012 \cdot (\text{temperature in degree})$$
$$+ \left[0.1 - 0.2 \left(0.35 - \frac{\text{temperature in degree}}{1000} \right)^2 \right]$$

2. Structures in a corrosive environment (including sea water), see Section 7.4
3. Structures subjected to single impact
4. Concrete reinforcement

7.3 Effect of residual stresses on the fatigue resistance of hollow section joints

Residual stress is set up, when a structural element is manufactured or when a joint in a

7.3 Effect of residual stresses on the fatigue resistance of hollow section joints

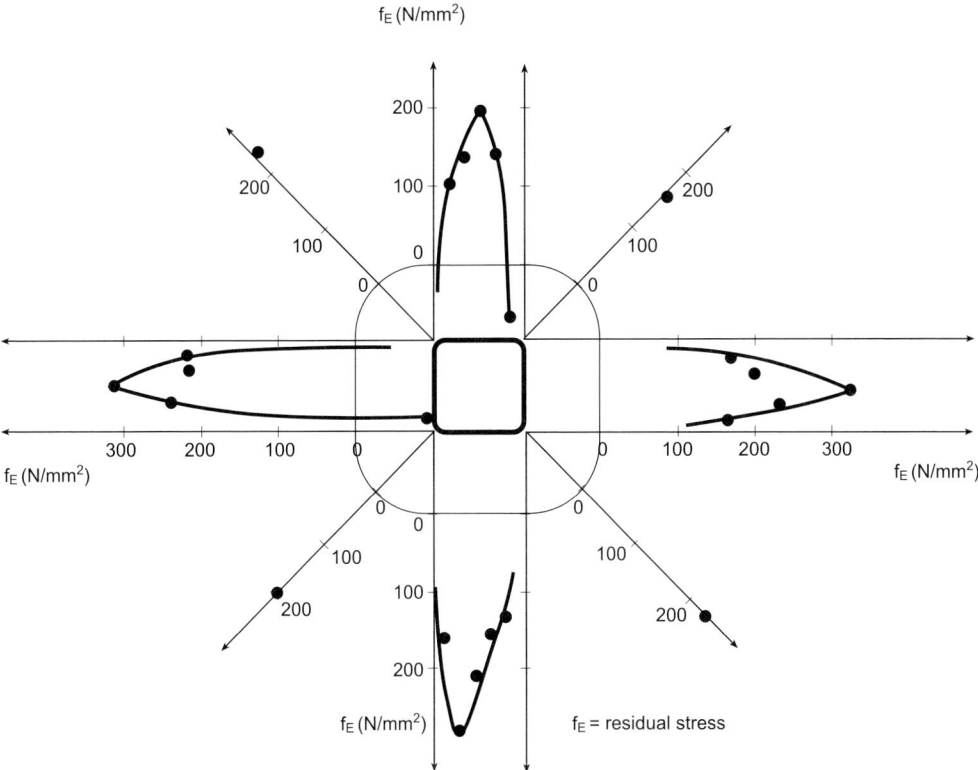

Fig. 7-13 State of residual stress in a cold formed square hollow section

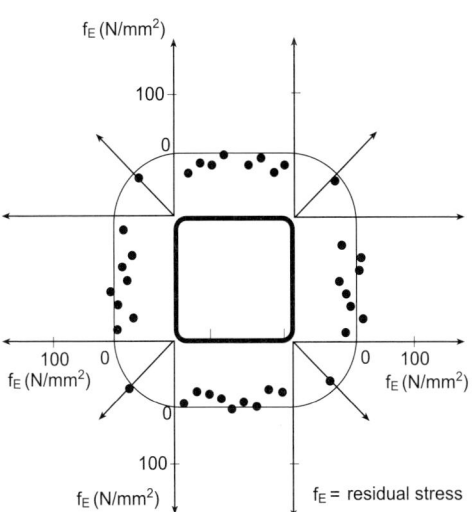

Fig. 7-14 State of residual stress in a hot finished square hollow section

structure is fabricated i.e. residual stress in weld. Figs. 7-13 and 7-14 illustrate the state of residual stress in cold formed and hot finished square hollow sections after they are shaped from plates or "coils". These stresses superpose external loads and can thereby reduce the fatigue life. The internal stresses can be lowered by means of stress relieving heat treatment. This is done by heating to a temperature between 530 to 580 °C according to the steelgrade used. Workpiece is then slowly cooled.

As regards welded structures, residual stresses due to shrinkage of welding (especially when oversized) induce a very different stress distribution in the structure – other than the general calculated stresses.

However, the tests on welded joints of relative thin hollow sections (cold formed or hot finished, with or without stress relieving) from various manufacturing mills (CHS diameter or

RHS width ≤ 400 mm, which are usually applied in steel structures), have demonstrated that the effect of the residual stresses on the fatigue behaviour of these joints is only small [45]. The dominant influence on the fatigue resistance of the hollow section joints is wielded by the weld notches e.g. undercut, crack etc. or the stress concentration due to the geometrical form.

Based on the occurrence of the weld residual stresses, the stress ratio $R = f_{min}/f_{max}$ need not be taken into consideration while designing as-welded hollow section constructions nowadays.

7.4 Effect of corrosive environment on the fatigue of hollow section joints

Corrosion exercises a significant influence on the fatigue behaviour of steel structures and plays a far more critical role than for the constructions under static load. In the static case, the rusting of steel threats a construction in the long run with the reduction of the cross-sectional area, while under fatigue load, the combined action of cyclic loading and corrosive environment often results in significant worsening of fatigue performance (as for example, of offshore structures exposed to seawater environment). Severe corrosion acts like sharp notches reducing the lifetime of the structure under fatigue considerably.

The fatigue life of a welded CHS joint under cyclic loading in seawater (corrosion fatigue) is about 40% of that subjected to "air fatigue" [46]. This subject will not be further dealt with closely in this book, as steel structures exposed to seawater or other aggressive mediums are seldom used in the conventional structural engineering.

It is to be mentioned here in this context that weathering steel can be classified into the same detail categories as slight corrosion notches, which are of less influence than the notches produced by welding. However, welding plus excessive corrosion is very detrimental.

7.5 Stress or strain distribution in hollow section joints

Bolted as well as welded joints are highly susceptible to the notch effects. Peak stresses occur at the notch locations due to local stress concentrations. Under cyclic loading, these locations are very critical, as, contrary to the static loading, the peak stresses cannot be reduced by plastic deformation because of the quick change of the loading cycle. The bolt holes weaken the bolted joints and the peak stresses at the hole edge affects the fatigue behaviour unfavourably. The threads of a bolt act as notches and besides, they are highly loaded in tension. Due to these reasons, these constructions are less suitable to withstand fatigue loading.

Local, enlarged stresses appear at the weld penetration notches e.g. undercut in a welded connection by stress concentration. Often an additional increase of stress takes place at the connection of the structural members with different rigidities (stiffness kink) or at the weld itself (due to the change of stiffness by the accumulation of excess weld metal). As special attention has to be paid to the fatigue of this type in welded hollow section joints, the fatigue investigations on these joints were carried out with priority in the last two decades [5, 9, 10, 31–33, 35, 36, 46, 47].

Fig. 7-15 shows two examples of non-uniform stress distribution in CHS and RHS welded joints, where the chord and the bracing are directly connected with each other. A local stress or strain concentration is initiated at the critical point of discontinuity in shape and cross section of members (usually at weld toe), which affects the fatigue behaviour of welded hollow section joints in a truss predominantly (Figs. 7-16 and 7-17). This causes a many times enlarged nominal stress or strain.

The local peak stress in a construction can be reduced by the redistribution of the stress pattern in the joint due to local plastic deformation. The effect of this process is markedly higher for static load than for the fluctuating one. In the latter case, a fatigue crack starts at the location of the highest (hot spot) stress or strain, which leads to ultimate failure e.g. crack through the wall thickness by further propagation. In order to estimate the fatigue behaviour, the knowledge of the hot spot stress or strain is therefore required.

7.5 Stress or strain distribution in hollow section joints

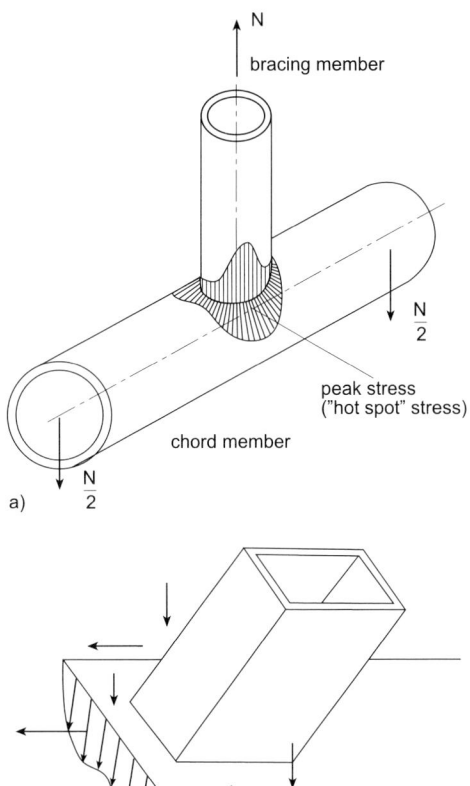

Fig. 7-15 Stress distribution in a) CHS joint, b) RHS joint

The considerations described above lead to the conclusion that the hot spot stress or strain range $S_{r,hs}$ or $\varepsilon_{r,hs}$, which is defined as the local maximum main stress or strain range in a section wall of the joint at the toe (crown) of the weld, is dependent on the following factors:

– Global geometry of the joint described by the geometrical ratios $2 l_0/d_0$ or b_0, d_i/d_0 or b_i/b_0, $d_0/2 t_0$ or $b_0/2 t_0$, t_i/t_0, g/d_0 or g/b_0, θ and joint configuration (see Figs. 7-16 to 7-18, $i = 1, 2$)
– Global geometry of the weld (weld thickness a, see Fig. 6-88; weld length l_W, see Section 6.6.3.3)
– Condition of the weld toe e.g. angle between weld and parent material (Fig. 7-19), undercut (Fig. 8-34, can be improved by grinding metal from the weld toe, see Fig. 7-20)
– Type of loading (tension, compression, in-plane and out-of-plane bending moment, see Fig. 7-21)

As has been already mentioned, the hot spot stress or strain is taken as the basis for designing welded hollow section joints under cyclic loading. In practically all new standards and recommendations as well as recent publications, "hot spot stress range $S_{r,hs}$ vs. loading cycles to failure N_F" lines (usually in logarithmic scale) are applied for this purpose. The existing residual stresses plus the peak stresses (hot spot stress) may lead to yielding ($f_{max} = f_y$) locally, so that the stresses during the applied loading cycles vary between the yield strength f_y of the material and a lower stress f ($f_y - f$). This justifies the application of stress range S_r as a decisive parameter.

In order to establish "nominal stress range $S_{r,nom}$ vs. loading cycles to failure N_F" lines by the experimental method, the nominal stresses in the joint members are measured, in which the effect of the weld form and conditions involving other irregularities is integrated. By this procedure, it is not possible to determine the influence of the weld conditions (notches) on the fatigue behaviour separately from that given by the geometry of the joint only. While establishing a design line "$S_{r,hs}$ or $\varepsilon_{r,hs}$ vs. N_F", the stress or strain measurements or calculations can be made, where the results are valid for the geometry of the joint and the loading type only excluding the effect of weld (flatness, concavity, convexity, undercut etc.). The justification for taking $S_{r,hs}$ or $\varepsilon_{r,hs}$ instead of $S_{r,nom}$ as the basis for the fatigue design of hollow section joints lies in the fact that the effect of the weld on the fatigue behaviour is very much dependent on the fabrication procedure, which may differ from workshop to workshop (welding equipment, qualification of welders etc.). Consequently it is not practicable to formulate a general fatigue design proposal based on nominal stress range.

In the framework of an European offshore research programme in the late seventies, a test procedure was developed for this purpose

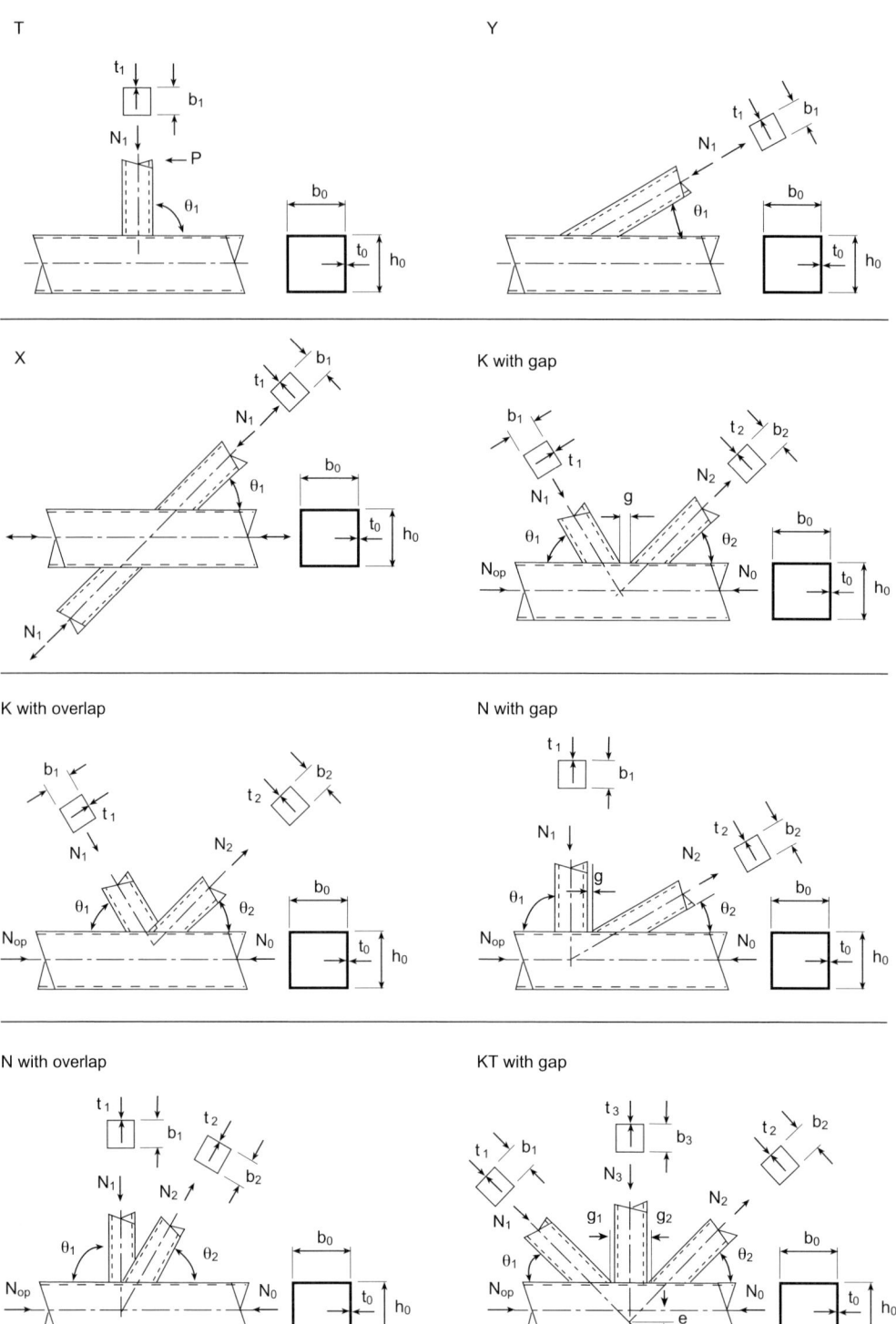

Fig. 7-16 Plane truss joints in RHS (square)

7.5 Stress or strain distribution in hollow section joints

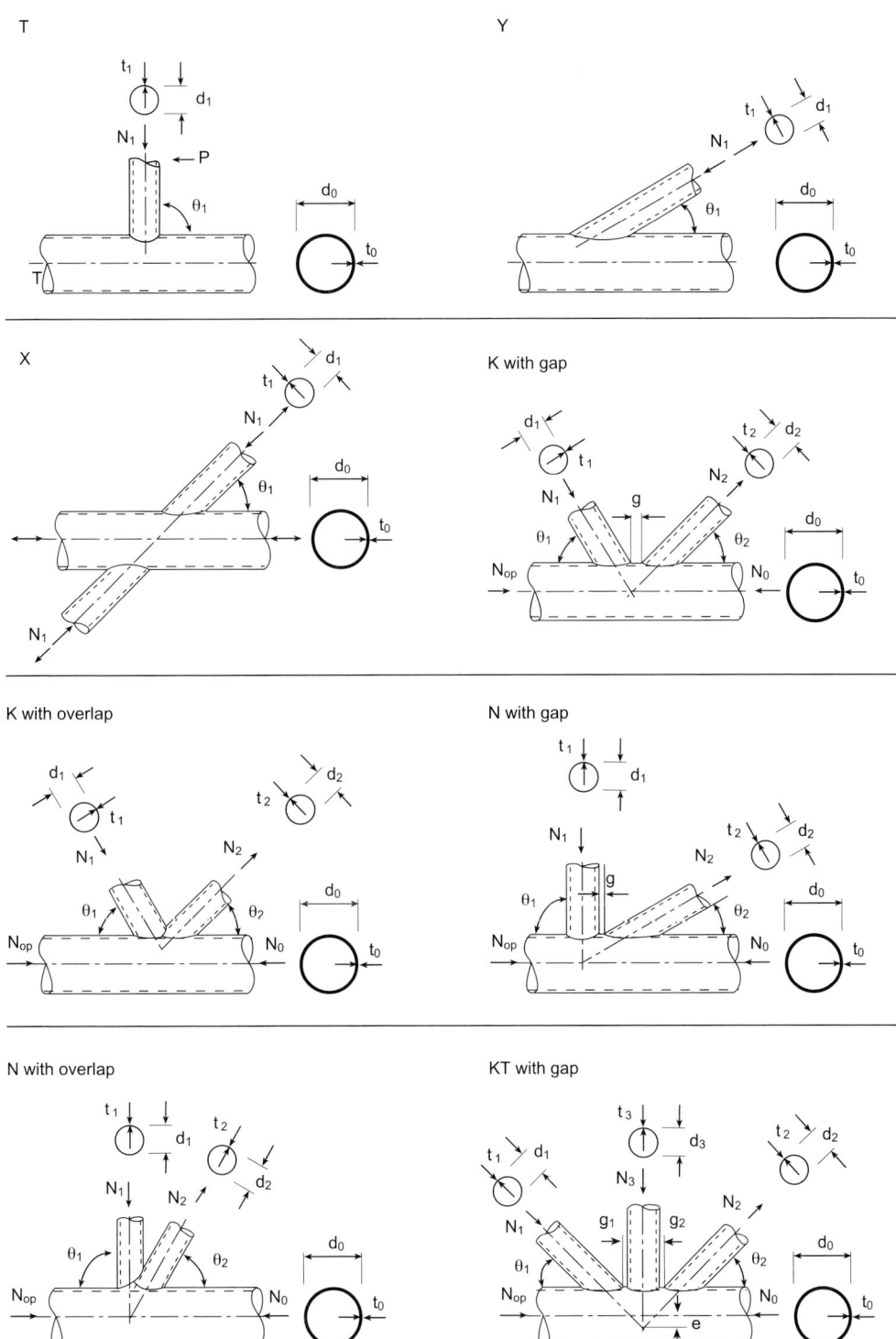

Fig. 7-17 Plane truss joints in CHS

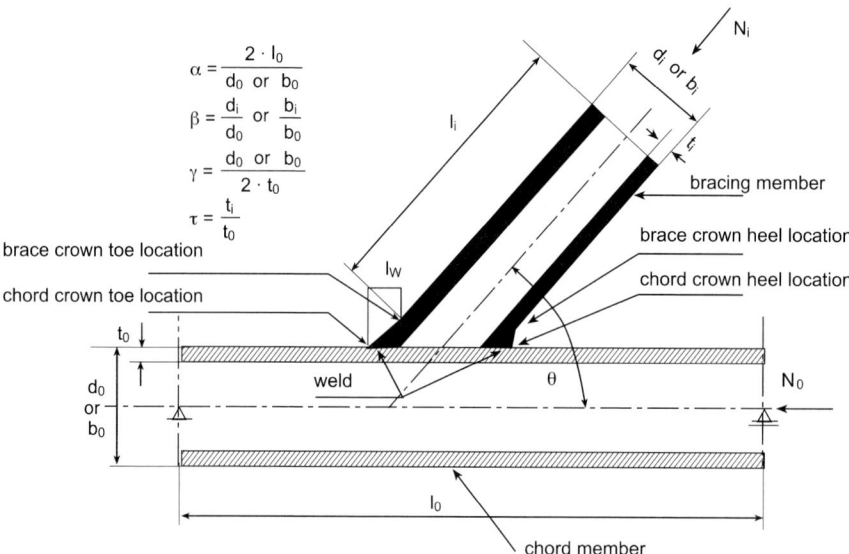

Fig. 7-18 Welded hollow section joint with notations

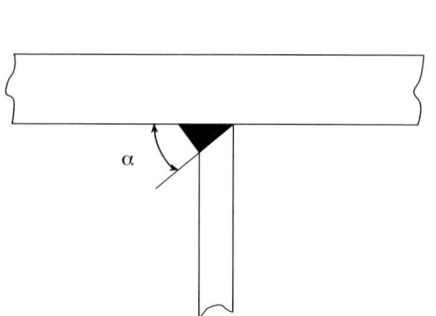

Fig. 7-19 Angle between hollow section and weld

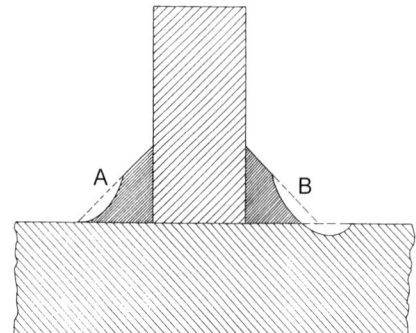

Fig. 7-20 Weld toe grinding – if grinding is performed tangentially to the surface of the hollow section (A), a small improvement of the fatigue strength can be achieved. Depth of grinding should be 0.5 to 0.8 mm below the bottom of the undercut (B) to eliminate its effect

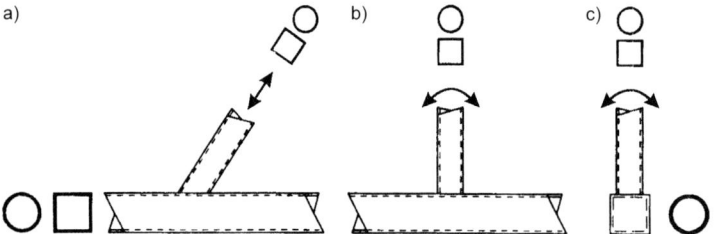

Fig. 7-21 Loading types. a) Axial forces, b) in-plane bending moment (M_{ip}), c) out-of-plane bending moment (M_{op})

7.5 Stress or strain distribution in hollow section joints

[47], which has been since used by most researchers to determine the hot spot stresses or strains in hollow section joints. This procedure as the state-of-the-art is described in Section 7.5.1.

7.5.1 Stress Concentration Factor SCF and Strain Concentration Factor SNCF

Hot spot stress range $S_{r,hs}$ adjacent to the weld toe of the members is the decisive value to construct and design hollow section joints. The hot spot stresses are determined in general by the following methods:

- Measurement of strain on the surface of the structural members with strain gauges at 1:1 scale
- Calculation of stress or strain theoretically by means of finite element analysis
- Photoelastic procedure with acrylic plastic models
- Method applying fracture mechanics

The first two are most suitable for practical application and are most frequently used by the researchers.

7.5.1.1 Experimental measurement of strains with strain gauges

The aim of this procedure is the measurement of hot spot strain on the surface of a welded hollow section joint taking only the "geometric" influence based on the relevant parameters into consideration. The influence of the weld notches at the weld toe is excluded.

The standard procedure recommended by ECSC WG III [47] consists of the extrapolation of the measured strains by strain gauges from a defined distance to the weld toe (Figs. 7-22 and 7-23). Principal strains are determined at each point using rosettes with two element 90° strain gauges and then the principal stresses are resolved assuming, the shear strain near the weld is negligible. It is recommended that the hot spot stresses are determined perpendicular to the weld toe, which can be calculated as follows:

$$f_x = E \cdot \varepsilon_x \cdot \frac{1 + v\,\varepsilon_y/\varepsilon_x}{1 - v^2}$$

where:
f_x is the calculated stress perpendicular to the weld, ε_x is the measured strain perpendicular to the weld and ε_y parallel to the weld, E and v are the modulus of elasticity and the Poisson's ratio respectively.

Assuming a similar degree of bi-axiality ($\varepsilon_y/\varepsilon_x$), this approach is simplified by measuring only ε_x with strip gauges and by resolving a fictitious hot spot stress based on assumed strain ratio.

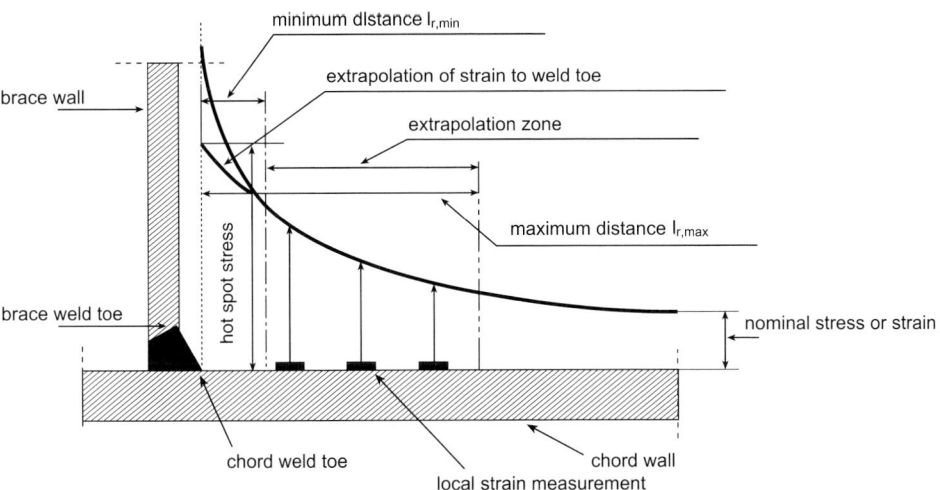

Fig. 7-22 Determination of hot spot strain by extrapolation of strain measurements by strain gauges

stress distribution in bracing member

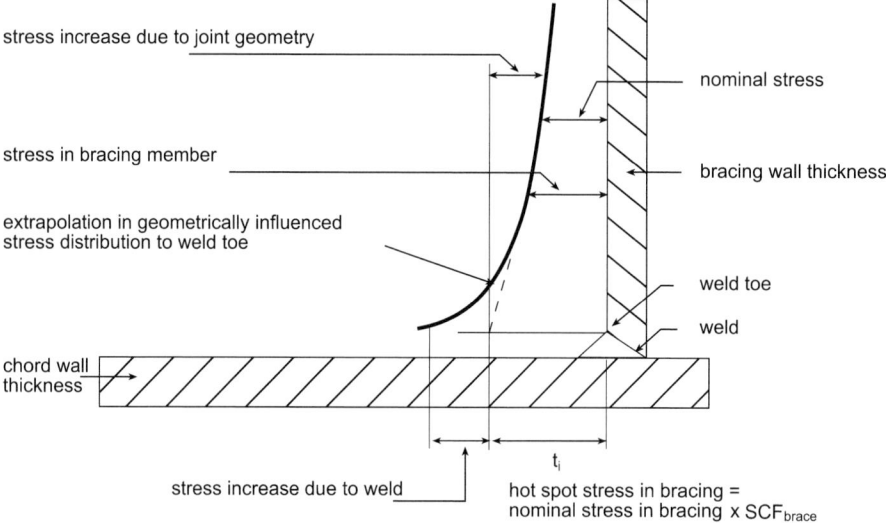

stress distribution in chord member

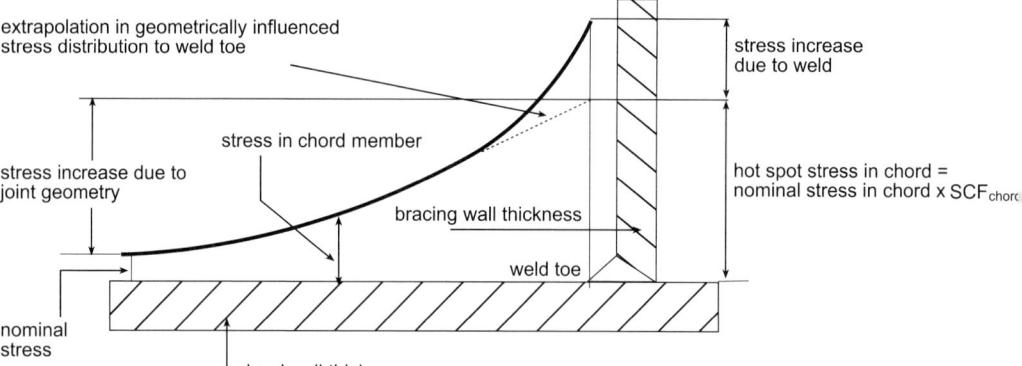

Fig. 7-23 Definitions for hot spot stress in welded hollow section joint

The first point of extrapolation is outside the influence area of weld. Further strip strain gauges are located perpendicular to the weld toe, so that only primary strains in this direction are considered. This is done mainly to avoid the use of strain gauge rosettes, because the length of the extrapolation region is insufficient for their accommodation. At the same time, this saves labour and expenditure.

Fig. 7-24 illustrates the linear and quadratic extrapolation methods used for welded RHS joints (these methods can be applied to welded CHS joints analogously).

For circular hollow section joints, the determination of the strains at the two specified points A and B is done basing on strain data or measurements by quadratic extrapolation (using least square method) to these points.

7.5 Stress or strain distribution in hollow section joints

The extrapolation to the weld toe C over A and B is based on linear extrapolation from these points leading to the hot spot strain range $\varepsilon_{r,hs}$ (compare with Fig. 7-24a). According to ECSC WG III [51], the maximum and minimum distances $l_{r,max}$ and $l_{r,min}$ measured from the weld toe location in a direction perpendicular to the weld toe for the chord member locations and parallel to the axis of the bracing member for brace member locations are given by:

On chord member:

$l_{r,min} = 0.4\ t_0$, minimum 4 mm (crown and saddle location)
$l_{r,max} = 0.4\ (r_1 \cdot t_1 \cdot r_0 \cdot t_0)^{0.25}$ (crown)
$l_{r,max} = r_0 \cdot \pi \cdot 5/180$

On bracing member:

$l_{r,min} = 0.4\ t_1$, minimum 4 mm (crown and saddle location)
$l_{r,max} = 0.55\ (r_1 \cdot t_1)^{0.5}$

r_0 and r_1 designate the external radii of the chord and bracing members respectively.

For joints in rectangular hollow sections, the geometrical strain from the maximum distance DF to the weld toe (Fig. 7-24b) can be strongly non-linear and therefore a quadratic extrapolation method should give more realistic value for the hot spot strain at the weld toe. All data points between the both limiting values at E ($FE = 0.4\ t_i$, minimum 4 mm) and at D ($ED = 1.0\ t_i$) are joined by means of a parabolic quadratic curve (using least square method). An extension of this curve to the weld toe F results in the hot spot strain range $\varepsilon_{r,hs}$.

Further, it is important to set up appropriate arrangements of strain gauges suited to the joint types and loading art. Fig. 7-25a, b shows a series of the strain gauge arrangements, which have been used in various research projects [9, 31, 33–36, 47-49]. The failure modes, which have been observed during the test execution, confirm the justification of these arrangements. Fig. 7-26 illustrates a number of these failure modes.

7.5.1.2 Theoretical determination of stresses or strains using "finite element" analysis

The finite element modelling of a joint consists of translating a joint and its loading into a mathematical model, the numerical solution of which can be achieved by the use of finite elements (shell, solid, beam etc.). Geometry and material properties, as well as load and boundary conditions assumed for the model, have to correspond to those of the joint as realistically as possible. The compatibility to the real joint behaviour and the accuracy of the numerical work depend largely on the type of element, mesh refinement, integration scheme and the weld shape considered. As found out by the research works already done [48, 50, 51, 53, 54, 66], the choice of element type and element size in the region of the intersection of the members is of major importance. Thick shell elements in combination with solid elements for modelling of the weld are to be recommended in general. The ambiguity of the application of shell elements at the location of weld toe can be overcome by using solid elements for the weld region, with proper modelling of the weld shape [87]. Mesh size in the hot spot region needs to be sufficiently small to allow extrapolation of stresses.

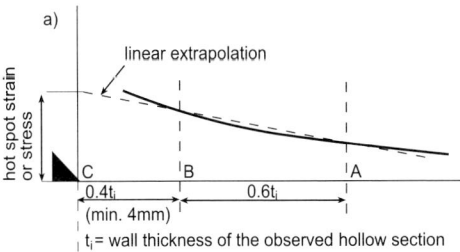

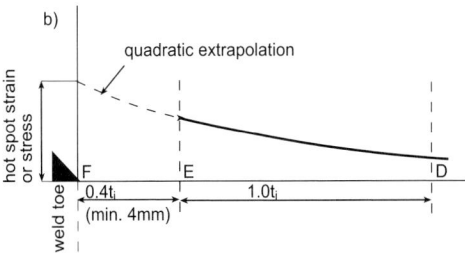

Fig. 7-24 Extrapolation methods. a) Linear extrapolation, b) quadratic extrapolation (RHS joints)

7 Design of welded hollow section joints subjected to fatigue loading

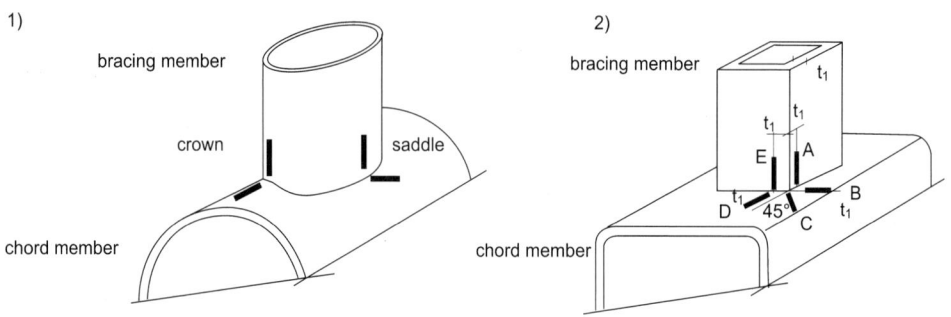

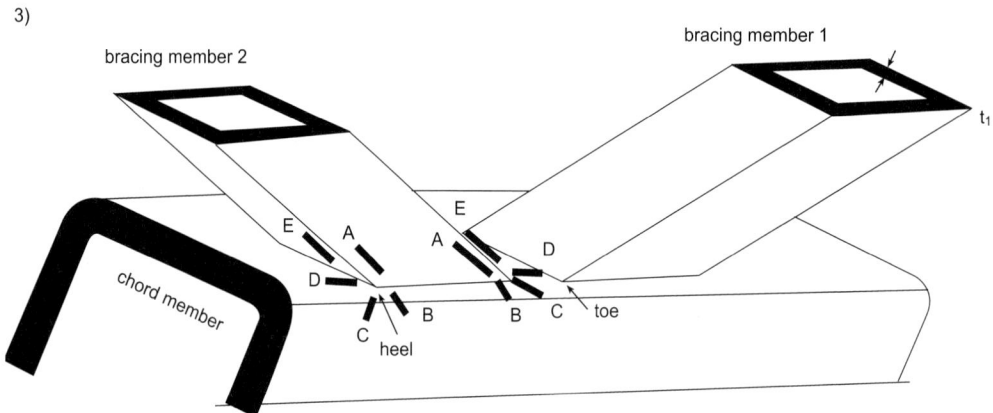

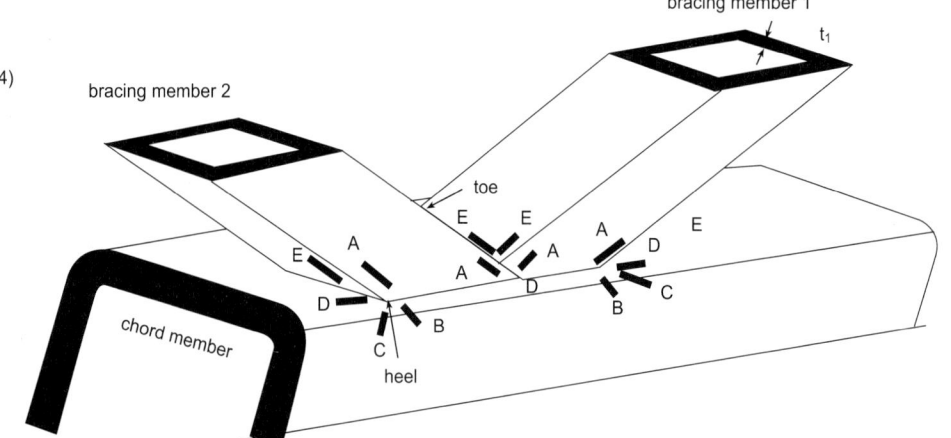

Fig. 7-25 a) Arrangements of strain gauges in uni-planar hollow section joints

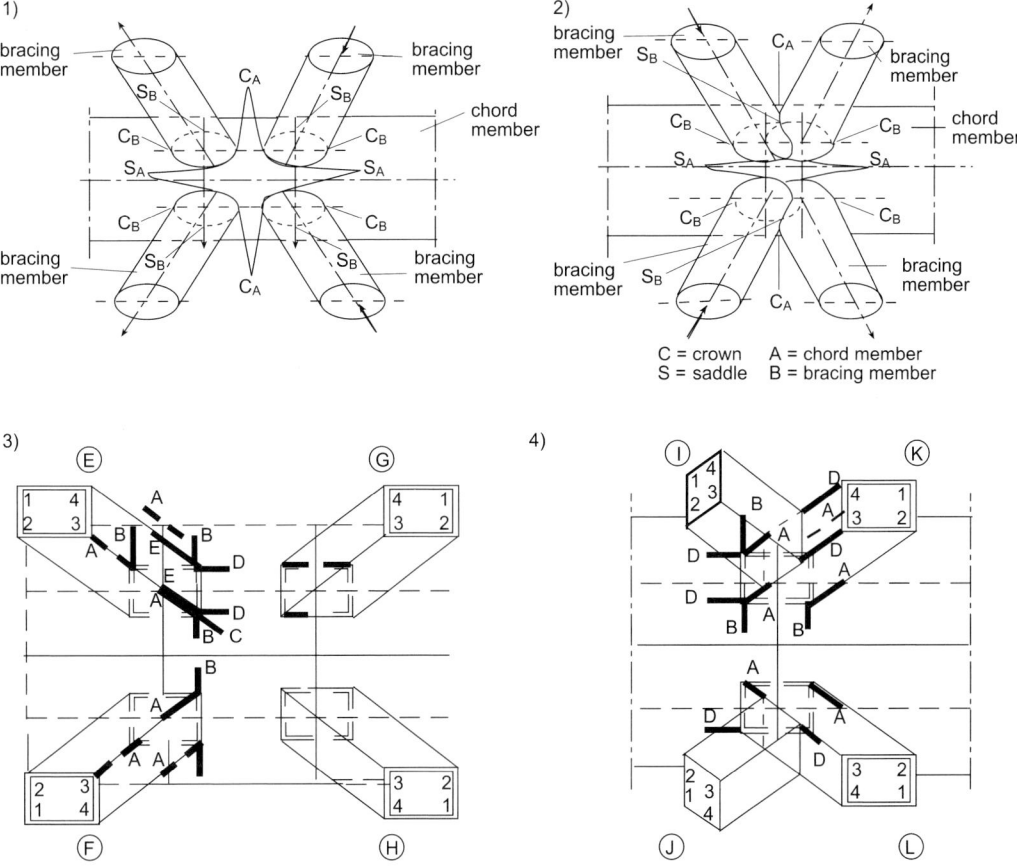

Fig. 7-25 b) Arrangements of strain gauges in multi-planar hollow section joints (KK type for a triangular girder)

The results of the experiments described in Section 7.5.1.1 should be compared with those by finite element analysis to come to the best possible finite element models. On the basis of the model with the best agreement, numerical work can be performed with extensive parametric variations. This enables to carry out parameter studies on a theoretical basis, which are less expensive than practical experiments.

7.5.1.3 Determination of stress or strain concentration factors (SCF or SNCF)

For the convenience of the calculation of hot spot stress range $S_{r,hs}$ in a hollow section truss joint, a relationship between the nominal stress range $S_{r,nom}$ in a bracing member, which is one of the initial inputs for the design calculation of a hollow section joint, and the hot spot stress in the joint has been established. This is defined as the stress concentration factor (SCF), which is the ratio of the geometrical hot spot stress range in the joint to the nominal stress range in the adjacent bracing member. In multi-planar joints, each bracing member has to be considered.

$$\text{SCF} = \frac{S_{r,hs,i,j,k} \text{ (in joint)}}{S_{r,nom} \text{ (in brace)}}$$

where:
i = chord or bracing member
j = location (e.g. crown, saddle)
k = loading type (ax = axial load, ip = in-plane bending, op = out-of-plane bending)

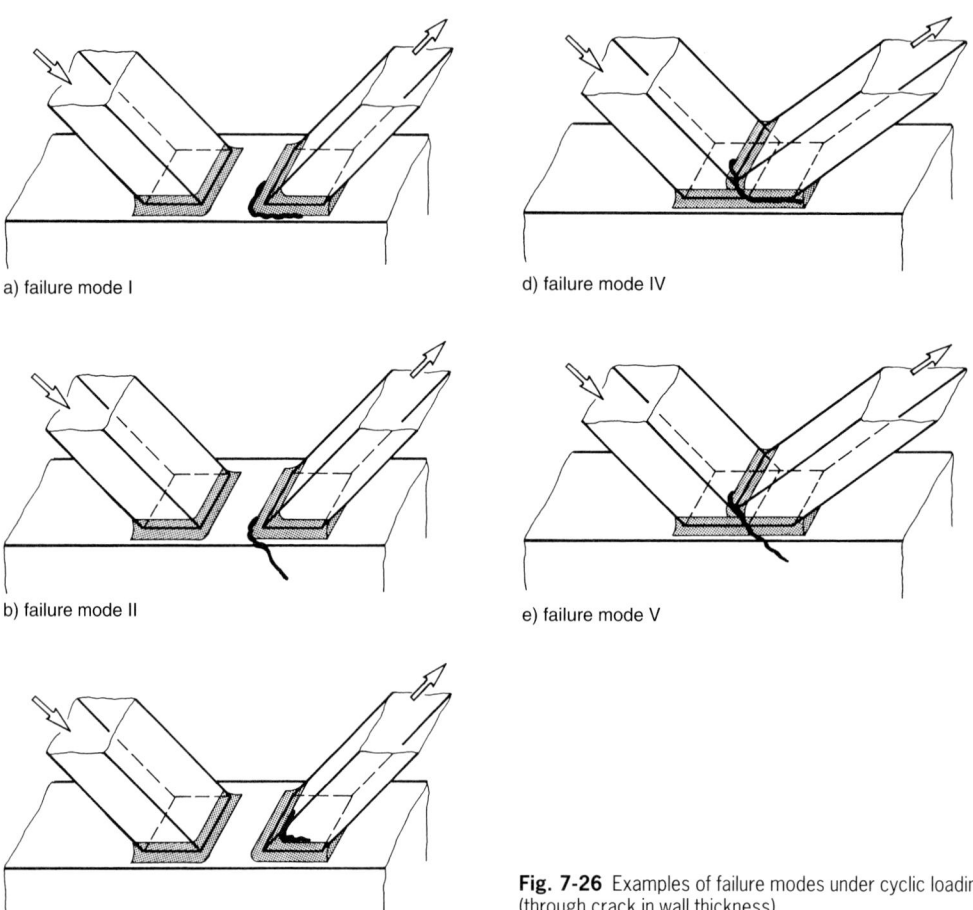

Fig. 7-26 Examples of failure modes under cyclic loading (through crack in wall thickness)

a) failure mode I
b) failure mode II
c) failure mode III
d) failure mode IV
e) failure mode V

Strain concentration factor SNCF is defined similarly:

$$\text{SNCF} = \frac{\varepsilon_{r,hs,i,j,k} \text{ (in joint)}}{\varepsilon_{r,\text{nom}} \text{ (in brace)}}$$

Although in most design recommendations, hot spot stress and stress concentration factors SCF are used as the basis for calculation, the design is really based on hot spot strain concentration factors SNCF. For experiments, this can be explained by the fact that the measurements are made by strain gauges.

The use of "strain" as the ultimate basis for measurement and calculation has one important advantage that the strains for exceeding the yield strain during a fatigue test, as can be the case for low cycle fatigue, show very little deviations in behaviour. Another problem may arise while determining SCF for hollow section joints, when the yield strength of the joint material is exceeded locally even by the application of a smaller load. It is therefore practicable to use SNCF instead of SCF as the basis for fatigue design, as the influence of the local plasticity is integrated in the strain itself. The other advantage is due to the easy measurement of strain by individual strip strain gauges in arbitrary directions, while strain gauge rosettes for various strain components have to be used to determine stresses.

However, the predominant use of SCF in the practical design necessitates to consider the conversion of strain into stress.

7.5 Stress or strain distribution in hollow section joints

The nominal strain and stress range can be easily converted:

$$S_{r,\text{nom}} = E \cdot \varepsilon_{r,\text{nom}}$$

where:
E = modus of elasticity (Young's modulus)

The nominal strain range in the tension bracing member of a CHS joint is defined as:

$$\varepsilon_{r,\text{nom}} = \varepsilon_{r,ax,\text{nom}} + \varepsilon_{r,res,\text{nom}}$$

where:
$$\varepsilon_{r,res,\text{nom}} = \sqrt{\varepsilon_{r,ip,\text{nom}}^2 + \varepsilon_{r,op,\text{nom}}^2}$$

and for a RHS joint as:

$$\varepsilon_{r,\text{nom}} = \varepsilon_{r,ax,\text{nom}} + \varepsilon_{r,ip,\text{nom}} + \varepsilon_{r,op,\text{nom}}$$

The nominal stress range can be calculated by multiplying the nominal strain range with the Young's modulus E for any arbitrary combination of loads.

Using the strain concentration factor SNCF and the nominal strain ranges $\varepsilon_{r,\text{nom}}$ described above, the hot spot strain range $\varepsilon_{r,hs}$ can thus be determined considering all actions on the joints, namely, axial loads and in-plane and out-of-plane bending moments:

$$\varepsilon_{r,hs} = \text{SNCF}_{ax} \cdot \varepsilon_{r,ax,\text{nom}} + \text{SNCF}_{ip} \cdot \varepsilon_{r,ip,\text{nom}} + \text{SNCF}_{op} \cdot \varepsilon_{r,op,\text{nom}}$$

The ratio between SCF and SNCF or the "so-called" conversion factor Snf = SCF/SNCF can vary substantially between 0.6 and 1.4. In the framework of the European Offshore Programme, van Delft et al. [50] found a mean Snf value of 1.15 for CHS joints. Romijn [51] proposed the following equation for Snf of CHS joints assuming a plane-stress condition and a fully isotropic behaviour of steel with $E = 2.068 \cdot 10^5$ N/mm² and $\nu = 0.3$ (ν = Poisson's ratio):

$$\text{Snf} = \frac{\text{SCF}}{\text{SNCF}} = 1.10 + 0.33 \frac{\varepsilon_y}{\varepsilon_{x,hs}}$$

where:
ε_y = strain measured in the direction parallel to the weld toe (chord member locations on outer surface) or along the bracing member surface perpendicular to the axis of the bracing member (bracing member locations on outer surface)

For K type RHS joints, Frater [52] determined Snf ratio between 1.091 and 1.146. Wingerde [53] proposed the mean Snf value for T and X joints in square hollow sections to be 1.1, while Panjeshahi [54] found this average Snf value for multi-planar KK joints in square hollow sections to be 1.12.

Based on these research works, the following conversion factors have been proposed by the recent researchers [48, 49, 51, 53, 54] to be used in general:

SCF = 1.2 · SNCF for CHS joints
SCF = 1.1 · SNCF for RHS joints

7.5.1.4 Determination of total hot spot stress range $S_{r,hs,\text{total}}$ using stress concentration factors

The total hot spot stress $S_{r,hs,\text{total}}$ at a particular location a around the bracing to chord connection (e.g. crown, saddle or in-between for CHS joints) is to be determined by the superposition of the hot spot stress range $S_{r,hs}$ in the individual joint members for a combination of various types of loads (axial load, in-plane and out-of-plane bending moments):

$$S_{r,hs,\text{total},a} = \\ S_{r,ax,br,\text{nom},a} \cdot \text{SCF}_{ax,br,a} + S_{r,ip,br,\text{nom},a} \\ \cdot \text{SCF}_{ip,br,a} + S_{r,op,br,\text{nom},a} \cdot \text{SCF}_{op,br,a} \\ + S_{r,ax,ch,\text{nom},a} \cdot \text{SCF}_{ax,ch,a} + S_{r,ip,ch,\text{nom},a} \\ \cdot \text{SCF}_{ip,ch,a} + S_{r,op,ch,\text{nom},a} \cdot \text{SCF}_{op,ch,a}$$

However, $S_{r,hs,\text{total},a}$ can be subdivided according to the share of the chord and the bracings.

This procedure is similar for RHS joints.

$S_{r,hs,\text{total}}$, as described above, are only taken, when the hot spot stress locations corresponding to various applied load types are not identified. Due to safety reason, all hot spot stresses due to axial load and in-plane and out-of-plane bending moment are added together.

In case of multi-planar joints, where bracing members lie in different planes along the chord axis, a distinction is made between reference loads and carry-over loads (see Fig. 7-27). For the XX joint with axial balanced bracing loading (Fig. 7-27b) as an example, the reference effect is caused by the loads on the reference bracing "a" due to $N_{ax,br,a}$ ($M_{ip,br,a}$ and $M_{op,br,a}$ if existing) and the loads on the chord member i.e. SCFs due to $N_{ax,ch}$, $M_{ip,ch}$ and $M_{op,ch}$ (if available). The carry-over effect on SCFs around the connection of the bracing member "a" to the chord member is caused by the loads on the other (carry-over) bracing members. Under general loading conditions, the hot spot stress range at any location, in the **chord** member, is given by:

- For CHS XX joints,

$S_{r,hs,ch} =$
$SCF_{ax,ref\text{-}br} \cdot S_{r,ax,ref\text{-}br} + SCF_{ip,ref\text{-}br} \cdot S_{r,ip,ref\text{-}br}$
$+ SCF_{op,ref\text{-}br} \cdot S_{r,op,ref\text{-}br} + SCF_{ax,ch} \cdot S_{r,ax,ch}$
$+ SCF_{ax,cov\text{-}br} \cdot S_{r,ax,cov\text{-}br} + SCF_{op,cov\text{-}br}$
$\cdot S_{r,op,cov\text{-}br}$

- For all other joints,

$S_{r,hs,ch} =$
$SCF_{ax,br} \cdot S_{r,ax,br} + SCF_{ip,br} \cdot S_{r,ip,br}$
$+ SCF_{op,br} \cdot S_{r,op,br} + SCF_{ax,ch} \cdot S_{r,ax,ch}$
$+ SCF_{ip,ch} \cdot S_{r,ip,ch}$

Under general loading conditions, the hot spot stress at any location, in the **bracing** member, is given by:

- For CHS XX joints,

$S_{r,hs,br} =$
$SCF_{ax,ref\text{-}br} \cdot S_{r,ax,ref\text{-}br} + SCF_{ip,ref\text{-}br} \cdot S_{r,ip,ref\text{-}br}$
$+ SCF_{op,ref\text{-}br} \cdot S_{r,op,ref\text{-}br} + SCF_{ax,cov\text{-}br}$
$\cdot S_{r,ax,cov\text{-}br} + SCF_{op,cov\text{-}br} \cdot S_{r,op,cov\text{-}br}$

- For all other joints,

$S_{r,hs,br} =$
$SCF_{ax,br} \cdot S_{r,ax,br} + SCF_{ip,br} \cdot S_{r,ip,br}$
$+ SCF_{op,br} \cdot S_{r,op,br}$

Efthymiou [59] introduced the concept of an influence function IF (see Fig. 7-30) for CHS joints, which is an expression for the hot spot stress at a certain location of a specific brace arising from a nominal stress of unit magnitude acting on another bracing of the joint. Once the influence functions are established, the hot spot stress at a given location can be obtained by multiplying each influence function with its respective nominal stress and superimposing the contributions from all bracings of the joint.

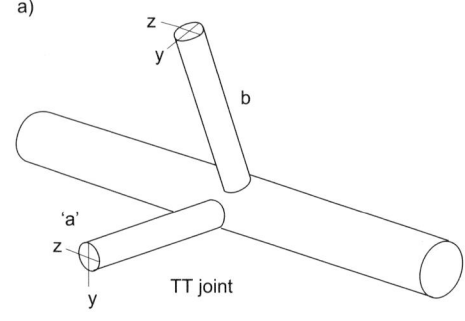
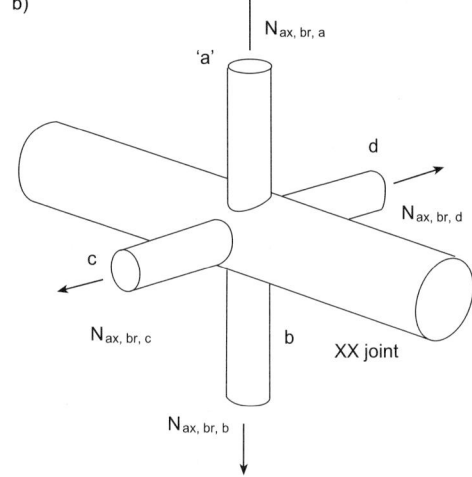
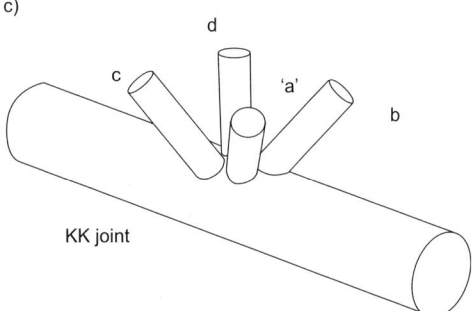

Fig. 7-27 Multi-planar CHS joints: a) TT, b) XX, c) KK

7.5.1.5 Parametric formulae for the determination of stress concentration factors for welded hollow section joints

In order to simplify the numerical calculation of the geometrical stress concentration factors, the parametric formulae have been established applying the following procedures, which are generally based on "best fit" analysis to reference data:

- Determination of SNCF or SCF experimentally with strain gauge measurements.
- Numerical determination of hot spot stresses and stress concentration factors by the "finite element" analysis. A calibrated FE model is developed, which has to result in the best possible agreement with the outcome of the experiments. The best FE models for different types of joints (T, X, K, KT, TT, XX and KK) and loading modes can be obtained by comparing the results of the FE analysis with the test results. The strain is taken as the basis for the FE analysis to be able to make a direct comparison.

The weld seam is modelled according to the measured dimensions.

- Investigation of the effect of the joint parameters α, β, γ, τ, g and θ (see Fig. 7-18) and the loading modes on SCF or SNCF by varying the parameters and then calculating the SCF and SNCF using the best possible FE models.
- Parameter studies based on the extensive test results lead to SNCF formulae worked out with the aid of the regression analysis. They are valid within the investigated parameter ranges. SNCF is transformed into SCF in order to make the application in practice more convenient.

It is required that SCFs derived from FE analysis are independent of the boundary conditions. Therefore, the balanced bending moments at the chord ends of a joint have to be taken into consideration in the case of loading in the bracings.

In general, the parametric equations of the following type are used:

$$SCF = K \cdot \alpha^{n1} \cdot \beta^{n2} \cdot \gamma^{n3} \cdot \tau^{n4} \cdot (\sin\theta)^{n5}$$

where K is constant, ni are exponents and α, β, γ, τ are the geometrical ratios of the joint parameters.

Parametric equations to determine SCF for CHS joints

In the late sixties, extensive research works on CHS joints were carried out in various industrial countries of the world. Numerous parametric formulae or diagrams for SCF have been established based on the results of these investigations. These research projects were restricted to simple T/Y, X, K, N and KT joints under various loading combinations [1, 2, 5, 6, 14–16, 18, 21–23, 55, 56]. However, the results obtained from these research works showed large scatter. The validity of these formulae for general application was therefore called into question.

Reliable research results have been first published in the last fifteen years [4, 51, 57–60]. Lit. [51, 59] deal with the multi-planar TT, XX and KK joints (Fig. 7-27).

The SCF equations for CHS T, Y, X, K and KT joints under axial forces and in-plane and out-of-plane bending moments were set up in the late eighties in the framework of an European Offshore Program carried out in the United Kingdom and then recommended for application [57]. They show good coincidence with the practical test results [61]. Tables 7-1a and b contain the formulae, which are often used for offshore structures in particular.

The above mentioned research works give the following indications regarding the effect of individual geometrical parameters of CHS joints on the SCF:

- Influence of $\alpha = \dfrac{2l_0}{d_0}$

The influence of α is very feeble and therefore in most cases neglected.

- Influence of $\beta = \dfrac{d_i}{d_0}$

For the bracing under axial force, the hot spot stress occurs at the lowest point (saddle) of

Table 7-1a Parametric formulae for the stress concentration factors SCF for welded, unstiffened CHS joints under axial load in the bracing [4, 57]

Load transmission	SCF formulae	Validity range
T/Y joint	Chord: (two controls: saddle (index s) and crown (index c)) $SCF_s = \gamma \cdot \tau \cdot \beta \cdot (6.78 - 6.42\beta^{0.5}) \cdot \sin^{(1.7+0.7\beta^3)}\theta$ or $SCF_c = K'_c + K_o \cdot K''_c$ $K'_c = [0.7 + 1.37 \cdot \gamma^{0.5} \cdot \tau \cdot (1-\beta)] \cdot (2\sin^{0.5}\theta - \sin^3\theta)$ $K_o = \dfrac{\tau\left(\beta - \dfrac{\tau}{2\gamma}\right) \cdot \left(\dfrac{\alpha}{2} - \dfrac{\beta}{\sin\theta}\right) \cdot \sin\theta}{\left(1 - \dfrac{3}{2\gamma}\right)}$ $K''_c = 1.05 + \left[\dfrac{30 \cdot \tau^{1.5}}{\gamma} \cdot (1.2-\beta) \cdot (\cos^4\theta + 0.15)\right]$	
	Brace: $SCF = 1 + 0.63\, SCF_s$ or $SCF = 1 + 0.63\, SCF_c$	$0.13 \leq \beta \leq 1.0$ $12 \leq \gamma \leq 32$ $0.25 \leq \tau \leq 1.0$ $30° \leq \theta \leq 90°$ $8 \leq \alpha \leq 40$
X joint	Chord: $SCF_s = 1.7\gamma \cdot \tau \cdot \beta(2.42 - 2.28\beta^{2.2})\sin^{\beta^2 \cdot (15-14.4\beta)}\theta$	
	Brace: $SCF = 1 + 0.63 \cdot SCF_s$	
K, KT joints	Chord: $SCF_s = [\gamma\tau\beta(6.78 - 6.42\beta^{0.5})] \cdot \left[(\sin^{(1.7+0.7\beta^3)}\theta_1)\right.$ $\left. -((0.012\gamma)^{(2\zeta/3+0.4)}\left(\dfrac{\sin\theta_1}{\sin\theta_2}\right)^{1.8}\sin^{(1.7+0.7\beta^3)}\theta_2\right]$ $SCF_c = 1.1\,\gamma^{0.65}\tau\left(\dfrac{\sin\theta_1}{\sin^{0.5}\theta_2}\right)(2\zeta)^{0.05/\beta} \cdot (1.5\beta^{0.25} - \beta^2)$	$0.13 \leq \beta \leq 1.0$ $12 \leq \gamma \leq 32$ $0.25 \leq \tau \leq 1.0$ $30° \leq \theta \leq 90°$ $\theta_1 \geq \theta_2$ $\theta \leq 90°$ $0.01 \leq \zeta \leq 1.0$ $8 \leq \alpha \leq 40$
	Brace: $SCF = 1 + 0.63\, SCF_s$ or $SCF = 1 + 0.63\, SCF_c$	

7.5 Stress or strain distribution in hollow section joints

Table 7-1 b Parametric formulae for the stress concentration factors SCF for welded, unstiffened CHS joints under in-plane and out-of-plane bending moment in the bracing [4, 57]

Joint type and loading mode	Chord, saddle SCF_s	Chord, crown SCF_c
In-plane bending		
T and Y joint		
X joint	SCF = 0	$0.75\gamma^{0.6}\tau^{0.8}(1.6\beta^{0.25}-0.7\beta^2)$ $\cdot \sin^{(1.5-1.6\beta)}\theta$
K, KT joints		
Out-of-plane bending		
T and Y joint	$\gamma\tau\beta(1.6-1.15\beta^5)\cdot \sin^{(1.35+\beta^2)}\theta$	SCF = 0
X joint	$\gamma\tau\beta(1.56-1.46\beta^5)\cdot \sin^{(\beta^2(15-14.4\beta))}\theta$	SCF = 0
K joint	$\left[\gamma\tau\beta(1.6-1.15\beta^5)\right]\left[(\sin^{(1.35+\beta^2)}\theta_1)+ (\sin^{(1.35+\beta^2)}\theta_2)\cdot((0.016\beta\gamma)^{(\zeta+0.45)})(\theta_1/\theta_2)^{0.3}\right]$ $\left[1-0.1^{(1+4\zeta)}\right]$	SCF = 0
KT joint	$\left[\gamma\tau\beta(1.6-1.15\beta^5)\right]\left[(\sin^{(1.35+\beta^2)}\theta_1)+ (\sin^{(1.35+\beta^2)}\theta_2)\cdot((0.016\beta\gamma)^{(\zeta+0.45)})2(\theta_1/\theta_2)^{0.3}\right]$ $\left[1-0.1^{(1+4\zeta)}\right]^2$	SCF = 0

the weld for β extending up to 0.8. A relative uniform stress distribution over the weld perimeter takes place, when β has a higher value (Fig. 7-28).

As β moves towards 1.0, the type of the stress distribution changes, where the location of the hot spot stress moves away from the "saddle" point depending on the reducing SCF.

In T/Y joints with larger $\gamma = d_0/2\, t_0$, this change is less distinct than in a joint with a more rigid chord. For a critical value of γ, the hot spot stress remains close to the "saddle" point due to a reduced chord rigidity (valid also for $\beta \to 1.0$).

- Influence of $\gamma = \dfrac{d_0}{2\, t_0}$

The deformation under load as well as the non-uniform stress distribution in the connection zone are enlarged with increasing γ ratio. Fig. 7-29 demonstrates the magnitude and distribution of SCF under axial force and in-

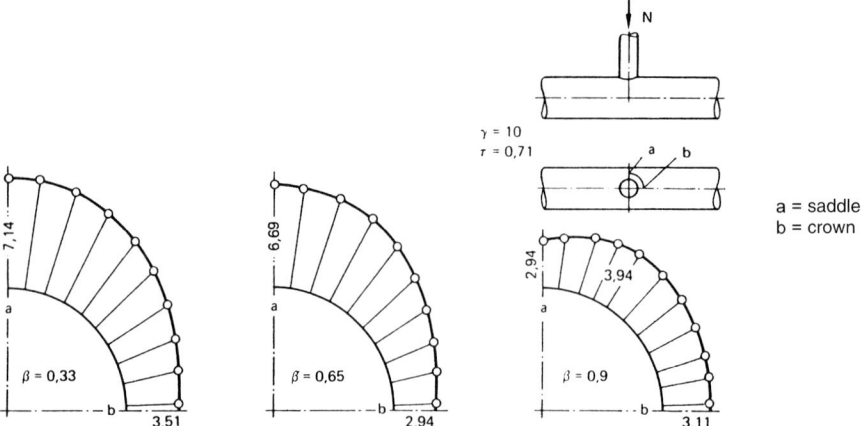

Fig. 7-28 Effect of $\beta = d_i/d_0$ on the SCF distribution on the chord (weld connecting the chord with the bracing) in a CHS T joint

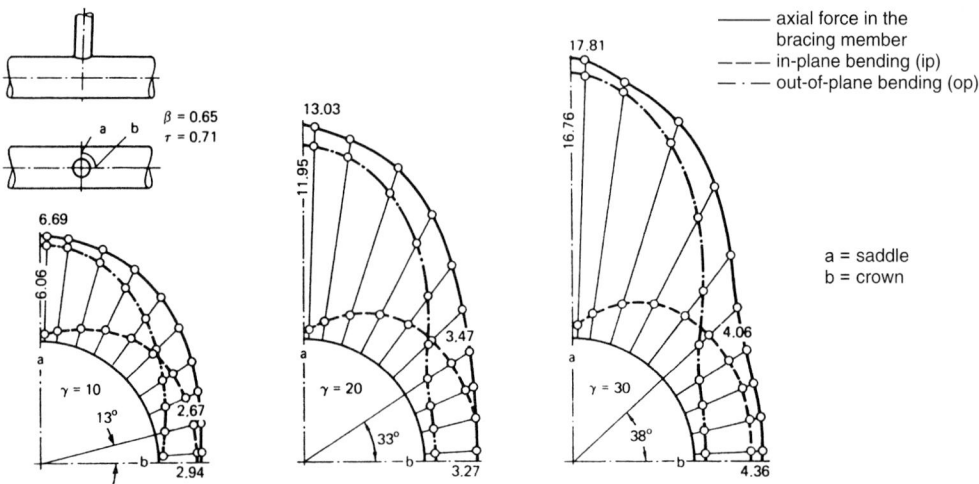

Fig. 7-29 Effect of $\gamma = d_0/2\, t_0$ on the SCF distribution on the chord (weld connecting the chord with the bracing) in a CHS T joint

7.5 Stress or strain distribution in hollow section joints

plane and out-of-plate bending moments for $\gamma = 10/20/30$.

Under in-plane bending moment, a larger lateral shift of the hot spot stress takes place from the central zone of the joint to the crown with increasing γ. The difference between the SCFs at the crown and the lateral location is so small that the shift of the hot spot stress can be neglected. In general, the dependence of SCF and γ is nearly linear (see Table 7-2).

- Influence of $\tau = \dfrac{t_i}{t_0}$

Theoretically, the bracing is the critical structural member for small τ ratio as the hot spot stress in it is larger than that in the chord in this case. Cracks in the bracing or the chord are initiated, mostly at the transition of the weld to the parent material, when a critical limit for τ is exceeded. SCF is enlarged with increasing τ, while the relationship of the both to each other is quasi-linear.

- Influence of the angle of inclination θ of the bracing to the chord

The location of the hot spot stress as well as the SCF value for a CHS joint with an inclined bracing is very susceptible to the loading types and the geometrical parameter ranges. In order to avoid a difficult execution of the weld, the included angle θ, has not to be less than 30°. This requirement corresponds to the lowest limit for the validity of the SCF formulae. The influence of θ is represented by a sine-function, which results the maximum value for SCF at $\theta = 90°$.

- Influence of $\zeta = \dfrac{g}{d_0}$

When ζ increases, the rigidity of K/N or KT joints is reduced insignificantly, while the SCF values become higher. However, this effect is so minor, that this can be neglected for K/N joints under in-plane bending moment and while determining SCF$_{chord}$ in KT joint under axial force.

A minimum value of SCF in K/N and KT joints can be achieved under the following conditions:

$\beta = \dfrac{d_i}{d_0}$ large as far as possible

$\gamma = \dfrac{d_0}{2 t_0}$ small as far as possible

$\tau = \dfrac{t_i}{t_0}$ small as far as possible

$\zeta = \dfrac{g}{d_0}$ small as far as possible

$30° < \theta < 45°$

These conditions are also valid for T/Y and X joints, as long as the unfavourable range with smaller β values is left out of consideration.

Table 7-2 Effect of $\gamma = d_0/2 t_0$ and $\tau = t_i/t_0$ on SCF$_{chord}$

Load type	τ	SCF		$\dfrac{SCF(\gamma = 20)}{SCF(\gamma = 10)}$
		$\gamma = 10$	$\gamma = 20$	
Axial force	0.47	3.75	8.0	2.13
	0.71	6.69	13.03	1.96
	1.0	10.72	21.1	1.96
In-plane bending (IPB)	0.47	1.77	2.37	1.34
	0.71	2.67	3.47	1.3
	1.0	3.79	5.08	1.34
Out-of-plane bending (OPB)	0.47	3.77	7.92	2.1
	0.71	6.06	11.94	1.97
	1.0	9.01	18.04	2.0

A further development was made by the numerical investigations leading to the SCF formulae for CHS joints carried out by Efthymiou [59] (see Tables 7-3a–d). General expressions and the so-called influence functions have been developed to calculate the hot spot stresses in simple as well as complex joints under various loadings in the bracing members. The derivation of the influence functions is based on the superimposing of linear elastic stress fields.

Fig. 7-30 illustrates the concept of the influence function IF. Using the superposition principle, the influence function is derived for a certain location in a CHS joint (e.g. chord saddle), which results in the connection of a specific bracing (i) arising from a nominal axial stress on another bracing (j) (Fig. 7-30a).

Fig. 7-30b shows the method to determine the hot spot stress at a certain location on the chord at the connection with the reference bracing i, which is obtained by superposition summing up the contributions from all other bracings j, k and l. In case the bracing i is also loaded, the hot spot stress due to this load has to be added to the superposed hot spot stresses.

Tables 7-3a–d list the formulae for stress concentration factors SCF of CHS T, Y, X, K and KT joints recommended by Efthymiou [59]. Fig. 7-31 contains the definitions for the geometrical parameter together with the validity ranges for the application of the SCF formulae proposed by Efthymiou.

Further, the influence functions (X, K and KT joints) for the superposition of the effects exercised by all member loads at a particular location are contained in Tables 7-4a–c. Table 7-5 describes the influence functions for the bracings in multi-planar joints. In addition, Table 7-6 shows the hot spot stresses for KT joints under out-of-plane bending moment (OPB).

Finally, Romeijn [51] in his dissertation determined the stress concentration factors for several common types of CHS, welded uni-planar (T, Y, X and K) and multi-planar (TT, XX and KK) joints (Fig. 7-27) with the aid of tests and finite element analysis. The basis of this research was mainly given by the experimental results described in [49, 57, 64–67] and the restricted (only for axial force) numerical studies on CHS multi-planar joints [59]. Figs. 7-32 to 7-34 show a number of examples for tests on joints in triangular girders [49]. However, no parametric formulae for SCF were developed, as the large number of loading and geometrical parameter was too numerous and confusing. Instead, substantial quantity of results have been stored in the data files, from which, for e.g. by the use of an input-file and a programme-file, the SCFs (SNCFs) and hot spot stresses (strains) can be easily obtained.

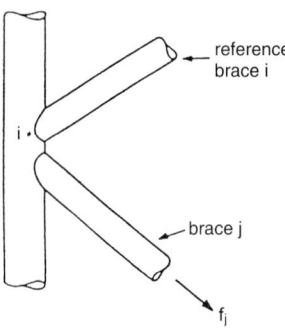

a) Uni-planar joint (uni-axial)
IF_{ij} = influence function at chord saddle location of brace i due to axial load on brace j
$f_{i,hs}$ = hot spot stress at chord saddle of brace i due to axial load on brace j
$f_{i,hs} = f_j \cdot IF_{ij}$ (Geometry)
f_j = nominal stress in brace j

b) Multi-planar joint (multi axial)
$f_{i,hs}$ = hot spot stress at chord saddle of brace i due to axial loads on braces $j, k,$ and $l = f_j \cdot IF_{ij} + f_k \cdot IF_{ik} + f_l \cdot IF_{il}$
f_j, f_k, f_l = nominal stress in the corresponding braces $j, k,$ and l

Fig. 7-30 Definition of influence function

7.5 Stress or strain distribution in hollow section joints

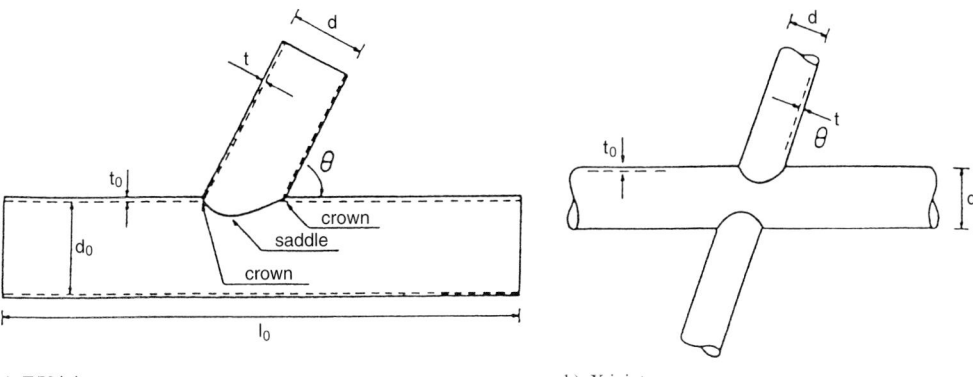

a) T/Y joints

b) X joint
$\beta = d/d_0$
$\gamma = d_0/2t_0$
$\tau = t/t_0$
$\alpha = 2l_0/d_0$

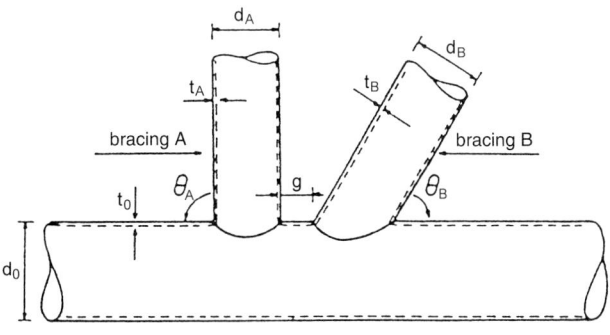

c) K joint
$\beta_A = d_A/d_0; \quad \beta_B = d_B/d_0$
$\tau_A = t_A/t_0; \quad \tau_B = t_B/t_0$
$\gamma = d_0/2t_0; \quad \zeta = g/d_0$

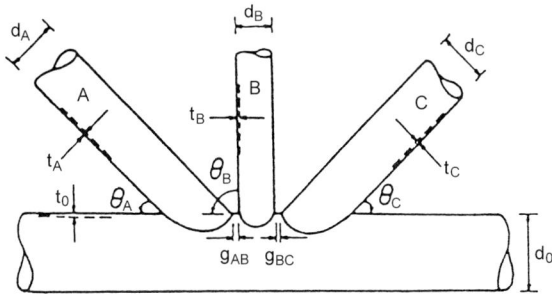

d) KT joint
$\beta_A = d_A/d_0; \quad \beta_B = d_B/d_0; \quad \beta_C = d_C/d_0$
$\tau_A = t_A/t_0; \quad \tau_B = t_B/t_0; \quad \tau_C = t_C/t_0$
$\zeta_{AB} = g_{AB}/d_0; \quad \zeta_{BC} = g_{BC}/d_0; \quad \gamma = d_0/2t_0$

validity ranges
$0.2 \leq \beta \leq 1.0$
$0.2 \leq \tau \leq 1.0$
$8 \leq \gamma \leq 32$
$4 \leq \alpha \leq 40$
$20° \leq \theta \leq 90°$
$\dfrac{-0.6\beta}{\sin\theta} \leq \zeta \leq 1.0$

Fig. 7-31 Definitions for the geometrical parameter and the validity ranges for Tables 7-3 to 7-6

Table 7-3a SCF formulae for T/Y CHS joints [59]

Loading types and fixity conditions	SCF formulae	Equation number	Correction factor for "short" chord*
Axial force Fixed chord ends	Chord saddle: $\gamma \tau^{1.1} [1.11 - 3(\beta - 0.52)^2] \sin^{1.6} \theta$	T1	F1
	Chord crown: $\gamma^{0.2} \tau [2.65 + 5(\beta - 0.65)^2] + \tau \beta (0.25 \alpha - 3) \sin \theta$	T2	none
	Bracing saddle: $1.3 + \gamma \tau^{0.52} \alpha^{0.1} [0.187 - 1.25 \beta^{1.1} (\beta - 0.96)] \cdot \sin^{(2.7 - 0.01 \alpha)} \theta$	T3	F1
	Bracing crown: $3 + \gamma^{1.2} [0.12 \exp(-4\beta) + 0.011 \beta^2 - 0.045] + \beta \tau (0.1 \alpha - 1.2)$	T4	none
Axial force General fixity conditions	Chord saddle: $[T1] + C_1 (0.8 \alpha - 6) \tau \beta^2 (1 - \beta^2)^{0.5} \sin^2 2\theta$	T5	F2
	Chord crown: $\gamma^{0.2} \tau [2.65 + 5(\beta - 0.65)^2] + \tau \beta (C_2 \cdot \alpha - 3) \sin \theta$	T6	none
	Bracing saddle: Eq. (T3)		F2
	Bracing crown: $3 + \gamma^{1.2} [0.12 \exp(-4\beta) + 0.011 \beta^2 - 0.045] + \beta \tau (C_3 \cdot \alpha - 1.2)$	T7	none
In-plane bending moment	Chord crown: $1.45 \beta \tau^{0.85} \gamma^{(1 - 0.68 \beta)} \sin^{0.7} \theta$	T8	none
	Bracing crown: $1 + 0.65 \beta \tau^{0.4} \gamma^{(1.09 - 0.77 \beta)} \sin^{(0.06 \gamma - 1.16)} \theta$	T9	none
Out-of-plane bending moment	Chord saddle: $\gamma \tau \beta (1.7 - 1.05 \beta^3) \sin^{1.6} \theta$	T10	F3
	Bracing saddle: $\tau^{-0.54} \gamma^{-0.05} (0.99 - 0.47 \beta + 0.08 \beta^4) \cdot [T10]$	T11	F3

* Correction factor for $\alpha < 12$:
$F1 = 1 - (0.83 \beta - 0.56 \beta^2 - 0.02) \gamma^{0.23} \exp[-0.21 \gamma^{-1.16} \alpha^{2.5}]$
$F2 = 1 - (1.43 \beta - 0.97 \beta^2 - 0.03) \gamma^{0.04} \exp[-0.71 \gamma^{-1.38} \alpha^{2.5}]$
$F3 = 1 - 0.55 \beta^{1.8} \gamma^{0.16} \exp[-0.49 \gamma^{-0.89} \alpha^{1.8}]$
where $\exp[x] = e^x$

Fixity parameter for chord ends
$C_1 = 2(C - 0.5)$
$C_2 = C/2$
$C_3 = C/5$
C = fixity parameter for chord ends
$0.5 \leq C \leq 1.0$. standard $C = 0.7$
$C = 0.5$ (fully fixed chord end)
$C = 1.0$ (pinned chord end)

7.5 Stress or strain distribution in hollow section joints

Table 7-3b SCF formulae for CHS X joints [59]

Loading types	SCF formulae	Equation number
Axial force (balanced)	Chord saddle: $3.87\,\gamma\tau\beta(1.10-\beta^{1.8})(\sin\theta)^{1.7}$	X1
	Chord crown: $\gamma^{0.2}\tau[2.65+5(\beta-0.65)^2]-3\tau\beta\sin\theta$	X2
	Bracing saddle: $1+1.9\,\gamma\tau^{0.5}\beta^{0.9}(1.09-\beta^{1.7})\sin^{2.5}\theta$	X3
	Bracing crown: $3+\gamma^{1.2}[0.12\exp(-4\beta)+0.011\beta^2-0.045]$	X4
	In joints with $\alpha<12$, the saddle SCFs can be reduced by the factor F1 (fixed chord ends) or F2 (pinned chord ends) where: $F1=1-(0.83\beta-0.56\beta^2-0.02)\gamma^{0.23}\exp[-0.21\gamma^{-1.16}\alpha^{2.5}]$ $F2=1-(1.43\beta-0.97\beta^2-0.03)\gamma^{0.04}\exp[-0.71\gamma^{-1.38}\alpha^{2.5}]$	
In-plane bending moment	Chord crown: Eq. (T8)	
	Bracing crown: Eq. (T9)	
Out-of-plane bending moment (balanced)	Chord saddle: $\gamma\tau\beta(1.56-1.34\beta^4)(\sin\theta)^{1.6}$	X5
	Bracing saddle: $\tau^{-0.54}\gamma^{-0.05}(0.99-0.47\beta+0.08\beta^4)\cdot[X5]$	X6
	In joints with $\alpha<12$, Eqs. X5 and X6 can be reduced by the factor F3, where $F3=1-0.55\,\beta^{1.8}\gamma^{0.16}\exp[-0.49\gamma^{-0.89}\alpha^{1.8}]$	

Table 7-3b (continued)

Loading types	SCF formulae	Equation number
Axial force on one bracing only	Chord saddle: $[T5] \cdot [1 - 0.26 \beta^3]$	X7
	Chord crown: Eq. (T6)	
	Bracing saddle: $[T3] \cdot [1 - 0.26 \beta^3]$	X8
	Bracing crown: Eq. (T7)	
	In joints with $\alpha < 12$, the saddle SCFs can be reduced by the factor F1 (fixed chord ends) or F2 (pinned chord ends) where:	
	$F1 = 1 - (0.83\beta - 0.56\beta^2 - 0.02)\gamma^{0.23} \exp[-0.21\gamma^{-1.16}\alpha^{2.5}]$	
	$F2 = 1 - (1.43\beta - 0.97\beta^2 - 0.03)\gamma^{0.04} \exp[-0.71\gamma^{-1.38}\alpha^{2.5}]$	
Out-of-plane bending moment on one bracing only	Chord saddle: Eq. (T10)	
	Bracing saddle: Eq. (T11)	
	In joints with $\alpha < 12$, Eqs. T10 and T11 can be reduced by the factor F3, where	
	$F3 = 1 - 0.55 \beta^{1.8} \gamma^{0.16} \exp[-0.49\gamma^{-0.89}\alpha^{1.8}]$	

7.5 Stress or strain distribution in hollow section joints

Table 7-3c SCF formulae for CHS K joints with gap or overlap [59]

Load types	SCF formulae	Equation number	Correction factor for "short" chord*
Balanced axial forces (crown, saddle)	Chord: $\tau^{0.9}\gamma^{0.5}(0.67-\beta^2+1.16\beta)\sin\theta \left[\dfrac{\sin\theta_{max}}{\sin\theta_{min}}\right]^{0.30} \left[\dfrac{\beta_{max}}{\beta_{min}}\right]^{0.30} \cdot$ $[1.64+0.29\beta^{-0.38}\,\text{ATAN}(8\zeta)]$	K1	none
	Bracing: $1+[\text{K1}](1.97-1.57\beta^{0.25})\,\tau^{-0.14}\sin^{0.7}\theta + C\cdot\beta^{1.5}\gamma^{0.5}\tau^{-1.22}$ $\sin^{1.8}(\theta_{max}+\theta_{min})\cdot[0.131-0.084\,\text{ATAN}(14\zeta+4.2\beta)]$	K2	none
	where $C=0$ for gap joints $C=1$ for the through bracing $C=0.5$ for the overlapping bracing Note that τ, β, θ and the nominal stress relate to the bracing under consideration. ATAN is arctangent evaluated in radians.		
Unbalanced in-plane bending moment	Chord crown: Eq. (T8) (for overlaps $\lambda_{ov}>30\%$ use $1.2 \cdot$ T8) Gap joint, bracing crown: Eq. (T9) Overlap joint, bracing crown: (T9) $\cdot$ $(0.9+0.4\beta)$	K3	
Unbalanced out-of-plane bending moment	Chord saddle adjacent to bracing A $[\text{T10}]_A\,[1-0.08(\beta_B\gamma)^{0.5}\exp(-0.8x)]+$ $[\text{T10}]_B\,[1-0.08(\beta_A\gamma)^{0.5}\exp(-0.8x)][2.05\,\beta_{max}^{0.5}\exp(-1.3x)]$ where $x=1+\dfrac{\zeta\sin\theta_A}{\beta_A}$	K4	F4
	Bracing A saddle $\tau^{-0.54}\gamma^{-0.05}(0.99-0.47\beta+0.08\beta^4)\cdot[\text{K4}]$	K5	F4

$F4 = 1-1.07\beta^{1.88}\exp[-0.16\gamma^{-1.06}\alpha^{2.4}]$

$[\text{T10}]_A$ is the chord SCF adjacent to the bracing A as calculated from Eq. (T10)

Note that the designation of the bracings A and B is not dependent on geometry. It is nominated by the user.

Table 7-3c (continued)

Load types	SCF formulae	Equation number	Correction factor for "short" chord
Axial force on one bracing only	Chord saddle: Eq. (T5)		F1
	Chord crown: Eq. (T6)		–
	Bracing saddle: Eq. (T3)		F1
	Bracing crown: Eq. (T7)		–
	Note that all geometric parameters and the resulting SCFs relate to the loaded bracing.		
In-plane bending moment on one bracing only	Chord crown: Eq. (T8)		–
	Bracing crown: Eq. (T9)		–
	Note that all geometric parameters and the resulting SCFs relate to the loaded bracing.		
Out-of-plane bending moment on one bracing only	Chord saddle: $$[T10]_A \, [1 - 0.08(\beta_B \gamma)^{0.5} \exp(-0.8x)]$$ where $x = 1 + \dfrac{\zeta \sin\theta_A}{\beta_A}$	K6	F3
	Bracing saddle: $$\tau^{-0.54} \gamma^{-0.05}(0.99 - 0.47\beta + 0.08\beta^4) \cdot [K6]$$	K7	F3

"Short" chord correction factors:

$F1 = 1 - (0.83\beta - 0.56\beta^2 - 0.02)\gamma^{0.23} \exp[-0.21\gamma^{-1.16}\alpha^{2.5}]$

$F2 = 1 - (1.43\beta - 0.97\beta^2 - 0.03)\gamma^{0.04} \exp[-0.71\gamma^{-1.38}\alpha^{2.5}]$

$F3 = 1 - 0.55\beta^{1.8}\gamma^{0.16} \exp[-0.49\gamma^{-0.89}\alpha^{1.8}]$

7.5 Stress or strain distribution in hollow section joints

Table 7-3d SCF formulae for CHS KT joints [59]

Load type	SCF formulae	Equation number
Balanced axial forces	Chord: Eq. (K1) Bracing: Eq. (K2) for the diagonal bracings A and C, use $\zeta = \zeta_{AB} + \zeta_{BC} + \beta_B$ for the central bracing B, use ζ = maximum of ζ_{AB} and ζ_{BC}	
In-plane bending (IPB)	Chord crown: Eq. (T8) Bracing crown: Eq. (T9)	
Unbalanced out-of-plane bending moment	Chord saddle adjacent to diagonal bracing A: $[T10]_A [1-0.08(\beta_B\gamma)^{0.5} \exp(-0.8 x_{AB})] \cdot [1-0.08(\beta_C\gamma)^{0.5} \exp(-0.8 x_{AC})]$ $+ [T10]_B [1-0.08(\beta_A\gamma)^{0.5} \exp(-0.8 x_{AB})] \cdot [2.05 \beta_{max}^{0.5} \exp(-1.3 x_{AB})]$ $+ [T10]_C [1-0.08(\beta_A\gamma)^{0.5} \exp(-0.8 x_{AC})] \cdot [2.05 \beta_{max}^{0.5} \exp(-1.3 x_{AC})]$ where $x_{AB} = 1 + \dfrac{\zeta_{AB} \sin\theta_A}{\beta_A}$ $x_{AC} = 1 + \dfrac{(\zeta_{AB} + \zeta_{BC} + \beta_B) \sin\theta_A}{\beta_A}$ Chord saddle adjacent to the central bracing B: $[T10]_B [1-0.08(\beta_A\gamma)^{0.5} \exp(-0.8 x_{AB})]^{(\beta_A/\beta_B)^2} \cdot$ $[1 - 0.08(\beta_C\gamma)^{0.5} \exp(-0.8 x_{BC})]^{(\beta_C/\beta_B)^2}$ $+ \quad [T10]_A [1-0.08(\beta_B\gamma)^{0.5} \exp(-0.8 x_{AB})] \cdot [2.05 \beta_{max}^{0.5} \exp(-1.3 x_{AB})]$ $+ \quad [T10]_C [1-0.08(\beta_B\gamma)^{0.5} \exp(-0.8 x_{BC})] \cdot [2.05 \beta_{max}^{0.5} \exp(-1.3 x_{BC})]$ where $x_{AB} = 1 + \dfrac{\zeta_{AB} \sin\theta_B}{\beta_B}$ $x_{BC} = 1 + \dfrac{\zeta_{BC} \sin\theta_B}{\beta_B}$	KT1 KT2
SCFs for the out-of-plane bending moment in the bracing	SCFs for the out-of-plane bending moment in the bracing are obtained directly from the adjacent chord SCFs using: $\tau^{-0.54} \gamma^{-0.05} (0.99 - 0.47\beta + 0.08\beta^4) \cdot SCF_{chord}$ where: SCF_{chord} = KT1 or KT2	

Table 7-3d (continued)

Load type	SCF formulae	Equation number
Axial force on one bracing only	Chord saddle: Eq. (T5) Chord crown: Eq. (T6) Bracing saddle: Eq. (T3) Bracing crown: Eq. (T7)	
Out-of-plane bending moment on one bracing only	Chord adjacent to the diagonal bracing A: $[T10]_A \, [1-0.08(\beta_B \gamma)^{0.5} \exp(-0.8 x_{AB})] \cdot [1-0.08(\beta_C \gamma)^{0.5} \exp(-0.8 x_{AC})]$ where $x_{AB} = 1 + \dfrac{\zeta_{AB} \sin \theta_A}{\beta_A}$ $x_{AC} = 1 + \dfrac{(\zeta_{AB} + \zeta_{BC} + \beta_B) \sin \theta_A}{\beta_A}$ Chord adjacent to the central bracing B: $[T10]_B \, [1-0.08(\beta_A \gamma)^{0.5} \exp(-0.8 x_{AB})]^{(\beta_A/\beta_B)^2} \cdot$ $[1-0.08(\beta_C \gamma)^{0.5} \exp(-0.8 x_{BC})]^{(\beta_C/\beta_B)^2}$ where $x_{AB} = 1 + \dfrac{\zeta_{AB} \sin \theta_B}{\beta_B}$ $x_{BC} = 1 + \dfrac{\zeta_{BC} \sin \theta_B}{\beta_B}$	KT3 KT4
SCFs for the out-of-plane bending moment in the bracing	$SCF_{op,br}$ are obtained directly from the adjacent chord SCFs using $\tau^{-0.54} \gamma^{-0.05} (0.99 - 0.47 \beta + 0.08 \beta^4) \cdot SCF_{chord}$	

7.5 Stress or strain distribution in hollow section joints

Table 7-4a Influence functions for CHS X joints under axial force and out-of-plane bending moment [59]

Loading type	Influence function for the bracing A	Equation number
Axial force	Chord saddle: $f_B \dfrac{A_B \sin\theta_B}{A_A \sin\theta_A} \left[[X1]_A - [X7]_A\right]$	IX 1
	Chord crown: $f_B \dfrac{A_B \sin\theta_B}{A_A \sin\theta_A} \left[[X2]_A - [T6]_A\right]$	IX 2
	Bracing saddle: $f_B \dfrac{A_B \sin\theta_B}{A_A \sin\theta_A} \left[[X3]_A - [X8]_A\right]$	IX 3
	Bracing crown: $f_B \dfrac{A_B \sin\theta_B}{A_A \sin\theta_A} \left[[X4]_A - [T7]_A\right]$	IX 4
	where: A_A = cross-sectional area of bracing A A_B = cross-sectional area of bracing B	
OPB	Chord saddle: $f_B \dfrac{W_B \sin\theta_B}{W_A \sin\theta_A} \left[[X5]_A - [T10]_A\right]$	IX 5
	Bracing saddle: $f_B \dfrac{W_B \sin\theta_B}{W_A \sin\theta_A} \left[[X6]_A - [T11]_A\right]$	IX 6
	where: W_A = sectional modulus of bracing A W_B = sectional modulus of bracing B	

Note: In the above expressions, the influence function is only the geometric part, i.e. without the nominal stress f_B. The total expression gives the hot spot contribution.

Table 7-4b Influence functions for CHS K joints under axial force and out-of-plane bending moment [59]

Loading type	Influence function for the bracing A	Equation number
Axial force	Chord saddle: $f_B \dfrac{A_B \sin\theta_B}{A_A \sin\theta_A} \left[[T5]_A - [K1]_A \right]$	IK 1
	Chord crown: $f_B \dfrac{A_B \sin\theta_B}{A_A \sin\theta_A} \left[[T6]_A - [K1]_A \right]$	IK 2
	Bracing saddle: $f_B \dfrac{A_B \sin\theta_B}{A_A \sin\theta_A} \left[[T3]_A - [K2]_A \right]$	IK 3
	Bracing crown: $f_B \dfrac{A_B \sin\theta_B}{A_A \sin\theta_A} \left[[T7]_A - [K2]_A \right]$	IK 4
	where: A_A or A_B = cross-sectional area of bracing A or B	
OPB	Chord saddle: $f_B \left[[K4]_A - [K6]_A \right]$	IK 5
	Bracing saddle: $f_B \left[[K5]_A - [K7]_A \right]$	IK 6

Note: In the above expressions, the influence function is only the geometric part, i.e. without the nominal stress f_B. The total expression gives the hot spot contribution.

7.5 Stress or strain distribution in hollow section joints

Table 7-4c Influence functions for CHS KT joints under axial force [59]

Loading type	Influence function for the bracing A	Equation number
	Chord saddle: $f_B \dfrac{A_B \sin\theta_B}{A_A \sin\theta_A}\left[[T5]_A - [K1]_{AB}\right] + f_C \dfrac{A_C \sin\theta_C}{A_A \sin\theta_A}\left[[T5]_A - [K1]_{AC}\right]$	IKT 1
	Chord crown: $f_B \dfrac{A_B \sin\theta_B}{A_A \sin\theta_A}\left[[T6]_A - [K1]_{AB}\right] + f_C \dfrac{A_C \sin\theta_C}{A_A \sin\theta_A}\left[[T6]_A - [K1]_{AC}\right]$	IKT 2
	Bracing saddle: $f_B \dfrac{A_B \sin\theta_B}{A_A \sin\theta_A}\left[[T3]_A - [K2]_{AB}\right] + f_C \dfrac{A_C \sin\theta_C}{A_A \sin\theta_A}\left[[T3]_A - [K2]_{AC}\right]$	IKT 3
	Bracing crown: $f_B \dfrac{A_B \sin\theta_B}{A_A \sin\theta_A}\left[[T7]_A - [K2]_{AB}\right] + f_C \dfrac{A_C \sin\theta_C}{A_A \sin\theta_A}\left[[T7]_A - [K2]_{AC}\right]$	IKT 4
	where: A_A = cross-sectional area of bracing A $[T5]_A$ = SCF equation (T5) evaluated using the geometric parameters of the bracing A $[K1]_{AB}$ = SCF equation (K1) evaluated using the bracings A and B with the bracing A acting as the reference bracing, i.e. τ, β and θ refer to the bracing A	

Table 7-5 Influence functions for CHS non-planar bracing [59]

Loading type	Influence function	Equation number
Axial force	Chord saddle: $\dfrac{P_2}{A_i \sin\theta_i} \left[[X1]_i - [K1]_i \right]$	IM 1
	Chord crown: $\dfrac{P_1}{A_i} \left[\dfrac{C}{2} \alpha \beta_i \tau_i \right]$	IM 2
	Bracing saddle: $\dfrac{P_2}{A_i \sin\theta_i} \left[[X3]_i - [T3]_i \right]$	IM 3
	Bracing crown: $\dfrac{P_1}{A_i} \left[\dfrac{C}{5} \alpha \beta_i \tau_i \right]$	IM 4

where:

$$P_1 = \sum_{j=1}^{n} f_j A_j \cos\phi_j \cdot \sin\theta_j$$

$$P_2 = \sum_{j=1}^{n} f_j A_j \cos 2\phi_j \cdot \sin\theta_j$$

i = suffix denoting the bracing under consideration
j = suffix denoting non-planar bracings
n = number of non-planar bracings
f_j = nominal axial stress on the bracing j
A_j = cross-sectional area of the bracing j
C = chord-end fixity parameter ($0.55 \leq C \leq 1.0$)
    $C = 0.55$ for fixed end; $C = 1.0$ for pinned end)

7.5 Stress or strain distribution in hollow section joints

Table 7-6 Hot spot stresses in CHS KT joints under out-of-plane bending moment [59]

Hot spot stress f_{hs} expression	Equation number
Chord saddle hot spot stress adjacent to the diagonal bracing A: f_A [T10]$_A$ $[1-0.08(\beta_B\gamma)^{0.5} \exp(-0.8 x_{AB})] \cdot [1-0.08(\beta_C\gamma)^{0.5} \exp(-0.8 x_{AC})]$ $+ f_B \cdot$ [T10]$_B$ $[1-0.08(\beta_A\gamma)^{0.5} \exp(-0.8 x_{AB})] \cdot [2.05 \beta_{max}^{0.5} \exp(-1.3 x_{AB})]$ $+ f_C \cdot$ [T10]$_C$ $[1-0.08(\beta_A\gamma)^{0.5} \exp(-0.8 x_{AC})] \cdot [2.05 \beta_{max}^{0.5} \exp(-1.3 x_{AC})]$ where: $x_{AB} = 1 + \dfrac{\zeta_{AB} \sin\theta_A}{\beta_A}$ $x_{AC} = 1 + \dfrac{(\zeta_{AB} + \zeta_{BC} + \beta_B) \sin\theta_A}{\beta_A}$	HSS1
Saddle hot spot stress in the diagonal bracing A: $\tau^{-0.54} \gamma^{-0.05} (0.99 - 0.47\beta + 0.08\beta^4) \cdot$ HSS1	HSS2
Chord saddle hot spot stress adjacent to the vertical bracing B: f_B [T10]$_B$ $[1-0.08(\beta_A\gamma)^{0.5} \exp(-0.8 x_{AB})]^{(\beta_A/\beta_B)^2} \cdot$ $[1-0.08(\beta_C\gamma)^{0.5} \exp(-0.8 x_{BC})]^{(\beta_C/\beta_B)^2}$ $+ f_A \cdot$ [T10]$_A$ $[1-0.08(\beta_B\gamma)^{0.5} \exp(-0.8 x_{AB})] \cdot [2.05 \beta_{max}^{0.5} \exp(-1.3 x_{AB})]$ $+ f_C \cdot$ [T10]$_C$ $[1-0.08(\beta_B\gamma)^{0.5} \exp(-0.8 x_{BC})] \cdot [2.05 \beta_{max}^{0.5} \exp(-1.3 x_{BC})]$ where: $x_{AB} = 1 + \dfrac{\zeta_{AB} \sin\theta_B}{\beta_B}$ $x_{BC} = 1 + \dfrac{\zeta_{BC} \sin\theta_B}{\beta_B}$	HSS3
Saddle hot spot stress in the vertical bracing B: $\tau^{-0.54} \gamma^{-0.05} (0.99 - 0.47\beta + 0.08\beta^4) \cdot$ HSS3	HSS4

Simplified methods of determining SCF for CHS joints recommended by IIW [88]

Based on the SCF calculation formulae proposed by Durkin and Efthymiou [59, 90] (see Tables 7-3 to 7-6), IIW published recently [88] the recommended fatigue design procedure for welded hollow section joints, which includes the numerical formulae and simplified design graphs to determine the SCFs for the following CHS joints:

1. Uni-planar CHS T and Y joints
2. Uni-planar CHS X joints
3. Uni-planar CHS K joints with gap
4. Multi-planar CHS XX joint
5. Multi-planar CHS KK joints with gap

The weld types for uni-planar joints are shown in Fig. 6-94 and 6-96 and those for multi-planar joint in Fig. 7-50.

The SCF formulae have been derived with the aim, so that the designer can use them simply, specially for computer-based calculations. However, it is even better to establish simple graphs and have the designer read the SCF, which is significantly more convenient.

In general, SCF is a function of β, γ, τ, θ and g' or λ_{ov}. It is obvious that plotting the SCF directly against 5 parameters is impossible and would lead to unreadable graphs. Therefore, the parameters are separated into 2 sets, of 2 and 3 parameters respectively, by defining the SCFs as:

$$\text{SCF} = f_1 \cdot f_2$$

Fig. 7-32 Tests on CHS joints in triangular girders [49]

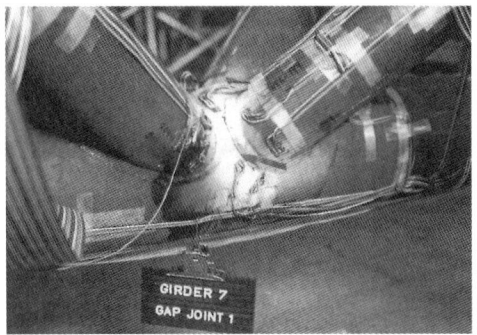

Fig. 7-33 Failure of a CHS joint with gap in a triangular girder [49]

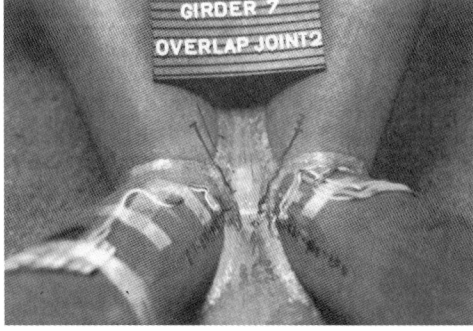

Fig. 7-34 Failure of a CHS joint with overlap in a triangular girder [49]

Uni-planar CHS T and Y joints

Hot spot locations are given in Fig. 7-31 a.

The SCF formulae are according to Table 7-3 a.

The following graphs can be read from [88].

Graphs for axial force ($\alpha = 12$; $C = 0.7$):

a1) $SCF_{br,s,ax}$ vs. β
a2) $SCF_{br,c,ax}$ vs. β
a3) $SCF_{ch,s,ax}$ vs. β
a4) $SCF_{ch,c,ax}$ vs. β

Graphs for in-plane bending moment ($\alpha = 12$):

a5) $SCF_{br,c,ip}$ vs. β
a6) $SCF_{ch,c,ip}$ vs. β

Graphs for out-of-plane bending moment ($\alpha = 12$):

a7) $SCF_{br,s,op}$ vs. β
a8) $SCF_{ch,s,op}$ vs. β

The related parameters are $2\gamma = 15, 30, 50$ and $\tau = 0.5$ and 1.0.

In the case of $\beta \geq 0.95$, SCF for $\beta = 0.95$ is to be used.

The validity range is as follows:

$0.2 \leq \beta \leq 1.0$
$16 \leq 2\gamma \leq 64$
$0.2 \leq \tau \leq 1.0$
$4 \leq \alpha \leq 40$
$30° < \theta < 90°$

The Factor C corresponds to chord-end fixity.

Uni-planar CHS X joints

The SCF formulae are according to Table 7-3 b.

Graphs for axial force ($\alpha = 12$; $C = 1.0$):

b1) $SCF_{br,s,ax}$ vs. β
b2) $SCF_{br,c,ax}$ vs. β (independent of τ)
b3) $SCF_{ch,s,ax}$ vs. β
b4) $SCF_{ch,c,ax}$ vs. β

Graphs for in-plane bending moment ($\alpha = 12$):

b5) $SCF_{ch,c,ip}$ vs. β
b6) $SCF_{br,c,ip}$ vs. β

Graphs for out-of-plane bending moment ($\alpha = 12$):

b7) $SCF_{ch,s,op}$ vs. β
b8) $SCF_{br,s,op}$ vs. β

The related parameters are $2\gamma = 15, 30, 50$ and $\tau = 0.5, 1.0$ except for b2, which is independent of τ.

In the case of $\beta \geq 0.95$, SCF for $\beta = 0.95$ is to be used.

The validity range is the same as that given for T and Y joints.

Uni-planar CHS K joint with gap

SCF for the chord under basic balanced axial forces:

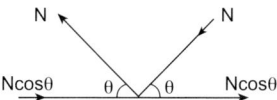

$SCF_{ch,ax} = SCF_{0,ch,ax} \times$ correction factor

Lit. [88] contains the following graphs.

Graphs:

c1) $SCF_{0,ch,ax}$ vs. β
c2) Correction factor vs. τ

The related parameters for c1 are $\theta = 30°, 45°, 60°$ and $2\gamma = 24$ as well as $\tau = 0.5$.

For c2, the related parameters are varied as follows: $2\gamma = 25, 35, 45$ and 55.

SCF for the bracings under basic balanced axial forces:

$SCF_{br,ax} = SCF_{0,br,ax} \times$ correction factor

Graphs:

c3) $SCF_{0,br,ax}$ vs. β
c4) Correction factor vs. τ

The related parameters for c3 are $\theta = 30°, 45°, 60°$ and $2\gamma = 24$ as well as $\tau = 0.5$.

For c4, the related parameters are varied as follows: $2\gamma = 25, 35, 45$ and 55.

The minimum values of $SCF_{br,ax}$ are 2.64, 2.30 and 2.12 for $\theta = 30°, 45°$ and $60°$ respectively.

SCF for the chord under chord loading (axial force and in-plane bending moment):

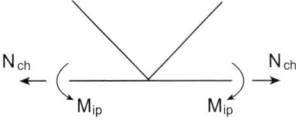

SCF$_{ch,(ax+ip)}$ is to be read from the graph.

c5) SCF$_{ch,(ax+ip)}$ vs. τ; the related parameters are $\theta = 30°$, $35°$, $40°$ and $\theta \geq 45°$

SCF for the bracings under chord loading (axial force and in-plane bending moment):

SCF$_{br,(ax+ip\ in\ chord)} = 0$ (negligible)

The range of validity for the graphs c1 to c5 is as follows:

$0.3 \leq \beta \leq 0.6$
$24 \leq 2\gamma \leq 60$
$0.25 \leq \tau \leq 1.0$
$30° \leq \theta \leq 60°$

Multi-planar CHS XX joint

Hot spot locations are shown below.

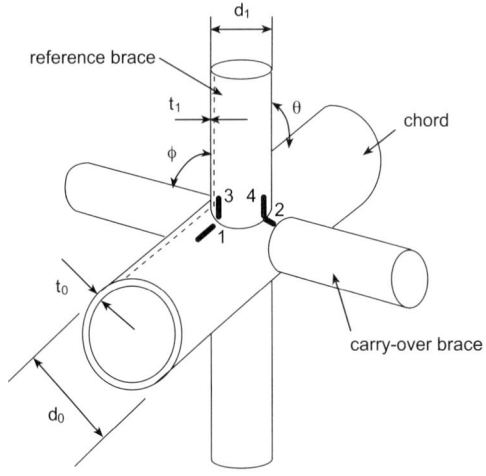

The following load conditions have been taken into consideration:

1. Axially balanced brace loading

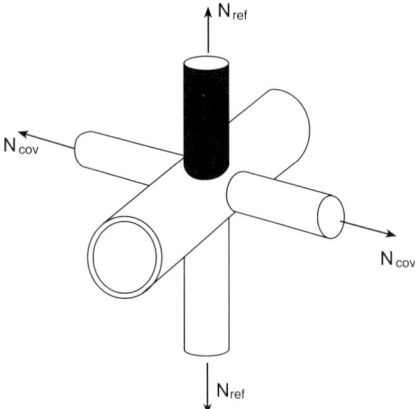

2. Balanced in-plane bending on braces

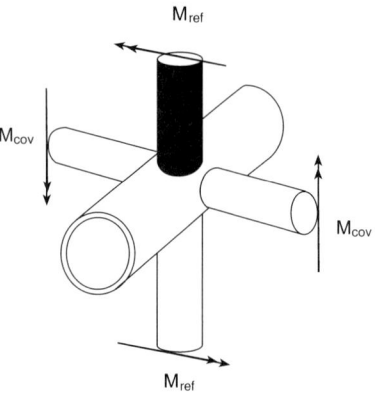

3. Balanced out-of-plane bending on braces

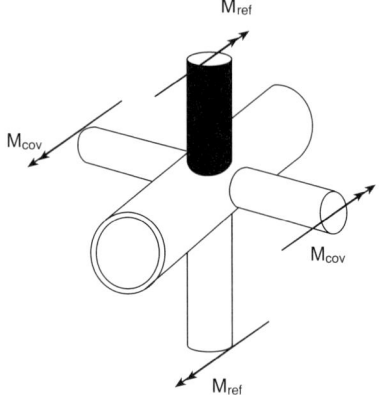

7.5 Stress or strain distribution in hollow section joints

4. Axially balanced chord loading

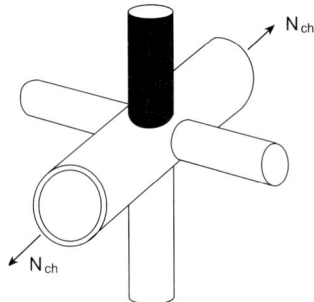

SCFs for the above mentioned loading conditions have been recommended with the aid of equations and diagrams. Effects of reference brace ($N_{br,ref}$, $M_{br,ref}$) and carry-over brace ($N_{br,cov}$, $M_{br,cov}$) have to be combined.

The range of validity is as follows:

No eccentricity
Equal braces

$0.3 \leq \beta \leq 0.6$
$15 \leq 2\gamma \leq 64$
$0.25 \leq \tau \leq 1.0$
$0.25 \leq \tau \leq 1.0$
$\theta = 90°$
$\phi = 90°$
$\psi = \phi - 2 \cdot \arcsin(\beta) \geq 16.2°$

Multi-planar CHS KK joints with gap

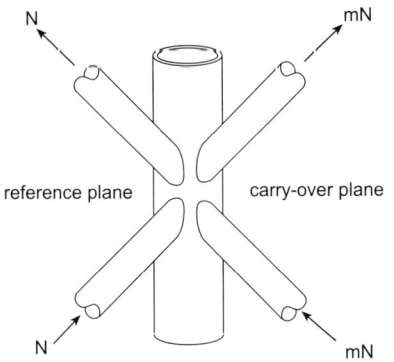

Axially balanced loading in CHS KK joints

Value of m	Referred to as
1	Symmetrical loading
0	Reference-plane loading
−1	Anti-symmetrical loading

Multi-planar correction factors (MCF) on SCF for CHS KK joints with gap ($\phi \leq 90°$)

Load Case	Chord			Brace		
	m = +1	m = 0	m = −1	m = +1	m = 0	m = −1
Axially balanced brace loading	1.0	1.0	1.25	1.0	1.0	1.25
Chord loading	1.0	1.0	1.0	1.0	1.0	1.0

The SCFs for multi-planar CHS KK joints with gap (SCF_{KK}) is to be calculated as follows:

$SCF_{KK} = MCF \cdot SCF_K$

where SCF_K is the SCF for uni-planar CHS K joints with gap and MCF is the multi-planar correction factor accounting for the effects of geometry and loading. The values of MCF for $\phi = 180°$ are 1.0 for all m values. The values of MCF for $\phi \leq 90°$ are given in the table above. Interpolation is allowed for m between 0 and −1 and for ϕ between 90° and 180°.

The range of validity is as follows:

No eccentricity
Equal braces

$0.3 \leq \beta \leq \cos(\theta)$
$24 \leq 2\gamma \leq 48$ $\quad 30° \leq \theta \leq 60°$
$0.25 \leq \tau \leq 1.0$ $\quad 60° \leq \phi \leq 180°$

Minimum SCF values

Uni-planar CHS joints: A minimum SCF value of 2.0 is recommended unless otherwise specified such as "negligible" or "no minimum SCF values required".

Multi-planar CHS joints: When using the Eq. $SCF_{KK} = MCF \cdot SCF_K$, the calculated SCF_K for uni-planar CHS K joints should be adopted even if it is less than 2.0. A minimum SCF value of 2.0 is recommended after applying the MCF to SCF_K.

Parametric equations to determine SCF for joints with square hollow sections

In the late eighties as well as early nineties, extensive numerical and experimental investigations were carried out to determine the stress concentration factors SCF for welded, unstiffened T, X and K joints with square hollow sections [53, 68, 69]. They were later extended to the multi-planar joints TT, XX and KK [54, 70]. The procedure to determine the SCF for square hollow section joints is similar to that for CHS joints, although the differences regarding the profile form have to be taken into consideration. This concerns mainly the locations of hot spot stresses, where the strain gauges must be placed (Fig. 7-25 a and b).

Based on the experimental results and parametric studies aided by FE analysis, Wingerde [53] established the SCF formulae for T and X joints with square hollow sections (see Table 7-7) under axial force and in-plane bending moment. The application of these formulae leads to the calculation of the SCF values at the locations A through E (Fig. 7-25, a2). They are however valid for groove welds with specific dimensions (Fig. 7-35c, $W_0=t_1/2$; $W_1=t_1+2$ and corner radii (r_i/t_i, r_0/t_0, see Fig. 7-36).

The investigation on the influence of the weld (see Fig. 7-35, a) fillet weld $W_0=W_1=t_i\cdot\sqrt{2}$, b) fillet weld with full wall penetration $W_0=W_1=t_i\cdot\sqrt{2}$, c) groove weld $W_0=t_1/2$, $W_1=t_1+2$) led to the proposal for the following correction factors, which have to be multiplied with the SCF in Table 7-7, when the fillet welds (Fig. 7-35 a and b) are applied instead of the groove weld (Fig. 7-35 c):

$SCF_{fillet\ weld} = 1.4\ SCF_{Table\ 7-7}$ for the bracing

$SCF_{fillet\ weld} = 1.0\ SCF_{Table\ 7-7}$ for the chord

Regarding the influence of the corner radii of the square hollow sections on the SCF, no correction measure has been proposed. This is due to the fact that this usually lies within the scatter band of the parametric formulae in Table 7-7 and hence can be left unconsidered. Otherwise, the formulae will be considerably complicated if all the influences including the manufacture and fabrication of the joints are incorporated.

In the framework of further research projects, the SCF formulae for uni-planar K joints with square hollow sections were developed [68, 69]. As this set of equations was very complex, a simplification was necessary for the purpose of application [91]. In this analysis,

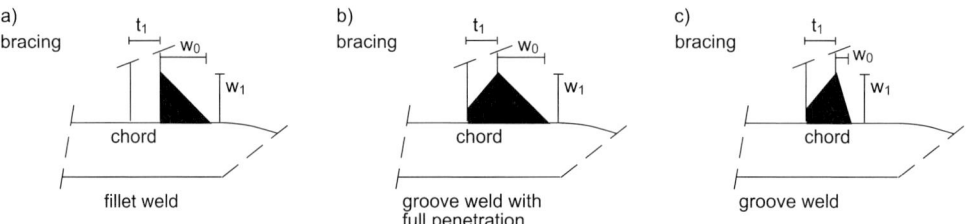

Bild 7-35 Types of weld

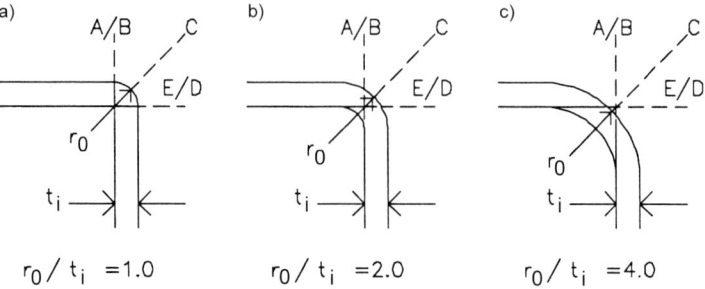

Fig. 7-36 Investigated corner radii

7.5 Stress or strain distribution in hollow section joints

Table 7-7 SCF formulae for T and X joints made of square hollow sections [53] (groove weld) (strain gauges, see Fig. 7-25 a 2)

Line	SCF under IPB on the bracing
B	SCF = $(-0.011 + 0.085 \cdot \beta - 0.073 \cdot \beta^2) \cdot 2\gamma^{(1.722+1.151 \cdot \beta - 0.697 \cdot \beta^2)} \cdot \tau^{0.75}$
C	SCF = $(0.952 - 3.062 \cdot \beta + 2.382 \cdot \beta^2 + 0.0228 \cdot 2\gamma) \cdot 2\gamma^{(-0.690+5.817 \cdot \beta - 4.685 \cdot \beta^2)} \cdot \tau^{0.75}$
D	SCF = $(-0.054 + 0.332 \cdot \beta - 0.258 \cdot \beta^2) \cdot 2\gamma^{(2.084-1.062 \cdot \beta + 0.527 \cdot \beta^2)} \cdot \tau^{0.75}$
A, E	SCF = $(0.390 - 1.054 \cdot \beta + 1.115 \cdot \beta^2) \cdot 2\gamma^{(-0.154+4.555 \cdot \beta - 3.809 \cdot \beta^2)}$

Line	SCF under axial force on the bracing
B	SCF = $(0.143 - 0.204 \cdot \beta + 0.064 \cdot \beta^2) \cdot 2\gamma^{(1.377+1.715 \cdot \beta - 1.103 \cdot \beta^2)} \cdot \tau^{0.75}$
C	SCF = $(0.077 - 0.129 \cdot \beta + 0.061 \cdot \beta^2 - 0.0003 \cdot 2\gamma) \cdot 2\gamma^{(1.565+1.874 \cdot \beta - 1.028 \cdot \beta^2)} \cdot \tau^{0.75}$
D	SCF = $(0.208 - 0.387 \cdot \beta + 0.209 \cdot \beta^2) \cdot 2\gamma^{(0.925+2.398 \cdot \beta - 1.881 \cdot \beta^2)} \cdot \tau^{0.75}$
A, E	SCF = $(0.013 + 0.693 \cdot \beta - 0.278 \cdot \beta^2) \cdot 2\gamma^{(0.790+1.898 \cdot \beta - 2.109 \cdot \beta^2)}$

Line	SCF under IPB on the chord ($SCF_{ip,ch}$) and under axial force on the chord ($SCF_{ax,ch}$)
C	SCF = $0.725 \cdot 2\gamma^{0.248 \cdot \beta} \cdot \tau^{0.19}$
D	SCF = $1.373 \cdot 2\gamma^{0.205 \cdot \beta} \cdot \tau^{0.24}$
B, A, E	negligible: SCF = 0

Range of validity:	$0.35 \leq \beta \leq 1.0$
	$12.5 \leq 2\gamma \leq 25.0$
	$0.25 \leq \tau \leq 1.0$
	$1.0 \leq r_0/t_i \leq 4.0$ (see Fig. 7-36)
Min. SCF for the bracing:	$SCF_{ax}, SCF_{ipb} \geq 2.0$
X joint, $\beta = 1.0$	line C $SCF_{ax} = 0.65 \cdot SCF_{formula}$
	line D $SCF_{ax} = 0.50 \cdot SCF_{formula}$
	lines A, E $SCF_{ax/ipb} = 1.40 \cdot SCF_{formula}$
Fillet welds: (if β i close to 1.0, line A cannot have a fillet weld)	

only symmetrical K joints with identical bracings and same angles of inclination of the both bracings to the chord have been considered. By this, it has been made sure that the β and τ values are valid for both bracings. A simplification of the complex formulae given in [69] is contained by Table 7-8 [82, 91], where only two sets of three equations cover gapped and overlapped K joints. The SCF equations for each set have been derived for the balanced axial loads in bracings and chord respectively, as well as a SCF equation for additional axial load in the chord. 96 SCF formulae corresponding to the measurement lines A through E have been dispensed with and only the maximum SCF values in the bracings and chord of the K type square hollow section joints with gap and overlap under axial forces and in-plane bending moments have been selected (Table 7-8).

Lit. [82] contains a comparison of the parametric equations proposed by a number of other researchers [70, 76, 83], which leads to the conclusion that the SCF formulae shown in [82] are best suitable as regards accuracy and user-friendliness.

In the early nineties, Panjeshahi [54] performed extensive experimental and numerical investigations on the fatigue behaviour of

Table 7-8 Simplified parametric SCF formulae for symmetrical K joints made of square hollow structural sections with gap and overlap (see Fig. 7-25 a ③ and ④)

K joint with gap

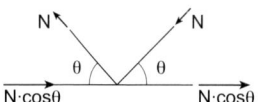

Basic balanced axial forces

Bracing SCF = $[-0.008 + 0.45\beta - 0.34\beta^2] \cdot 2\gamma^{+1.36} \cdot \tau^{-0.66} \cdot \sin^{1.29}\theta$

Chord SCF = $\left[+0.48\beta - 0.5\beta^2 - \dfrac{0.012}{\beta} + \dfrac{0.012}{g'}\right] \cdot 2\gamma^{+1.72} \cdot \tau^{+0.78} \cdot g'^{+0.2} \cdot \sin^{2.09}\theta$

use SCF ≥ 2.0

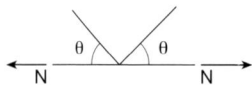

Axial force on the chord

Chord SCF = $[2.45 + 1.23 \cdot \beta] \cdot g'^{-0.27}$ Bracing SCF = 0 (negligible)

K joint with overlap

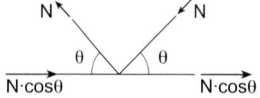

Basic balanced axial forces

Bracing SCF = $\left[0.15 + 1.1\beta - 0.48\beta^2 - \dfrac{0.14}{\lambda_{ov}}\right] \cdot 2\gamma^{+0.55} \cdot \tau^{-0.3} \cdot \lambda_{ov}^{-0.271+1.62\beta^2} \cdot \sin^{0.31}\theta$

Chord SCF = $[0.5 + 2.38\beta - 2.87 \cdot \beta^2 + 2.18\beta \cdot \lambda_{ov} + 0.39 \cdot \lambda_{ov} - 1.43 \cdot \sin\theta]$
$\cdot 2\gamma^{+0.29} \cdot \tau^{+0.7} \cdot \lambda_{ov}^{+0.73-5.53\cdot\sin^2\theta} \cdot \sin^{-0.4-0.08\lambda_{ov}}\theta$

use SCF ≥ 2.0

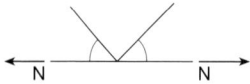

Axial force on the chord

Chord SCF = $[1.2 + 1.46\beta - 0.028\beta^2]$ Bracing SCF = 0 (negligible)

multi-planar, welded square hollow section joints (KK, XX, KY and YY joints, see Fig. 7-37), which were carried out jointly in Delft (the Netherlands) and Karlsruhe (Germany). The research programme also included FE analysis and parameter studies to determine SCF (SNCF) factors. The lines A through E in Fig. 7-25 b show the strain gauge arrangements, where the SCF values were determined.

The numerical work is based on the results of the practical fatigue tests on triangular girders [49] (Fig. 7-38).

The following conclusions regarding the SCF behaviour of the multi-planar square hollow

7.5 Stress or strain distribution in hollow section joints

Table 7-8 (continued)

K joint with gap or overlap

In-plane bending moment
Identical SCF formulae as for basic balanced axial forces are used

Validity ranges

$0.35 \leq \beta \leq 1.00$
$10 \leq 2\gamma \leq 35$ (for smaller 2γ, use $2\gamma = 10$)
$0.25 \leq \tau \leq 1.00$
$30° \leq \theta \leq 60°$
$2 \cdot \tau \leq g'$ $(g \geq 2 t_i)$
$0.50 \leq \lambda_{ov} \leq 1.00$ (= 50% bis 100% overlap)
$-0.55 \leq \dfrac{e}{b_0} \leq 0.25$

Note:
1. The formulae give the maximum SCFs in the both bracings and the chord.
2. They are applicable only to square hollow section bracings of identical dimensions.
3. The hot spot stress ranges $S_{r,hs,ch\ or\ br}$ in the chord or the bracings are calculated by multiplying $SCF_{formula}$ (for the chord and the bracings) with the nominal stress range in the bracing $S_{r,nom,br}$ for the chord as well as the bracings. $S_{r,nom,br}$ is determined as follows:

    $S_{r,nom,br}$ = maximum sum of the absolute axial stress range $S_{r,nom,ax,br}$ and the absolute in-plane bending stress range $S_{r,nom,ip,br}$ in either of the bracings. This maximum can occur at heel or toe of either of the bracings.

4. The additional hot spot stress range in the chord $S_{r,hs,ch,additional}$ is calculated by multiplying $SCF_{formula}$ (axial force on the chord) with the nominal stress range in the chord $S_{r,nom,ch,additional} \cdot S_{r,nom,ch}$ is determined as follows:

    $S_{r,nom,ch}$ = maximum sum of the absolute axial stress range $S_{r,nom,ax,ch}$ and the absolute in-plane bending stress range $S_{r,nom,ip,ch}$

    Since the balanced load case already includes some chord load, the additional nominal stress range in the chord is found by: $f_{0,nom,additional} = f_{0,nom} - f_{1,nom} \cdot \cos\theta \cdot A_1/A_0$

    $f_{1,nom}$ is the value, which is used for $S_{r,nom,br}$ in Pos. 3.

5. In order to obtain the hot spot stress range in the chord including the additional hot spot stress range in the chord, the following values have to be added:

    $S_{r,hs,ch}$ (from Pos. 3) + $S_{r,hs,ch,additional}$ (from Pos. 4)

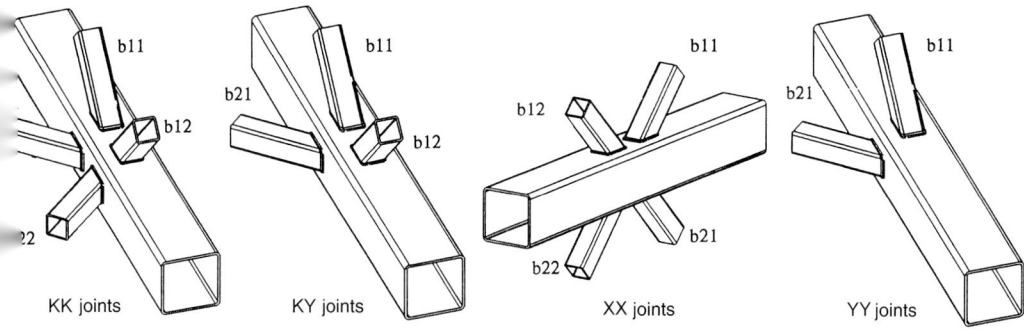

Fig. 7-37 Investigated multi-planar joints in square hollow sections [49]

section joints can be drawn from this research:

- The dependence of the locations of the hot spot stresses on the joint types and parameters as well as on loads and their combinations (Fig. 7-39) is very significant. Hot spots can shift from one line to the other, when the loads undergo variations.
- Highest SCF values occur when the bracings are loaded by axial forces. Significantly large SCFs are also produced by in-plane bending moments (IPB) in the bracings. The SCF values due to the out-of-plane bending moment in the reference bracing are generally small.
- The direction of the bending moment in the bracings plays an important role for the location of maximum hot spot stress (hence for maximum SCF).
- The influence of the axial force, in-plane and out-of-plane bending moments on bracings other than the reference bracing on

Fig. 7-38 Tests on square hollow section joints in a triangular girder, Delft [49]

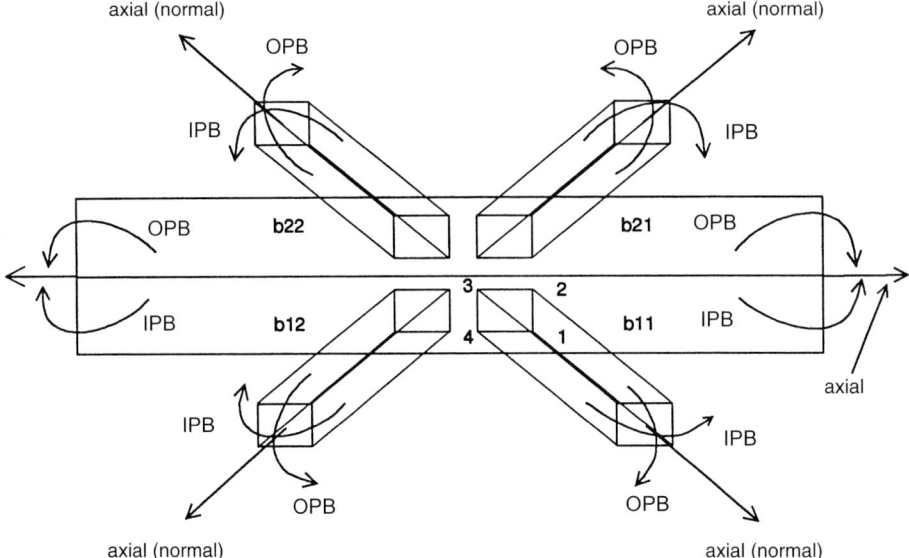

Fig. 7-39 Load types on KK joints in square hollow sections

the SCFs near the weld toes is significant. This is demonstrated by the results showing that a K joint with balanced axial loads in the bracings has lower SCFs compared to a Y joint. A KK joint with balanced axial loads in the bracings has SCFs even lower than a K joint with balanced axial load (Fig. 7-40). The presence of an unloaded bracing may have large influence on reducing the SCFs of the loaded reference bracing.

- As the angle of inclination θ between the bracing and the chord increases, the SCFs also increase in value.

Similar to Romeijn [51] for multi-planar, welded CHS joints, Panjeshahi [54] too refrained from giving parametric formulae for SCFs in multi-planar, welded square hollow section joints. In this case, the abstention is based on the following arguments:

A parametric formula is usually made on the basis of regression analysis of a parametric study for one line location and one load combination at a time. For a square hollow section joint, there are 20 possible locations, where the maximum hot spot stress can occur on the basis of the joint geometry and the loading parameters. Parametric formulae need to be given for every joint type with a given load combination. In practice, the ratios between the loads in the bracings and the bending moments can be of too many combinations. For a multi-planar joint, it is not practical to give parametric formulae for SCF, because their number will be too large and confusing to be accepted and used by the designers.

That is why, Panjeshahi developed a FORTRAN program "HOTSHS", which calculates the hot spot stresses for the 20 line locations around the reference bracing for a given joint geometry and given load combinations, on the basis of the large data base obtained by the parametric study. They are stored in the data bank of the Delft University of Technology in the Netherlands.

The application facilities for the program "HOTSHS" are as follows:

Joint type: KK, XX, KY, YY, Y, X, K

Validity ranges: $0.25 \leq \beta \leq 0.6$
$12.5 \leq 2\gamma \leq 25$
$0.5 \leq \tau \leq 1.0$
$30° \leq \theta \leq 60°$

Weld: see Fig. 7-35

However, Herion [70, 72] recommended the parametric SNCF formulae (see Table 7-9) for multi-planar, welded KK joints with square hollow sections (see Fig. 7-41) under the following assumptions:

- Maximum SNCFs occur depending on loadings (axial force, IPB and OPB) mainly in one corner of the bracing or chord.
- The SNCF in the bracing is always lower than 5.0.
- The SNCF in the chord under axial force and IPB can reach 11.0; under OPB, SNCF remains lower than 5.0.

The advantage of the application of Table 7-9 consists of the fact that only one SNCF formula for the chord and the bracing is given for each loading case. The results lie on the safe side, as only the maximum SNCF values are selected for all load combinations.

In order to calculate SCF from SNCF, SNCF is multiplied by 1.1.

Simplified methods of determining SCF for the welded, unstiffened RHS (square) joints [88]

Numerical formulae and simplified design graphs for the following welded, square hollow section joints have been recommended by IIW [88] recently for the easy use by the designers:

1. Uni-planar RHS T and X joints
2. Uni-planar RHS K joints with gap
3. Uni-planar RHS K joints with overlap
4. Multi-planar RHS KK joints with gap

Apart from the results of the numerous research works, these design recommendations are mainly based on the evaluations given in Lit. [69, 82, 91–96]. The weld types for uni-planar joints are shown in Figs. 6-95 and 6-97 and those for multi-planar joints in Fig. 7-51.

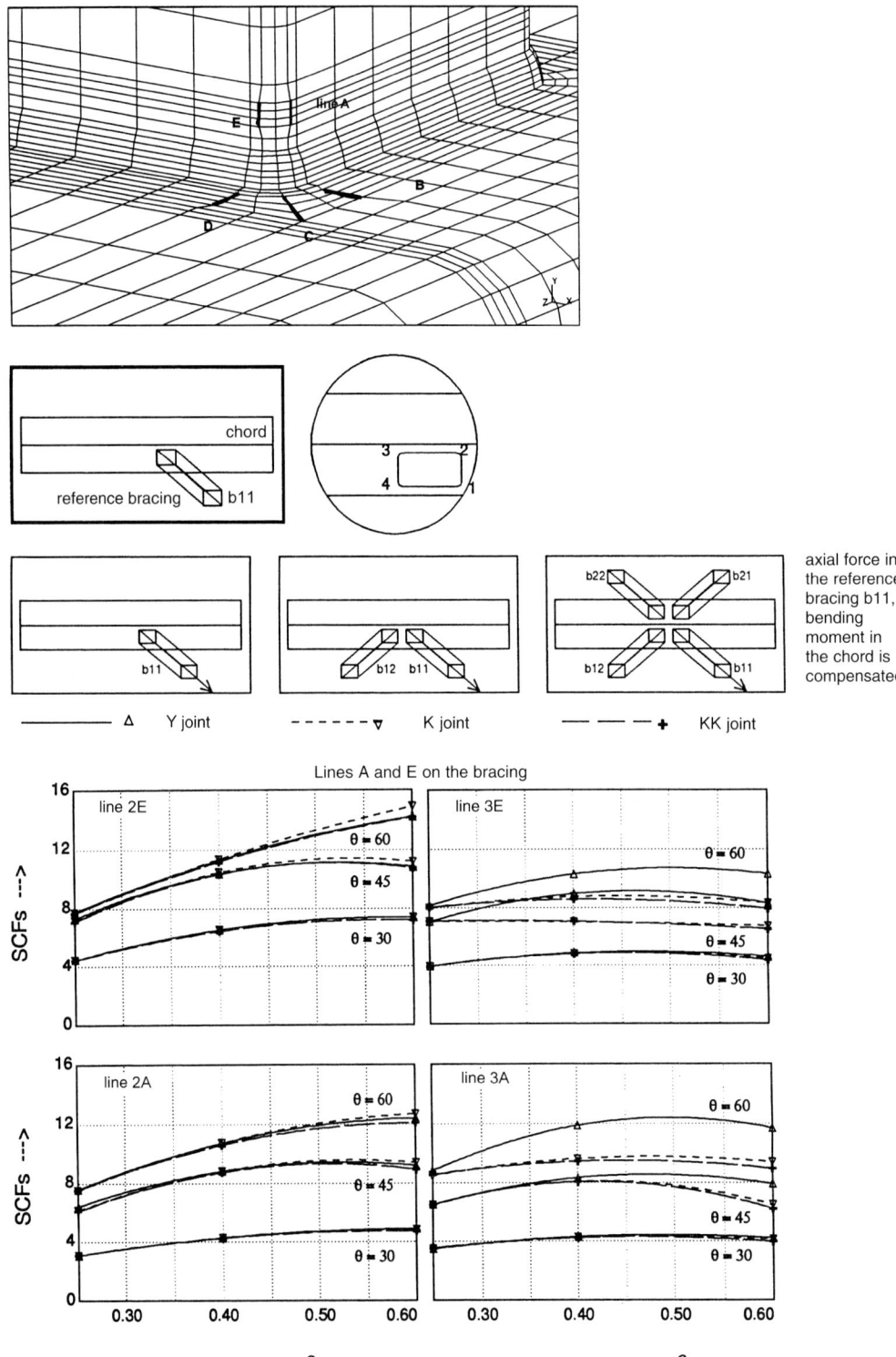

Fig. 7-40 Influence of the other unloaded bracing on the SCFs in the reference bracing of Y, K and KK joints; variable: β and θ; constant parameter: $2\gamma = 25$, $\tau = 0.5$ [49]

7.5 Stress or strain distribution in hollow section joints

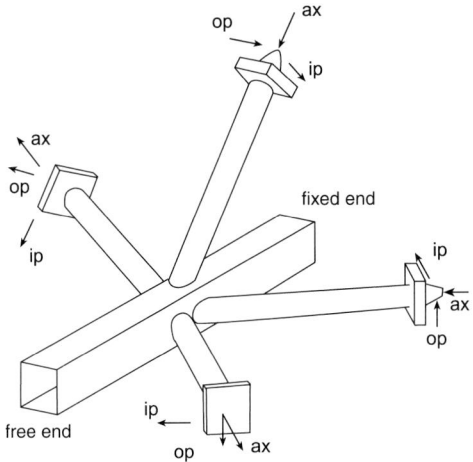

Fig. 7-41 Applied loads and the boundary conditions

Uni-planar RHS T and X joints

Hot spot locations and the corresponding strain gauge arrangements are shown in Fig. 7-25 a(2).

The SCF formulae are according to Table 7-7 ($\theta = 90°$) with the following simplifications:

Load condition: in-plane bending (IPB) on the bracing

$SCF_{ip,ch}$ for the chord lines B, C and D and $SCF_{ip,br}$ for the bracing lines A and E have to be checked.

For joints with fillet welds, $SCF_{A,ip,br}$ and $SCF_{E,ip,br}$ have to be multiplied by a factor of 1.40.

Load condition: axial force on the bracing

$SCF_{ax,ch}$ for the chord lines B, C and D and $SCF_{ax,br}$ for the bracing lines A and E have to be checked.

For joints with fillet welds, $SCF_{A,ax,br}$ and $SCF_{E,ax,br}$ have to be multiplied by a factor of 1.40.

For X joints with $\beta = 1.0$:

$SCF_{C,ax,ch}$ is to be multiplied by 0.65 and $SCF_{D,ax,ch}$ by 0.50.

Load condition: IPB or axial force on the chord

$SCF_{ax,ch}$ or $SCF_{ip,ch}$ for the chord lines C and D have to be checked.

Table 7-9 SNCF formulae for square hollow section KK joints [70, 72]

	Axial force	Validity range
Chord	SNCF = $-0.4961 + 38.368 \cdot \xi \cdot \tau^2 + 33.743 \cdot \xi \cdot \beta^3 - 31.436 \cdot \xi \cdot \tau^3 + 0.0543 \cdot \beta \cdot \gamma^2$ $+ 0.020768 \cdot \gamma^2 \cdot \tau^3 + 0.002728 \cdot \xi^2 \cdot g'^3 - 4.1366 \beta^3 \cdot g' - 0.004644 \xi^3 g'^3$	$0.25 \leq \tau = \dfrac{t_i}{t_0} \leq 1.14$
Bracing	SNCF = $1.329 - 4.31 \cdot \tau^2 + 4.274 \cdot \beta \cdot \tau^2 - 1.31732 \cdot \gamma \cdot \beta^3 + 3.472 \cdot 10^{-4} \cdot \tau \cdot \gamma^3$ $+ 0.1651 \cdot \tau \cdot g' + 1.1478 \cdot \beta \cdot \gamma - 7 \cdot 10^{-4} \tau \cdot g'^3$	$0.40 \leq \beta = \dfrac{b_i}{b_0} \leq 0.80$
	IPB (in-plane bending)	$12.5 \leq \gamma = \dfrac{b_0}{2t_0} \leq 25$
Chord	SNCF = $-0.396 + 81.5278 \cdot \beta \cdot \xi^2 + 0.032952 \cdot \beta \cdot \gamma^3 - 7.5135 \cdot \dfrac{\xi^3}{\tau} - 1.496 \cdot 10^{-3} \cdot \dfrac{\gamma^3}{\tau}$ $- 69.206 \cdot \beta^2 \cdot \xi^3 - 0.030184 \cdot \beta^2 \cdot \gamma^3 - 1.7706 \cdot g' \cdot \gamma + 2.508 \cdot 10^{-3} \cdot \dfrac{g'^3}{\tau^2}$	$2.03 \leq g' = \dfrac{g}{t_0} \leq 10.88$ $0.32 \leq \xi = \dfrac{g}{b_i} \leq 1.09$
Bracing	SNCF = $1.4338 + \tau(-4.8522 \cdot 10^{-4} \cdot \gamma^3 - 2.733 + 3.297 \cdot \beta)$ $+ g'^2 (0.01362 - 1.289 \cdot 10^{-3} \cdot g') + \beta \cdot \gamma^2 (0.074184 - 0.10126 \cdot \beta^2)$	$\theta_l = \theta_q = 90°$ Konst. θ_l = angle between the diagonals in plane parallel to the longitudinal axis of the chord
	OPB (out-of-plane bending)	
Chord	SNCF = $-0.3893 + \beta^3 [\xi(46.02 - 15.252 \cdot \xi) + 1.677 \tau - g'(3.5973 - 0.0873 \cdot g')]$ $+ 8\gamma^3 [1.18 \cdot 10^{-4} \cdot \beta + \tau(4.32 \cdot 10^{-4} - 2.09 \cdot 10^{-4} \cdot \tau)]$	θ_q = angle between the diagonals in plane transverse to the longitudinal axis of the chord
Bracing	SNCF = $2.012 + \tau^3 (5.7727 + 1.475 \beta - 1.66162 \cdot \gamma)$ $- \gamma^2 (8.64 \cdot 10^{-4} \cdot \gamma - 0.069264 \cdot \beta + 0.074696 \cdot \beta^3 - 0.086876 \cdot \tau^3) + 0.222 \cdot \zeta \cdot \beta$	

$SCF_{ax,ch}$ and $SCF_{ip,ch}$ for the chord line B and $SCF_{ax,br}$ and $SCF_{ip,br}$ for the bracing lines A and E can be neglected.

SCF can be read from the graphs given in [88].

Graphs for axial force on the bracing ($\theta = 90°$):

a1) $SCF_{ax,ch,B}$ vs. β
a2) $SCF_{ax,ch,C}$ vs. β
a3) $SCF_{ax,ch,D}$ vs. β
a4) $SCF_{ax,br,A \text{ or } E}$ vs. β (for all τ)

The related parameters are $2\gamma = 12.5, 16, 25$ and $\tau = 0.5$ and 1.0.

Graphs for in-plane bending moment on the bracing ($\theta = 90°$):

a5) $SCF_{ip,ch,B}$ vs. β
a6) $SCF_{ip,ch,C}$ vs. β
a7) $SCF_{ip,ch,D}$ vs. β
a8) $SCF_{ip,br,A \text{ or } E}$ vs. β (for all τ)

Graphs for axial force or in-plane bending moment on the chord ($\theta = 90°$):

a9) $SCF_{ax,ch,C}$ or $SCF_{ip,ch,C}$ vs. β
a10) $SCF_{ax,ch,D}$ or $SCF_{ip,ch,D}$ vs. β

For $\theta = 90°$, the range of validity for RHS T and X joints is as follows:

$0.35 \leq \beta \leq 1.0$
$12.5 \leq 2\gamma \leq 25.0$
$0.25 \leq \tau \leq 1.0$

For RHS X joints with $40° \leq \theta \leq 80°$, SCF is to be determined using for $\theta = 90°$ with the correction factors below:

For the chord lines B, C and D, $SCF_\theta = 1.2 \cdot SCF_{\theta=90°} \cdot \sin^2\theta$

For the bracing lines A and E, $SCF_\theta = 1.2 \cdot SCF_{\theta=90°} \cdot \sin\theta$

Uni-planar RHS K joints with gap

SCF for the chord under basic balanced axial forces:

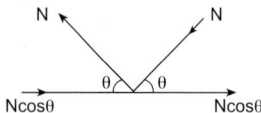

$SCF_{ch,ax} = SCF_{0,ch,ax} \times$ correction factor

Graphs:

b1) $SCF_{0,ch,ax}$ vs. β ($g' = 1.0$)
b2) $SCF_{0,ch,ax}$ vs. β ($g' = 2.0$)
b3) $SCF_{0,ch,ax}$ vs. β ($g' = 4.0$)
b4) $SCF_{0,ch,ax}$ vs. β ($g' = 8.0$)
b5) Correction factor on $SCF_{0,ch,ax}$ vs. τ

The related parameters for b1 to b4 are $\theta = 30°, 45°$ and $60°$ and those for b5 are $2\gamma = 10, 15, 20, 25, 30$ and 35. Interpolation is allowed between the lines for other angles and between the graphs for other g' and 2γ values.

SCF for the bracings under basic balanced axial forces:

$SCF_{br,ax} = SCF_{0,br,ax} \times$ correction factor

Graphs:

b6) $SCF_{0,br,ax}$ vs. β (all g' values)
b7) Correction factor on $SCF_{0,br,ax}$ vs. τ

The related parameters for b6 are $\theta = 30°, 45°$ and $60°$ and those for b7 are $2\gamma = 10, 15, 20, 25, 30$ and 35. Interpolation is allowed between the lines for other angles and for other 2γ.

SCF for the chord under the chord loading (axial force and in-plane bending moment):

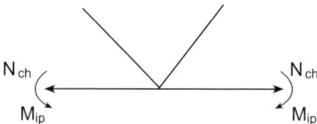

$SCF_{ch,(ax+ip)}$ is to be read from the following graph:

b8) $SCF_{ch,(ax+ip)}$ vs. β

The related parameters are $g' = 1, 2, 4$ and 8; for other g' values, interpolation is allowed.

SCF for the bracings under the chord loading (axial force and in-plane bending moment):

$SCF_{br,(ax+ip \text{ in chord})} = 0$ (negligible)

The range of validity for the graphs b1 to b8 is as follows:

Equal braces

$0.35 \leq \beta \leq 1.0$
$10 \leq 2\gamma \leq 35$
$0.25 \leq \tau \leq 1.0$
$30° \leq \theta \leq 60°$
$2\tau \leq g'$
$-0.55 \leq e/h_0 \leq 0.25$

Uni-planar RHS K joints with overlap

SCF for the chord under basic balanced axial forces:

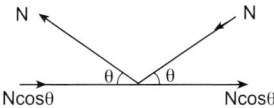

$SCF_{ch,ax} = SCF_{0,ch,ax} \times$ correction factor

Graphs:

c1) $SCF_{0,ch,ax}$ vs. β ($\lambda_{ov} = 50\%$)
c2) $SCF_{0,ch,ax}$ vs. β ($\lambda_{ov} = 75\%$)
c3) $SCF_{0,ch,ax}$ vs. β ($\lambda_{ov} = 100\%$)
c4) correction factor on $SCF_{0,ch,ax}$ vs. τ

The related parameters for c1 to c3 are $\theta = 30°, 45°$ and $60°$ and those for C4 are $2\gamma = 10, 15, 20, 25, 30$ and 35. Interpolation is allowed between the lines for other angles and between the graphs for other overlap percentage λ_{ov} and 2γ values.

SCF for the bracings under basic balanced axial forces:

$SCF_{br,ax} = SCF_{0,br,ax} \times$ correction factor

Graphs:

c5) $SCV_{0,br,ax}$ vs. β ($\lambda_{0v} = 50\%$)
c6) $SCV_{0,br,ax}$ vs. β ($\lambda_{0v} = 75\%$)
c7) $SCV_{0,br,ax}$ vs. β ($\lambda_{0v} = 100\%$)
c8) correction factor on $SCF_{0,br,ax}$ vs. τ

The related parameters for c5 to c7 are $\theta = 30°, 45°$ and $60°$ and those for c8 are $2\gamma = 10, 15, 20, 25, 30$ and 35. Interpolation is allowed between the lines for other angles and between the graphs for other overlap percentage λ_{ov} and 2γ values.

SCF for the chord under the chord loading (axial force and in-plane bending moment):

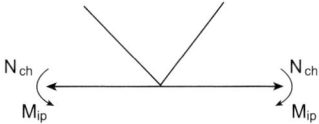

$SCF_{ch,(ax+ip)}$ is to be read from the following graph:

c9) $SCF_{ch,(ax+ip)}$ vs. τ (for all overlaps)

SCF for the bracings under the chord loading (axial force and in-plane bending moment:

$SCF_{br,(ax+ip\,in\,chord)} = 0$ (negligible)

The range of validity for the graphs c1 to c9 is as follows:

Equal braces

$0.35 \leq \beta \leq 1.0$
$10 \leq 2\gamma \leq 35$
$0.25 \leq \tau \leq 1.0$
$30° \leq \theta \leq 60°$
$50\% \leq \lambda_{ov} \leq 100\%$
$-0.55 \leq e/h_0 \leq 0.25$

Multi-planar RHS KK joints with gap

The procedure to determine SCF_{KK} for RHS joints is the same as for that for CHS joints, which has been described previously. However, the range of validity is as follows:

Equal bracings

$0.25 \leq \beta \leq 0.60$
$12.5 \leq 2\gamma \leq 25.0$
$0.5 \leq \tau \leq 1.0$
$30° \leq \theta \leq 60°$
$2\tau \leq g'$
$-0.55 \leq e/h_0 \leq 0.25$
$60° \leq \phi \leq 180°$

7.6 Effect of secondary bending moments on the fatigue strength of RHS and CHS truss joints of K and N type

In order to avoid extensive stress analysis and joint modelling, the distribution of axial forces in a truss is determined under the assumption that all structural members are pin connected (Fig. 6-73). Resulting from the really existing end fixity of the web members to the flexible chord wall, secondary bending moments are however initiated (besides the primary bending moments due to eventual nodal eccentricity depending on the joint configuration), which have to be taken into account when calculating the fatigue strength. It is therefore recommended to multiply the existing hot spot stress ranges due to the axial load by the given factors in Table 7-10 (for square hollows), so that complicated numerical idealization can be avoided [68].

The following minimum value has been proposed:

$$S_{r,hs,\text{total}} = M \cdot \sum_{i=1}^{n} \text{SCF}_{i,\text{ax}} \cdot S_{r,\text{nom,ax}}$$

This is valid independent of the procedure for the numerical idealization as well as the location of hot spot stress.

EC 3 [42] recommends the multiplication factors for K and N type CHS or RHS joints for general application according to Table 7-11 or 7-12.

The recommendation of IIW [88] regarding the multiplication factor M accounting for secondary bending moments deals only with K joints in CHS with gap and in RHS with gap and overlap, which have been taken from Tables 7-10 to 7-12.

7.7 Basic "$S_{r,hs}$ vs. N_F" lines for welded, uniplanar CHS and RHS (square) section joints (T, X, K, N and KT)

The experimental investigations on square hollow sections [49, 68] as well as the research works by DEn [38, 39, 73, 74] and AWS [13] on CHS joints were carried out to establish "hot spot stress range $S_{r,hs}$ vs. number of loading cycles to failure N_F" lines and to determine SCF values in addition, which serve as the basis for the design of hollow section joints under fatigue loading.

In these tests, the hot spot stress ranges $S_{r,hs}$ were measured with strain gauges on the test specimens (T, X and K joints) and recorded against the corresponding number of loading cycles to failure N_F in a diagram.

The life of a hollow section joint with regard to fatigue strength can be in general represented by the number of loading cycles to failure N_F (see Figs. 7-42 to 7-47), which corresponds to the following failure modes:

Table 7-10 Multiplication factor M to account for secondary bending moments in K type square hollow section joints under fatigue load

K joint type	Multiplication factor M	
	chord	diagonal
with gap	1.5	1.5
with overlap	1.5	1.3

Table 7-11 Multiplication factor M to account for secondary bending moments in K and N type CHS joints under fatigue load

Joint type			Multiplication factor M	
		Chord	Vertical	Diagonal
with gap	K	1.5	–	1.3
	N	1.5	1.8	1.4
with overlap	K	1.5	–	1.2
	N	1.5	1.65	1,25

Table 7-12 Multiplication factor M to account for secondary bending moments in K and N type RHS joints (in the range $0.5 \leq h_0/b_0 \leq 2$) as well as square hollow section joints under fatigue load

Joint type			Multiplication factor M	
		Chord	Vertical	Diagonal
with gap	K	1.5	–	1.5
	N	1.5	2.2	1.5
with overlap	K	1.5	–	1.3
	N	1.5	2.0	1.4

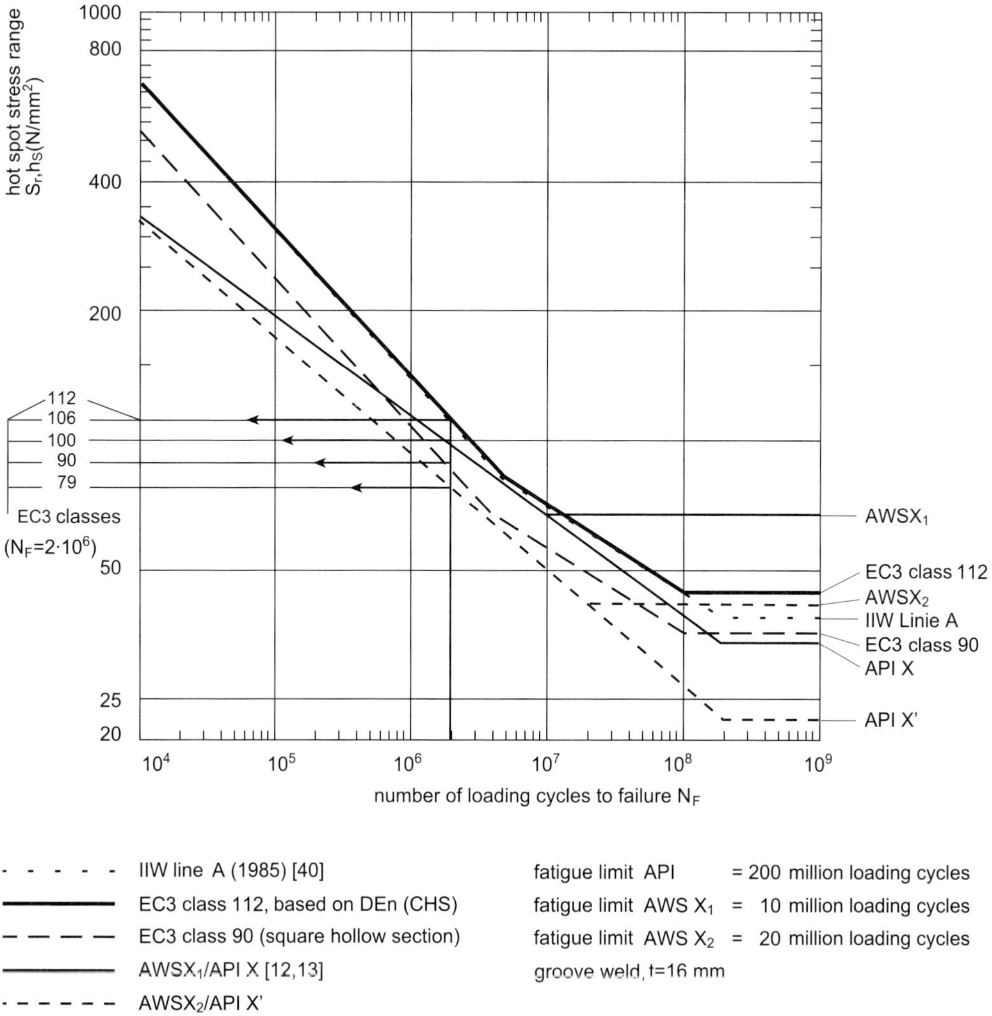

7.7 Basic "$S_{r,hs}$ vs. N_F" lines for welded, uni-planar CHS and RHS (square) section joints (T, X, K, N and KT)

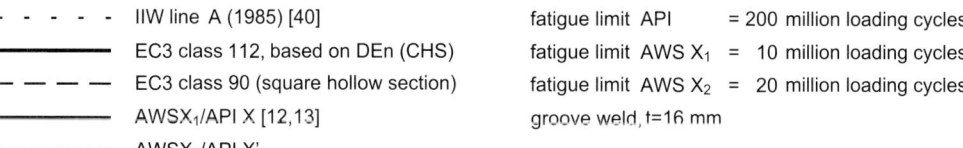

- - - - - IIW line A (1985) [40]
────── EC3 class 112, based on DEn (CHS)
── ── ── EC3 class 90 (square hollow section)
────── AWSX$_1$/API X [12,13]
- - - - - AWSX$_2$/API X'

fatigue limit API = 200 million loading cycles
fatigue limit AWS X$_1$ = 10 million loading cycles
fatigue limit AWS X$_2$ = 20 million loading cycles
groove weld, t=16 mm

Fig. 7-42 Major "$S_{r,hs}$ vs. N_F" lines for welded, uni-planar hollow section joints

1. First visible crack
2. Reaching large crack length
3. Through crack in the profile wall
4. Total loss of joint strength (end of test)

The failure mode 3 is at present accepted as the design criterion for hollow section joints under fatigue load. This corresponds to about 80% of the total life (failure mode 4) of the joints.

Another possibility to determine $S_{r,hs}$ is to multiply the nominal stress ranges $S_{r,nom}$ with the SCF factors, which can be calculated with the aid of Tables 7-1 to 7-8. The $S_{r,hs}$ values are then drawn against the loading cycles to failure N_F.

As these tests were conducted with the steel grades S235 and S355, the application of the "$S_{r,hs}$ vs. N_F" or the "$S_{r,nom}$ vs. N_F" lines are limited to S 355.

Fig. 7-42 shows a comparison of the existing major "$S_{r,hs}$ vs. N_F" lines [75] given by various design guidelines together with the classified "EC 3-90" and "EC 3-112" lines, which have been proposed as the design basis finally (Fig. 7-43). The "EC 3-112" is valid for welded

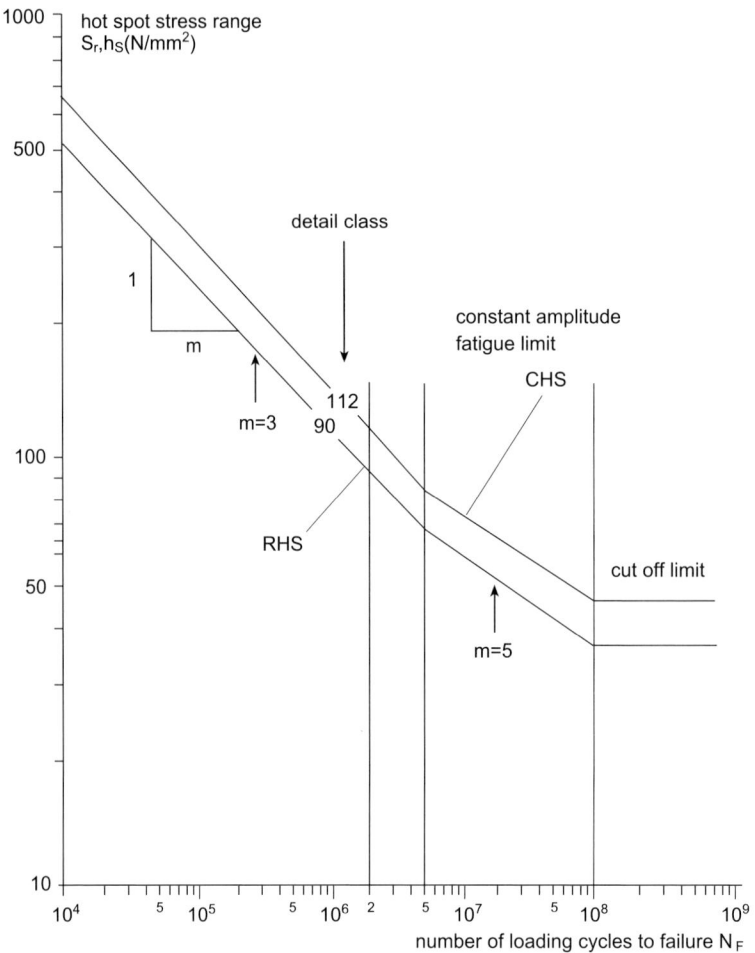

Fig. 7-43 "EC 3-90" and "EC 3-112" lines for welded RHS and CHS joints (T, X, K and N) for fatigue design [75]

Joint	$S^*_{r,hs}$ [Mpa = N/mm²]
Square hollow section	90 (EC3–90)
CHS	112 (EC3–112)

* $t = 16$ mm

CHS joints, while the "EC 3-90" line is to be used for square or rectangular (in the range $0.5 \leq h_0/b_0 \leq 2$) hollow section joints. These lines make it possible to take the hot spot stresses in hollow section joints for the design recommendation according to EC 3 [42] into consideration. The lines shown in Fig. 7-42 are for groove welded connections with wall thickness $t = 16$ mm.

7.7.1 Correction factors accounting for the wall thickness of the applicable members (chord and bracings) being checked for the fatigue cracking and the fatigue lines "$S_{r,hs}$ vs. N_F"

It has been demonstrated by the investigations that the wall thicknesses of structural members exercise a significant influence on the fatigue strength, which has to be taken into account apart from other effects. This phenomenon is based on the following reasons:

7.7 Basic "$S_{r,hs}$ vs. N_F" lines for welded, uni-planar CHS and RHS (square) section joints (T, X, K, N and KT)

Table 7-13 Equations for the "$S_{r,hs}$ vs. N_F" lines accounting for wall thicknesses of the members of the welded hollow section joints and the ranges of the number of the loading cycles to failure

t	N_F	$S_{r,hs}$
t<16 mm CHS	$10^3 < N_F < 5 \cdot 10^6$	$\log(S_{r,hs}) = \frac{1}{3} \cdot (12.476 - \log(N_F)) + 0.11 \cdot \log(N_F) \cdot \log\left(\frac{16}{t}\right)$
	$5 \cdot 10^6 < N_F < 10^8$ (for variable amplitude only)	$\log(S_{r,hs}) = \frac{1}{5} \cdot (16.327 - \log(N_F)) + 0.737 \cdot \log\left(\frac{16}{t}\right)$
t ≥ 16 mm CHS	$10^3 < N_F < 5 \cdot 10^6$	$\log(S_{r,hs}) = \frac{1}{3} \cdot (12.476 - \log(N_F)) + 0.30 \cdot \log\left(\frac{16}{t}\right)$
	$5 \cdot 10^6 < N_F < 10^8$ (for variable amplitude only)	$\log(S_{r,hs}) = \frac{1}{5} \cdot (16.327 - \log(N_F)) + 0.30 \cdot \log\left(\frac{16}{t}\right)$
t<16 mm RHS	$10^3 < N_F < 5 \cdot 10^6$	$\log(S_{r,hs}) = \frac{1}{3} \cdot (12.151 - \log(N_F)) + 0.11 \cdot \log(N_F) \cdot \log\left(\frac{16}{t}\right)$
	$5 \cdot 10^6 < N_F < 10^8$ (for variable amplitude only)	$\log(S_{r,hs}) = \frac{1}{5} \cdot (15.786 - \log(N_F)) + 0.737 \cdot \log\left(\frac{16}{t}\right)$
t ≥ 16 mm RHS	$10^3 < N_F < 5 \cdot 10^6$	$\log(S_{r,hs}) = \frac{1}{3} \cdot (12.151 - \log(N_F)) + 0.30 \cdot \log\left(\frac{16}{t}\right)$
	$5 \cdot 10^6 < N_F < 10^8$ (for variable amplitude only)	$\log(S_{r,hs}) = \frac{1}{5} \cdot (15.786 - \log(N_F)) + 0.30 \cdot \log\left(\frac{16}{t}\right)$

- The stress gradient at the weld notch is less steep for larger wall thickness. As a consequence, the stresses at the crack tips are higher and lead to the enlargement of cracks.
- The probability of a larger notch becomes more with increasing wall thickness (larger volume); the fatigue is reduced as the weld defect grows.
- The following occurrences become more probable with increasing wall thickness: larger grain size, lower yield strength, higher residual stress, lower ductility and higher probability for hydrogen brittle fracture.

The proposed fatigue design lines "$S_{r,hs}$ vs. N_F" in Fig. 7-43 correspond to the basic wall thickness $t = 16$ mm of the members of uni-planar hollow section joints. In order to account for the wall thicknesses smaller or larger than 16 mm, the correction factors have been proposed, which lead to the equations for various ranges of the numbers of loading cycles to failure N_F (see Table 7-13) [75]. The design of the welded hollow section joints has been further made easier for the constructors and calculators by means of "$S_{r,hs}$ vs. N_F" lines for the wall thicknesses between 4 and 25 mm, illustrated in Figs. 7-44 and 7-45.

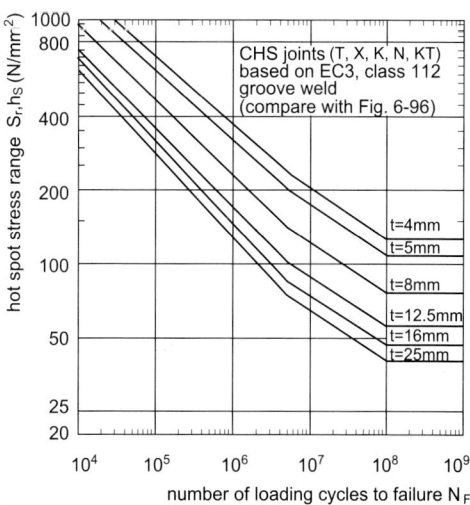

Fig. 7-44 Design lines "$S_{r,hs}$ vs. N_F" according to "EC 3-112"

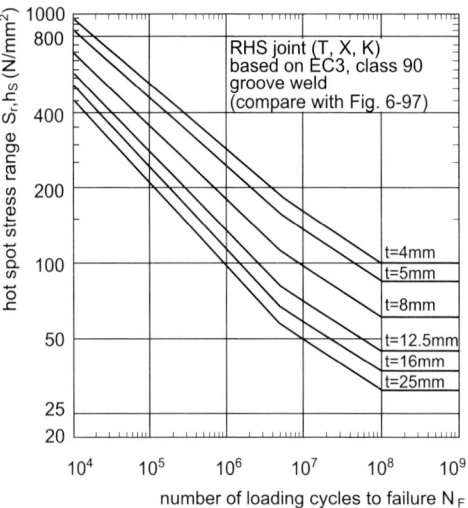

Fig. 7-45 Design lines "$S_{r,hs}$ vs. N_F" according "EC 3-90"

7.7.2 Application of high strength steels

The static as well as fatigue design methods of hollow section joints described up to now [42, 78] deal with structural steels S 235, S 275 and S 355. The tendency in the last years goes strongly in the direction of using steel grades of higher strengths. In order to meet this demand, EC 3 [42], Appendix D proposes the application of S 420 and S 460 also for the structural purpose. Besides, the national German guideline DASt 011 [79] "Application of weldable high strength fine grain steels St E 460 and St E 690 in steel structures" take the rising use of high strength steels into account.

As has been already described in Chapter 3, the higher strength in steelgrade St E 460 is produced by the addition of the alloy elements e. g. Nb, V, Ti, Cr, Ni, MO and Cu as a rule. St E 690 is predominantly quenched and tempered. Another procedure to achieve higher yield strengths is to apply thermo-mechanical rolling (TM steels), where the final shaping of the product takes place in a certain temperature range. However, a reduction of this higher strength occurs by a post-heat treatment at more than 580 °C.

The advantages of hollow sections made of high strength steels consist of the smaller, required cross-sectional area under load and the resulting saving of material. As long as the structural element is not welded or the welded joint under static load can attain the full plasticity exploiting the higher yield strength, this advantage can be quite significant. Unfortunately, the welded, hollow section joints and connections of high strength steels under static load have not been amply investigated until today. Therefore, the existing formulae for the design resistances and the corresponding design rules (see Sections 6.6.4.1 to 6.6.4.5) are not applicable to higher strength steel grades. Further research works are necessary in this field.

A number of tests on welded hollow section joints and connections made of S 460 under fatigue load has been carried out, which gives indications regarding the fatigue behaviour of these joints [80, 81].

Figs. 7-46 and 7-47 show the comparisons of "$S_{r,nom}$ vs. N_F" lines for X joints of S 235 and S 460, which demonstrate that the joints of higher strength steel grades deliver larger fatigue resistance values in the case of high or average stress concentration in the connection zone.

Further investigations on K joints lead to the conclusion that the higher susceptibility of the weld notches in a hollow section joint of S 460 influences the fatigue behaviour decisively. For existing sharp notches, the fatigue strengths of K joints of S 235 and S 460 are about equal. If the weld notches lack sharpness, the higher yield strength of the high strength steels can be exploited. The joint made of S 460 with a smooth transition of the weld to the parent metal shows a more favourable fatigue behaviour than the joint of S 235. The lessening of the weld notches by grinding or TIG or plasma dressing increases the fatigue strength of a joint of higher strength steel markedly. Although the reduction of the weld notch also occurs in the joints of the general structural steels with lower yield strengths, the enlargement of the stress amplitude in this case causes plasticity quickly and leads further to crack. As the yield strength of a higher strength steel is larger, the hot spot stress remains still below the elasticity limit.

7.7 Basic "$S_{r,hs}$ vs. N_F" lines for welded, uni-planar CHS and RHS (square) section joints (T, X, K, N and KT)

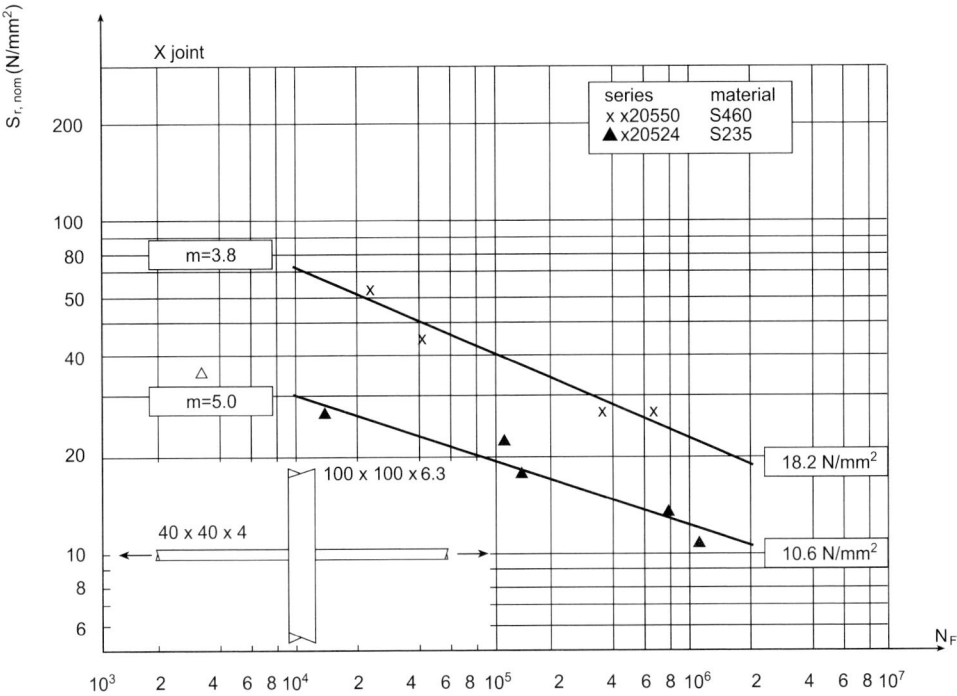

Fig. 7-46 X joints ($b_1/b_0 = 0.4$; $t_1/t_0 = 0.63$); comparison of S 235 and S 460

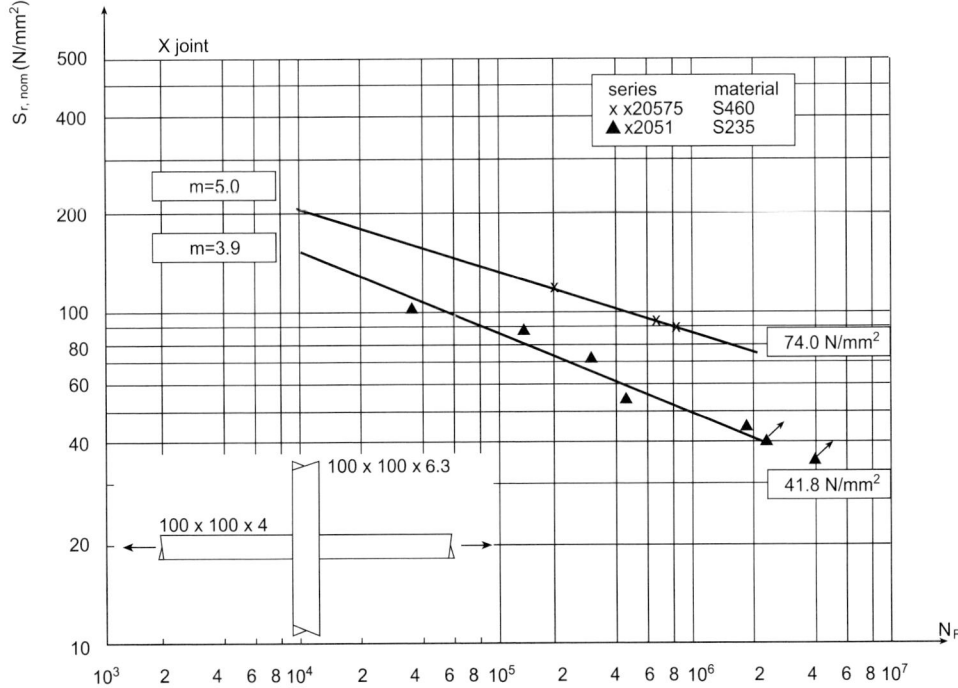

Fig. 7-47 X joints ($b_1/b_0 = 1.0$; $t_1/t_0 = 0.63$); comparison of S 235 and S 460

The effect of wall thickness on the fatigue resistance of T, X, N, K and KT joints of general structural steels S 235 and S 355 is known (see Section 7.7.1). This can have more significance for joints made of higher steel grades. For a thickwalled hollow section, a tri-axial stress condition occurs, which can cause brittle fracture. As the higher strength steels are also loaded by a larger mean stress f_m, the brittle fracture exercises a higher influence.

Finally, this has to be mentioned that further investigations on the static as well the fatigue behaviour of hollow section joints and connections of higher strength steels are required.

7.8 Basic "$S_{r,hs}$ vs. N_F" lines for welded, multi-planar CHS and square hollow section joints (TT, XX and KK)

Extensive practical tests and numerical analysis were performed to establish the basic "$S_{r,hs}$ vs. N_F" lines for KK type welded CHS and square hollow section joints in the framework of an European research programme [49] (see Figs. 7-48 and 7-49). They are also valid for the TT and XX type joints (Fig. 7-27).

Figs. 7-50 and 7-51 illustrate the recommendations for the execution for the multi-planar joint welds.

The design line for the multi-planar CHS joint is compared with the DEn recommendation

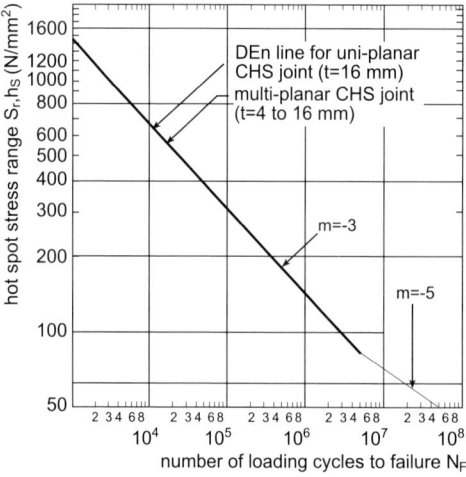

Fig. 7-48 Design line "$S_{r,hs}$ vs. N_F" for welded, multi-planar CHS joints (also DEn line for uni-planar CHS joints)

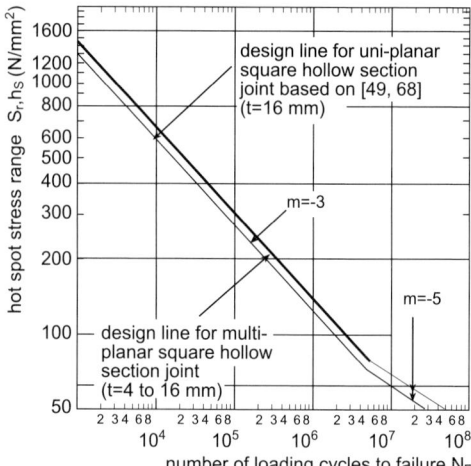

Fig. 7-49 Design line "$S_{r,hs}$ vs. N_F" for welded, multi-planar square hollow square hollow section joints

[73] for uni-planar CHS joints, which is valid for $t = 16$ mm (see Fig. 7-48). Both lines are practically identical. The difference lies in the fact that the correction for the wall thickness of the uni-planar joint (DEn) for the wall thickness more and less than 16 mm must be carried out. Contrary to this, no correction is necessary for the wall thicknesses between 4 and 16 mm in the case of the multi-planar CHS joint. Fundamentally, this statement is also valid for square hollow section joints.

If no detailed analysis be carried out to determine the secondary bending moments in the members of the multi-planar joints, the similar measures as for uni-planar joints has to be taken (see Tables 7-11 and 7-12), which means, the axial hot spot stress range $S_{r,hs,ax}$ has to be multiplied by a multiplication factor M.

$$S_{r,hs} = M \cdot S_{r,hs,ax}$$

Table 7-14 Multiplication factor M accounting for the secondary bending moments in KK type CHS and square hollow section joints

KK joint	Multiplication factor M to axial hot spot range	
	chord	bracings
with gap	1.5	1.5
with overlap	1.5	1.5

7.9 Design procedure for welded, uni-planar or multi-planar CHS or RHS (square) truss joints

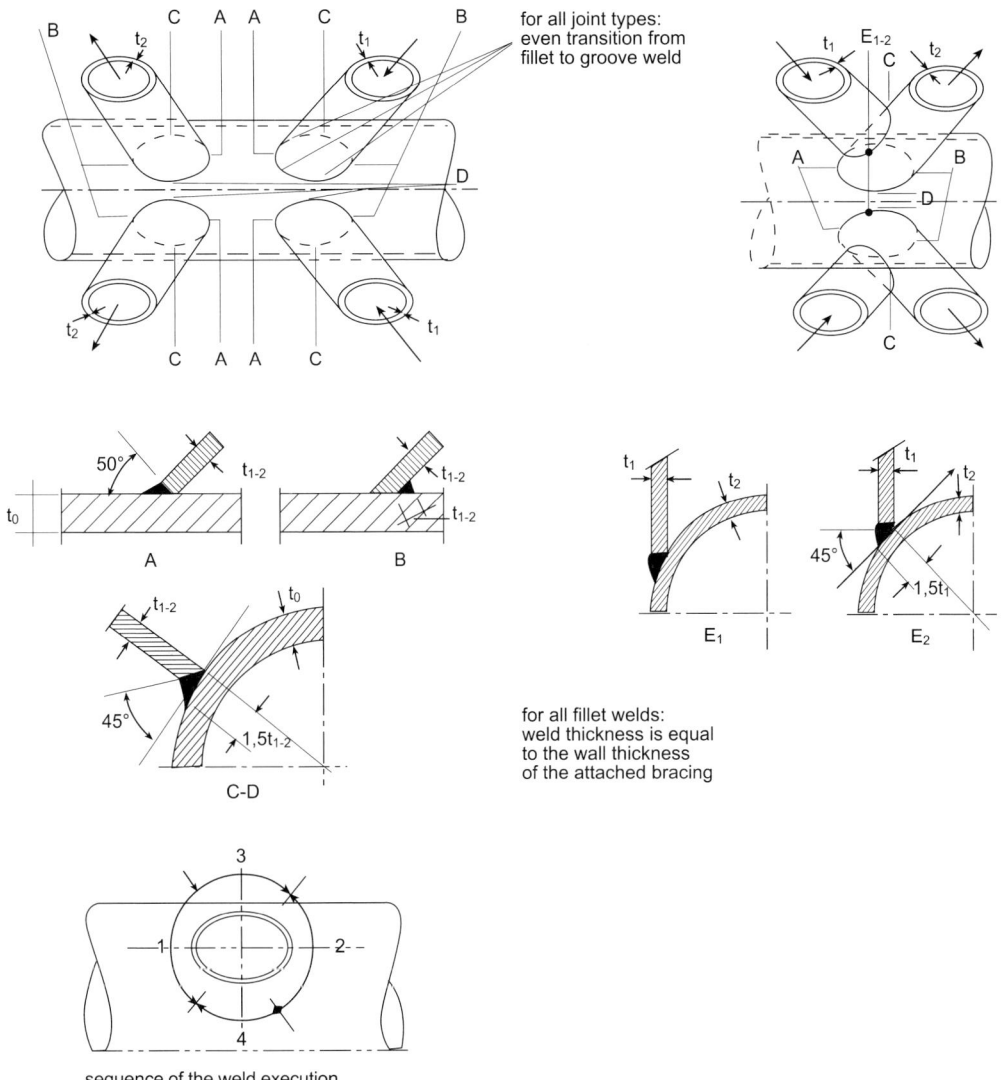

Fig. 7-50 Recommended weld execution and sequences for a multi-planar CHS joint (cf. Fig. 7-48)

These factors for KK type, welded, multi-planar CHS and square hollow section joints with gap as well as overlap are listed in Table 7-14.

7.9 Design procedure for welded, uni-planar or multi-planar CHS or RHS (square) truss joints subjected to fatigue loading

This method to design welded CHS and RHS joints refers directly to the stress concentration factor SCF.

Fig 7-52 shows a flow diagram of the design procedure; the graphical illustration is given in Fig. 7-53.

The life of each single joint must be at least equal to the service duration of the total structure. For the critical joint, the failure of which may lead to the destruction of a structure, EC 3 [42] has suggested an additional partial safety factor γ_F for the nominal stress range $S_{r,\text{nom}}$. They are listed in Table 7-15.

292   7 Design of welded hollow section joints subjected to fatigue loading

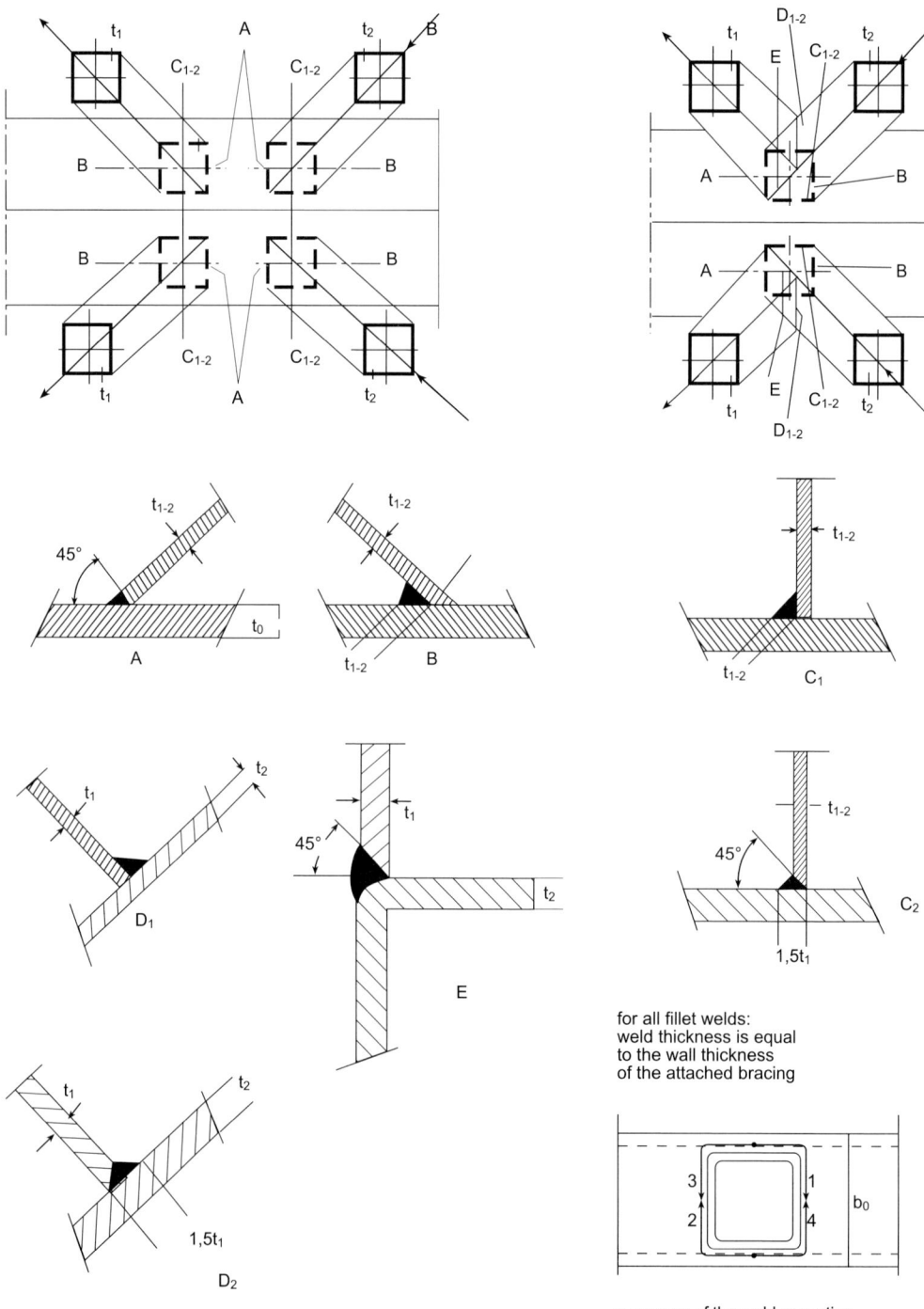

Fig. 7-51 Recommended weld execution and sequence for a multi-planar square hollow section joint (cf. Fig. 7-49)

7.10 Fatigue strength design procedure for hollow section connections and truss joints

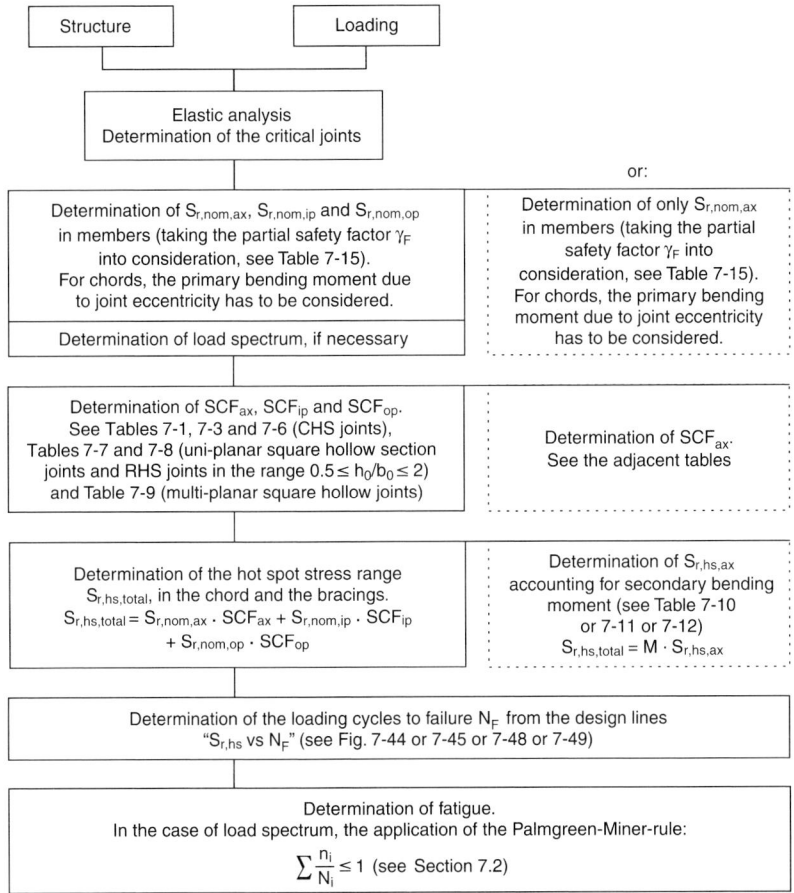

Fig. 7-52 Flow chart of design procedure for hollow section joints under fatigue load with the aid of SCF

Table 7-15 Partial safety factor γ_F for the fatigue load

Inspection and access	"Fail-safe"* components	Non "fail-safe" components
Periodic inspection and maintenance; accessible joint detail	1.00	1.25
Periodic inspection and maintenance; poor accessibility	1.15	1.35

* Structural components with reduced consequences of failure, such that the local failure of one component does not result in failure of the structure

The design verification of a hollow section truss does not necessitate that all joints have to be checked. It is sufficient to select one or two critical joints in accordance with structural dimensions and geometry as well as design loads and verify their fatigue resistances and life.

7.10 Fatigue strength design procedure for hollow section connections and truss joints using the "classification" method

The Committee TC 6 "Fatigue" of the European Convention for Constructional Steelwork (ECCS) published in 1985 the fatigue design recommendations for steel structures [41] in order to standardize the technical design and calculation procedure of steel constructions under fatigue load. This was later modified by IIW [40], which has been then integrated in EC 3 [42]. Besides the joints of the conventional rolled steel sections, also

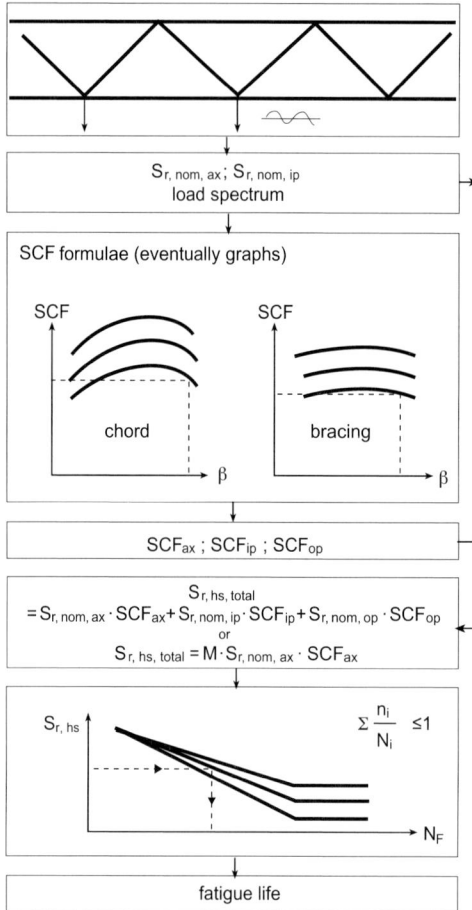

Fig. 7-53 Graphical presentation of the fatigue design procedure for hollow section joints aided by SCF

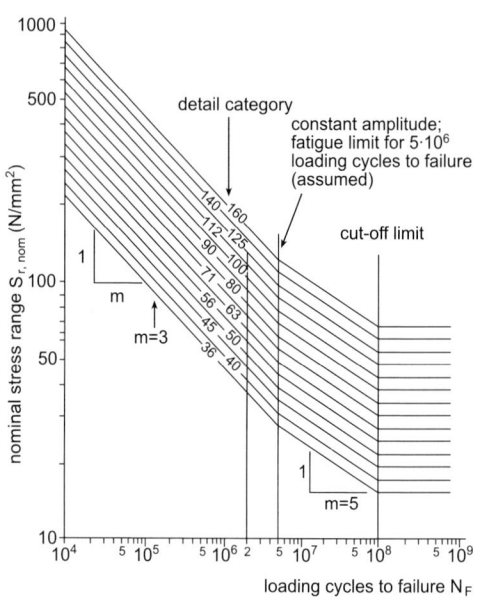

Fig. 7-54 "$S_{r,\text{nom}}$ vs. N_F" lines for the fatigue design following the "classification" method (compare with Table 7-16) [42]

those made of CHS and RHS have been dealt with here. Effort was made to harmonize the design of the joints and connections with hollow sections and conventional rolled steel sections, which however led to very conservative design of the hollow section joints from case to case. The "classification" method, which is described in the following, is very simple to apply.

The concept of design according to the "classification" method is based on the nominal stress range $S_{r,\text{nom}}$. The uniform design lines "$S_{r,\text{nom}}$ vs. N_F" with the inclination $m = 3$ up to $N_F = 5 \cdot 10^6$ and then with $m = 5$ up to $N_F = 10^8$ have been proposed for use (see Fig. 7-54).

While taking the nominal stress range $S_{r,\text{nom}}$ as the basis for design, it has been assumed that the test values of $S_{r,\text{nom}}$ to establish the Woehler curves "$S_{r,\text{nom}}$ vs. N_F" already integrate the effect of the weld, the residual stress, the joint eccentricity and the secondary bending moment due to the end fixity of the joint members.

Adapted to Fig. 7-54, the fatigue resistances of the joints and connections have been classified in "detail categories" (see Table 7-16). A higher detail category designates a larger serviceability resistance compared to that with a lower detail category.

Detail category =
nominal stress range $S_{r,\text{nom}}$ in N/mm² at $2 \cdot 10^6$ loading cycles to failure

Table 7-17 contains the numerical $S_{r,\text{nom}}$ values for 10^5, $2 \cdot 10^6$, $5 \cdot 10^6$ and 10^8 loading cycles to failure. For hollow section joints, the inclination of the $S_{r,\text{nom}}$ vs. N_F lines is $m = 5$ and extends over the total range up to $N_F = 10^8$ (see Fig. 7-55 and Table 7-18). Table 7-19 shows the numerical $S_{r,\text{nom}}$ values for $N_F = 10^5$, $2 \cdot 10^6$ and 10^8 corresponding to Fig. 7-55.

7.10 Fatigue strength design procedure for hollow section connections and truss joints

Table 7-16 Hollow sections* and hollow section connections with the detail categories [42]

Detail category	Constructional detail	Description	Requirement
160	1	Rolled and extruded products ① non-welded elements	① Sharp edges and surface flaws to be improved by grinding
140	2	Continuous longitudinal welds ② automatic longitudinal weld seams	② No stop/start positions, and free from defects outside the tolerances
71	3	Transverse butt welds ③ groove-welded end-to-end connection of CHS	③ and ④ – Height of the weld convexity less than 10% of the weld thickness with smooth transitiion to the plate surface – Welds made in flat position and found by inspection free from defects outside the tolerances – Details with wall thickness larger than 8 mm may be classified two categories higher
56	4	④ groove-welded end-to-end connection of RHS	
71	5	Non-bearing welded connections ⑤ circular or rectangular hollow section, fillet welded to another section	⑤ – Non-load-carrying welds – Section width parallel to stress direction ≤ 100 mm
50	6	Welded flanges ⑥ CHS groove-welded end-to-end connection with a flange plate	⑥ and ⑦ – Load-carrying welds – Welds inspected and found free from defects outside the tolerances – Details with wall thickness larger than 8 mm may be classified one category higher
45	7	⑦ RHS groove-welded end-to-end connection with a flange plate	
40	8	⑧ CHS fillet-welded end-to-end connection with a flange plate	⑧ and ⑨ – Load-carrying welds – Wall thickness less than 8 mm
36	9	⑨ RHS fillet-welded end-to-end connection with a flange plate	

* t ≤ 12.5 mm

Table 7-17 Numerical nominal fatigue stress ranges $S_{r,nom}$ (N/mm^2) for hollow sections and their welded joints according to Fig. 7-54 [42]

	Detail category $2 \cdot 10^6$	Fatigue limit $5 \cdot 10^6$	Cut-off limit 10^8
10^5 loading cycles	loading cycles	loading cycles	loading cycles
434	160	118	65
380	140	103	57
339	125	92	51
304	112	83	45
271	100	74	40
244	90	66	36
217	80	59	32
193	71	52	29
171	63	46	25
152	56	41	23
136	50	37	20
122	45	33	18
109	40	29	16
98	36	26	15

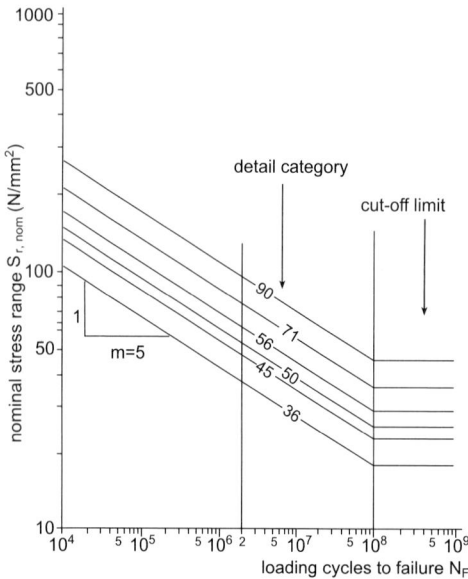

Fig. 7-55 "$S_{r,nom}$ vs. N_F" lines for the fatigue design of the hollow section joints (compare with Table 7-18) [42]

ondary bending moments due to the end fixity of the members have to be taken for the calculation of the joint into account and the primary bending moment due to joint eccentricity goes predominantly into the chord (see Section 6.6.4), which is a part of the preliminary calculation for the chord. Accounting for the secondary bending moment, the axial forces in the truss members are simply multiplied by a factor M given in Table 7-11 or 7-12 according to EC 3 [42]. Axial forces are further to be multiplied by the partial safety factor γ_F listed in Table 7-15 [42].

The fatigue life of a truss joint in hollow sections can be determined as follows using the "classification" method:

- Calculation of the nominal stress ranges $S_{r,nom}$ in the critical joint members assuming pinned connections.

- Taking the secondary bending moments and additional safety factor into account using the following equation:

$$S_{r,nom,existing} = S_{r,nom,calculation} \cdot M \cdot \gamma_F$$

- Determination of the fatigue life N_F of the joint under $S_{r,nom}$ vs. N_F" in Fig. 7-55.

- Application of the "Palmgren-Miner" rule in the case of load spectrum:

$$\sum_{n}^{i=1} \frac{n_i}{N_i} \leq 1.0$$

7.10.1 CHS connection with splice plates

Fig. 7-56 illustrates a number of the investigated splice plate connections with CHS, where the details of the weld are also shown. The applied steel grades in the fatigue tests are as follows: S 235, S 355 and St E 690 (see Chapter 3).

The "$S_{r,nom}$ vs. N_F" lines obtained by the tests are the basis for the extension of the design recommendations according to the "classification" method (see Table 7-20) [77]. Fig. 7-54 is used for the fatigue design, whereas the construction of the connection has to lean against those shown in Fig. 7-56.

Usually, the calculation of the design resistances of truss members is made assuming pinned joint (free of moments), which allows to avoid extensive numerical analysis. This makes only the calculation of axial forces in the structural members necessary. The sec-

7.10 Fatigue strength design procedure for hollow section connections and truss joints

Table 7-18 Lattice girder joints in hollow sections ($m = 5$) [42]

Detail category	Constructional detail	Description	Requirement
90	$t_0/t_i \geq 2.0$	Joints with gap* ① CHS K and N joints ② RHS K and N joints	① $0.5(b_0-b_i) \leq g \leq 1.1(b_0-b_i)$ $g \geq 2\,t_0$ g: gap e: positive eccentricity
45	$t_0/t_i = 1.0$		
71	$t_0/t_i \geq 2.0$		
36	$t_0/t_i = 1.0$		
71	$t_0/t_i \geq 1.4$	Joints with overlap* ③ K joint	③ and ④ Overlap between 30% and 100% $(q/p)\cdot 100$: overlap e: negative eccentricity
56	$t_0/t_i = 1.0$		

Table 7-18 (continued)

Detail category	Constructional detail	Description	Requirement
71	$t_0/t_i \geq 1.4$	Joints with overlap* ④ N joint	① through ④ – $t_0, t_i \leq 12.5$ mm – $35° \leq \theta \leq 50°$ – $b_0/t_0 \leq 25$ – $d_0/t_0 \leq 25$ – $0.4 \leq b_i/b_0 \leq 1.0$ – $0.4 \leq d_i/d_0 \leq 1.0$ – $b_0 \leq 200$ mm – $d_0 \leq 300$ mm – $-0.5 h_0 \leq e \leq 0.25 h_0$ – $-0.5 d_0 \leq e \leq 0.25 d_0$ – Out-of-plane eccentricity $\leq 0.02 b_0$ or $\leq 0.02 d_0$ – Fillet welds are permitted in bracing with wall thickness ≤ 8 mm. – For wall thickness larger than 12.5 mm, see clause 9.6.3 of EC3 [42]
50	$t_0/t_i = 1.0$		

* For intermediate t_0/t_i values, linear interpolation between nearest detail categories can be used. The bracings and the chords require separate fatigue assessments.

7.10 Fatigue strength design procedure for hollow section connections and truss joints

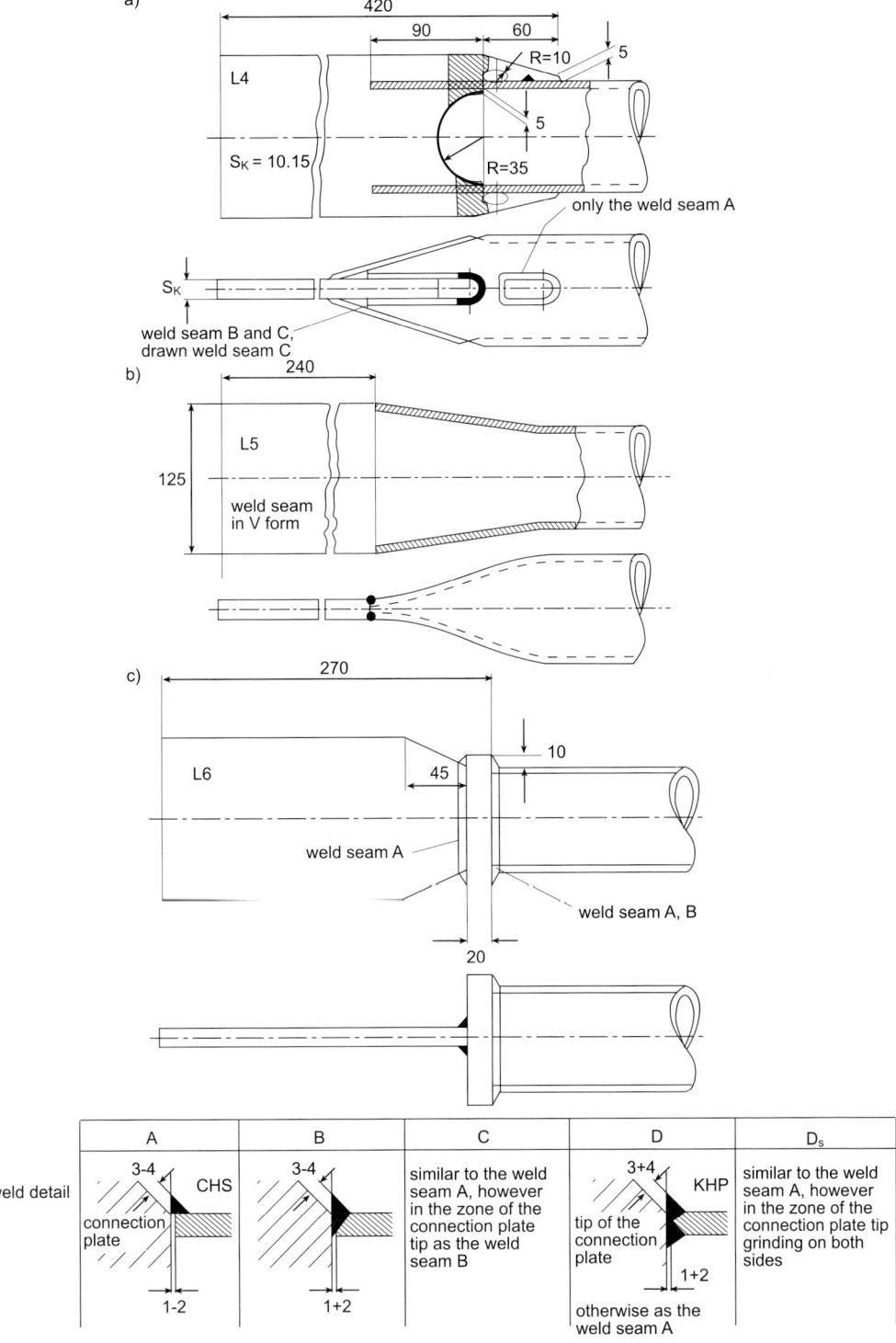

Fig. 7-56 Test specimens for splice plate connection with CHS [5]

Table 7-19 Numerical nominal stress ranges $S_{r,nom}$ (N/mm²) for hollow section joints according to Fig. 7-55 [42]

10^5 loading cycles	Detail category $2 \cdot 10^6$ loading cycles	Cut-off limit 10^8 loading cycles
164	90	41
126	71	32
102	56	26
91	50	23
82	45	21
66	36	16

7.10.2 Recommendation for the modification of the detail categories for the hollow section joints according to EC 3 [42]

Because of the effort to harmonize the fatigue design of the connections and joints in conventional rolled profiles and hollow sections using the "classification" method according to EC 3 [42], the detail categories for the hollow section joints are recommended in a generalised manner on a safe (conservative) side (see Fig. 7-55 and Table 7-18). This version of the recommended categories is often too safe and at the same time uneconomical, which may be detrimental to the competitiveness of the K and N type hollow section joints while applying them from case to case.

There are proposals to modify the detail categories [77] for the K and N type hollow section joints according to EC 3 [42], which specify the constructional details more precisely (e. g. giving ranges not only for t_0/t_i, but also for d_i/d_0). They lead to significantly higher categories in some ranges. The evaluations have shown very good agreement with the test results.

These proposals with the precisely evaluated detail categories for welded K and N type hollow section joints are presented in Tables 7-21 and 7-22 [77] and can be used together with Fig. 7-55 ($m = 5$). Table 7-22 contains transformation factors, with which the detail categories of other joint types can be calculated

Table 7-20 Splice plate connections with CHS together with the detail categories

Detail category	Constructional detail	Description and requirements
72		CHS-plate connection with flattened CHS end and butt weld (X form) ranges: plate thickness ≤ 20 mm CHS outer diameter ≤ 200 mm
64	groove weld / fillet weld / groove weld	CHS-plate connection, CHS slotted and welded to the plate ranges: CHS outer diameter ≤ 200 mm plate thickness ≤ 20 mm, rounded
45	groove weld	In-line CHS-plate connection with a plate in-between, welded by groove weld ranges: CHS outer diameter ≤ 200 mm plate thickness ≥ 20 mm
45	groove weld / fillet weld	CHS-plate connection slotted CHS end welded to the plate

7.10 Fatigue strength design procedure for hollow section connections and truss joints

Table 7-21 Modified detail categories for welded CHS* joints according to the "classification" method (use Fig. 7-55) [77]

Detail category	Constructional details		Description and requirement
		The diagonal members are to be dimensioned	Truss joints in CHS and RHS
36	$1.0 \leq t_0/t_i < 1.5$	$0.28 \leq d_i/d_0 < 0.75$	1. Joint with a gap $g = 25.4$ mm between the intersection lines of the diagonal members with the chord (K joint) or between the intersection lines of the diagonal and the vertical member with the chord (N joint)
64	$1.0 \leq t_0/t_i < 1.5$	$0.75 \leq d_i/d_0 \leq 1.00$	
	$1.5 \leq t_0/t_i < 2.0$	$0.28 \leq d_i/d_0 < 0.75$	
80	$1.5 \leq t_0/t_i < 2.0$	$0.75 \leq d_i/d_0 \leq 1.00$	2. Joint with gap = 0 (this case is taken for the sake of simple interpolation)
	$t_0/t_i > 2.0$	$0.28 \leq d_i/d_0 < 0.75$	
101	$t_0/t_i \geq 2.0$	$0.75 \leq d_i/d_0 \leq 1.00$	3. Joint with 50% overlapped bracing
		1. joint with 25.4 mm gap	4. Joint with 100% overlapped bracing
45	$1.0 \leq t_0/t_i < 1.5$	$0.28 \leq d_i/d_0 < 0.75$	For the values of t_0/t_i in-between, a linear interpolation is possible
72	$1.0 \leq t_0/t_i < 1.5$	$0.75 \leq d_i/d_0 \leq 1.00$	Ranges:
	$1.5 \leq t_0/t_i < 2.0$	$0.28 \leq d_i/d_0 < 0.75$	$b_0 \leq 200$ mm $d_0 \leq 300$ mm
90	$1.5 \leq t_0/t_i < 2.0$	$0.75 \leq d_i/d_0 \leq 1.00$	
	$t_0/t_i \geq 2.0$	$0.28 \leq d_i/d_0 < 0.75$	$0.4 \leq \dfrac{b_i}{b_0} \leq 1.0$ $0.25 \leq \dfrac{d_i}{d_0} \leq 1.0$
114	$t_0/t_i \geq 2.0$	$0.75 \leq d_i/d_0 \leq 1.00$	
		2. joint with 0.0 mm gap	$\dfrac{b_0}{t_0}$ or $\dfrac{d_0}{t_0} \leq 25.0$
80	$1.0 \leq t_0/t_i < 1.5$	$0.28 \leq d_i/d_0 \leq 0.50$	$35° \leq \theta \leq 55°$ (for joint type 1 and 2)
72	$1.0 \leq t_0/t_i < 1.5$	$0.50 < d_i/d_0 \leq 1.00$	$45° \leq \theta \leq 55°$ (for joint type 3 and 4)
101	$1.5 \leq t_0/t_i < 2.0$	$0.28 \leq d_i/d_0 \leq 1.00$	The values in this table are valid for K type joints.
114	$t_0/t_i \geq 2.0$	$0.28 \leq d_i/d_0 \leq 1.00$	For N joints of CHS and K and N joints of RHS, the values given in this table have to be multiplied by the reduction factors according to Table 7-22.
64	$1.0 \leq t_0/t_i < 1.5$	$0.28 \leq d_i/d_0 \leq 0.50$	
51	$1.0 \leq t_0/t_i < 1.5$	$0.50 \leq d_i/d_0 \leq 0.75$	
36	$1.0 \leq t_0/t_i < 1.5$	$0.75 < d_i/d_0 \leq 1.00$	For the chord design for $N_F = 2 \cdot 10^6$, the "detail category" of 127 has to be taken.
80	$1.5 \leq t_0/t_i < 2.0$	$0.28 \leq d_i/d_0 \leq 0.75$	
64	$1.5 \leq t_0/t_i < 2.0$	$0.75 < d_i/d_0 \leq 1.00$	
80	$t_0/t_i \geq 2.0$	$0.28 \leq d_i/d_0 \leq 1.00$	
		4. joint with 100% overlap	

* For RHS joints, d_i or d_0 has to be substituted by b_i or b_0.

Table 7-22 Reduction (transformation) factors of detail categories for other joint types (N joint in CHS and N and K joints in RHS)

N joint in CHS		K joint in RHS	N joint in CHS	N joint in RHS
	gap 25.4 mm	0.8	0.9	0.4
	gap 0.0 mm	0.8	0.9	0.4
	overlap 50%	0.8	1.0	0.7
	overlap 100%	0.8	1.0	0.7

Table 7-23 Detail categories for cross joints in RHS

Detail category	Constructional detail *			
	k	b/b_0	t_0/t	
k.51	0.15	$0.4 \leq b/b_0 < 0.6$	$t_0/t \geq 1.0$	
	0.15	$0.6 \leq b/b_0 < 0.7$		
	0.20	$0.7 \leq b/b_0 \leq 0.8$	$1.0 \leq t_0/t \leq 1.5$	
	0.40	$0.8 < b/b_0 < 1.0$		
	0.40	$0.6 \leq b/b_0 < 0.8$	$t_0/t > 1.5$	
	0.50	$0.8 \leq b/b_0 < 1.0$		
	0.70	$b/b_0 = 1.0$	$t_0/t \leq 1.5$	
	1.10		$t_0/t > 1.5$	

* Valid for an angle of inclination $\theta = 90°$ and for axial load with the aid of the detail categories of the K type CHS joints given in Table 7-21.

The detail categories for the X type RHS joints with the included angle $\theta = 90°$ (cross joint) have also been proposed in [77], which may be taken as a further extension of the EC 3 recommendations (see Table 7-23).

7.11 Effect of the local reinforcement by plates on the fatigue strength of RHS joints

A reinforcement of the RHS joints by welding a plate to the connecting face of the chord member (Fig. 7-57, form 1) or by setting an welded transversal plate (Fig. 7-57, form 2) improves the fatigue strength in general. A combination of the forms 1 and 2 is also possible (Fig. 7-57, form 3).

A significant rise of f_{max} ($N_F = 2 \cdot 10^6$) can be achieved by the reinforcement of the form 1 with increasing plate thickness for small b_i/b_0 ratios (see Fig. 7-58). This applies also to the fatigue strength for finite life.

The influence of the width of the reinforcement plate on the ultimate maximum tensile stress f_{max} is illustrated in Fig. 7-59.

It can be observed from Fig. 7-60 that a short plate length in the case of a joint with gap is advantageous. However no effect could be seen beyond a certain plate length.

For a joint with overlap, which in its non-reinforced form fails by a chord crack transverse to the longitudinal axis of the chord, an increase of the fatigue strength can be attained by using this reinforcement type and a larger plate length. Hereby, a continuous longitudinal weld lies in the critical zone instead of a transversal weld and the decisive loads in the connection zone are reduced by the additional reinforcement plate.

A further increase of the fatigue strength applying the reinforcement form 3 is visible in Fig. 7-61, where a comparison of the data obtained from the tests on the form 1 with those on the form 3 has been made.

Fig. 7-62 demonstrates that the same tendency prevails for the reinforced as well as for the non-reinforced joints regarding the effect of overlaps. Fundamentally, a 70% overlap is not more favourable than a 37% overlap.

7.12 Repair and reconstruction of hollow section joints in the critical zones susceptible to ruptures by cracks

As a rule, the life of a steel structure is set to be between 30 and 70 years. Damages on single structural members or joints can however occur before the expiry of the set time and must then be repaired. The failure modes and the crack figures (see examples in Figs. 7-26, 7-33 and 7-34) determine the practical reinforcement and repairing measures. Investigations on the repair and reconstruction of hollow section joints under fatigue load were carried out within a research work sponsored by the European Community [49].

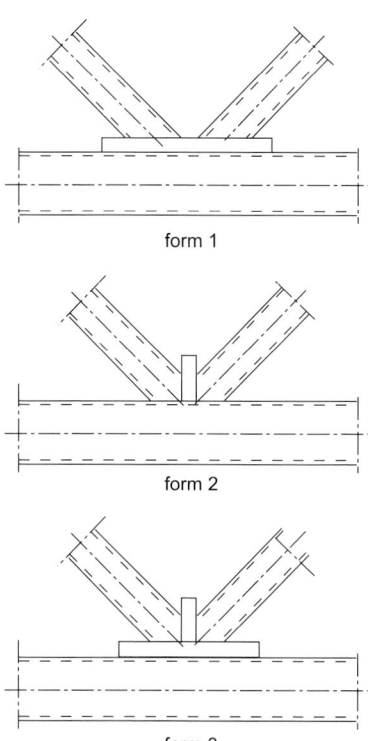

Fig. 7-57 Types of reinforcement for RHS joints

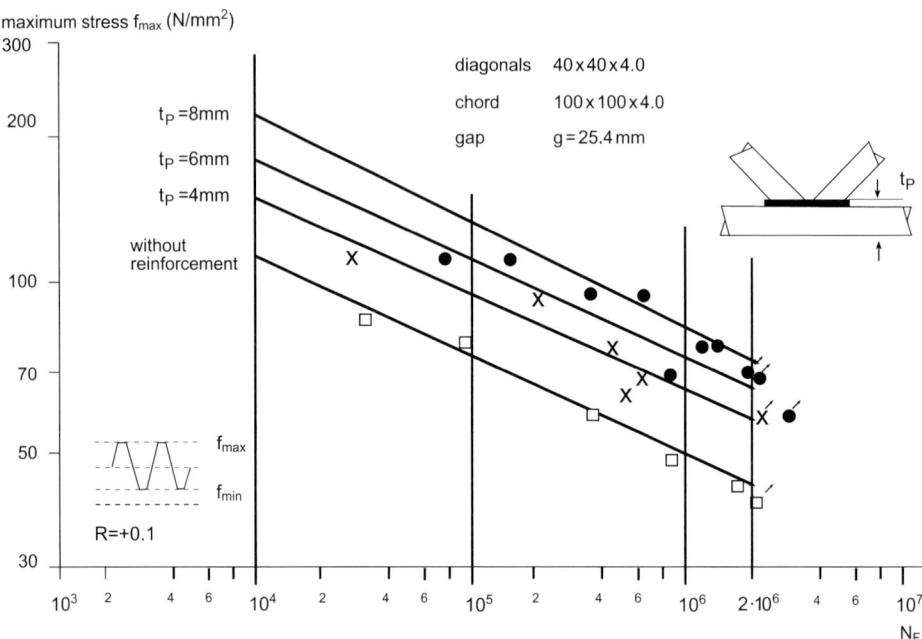

Fig. 7-58 Effect of the reinforcement plate thickness on the fatigue strength of the K type RHS joints

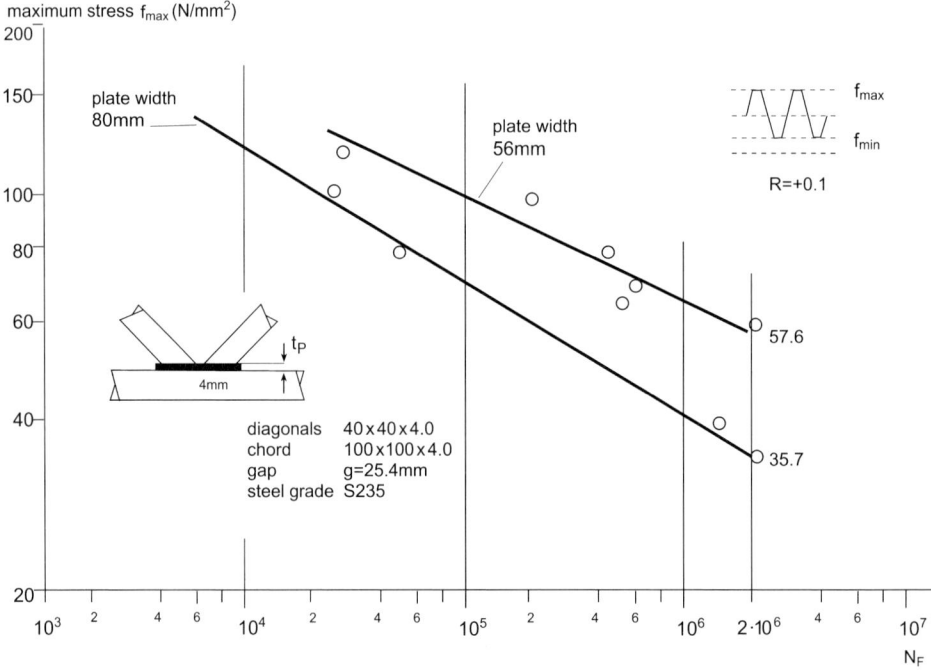

Fig. 7-59 Effect of the width of the reinforcement plate on the fatigue strength of the K type RHS joints

7.12 Repair and reconstruction of hollow section joints in the critical zones susceptible to ruptures by cracks

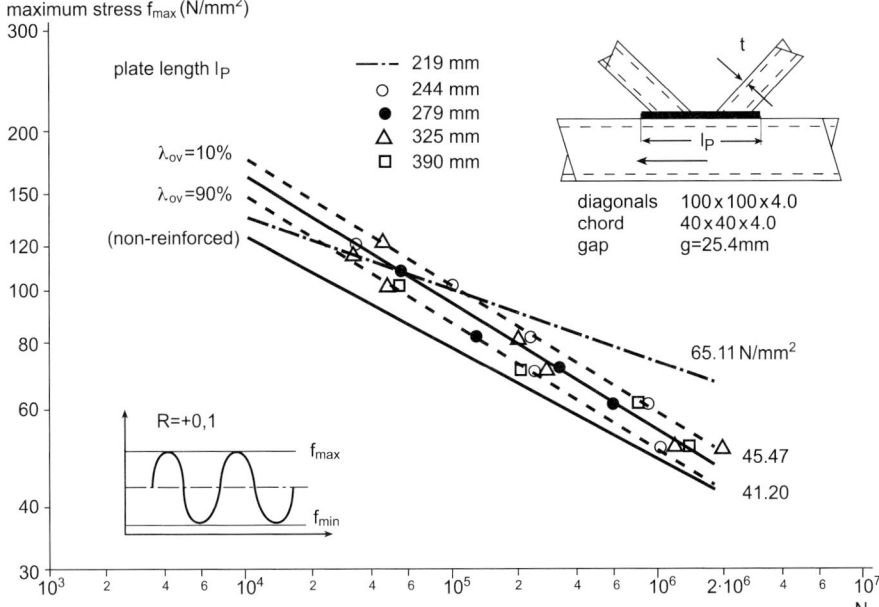

Fig. 7-60 Effect of the length of the reinforcement plate on the fatigue strength on the K type RHS joints

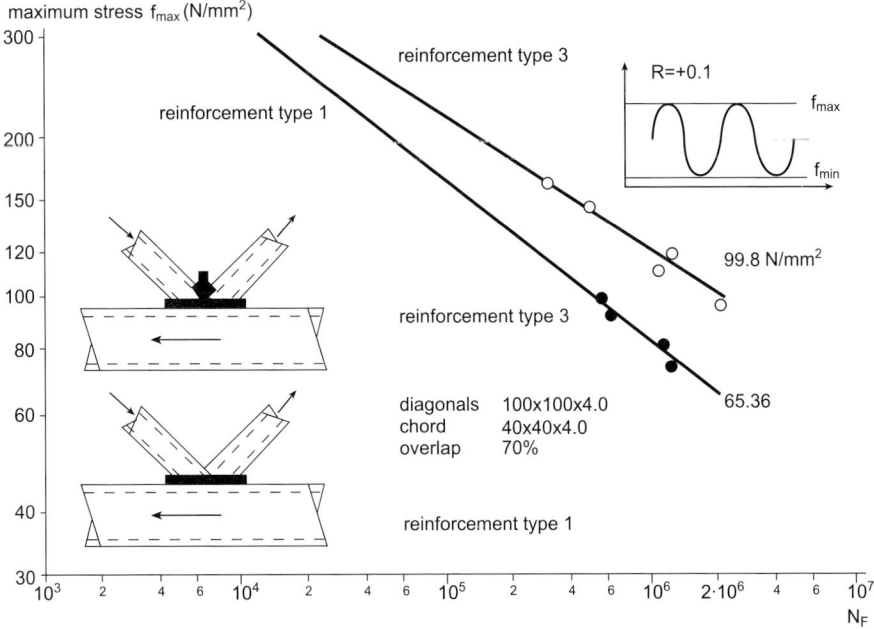

Fig. 7-61 Comparison of the reinforcement types 1 and 3

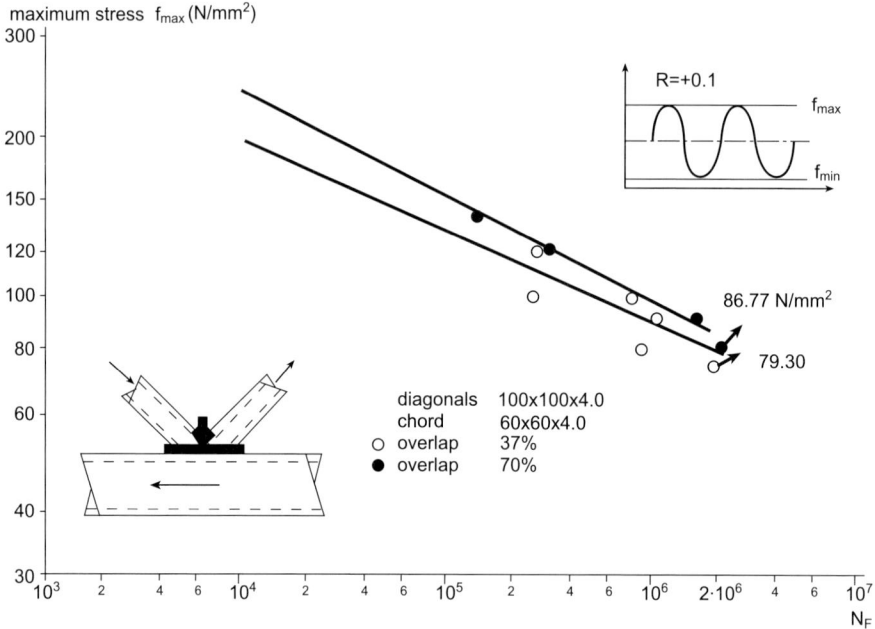

Fig. 7-62 Effect of overlap on the fatigue strength of the K type RHS joints with reinforcement type 3

Strengthening of a structure requires the following considerations:

- The structure under load has a certain amount of stress in it; any additional member (plate, profile) is under zero stress; the consequence is a very different level of stresses.
- Strengthening and stiffening of a structure often shift the problems to the spot next to the strengthening part.

In the following, the involved repairing measures for the hollow sections and their joints have been described.

7.12.1 Application of a crack stopper in the form of a drilled hole

The occurred cracks are not repaired in this procedure. But a hole is drilled as a crack stopper at the end of the crack. The hole is then loaded with a high strength prestressed bolt so that the crack can not propagate further. This measure is only possible for constructions with large hollow section dimensions.

7.12.2 Grinding or gouging of the crack and rewelding

By a simple welding shutting off a crack, the initial life can increase by about 0.25 to 0.50 times. This means that this simple repair can bring about a short-lived bridging of the standstill period.

Figs. 7-63 and 7-64 show a suitable repairing possibility to reach a long life (particularly for hollow section joints), which consists of gouging the crack and rewelding.

The above mentioned investigation [49] on the triangular girders in hollow sections has shown that this repairing measure can lead to a further rest life of 40% of the design life. After gouging, welding should be done with the basic coated electrodes and the last pass with the rutile coated electrodes. As cracks in hollow section joints are initiated at the weld toe and the repair is carried out adjacent to the weld, a smooth transition must be made from the old to the new weld.

7.12 Repair and reconstruction of hollow section joints in the critical zones susceptible to ruptures by cracks 307

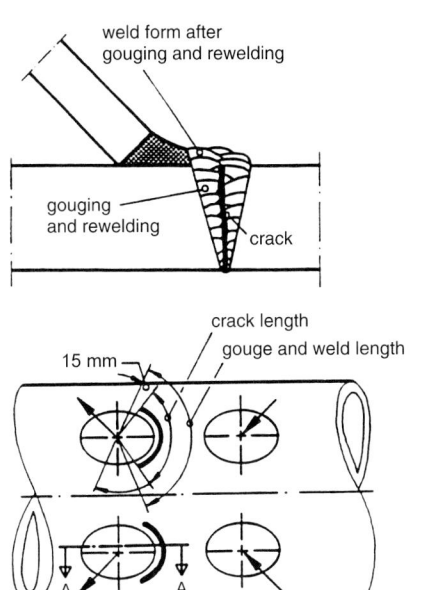

Fig. 7-63 Gouging and rewelding the crack in a CHS joint

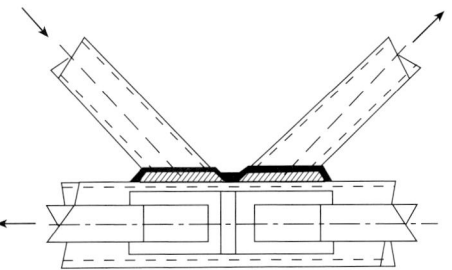

Fig. 7-65 Repairing a RHS joint by mounting a plate on the chord flange

Fig. 7-66 Elevating the load-bearing section in the butt-welded zone (lack of fusion in weld) by welding a half-shell (predominantly static construction)

Fig. 7-64 Gouging and rewelding the crack in a CHS joint with overlap

7.12.3 Mounting a plate or a shell on the chord flange (crack in the chord flange)

In this procedure, the weld is gouged at the crack location, then rewelded and an additional plate or shell is mounted on the chord flange as a reinforcement (Figs. 7-65 and 7-66).

In case a crack occurs in the bracing, a part of it must be cut off and then replaced by a new section (Fig. 7-67).

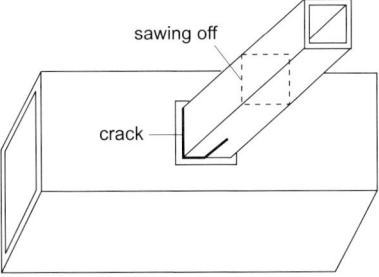

Fig. 7-67 Repair by replacing a bracing part (crack in bracing)

Fig. 7-68 Repair of the RHS joint with gap by welding RHS angle pieces

7.12.4 Fitting RHS angle pieces

Although more expensive than other repairing measures, the fitting of RHS angle pieces in the crack location shows very good results (Fig. 7-68). In addition to the bridging of the cracks, a reinforcement and stiffening of the joint can also be attained by this measure. After experimental investigations [49], an increase of the fatigue life by twice the initial life can be expected after repair.

7.13 Improvement of the fatigue strength applying mechanical and thermal processes

The improvement of the fatigue strength of the welded steel joints susceptible to fatigue rupture in the transition zone from the weld to the base metal involves mainly the following two points:

- Creation of favourable residual stress condition in the weld transition zone.
- Execution of notch-free weld transition as far as possible where the weld metal blends smoothly with the base metal. This concerns weld profiling and the use of special electrodes, which are integral part of the welding process itself.

The methods for the fatigue strength improvement in steel constructions are classified in the following main groups:

– Residual stress technique
– Weld toe geometry modification methods

Apart from the above mentioned improvement measures, a better fatigue strength of an welded hollow section joint can already be obtained by choosing special electrodes with a suitable coating, which enables a smooth geometric transition at the weld toe as a result of an improved filler metal quality [97].

The objective of the residual stress technique is in general the creation of compressive residual stresses where fatigue cracks are likely to be initiated. A totally tensile stress cycle can even be made at least partly compressive applying this method. This creates residual stress conditions, which reduce the stress ratio R ($= f_{min}/f_{max}$) (this controls the growth of the fatigue cracks). The residual stress technique comprises the following processes:

– Post-weld heat treatment
– Shot peening or hammer peening
– Overloading
– Stress relieving vibration

The aim of the weld toe geometry modification is to remove the crack-like defects at the weld toe and thereby reduce the stress concentration factor of the weld. There is substantial scope of fatigue life increase by making the weld transition smooth, which delays the crack initiation. The modification of the weld toe geometry is carried out applying the following processes:

– Grinding of weld transition zones
– TIG or plasma post-treatment

7.13.1 Post-weld heat treatment

Unfavourable residual stress condition in weld e.g. tensile residual stress can be removed by post-weld heat treatment, especially if the applied load cycle is wholly or partly in compression. However stress relief does not introduce compressive residual stresses but relies on the removal of tensile residual stresses so that compressive stresses may be experienced at the weld toe.

A careful check of the residual stress condition is necessary, as there may be cases, where the stress relieving heat treatment brings about no change in the fatigue behaviour or even lowers the fatigue load. This is due to the fact that the welding may result in favourable compressive stresses at the locations susceptible to fatigue cracks (weld transition or weld root), which are then reduced by the heat treatment.

The process consists of heating the connection zone with special heating equipments (gas burner or induction coil) to about 500 °C. The aim is to reduce the residual stresses by local plastification.

Special advantage is offered to the higher strength steels with higher yield strengths by this method, as the degree of improvement for higher strength steel is generally larger than for mild steels.

Another procedure of heat treatment is to apply hot point or heated wedge depending on the type and form of the structural element. Hereby, compressive residual stresses are introduced in the critical locations in order to reduce existing tensile residual stresses.

It is to be mentioned here specially that the post-weld heat treatment is less expensive than grinding the weld.

7.13.2 Shot peening and hammer peening

The mechanical methods e.g. shot peening and hammer peening, are cold working processes exercising two effects:

- Creation of residual compressive stresses in the surface layer of the construction; local peak stresses are reduced by the superposition of compressive stresses.
- Improvement of the weld profile by cold plastic deformation and smoothening of the under-cut.

In shot peening, the surface of the component is deformed by a high velocity stream of small shot (either cast iron spheres of small diameter or pieces of high tensile steel wires). The grade of the shot peening is determined by two parameters:

- Degree of surface plastic deformation.
- Surface area coverage by the produced dimples. 100% coverage is obtained when visual examination at 10x magnification of the surface shows that all dimples just overlap.

The effectivity of the shot peening depends on the static strength of steel among others. In particular, it renders improvement in the range higher than 10^6 loading cycles. Hammer peening is carried out either with a solid tool with a rounded tip of 6–14 mm radius or with a tool consisting of a bundle of wires or rods with the tip of each wire rounded. Both tools are normally pneumatically operated. The required air pressure is between 5 and 6 bar. Optimum results for hammer peening are obtained after four passes, giving a severely deformed weld toe with an indentation depth of about 0.6 mm, providing a simple inspection criterion. Hammer peening is a noisy and tedious operation; but the fatigue strength improvements are significantly high, also for high strength steels.

7.13.3 Overloading of a structural element

This method consists of mechanical overloading of a structural element to attain a planned plastification of the critical fatigue zones, where compressive residual stresses are initiated after removal of the load at least close to the surface.

In the local compression technique, a part of the structure is made to yield by local compression between dies, usually of circular cross-section. A sufficiently large load is applied to the dies to induce a state of general yield, so that an indentation is formed and the material is squeezed out radially. When the load is removed, the indentation remains and compressive residual stresses exist in the region of the indentation. This technique is suited to geometries, in which the expected fatigue initiation site is localized, such as spot welds or the ends of stiffeners.

7.13.4 Stress reliving vibration

It is possible to reduce the residual stresses in a welded joint by inducing vibrations. The research works done in this field demonstrate however large scatter of the results. It is therefore not sufficiently sure whether the unfavourable residual stresses can be removed or reduced by this treatment. Due to this reason, the application of this method can only then be recommended when the reduction of the residual stresses is confirmed by the tests.

Another disadvantage of this process lies in the fact that vibrations may use up a considerable proportion of the fatigue life of a structure themselves.

7.13.5 Grinding of weld transition zones

Grinding of the weld by a rotary burr grinder or a disc grinder removes under-cuts and thereby also improves the weld profile. It has to be taken into consideration that grinding has to be extended to a depth of minimum 0.5 mm below the weld notches to ensure the removal of slag intrusions. An inexperienced operator may inadvertently remove too much material reducing the cross-section of the structure in this procedure. This can even worsen the fatigue behaviour of a joint.

Grinding is mainly carried out to avoid crack initiation by removing metal from the weld toe. In order to eliminate the locations of initial cracks, this has to be done with a larger radius and an adequate depth (0.5–1 mm). Fig. 7-69 illustrates the AWS recommendation for the grinding to an improved weld profile [13]. In this code, a low stress concentration factor is sought by controlling the overall shape of the weld to obtain a concave profile and requiring a gradual transition at the weld toe. The "disc test" or "dime test" is specified by AWS to ensure an acceptable weld.

Grinding process is relatively simple as well as priceworthy. For the handling of larger areas, this offers particular advantages. However, grinding is often not intended for initial de-

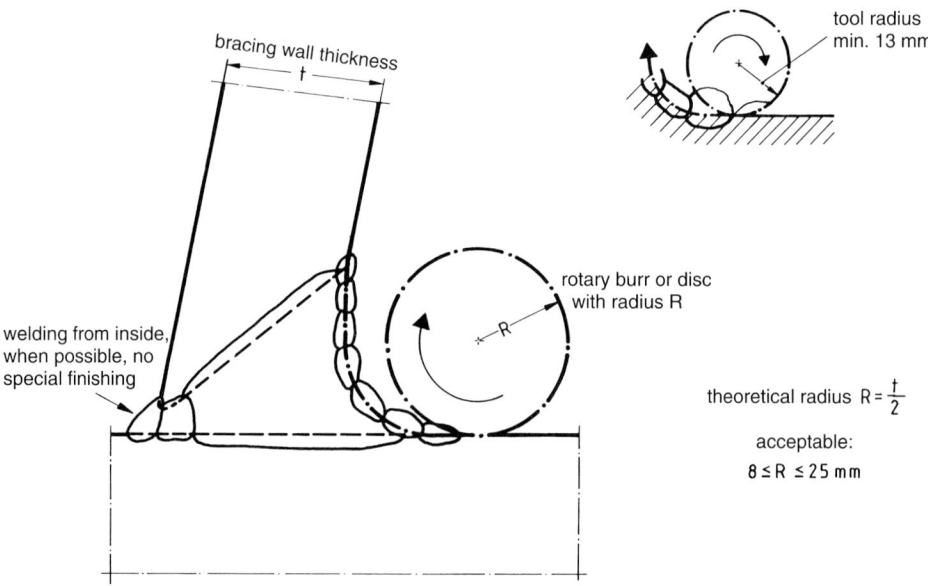

Fig. 7-69 AWS recommendation [13] for grinding the weld profile

sign; it is instead used as a remedial measure if the design life is shown to be inadequate at a later stage during design or construction.

Grinding is also recommended in combination with other methods; common combinations are grinding and shot peening or grinding according to AWS weld profile control and shot peening [98]. This enlarges the fatigue values for an welded joint by a factor of about 2.5.

7.13.6 TIG or plasma post-treatment

Tungsten-inert-gas (TIG) or plasma heat treatment consists of remelting the weld toe region, which results in large gains in fatigue strength due to the following reasons:

- Weld toe transition becomes continuous and smooth and the weld metal blends smoothly with the base plate. This reduces the stress concentration factor.
- Slag inclusions and undercuts are removed.
- Higher hardness in the heat affected zone contributes to higher fatigue strength.

Scale deposits on the surface to be treated may cause small notches or undercuts in the remelting zone. The weld and the base metal should therefore be cleaned using metal brushes in order to remove scale, slag or other impurities before TIG dressing. Sand blasting or light grinding may also be used.

Plasma dressing is similar to TIG dressing, the main difference being the higher heat input, about twice that used in TIG dressing and a wider weld pool. The resulting improvements in fatigue strength are generally higher than for TIG dressing, particularly for higher strength steel.

A special problem attributed to the TIG and plasma remelting procedures involve the stop-start locations, which may lead to irregularities in weld bead contours. It is therefore suggested to stop and start the weld curve on the weld surface.

7.14 Fatigue of bolted flange connections

Fatigue tests on bolted flange connections loaded in tension [26] have demonstrated that a connection can be structurally designed particularly with regard to the location of the contact faces in a manner, that it is possible entirely to eliminate bolt fatigue in tension, thereby dispensing with an elaborate analysis of the fatigue strength of the bolts.

Fig. 7-70 shows some favourable and not-so-favourable forms of bolted connections for CHS members with blind flanges under bolt fatigue.

The diagram in Fig. 7-71 is often used to determine the variation in bolt tension (ΔF_b):

F_v = clamping force (prestressing load by bolt tightening with controlled torque)
F_t = external tensile force
ΔF_b = change in bolt tension (F_b)
ΔF_c = change in contact force (F_c)

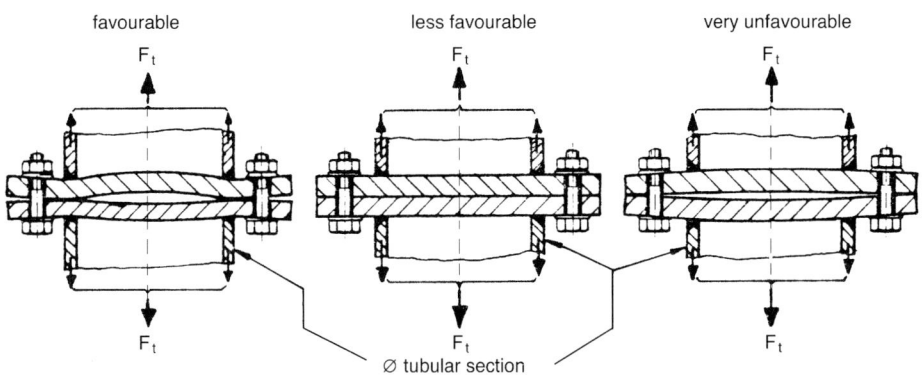

Fig. 7-70 Bolted CHS blind flange connections

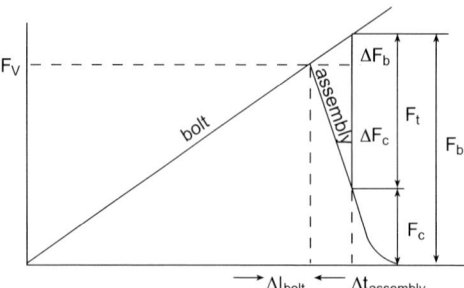

Fig. 7-71 Diagram to illustrate the variation in bolt tension

When the bolt is tightened to F_v, this force will be acting in the bolt (bolt tension) and also in the assembly. From the equilibrium of the flange plate, it follows:

$$F_b = F_t + F_c$$

If the external tensile force is marked off between the "bolt" line and the "assembly" line in Fig. 7-71, it is found that the relationship $F_b = F_t + F_c$ is satisfied. If the external tensile force varies between 0 and F_t, the bolt tension will vary by an amount ΔF_b.

This conception considering only the stiffness of the bolt and the assembly has been however found inadequate and a more comprehensive modification has been proposed in Fig. 7-72.

This modification considers the following additionally:

- The overall deformation i.e. deformation of bolt + deformation of flange occurs if the transfer of force takes place via the bolts.

- The deformation due to the transfer of force affected by the reduction of contact force has to be considered. In this case, the location of the contact force related to the location of the external tensile force must be taken into account.

The following conclusions have been derived from the evaluation of the results of the tests [26] on bolted flange connections loaded in tension in general:

- The influence of the location of the contact force produced by the tightening of the bolts on the bolt fatigue is of major importance. This should be so located that the least deformation occurs in the case of load transfer through reduction of contact force.

- For favourable locations of contact faces, there will be no risk of bolt fatigue as long as the tensile forces are smaller than the contact force. This being so, an elaborate analysis of the bolt fatigue can be dispensed with in many cases (however, any weld for transmitting the tensile force to the flange will have to be checked for fatigue).

- The clamping forces should be as large as possible to improve the fatigue behaviour. It is advisable to tighten the bolts within the plastic range for the sake of obtaining adequate certainty as to the magnitude of the clamping force.

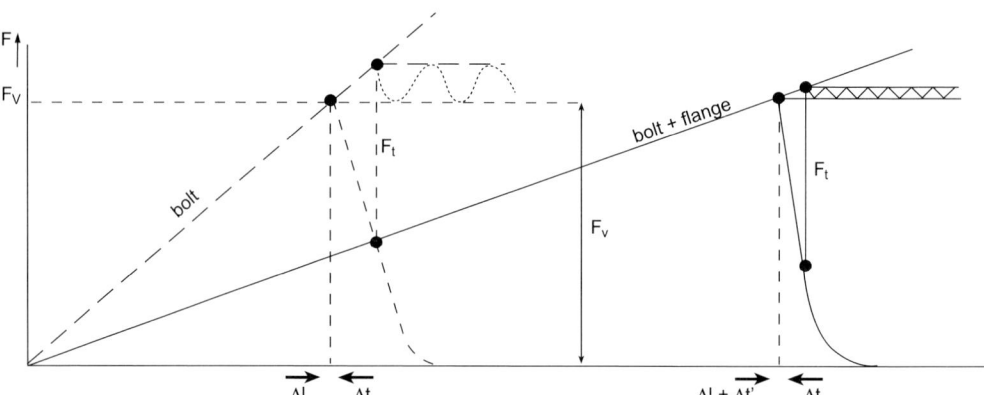

Fig. 7-72 Modified diagram to show the variation in bolt tension

7.14 Fatigue of bolted flange connections

- The distance of the bolts from the component, in which the tensile force acts, should be as small as possible.

As is apparent from the illustrations in Fig. 7-70, the contact face should be located as close as possible to the components in which the tensile force acts. In the case of CHS, it is the wall, not the central axis of the CHS.

Fig. 7-73 shows the connections, where the tensile forces do not act in line with each other. A favourable situation for the bolts can enable the external tensile forces to follow as straight a path as possible. The location of the contact face is best suited where the components, in which the external tensile forces act, cross each other. Thin packing plates (shims), may, for example, be inserted at these points, thus ensuring the most favourable location of the contact force (see the view in plan, Fig. 7-73).

Favourable location of the contact pressure can often be ensured at the design stage. An example with blind flanges is given in Fig. 7-74. If the flanges are not machined, a thin packing ring (e.g. 1 mm thick) may alternatively be inserted so as to form a continuation of the CHS wall across the joint.

An open-flange bolted CHS connection can be detailed as shown in Fig. 7-75. If the CHS wall is relatively thin, this connection is liable to give rise to deformations during the tightening of the bolts. In order to prevent such deformations, arrangements as illustrated in Fig. 7-76 may be adopted.

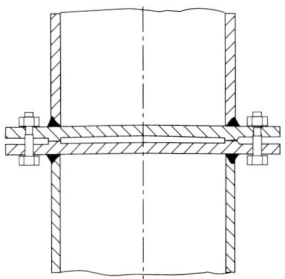

Fig. 7-74 Bolted CHS blind flange connection with favourable contact face location

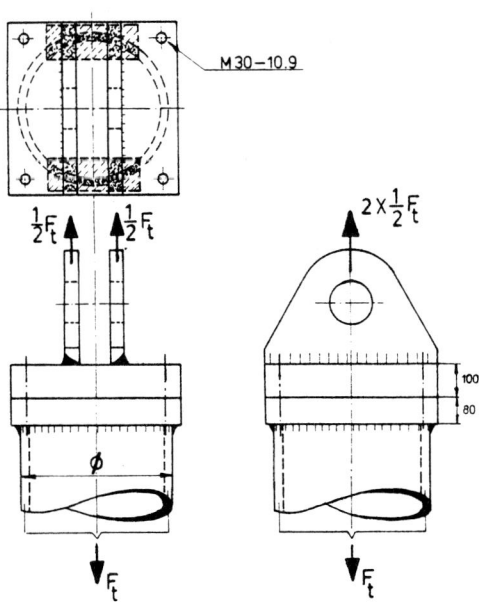

Fig. 7-73 CHS flange connection with components, in which the tensile forces act, are not completely aligned with each other

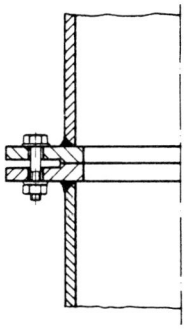

Fig. 7-75 Bolted open-flange CHS connection with favourable contact face location

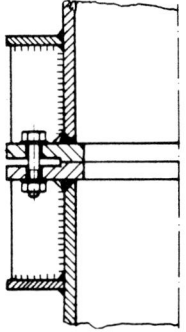

Fig. 7-76 Modification of the bolted open-flange CHS connection to prevent CHS wall deformation

7.15 List of symbols

A	cross-sectional area
A_i	cross-sectional area of a member i ($i = 0, 1, 2, 3$)
D	damage ratio
E	modulus of elasticity
M	moment or multiplication factor accounting for secondary bending moment
N, n	number of loading cycles
N_F	number of loading cycles to failure
N_i	applied axial force in a member i ($i = 1, 2, 3$)
N_0	design value of the axial load in chord member
N_{op}	chord "preload" (additional axial force in the chord member at a connection which is not necessary to resist the horizontal components of the bracing member forces
P	failure probability
R	stress ratio (f_{min}/f_{max})
S_r	stress range ($f_{max} - f_{min}$)
$S_{r,F}$	stress range as fatigue strength (resistance)
$S_{r,nom}$	nominal stress range
$S_{r,hs}$	hot spot stress range
W	section modulus of a member
a	throat thickness of a weld
b	width
b_i	external width of a square or rectangular hollow section (RHS) member i (90° to the plane of a truss) ($i = 0, 1, 2, 3, j$)
d	external diameter
d_i	external diameter of a member i ($i = 0, 1, 2, 3$)
e	noding eccentricity for a joint (see Fig. 6-101)
f	stress
f_a	stress amplitude $\left(= \dfrac{f_{max} - f_{min}}{2}\right)$
f_F	fatigue stress (strength)
f_E	residual stress
f_{hs}	hot spot stress
f_i	stress in a member i ($i = 0, 1, 2, 3$)
f_m	average stress $\left(= \dfrac{f_{max} + f_{min}}{2}\right)$
f_{max}	maximum stress
f_{min}	minimum stress
f_y	tensile yield strength
g	gap between the bracing member toes on the mantle (CHS) or the flange (RHS) ignoring welds of a K, N or KT joint
g'	gap divided by the chord wall thickness t_0 (g/t_0)
h_i	external depth of a square or rectangular hollow section member i ($i = 0, 1, 2, 3$)
i	index to designate a member in a joint $i = 0$ designates a chord; $i = 1$ refers in general to the bracing for T, Y and X joints or it refers to the compression bracing member for K, N and KT joints; $i = 2$ refers to the tension bracing member for K, N and KT joints; $i = 3$ refers to the vertical member for KT joint
L, l, l_0	length
l_P	plate length
l_w	weld length
m	inclination of a design line
r	radius
r_0	external corner radius of RHS
r_i	internal corner radius of RHS
t	wall thickness
t_i	wall thickness of a hollow section member i ($i = 0, 1, 2, 3$)
t_P	thickness of a reinforcement plate
α	$2\, l_0/d_0$ or $2\, l_0/b_0$
β	external width or external diameter ratio between bracing member(s) and chord

$$\beta = \dfrac{d_1}{d_0}; \dfrac{d_1}{b_0}; \dfrac{b_1}{b_0} \quad (T, Y, X)$$

$$\beta = \dfrac{d_1 + d_2}{2\, d_0}; \dfrac{d_1 + d_2}{2\, b_0};$$

$$\dfrac{b_1 + b_2 + h_1 + h_2}{4\, b_0} \quad (K, N)$$

$$\beta = \dfrac{d_1 + d_2 + d_3}{3\, d_0}; \dfrac{d_1 + d_2 + d_3}{3\, b_0};$$

7.15 List of symbols

γ	$\dfrac{b_1+b_2+b_3+h_1+h_2+h_3}{6b_0}$ (KT) half diameter or width to thickness ratio of the chord $\left(\dfrac{d_0}{2t_0}\text{ or }\dfrac{b_0}{2t_0}\right)$		tion Tubulaire (International Committee for the Development and Study of Tubular Structures)
		EC	Eurocode
		ECSC	European Committee for Steel and Coal
γ_F	partial safety factor for load (action)	ECCS	European Convention for Constructional Steelwork
γ_M	partial safety factor for resistance	FE	Finite Element
ε	strain	IF	influence function
ε_r	strain range ($\varepsilon_{max} - \varepsilon_{min}$)	IPB	In-plane bending moment
$\varepsilon_{r,nom}$	nominal strain range	OPB	Out-of-plane bending moment
$\varepsilon_{r,hs}$	hot spot strain range	SCF	Stress concentration factor
ζ	g/d_0 or g/b_0	SNCF	Strain concentration factor
ξ	g/b_i	Snf	Conversion factor SCF/SNCF
θ_i	included angle between bracing member i (i = 1, 2, 3) and the chord	TC	Technical Committee
ν	Poisson's ratio		
τ	shear stress	**Subscripts:**	
τ	$t_{1,2}/t_0$ ratio		
λ_{ov}	overlap (see Fig. 6-82, $\dfrac{q}{p}\cdot 100$ in %)	ax	axial force
		br	bracing
ϕ	angle between the bracing axes in a multi-planar joint	ch	chord
		cov	carry-over
CHS	circular hollow section	c	crown
RHS	rectangular (incl. square) hollow section	hs	hot spot
		ip	in-plane bending
API	American Petroleum Institute	op	out-of-plane bending
AWS	American Welding Society	ref	reference
CIDECT	Comité International pour le Developpement et l'Etude de la Construc-	s	saddle
		w	weld

8 Fabrication, assembly and transport of hollow section structures

8.1 General

In principle, the fabrication, the assembly and the transport of hollow section structures do not differ from those of general, conventional steel constructions. However, the following aspects specific to the structural engineering related to hollow sections have to be accounted for additionally:

- As a closed profile, a hollow section is accessible externally, however only with restrictions to the inside e.g. by means of hand access holes or for locations close to the open ends. Due to the accessibility from one side, the joints in hollow sections can be easily manufactured by welding the members directly to one another, while a direct joining by bolting represents a problem, as the nuts of the bolts cannot be locked on account of the inaccessibility to the inside of the hollow section. Therefore, for connections of hollow sections made by bolting, predominantly the members are indirectly joined using expensive application of gusset or head plates. In the jointing technique, the welding of hollow sections has the position of prime importance. In the recent time, a number of blind bolting types e.g. Flowdrill [1, 2] and Lindapter [3, 4] are however available, which have made an easy direct bolting of hollow sections possible. A further possibility is the application of "through" bolts, as far as it is facilitated by the given hollow section sizes.

- Holes in hollow sections can be made by penetration drilling. Through punching is only possible with the aid of a die plate.

- Although indirectly bolted joints are more labour-intensive and expensive than directly welded joints, they are very often used for joint assembly on site. Site welding is far more costly than site bolting and the susceptibility to weld defects due to ambient adversities is higher.

Therefore, welded sub-assemblies in hollow sections, prefabricated in the workshop are in general assembled on site, preferably by bolting.

A careful planning and execution belong to the fabrication of hollow section constructions, which consists of the following jobs in the workshop:

- Cutting (sawing and flame cutting, also laser and plasma cutting)
- Slotting
- Flattening and cropping of hollow section ends
- Bending (arching)
- Bolting
- Welding
- Nailing

Before the fabrication procedure starts, the fabrication drawings have to be prepared, which are based on the design drawings by the designer and draughtsman. Corresponding to the existing technological capability as well as the available equipments and machines in the workshop, each structural element and joint detail shall be modified and drawn. It is often recommended to draw full-scale drawings of the complicated parts in order to demonstrate precise details of constructions and the fitting of the elements to one another.

Dimensions of structural components, the selected steel grades and the manufacturing processes shall be carefully recorded so that any discrepancy between the fabrication and the design drawing can be avoided. As it is difficult to check the wall thickness of a hollow section during the manufacturing phase, the size has to be labelled on the outside surface of the hollow section. It is necessary to engage an inspector-in-charge in the workshop, who supervises and coordinates the above mentioned jobs.

As for all other steel constructions, the fabrication procedure of hollow section structures in the workshop should be so arranged that the material goes through a one-way system from the entry to the final delivery. Before the actual fabrication starts, the hollow sections are to be kept in a stock room for a short period, so that they can be easily identified and moved. In a modern workshop, the sizes, the lengths and the steel grades of the hollow sections for a certain project shall be recorded in a list. Hollow sections are then marked with identification labels.

Following the short-time storage the structural elements are transported by a conveyor belt or a lifting equipment to the workshop, where the following steps are undertaken in general:

a) Marking the structural members

b) Cutting the structural elements to length by sawing or flame cutting
In case flame cutting is applied, a) and b) can be combined. Cut-outs can also be executed together with cutting to length.

c) Flattening or cropping of the member ends (if planned)
In light constructions, a simultaneous cutting and crimping is carried out in one operation by means of a punching press.

d) Bending (if planned)

e) Edge preparation of the member ends for welds in welded structures
This can be done together with b). The measurement of the real member dimensions (tolerances) is compulsory in order to obtain precise required cut-outs and weld bevelling.

f) Drilling holes in bolted structures

g) Assembling the members or constructional parts by welding or bolting or a combination of both. This process is often executed in two steps:
– Provisional assembly e.g. by tack welding
– Final assembly

h) Shot blasting (removing rust and descaling)
This can be done from case to case also prior to g), as shot blasting may be difficult after the assembly, especially for large structures.

i) Finishing with primer coating (two or more layers depending on requirements)

j) Top coating for protection against external corrosion or with intumescent paints for protection against fire

In case head plates, ribs or cleats have to be welded to beams or columns, a second line (e.g. conveyor system) parallel to the first for the fabrication flow has to be installed. The process may be similar, if the hollow sections have to be bent or straightened.

8.2 Cutting

The preparation for the assembly of a hollow section structure by welding and bolting starts with flame cutting or sawing. The following types of cuts are commonly used: plane, straight cuts, square cuts or cuts at an angle, mitre cuts, profile cuts or saddling (multi-planar), cropping (see Section 6.6.8).

8.2.1 Flame cutting

Flame cutting is a method for thermal separation of workpieces, where the oxyfuel gas flame is used as the heating source. This is mainly applied to non-alloy and low-alloy steels. In the case of high-alloy steels and cast iron, the plane of separation must be smelted by a stronger heating source e.g. by electric arc. The direct joining of circular hollow sections often applied in steel structures, necessitates a multi-planar, "profiled" cut, referred to as a "saddle", where the angle of intersection between the bracing and the chord changes from point to point around the perimeter of the bracing (see Fig. 8-1). In general, the flame cutting is done with a burner

8.2 Cutting

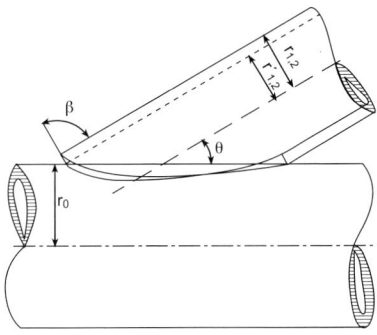

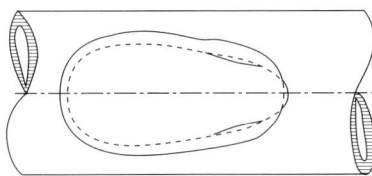

Fig. 8-1 Penetration of a CHS bracing to a CHS chord
r_0 = outer radius of chord
$r_{1,2}$ = outer radius of bracing
$r'_{1,2}$ = inner radius of bracing
θ = included angle between chord and bracing axes
β = angle of chamfer

nozzle, which can be performed manually as well as by means of automatic, coordinate-controlled machines. The flame cutting process is illustrated schematically in Fig. 8-2.

8.2.1.1 Manual flame cutting

This method is mainly used for cutting on site and also for cutting large sized sections (Fig. 8-2).

Manual flame cutting involves the following working steps:

- Making a template for marking-off for profile-shaping the end of a CHS bracing (see Fig. 8-5)

- Marking a tube with the help of the template

- Flame cutting by hand
  The flame cutting torch is held by hand and follows the line of the cut on the CHS with or without a guide. The path of the cut can be marked directly on the hollow section or on a template made of a thin metal sheet or oiled paper or cardboard depending on the degree of permanence, which is required.

- Correction of weld bevel
  This is not necessary for fillet welds, when the bracing wall thickness is $t_{1,2} \leq 5$ mm. The ends of the thicker sections have to be chamfered if a groove welded connection is required (Fig. 8-4).

- Correction of the shift zone (contact line from the inner to the outer diameter)

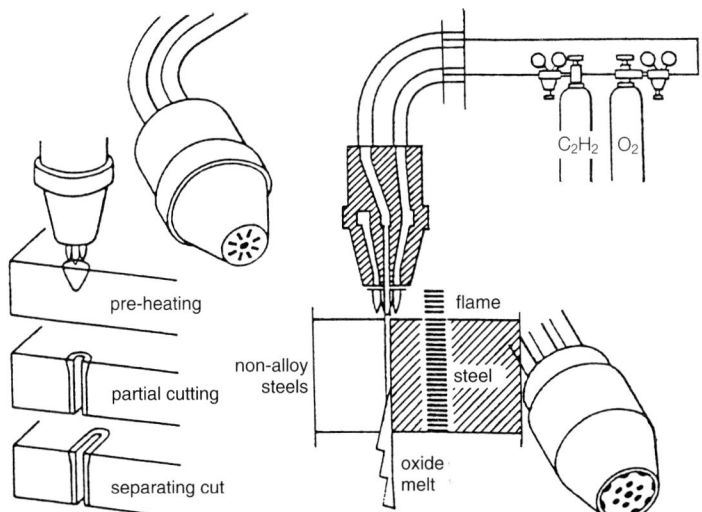

Fig. 8-2 Flame cutting process

Fig. 8-3 Offshore structural part in large sized CHS

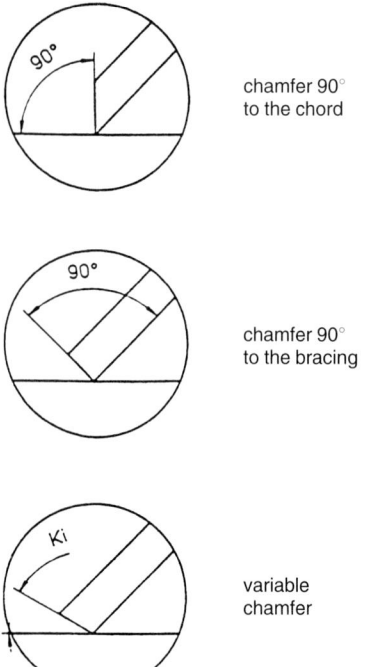

Fig. 8-4 Welding chamfer at the bracing inside saddle

The details of the weld preparation at the penetration of bracing to chord are illustrated in Figs. 6-94 and 6-96.

- Final correction by grinding
  This is required as manual flame cutting cannot be executed very precisely.

8.2.1.2 Automatic flame cutting by machines

Today, multi-planar intersection curves for CHS members are mostly cut with the aid of the automatic flame cutting machines, which were introduced by the firm Müller, Opladen, Germany in the fifties and have been further modified by the firm AGT, Geesthacht to a more efficient computer controlled machines (Fig. 8-6).

These machines execute multi-planar intersection curves simultaneously with the bevelling for the weld. The cutting speed and the reproducibility as well as the repeatability is excellent compared to the manual flame cutting and is very favourable for weld execution. The handling of these machines is simple and can be conducted by the trained personnel. The use of the automatic machines is therefore more economical than that by manual flame cutting. The running procedure is in general as follows:

- The following inputs are read by the machine:
  - Outside diameter of the chord member
  - Outside diameter of the bracing member
  - Inside diameter of the bracing member
  - Opening angle of chamfer of weld
  - Included angle between the axes of the bracing and the chord

- Lighting the flame is followed by the preheating of the workpiece at first and then by the plunge cutting, when the flame cutter moves automatically a few centimeters beyond the line of intersection into the waste metal.

- After plunge cutting, the cutter takes the previously selected cutting direction automatically. In this process, the flame cutter comes out of the waste part and continues the cutting operation in the given cutting curve. In case the plunge cutting does not

8.2 Cutting

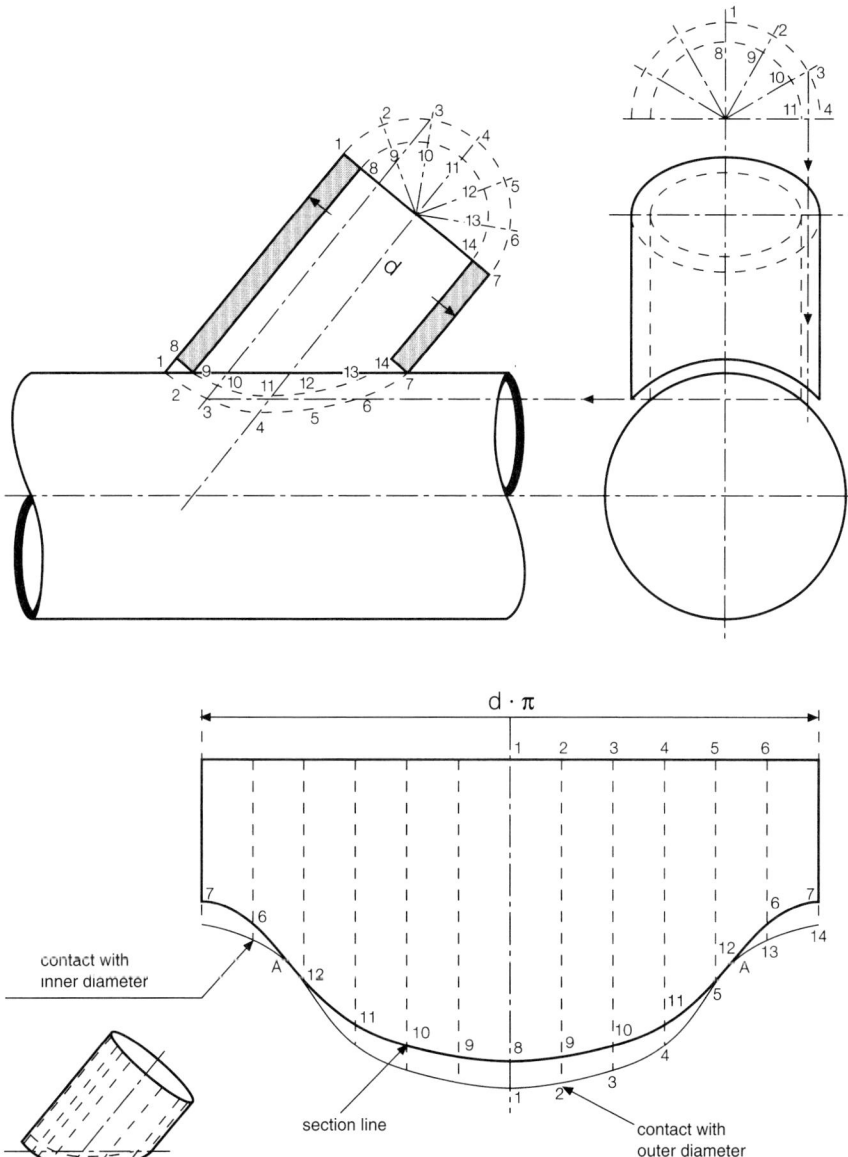

Fig. 8-5 Making a template for "profile" cut of CHS bracings

take place beyond the prescribed intersection curve, defects may occur causing weld problems later.

- After a bit more than 360° rotation, the cutting procedure is completed with the removal of the fuel gas.

Today, various computer-controlled flame cutting machines are available on the market. The modern machines offer the following facilities:

– Single or double mitres
– Concentric or eccentric adjustment

Fig. 8-6 Automatic CHS flame cutting machine

- Concentric or eccentric penetration
- Automatic bevelling for weld with various penetrations
- Various variants of cuts
- Other special adjustments

8.2.2 Sawing

Sawing is mainly applied for the end preparation of particularly RHS members to fit into planar intersections, where cuts may be square or at an angle (Fig. 8-7). Usually, the cutting tool is either a heavy duty circular saw with hydraulic feed or a heavy band saw or a power hacksaw. Depending on the required quality and accuracy of the cut, the following equipments are used:

- Milling machine
  Relatively low speed of operation, very high precision, no burrs, working rate satisfactory

- Grinding wheel
  Rapid process, however not accurate, with sharp burrs left

- Friction-toothed machine
  Rapid process, good finish relatively free from burrs, usually limited to the smaller sections, although machines are available for cutting heavy sections, widely used

- Abrasive disc cutting machine
  Fast in operation, relatively free from burrs, usually limited to the smaller sections, widely used although capacity not as wide as for friction-toothed machine

- Bandsaws
  Relatively low speed of cutting, general use capable of tackling a larger range of sizes than the disc cutters, blades relatively cheap, fairly long lasting and easy repairing if broken

- Power hacksaws
  Relatively low speed of operation, very useful for small quantity production, can be used for larger sizes, available in most workshops

Double cuts can be carried out simultaneously with a swivel cutting head (Fig. 8-8).

A considerable progress with the time has been made in the development of the ma-

Fig. 8-7 RHS joint with plane cuts at the bracing ends

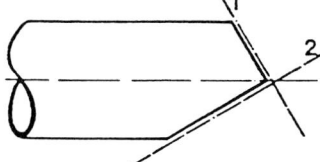

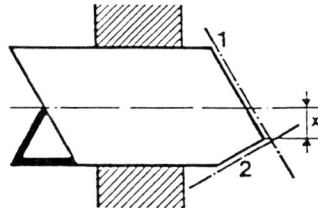

Fig. 8-8 Double cut at a hollow section end (concentric and eccentric)

chines with regard to the cutting quality and the optimisation of fabrication flow:

- Manually fed cutting machine
- Semi-automatic cutting installation
- Automatic cutting installation
- Workpiece feeding cutting machine

Cutting plants are also available, where a hollow section can be cut at both ends simultaneously.

The investment costs for the automatic flame cutting machines to execute "profile" cuts of CHS members are high, which prevent many small and medium sized firms from purchasing them. This means, they cannot manufacture CHS structures economically, although they offer architectural and static advantages compared to the constructions with conventional open profiles. Therefore, the question arises whether a multi-planar intersection curve for a CHS joint can be approximately substituted by a number of plane cuts using the sawing procedure, which depend on the relative CHS diameters in the joint. The minimum condition that has to be fulfilled is that a joint manufactured by this process must have the same strength as a CHS joint with "profile" cut members. At the same time, it is necessary to find out the gaps between the bracing ends and the chord, so that these can be bridged by welding.

Tests and numerical investigations [5] led to the following parameters, on which the weld gap depends:

- Number of plane cuts (one, two or three) (see Fig. 8-9)
- Diameter ratio $\left(\dfrac{\text{bracing diameter } d_{1,2}}{\text{chord diameter } d_0}\right)$
- Wall thickness of the bracing member $t_{1,2}$
- Angle of inclination θ of the bracing axis to the chord axis

The simplest procedure is the one with a single plane cut, which can be applied to joints with very small $d_{1,2}/d_0$ (Fig. 8-10). The following condition has to be fulfilled:

$g_1 \leq t_r$

where:
t_r is the smaller of the two values t_0 and t_1

A further condition is of a more general nature:

$g_2 \leq 3$ mm

Table 8-1 lists the combinations of the bracing and the chord diameters, which fulfil the condition $g \leq 3$ mm.

Fig. 8-9 CHS bracing ends with single, double and triple plane cuts

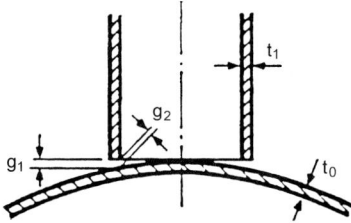

Fig. 8-10 CHS joint with a single plane cut at the bracing end

Table 8-1 Combinations of the bracing (d_1) and chord (d_0) diameters

d_0 mm	d_1 mm	d_0 mm	d_1 mm
33.7	26.9	88.9	33.7
42.4	26.9	101.6	42.4
48.3	26.9	114.3	42.4
60.3	33.7	139.7	48.3
76.1	33.7	168.3	48.3

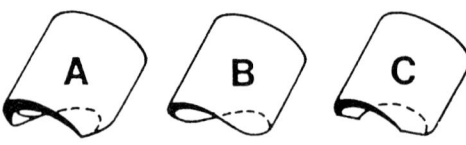

Fig. 8-11 Smoothing down the cut areas.
A) Grinding of the internal angles, B) profile grinding, C) shearing off

Large d_1/d_0 ratios lead to large weld gaps, which necessitate more weld metal and consequently more labour costs for welding. In these cases, the weld gap can be reduced by the following two methods; the larger the number of the plane cuts applied, the more precise is the adaptation to the weld gap:

– Two plane cuts, after which, the areas shown in Fig. 8-11 are smoothed by grinding or shearing off

– Two or three plane cuts with the cutting angles β_g and β_d, which can be determined according to the equations given with Fig. 8-12

Method A) is illustrated in Fig. 8-13.

The value of "a" shall be determined according to the following equation:

$$a = \frac{r_1^2}{2r_0} - r_1 \tag{8-1}$$

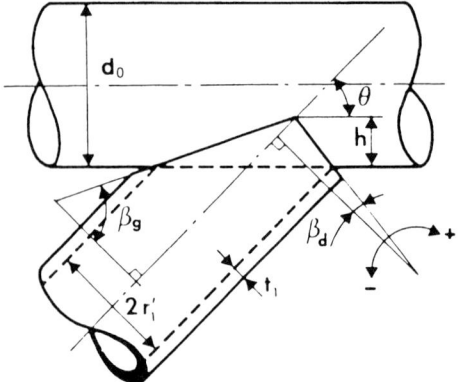

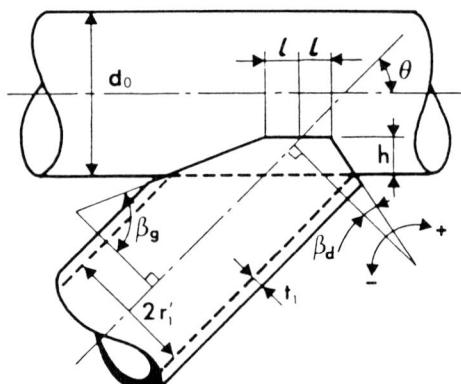

Two plane cuts

$$h = \frac{d_0}{2} - \sqrt{\frac{d_0^2}{4} - r_1'^2} \tag{8-2}$$

$$\alpha_g = \arctg\left(\frac{h \sin\theta}{r_1' + h\cos\theta}\right) \tag{8-3}$$

$$\alpha_d = \arctg\left(\frac{h \sin\theta}{r_1' - h\cos\theta}\right) \tag{8-4}$$

Three plane cuts

$$l = \sqrt{r_1'^2 - (r_1' - t_1)^2} \tag{8-5}$$

$$h = \frac{d_0}{2} - \sqrt{\frac{d_0^2}{4} - (r_1' - t_1)^2} \tag{8-6}$$

$$\alpha_g = \arctg\left(\frac{h \sin\theta}{r_1' + h\cos\theta - l\sin\theta}\right) \tag{8-7}$$

$$\alpha_d = \arctg\left(\frac{h \sin\theta}{r_1' - h\cos\theta - l\sin\theta}\right) \tag{8-8}$$

$$\beta_g = 90° - \theta + \alpha_g \tag{8-9}$$

$$\beta_d = -90° + \theta + \alpha_d \tag{8-10}$$

Fig. 8-12 Two or three plane cuts at the bracing end of a CHS joint (method B)

8.2 Cutting

where:

$$r_1 = \frac{d_1 - 2t_1}{2}$$

$$r_0 = \frac{d_0}{2}$$

The value of "a" is constant for all values of the included angle θ.

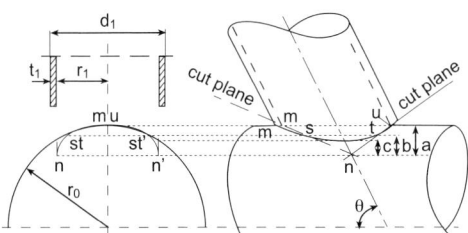

Fig. 8-13 Approximate "profile" cut by means of two plane cuts at the bracing end of a CHS joint (method A)

Starting from the point "n" determined by the above value "a", the lines "$n-m$" and "$n-u$" are drawn. These define the cutting planes, whose inclination must now be measured. When both cuts are made, the edges are trimmed so that the bracing fits neatly on the chord. The point area $t-n-s$ is then smoothed down by the methods shown in Fig. 8-11.

Method B), illustrated in Fig. 8-12, is applied for two or three cuts. "h" in Fig. 8-12 is constant irrespective of the value of the included angle θ and can be calculated using Eqs. (8-2) or (8-6). After the calculation of the intermediate values α_g (Eqs. 8-3 or 8-7) and α_d (Eqs. 8-4 or 8-8), the required cutting angles β_g (Eq. 8-9) and β_d (Eq. 8-10) can be determined.

The maximum weld gap J_m for two or three plane cuts is calculated as follows [6]:

Double cuts:

$$J_m = \sqrt{\frac{1}{2}\left(\frac{d_0^2}{4} + r_1'^2 + \frac{d_0}{2} \cdot \sqrt{\frac{d_0^2}{4} - r_1'^2}\right)} - \frac{d_0}{2} \quad (8\text{-}11)$$

Triple cuts:

$$J_m = \sqrt{\frac{d_0^2}{4} + r_1'^2 \cos^2\phi} - \frac{d_0}{2} \quad (8\text{-}12)$$

where:

$\phi = \arccos$

$$\left[\frac{1}{2}\sqrt[3]{\frac{d_0^2}{4r_1'^2} - 1}\left(\sqrt[3]{\frac{d_0}{2r_1'} + 1} - \sqrt[3]{\frac{d_0}{2r_1'} - 1}\right)\right] \quad (8\text{-}13)$$

8.2.3 Plasma cutting

In plasma cutting, the workpiece is hit by a thin stream of a gas (Ar, N_2 or Ar + N_2 or N_2 + H_2) heated by a concentrated electric arc with high speed. The material is cut by melting. Due to the high concentration of energy, the cut can be executed quicker than by other existing methods and in most cases without any distortion of the material. The most important condition for a flawless cut is a stable, concentrated electric arc as well as the adaptation of the capacity of the heating energy to cutting jobs. As a rule, high concentration is achieved by the double stabilized plasma gas pressure system. The cutting range amounts to 4 to 35 mm for quality cuts and 6 to 45 mm for simple separation cuts. Small compact transportable machines to high capacity installations are available on the market. The electric current required for the installations lies between 40 to 12 Amperes at a service voltage between 80 to 125 Volts.

8.2.4 Laser cutting

Due to the greater advantages of the laser cutting (higher quality, accurate performance or precision, more flexibility, lower fabrication costs, substitution of expensive manual work or of eventually required post treatment), this is gaining more and more importance in practice. An increasing number of small and medium-sized firms carries out this job on commission.

Now-a-days, plane cuts can be made on the following thickness without any problem with normal, priceworthy laser cutting systems:

– Non-alloy steels up to about 16 mm
– Stainless steels up to about 10 mm
– Aluminium up to about 6 mm

The cutting rate can reach up to 10 m/min with a very low size tolerance of ± 0.1 mm.

This can be of high significance for further fabrication steps e.g. welding with robots.

The heat affected zones (HAZ) on the cutting edges are very small. As the cutting quality is excellent, the costs for further improvement is reduced. The post-treatment is not required because the distortion is very little. A high repeatability is possible by means of CNS controlling device.

With the development of the new high capacity laser equipments, this cutting method is applied increasingly to cut thickwalled tubes. The cutting device is nearly identical to that of the computer controlled flame cutting machines; only the cutting medium is laser beam instead of gas.

However, high investment costs represent the largest obstacle against the wider application of the laser cutting.

8.3 Slotting

A series of end-to-end connections made by means of welded fittings, which are inserted through slots cut out in the hollow section members, are shown in Fig. 6-36e and Table 6-3. In Fig. 8-14, a gusset plate is inserted

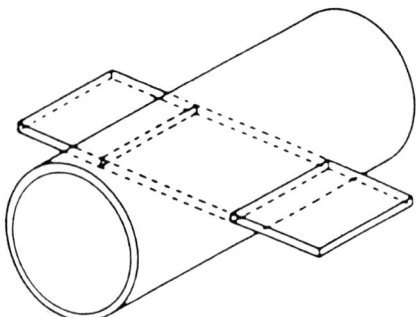

Fig. 8-15 Slots in the body of a hollow section

into the slot and welded, where the plate may be flat or bent (producing a fork or clevis that will give double shear in pins or bolts).

Fig. 8-15 demonstrates an other variant, where the slots are cut in the body of the members, away from the ends, to allow a connection plate to pass completely through them. This ensures the continuous transmission of loads without any risk of deformation.

The slots are, as a rule, entirely closed by weld seams to seal with a view to prevent internal corrosion. In some cases, the ends of the hollow section members are sealed by inserting and welding a semi-circular plate for additional safety. However, it is not permissible to close the openings for hot dip galvanized parts to avoid explosion by the immersion of closed volumes into molten zinc baths registering 450 °C.

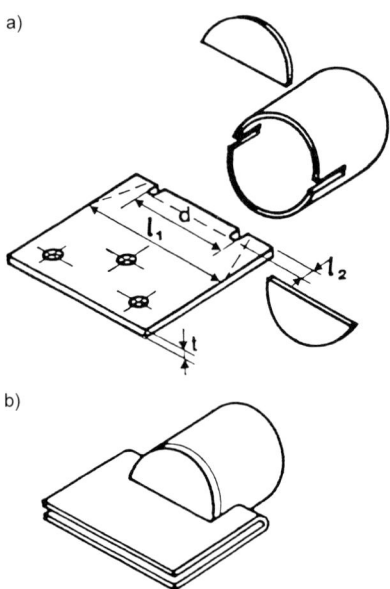

Fig. 8-14 Slotted hollow section connection with end gusset plates. a) Flat plate, b) bent plate

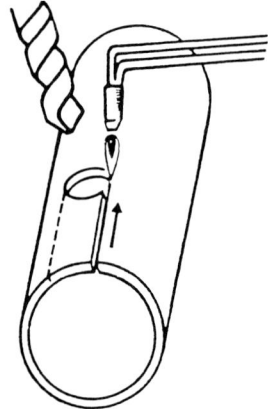

Fig. 8-16 Cutting a slot with a manual flame cutter

8.5 Bending (arching) of hollow sections

Slots are cut by the following methods:

- Notching using special blades
- Flame cutting with semi-automatic machine or manually
- Milling tools
- Abrasive discs

Fig. 8-16 illustrates slotting by manual flame cutting schematically. At first, a hole is drilled in the hollow section wall at the end of the intended slot, where the hole diameter should be a bit larger than the slot width. Finally, the slot is cut with the flaming torch starting from the ends towards the hole.

In case a full-automatic or semi-automatic flame cutting machine is available, two parallel cuts can be executed simultaneously by synchronizing the movements of both cutters (Fig. 8-17).

Further, it is possible to make a slot by using a milling saw, whose head is set parallel to the workpiece (Fig. 8-18). This is often the most rational process.

The following points have to be specially accounted for, when slotting is carried out with a flame cutter:

- Special care has to be taken for manual flame cutting as weld problems may arise later due to the lack of accuracy in the cut end or in the slit width or length. Slit length

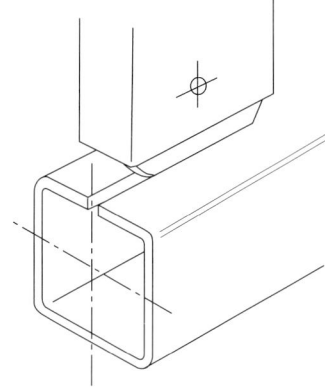

Fig. 8-18 Slotting with a milling cutter

  is a critical location for stress concentration in a joint. The flame cutter has to be guided in order to obtain a particularly accurate slotting.

- Stress relief due to heat input and cutting by a flame cutter can lead to the distortion of a section. This can be avoided by leaving the head end of the slit partially uncut and letting it get cooled.

8.4 Flattening of hollow section ends

This subject has been dealt with in detail in Section 6.6.8 and is mentioned here once again as a processing type.

8.5 Bending (arching) of hollow sections

Circular as well as rectangular hollow sections are used in bent form in various diverse fields, such as for architecturally interesting structures with vaults and domes (see Section 6.3) and also applied in mechanical engineering and for manufacturing pressure vessels and pipe lines.

Bending consists of a deformation in the plastic zone to give the section a permanent curvature. This results in an extension of the outer wall and a contraction of the inner one. In this process, an weakening of the wall thickness on the external side leading to the crack initiation by tension and a corrugation with the increase of the wall thickness of the internal side by compression take place.

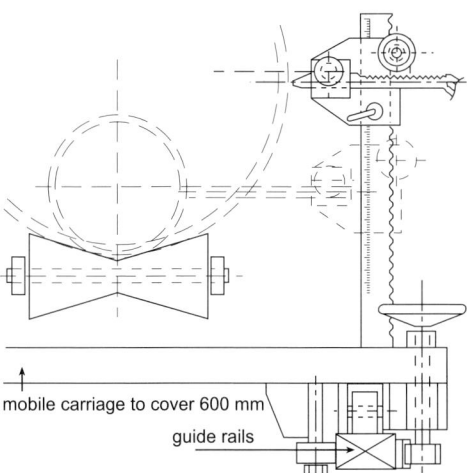

mobile carriage to cover 600 mm
guide rails

Fig. 8-17 Slotting with a semi-automatic flame cutter

These changes in wall thickness as well as the possible ovality of a circular hollow section have to be kept as small as possible.

The following parameters are decisive for the bending behaviour of hollow sections:

- $\dfrac{\text{Diameter of CHS or depth of RHS}}{\text{Wall thickness of hollow section}}$ – ratio

- $\dfrac{\text{Bending radius of curvature (Fig. 8-19)}}{\text{Diameter of CHS or depth of RHS}}$ – ratio

- Yield strength f_y of the hollow section steel. Bending is easier if the yield strength is lower.

- Ultimate tensile strength f_u of the hollow section steel.
  High ultimate tensile strength prevents early initiation of crack.

- Elongation of the hollow section steel. Adequate elongation property (minimum 20%) exercises a significant influence on the bending behaviour of hollow sections.

- Microstructure of the hollow section steel. Finer microstructure is favourable for bending.

Hollow sections can be bent in the hot as well as in the cold condition. Cold bending is more often used, as it is less expensive than hot bending. However, a heat treatment after cold bending is necessary corresponding to the steel grade used in order to obtain the initial microstructure of the material [7]. This however increases the manufacturing costs. There is a large overlapping zone, where both hot and cold bending are applied.

Hollow sections are bent in the workshops in general. However, this is done also on site, particularly when the dimensions of the workpieces are small.

In the following, a number of bending processes is described.

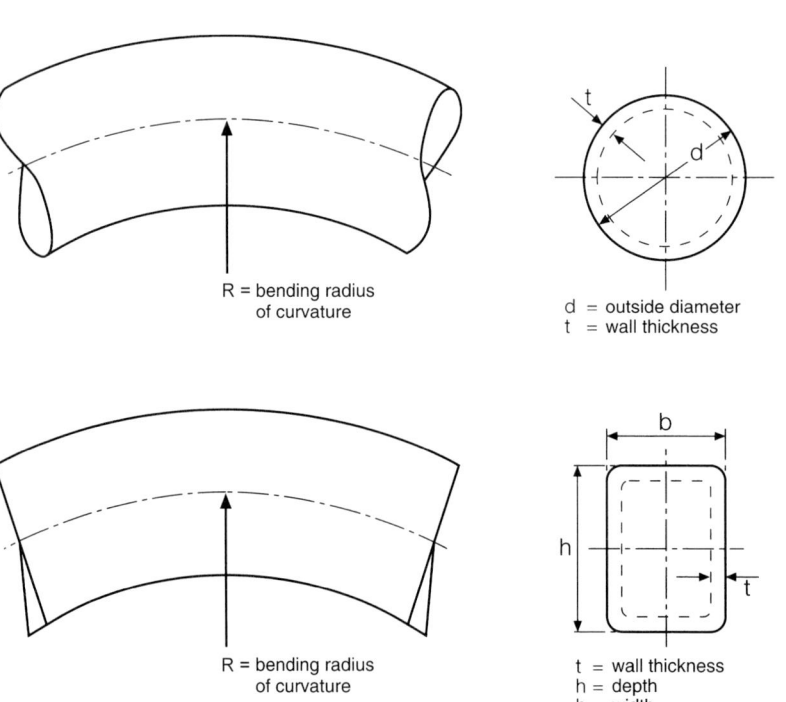

Fig. 8-19 Radius of curvature for the bending in a hollow section

8.5.1 Cold bending of CHS

8.5.1.1 Cold bending by pressing

The hollow section is set between two fixed rollers and bent by the displacement of a central former usually connected to a hydraulic actuator (Fig. 8-20).

The stamp encircles the CHS laterally to some extent and reduces thereby the initiated ovality. The operation may also be performed by holding the former motionless in the centre and pushing both side rollers simultaneously by connecting them to a press.

This process is usually used for 180° arches with a wide range of dimensions. However, localized pressing causes lower accuracy and poorer appearance than by other processes.

8.5.1.2 Cold bending using a "former" box

In principle, the hollow section is forced into a preshaped "former" box (Fig. 8-21). This is fixed to a horizontal base and a straight guide box is placed in front of the bent former. The hollow section is forced into the guide box and then into the former box by means of the actuator. The end of the section to be bent must be provided with a guide plug in order to avoid damaging of the tool walls. Further, lubrication is essential. This process is only then economical if many bends of identical sizes are manufactured.

8.5.1.3 Cold bending with a three-roller bender

According to this process, bending is obtained by passing the hollow sections through three rollers (Fig. 8-22). All three may be driven, but the central one, which determines the bending radius and the grade of deformation, can be idle.

The rollers have to be adapted to the shape and size of the hollow sections to be bent. Corresponding to the profile groups, the sizes of the rollers are different. For cold bending with roller bender, a bending radius of curvature of about five times the outside diameter of CHS is used in practice.

Some machines have four rollers, two of which are idle (Fig. 8-23). Roller A is idle, B and C are driven and compress while D downstream is also idle but adjustable in a transverse direction.

This deformation method is often favoured by the steel constructors. The smaller roller

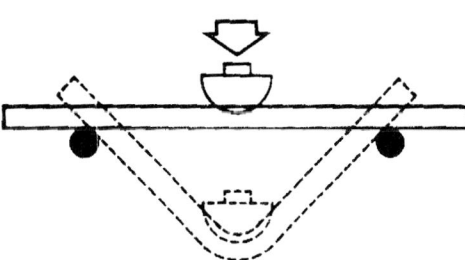

Fig. 8-20 Cold pressing, moving part – white, fixed part – black

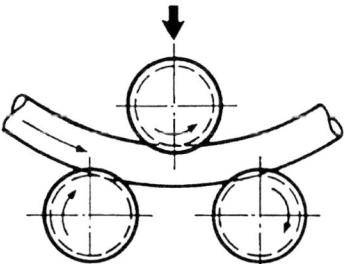

Fig. 8-22 Thee-roller bending

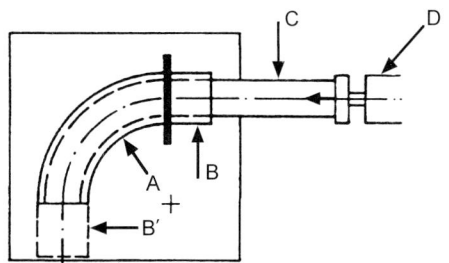

Fig. 8-21 "Former" box

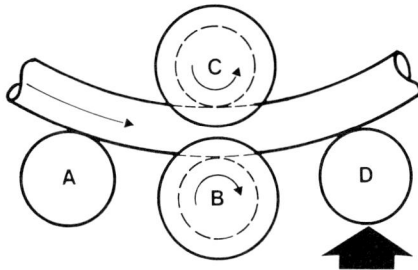

Fig. 8-23 Four-roller bending

benders are usually mounted vertically, while the larger parts are commonly bent on horizontal machines. Various improvements lead to further working facilities:

- A reversible motor with magnetic brake for bending sections over several passes and for producing elliptical or otherwise curved parts
- Independent rollers with hydraulic adjustments, which make it possible to reduce to a minimum the straight length (i.e. the offcut) remaining at the ends after bending
- A portable control desk, which enables the operator to place himself in the best position to conduct his work.

8.5.1.4 Arches by means of mitre cuts

In this manufacturing process, a bending in its true sense is dispensed with (Fig. 8-24).

Hollow section ends are cut with appropriate slants and an approximate curve is obtained by welding the plane sections end-to-end to one another (Fig. 8-24a). This method is mainly used for large radius bends.

8.5.2 Cold bending of RHS

8.5.2.1 Cold bending by pressing

This process is similar to that for CHS, which has been described in Section 8.5.1.1.

8.5.2.2 Arches by means of a) mitre cuts, b) "V" cutouts

The process a) is similar to that dealt with in Section 8.5.1.4.

For small size RHS, it is possible to carry out "V" cutouts on three of the faces and obtain the bend by folding the part on the remaining face so as to close the notches and then weld the edges (Fig. 8-24b).

8.5.2.3 Cold bending with a three-roller bender

The working principle is similar to that for CHS as shown in Fig. 8-22. The results of the bending are however different from CHS due to the difference in shape and the flat faces. The inside face is compressed and the outside face is stretched, while the side walls are subjected to both compression and tension. A strong curvature of the RHS with a large "width or depth to wall thickness" ratio leads to the convexity of the side faces and the concavity of the inside face. A reduction of the depth takes place in the curved zone, while the width is enlarged due to convexity (Fig. 8-25). The wall thickness or the cross-sectional area of RHS shall be reduced by this process.

As adequate knowledge regarding the bending behaviour of RHS in a three-roller machine is not available at present, it is recommended to perform a number of pretests before starting a project in order to determine the roller dia-

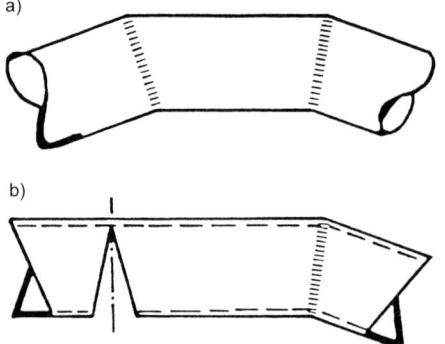

Fig. 8-24 a) Mitre cuts, b) "V" cutouts

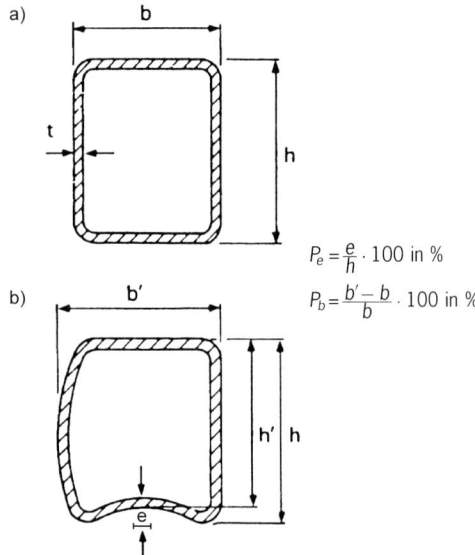

$P_e = \frac{e}{h} \cdot 100$ in %

$P_b = \frac{b' - b}{b} \cdot 100$ in %

Fig. 8-25 Typical RHS before and after three-roller bending, a) before bending, b) after bending

meter, the distance between the vertical axes of the rollers, the radius of curvature corresponding to the RHS sizes and the steel grades used. Thereby the allowable deformations P_e and P_b have to be accounted for (see Fig. 8-25).

In a research work [8, 9], a series of tests has been carried out to determine the minimum bending radius of curvature for RHS (including square) in cold bending by a three-roller machine as well as the deformations P_e and P_b. As the results of these tests depend on the specific details of the test conditions (e.g. outer diameters of the fixed and moving rollers, distance between the vertical axes of the rollers, RHS size and material properties), a general applicability of these test results is called into question. However, they give indications for a pre-estimation.

8.5.3 Hot bending of hollow sections

While bending hollow sections in hot condition, special attention has to be given to the aspect that changes in mechanical properties and microstructures of cold formed sections can occur due to the heat input during hot bending.

8.5.3.1 Hot bending of hollow sections filled up with sand

Although this process is applicable for both CHS and RHS, thickwalled CHS with a large diameter is in general bent using this method. The hollow section is filled up with dry sand before heating. The sand is at first compressed in order to avoid corrugations on the inside surface of the hollow section and to keep the ovality small. The bending zone is then heated to a temperature of 850–1100 °C and the workpiece is fixed at one end while the other end is pulled about a template. This procedure is repeated stepwise starting from the fixed location. After the first bending zone is cooled down, the mechanical working of the neighbouring zone is started and so on, till the required total bending is completed.

8.5.3.2 "Hamburger arch" (CHS only)

The firm "Hamburger Rohrbogenwerk" is the inventor and the holder of patent of this classical hot bending process for manufacturing foldfree bends in welded or seamless CHS. The CHS is heated to a temperature of about 850–1100 °C and then pushed on to a conical internal mandrel. Thereby, the bending to a desired radius of curvature takes place simultaneously with the enlargement of the diameter.

Even for very small bending radii, the wall thickness on the outer side is not reduced by using this process.

Manufacturing ranges:

– Up to 914.4 mm outer diameter (bending radius $R = 1.5 \times$ CHS outer diameter)
– Up to 609.9 mm outer diameter (bending radius $R = 2.5 \times$ CHS outer diameter)

8.5.3.3 Bending with inductive heating

Hot bending of CHS as well as RHS can be performed in this installation.

The working principle of this machine is based on the induction heating of a short length under precise temperature control by a ring-shaped inductor. The bending takes place in a small heated zone. The hollow section is then pushed forward into the inductor and the following zone is heated and so on, so that the total arch with the required bending radius is finally manufactured (Fig. 8-26).

As the bending installation is provided with two bending arms, it permits a higher flexibility regarding the diameter and the wall thickness as well as the bending radius and the angle to be accommodated. Very small radius of curvature can be achieved, even when the diameter and wall thickness of a CHS is quite large (Fig. 8-27).

Multiple bends, arches etc. with long legs can also be manufactured saving the number of welds.

It is recommended to bend a RHS with the bending radius $R = 10 \times$ depth of the RHS.

8.5.3.4 Hot bending with a three-roller bender

This method can be applied to both CHS and RHS.

For hot bending, the bending radius ($R = 3 \times$ outer diameter) of CHS is smaller than that for cold bending.

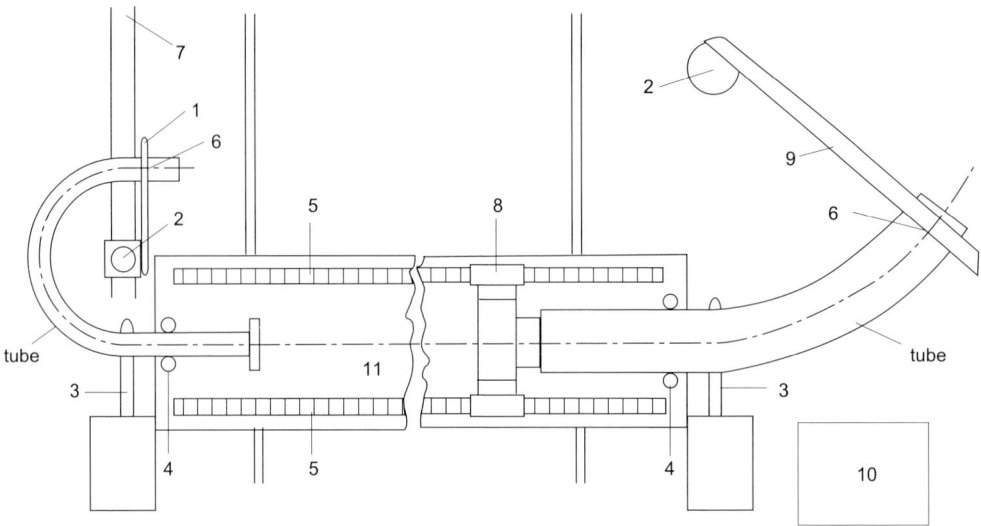

Fig. 8-26 Sketch of a bending installation with inductive heating
1. bending arm (light), 2. pivot point, 3. inductor, 4. guide roller, 5. drive chain, 6. stretching device, 7. rails, 8. feeder, 9. bending arm (heavy), 10. source of electricity, 11. machine bench

Fig. 8-27 Installation for induction bending

8.5.3.5 Cambering

In some steel structures, relatively small bending curvatures (large bending radius) e.g. precambered girders are required to ensure that they do not sag under load. Cambers can be produced in the cold condition by pressing (see Section 8.5.1.1) or by three-roller bending (see Section 8.5.1.3).

However, another method can also be applied, which uses no special equipment other than a heating torch. By heating just one side of a hollow section and allowing it to cool, the hollow section bends while cooling occurs in the direction of the preheated side. This can be conducted quite accurately with increasing experience considering also that the bending curvature depends on the heat input, the section size and the manufacturing process of the hollow section (cold or hot forming).

8.6 Bolting

Bolted connections are indispensable on account of the simplification of the assembly operation and for the purpose of transport. Prefabricated, welded subunits are bolted together on site. In Chapter 6, a number of constructional detail of the indirect, bolted connections in hollow sections has been described, e.g. column-to-truss connections (see Section 6.2.1), beam-to-column connections (see Section 6.2.2), knee joints (see Section 6.4) and end-to-end connections (see Section 6.5).

In principle, the bolting of hollow sections does not differ from that used for the conventional steel constructions. However, holes are normally made in structural hollow section only by drilling. It is not possible to do the holing by punching due to its hollow shape unless an internal support is used. The calculation procedures for the application of the ordinary and high strength bolts with or without prestress have been recommended by various national and international standards [10–13], see also Table 6-4. In general, bolts have to be checked for the resistances against shear, bearing pressure and tension.

Ordinary bolts are used to make indirect hollow section connections via splice or gusset plates or open profiles. The application of prestressed high strength bolts is favourable for connections under fatigue load (see Section 7.14).

Simple bolted connections are usually applied for the mechanical engineering purpose, where a good architectural appearance of the structure is not essential. However, they also must possess adequate design resistances and be economical in their application.

For hollow section joints, where the section walls are penetrated, care has to be taken regarding the prevention of water inlet, which may produce internal corrosion of the hollow sections subsequently. The structures are in danger, which are exposed to open atmosphere. In this case, the water assembled within the section may freeze and bring the section to explode.

A crowding of bolted joints in a location should be avoided as far as possible, as they act as water or snow traps, which promote external corrosion.

8.6.1 Blind bolting

At the beginning of this chapter, one-sided access to the hollow section has already been described as a disadvantage inherent to the indirect bolting for hollow section connections. Blind bolts are special bolts or bolting systems, which invalidate this problem. They allow bolting to take place from one side of the structural element only removing the need to get to both sides.

At present, a number of blind bolting systems for hollow section joint is available on the market. The following systems are described in this Section:

– Flowdrill
– Lindapter

8.6.1.1 Flowdrill

The flowdrill system is a patented method for making specially extruded and metric threaded holes for bolts in sheets and plates. This is a thermo-mechanical plastification process by the friction-initiated heat to produce holes in the walls of the hollow steel elements (CHS, RHS etc.) (Fig. 8-28). The system is based essentially on the heat produced by a rotating tool of tungsten carbide, which has the high qualities of hardness and resistance to high temperature. The tool is four lobed shaped with a straight body and a conical point. When the tool, rotating at sufficient speed, comes into contact with a metal plate, it produces a rapid heating, which softens the material in a local area. The tool is then forced through the metal plate by means of an axial load producing a hole by material extrusion. The material displaced by the tool forms a truncated hollow cone of the depth of one or two times wall thickness on the far side of the hollow section wall and a small upset on the near side. The upset can be removed directly in the drilling phase, while metal is still soft, using a tool provided with a milling cutter collar. The hole, in the second phase, is then threaded with a rolling tapping tool. The result of the entire process is a hole, which has a kind of thin-walled nut attached to the metal plate.

Regarding the performance of this bolting system, the test results have shown that

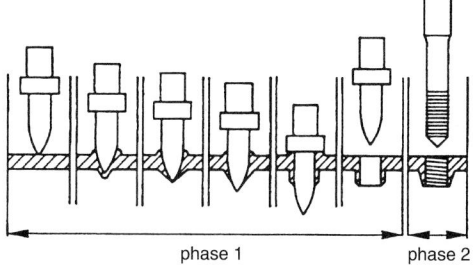

Fig. 8-28 Working principle of the flowdrill system

threaded holes can be produced in both hot and cold formed RHS in thicknesses from 5 to 12.5 mm with M 16, M 20 and M 24 ISO course thread profiles. The test steel grades are S 275 and S 355.

8.6.1.2 Lindapter Hollo-Fast

The Hollo-Fast is a component which expands behind steelwork to form a flush threaded connection point. From this, secondary steelwork is fixed. It is best applicable when connections need to be made in hollow section steelwork and where access from behind is not possible.

This blind bolting system consists of a standard bolt and a special steel insert to be introduced into a hole through the wall thickness into the hollow section. Once the insert is pushed into the hole with appropriate diameter using light hammer blows, the action of the tightening of the bolt in the truncated cone thread makes it separate the cone and pull it inside the Hollo-Fast body. The consequent deformation and expansion of the cylindrical part create four fins. They provide the mechanical interlock, which is necessary to prevent the pull-out of the bolt.

The steel insert is capable of accepting M 8, M 10, M 12 and M 16 bolts. Fixing can be made using lightly oiled grade 8.8 setscrews.

Fig. 8-29 illustrates the bolting steps for Hollo-Fast system. Ref. [3, 4] contain the results of the extensive investigations on these blind bolts.

When subjected to both shear and tension, Hollo-Fast must fulfil the following condition:

$$\frac{F_S}{P_S} + \frac{F_t}{P_t} \leq 1.4$$

where:
F_S = applied shear
P_S = shear capacity
F_t = applied tension
P_t = tension capacity

8.6.1.3 Huck Ultra-Twist

These fasteners developed by the Huck International Inc. in Utah, U.S.A. have tensile

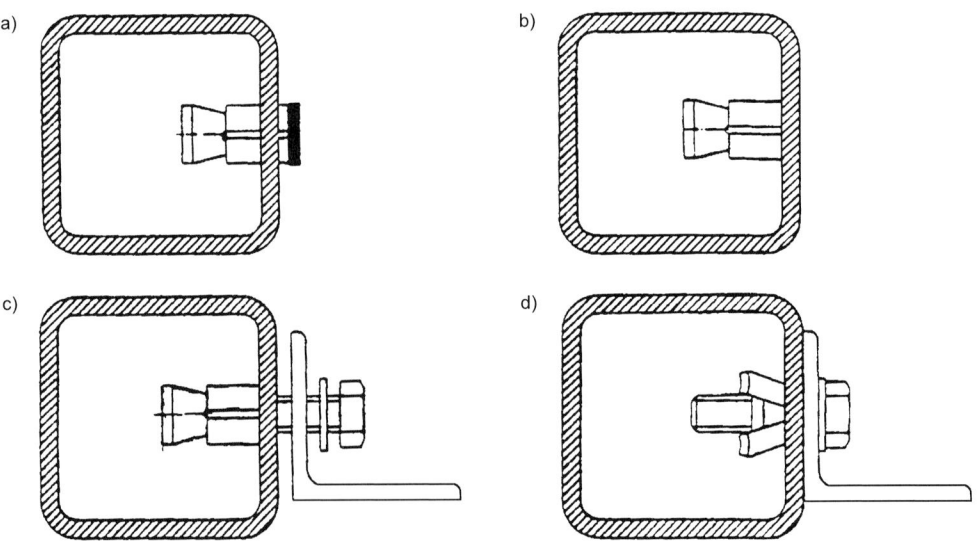

Fig. 8-29 Working principle of Lindapter-Hollo-Fast blind bolts.
a) After drilling hole in required position, Hollo-Fast body is inserted into hole, threaded cone end first. b) Hollo-Fast is pressed into hole until the knurled face is flush with the hollow section surface. c) Setscrew is passed through hole in the fixture and screwed into the threaded cone. d) The setscrew is further tightened. The action of tightening separates the cone and pulls it inside the Hollo-Fast body, which expands to form a secure threaded fixing

strengths and installed tensions meeting those specified for ASTM A325 bolts (equivalent to grade 8.8 bolts) of 3/4 inch (19 mm), 7/8 inch (22 mm) and 1 inch (25.4 mm) diameter. Fig. 8-30 shows an exploded view of an Ultra-Twist fastener together with the installation procedure using an electric bolting wrench.

If Ultra-Twist fasteners are used in a RHS column face and are loaded in tension, a potential failure mode is punching shear of the fastener through the column face, in which case the column thickness becomes a critical parameter. In order to avoid this, the following condition is to be fulfilled:

$t > (0.8\, T_r)/(d'_f \cdot f_u)$

where:
T_r = tension resistance of the fastener
d'_f = diameter of the Ultra-Twist fastener + 6 mm (estimated effective bulb diameter)
t = wall thickness of hollow section
f_u = specified minimum tensile strength of the RHS material

Another critical failure mode for an unstiffened RHS column face loaded by point tension load at the fastener positions is yielding of the RHS connecting face. The limit states resistance for this failure mode can be calculated by assuming that the column wall like a 90° T-Type joint with the bracing member in tension (see Eq. 6-73 or 6-78).

8.7 Welding

8.7.1 Hollow section steel grades and their weldability

Principally, the weldability of steels depends on their chemical compositions. They are specified in many national and international standards [14–27]. Today, steel is manufactured using modern melting processes, where the alloy contents of the steels e.g. phosphorus, sulphur and nitrogen, are low. In general, they are killed by the addition of Aluminium and have a high degree of oxidation purity and large ductility.

Decisive for the weldability of low-alloy steels (see Table 3-5) are their Carbon con-

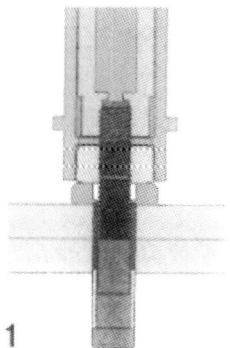 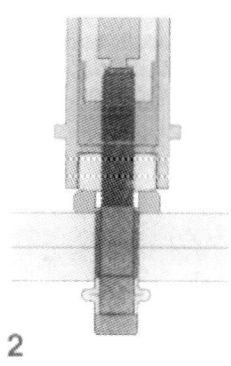 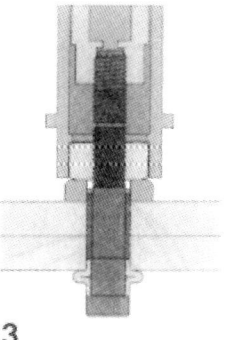 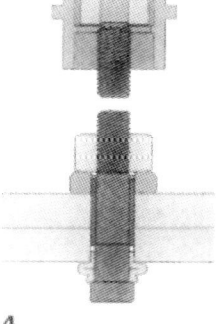

1
The ULTRA-TWIST blind bolt is installed from one side of the structure by a single operator.
The installation tool is the standard electric shear wrench tooling used for installation of Twist-Off Control (T-C) type fasteners. The fastener is inserted and the tool engaged.

2
The backside bulb is fully formed in the air to a uniform diameter regardless of grip.

3
As the installation load increases, a special internal washer shears allowing the backside bulb to come into contact with the work surface and for all clamp load to go into the work structure.

4
Continued torquing of the unit develops the required clamp and the torque pintail shears off, completing the installation.
Using a standard S60EZ shear wrench, installation time for a 3/4" fastener is approximately 30 seconds.

Fig. 8-30 Installation of a Huck Ultra-Twist fastener

tents (C ≤ 0.22%) and their purity shown by low Sulphur (S ≤ 0.045%), Phosphorus (P ≤ 0.045%) and Nitrogen (N$_2$ ≤ 0.009%) contents.

The fine grain steels (see Table 3-7) obtain good weldability and increased strength by relatively low Carbon content and with alloys e. g. Manganese, Silicon, Niobium, Vanadium, Aluminium, Titanium, Chromium, Nickel and Molybdenum. Low percentage of Carbon (C ≤ 0.20%) improves not only the weldability but also promotes the fine grain microstructure, which reduces the susceptibility to brittle structure. Higher strength is mainly attained by the Manganese alloy together with the low Carbon content.

The susceptibility to cold cracking in the heat affected zone (HAZ) is influenced by the chemical composition of the material. Cold cracks can occur, when one of the three following factors attains the critical value:

– Hardness of the heat affected zone adjacent to the weld
– Amount of diffusive hydrogen in the weld metal
– Tensile residual stress in the weld

Pre-heat treatment is the most effective measure to prevent cold cracking. As the low hydrogen gas shielded weld technology is preferably applied to weld steel structures at present, pre-heating is only required for high strength steels and here also for the relatively thick wall thicknesses. Informations regarding the prevention of cold cracks in welds of high strength steels are given in Lit. [28].

According to Ref. [33], any consideration regarding pre-heating should only be made, if the wall thicknesses of the structural members of a joint differ from one another by about 10 mm. Also for connections between hollow and solid sections, pre-heat treatment may be required.

Welding in humid weather, at low external temperature and also for thick structural members can lead to cold cracks. A pre-heating to a temperature of 50–150 °C depending on the steel grades can prevent this effectively.

For S 355 (see Table 3-5 and 3-9)

Up to 13 mm wall thickness (fillet weld)
Up to 20 mm wall thickness (groove weld)
} Pre-heating is not required

Wall thicknesses beyond the above mentioned necessitate pre-heat treatment.

S 460 (see Table 3-10):

Up to 8 mm wall thickness (fillet weld)
Up to 12 mm wall thickness (groove weld)
} Pre-heating is not required

Pre-heating to a temperature from about 80 to 150 °C is to be carried out for the wall thicknesses beyond the above mentioned.

The steel grades in general favour all fusion and pressure welding processes.

It is usual in practice to take the Carbon Equivalent Value CEV as the criterion for the weldability. This is defined as follows:

$$CEV = C + \frac{Mn}{6} + \frac{Cr + Mo + V}{5} + \frac{Ni + Cu}{15}$$

The smaller the value of CEV, the better is the weldability of a steel grade, particularly in combination with very low Carbon content, which is necessary to reach the required design strength.

No cold crack occurs in general for the wall thicknesses smaller than 16 mm, as far as CEV < 0.40. Precautions have to be undertaken with regard to the welding methods, when 0.40 < CEV < 0.45. Pre-heating is generally needed for CEV > 0.40.

8.7.2 Methods for welding hollow section joints

The same welding processes are used for hollow sections as for conventional plates and profiles. They are as follows:

– Shielded metal arc welding (SMAW)
– Gas metal arc welding (GMAW): Metal inert gas (MIG) and Metal active Gas (MAG)
– Flux cored arc welding (FCAW)
– Submerged arc welding (SAW)

Metal arc welding is commonly used for site welding and when restricted access prevails in the workshop. Gas metal arc welding with inert gas shield is as a rule popular for workshop fabrication. In principle, they belong to the group "fusion welding" [29], whose characteristic feature is the jointing in the fused mass. Thereby, the base metal melts by the heat input due to weld and also the filler metal as well as the electrode, the welding wire or rod are nearly always melt together with it.

Fundamentally, the application of the weld equipments and machines can be sub-divided in three groups: manual, semi-automatic machine weld and automatic machine weld. In common, the first two methods are used for steel structures in hollow sections. Generally, full automatic welding is not usual; it is however applied when necessary.

Besides the "fusion welding", the "friction welding" belonging to the group of "pressure welding" comes also into operation, in case the quantity required is large, i.e. head plate joints with hollow sections. Friction welding is mainly used for joining the members of the mechanical constructions e.g. axles for lorries. In this case, the stub shaft is joined to CHS by friction welding. Both parts are rubbed with each other. The heat produced is then used for welding. After reaching the required jointing temperature, both parts are pressed together.

8.7.2.1 Manual shielded metal arc welding with stick electrodes coated with shielding flux material [30, 31]

This welding process effects the fusion welding by the energy of the electric arc. This is particularly applied, when the welding locations allow access with difficulty or the welding positions are disadvantageous e.g. overhead (Fig. 8-31).

An electric arc burns between the fusing electrode and the workpiece. Electric arc and melt weld both are shielded from the detrimental atmospheric effects by means of slag or gas produced from the electrode coating, thereby preventing the undesired oxidation.

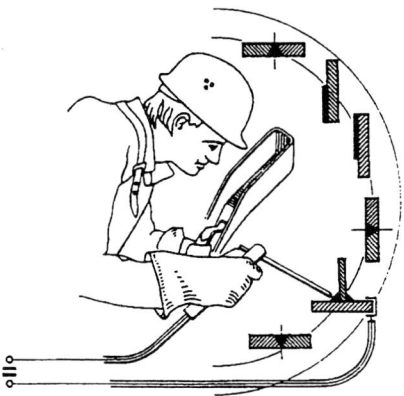

Fig. 8-31 Manual metal arc welding with various welding positions

Stick electrodes with a coating layer of shielding flux chemicals:

Decisive criteria for the selection of electrodes are as follows:

- Steel grade
- Ambient temperature
- Humidity
- Joint form
- Geometrical configuration and details
- Welding position

When workpieces of two different steelgrades are to be welded to each other, the welding process and the electrodes suited to the higher steel grade should always be adopted.

For metal arc welding, the electrodes with coatings of rutile acidic or alkaline flux chemicals are applied. In non-alloy and low alloy steels, the root welding and the welding of the filling layers are carried out by the rutile electrodes due to their easy handling and better gap bridging capacity for root welding. It is also worth mentioning that the slags produced by the rutile electrodes can be removed more easily than those produced by the alkaline electrodes, particularly from the narrow weld seams.

For more stringent ductility requirements, and in structures subjected to dynamic loads as well as to weld high strength fine grain steels, the hydrogen controlled electrodes with alkaline coating are used. They must be protected from humidity during storage and dried in a oven prior to the welding operation (for several hours at about 250–400 °C as a rule) according to the recommendations of the electrode manufacturers in order to avoid hydrogen embrittlement. The instructions of the manufacturer regarding current (AC or DC) and polarity should be followed.

In Ref. [32], the electrode types are recommended depending on the steel grades, wall thicknesses and weld forms:

For S 235 and S 275 (Table 3-5 and 3-9)

For wall thickness ≤ 16 mm (groove weld)
For wall thickness ≤ 30 mm (fillet weld) } Rutile or alkaline, hydrogen controlled electrodes

For wall thickness > 16 mm (groove weld) } Alkaline, hydrogen controlled electrodes

For S 355, S 460 (Table 3-10) and weathering steels

For all wall thicknesses } Alkaline, hydrogen controlled electrodes

Where there are several steel grades in a workshop, it is advisable to use only hydrogen controlled electrodes to avoid errors.

The root welding using alkaline electrodes is more difficult than with rutile electrodes, especially when the root welding or the welding with a disadvantageous position using low current has to be performed with electrodes of small diameters. The operation with this type of electrode demands adequate skill and experience from the welder.

8.7.2.2 Gas metal arc welding GMAW

This is a semi-automatic variant of the metal arc welding, where the arc and the molten metal are protected by the gas shield from the harmful effects of oxygen and nitrogen. In this process, the welding machine is fed by a continuous, solid wire electrode and a shielding gas through a nozzle (Fig. 8-32).

Gas metal arc welding consists of two variants:

1. The welding for MIG (metal inert gas) and MAG (metal active gas) processes is carried

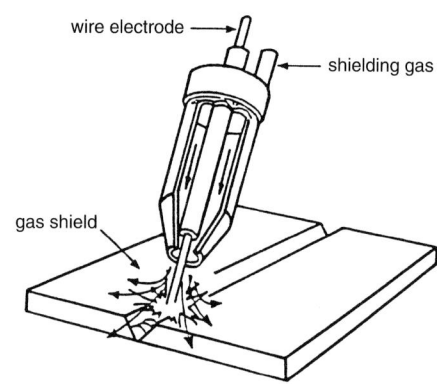

Fig. 8-32 Semi-automatic GMAW. Metal inert gas MIG with Argon or Helium or Metal active gas MAG with CO_2 or (Argon + CO_2) or a mixture (80% Ar + 15% CO_2 + 5% O_2)

out with the fusion of an wire electrode (the wire diameter ranges between 0.45 and 3.2 mm; the usual diameter is between 1.0 and 1.2 mm).

The difference between MIG and MAG processes lies in the shielding gas types, Argon or Helium partly with small content of O_2 for MIG and CO_2 or a gas mixture (80% Ar + 15% CO_2 + 5% O_2) for MAG. The latter is less expensive than the former one. Both processes necessitate an expensive special electric control, an wire feeder and a special shielding gas supplier.

The MIG process with high fusion capacity is applied to high alloy steels and non-iron metals, while the MAG process is used for welding non-alloy and low alloy steels. The MAG process has shown to be specially efficient for welding hollow sections in disadvantageous welding positions and for bridging wide weld gaps.

2. The tungsten electrode, which is not in the process of fusion, is used for the TIG (Tungsten inert gas) method. This consists of separate feeds from the heating source (electric arc burner) and from the weld electrode wire spools. The torch contains both the current supplying tungsten electrode and the shielding gas nozzle.

The TIG method is applied preferably for welding thinwalled hollow sections of alloy steels. Joint seams and throats with usual undercuts can be filled up by means of this method. The fusion capacity is however small.

8.7.2.3 Flux cored arc welding FCAW

This is a semi-automatic process using electrodes as continuous hollow wire fed in from a spool on the welding machine. The wire contains flux chemicals, which provide protection of the arc and the molten metal from the detrimental effects by O_2 and N_2. Over and above, shielding gases are also delivered to the operator's gun.

The system demands higher investments for the diverse equipments. However, this is compensated by a higher working rate i.e. two or three times faster than shielded metal arc welding and consequent saving of time.

8.7.2.4 Submerged arc welding SAW

Among various processes to protect an wire electrode by slags while welding, the full automatic welding under powder has proved to be most satisfactory. The metal arc burns buried under the flux powder, which piles up on the weld seam about the end of the electrode. Submerged arc welding machine consists of equipments for wire feed, powder supply, weld movement and control. However, the application of this process is restricted to horizontal weld seams and the welding has to take place under roof. That is why this welding process is applied in special cases e.g. offshore constructions.

8.7.3 Preparation of the welds in hollow section structures

An economical manufacture of structures in hollow sections demands not only the appropriate selection of the welding method but also a precise planning and execution of the jobs in the workshop. Proper end preparation of the structural elements as well as the adequate welding performance belong to this procedure. In Chapter 6, they have been described for various constructional details:

- Knee joints (see Section 6.4)
- End-to-end connections (see Section 6.5)
- Indirectly joined welded connection (see Section 6.5.2)
- Directly joined welded connections (see Section 6.5.3)
- Lattice girders (trusses) (see Section 6.6)
- Truss joints (see Section 6.6.2)

In some cases, the ends of the hollow section members are kept in their initial condition while in others, they are fully or partially flattened (see Section 6.6.8), see Fig. 8-33 for fillet welds.

The lack of proper care is mostly responsible for wrong weld preparations, which make the welding operation difficult. Above all, it may cause great problems to carry out root welding. For large section depth and too small angle of inclination, the risk exists that the sides of the joint members adjacent to the root are not fused by the metal arc. This may initiate an weld root with defect (lack of fusion). Problems may often arise while fully penetrating the root in groove welds joining offset plates.

8.7.4 Welding positions and sequences

Fig. 8-34 shows four principal welding positions together with the sequence of the welding operation for a hollow section construction. They can be regulated by the movement of the structural members.

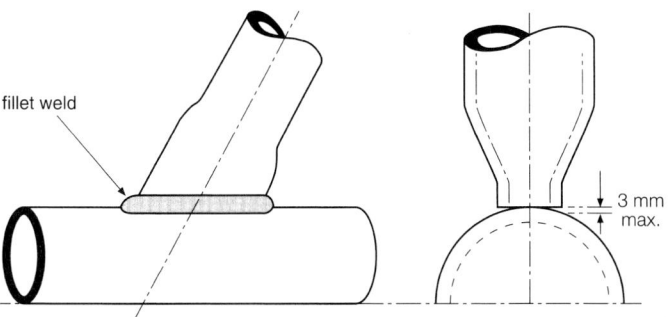

Fig. 8-33 Fillet welds in a lattice girder joint in CHS with flattened ends

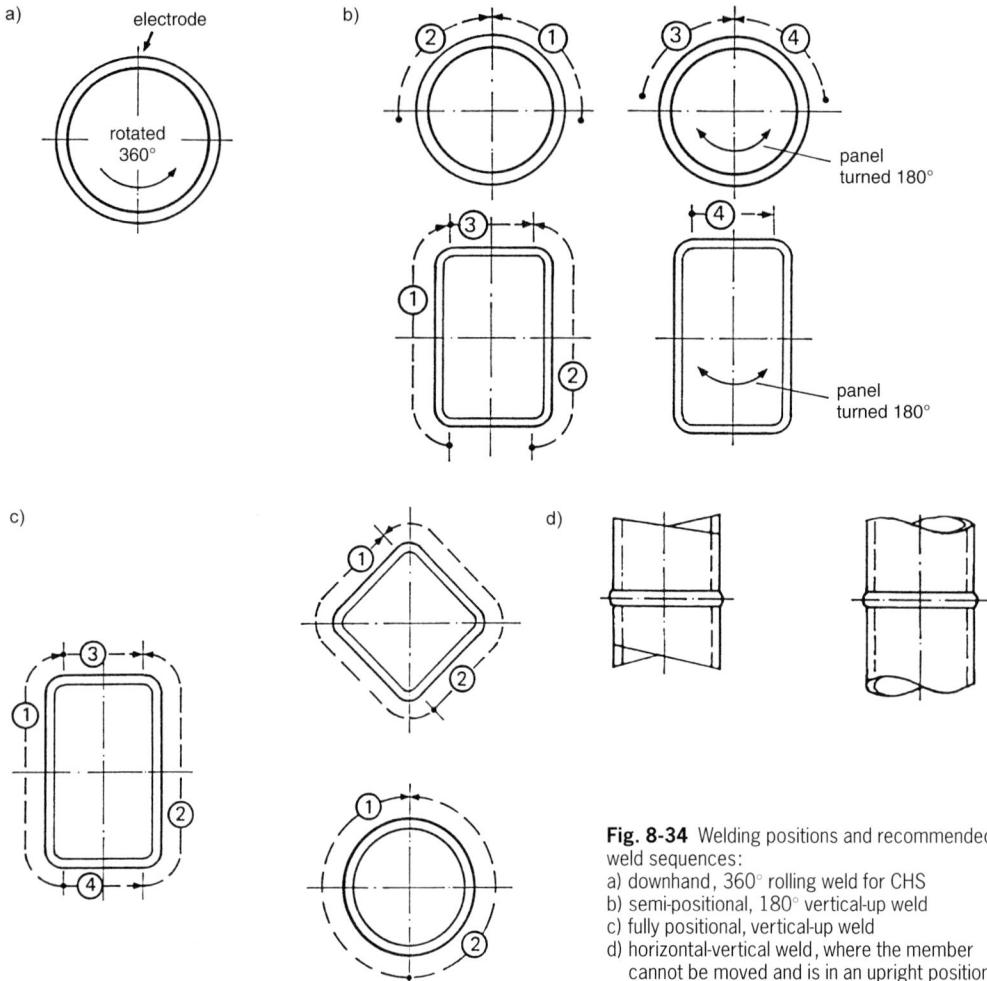

Fig. 8-34 Welding positions and recommended weld sequences:
a) downhand, 360° rolling weld for CHS
b) semi-positional, 180° vertical-up weld
c) fully positional, vertical-up weld
d) horizontal-vertical weld, where the member cannot be moved and is in an upright position

While fixing the welding sequence, the following points have to be particularly attended to:

– Welds must not start or stop on a corner of RHS.
– For the execution of welding in a hollow section construction, the proper selection of weld sequence is of high importance, as it affects the shrinkage, residual stresses and deformations significantly.

The sequence of welding in a lattice girder joint has already been described in Section 6.6.3.2.

Further, multi-pass welds should be avoided as far as possible, especially for smaller wall thicknesses and also when a large number structural elements are to be welded in a joint.

8.7.5 Tack welding

Tack welding consists of short welds for the preliminary joining of structural members before the subsequent final welding. This is a weld to hold parts of a weldment in proper alignment until final welds are made. In this case, the throat thickness of the tack weld has to be in accordance with the root position. They should guarantee a clean connection at the weld root. The beginning and the end of the weld seam has to be fixed following

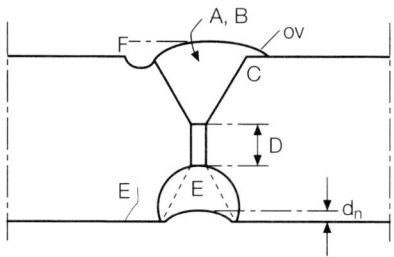
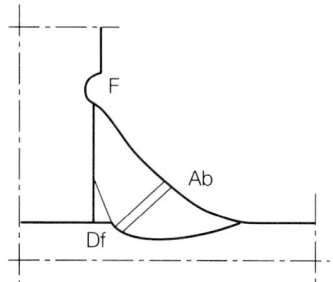

E = crack
C = lack of fusion
A = gas cavities
Ab = worm holes
B = slag inclusions

Df = incomplete root penetration in fillet weld
D = incomplete root penetration in groove weld
d_n = insufficient throat
ov = overlap
F = undercut

Fig. 8-35 Types of weld defects

Fig. 8-34. The ends of the tack welds should be dressed to permit proper fusion into the root run.

As tack welds become a part of the final weld itself, they have to be carried out free of defects. They have to be subjected to the same quality requirements as the final welds. Fig. 8-35 shows the possible defects in butt or fillet welds. That is why the welders need special certificate of competency to conduct tack welding [16].

While tack welding a CHS joint, welding has to be avoided at the symmetrical locations A shown in Fig. 8-36, as local stress concentration takes place there.

Tack welds, which are not incorporated into final welds, shall be removed, except that, for statically loaded structures, they need not be removed unless required by the operator.

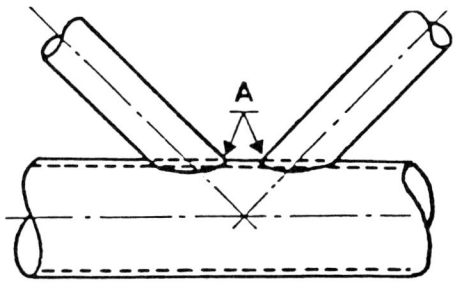

Fig. 8-36 Tack weld has to be avoided at A

CHS tacking is done by circumference welding, when the CHS diameter is small. The problems due to eventual large tack welding can be easily avoided by reducing the tacking length to a minimum of 1/10 of the CHS circumference.

In RHS joints, the straight parts are first aligned by tack welding. The length of the tack weld should be four times the thickness of the thicker part being joined, but not more than 50 mm.

Tack welds must not be applied at corners.

Backing members must always be tack welded to the root face, never internally.

8.7.5.1 Plug and slot welding

Plug weld is made in a circular hole in one member of a joint fusing that member to another member (see Figs. 6-62 and 6-63). Slot weld is the same for an elongated hole, where the hole may be open at one end. SMAW, GMAW or FCAW are used fusing and depositing successive weld layers to fill the hole to required depth. The slag covering the weld metal should be kept molten until the weld is finished. The slag must be completely removed after it is allowed to cool. Plug and slot welds are not permitted in quenched and tempered steels. They are also prohibited when the members are subjected to tension primarily.

8.7.6 Post-heat treatment of welded constructions of hollow sections

Stress relieving heat treatment after welding is only then performed, when the residual stresses due to welding have to be reduced on account of special considerations. Usually, the required temperature lies between 530 and 580 °C. For high strength steels e.g. S 460, this is about 30 to 50 °C lower than the annealing temperature of the material.

8.7.7 Residual stress and deformation due to welding and their reduction measures

In the process of the cooling of the welds, stress due to shrinkage occurs in the immediate, adjacent area to the weld. This takes place also by the contraction of the total welded construction. This stress leads either to deformations or to twists or remains in the welded part as residual stress initiated by weld. In rigid constructions, the deformation due to shrinkage by cooling after the welding operation is strongly prevented. The residual stresses due to welding on the other hand, increase significantly in the constructions. The capacity of a designer is restricted to building a structure, where the deformations and residual stresses due to welding are as small as possible. As these two effects are directed against each other, the design can only be made based on a compromise considering both effects.

The following parameters determine the deformation as well as the welding residual stress, which give the constructor the possibility to control them:

- Weld thickness
- Number of weld passes
- Distance of the weld to the neutral axis of the structural element
- Restraint of the welded structural elements in a joint
- Rigidity of the structural elements in a welded structure
- Welding process
- Welding sequence

In order to reduce the difficult and expensive straightening and aligning job after welding, the distortions due to shrinkage are sometimes compensated by pre-deformation of the

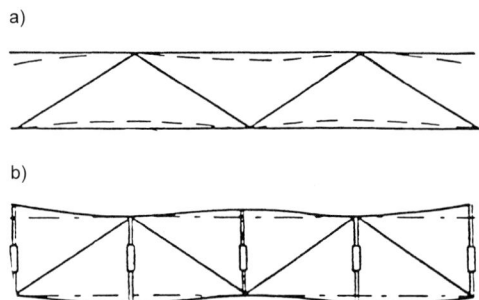

Fig. 8-37 Pre-deformation of a welded lattice girder

structure. Figure 8-37 illustrates an example with a lattice girder. After estimating the deformations of the lattice girder due to welding (Fig. 8-37a), chord members in a tack welded girder are preset by means of jack, screws or winches (Fig. 8-37b).

Weld arrangement and sequence of the welding operation determine the residual stress and the shrinkage proportionally. Tack welds must be so numerous and have such high strength that they are able to withstand all forces occurring after the final welding.

Groove welds should be carried out before the fillet welds and the longitudinal welds prior to the transverse ones in order to reduce the residual stresses.

The weld sequences for truss joints, already recommended in Section 6.6.3.2, demonstrate that the welding shall always proceed from the inside to the outside direction. This allows the free movement of the structural parts to one another due to shrinkage resulting in small deformation and low residual stress.

In the following, a number of measures are described to reduce the deformations and residual stresses due to welding.

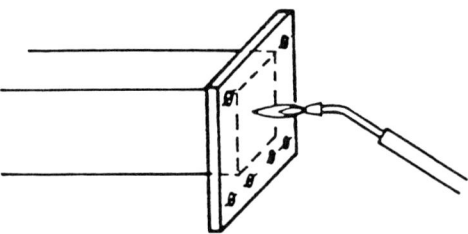

Fig. 8-38 Local circular corrective post-heat treatment

- One-sided local corrective circular heating (Fig. 8-38)
  One-sided heating brings about local shrinkage, which cancels out the distortion due to welding. A circular application of the corrective heat contracts the adjoining metal. It is mainly used to flatten out swellings.

- One-sided local corrective linear heating (Fig. 8-39)
  A linear type of corrective heating is applied to avoid shrinkage on seams of large length. The straightening action is obtained by shrinkage resulting from compensating on the opposite side.

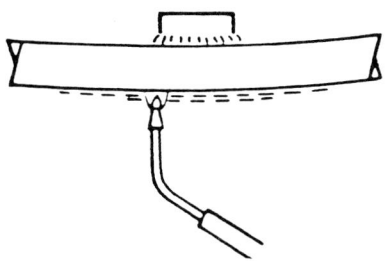

Fig. 8-39 Application of the corrective linear heating

- One-sided local corrective triangular heating (Fig. 8-40)
  In order to suppress angular distortion due to fillet welding, corrective heating in a triangular manner on the side opposite to the weld is applied. This heating process must start from the apex of the triangle. Subsequent cooling produces a shrinkage in the opposite direction.

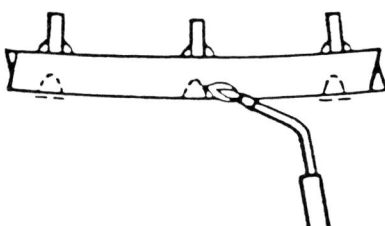

Fig. 8-40 Application of the local corrective triangular heating

- Further methods
  The straightening up of structural units is also carried out by cold bending, particularly for light structures and by hammering of the welds (seldom used).

The location of corrective heat treatment must be precisely determined beforehand, so that any torsional distortion of the structural element following an unregulated course of the flame can be avoided. The global effects must always be considered alongside the local ones.

8.7.8 Weld defects and their repairs

Fig. 8-35 shows various types of defects in groove or fillet welds. Lit. [16] presents the undesired as well as unacceptable weld profiles of groove and fillet welds (see Fig. 8-41).

The weld shall have a gradual transition to the surface of the structural members and be free of the discontinuities as excessive convexity, undersize welds, large undercuts and excessive weld metal deposits (overlap). Weld defects can be repaired by the removal of the weld metal or portions of base metal by means of machining, grinding, chipping or gouging.

Excessive concavity of weld, insufficient weld throat and large undercuts are compensated by depositing additional weld metal. Lack of fusion, gas cavities and slag inclusions have to be removed and rewelded. Cracks in weld metal or base metal shall be removed 50 mm beyond each end of a crack and rewelded.

8.7.9 Inspection of welds

Welds in all steel structures can be checked either by destructive (in laboratory) or by non-destructive tests. There are various methods of inspection belonging to each of these groups with their special merits and demerits and their specific fields of application.

Destructive tests (to determine ultimate tensile strength, yield strength, elongation, ductility values using impact bend or drop-weight tear or COD (crack opening displacement) method, hardness and deformability using bending, folding or flattening tests as well as fatigue resistance) are usually made prior to the final welding of a structure or when new

Fillet welds:

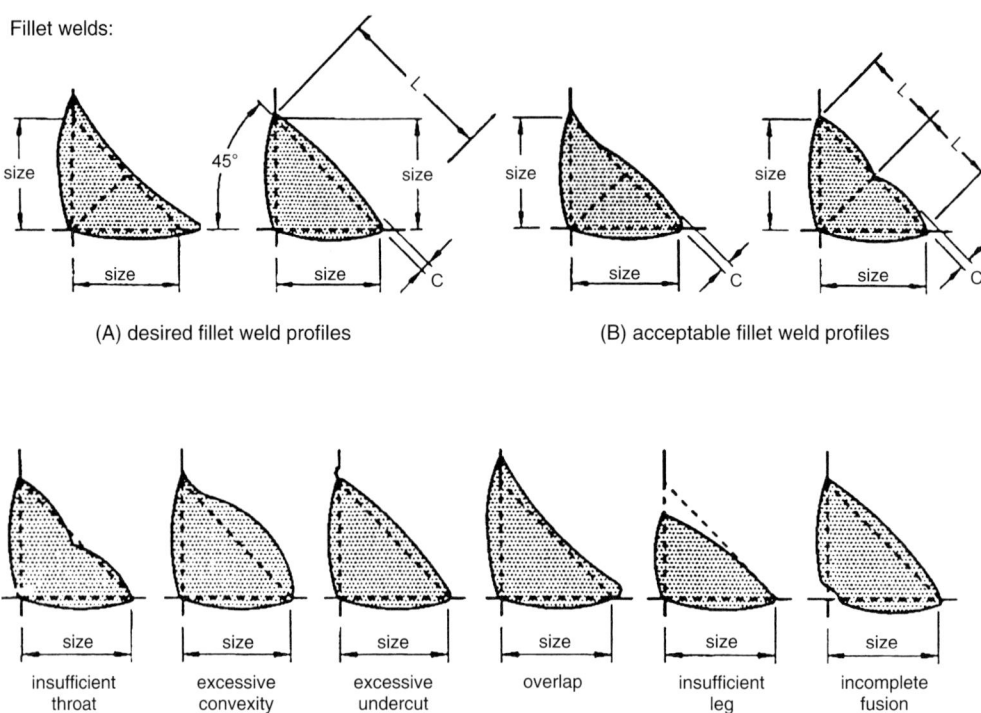

Fig. 8-41 Acceptable and unacceptable weld profiles [16]

materials, new construction types and welding processes have to be investigated. Further, they are also taken up to examine the competency of the welders. Mostly these tests are carried out in the laboratories or workshops.

Non-destructive tests are listed in the following five groups:

– Visual inspection
– Magnetic particle test
– Dye penetration test
– Ultrasonic inspection
– Radiographic inspection by X or γ rays

8.7.9.1 Visual inspection

It is of vital importance to make a very precise visual inspection of the weld seam as well as of the weld vicinity before and after welding. Specially for welded hollow section structures, this is the most practical and effective technique, as other methods mentioned above have only little or restricted applications.

It is recommended to check prior to welding the root gap between the components to be welded together, the included angle, the uniformity of the weld edge or bevelling, as well as the complete removal of oil, grease, slag etc. from the weld location. Besides, competent (certified) welders shall be commissioned to perform the required welding job. As a rule, a weld can be executed without defects, if the above mentioned conditions are fulfilled.

After welding, the welds have to be scrutinized by visual inspection to find out the surface flaws e.g. cracks, undercuts, overlap and excessive or insufficient weld thickness. Checking the weld appearance (roughness of bead surface, bead width etc.) is also important. Finally, the measurement of the weld throat thickness by means of the gauges developed for this purpose (see Section 6.6.3) and that of the transition of the weld to the base metal (specifically of importance for constructions under fatigue load) have to follow.

8.7.9.2 Magnetic particle test

This is a simple and quick method to discover surface defects like fine cracks, which are not obviously visible with eyes. The main area of application is to find weld defects in nodal joints, which are very difficult or not possible to determine by using other techniques e.g. ultrasonic or radiographic tests. It is also useful for checking routine spot welds.

Fine magnetic particles are sprayed on the location to be checked and a high current magnetic field is created by means of a magnetic yoke or coil. The magnetic field passes through the test piece when a crack distorts or interrupts this magnetic field, the magnetic particles line up along the cracks indicating even the finest ones (up to 1/10 000 mm). The measurement is recorded by photographs.

8.7.9.3 Dye penetration test

This testing method (Fig. 8-42) explores the weld defects rising to the surface of weldments. The process, although slow, is especially useful to find out the extent of the defects and applied particularly when gouging and grinding are done to repair a defect.

The procedure starts with cleaning the surface of the workpiece thoroughly. Subsequently, a penetrating red dye solution is put on the surface to be checked by means of a brush or a spray. The solution is allowed to act for about 5 to 10 minutes, during which the dye is drawn into the cracks. The red solution must therefore have low surface tension and capil-

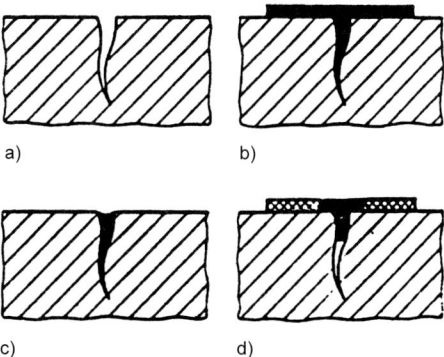

Fig. 8-42 Steps of the dye penetration process: Crack a) after cleaning the surface, b) after applying the penetration dye solution, c) after wiping off and cleaning the surface with water or a solvent, d) after applying white powder or a quick drying of white developer solution, the outline of the crack is visible

larity. After the solution is activated, the dye is wiped off with a cloth at first and the surface is then cleaned with water or a solvent specially developed for this purpose. When the surface is dry, either a thin layer of white powder is applied to it or a quick drying white developer solution is sprayed on the surface. In case cracks exists, the powder sucks the dye from the defects, into which it has been drawn making the cracks on the background visible. A clear outline of the defect can be documented by photographs.

8.7.9.4 Ultrasonic inspection

This is a simple and quick method, which however requires highly qualified and experienced inspectors.

An acoustic transmitter generates ultrasonic sound beams – generally at frequencies between 2 and 4 MHz and sends them into the weldment using, in most cases, water as a couplant. Imperfections in the weld, when being struck by the beam, reflect part of the sound energy. The reflected sound is partly received by a receiver, which is an oscilloscope displaying the echo electronically. The exact location and the approximate size of the defects can be determined by measuring the time for the sound wave to cover the known distance. It is however difficult to determine the type of the defect precisely. Only defects perpendicular to the wave direction can be found out using this method. In order to determine the lack of fusion i.e. in the groove weld flanks, angular transmitter and receiver are used, which sends and receives the sound waves at a required angle.

Main areas of the application of the ultrasonic method are as follows:

– Measurement of wall thickness from one side
– Proof of laminations in plates or shells
– Testing the manufacturing defects in bars and hollow sections
– Determination of defects e.g. slag inclusions, gas cavities and cracks in forgings, castings, welds etc.
– Testing of hot workpieces in manufacturing plants

In most cases, the ultrasonic method is more economical than the radiographic one. However, it demands inspectors with special knowledge and experience. It is not always without problems to decipher ultrasonic readings on the screen.

For the fabrication of hollow section joints, this method plays a limited role. Small fillet welds and partial penetration groove welds cannot be checked reliably, as the signals are difficult to interpret unambiguously due to the large variations of the joint geometry. Differing interpretations of the readings are also possible for the complete joint penetration groove welds with or without backing bars.

8.7.9.5 Radiographic inspection by X or γ rays

Radiographic tests consist of directing either X-rays or γ (gamma) rays from Cobalt or Iridium source through a weldment and producing a film showing weld defects (Fig. 8-43). Thereby, the test results can be well documented. This method is particularly suitable to find out lack of fusion, gas cavities and slag

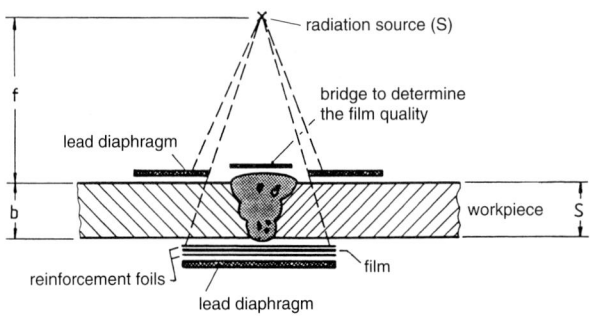

Fig. 8-43 Radiographic test (sketch)
f = distance between the radiation source and the workpiece surface
b = distance between the workpiece surface and the film

inclusions. Due to the irregular shape of the joints and the resulting variations in thickness of the material as projected on the film, this method plays a limited role to check welds in hollow section structures. However, nearly all critical butt welds can be checked using this method. As regards the application to butt welds in hollow sections, the ability or inability to place either the source of radiation or the film inside the hollow section determines whether single or double wall exposure has to be made. A minimum of three exposures is recommended to cover the complete circumference [16].

X-ray test is applicable for wall thicknesses from 2 to 50 mm. For inspection on site and for larger wall thicknesses, the application of γ-rays is recommended.

As long-time exposure to radioactive rays is detrimental to health, the inspection has to be carried out in a closed room or space.

Summarizing, the following recommendations can be made with regard to the weld inspection methods described above:

- X-ray tests can be applied most effectively to butt welds.
- Application of γ-ray tests is recommended for inspection on site.
- Radiographic tests deliver reliable values for object thicknesses of about 16–20 mm. Beyond that, the results supplied by the ultrasonic tests are more reliable for the inspection carried out both in workshop or on site. It is important to know the feasibility of the ultrasonic inspection on a joint, before it is applied to a especially critical joint.
- Due to their unreliability, the fillet welds should not be tested radiographically.
- Magnetic particle or dye penetration tests are mainly used to determine surface cracks.
- Visual inspection within its applicability range is the most practical, the cheapest and the most effective among all weld inspection methods for welded hollow section structures. This however requires experienced and preferably certified inspectors.

8.7.10 Performance qualification tests for welders and welding workshops

In various codes and standards [16, 34–37], it is prescribed that the welders and the welding workshops must be in possession of valid certificates of competency, which allow them to work on different types of welded constructions. This is especially to prevent non-qualified personnel to weld safety relevant structural parts. As an example, Table 8-2 according to [37] lists the test joints with RHS (group B) or CHS (group R) required for examining the quality of a welder. Further, a welder, who welds hollow section lattice structures, shall be approved by means of an weld test on an additional joint shown in Fig. 8-44 (see Table 8-2, line 2). This subject has also been dealt with in [16].

Welding jobs done in the workshop determine the type and the extent of the performance tests for the welder's qualifications. Ten different criteria to be taken into consideration are given by [35]:

- Welding method
- Form of structural element and joint

Table 8-2 Assigning test joint types to welder performance qualifying tests [37]

	Joint type	Required welder performance qualification tests
1		R I, R II[a]
2		R I and additional performance qualifying test
3		B I, B II[a]
4		B I
5	$\theta<90°$	B I and additional performance qualifying test

[a] The tensile resistance depends on the quality of the weld execution.

- Weld type
- Steel grade
- Electrode and weld metal
- Specimen thickness
- CHS diameter
- Groove weld detail
- Groove weld root execution
- Welding position

Approved welding research centres, technical supervision authorities or material testing institutes are responsible to examine the welder's performance.

8.7.11 Welding of cold formed hollow sections

The welding of cold formed hollow sections has already been dealt with in detail in Section 3.1 (compare with Table 3-15).

8.7.12 Stud welding

Welding of studs to hollow sections is done by the arc welding with the use of automatically timed stud welding equipment (stud gun) connected to a suitable source of direct current. The process produces coalescence of metals by heating them with an arc between a metal stud or a similar part and the workpiece. When the surfaces are properly heated, they are brought together under pressure. Studs are furnished with surrounding ceramic ferrule deoxidizing gas or flux for shielding the arc.

At the time of welding, the studs and the areas, to which the studs are to be welded, shall be cleaned from scale, rust, moisture or other injurious materials by means of wire brushing, scaling, prick-punching and grinding. The stud bases shall not be painted, galvanized or cadmium-plated prior to welding. The arc shields or ferrules shall be kept dry or be dried in oven if necessary.

Stud application qualification requirements are met applying bending, torque and tension test methods. If a visual inspection reveals any stud that does not show a full 360° flush or any stud that has been repaired by welding, such stud shall be bent to an angle of approximately 15° from its original axis. Threaded studs shall be torque tested.

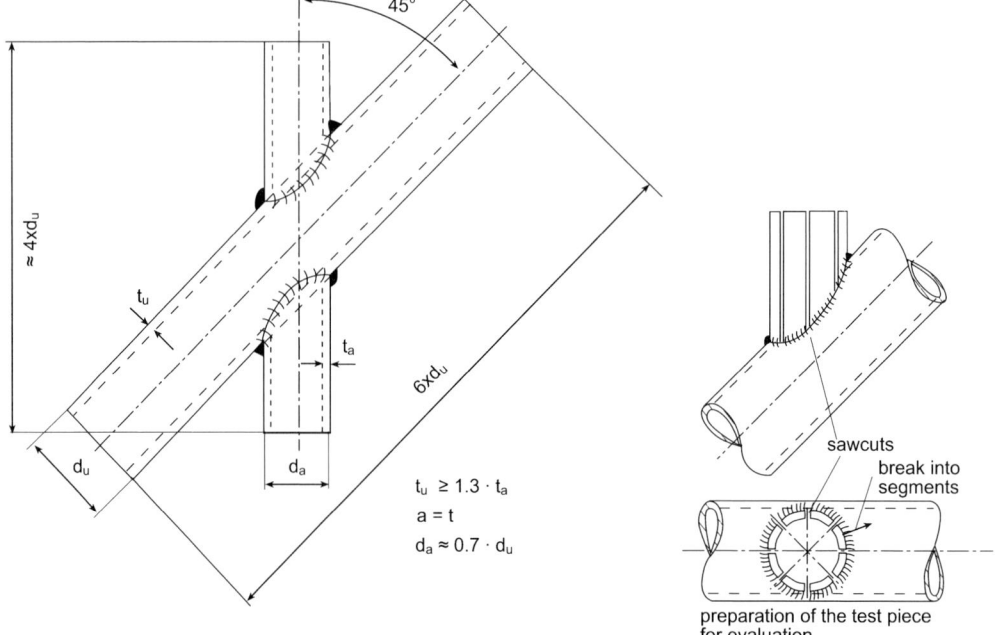

Fig. 8-44 Specimen for additional performance qualifying tests for the welders

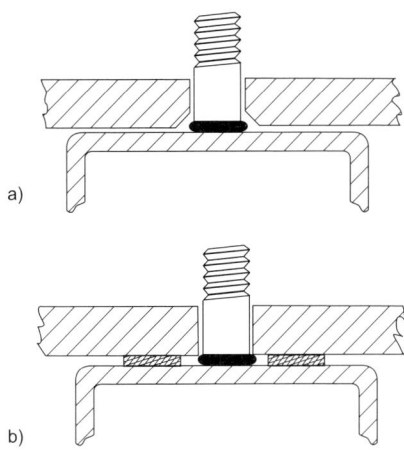

Fig. 8-45 Stud welding with a) recess at the bolt hole in the connecting plate, b) a fitted clearance washer

Fig. 8-45 shows two applications of studs welded onto the face of RHS, where the stud welding leaves a collar at the root (where the stud meets the section). In these cases, the bolt hole in the plate to be connected is chamfered clearing the collar (Fig. 8-45a) or a clearance washer is fitted (Fig. 8-45b).

For light weight fixings such as claddings fixed directly to the hollow section purlins, stud and self tapping screws are employed.

8.7.13 Laser welding

Due to the progressive development of the laser equipment with high efficiency, the laser welding is applied more and more, preferably in the sector of thin plate welding. The large concentration of energy at the focal point of the laser beam leads to a high welding speed. The lower heat input minimises the residual stresses and the shrinkage or distortion of the structural parts. The welding of the coated structural elements or workpieces is possible due to narrow weld seams and heat affected zones. The coating can only be impaired or destroyed in a small zone.

In the industrial manufacture, CHS welded to a head plate by the laser beams is an example of the structural applications. The following advantages can be exploited with the aid of the constructional technique using the laser welding method:

- The constant quality of the weld is guaranteed due to the use of an automatic welding process.
- Inspite of high investment cost, lower production time makes the laser method economical in the long run.
- On the one hand, the application of the electrode wire enlarges the gap bridging capacity, which reduces the expenditure for the weld preparation. On the other hand, the mechanical properties of the weld seam can be improved by using suitable flux material.

8.7.14 General recommendations for welding

- A thorough cleaning of the weldment is of extremely high importance, as contaminants like rust, slag, milling scale etc. lead to slag inclusions in the weld.

- The easy accessibility to the weld location is of high significance for carrying out welding. There must be adequate space for welding torch, shielding gas nozzle and electrode clamp to conduct welding conveniently.

- The fabricators are often inclined to specify and carry out more extensive and larger welds with higher throat thicknesses (over size weld) than is required for the structural function (Fig. 8-46). This is not only costly but also wasteful and possibly harmful as the shrinkages and distortions of a structure become larger due to overlap and excessive weld deposits. Over and above, additional heat input causes a change of the microstructure of the base metal in the heat affected zone. The resistance of the weld is not very much affected by these additional stresses when the structure is subjected to static loading. Under cyclic loading however, the hot spot stresses cannot be re-

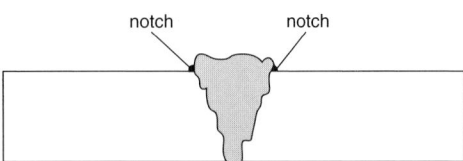

Fig. 8-46 Overlap of weld

duced by local plastification, so that a full exploitation of the mechanical properties of the base metal is not possible. The welds belonging to dynamically loaded constructions can undergo an improvement by grinding off excessive weld metal.

- While welding stiffeners, welding shall not be performed up to the edge of the stiffening plate keeping the weld root at the corner free. That means that the corner of the stiffener must be taken off (Fig. 8-47).

Fig. 8-48 Example of a welded joint clustered with too many CHS components

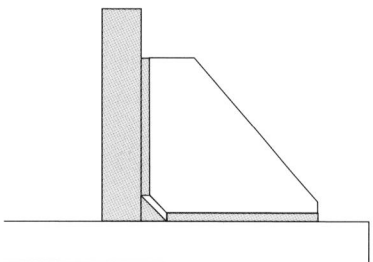

Fig. 8-47 Taking off the corner of a stiffening plate

- Fillet welds are preferred to groove welds as long as they do not become too thick. They are applied for a leg length equal to or smaller than 16 mm. Beyond this limit or when fillet welds are not feasible due to other reasons, partial joint penetration groove welds should be used, as they are less expensive than complete joint penetration groove welds. The latter can however be applied for butt splices on to the backing bars for similar size sections. It is recommended to avoid their use without backing bars.

- In order to assure adequate weld penetration at the heel of a branch member in a lattice joint of hollow sections, the angle of intersection between the bracing member and the chord should be 30° or more.

- A joint should not be locally crowded by welding too many plates and sections, as this makes not only the accessibility to the weld location difficult but also promotes corrosion by creating water or snow traps (Fig. 8-48).

- Distortion of a structure in hollow sections can be reduced by following the welding sequences shown in Fig. 6–100 (lattice joints) and Fig. 8-34. They also avoid starts and stops at the corners of RHS, which is particularly important for critical joints under fatigue loading. This is because the eventual weld flaws initiate premature cracks in the RHS corners.

- It is not necessary to apply pre-heat treatment for general structural steels used in hollow section structures (see Section 8.7.1). However, when the surfaces are wet or when the ambient temperature is 5° or less, the workpieces have to be preheated until they are warm to the touch for a distance of not less than 75 mm on either side of the weldment.

- Welding of galvanized or zinc-coated hollow sections can be done without any problem. However, one has to take care of proper ventilation to remove toxic gases and fumes, when the zinc volatilises. The proper and quick method of welding consists of volatilising the coating with a lengthened arc and executing the weld finally. Following, the zinc coating can be renewed by spray metallization or zinc rich painting.

- Visual inspection of welds in a hollow section structure is performed most frequently, as it is most economical and technically without any problem in the standard cases, where it is sufficient to find out the weld defects on the surface. It is however necessary that the inspectors possess adequate welding knowledge and experience. Other testing methods e.g. ultrasonic, radiographic etc. are only used to check the critical joints depending on their feasibility and only when they are capable of supplying reliable results.

- Welds in combination with bolts and rivets must carry the entire stress in the connection alone. It is however permitted to have connections that are welded to one member and bolted or riveted to another member. High strength bolts as a friction type connection installed prior to welding may share stresses with the welds.

- Molten metal, sparks, slag etc. produced by welding and cutting can cause fire or explosion if precautionary measures are not used.

- Welding and cutting may produce radiation harmful to health. Most arc weldings except SAW, laser welding and torch welding can produce quantities of non-ionizing radiation such that precautionary measures are necessary.

8.8 Nailing

As an alternative to the splice connections between CHS members by groove welding from one side (see Section 6.5.3) or using blind bolting systems (see Section 8.6.1), the nailing of one tube to another coaxial tube within it is a relatively simple and more economical jointing method. In this process, one tube is inserted inside another (the inside diameter of the larger equals the outside diameter of the smaller) to achieve a reasonably snug fit. High strength ballistic point nails are then shot fired with an independent powder actuated tool (or gun) penetrating through the wall thicknesses of both tubes. They are arranged symmetrically around the tube perimeter (Fig. 8-49). Another constructional possibility may be to join two tubes of the same outside diameter by means of a tubular collar over both tube ends. It is worth mentioning that the involved procedure requires minimal training and can be performed by relatively unskilled workers on site at an extremely fast rate.

Fig. 8-49 Nailing process

A special type of nails was checked successfully [40] to connect the tubes in an experimental research programme [38], which involved tests with tubes ranging up to 400 mm outside diameter investigating into the following parameters:

- Diameter to thickness ratio of tube
- Wall thickness of tube
- Degree of "Lack-of-fit" (tight or loose)

The nails used in these tests have two slightly cupped washers located on the nail shank, which move up towards the head of the nail when the nail is driven into the steel. These washers provide a good clamp between the tube walls.

The tests demonstrated that the combined steel thicknesses of up to 13 mm could be easily connected.

Due to the closed connection, it is not possible to visually examine the inside of the tube in order to confirm adequate penetration of the nail through both CHS walls. Proper penetration depth and consequently the connectivity can be determined by the "stand-off height" of the nail head from the outer CHS wall and the measurement of sufficient insertion of one CHS into the other.

Two potential failure modes were identified by the tests:

- Shear failure of the nails
- Bearing/shear-out failure of CHS wall

Simple prediction formulae, derived from the bolted connections [41], have been verified against the test results for the ultimate nailed connection strength and found to be adequate.

In order to calculate the ultimate connection strength R_n, the following formulae are recommended:

Nail shear failure

$R_{n,\text{nom}} = (\text{single shear strength of a nail}^{1)}) \cdot n$

(8-14)

Bearing/shear-out failure (in this case, the deformation around the "bolt hole" is a design consideration. The design is based on AISC LRFD [42] criteria)

If $L_e \geq 1.5\, d_n$ and $s \geq 3\, d_n$,

$R_{n,\text{nom}} = 2.4\, d_n \cdot t \cdot n \cdot f_u$ (8-15)

If $L_e < 1.5\, d_n$ or $s < 3\, d_n$,

$R_{n,\text{nom}} = L_e \cdot t \cdot n \cdot f_u$ (8-16)

but $R_{n,\text{nom}} \leq 2.4 \cdot d_n \cdot t \cdot n \cdot f_u$ for the nail row nearest the free edge and

$R_{n,\text{nom}} = \left(s - \dfrac{d_n}{2}\right) \cdot t \cdot n \cdot f_u$ (8-17)

but $R_{n,\text{nom}} \leq 2.4\, d_n \cdot t \cdot n \cdot f_u$ for the remaining nail rows

In the foregoing equations,
L_e = edge distance measured along the axis of CHS from the centre of the nail
n = number of nails
d_n = diameter of the nail

---

1) The unregulated control of the properties of nails from manufacturer to manufacturer requires that a nail be independently tested by a third party and the results be used to predict nail shear failure of a nailed connection. This would then result in Eq. (8-14) for the nominal ultimate **shear strength** of a nailed connection ($R_{n,\text{nom}}$).

s = pitch or spacing between the rows of nails measured along the axis of CHS
t = CHS wall thickness
f_u = ultimate tensile stress of CHS material
$R_{n,\text{nom}}$ = nominal ultimate **bearing strength** for a nailed connection

Eqs. (8-15), (8-16) and (8-17) represent reasonable predictors for the bearing/shear-out failure mode and give a predicted connection factor resistance implementing a resistance factor ϕ of 0.75 as recommended by AISC LRFD [42] (equivalent to the partial safety factor $\gamma_M = 1.33$ according to Eurocode 3 [43]).

Besides simple application of nailing to make lapped splice connections for manufacturing tubular end-to-end joints, as for example for electricity transmission line poles, this can also be used as a mechanical shear connector in concrete filled hollow sections. In this case, the nail is driven through the steel wall from the outside surface to penetrate into the concrete. Analytical and experimental investigations have been carried out to determine the joint behaviour, which involves the prevailing mode of failure by nail-pushout after extensive nail plastic deformation [39].

8.9 Application of cast steel elements in hollow section structures

Iron casting is one of the very early developments to be used as structural elements. The examples of cast iron columns used in the early nineteenth century are existing till today. The type of cast iron applied to structures is generally spheroidal graphite or laminated graphite iron, because of their improved ductility values compared to other special cast irons. These irons, depending on their grade, have a 0.2% proof strength between 200 and 700 N/mm^2 (MPa) in tension. The cast irons are not easily weldable, where the welding technology is subdivided into three types e.g. hot welding (welding temperature $t_W \geq 500\,°C$), semi-hot welding ($t_W \approx 150\,°C$) and cold welding (t_W = room temperature) corresponding to their effects on the heat affected zone and the weld metal compared to the basic material.

Welding is not usually applied to cast irons; they are more suitable for bolted connections.

With the progress of the steel manufacturing processes, the cast iron became practically without any significance in the structural and mechanical engineering. It is substituted by cast steels, non-alloyed, low-alloyed, high-alloyed and stainless, which can be produced with comparable ductility and mechanical properties to those of the rolled steels used for structural hollow sections. They are listed in Table 8-3, measured in the framework of a recent research [43].

Welding of cast elements

Cast steels have also similar welding characteristics to rolled steels for structural hollow sections with about the same carbon equivalent value. Suitable welding processes for cast steels belong to the group "fusion welding" differing from one another with regard to flux materials. They are of two categories: the alien and the same type. While the welding with the flux material of the same type leads to an adequate cross-sectional area with an allowable stress equal to ca. 60% of the yield strength of the base material, the bearing strength of the joint by welding with the low strength flux material of the nickel-based alien type is so small that they should not be loaded by tension or bending. The welding processes generally applied to cast steels in practice are manual metal arc (MMA), metal inert gas (MIG), tungsten inert gas (TIG) and metal argon gas (MAG).

Further the malleable cast iron is a special product, which is manufactured similarly to general cast iron and then obtains the composition close to that of cast steel by specific heat treatment. The demerit lies in the limitation of production up to 10 mm wall thickness and in high costs for the manufacture.

An overview of the structural application behaviour of fusion welding in cast iron or cast steel is given in Table 8-4 [45].

Quality assurance of cast elements and their welded joints

The quality assurance of the cast steel or iron parts as well as their welded connections is of outstanding importance for their application in structures. The primary responsibility lies with the supplier of the cast products, whose job is to find out the defects by means of the radiation and the ultra-sonic tests (see Sections 8.7.9.4 and 8.7.9.5). An independent inspector checks the results later by repeating these tests with random sampling. The reliability of the mentioned non-destructive test methods to make proper defect indications shall be once again confirmed by sawing the cast specimen. Defective locations are then made visible by cutting the specimen further

Table 8-3 Material properties of rolled steel and cast steel

Material	Thickness	Yield strength	Ultimate tensile strength	Elongation A_5	Impact energy (ISO-V)	
	mm	N/mm^2	N/mm^2	%	Test temp.	Joule (min)
Cast steel 20 Mn 5 V quenched and tempered	$t \leq 50$	360	500–650	24	−30 °C 20 °C	40 70
	$50 \leq t \leq 100$	300	500–650	24	−30 °C 20 °C	40 50
	$100 \leq t \leq 160$	280	500–650	22	−30 °C 20 °C	40
St 52-3 (S 355 J2, see Table 3-9)	$16 \leq t = 40$	345	490–630	22	−20 °C	27
	$40 < t \leq 63$	335	490–630	21	−20 °C	27
	$80 \leq t \leq 100$	315	490–630	20	−20 °C	27

Table 8-4 Structural application of fusion welding in cast iron or cast steel

Material	Flux material	Weldability	Load bearing possible	Pre-heat treatment temperature
Cast steel CEV ≤ 0.25	Same type, permissible stress is equal to about 60% of the yield strength of the base material	good	yes	room temp.
Cast steel CEV > 0.25		not adequate	with restriction	~200 °C
Malleable cast iron, white		good for wallthickness ≤ 6 mm	yes	room temp.
Spheroidal graphite iron		not adequate	with restriction	~500 °C
Malleable cast iron, black		very little	no	~500 °C
Laminated graphite iron		very little	no	~500 °C
Cast steel CEV ≤ 0.25	Alien type, permissible stress is due, to nickel-based flux material very low	good	yes	room temp.
Cast steel CEV > 0.25		good	yes	room temp.
Malleable cast iron, white		good	yes	room temp.
Spheroidal graphite iron		good	yes	room temp.
Malleable cast iron, black		good	yes	room temp.
Laminated graphite iron		good	yes	room temp.

and subjecting part of it to a mechanical process or etching it.

Defects close to the surface are to be determined by the magnetic particle test (see Section 8.7.9.2) or the dye penetration tests (see Section 8.7.9.3). In case defects are found in castings and welds in the process of quality control, they have to be removed as follows:

– By grinding-off the defects
– By reinforcing the casting (cast steel spheres or joints with welded ribs)
– By replacing the defected part by a new one

Summarizing, this can be stated that cast steels can fulfil today all the structural requirements regarding strength of material, ductility, fracture toughness, weldability and corrosion resistance.

Cast elements in structures

As has been already described, structural parts in cast steel have been playing an innovative and at the same time extensive role during the last decade in the field of structural engineering, particularly in bridge constructions, multi-planar trusses, offshore constructions, heavy duty cranes, wind bracing connections, column heads as rope guides etc. It is important to note that the cast steel constructions can be a very good substitute to the welded ones in view of technical integrity as well as of economy. Especially, when the joints in a structure are so complex that the fabrication by welding becomes difficult and when many identical joints are required, it can be economic to use castings instead of welded joints. Due to constructional reasons (i. e. when a number of structural elements meet at a nodal point in a space structure and the accessibility necessary for jointing is difficult) or on account of architectural aesthetic or for a meaningful load transmission or load reorientation in design and statics, it can be required to manufacture the nodes with castings and then join them with a large number of single members by welding or bolting. Fig. 8-50 shows a nodal sphere in cast steel for a tubular space frame, where rolled CHS members are welded directly to the sphere. Fig. 8-51 illustrates a cast steel node in the form of a bolted construction. Sometimes, forged parts may also be used instead of cast steels as shown in Fig. 8-52.

As regards cast steel nodes, the capability to exploit the material qualities irrespective of the loading direction is of high significance. This is supported by the feasibility to vary the

8.9 Application of cast steel elements in hollow section structures

Fig. 8-50 Sphere in cast steel welded directly to a number of rolled CHS members in a space frame (Okta-node, sphere outer diameter 900 mm) [45]

Fig. 8-51 Nodus-node, bolted construction

Fig. 8-52 Forged semi-spheres welded together with a plate in-between to form a spherical node [46]

casting thickness to accommodate any area of high stress.

The joint forms made by cast steels and their wallthicknesses can be optimally adapted to the flow of forces transmitted by the incoming members (mostly tubes) to be joined. Cast steel joints with smoothed-out transitions do not require additionally to be calculated and dimensioned if the connected rolled steel tubes or the butt joints between the castings and the tubes are properly designed and sized.

Cast steel makes it possible to manufacture a rounded-off joint without any sharp edges, also in re-entrant angles. Thereby, local stress concentrations and notch effects do not occur and problematic welds producing discontinuities are avoided. Fig. 8-53 illustrates the continuous smooth transition in a cast steel joint avoiding any abrupt change of cross section and inclination.

Due to the smooth and rounded-off corners in a cast steel joint, it is less susceptible to atmospheric corrosion; rain water drains away easily and the joint is thoroughly aired. Over and above, the accessibility for inspection and maintenance is adequately given.

The most innovative and modern area of application of cast steel joints is the truss construction in hollow sections, particularly under fatigue loading. Fatigue resistances of various welded hollow section joint types (uni-planar, multi-planar, see Figs. 7-16, 7-17, 7-21, 7-37, 7-50 and 7-51) have been extensively dealt with in Chapter 7, where the up-to-date dimensioning methods for their application in atmosphere (not in the corrosive medium) have been described. Although extensive research works have been performed on directly welded truss joints in the last twenty years throughout the world (including also the tubular joints for offshore platforms) and even if large improvements have been achieved meanwhile in the cutting and welding technique, there are apparently still big gaps in the knowledge regarding the fatigue design of welded hollow section joints in rolled steels. This concerns mainly the restricted validity ranges of the application parameters i.e. α, β, γ, τ, θ etc. (see Fig. 7-18, and the list of symbols for Chapter 7) and also numerous additional possibilities for joint forms and configurations, which cannot be designed and dimensioned using the proposed formulae in Chapter 7.

Another special aspect, which have to be taken into consideration for the fatigue design of the welded hollow section joints, is the oc-

Fig. 8-53 Air terminal building in Stuttgart. The transition elements in the tree-shaped columns are made of cast steel; they are welded to the rolled tubes

currence of local "hot spot" stress, where the crack leading to ultimate failure is initiated. Hot spot stress can be calculated by means of the stress concentration factor SCF depending on the above mentioned geometric parameters (see Section 7.5.1.5) indicating the discontinuity in the joint stiffness. Further the weld notches and the welding residual stresses play a significant role in fatigue design.

As the above mentioned SCF can at present be calculated within the existing limited validity range only, directly welded, hollow section truss joints cannot be designed beyond this range. A further possibility is to determine hot spot stresses in individual joints with the aid of finite element analysis or to carry out tests to find out the fatigue resistance for each joint. Both of them are time-consuming and expensive. Depending on the complexity of a joint, they are sometimes not practicable.

All these problems inherent to the welded hollow section joints lose their significance by replacing them with cast steel joints, which offer a homogenous integral component with low residual stresses and small stress concentrations compared to welded joints (see Fig. 8-54). This is due to the fact that the defects in cast steel have larger notch radii than those in the welds of directly welded joints. Consequently, the fatigue performance of cast steel joints is specifically better. Over and above, the nominal stresses can always be kept low by simply increasing the thicknesses of the parts of a cast steel joint under high load; this is feasible without any problem. Considering the possibility of defects in a casting, the wallthicknesses of castings should always be made adequate on the inside with reserves from 20% to 50% [44].

Usually, structures with cast steel joints are manufactured by welding the cast steel parts to other rolled profiles (mostly tubes). A smooth transition from the casting to the rolled tube, which allows a uniform flow of forces, is decisive for a proper design of

8.9 Application of cast steel elements in hollow section structures

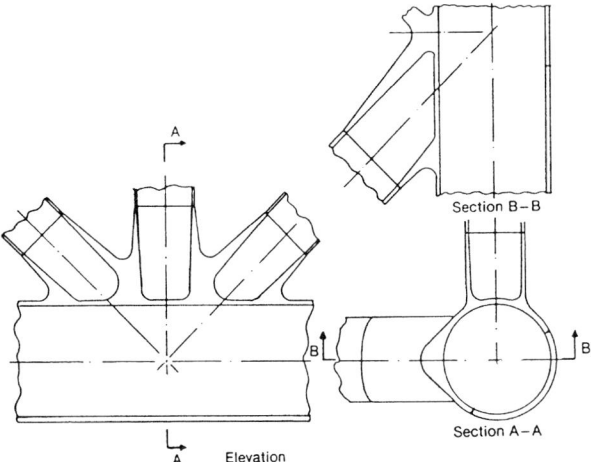

Fig. 8-54 KT type cast steel joint

weld. An abrupt discontinuity in thickness between the casting and the tube shall lead to undesired stress increase. It is recommended to place the weldment between the cast steel and rolled steel at a certain distance from the joint in a lower loaded zone, thereby also avoiding the secondary stresses in the joint due to welding (Fig. 8-55). It has to be attended to that the welds are simple and accessible.

As well as different casting materials, there are also many different casting methods. Their selection depends on the material type, the number of the castings to be manufactured, their weight and size, dimensional tolerances and surface finish. If it is decided to use castings in a structure, the casting company should be contacted at an early stage in order to discuss all these items extensively and make a reliable pre-estimation of the costs. The company should have sufficient experience and be in a position to give important related informations with the aid of solidification simulations to follow up the internal flow of forces in the castings by the corresponding forces through the incoming members and then to design the members in compression or tension adequately.

Fig. 8-55 Cast steel joints in the main arch and the abutment of the Humboldt-hafen-bridge in Berlin

Static and fatigue behaviour of welded steel-cast steel connections

Although in the last decades, quite a number of projects has been accomplished with the application of cast joints [44, 46, 48–53], there has been not enough information available regarding the investigations into the static and fatigue behaviour of the rolled steel-cast steel joints.

In the eighties offshore nodal joints in rolled steel and cast steel (material 8Mn7, Fig. 8-54) were tested under fatigue loading, where the composite cast steel joints delivered higher fatigue results than the joints in rolled steel (Fig. 8-56). The chemical composition and the mechanical properties of the cast steel 8Mn7 used are given in Table 8-5.

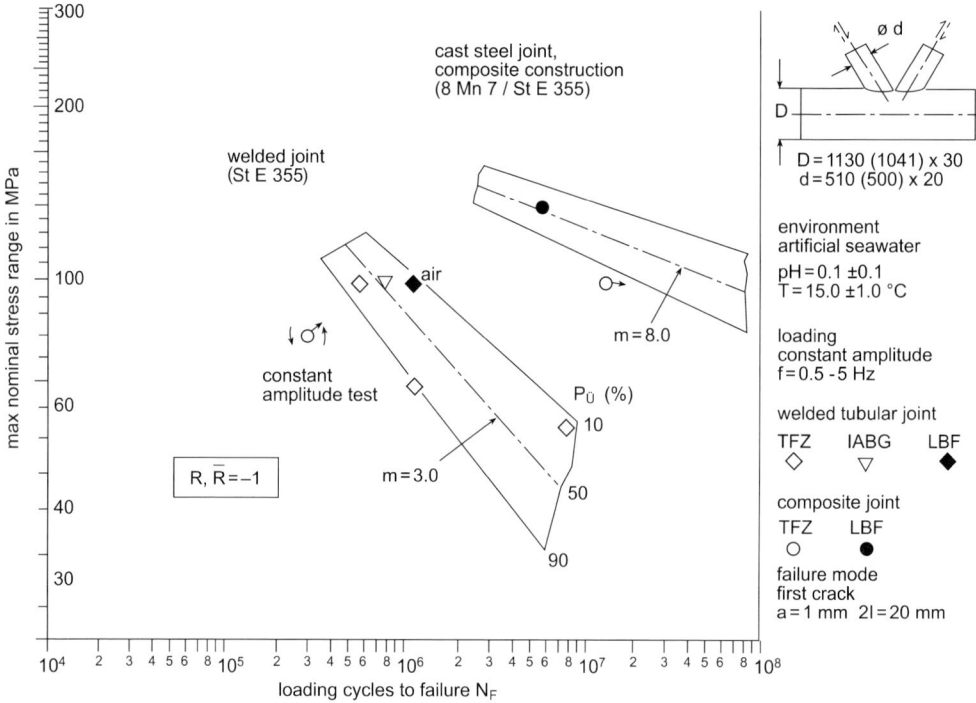

Fig. 8-56 Results of the fatigue tests on cast steel and rolled steel tubular joints for offshore constructions [56, 57]

Table 8-5 Chemical composition and mechanical properties of the cast steel 8Mn7

Chemical composition (mass in %)

C	Si ≤	Mn	P ≤	S ≤	Cr ≤	Mo	Ni	Others
0.06 to 0.10	0.60	1.50 to 1.80	0.020	0.015	0.20	–	–	Nb ≤ 0.05 V ≤ 0.10 N ≤ 0.02

Mechanical properties

Heat treatment	Wallthickness (mm)	Yield strength min. (N/mm^2)	Ultimate tensile strength (N/mm^2)	Elongation min. (%)	Impact energy (ISO-V) min. (Joule)
Quenched and tempered	≤ 60	350	500–600	22	80

8.9 Application of cast steel elements in hollow section structures

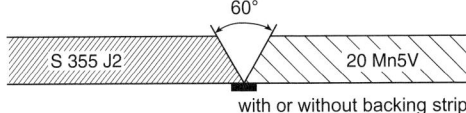

Fig. 8-57 Small plate specimen

Recently, the weldments between cast steel (20 Mn5V, see Table 8-3) and rolled steel (S 355 J2) were tested under static and fatigue loading more extensively to check the safety aspect of a project [44, 45, 47, 53] relevant to those joints:

- The static tests on small specimens of butt-welded connections with or without backing strips (40 to 50 mm wide and 3 to 5 mm thick) (see Fig. 8-57) under tensile load showed excellent load bearing strength; ultimate rupture took place at the location, where strong yielding was initially observed.

- The results of the fatigue tests on the above mentioned small specimens (Fig. 8-57) are shown in Fig. 8-58, where the lines (50% failure probability) are designated as follows:

  a) flawless butt-weld without backing strip
  b) flawless butt-weld with backing strip
  c) butt-weld with defects (lack of fusion up to 6 mm depth)

They lead to the following conclusions:

- Thicker welds ($t = 40$ mm) result a shorter life time than thinner welds ($t = 25$ mm).
- Butt-welds without backing strips show slightly higher fatigue strength than those with backing strip.
- Welds with defects lead to significantly shorter life corresponding to the expectation.

Table 8-6 contains a short evaluation of the fatigue results for the small specimen tests according to EC 3 [43].

- Further fatigue tests were carried out on large specimens (see Fig. 8-59) made of

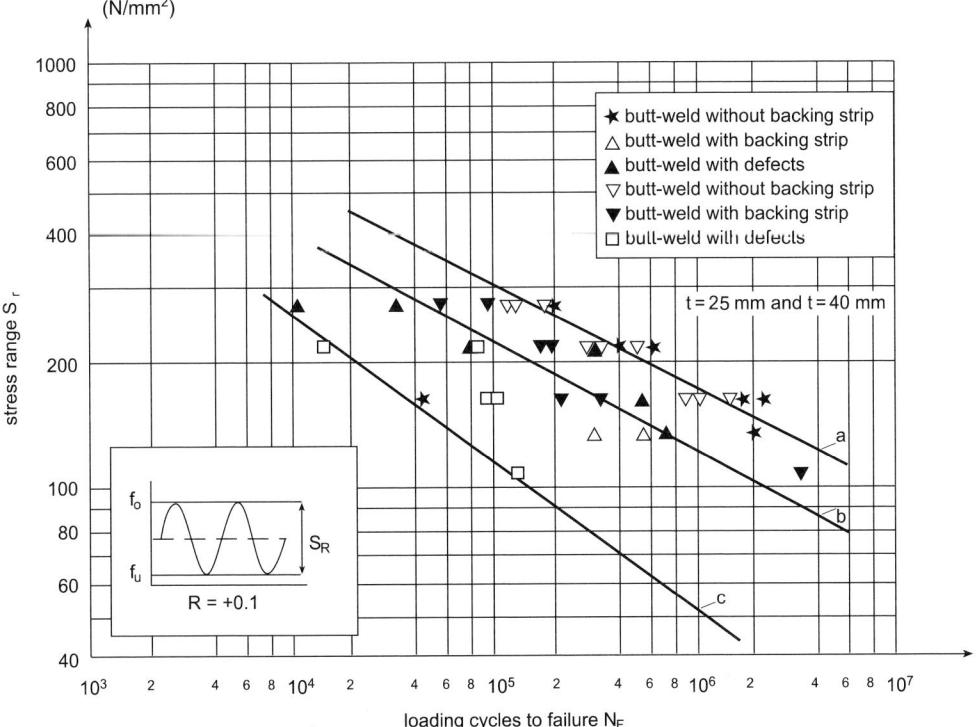

Fig. 8-58 Fatigue test results on small butt-welded plate specimens (20 Mn 5 V – S 355 J2)

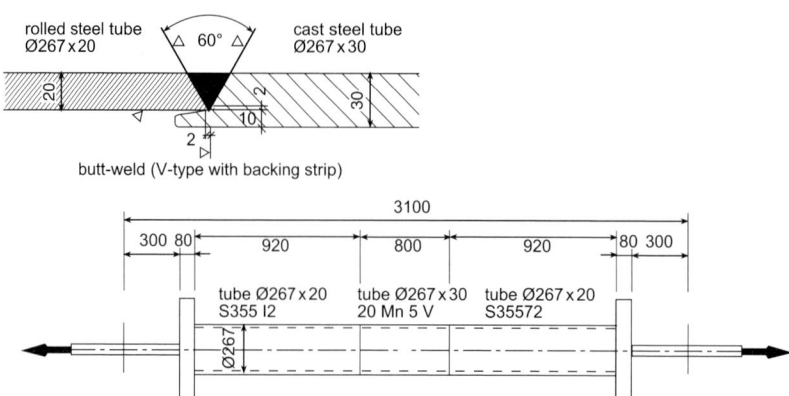

Fig. 8-59 Large tubular specimen

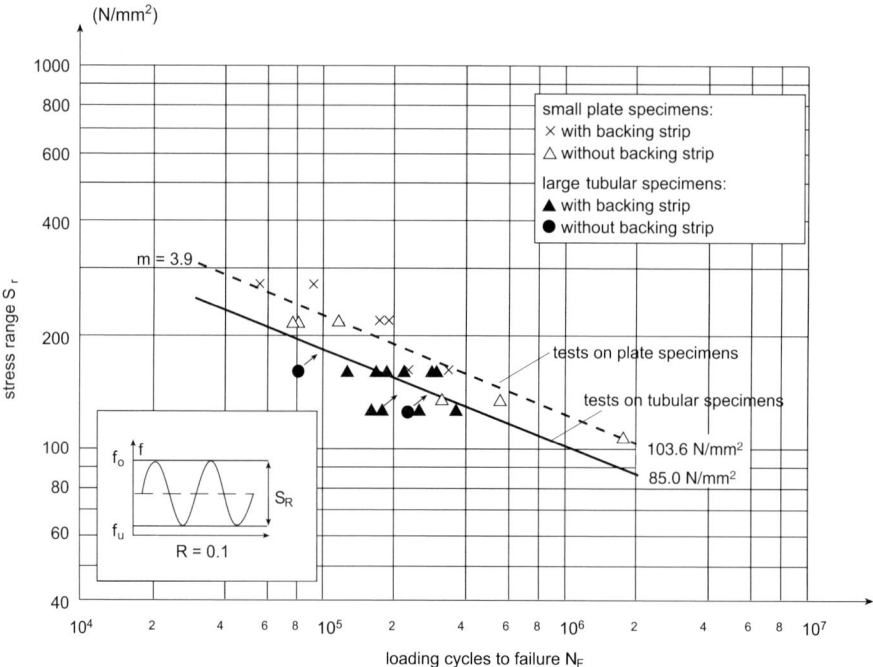

Fig. 8-60 Fatigue test results on butt-welded tubular and plate specimens (20 Mn 5 V – S 355 J2)

Table 8-6 Evaluation of the fatigue tests on the small specimens (Fig. 8-57)

Thickness of the plates (mm)	Stress range S_r (N/mm²) determined for $2 \cdot 10^6$ loading cycles			
	With backing strip, failure probability =		Without backing strip, failure probability =	
	50%	96.5%	50%	96.5%
25	112.7	87.5	146.1	123.5
40	96.9	78.8	154.0	135.1

tubes (267 mm outer diameter and 20 mm wall thickness) in cast steel 20 Mn5V and rolled steel S 355 J2. The diagrams in Fig. 8-60 show the results for the large specimens together with the small ones. They illustrate that the fatigue strength values for large tubular specimens lie in the lower scatter zone for those determined for smaller specimens.

8.10 Assembly

The key to a technically efficient and economical manufacture is the rationalised assembly, which is generally subdivided into the preassembly and the final assembly.

The decisive influence on the assembly of a steel structure is exercised by the techniques to fabricate joints (particularly welding or bolting), the sizes of the preassemblies and of the fully assembled units. Further, the following factors are of importance:

- Well-trained, skilled personnel
- Machine tools and lifting equipments or cranes in the workshop
- Working space available in the workshop
- Number of assembly units and subassemblies to be made
- Transport facilities and the distance between the storage and stacks for the structural members or components and the workshop and that between the workshop and the installation site

During the preassembly, all single constructional components are arranged on the jigs according to the given drawings and then tack-welded or bolted to one another. The assembly and rotation jigs facilitate the tack welding and then the final welding in a favourable working position. Welding in a difficult position has to be avoided. Time, labour and money can thus be saved.

The shrinkage and distortion, which evidently takes place during welding are preferably compensated for by jig constraining.

In the case of welded subassemblies, two working methods are used depending on the available jig types.

First method:

- The positions of the various components of the subassembly are marked off.
- As a first operation, the marked components are tack-welded to one another.
- The tack-welded subassemblies are transported to the welding shop, where the welders carry out the final welding in a pre-established sequence, which enables reduction of distortions.

Second method:

Preassembly by tack welding as well as the final welding are performed on the jig in the same workshop. Welders can perform preassembly and final assembly directly one after another. The jigs are so constructed that they can keep the shrinkages and deformations as small as possible and the structural tolerances can also be observed.

Various devices are described in the following:

- Assembly frame with cradles

Fig. 8-61 shows a sketch of the device. This consists of a main frame set at bench level and is provided with cradles (A and B), which hold the members of a lattice in their proper positions relative to one another. If the members are already fitted with attachment faces (flanges, gussets etc.), welded at a preassembly stage, the cradles can be replaced by corresponding plates or brackets (C) for bolting up.

- Marking-off slab

This procedure consists of simply marking a concrete floor in the workshop suitably. The area of application is preferably very large subassemblies or even complete assembly units of ordinary size. This device is also used, if the number of units to be assembled does not justify the manufacture of an assembly frame.

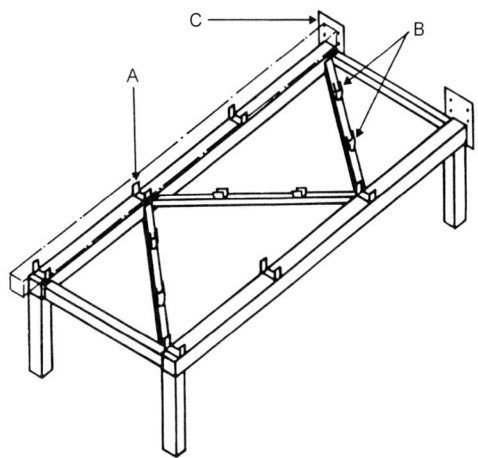

Fig. 8-61 Assembly frame with cradles

- Rotation frame

This device allows to clamp down the structural parts to be welded or already tack-welded preassemblies. It has a pivot line, about which the whole workpiece can be rotated. This offers the decisive advantage that the welding can be always done in the most favourable position.

The assembly of uni-planar and multi-planar trusses is as a rule conducted in a workshop, if the assembly units are relatively small and the transport facilities are available. In other cases, the trusses can be welded together on site or the subassemblies of the trusses are bolted together on site by means of flange connections (see Fig. 8-62).

In case welding is done on site, the assembly at a higher level of 4–5 m above the ground has proved to be practicable, as single structural members can be brought to their final positions with the aid of small mobile cranes or other lifting devices and there is enough free space under the construction for their easy movement. With this procedure, it is feasible to carry out the assembly and the erection simultaneously. Thereby, the panel nodes in the lower chords are placed on column heads and adjusted. Subassemblies of a truss are made by welding the other components (diagonals, upper chords) stepwise to the lower chords, which are then connected to one another by butt welds.

More frequently, subassembly units are welded in the workshop, transported to the site and then joined to one another on the ground through bolted flanges. Then they are hoisted by a mobile crane, set on the column heads and adjusted. Large units can only be handled if powerful cranes are available. The application of the structures in hollow sections is favoured for their relative lightness, extreme stiffness and low wind resistance (especially for CHS constructions).

Fig. 8-63 Assembly on site and erection of a lattice girder

Fig. 8-62 Triangular girder assembled on ground

Fig. 8-64 Quadrangular lattice girder during the erection on site

For site welding, special attention has to be given to the weather conditions e.g. wind, rain, ambient temperature and humidity. Gas metal arc welding (GMAW) requires protection against the wind especially. Recently, the semi-automatic self-shielded arc welding, also with robots, is used increasingly; thus the welding problems due to adverse weather conditions can be avoided.

As the welding power source and the welding location are often far from each other on site, a remote control function is required enabling the welder to adjust the welding condition at hand.

8.11 Transport of hollow sections and their structures

In order to lower the total budget for a structural project, it is important to take care that the subassembly or assembly units coming out of the workshops should be selected as large as possible without making the transport costs too high. The means of transport (road, rail, river, canal or sea) have a significant effect on the choice of a structure, the size of the units and thus upon the number and locations of the (butt) joints to be made on site.

A special advantage of RHS regarding their transport is given by their ability for favourable stacking. Further, the mechanical properties of CHS and RHS e.g. high buckling and torsional stiffness and multi-axial bending resistance are also very favourable for the transportation of the single elements as well as the large fabricated assemblies.

- Road transport (see Fig. 8-65)

This transport facility is the most convenient and the most frequently used, when the job site is relatively near the workshop. The maximum heights, widths and lengths of the transport objects authorised on the road by the national standards vary from country to country. Further, care must be taken regarding local restrictions imposed by the height limits for the bridges, direct accessibility to the job site etc.

Configuration and construction of certain types of subassemblies, such as triangular sections exercise a positive influence on the transport behaviour; they can be stowed quite economically on a truck.

- Rail transport

This type is the most preferred (often also the cheapest) transport means, when the workshop and the job site are directly linked by a railway line. The railway authorities also prescribe size and weight limits, which may be different in different countries.

- Water transport

This is a very economical means of transport and often offers the facility to transport exceptionally large assembly units. Ships and sometimes cargo boats (Fig. 8-66) are used for overseas locations and also when the site and the workshop are on – or can be brought

Fig. 8-65 Road transport of a triangular assembly unit

Fig. 8-66 Transport of subassemblies for offshore platform by a cargo boat

Fig. 8-67 Offshore structure floated to the location of installation by a tug

Fig. 8-68 Mobile crane erecting a pipe line bridge

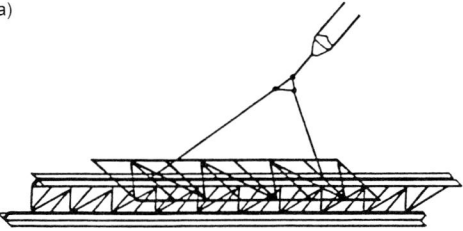

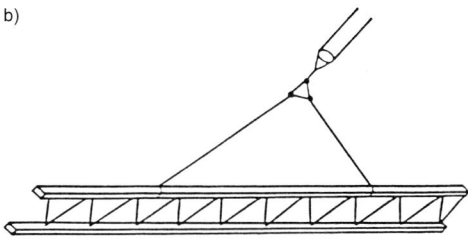

Fig. 8-70 Hoisting a lattice girder with a crane.
a) Girder in open sections are laterally stiffened by an auxiliary construction, b) girder in hollow sections without any auxiliary construction

Fig. 8-69 Mobile crane erecting an electric mast

to – a major waterway. Structures (particularly in hollow sections) can float themselves in some cases e.g. offshore platforms, where they are towed away to the locations of installation or erection (Fig. 8-67).

- Cranes as aid to assembly and transport (Figs. 8-68 and 8-69)

Mobile cranes are often used to lift and transport hollow section constructions because of their light weight. However, fixed cranes come also to application on site.

Due to the high torsional rigidity of hollow sections and consequently their reserve resistance against torque and lateral-torsional buckling, lifting and handling operations of hollow section assemblies are greatly facilitated. The lateral rigidity of structures in open rolled sections must be secured by means of auxiliary constructions while they are hoisted and transported (see Fig. 8-70a). These precautionary measures are not required for hollow section constructions (Fig. 8-70b).

8.12 List of symbols

CHS circular hollow section
RHS rectangular (incl. square) hollow section

9 Space structures

9.1 General

Grid frameworks are typical examples of interconnected systems made by the combination of single bars, which meet at nodal points exercising the statical load bearing function. Space structures are by definition, "spatial" or "tridimensional", whose constituent elements, generally as a lattice, can be oriented in the varying directions of the tridimensional space they generate. A chief property of all space structures is the uniformity of their elements or parts. The design concept is that of identical interlinked components having the same inertia and ensuring a uniform distribution of loads which are uniformly transferred along the members towards the support areas. The resulting system of intersecting girders pointing in several directions determine intercrossing lattice planes, vertical or inclined, wherein the connections are made between the upper and lower chords of these girders.

The efficiency of any grid structure is demonstrated by its ability to distribute the applied loads as widely as possible, where the loads are shared among many members due to their interconnections. This decreases the high stresses at the location of direct load transmission, peak stresses are avoided and a fairly even stress distribution can be achieved over the whole framework. The stress distribution in a grid varies according to the type of bracing, location of supports and type of applied loading.

The type of loading depends also on the external form of a load bearing structure:

– Bending stress is initiated in a plane plate (Fig. 9-1 a).
– Bending stress does not occur in a arched plate or shell, where only axial and shear stresses or membrane stresses are initiated (Fig. 9-1 b).

In order to be loaded by bending moment, space frames with two or more layers have to be used (Fig. 9-1 c). A frame with a single layer is adequate for membrane action (Fig. 9-1 d). Whilst single layer space structures consist essentially of reticulated shell e.g. barrel vault (see Fig. 9-2), spherical dome, hyperbolic-paraboloid shell, twin layer ones can be simulated to reticulated plate, i.e. three-dimensional framework produced by joining up units or modules having identical lattices (Fig. 9-3).

The geometry of the space frame can be a uni-planar plate (Fig. 9-1 e) or a shell with single or double curvature (Fig. 9-1 f) or a foldwork consisting of plane plates (Fig. 9-1 g).

Depending on the number of the chord directions in the parallel layers, the meshes are built up according to Fig. 9-1 h–k. If the chord members run in two directions, one refers to a two-way frame. It is also possible to design a three or more way frame. In systems with a number of layers, the meshes in different planes can be fully congruent or be constructed offset to one another.

Fig. 9-4 shows a few elementary structural modules, which can be combined to design a space frame. The stability of the instable modules can be achieved, if necessary, by means of stiffening diagonals in the plane rectangular fields at the edges of the frame or by stabilizing the support.

In the buildings, the use of plate space frame with rectangular or specially square plan is predominant, where the supports are placed at the corners or on the edges in general.

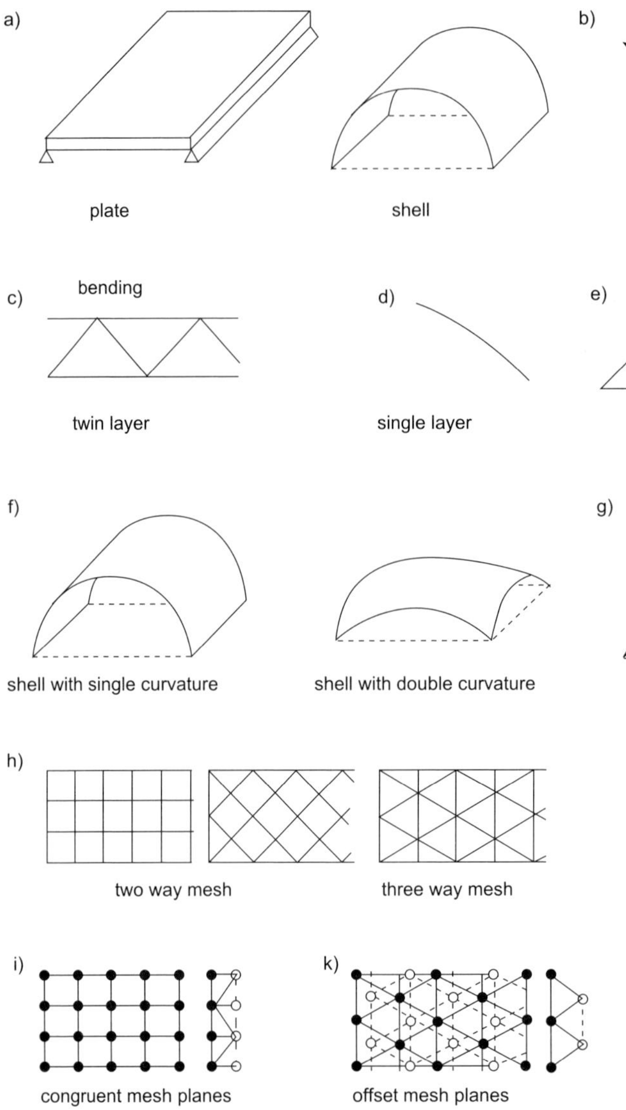

a) plate	shell	b) membrane
c) bending — twin layer	d) single layer	e) plate
f) shell with single curvature	shell with double curvature	g) foldwork
h) two way mesh	three way mesh	
i) congruent mesh planes	k) offset mesh planes	

Fig. 9-1 An overview of space structures

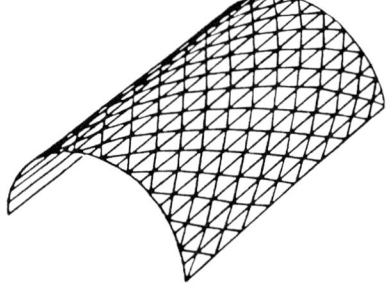

Fig. 9-2 Single layer space frame – barrel vault

The external loads e.g. dead weight of the roof construction are generally transmitted through the nodal joints (Fig. 9-9).

9.2 Constructional elements of space structures

The members of a space structure are often in an isotropic state as regards buckling and bearing loads, which are either in axial tension or compression. Due to the excellent buckling and torsional behaviour of hollow sections,

9.2 Constructional elements of space structures

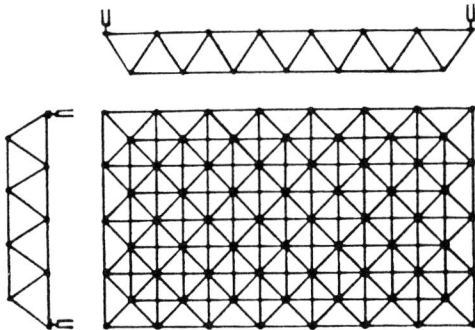

Fig. 9-3 Twin layer, three dimensional frame featuring an upper layer, a lower layer and an intermediary lattice

their higher permissible load than for open sections and the reduced surface area meaning less paint and maintenance work as well as their aesthetic appeal, optimisation leads naturally to choosing hollow sections. They are mostly circular, in some cases also square or rectangular. It is also possible to use open sections as bars, which is however seldom.

Another characteristic element of all space structures is the nodal joint or connector. In most systems, the convergent members are joined through the nodal joint to one another by means of bolting (see Fig. 9-5) and welding (Fig. 8-50), seldom by screw fastening. The total joint, i.e. the manufacture and the fabrication of the nodes and the member ends, exercises significant influence on the economy of the space structures.

Using the example of the "Okta-S" system (Fig. 9-6), the construction of the components of a space structure in general has been described below:

– The hollow steel sphere consists of two hot pressed half shells and an internal diaphragm plate, all welded together.
– Using appropriate jigs, the flanged stubs are then welded to the steel spheres after the sleeved nuts have been slipped in.
– Screwed trunnions are welded on both sides of CHS members to be screwed with sleeved nuts.
– During the assembly, the sleeved nuts are screwed on to the trunnions till a tight junction between the sleeved nut and the shoulder of the flanged stub as well as between the trunnion and the face of the flanged stub takes place. The sleeved nuts are then tightened with a preset torque thus ensuring a specific prestressed junction between trunnions and stubs.

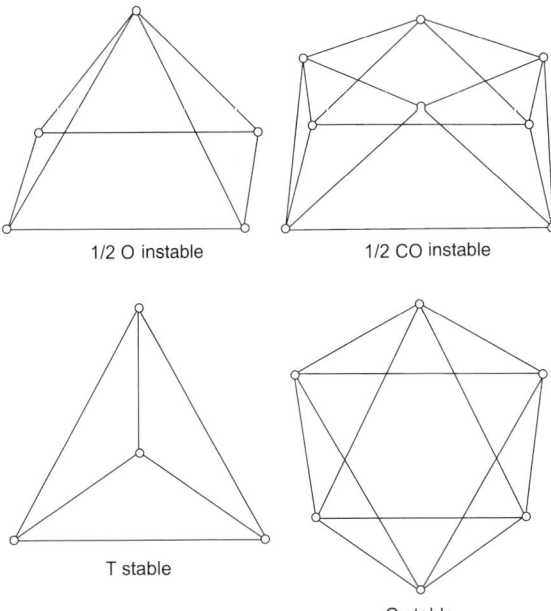

Fig. 9-4 A number of elementary modules
O = octahedron, CO = cube and octahedron, T = tetrahedron

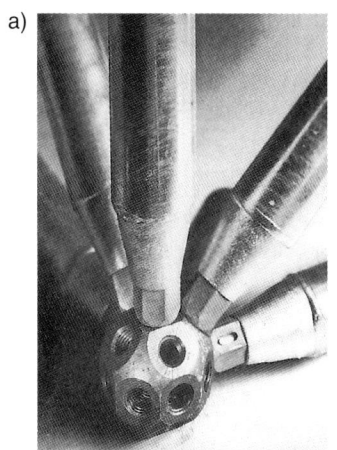

Fig. 9-5 A number of nodal connectors for space frames. a) Mero, b) Nodus, c) Okta-S

9.3 Assembly and erection of space frames

In regular space frames of medium or large dimensions, the number of single elements – open profiles or hollow sections and nodal joints – of identical sizes can be so large that it is worthwhile to standardize them. This offers the feasibility for an economical manufacture.

The assembly of space structures can be done following one of the methods below:

1. Industrially prefabricated subassemblies welded and/or bolted together in the workshop are transported to the site by lorries, railways or boats/ships (whichever is cheapest and most convenient) and then fi-

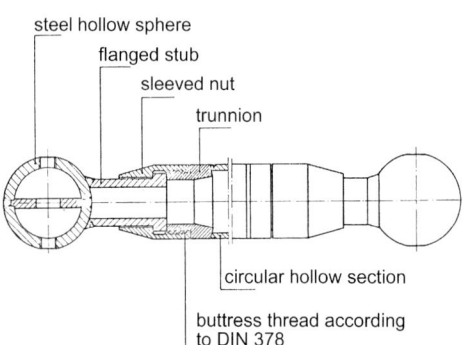

Fig. 9-6 Nodal connector "Okta-S"

nally assembled (site welding or preferably bolting) on the ground there. The whole operation is obviously governed by transport limitations.

2. All single components are prefabricated in the workshop and then transported to the job site by road, waterway or railway. Transport in this case is very economical, as the space in the vehicles can be used to nearly full capacity. The components are then fully assembled on the ground by joining them preferably with bolting (Fig. 9-7) (site welding should be avoided, if possible, as it is more expensive and also requires precautionary measures).

In both methods, the assembly works on site are carried out without scaffolding. They are normally assembled on the ground in units as large as the available space and the lifting equipment will allow.

The assembly on the ground offers the following advantages:

– It is possible to engage relatively large number of workers to finish the job quickly in order to keep the site occupation as short as possible.

– The supervision is more simple, it is sufficient to have one skilled person or a small number of experts for a larger project for the supervisory job.

As regards prefabrication of the single elements or the subassemblies in the workshop, high precision is a fundamental aspect, as the observance of precise tolerances in the manufacture of the single components or of the subassemblies (all units are built in special jigs) makes the individual pieces easy to install. This is clear that the joining of a large number of similar pieces may accumulate the manufacturing tolerances leading to unacceptable dimensional and angular deviations for a whole structure.

Erection can be performed very quickly after the columns or other supporting devices are installed. In usual cases, the assembly units are hoisted from the ground by means of mobile cranes and set upon the supports (Fig. 9-8).

Another special method of assembly and erection consists of carrying out the operation with the components one after another at the final height level. The advantage of this procedure lies in the aspect that this does not require high strength lifting devices or supporting constructions.

Fig. 9-7 Space frame "Okta-S" is assembled on the ground at the site

Fig. 9-8 Erection of a space frame

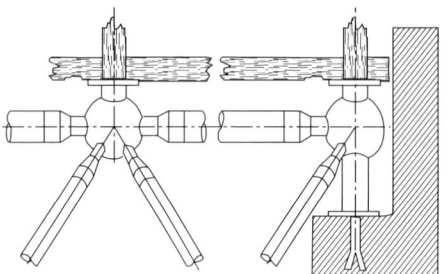
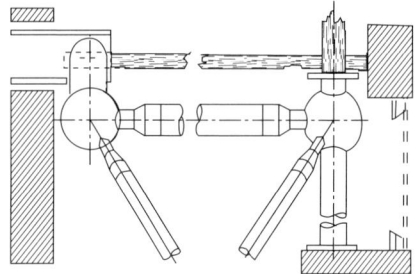

Fig. 9-9 Transmission of the roof weight through the nodes

9.4 Calculation of space structures

Considering the space structure as a planar construction, it represents an ideal case for the finite element analysis, as each of its members is in fact a finite element. So the calculation can be made using this method, although the work involved to calculate the action and resistance in the members may be time and labour consuming and consequently expensive. However, computer programs are available to carry out this job, which make it possible to determine the forces in all members on the basis of linear statics and elastic behaviour of materials. These assumptions apply to a structure with predominantly axially loaded members. The solution is achieved based on the displacement method. The members are usually considered as being hinged at both ends. For some joint systems, it is, on the other hand, more realistic to consider that there is at least a partial fixity for members located in the upper or lower chord layers [12]. As it is necessary to supply, for the related computer programs, all the mechanical properties of all the members (in order to derive the rigidity matrix for the members required for the finite element calculation), a preliminary calculation may well be necessary.

The methods for the approximate calculations of the space structures with hollow section members in analogy to the plate theory have been described in [2].

The cross sections of the members are calculated in accordance with the member forces.

Basing on an economical serial manufacture, it must be considered how far the required cross sections can be standardised to limit the number of member sizes.

It is not necessary to have the knowledge of all the member forces in a space structure already in the phase of design conception. Mostly it is adequate to determine the maximum member forces in a number of defined zones. They deliver the solutions for the theories of plate or shells valid for the corresponding structural system.

For the pre-calculation, the operation can be simplified by replacing the space structure by an equivalent continuous medium, such as a plate or possibly a shell. This "membrane analogy" has been described in [3] and a number of pre-calculation examples are given there.

9.5 Economic optimization of space structures

A study on the economic optimization of the space structures related to the plan sizes, the supports and the loadings has been published in [4]. The open parameters are the structure type, the raster size and the depth of the structure. Table 9-1 contains the list of investigated parameters. Space structure plates under consideration are shown in Table 9-2. Fig. 9-10 illustrates one example of the results in graphical form. Design aids are presented in these diagrams in order to be able to compare different structures without doing extensive design calculations. Final cost estimation can be made considering the applied structural system and the individual cost structure of a structural engineering firm.

9.6 Further remarks to the design of space structures

- Space structures are statically indeterminate systems and buckling of a compression member under heavy concentrated load does not lead to collapse of the whole framework. This has further the advantage that the load transmission does not take place with a jerk but occurs continuously. Local peak loads can be distributed among the neighbouring members due to the plastic behaviour of steel. A construction does not collapse even when one or more members fail.

- The depth of a tri-dimensional structure is much smaller than that of independent parallel girders of equal span with the same deflection limit (about 1/30 of the span for

Table 9-1 List of the parameters for the economic optimization of plate space structures in [4]

Parameter designation	Parameter symbol	Parameter valence	Parameter evaluation			
Plan view	G	3 in general	In the range of possible spans			
Support	L	2	Edge and corner support			
Loading	Q	3 in general	[kN/m²]	1.35	2.50	3.50
Load transmission	E	1	Nodal load (no transversal load on the members)			
Structure type	S	4 + (1)	R3A, R3B, (R3C), R4B, R75 (see Tabel 9-2)			
Raster value	R	3–4	2.0 m/3.0 m/4.0 m/5.0 m			
Structure depth	H	1	(usual depth)			
Member cross sections	T	2	Tubes and ⌐⌐ angles in S235			
Joint cost	A	1	Represented by modified total mass in diagram			
Structural system	B	1				
Span ratio	Ly/Lx	2 in general	Square 1:1 Rectangular 1:1.7			

Table 9-2 List of the investigated plate space structure types according to Table 9-1 [4]

	Index	Classification	Symbol	Chord orientation
Structure type	R3A	Half octahedron and tetrahedron package	1/2 O + T	Upper and lower chord parallel to the edges
	R3B			Upper and lower chord not parallel to the edges
	R4A	Half octahedron and half cube package	1/2 O + 1/2 CO	Upper chord parallel to the edges and lower chord not parallel to the edges
	R75	Octahedron and tetrahedron package	O + T	Upper and lower chord not parallel to the edges

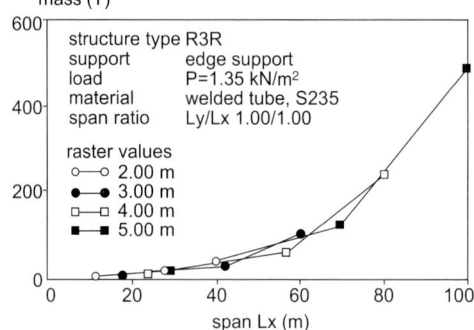

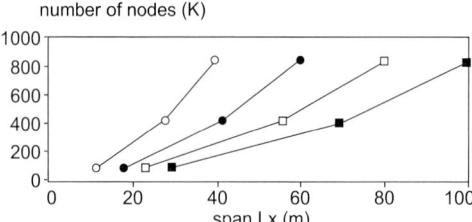

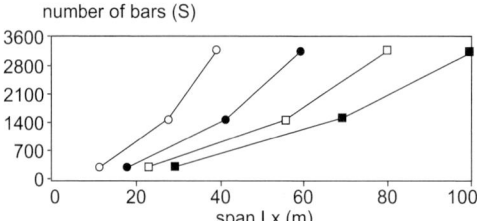

Fig. 9-10 Example of an economical optimization of a plate space structure

a parallel to side layout and about 1/40 of the span for a diagonal layout of the space frame compared to 1/10 to 1/15 of the span of the uni-planar girder, as the deflection is reduced significantly by the large rigidity of the space structure. The free space between the roof and the lower chord can thereby be smaller.

- Space structures are exceptionally light per unit area. This may sometimes be a decisive quality for the selection of a space structure.

- In most cases, it is advantageous to select a mesh or module as large as possible in order to reduce the labour and fabrication costs.

- Space frames are often applied to roof over areas with large spans without intermediary supports. The larger the span, the more justified is the adoption of a space structure. This can often be the case for sport stadia, swimming halls, exhibition centres, churches, theatres, airport hangars, warehouses, factories etc. Further, they can also be used as erecting scaffolding in bridge constructions.

- The plan view of a building is a further criterion for the selection of a space structure. A circular arrangement can result in the roof construction of i.e. a spherical dome, including also pentagonal, hexagonal or octagonal forms. Here it is justified to use a three-dimensional frame with meshes in three directions in one layer or mostly two layers of planes.

Very economic solutions for flat space structures can be obtained, when nearly equal spans exist in all directions (quadrangular, pentagonal, hexagonal etc.), as the applied loads are shared by the structural components more uniformly in this case.

- Space structures offer the architects a great diversity of possibilities for design. In many cases, aesthetics plays a decisive role. The space frame can be included in the visible part of the construction without any problem and thereby an additional ceiling to conceal it can be dispensed with. These constructions are specially preferred for cultural and sports buildings (see Fig. 1-21).

- Specially in double-layer grid structures, the space between the top and bottom layers can be used for location of mechanical and electrical services, such as heating, cooling and ventilation with convenient access for maintenance.

10 End fixity of rectangular hollow section columns in a concrete fundament

Steel columns have often to be bedded in concrete as elements fully resistant to bending. This is in particular the case for single storeyed structures such as railway stations or other roof constructions and masts for various purposes. Besides the standard application of plates, anchoring bolts or bars, the direct embedding of columns in concrete presents a very economical solution. CIDECT looked at the hollow section aspect of this application and sponsored a research programme to clarify the fixity behaviour of RHS columns embedded in concrete. The testing centre for steel, timber and stone of the University of Karlsruhe in Germany carried out the research project; the report is given in [5].

Regarding the end fixity of columns in concrete and its calculation, a number of other publications are also available [1–4].

Figs. 10-1 to 10-3 illustrate the assumptions related to the pressure distribution p between the column and the concrete. While Lit. [1, 2] use the linearly varying pressure according to Fig. 10-1, Fig. 10-2 illustrates another assumption, where the reaction forces are applied in a concentrated form at two locations. In [3], the depth of fixity has been determined as a function for the pressure depending on the "embedment modulus". It has been demonstrated in [4] that the maximum edge pressures do not change beyond a certain depth of fixity.

The publications [1–4] do not take the particular behaviour of RHS into consideration. This was made up by the investigations reported in [5].

The results can be summarized as follows:

1. The reference cross section of end fixity is that at the upper edge of the concrete funda-

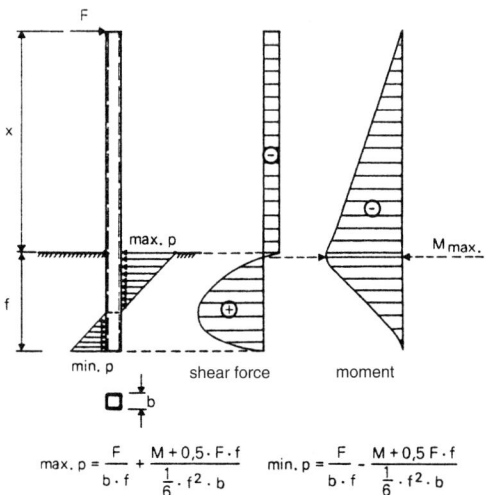

Fig. 10-1 Pressure distribution between the column and the concrete, assumption: linearly varying pressure

$$\text{max. } p = \frac{F}{b \cdot f} + \frac{M + 0{,}5 \cdot F \cdot f}{\frac{1}{6} \cdot f^2 \cdot b} \qquad \text{min. } p = \frac{F}{b \cdot f} - \frac{M + 0{,}5 \, F \cdot f}{\frac{1}{6} \cdot f^2 \cdot b}$$

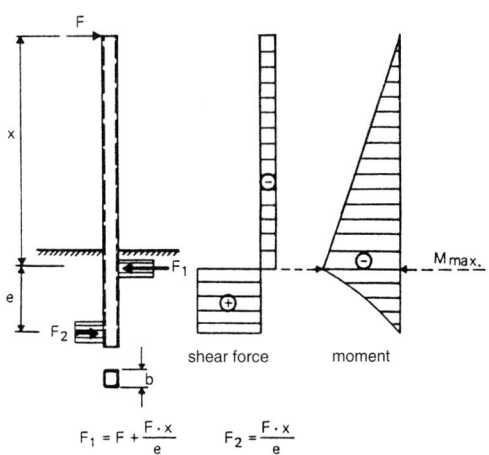

Fig. 10-2 Pressure distribution between the column and the concrete, assumption: reaction forces applied in a concentrated form

$$F_1 = F + \frac{F \cdot x}{e} \qquad F_2 = \frac{F \cdot x}{e}$$

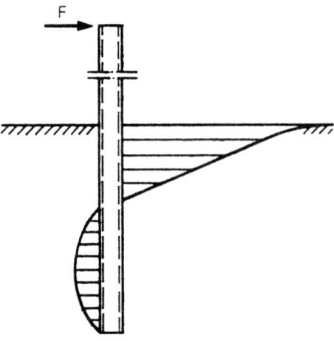

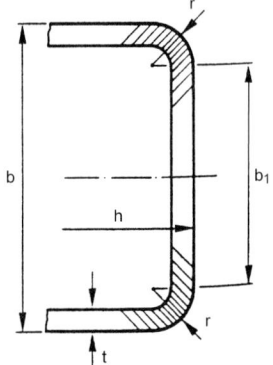

Fig. 10-3 Pressure distribution between the column and the concrete, assumption: depth of fixity as a function of the pressure dependent on "embedment modulus"

ment. Bending moment, shear and axial force on the hollow section have to be determined at this location.

The concrete enclosure can be taken as a local stiffening of the RHS column. This is expressed among others by the shift of the cross section of failure beyond the concrete enclosure.

Adequate strength and rigidity of the enclosing concrete and the steel inserts are required. Guidelines for the calculation are described in the cipher 4.

2. The deflections occurring in the statically determined tests (cantilever arm) are 20–30% higher than the calculated ones depending on the hollow section sizes. For the profiles with $b/h > 1$, the deviations were measured up to 50%. This means, depending on the type of end fixity the rigidity for the static indeterminate systems can be 20–30% lower (in case $b/h > 1$, up to 50% lower) than that for the full end fixity. For CHS and RHS with large corner radii, the decrease in rigidity can be less, as the distribution of pressure concrete/steel is more uniform.

3. Following conditions have to be fulfilled in order to avoid the local buckling of the hollow sections:

S 235: $b/t \leq 43$ or $b_1/t \leq 39$
S 355: $b/t \leq 36$ or $b_1/t \leq 32$

$b_1 = b - 2r$ where r = corner radius of RHS

4. According to [1, 3] the depth of fixity f (see Fig. 10-4) can be determined with the aid of the allowable concrete compression.

$$f \geq \frac{2H}{b \cdot \beta_R} \cdot \left(1 + \sqrt{1 + \frac{3M \cdot b \cdot \beta_R}{2H^2}}\right) \quad (10\text{-}1)$$

where:
H = the resultant horizontal force
M = the resultant moment
β_R = the calculation strength of concrete

According to [3], for common hollow sections and concrete strengths, the depth of fixity f shall be within the following range:

$1.5\,b \leq f \leq 2.5\,b$

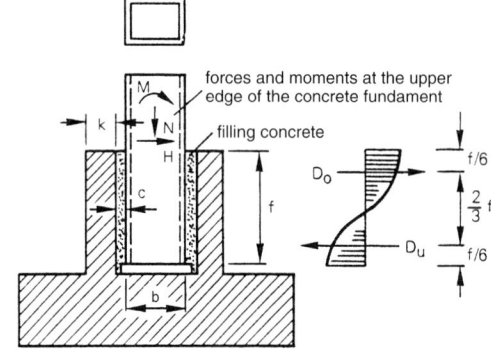

Fig. 10-4 Depth of fixity and the forces acting in concrete

10 End fixity of rectangular hollow section columns in a concrete fundament

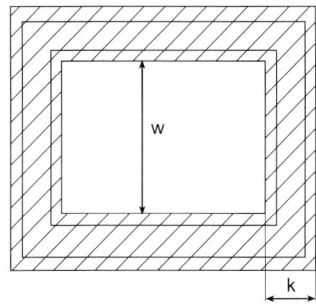

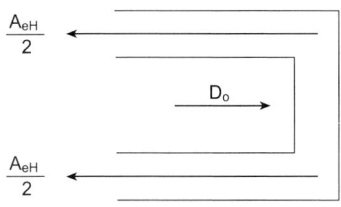

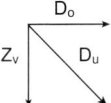

Fig. 10-5 Forces in concrete with reinforcement

$$D_o = \frac{3}{2}\frac{M}{f} + \frac{5}{4}H \qquad (10\text{-}2)$$

$$D_u = \frac{3}{2}\frac{M}{f} + \frac{1}{4}H \qquad (10\text{-}3)$$

Hereby, the friction between steel and concrete is neglected.

The maximum compressive stress $f_{B,\max}$ in concrete are determined as follows:

$$f_{B,\max} = \frac{4D_o}{b \cdot f} \qquad (10\text{-}4)$$

The thickness of the concrete filling c fits in with the vibrator type. The filling concrete must have the same grade as the concrete sleeve and can be compressed perfectly with a vibrator.

The force D_o is transmitted to the longitudinal walls through a horizontal ring reinforcement in the upper fundament zone. The reinforcement is to be dimensioned for each $D_o/2$.

The force D_u is transmitted to the fundament plate without any reinforcement.

The thickness k of the concrete sleeve should be $> 1/3\ w$, but equal to 10 cm at the least (w = smaller hole width, see Fig. 10-5).

The calculation of the concrete fundament can be made based on the following forces (see Fig. 10-4):

The longitudinal wall in the direction of D_o acts as the consoles fixed in the fundament, which transfers the force D_o with the triangle of forces Z_v and D_u in the fundament plate (see Fig. 10-5). The tensile force Z_v is reacted by the stirrups.

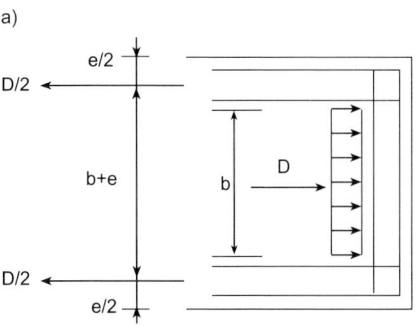

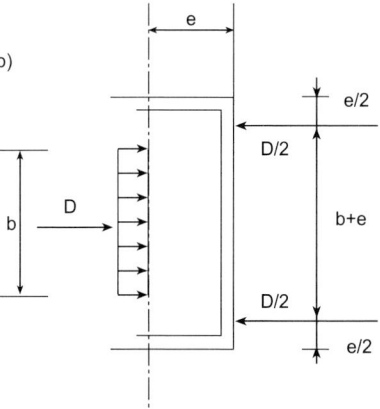

Fig. 10-6 Forces in the horizontal reinforcement. a) Longitudinal looped bars, b) transversal looped bars

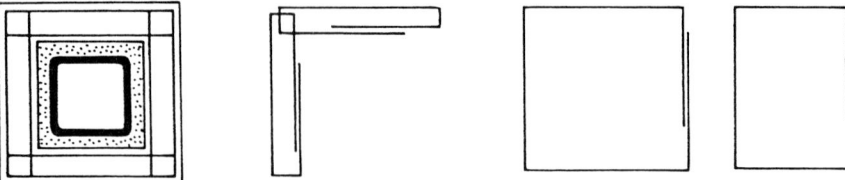

Fig. 10-7 Arrangement of horizontal reinforcement bars

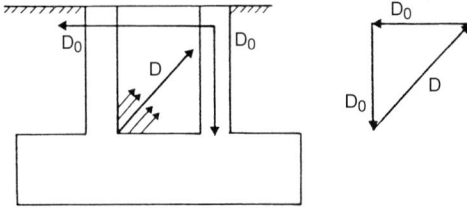

Fig. 10-8 Forces in the vertical reinforcement bar system

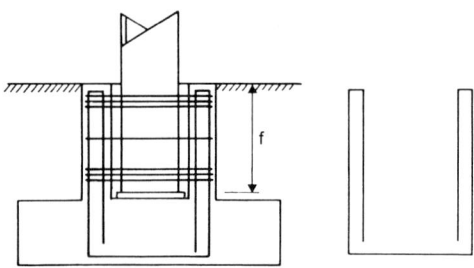

Fig. 10-9 Arrangement of vertical reinforcement bar system

The fundament plate is to be dimensioned to withstand the bending moment in the cross section at the concrete foot (Fig. 10-4). For smaller thickness of the fundament base, it is necessary to check the strength of the base against penetration by the faces of the sleeve.

For $\dfrac{M}{N \cdot b} > 0{,}15$, internal and external ring reinforcements are to be arranged in the longitudinal and transverse walls.

For $\dfrac{M}{N \cdot b} \leq 0{,}15$ and smaller dimensions, the closed rings in the outside zone of the walls shall be sufficient.

Fig. 10-6 shows the forces in the horizontal reinforcement bars; their arrangement is given in Fig. 10-7.

The forces in the vertical reinforcement bars and their arrangements are illustrated in Figs. 10-8 and 10-9 respectively.

List of symbols

CHS   circular hollow section
RHS   rectangular hollow section
CIDECT Comité International pour le Développement et l'Etude de la Construction Tubulaire (International Committee for the Development and Study of Tubular Structures)

11 Hollow sections in composite construction

11.1 Hollow section composite columns

Composite column in steel hollow section filled up with concrete is a structural element, the application of which has increased substantially in the recent time. However, already in the fifties, a number of electrical masts was installed, where the hollow angular joints were filled up with concrete in order to increase the allowable compressive strength. This led to the first investigations into the strength behaviour of concrete filled tubes [1].

Hollow section composite columns offer the architects the possibility to let a steel construction remain visible irrespective of fire protection requirements and make the optical design accordingly. This aspect will be discussed in detail in Chapter 13.

The characteristic features of hollow composite columns are described in the following:

- Larger strength and rigidity by filling with concrete, higher loads with slender columns, smaller external dimensions and reduction of the effective floor space occupied by the columns (Fig. 11-1) in a building.
- The structure can be visible and transparent. This allows a pretentious architectural design with various colourings.
- Standard connections for conventional steel structures can also be applied here. There is seldom any problem due to the highly developed joint technique of today.
- Prefabrication in the workshop and a short time for dry assembly and erection are possible.
- Hollow section is the casing and the reinforcement of the concrete at the same time.
- With corresponding percentage of reinforcements, concrete-filled hollow section columns can reach more than 90 minutes fire resistance time, that means a fire resistance class of R90. Further external fire insulation is not necessary.
- Special equipments are not required for the concrete-filling of hollow sections and can be included in other usual concreting jobs.
- The hardening of concrete is adequately short, so that the building progress is not prevented.
- Concrete-filling protects a column against mechanical damage.

Fig. 11-2 shows the possible cross sections of composite columns in hollow sections filled up with concrete. CHS and RHS with or with-

Fig. 11-1 Concrete-filled CHS in a multi-storey building

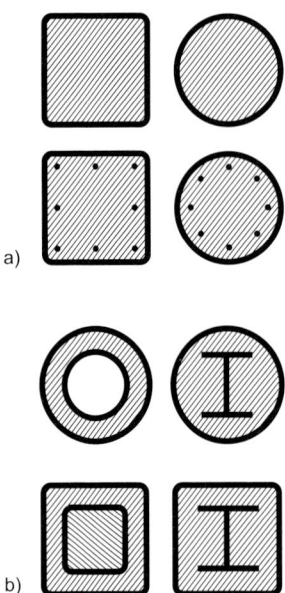

Fig. 11-2 Hollow section composite columns.
a) Regular cross sections, b) special cross sections

out additional reinforcements are the regular forms (Fig. 11-2 a). A number of special designs is illustrated in Fig. 11-2 b. RHS is also used, when subjected to bending moment; but the increase of the load bearing capacity is small, so that high manufacturing cost can be hardly justified. Therefore, the main field of application involves predominantly concentrically or light eccentrically loaded columns. These are, as for example, hinged columns (also continuous through columns) in multi-storey buildings and highly loaded columns in traffic tunnels. The concreting is particularly profitable for thin walled hollow sections of S 235 with high grade concrete e.g. C 45.

Investigations into the strength behaviour of the concrete-filled hollow section columns were carried out in Belgium, Germany, United Kingdom, Japan and the U.S.A. The results have been used to formulate various national and international standards [2–5].

In Europe, three calculation methods have been developed for composite columns, which make the basis for the European standard Eurocode 4 [5]:

– In Belgium and France, the method proposed by Guiaux and Janss [6–10]
– In United Kingdom, the method proposed by Dowling and Virdi [11–16]
– In Germany, the method proposed by Roik, Bode and Bergmann [17–22]

11.1.1 Design strength (resistance) of hollow section composite columns

11.1.1.1 General

An adequate load bearing safety is assured if the occurring axial, shear and bending stresses under design loading (= γ times serviceability loads) at any location of the construction taking also the deformations into consideration (theory of the second order) are smaller than or equal to those for the ultimate limit state. A stable equilibrium must exist in the total system.

An exact calculation of the ultimate load of a composite column is very extensive. If an adequate computer program is a available [22], it is possible to derive the design curves as shown in Fig. 11-3. Such design methods take account of the material and geometrical imperfections, the non-linear material behaviour and the effect of the deformation on the equilibrium of forces (theory of the second order).

Instead of calculating exactly, the approximation methods described in the following Sections can also be used.

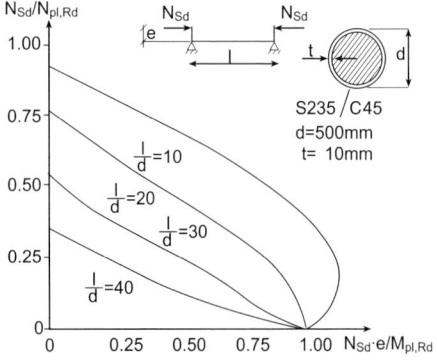

Fig. 11-3 Design (interaction) curves based on the computer calculations

11.1.1.2 Methodically concentric axial compression

For concrete-filled hollow section columns under concentric compressive load, the following simplified calculation method was developed by Roik, Bergmann and Bode based on numerous comparison calculations, which were confirmed by the tests [4, 21, 24, 25]. The procedure has also been recommended by Eurocode 4 [5]:

$$N_{Sd} \leq N_K \qquad (11\text{-}1\text{a})$$

with

$$N_K = \chi \cdot N_{pl,Rd} \qquad (11\text{-}1\text{b})$$

where:
N_{Sd} = design compressive load (= γ times serviceability load)
N_K = calculated ultimate load
χ = reduction factor depending on the non-dimensional slenderness $\bar{\lambda}$ and the European buckling curve "a" (compare Section 5.3.3)

$$N_{pl,Rd(\text{squash load})} = A_a \cdot \frac{f_y}{\gamma_a} + A_b \cdot \frac{f_{bk}}{\gamma_b} + A_s \cdot \frac{f_{sk}}{\gamma_s}$$

$$= N_{pl,a,Rd} + N_{pl,b,Rd} + N_{pl,s,Rd}$$
$$(11\text{-}2)$$

where:
A_a, f_y = cross-sectional area, yield strength of steel hollow section
A_b, f_{bk} = cross-sectional area, cylinder compressive strength of concrete (cylinder compressive strength = 0.83 × cube compressive strength)
A_s, f_{sk} = cross-sectional area, yield strength of reinforcement steel
γ_a = 1.1 for steel hollow section
γ_b = 1.5 for concrete
γ_s = 1.15 for reinforcement steel
} partial safety factors [5]

The non-dimensional slenderness $\bar{\lambda}$ can be calculated using

$$\bar{\lambda} = \sqrt{\frac{N_{pl,Rd}}{N_{Ki}}} \qquad (11\text{-}3)$$

Hereby, the cross-sectional resistance $N_{pl,Rd}$ is determined taking $\gamma_a = \gamma_{bk} = \gamma_{sk} = 1.0$.

$$N_{Ki} = \frac{\pi^2}{l_K^2}(E_a \cdot I_a + E_{bi} \cdot I_b + \sum E_s \cdot I_s) \qquad (11\text{-}4)$$

where:
l_K = buckling length of column
E_a, I_a = modulus of elasticity, second moment of area of steel hollow section
E_s, I_s = modulus of elasticity, second moment of area of reinforcement steel
I_b = second moment of area of concrete
E_{bi} = ideal calculation value for the modulus of elasticity of concrete

$$E_{bi} = 600 \cdot f_{bk} \quad (\text{see also Table 11-1}) \quad (11\text{-}5)$$

Without the additional reinforcement steel in hollow section, the share of $\sum A_s \cdot f_{sk}$ in Eq. (11-2) and $\sum E_s \cdot I_s$ in Eq. (11-4) is equal to zero.

For the application of the simplified design procedure, the following limits are valid:

- $\bar{\lambda} \leq 2.0$; the range $\bar{\lambda} > 2.0$ is not covered by tests. This does not mean any restriction for the practice, as $\bar{\lambda} > 2.0$ should not or seldom occur in any case.

- The applied longitudinal reinforcement cross-sectional area A_s shall be considered in the calculation up to a maximum value of

$$\left(\frac{A_s}{A_s + A_b}\right)_{\text{calculation}} = 3\%$$

For end-to-end connections, the following is valid:

$$\left(\frac{A_s}{A_s + A_b}\right)_{\text{calculation}} \leq 6\%$$

For the theoretical verification of ultimate load in the case of fire, also a higher share of the reinforcement than 3% or 6% can be taken into account.

- $0.2 \leq \dfrac{N_{pl,a,Rd}}{N_{pl,Rd}} \left(= \dfrac{A_a \cdot f_y}{\gamma_a}\right) \leq 0{,}9$; the share of the steel hollow section in the total ultimate load shall lie within the mentioned

Table 11-1 Cylinder compressive strength and ideal modulus of elasticity for concrete

Strength grade of concrete	C20	C25	C30	C35	C40	C45	C50	C55	C60
Cylinder compressive strength f_{bk} (kN/cm²)	2.0	2.5	3.0	3.5	4.0	4.5	5.0	5.5	6.0
Modulus of elasticity $E_{bi} = 600\, f_{bk}$ (kN/cm²)	1200	1500	1800	2100	2400	2700	3000	3300	3600

limits; $N_{pl,a,Rd}$ and $N_{pl,Rd}$ according to Eq. (11-2).

- Hollow section must be filled with concrete over its total length.

- The limiting ratios d/t or b/t or h/t against local buckling of hollow sections (see Tables 5-17 and 5-24) have to be observed.

11.1.1.3 Effect of long-term behaviour of concrete on the design strength (resistance) of slender columns

For slender columns under long-term loading, the effect of creep and shrinkage has to be taken into consideration and precisely for:

1 a) $\bar{\lambda} > \dfrac{0.8}{1-\delta}$ for columns in braced and non-sway systems

1 b) $\bar{\lambda} > \dfrac{0.5}{1-\delta}$ for columns in unbraced or sway systems

where:
$$\delta = \dfrac{N_{pl,a,Rd}}{N_{pl,Rd}} = \dfrac{A_a \cdot f_y}{N_{pl,Rd} \cdot \gamma_a}$$

2) $\dfrac{M_{Sd}}{N_{Sd}} = e$

where: M_{Sd} = Moment according to theory of the first order, $e < 2d$ or $e < 2h$

In a hollow section, the complete drying of concrete does not take place. The lower long-term effect due to this is compensated partially by the increasing strength with the growing age of concrete. The long-term effect is ascertained by a reduction of the ideal modulus of elasticity of concrete E_{bi}:

$$E_{bi,\infty} = E_{bi}\left(1 - 0.5\,\dfrac{N_{permanent}}{N_{Sd}}\right) \quad (11\text{-}6)$$

where:
N_{Sd} = axial design load (γ times increased)
$N_{permanent}$ = permanently acting share of N_{Sd}

11.1.1.4 Increased design strength (resistance) for compact CHS columns filled up with concrete

For compact CHS composite columns, the increased design strength due to the confinement effect of concrete can be calculated [4, 16, 24] making the following substitutions in Eq. (11-2):

$$f_y \text{ by } \eta_2 \cdot f_y \quad (11\text{-}7)$$

and

$$f_{bk} \text{ by } f_{bk}\left(1 + \eta_1\,\dfrac{t}{d}\cdot\dfrac{f_y}{f_{bk}}\right) \quad (11\text{-}8)$$

where:
d = CHS outer diameter
t = CHS wall thickness

Conditions: $e \leq d/10$
$\bar{\lambda} \leq 0.5$

For a loading eccentricity $\dfrac{M_{max,Sd}}{N_{Sd}} = e > d/10$, one has to set $\eta_1 = 0$ and $\eta_2 = 1$. Between $0 < e < d/10$, a linear interpolation is allowed. The increased design strength shall not be considered in the calculation, if by the constructional design of the area of force application, only the steel tube is loaded and not the concrete cross section at the same time.

The coefficients η_1 and η_2 are determined as follows:

$$\eta_1 = \eta_{10}\left(1 - 10\,\dfrac{e}{d}\right) \geq 0 \quad (11\text{-}9)$$

$$\eta_2 = \eta_{20} + (1 - \eta_{20}) \cdot 10\,\dfrac{e}{d} \leq 1{,}0 \quad (11\text{-}10)$$

where:

$$\eta_{10} = 4.9 - 18.5\,\bar{\lambda} + 17\,\bar{\lambda}^2 \geq 0 \qquad (11\text{-}11)$$

$$\eta_{20} = 0.25\,(3 + 2\,\bar{\lambda}) \leq 1.0 \qquad (11\text{-}12)$$

11.1.1.5 Design strength (resistance) of composite hollow section columns under compression and uni-axial bending

The ultimate load bearing strength of a composite hollow section column under axial compression and uni-axial bending is determined through the failure of the cross section taking the slenderness of the column (global buckling) and the increase of the bending moment (theory of second order) into account.

The design verification is to be carried out with the aid of M-N-interaction curves (see Fig. 11-4). The calculation of $N_{pl,Rd}$ can be done using Eq. (11-2). For the increase of bending moments, see Section 11.1.1.6.

Interaction curves for circular, square and rectangular hollow sections are shown in Figs. 11-5 to 11-8. They have been calculated electronically considering fully plastic stress distribution in the composite cross section [21, 25]. An approximation procedure to calculate the interaction curves for composite cross sections has been described in Section 11.1.1.8.

First of all, the ultimate load N_K for concentric compression and the squash load $N_{pl,Rd}$ are calculated according to Section 11.1.1.2; one obtains the points E and F in Fig. 11-4a. Joining the point F with the origin 0, the zone to the left of the straight line $\overline{OF}$ can be seen as belonging to the moment due to the geometrical imperfections and residual stresses for methodically concentric load. The zone to the right of $\overline{OF}$ (distance S) takes up the methodical bending moment (load bearing capacity of the moment). This verification starts from the presumption that the methodical moment (according to the theory of the second order) is always added to the imperfection moment or in other words, the failure occurs in the middle of the column. If the maximum moment due to external loading does not occur in the middle of the column, the edge cross section can fail when subjected to large edge moments and small axial loads. With the straight line $\overline{OF}$, one is on a very safe side. Therefore Bergmann [21] proposes to join the point F with the point χ_n on the ordinate (Fig. 11-4b), which is dependent on the course of the moment (see Table 11-2).

This modification is adopted also by DIN 18806 [4]. $\chi_n = 0$ is to be used for other moment areas.

If a methodical bending moment M_{Sd} (according to the elastic theory of the second order) is applied besides axial load N_{Sd} as γ times load-

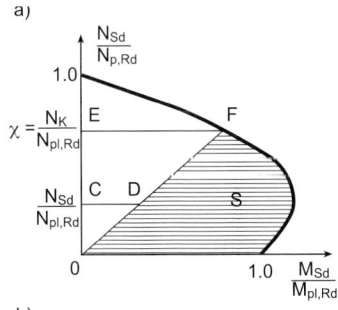

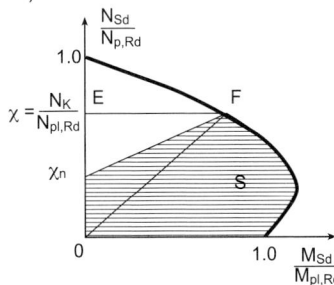

Fig. 11-4 a) Design principle for composite hollow section column under compression and uni-axial bending moment, b) modified method compared to a)

Table 11-2 Factors for various moment areas [21]

Moment area	χ_n	χ_n in general
	$0.50 \cdot \chi$	
	$0.25 \cdot \chi$	$\chi_n = \chi \cdot \dfrac{1-\psi}{4}$
	0	

ψ is the ratio of the larger end moment to the smaller one.

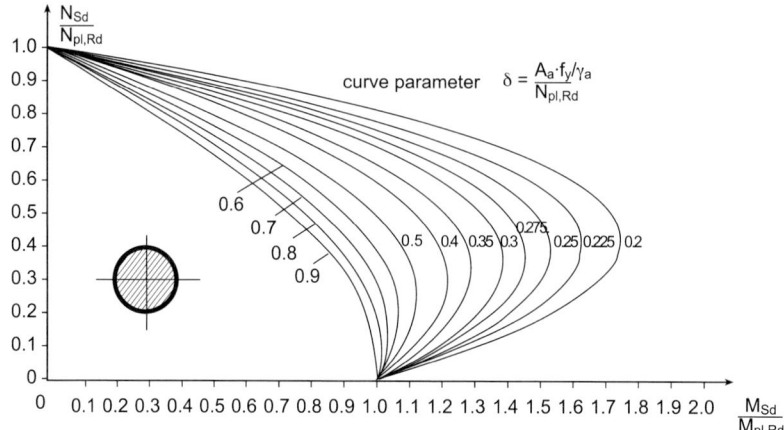

Fig. 11-5 M-N-interaction curves, concrete-filled CHS

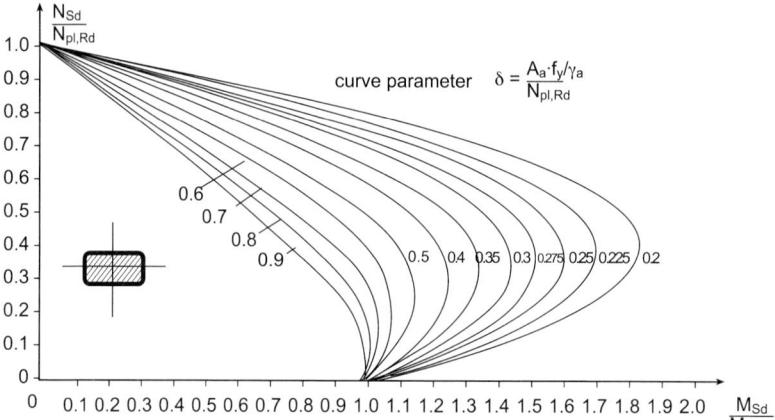

Fig. 11-6 M-N-interaction curves, concrete-filled RHS, flat lying

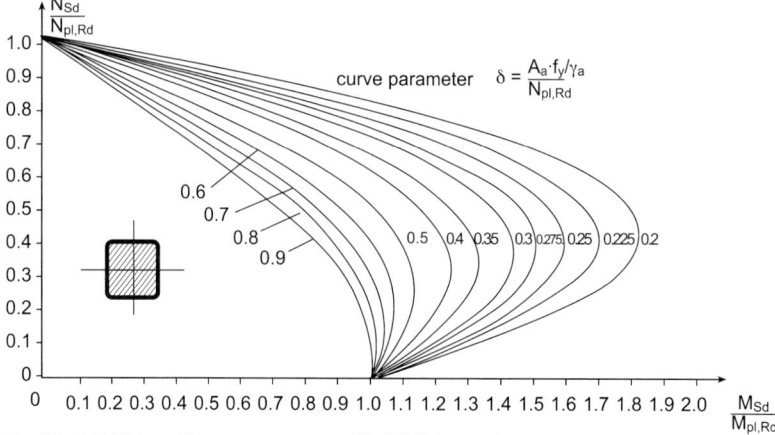

Fig. 11-7 M-N-interaction curves, concrete-filled RHS (square)

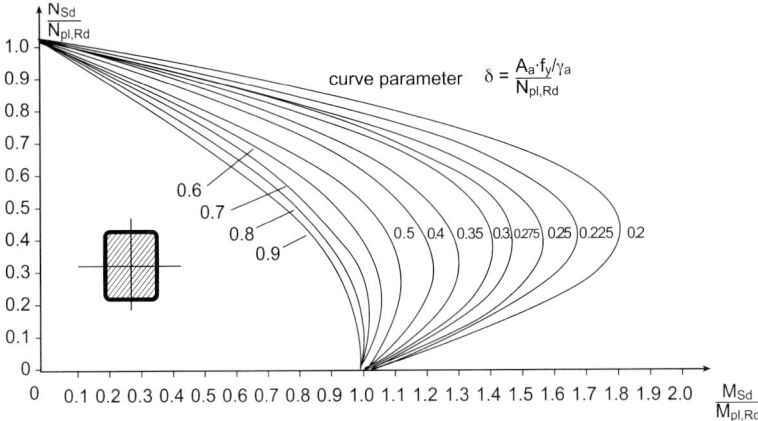

Fig. 11-8 M-N-interaction curves, concrete-filled RHS, vertical standing

ing, this must be restricted to the following limit:

$$M_{Sd} \leq 0.9 \cdot S \cdot M_{pl,Rd} \quad (11\text{-}13)$$

where:
$M_{pl,Rd}$ = fully plastic moment of the composite cross section
S = non-dimensional factor according to Fig. 11-4, to be taken from the interaction diagrams from the Figs. 11-5 to 11-8

The reduction of the factor S by 10% covers the simplifications in the stress-strain-law for concrete as well as the assumption of fully effective concrete cross section (condition I) for the rigidity values (E_{bi}, I_b) to determine the force and bending moment according to the theory of the second order [21].

The bending moment M according to the theory of the second order can be determined simply from the calculated bending moment M_0 according to the theory of the first order with the aid of an enlargement factor:

$$M = M_0 \cdot \frac{1}{1 - \dfrac{N_{Sd}}{N_{Ki}}} = M_0 \cdot \frac{1}{1 - \bar{\lambda}^2 \dfrac{N_{Sd}}{N_{pl,Rd}}} \quad (11\text{-}14)$$

where:
N_{Ki} according to Eq. (11-4)
$\bar{\lambda}$ according to Eq. (11-3)
$N_{pl,Rd}$ according to Eq. (11-2)

11.1.1.6 Ultimate strength (resistance) of a hollow section cross section under compression and bending

Under sole compression, $N_{pl,Rd}$ according to Eq. (11-2) gives the squash resistance.

Eqs. (11-7) and (11-8) are valid for compact CHS columns, as long as a correct load introduction (not only on the steel hollow section predominantly) is considered.

If high strength steels with yield strengths beyond that of S 355 are applied, the yield strength f_y of the hollow section (and the reinforcement bars) set in the calculation shall not exceed the value $21\,000 \cdot 0.002 = 420$ N/mm² taking the ultimate elongation (−2‰) of concrete under concentric compression into account.

The full plastic moment resistance $M_{pl,Rd}$ of the composite cross section can be calculated with the aid of the stress distribution shown in Fig. 11-9. As concrete cannot be loaded by tension, the formulae to calculate $M_{pl,Rd}$ are relatively complicated.

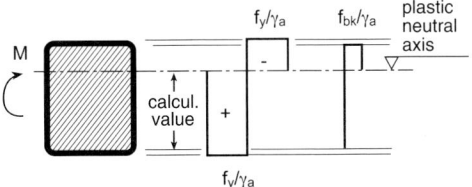

Fig. 11-9 Fully plastic stress distribution in RHS composite cross section for sole moment resistance $M_{pl,Rd}$

Eqs. (11-16) to (11-18) for $M_{pl,Rd}$ in Table 11-3 are taken from [21]. The coefficient $\bar{m}$ represents the relation for the bending capacities of composite hollow sections to steel hollow sections without concrete filling. They can be read from Table 11-3, the position of the neutral axis as well. For the rectangular and square hollow sections, an external corner radius of twice the wall thickness ($r_{ext} = 2t$) has been assumed. The values for fully plastic moments given by the Eqs. (11-16) to (11-18) do not consider any reinforcement in the concrete. For the reinforcement bars arranged symmetrical to the bending axis, this can be included simply by adding the plastic bending resistance of the reinforcement alone:

$$M_{pl,s,Rd} = \sum A_s \cdot \frac{f_{sk}}{\gamma_s} \cdot a_s \qquad (11\text{-}15)$$

where:
A_s = cross-sectional area of the reinforcement bar
f_{sk} = yield strength of the reinforcement steel
a_s = distance of the bar to the relevant bending axis

The influence of the reinforcement on the position of the neutral axis is in general relatively small and can be neglected mostly.

When axial force and bending moment act simultaneously, the ultimate resistance of the cross section can be determined using the interaction curves in Figs. 11-5 to 11-8. They are calculated electronically based on the fully plastic stress distribution similar to that shown in Fig. 11-9, however considering axial forces in addition. These curves are established for cross sections without reinforcement. If high percentage of reinforcement is applied, the use of these curves can lead to mistakes. In an approximate calculation, the parameter

$$\delta = \frac{A_a \cdot f_y}{\gamma_a \cdot N_{pl,Rd}} \text{ is to be substituted by}$$

$$\delta^* = \frac{A_a \cdot f_y/\gamma_a + A_s \cdot f_{sk}/\gamma_s}{N_{pl,Rd}} \qquad (11\text{-}19)$$

when symmetrical reinforcements exist on the sides subjected to tension and compression.

(Designations as given in Section 11.1.1.2, A_s is the total cross-sectional area of all the reinforcements.)

11.1.1.7 Design strength (resistance) of composite hollow section columns under compression and bi-axial bending

The interaction relationship for the ultimate resistance of a composite hollow cross section under axial compression and bi-axial bending moment is illustrated in Fig. 11-10. The (spatial) M_y-M_z-curve has been approximated by a straight line in the calculation.

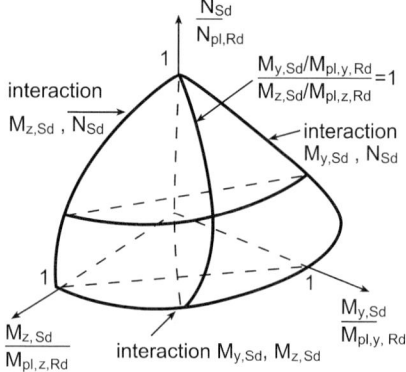

Fig. 11-10 M-N-interaction surface between compression and bi-axial bending moment

Leaning against the approximation procedure for axial force and uni-axial bending according to Section 11.1.1.5, the following method can be used for bi-axial bending:

- Calculation of the plastic bending resistances $M_{pl,y,Rd}$ and $M_{pl,z,Rd}$ separately for each main axis (see Section 11.1.1.6).
- Determination of ultimate uni-axial bending resistances = distance S_y and S_z in Fig. 11-11 a and b.
- The ultimate bending moment resistances are the boundary values for a new interaction curve made by joining S_y and S_z on the two axes (see Fig. 11-11 c).

11.1 Hollow section composite columns

Table 11-3a Coefficient $\bar{m}$ for RHS ($h_a/b_a = 0.5$) to calculate $M_{pl,Rd}$ and k_x to determine the position of the neutral axis of hollow section composite columns

	$f_y = 240$ N/mm² (Fe 240 ≈ S235)							
	C25		C35		C45		C55	
h_a/t	$\bar{m}$	k_x	$\bar{m}$	k_x	$\bar{m}$	k_x	$\bar{m}$	k_x
10	0.9784	0.402	0.9885	0.375	0.9970	0.352	1.0041	0.333
15	1.0190	0.354	1.0322	0.320	1.0425	0.293	1.0509	0.271
20	1.0443	0.316	1.0589	0.279	1.0699	0.251	1.0784	0.229
25	1.0625	0.285	1.0776	0.247	1.0887	0.219	1.0971	0.198
30	1.0765	0.260	1.0917	0.222	1.1025	0.194	1.1106	0.174
40	1.0970	0.221	1.1117	0.184	1.1218	0.159	1.1290	0.140
50	1.1114	0.192	1.1254	0.157	1.1346	0.134	1.1411	0.117
60	1.1223	0.170	1.1353	0.137	1.1438	0.116	1.1497	0.101

	$f_y = 360$ N/mm² (Fe 360 ≈ S355)							
	C25		C35		C45		C55	
h_a/t	$\bar{m}$	k_x	$\bar{m}$	k_x	$\bar{m}$	k_x	$\bar{m}$	k_x
10	0.9683	0.429	0.9765	0.407	0.9837	0.388	0.9900	0.371
15	1.0050	0.390	1.0165	0.361	1.0260	0.336	1.0341	0.315
20	1.0280	0.358	1.0414	0.324	1.0522	0.296	1.0609	0.273
25	1.0448	0.331	1.0594	0.293	1.0707	0.265	1.0797	0.242
30	1.0580	0.307	1.0733	0.268	1.0848	0.239	1.0938	0.216
40	1.0781	0.269	1.0938	0.229	1.1051	0.201	1.1137	0.179
50	1.0929	0.239	1.1084	0.200	1.1192	0.173	1.1272	0.153
60	1.1043	0.215	1.1193	0.177	1.1296	0.152	1.1370	0.133

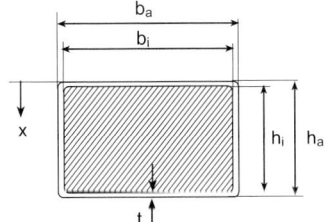

$$M_{pl,Rd} = \bar{m} \cdot \frac{1}{4}(h_a^2 \cdot b_a - h_i^2 \cdot b_i) \cdot f_y/\gamma_a = M_{pl,o,Rd} \qquad (11\text{-}16)$$

neutral axis: $x = k_x \cdot h_a$

$M_{pl,o,Rd}$ = plastic moment resistance of the composite cross section without longitudinal reinforcement

Table 11-3b Coefficient $\bar{m}$ and k_x for RHS ($h_a/b_a = 2.0$)

	$f_y = 240$ N/mm² (Fe 240 ≈ S235)							
	C25		C35		C45		C55	
h_a/t	$\bar{m}$	k_x	$\bar{m}$	k_x	$\bar{m}$	k_x	$\bar{m}$	k_x
10	0.8592	0.480	0.8666	0.473	0.8738	0.465	0.8807	0.459
15	0.9403	0.461	0.9542	0.447	0.9672	0.435	0.9794	0.423
20	0.9861	0.443	1.0053	0.424	1.0228	0.407	1.0387	0.391
25	1.0186	0.426	1.0422	0.403	1.0631	0.383	1.0817	0.364
30	1.0443	0.411	1.0714	0.384	1.0950	0.361	1.1156	0.340
40	1.0845	0.383	1.1170	0.351	1.1441	0.324	1.1671	0.301
50	1.1161	0.358	1.1521	0.323	1.1812	0.294	1.2053	0.270
60	1.1422	0.337	1.1806	0.299	1.2108	0.269	1.2352	0.244

	$f_y = 360$ N/mm² (Fe 360 ≈ S355)							
	C25		C35		C45		C55	
h_a/t	$\bar{m}$	k_x	$\bar{m}$	k_x	$\bar{m}$	k_x	$\bar{m}$	k_x
10	0.8527	0.486	0.8579	0.481	0.8629	0.476	0.8678	0.471
15	0.9279	0.473	0.9379	0.463	0.9474	0.454	0.9565	0.445
20	0.9685	0.460	0.9827	0.446	0.9960	0.433	1.0084	0.421
25	0.9965	0.448	1.0144	0.431	1.0307	0.414	1.0458	0.400
30	1.0182	0.436	1.0393	0.416	1.0583	0.397	1.0756	0.380
40	1.0521	0.415	1.0785	0.389	1.1015	0.366	1.1218	0.346
50	1.0788	0.395	1.1092	0.365	1.1351	0.339	1.1574	0.317
60	1.1013	0.377	1.1348	0.344	1.1626	0.316	1.1861	0.293

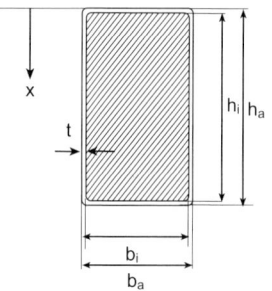

$$M_{pl,Rd} = \bar{m} \cdot \frac{1}{4} (h_a^2 \cdot b_a - h_i^2 \cdot b_i) \cdot f_y/\gamma_a = M_{pl,o,Rd}$$

neutral axis: $x = k_x \cdot h_a$

$M_{pl,o,Rd}$ = plastic moment resistance of the composite cross section without longitudinal reinforcement

11.1 Hollow section composite columns

Table 11-3c Coefficient $\bar{m}$ and k_x for RHS (square)

	\multicolumn{8}{c}{$f_y = 240$ N/mm² (Fe 240 ≈ S235)}							
	C25		C35		C45		C55	
h_a/t	$\bar{m}$	k_x	$\bar{m}$	k_x	$\bar{m}$	k_x	$\bar{m}$	k_x
10	0.9310	0.450	0.9418	0.433	0.9516	0.418	0.9605	0.404
15	0.9905	0.417	1.0068	0.393	1.0210	0.371	1.0334	0.352
20	1.0268	0.389	1.0470	0.359	1.0638	0.333	1.0781	0.312
25	1.0534	0.364	1.0761	0.330	1.0945	0.302	1.1097	0.280
30	1.0745	0.342	1.0990	0.306	1.1182	0.277	1.1337	0.253
40	1.1070	0.306	1.1332	0.266	1.1530	0.237	1.1684	0.213
50	1.1314	0.276	1.1582	0.236	1.1777	0.207	1.1926	0.184
60	1.1507	0.252	1.1774	0.212	1.1963	0.183	1.2105	0.162

	\multicolumn{8}{c}{$f_y = 360$ N/mm² (Fe 360 ≈ S355)}							
	C25		C35		C45		C55	
h_a/t	$\bar{m}$	k_x	$\bar{m}$	k_x	$\bar{m}$	k_x	$\bar{m}$	k_x
10	0.9211	0.465	0.9291	0.453	0.9365	0.441	0.9435	0.430
15	0.9748	0.441	0.9875	0.422	0.9990	0.405	1.0093	0.389
20	1.0068	0.419	1.0231	0.395	1.0374	0.373	1.0500	0.354
25	1.0299	0.400	1.0491	0.371	1.0654	0.346	1.0795	0.325
30	1.0485	0.382	1.0698	0.350	1.0875	0.323	1.1025	0.300
40	1.0775	0.350	1.1017	0.314	1.1211	0.285	1.1369	0.261
50	1.1000	0.323	1.1259	0.284	1.1460	0.254	1.1619	0.230
60	1.1184	0.300	1.1452	0.260	1.1653	0.230	1.1810	0.206

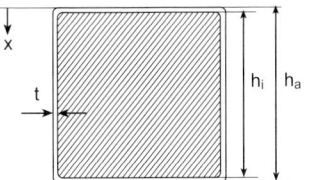

$$M_{pl,Rd} = \bar{m} \cdot \frac{1}{4}(h_a^3 - h_i^3) \cdot f_y/\gamma_a = M_{pl,o,Rd} \qquad (11\text{-}17)$$

neutral axis: $x = k_x \cdot h_a$

$M_{pl,o,Rd}$ = plastic moment resistance of the composite cross section without longitudinal reinforcement

Table 11-3d Coefficient $\bar{m}$ and k_x for CHS

	$f_y = 240$ N/mm² (Fe 240 ≈ S235)							
	C25		C35		C45		C55	
d_a/t	$\bar{m}$	k_x	$\bar{m}$	k_x	$\bar{m}$	k_x	$\bar{m}$	k_x
10	1.0337	0.460	1.0452	0.447	1.0556	0.435	1.0653	0.424
15	1.0560	0.435	1.0733	0.415	1.0886	0.399	1.1022	0.384
20	1.0757	0.413	1.0974	0.389	1.1160	0.369	1.1322	0.352
25	1.0933	0.394	1.1182	0.367	1.1391	0.345	1.1569	0.327
30	1.1091	0.377	1.1365	0.348	1.1590	0.325	1.1779	0.306
40	1.1363	0.348	1.1671	0.316	1.1915	0.292	1.2117	0.273
50	1.1591	0.324	1.1919	0.292	1.2174	0.267	1.2380	0.248
60	1.1785	0.305	1.2126	0.272	1.2385	0.248	1.2593	0.229
80	1.2103	0.274	1.2454	0.241	1.2715	0.218	1.2921	0.200
100	1.2353	0.250	1.2706	0.219	1.2964	0.196	1.3164	0.179

	$f_y = 360$ N/mm² (Fe 360 ≈ S355)							
	C25		C35		C45		C55	
d_a/t	$\bar{m}$	k_x	$\bar{m}$	k_x	$\bar{m}$	k_x	$\bar{m}$	k_x
10	1.0234	0.472	1.0317	0.462	1.0396	0.453	1.0470	0.445
15	1.0396	0.454	1.0528	0.438	1.0649	0.425	1.0760	0.412
20	1.0545	0.437	1.0717	0.417	1.0870	0.400	1.1007	0.386
25	1.0683	0.421	1.0886	0.399	1.1063	0.379	1.1220	0.363
30	1.0809	0.407	1.1039	0.382	1.1235	0.361	1.1405	0.343
40	1.1035	0.383	1.1303	0.354	1.1526	0.331	1.1715	0.312
50	1.1231	0.362	1.1527	0.331	1.1766	0.307	1.1966	0.287
60	1.1403	0.344	1.1718	0.312	1.1968	0.287	1.2173	0.267
80	1.1694	0.314	1.2031	0.281	1.2293	0.256	1.2502	0.237
100	1.1931	0.290	1.2280	0.257	1.2545	0.233	1.2754	0.214

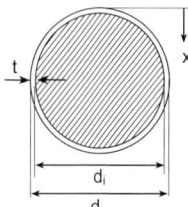

$$M_{pl,Rd} = \bar{m} \cdot \frac{1}{6}(d_a^3 - d_i^3) \cdot f_y/\gamma_a = M_{pl,o,Rd} \qquad (11\text{-}18)$$

neutral axis: $x = k_x \cdot d_a$

$M_{pl,o,Rd}$ = plastic moment resistance of the composite cross section without longitudinal reinforcement

It has to be verified that

$$\frac{M_{y,Sd}}{S_y \cdot M_{pl,y,Rd}} + \frac{M_{z,Sd}}{S_z \cdot M_{pl,z,Rd}} \leq 1 \quad (11\text{-}20)$$

where the single values have to fulfil the following condition in order to avoid tri-dimensional interaction surfaces:

$$\frac{M_{y,Sd}}{S_y \cdot M_{pl,y,Rd}} \text{ bzw. } \frac{M_{z,Sd}}{S_z \cdot M_{pl,z,Rd}} \leq 0{,}9 \quad (11\text{-}21)$$

M_y, M_z = moments according to the theory of the second order

For the determination of S_y and S_z, the modified procedure according to Fig. 11-4b can be used (see also [4]).

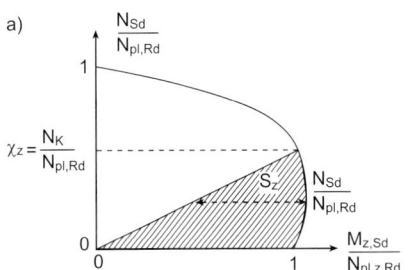

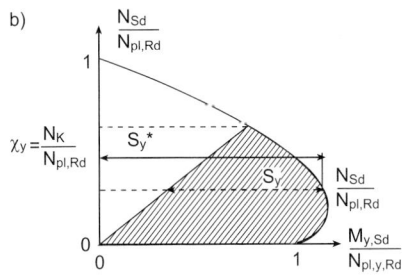

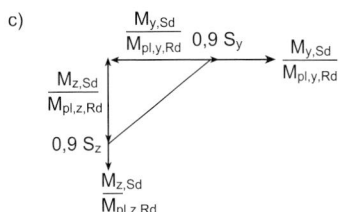

Fig. 11-11 Calculation procedure for composite hollow section column under compression and bi-axial bending moment

As is proposed by DIN 18800, part 2 [23], only a imperfection for the decisive axis has to be considered in the case of axial compression and bi-axial bending. This proposal has also been adopted by DIN 18806 [4]. In Fig. 11-11b, S_y^* can substitute S_y or (if unfavourable) S_z^* can substitute S_z, however keeping S_y as it is (see Eq. 11-20).

11.1.1.8 Approximate calculation for M-N-interaction in composite hollow section columns

When four points on the M-N-interaction curve are known, the curve can be represented by a polygonal course (Fig. 11-12) [21].

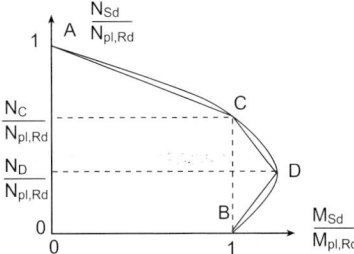

Fig. 11-12 Approximation of a M-N-interaction curve by a polygonal course

The point A, B, C and D result from the following considerations:

- The point A can be obtained using the plastic axial forces (Fig. 11-13a):

$N_A = N_{pl,Rd}$ (see Eq. 11-2)
$M_A = 0$

- The point B is given by the plastic bending moment (Fig. 11-13b):

$N_B = 0$

$$M_B = M_{pl,Rd} = \\ M_{pl,o,Rd} + A_s \cdot c \cdot \left(2 \cdot \frac{f_{sk}}{\gamma_s} - \frac{f_{bk}}{\gamma_b}\right) \quad (11\text{-}22)$$

where:

$M_{pl,o,Rd}$ = plastic moment of the composite column without reinforcement, see Eqs. (11-16) to (11-18)

A_s = cross-sectional area of the total reinforcement lying symmetrical to the central axis

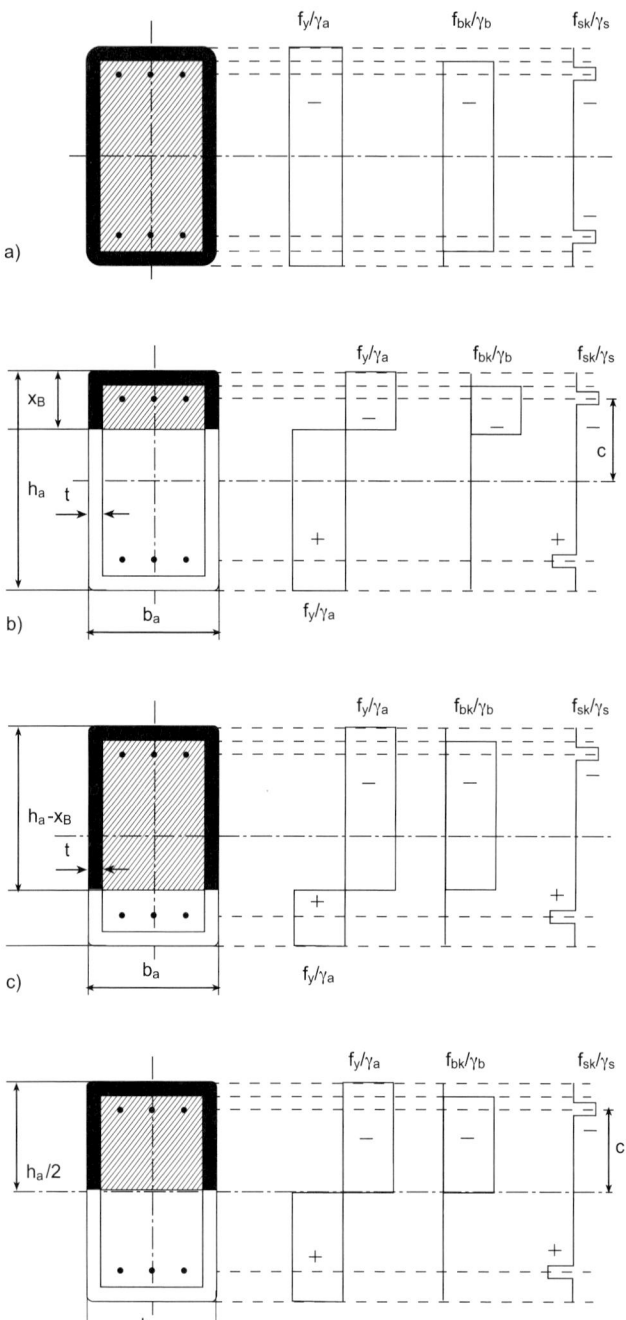

Fig. 11-13 Stress distributions for selected positions of the neutral axis (points A through D in Fig. 11-12)

f_y = yield strength of steel hollow section
f_{sk} = yield strength of reinforcement
f_{bk} = compressive strength of concrete

- The neutral axis of the point C is the reflected image of the plastic neutral axis (point B) about the central axis (Fig. 11-13c). Therefore, the plastic bending moment $M_{pl,Rd}$ occurs at this location similar to the point B, but also an axial compression N_C simultaneously. One obtains N_C by the concentric compression of the stress distribution at the point B on that at the point C.

$$N_C = (h_a - 2 \cdot x_B)[4 \cdot t \cdot f_y/\gamma_a + (b-2t)f_{bk}/\gamma_b] \quad (11\text{-}23)$$

However, N_C can be simply determined using

$$N_C = 2 \cdot N_D \quad (11\text{-}24)$$

$$M_C = M_B = M_{pl,Rd} \quad (11\text{-}25)$$

- The point D is characterized by the maximum bending moment, which is obtained when the neutral axis coincides with the centre line of the cross section (Fig. 11-13d):

$$N_D = \frac{A_b}{2} \cdot f_{bk}/\gamma_b \quad (11\text{-}26)$$

$$M_D = M_{pl,a,Rd} + N_D \cdot \frac{h_a/2 - t}{2} + A_s \cdot c \cdot \left(2 \cdot \frac{f_{sk}}{\gamma_s} - \frac{f_{bk}}{\gamma_b}\right) \quad (11\text{-}27)$$

where:
A_b = cross-sectional area of concrete
$M_{pl,a,Rd}$ = plastic moment resistance of hollow section ($= f_y/\gamma_a \cdot W_{pl}$)
W_{pl} = plastic section modulus of hollow section

11.1.1.9 Influence of shear forces [31]

It can be assumed that the shear forces in a composite column can be either assigned to the steel hollow section alone or they can be divided into a steel and a reinforcement concrete component.

The calculation following the former assumption is simpler, where the reduction of the axial stresses due to shear stresses is taken into account. In this case, the yield strength f_y of the hollow section is reduced in those parts of the steel profile, which are able to bear the shear, when shear force V_{Sd} exceeds the value of $V_{pl,Rd}/2$ [32]:

$$f_{y,red} = f_y \left[1 - \left(\frac{2 V_{Sd}}{V_{pl,Rd}} - 1\right)^2\right] \quad (11\text{-}28)$$

where:
V_{Sd} = applied shear force
$V_{pl,Rd}$ = plastic shear resistance

$$V_{pl,Rd} = A_v \cdot \frac{f_y}{\gamma_a \cdot \sqrt{3}} \quad (11\text{-}29)$$

where:
$A_v = 2 (d_a - t) \cdot t$ for CHS
$A_v = 2 (h_a - t) \cdot t$ or $2 (b_a - t) \cdot t$ for RHS

The value of reduced A_v has to be determined by diminishing the width of the relevant areas.

In case $V_{Sd} > 0.5 V_{pl,Rd}$, a reduced wall thickness t_{red} of the steel parts loaded by shear can simply substitute the reduced yield strength $f_{y,red}$ in Eq. (11-28).

$$t_{red} = t \left[1 - \left(\frac{2 V_{Sd}}{V_{pl,Rd}} - 1\right)^2\right] \quad (11\text{-}30)$$

Although the reduction of only the hollow section webs is required, the wall thickness of the whole cross section is reduced to simplify the calculation.

The effect of shear stresses on the axial stresses can be neglected for $V_{Sd} \leq 0.5 V_{pl,Rd}$ [32].

11.1.1.10 Load introduction

The full bondage between steel and concrete in hollow section composite columns, which is presumed for the design, must be guaranteed by the construction for load introduction. It was determined by tests that a maximum transferable bond stress $\tau_{Rd} = 0.4$ N/mm² has to be observed for hollow sections (CHS, RHS). When this exceeds the admissible value, mechanical shear connectors have to be applied.

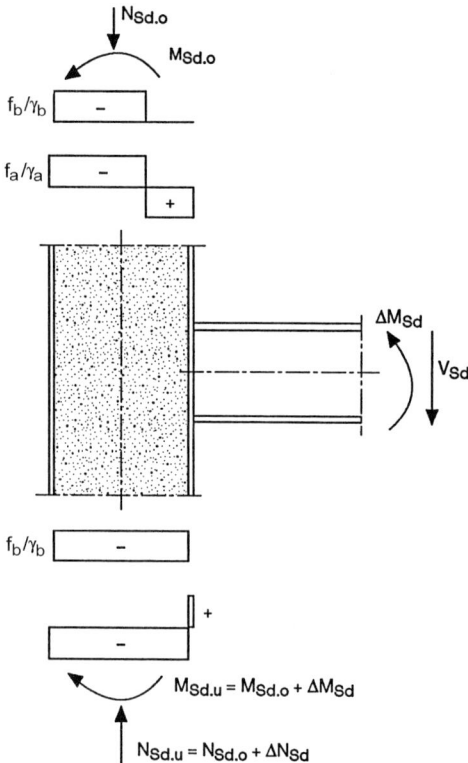

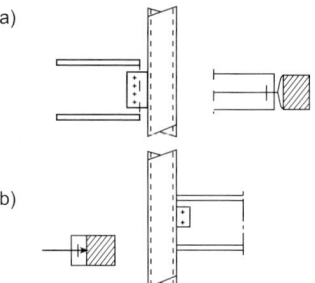

Fig. 11-15 Beam-to-column connections with concrete-filled square hollow section columns

Fig. 11-14 Difference of forces and moments in a load introduction region – fully plastic stress distribution [31]

In order to determine the bond stress, the forces in the bond area can be derived through the difference of the forces and the moments between the critical locations [31] (see Fig. 11-14).

It has to be assured that the concrete gets its share in the shortest way for shear transfer in the load introduction zone.

Figs. 11-15 and 11-16 show the connections, where the tensile forces from the bending moments in the connections are transmitted to the side walls and the compressive forces to the back side of the column. The compression is transmitted directly to the steel mantle with the concrete core behind it. The shear force is transferred only to the steel mantle. A theoretical calculation of the shear further transferred into the concrete core is hardly possible.

Fig. 11-16 Simple beam-to-column connections with concrete-filled CHS column (effective length for load introduction $l_c \leq 2\,d$)

Fig. 11-17 illustrates the application of dowel pins, which are fixed to the connecting plate or the profile walls and inserted into the inside of the profile through a drilled hole in the profile wall. The hollow section column is subsequently filled up with concrete. This type of connection is suitable to transmit the tensile component of the bending moment in the connection into the concrete core.

Lit. [31] proposes the calculation of the connections with inserted plates (see Fig. 11-18):

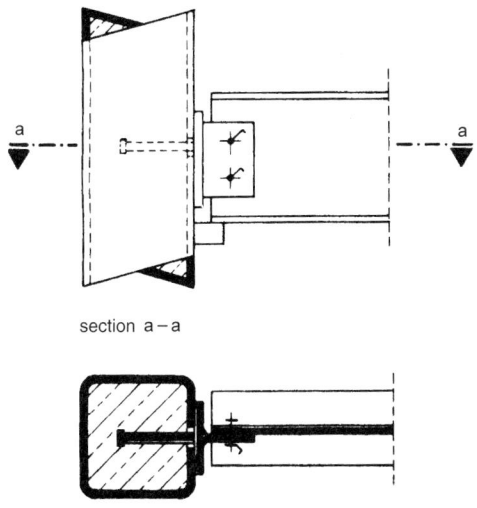

Fig. 11-17 Load transmission into the concrete core with inserted dowel pin or stud

$$f_{u1,Rd} = (f_{bk} + 35.0) \cdot \frac{1}{\gamma_b} \cdot \sqrt{\frac{A_b}{A_1}} \quad (11\text{-}31)$$

where:
A_b = cross-sectional area of the total concrete in a column
A_1 = cross-sectional area under the insertion
f_{bk} = characteristic compression strength of concrete in N/mm^2
γ_b = partial safety factor for concrete ($= 1.5$ according to [5])

$$f_{u1,Rd} < \frac{N_{pl,b,Rd}}{A_1}$$

$$\frac{A_b}{A_1} \leq 20$$

Columns between the floors of a building do not require any special device for load introduction, as the headplates act as dowels and functions as load transmitting elements (Fig. 11-19).

No cavity shall exist under the contact area between the head plate and the concrete. For this purpose, the column is overfilled and the surplus concrete is levelled off with a stickle before hardening. An alternative is to level off the ends by means of compensated setting mortar (Fig. 11-20).

Fig. 11-21 shows two design solutions for the connections between concrete floors and con-

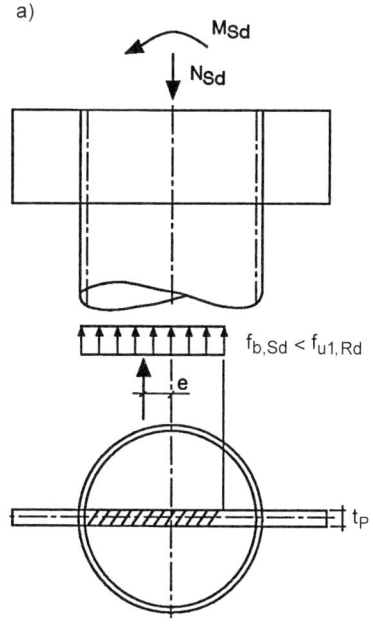

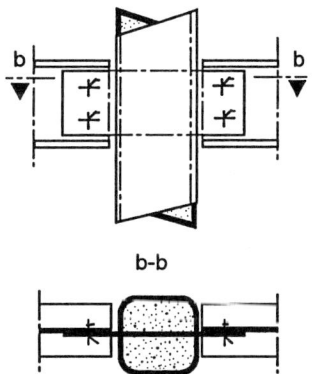

Fig. 11-18 Beam-to-column connection with inserted plate

crete-filled hollow section columns in a building.

11.2 Truss joints in RHS with concrete-filled chord members

The design resistance of hollow section joints in a truss can be improved by filling the chord members with concrete either over the whole length or in the critical chord zone of the joints. Experimental investigations on joints with concrete-filled RHS chord members un-

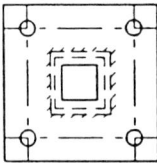

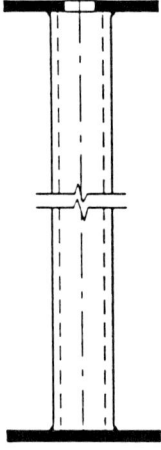

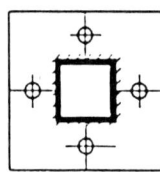

Fig. 11-19 Concrete-filled hollow section column with head and foot plate

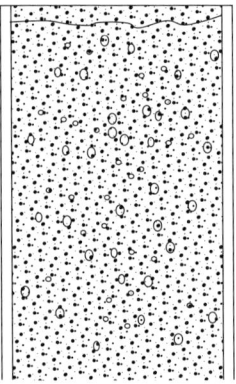

Fig. 11-20 Levelling of the column ends with setting mortar

der transversal compressive loads have been reported in [33].

The results of this research work can be summarized as follows:

- X type joints with the bracing members loaded in compression benefit most from the concrete filling of the chord members, because the vertical compression can be transferred directly through the concrete filling in the chord showing a higher resistance.

- For limit states design, the compressive resistance $N_{1,Rd}$ of RHS X joints with concrete-filled chord is given by:

$$N_{1,Rd} = (A_1 \cdot f_{bk}/\sin\theta_1) \cdot \sqrt{A_2/A_1}/\gamma_b \quad (11\text{-}32)$$

where:
f_{bk} = characteristic cylinder compressive strength of concrete
A_1 = load transmitting area over which the transverse load is applied = $h_1 \cdot b_1$ (see Fig. 11-22)
A_2 = dispersed load transmitting area at a slope 2 : 1 longitudinally along the chord member = $(h_1 + 2\,W_s) \cdot b_1$ (see Fig. 11-22)
γ_b = partial safety factor for concrete compressive strength (= 1.5 according to [5])

It is recommended to use:

L_b = length of concrete in RHS chord member
$\geq (h_1/\sin\theta_1) + 2\,h_0$

$$\frac{h_o}{b_o} \leq 1{,}4$$

h_0 = depth of the chord member
b_0 = width of the chord member
h_1 = depth of the bracing member
b_1 = width of the bracing member

- Tests have shown that the shrinkage in concrete away from the RHS inside side walls does not exercise any detrimental effect on the load bearing resistance of the joint.

Concrete-filling of chord members in the above mentioned tests was done in the workshop by tilting the truss and using concrete or grout with a high water-to-cement ratio.

11.3 Fabrication of concrete-filled hollow section columns

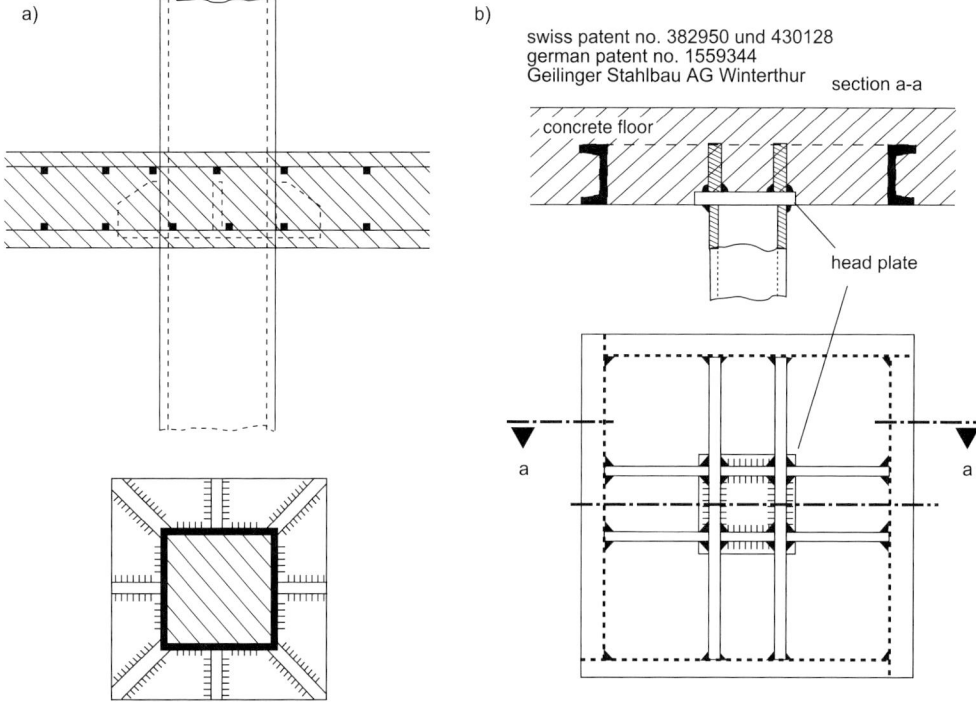

Fig. 11-21 Connection of a concrete-filled column to the concrete floor.
a) Through column, b) discontinuous column, load transmission with steel collar

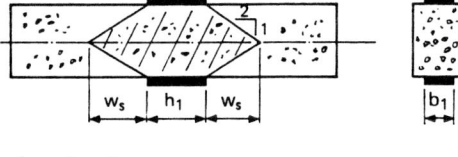

$A_1 = h_1 \cdot b_1$

$A_2 = (h_1 + 2w_s) b_1$

Fig. 11-22 Recommended method to determine the design resistance of concrete-filled RHS loaded in transverse compression

11.3 Fabrication of concrete-filled hollow section columns

11.3.1 Structural components

11.3.1.1 Hollow sections

Vent holes, preferably in pairs, should be drilled at the locations between 10 and 20 cm from the head and foot ends closed with steel plates in order to prevent the column from bursting under the steam pressure generated by the vaporisation, in case of fire, of the dehydration water locked in the concrete fill. The hole diameter shall be equal to 20 mm at the least. In case, the column length is larger than 5 meter, further intermediate holes have to be drilled.

Prior to the filling with concrete, the internal surface of the hollow section must be cleaned; it must be particularly free from oil, grease or excessive rust.

11.3.1.2 Concrete

Generally, the artificial Portland cement is used for filling the hollow section with concrete. The most commonly encountered concrete strength classifications are C35, C45 and C55 (compression strength f_{bk} after 28 days i.e. 45 MPa for C45).

Considering the relatively small sizes of the hollow sections used for structural purpose and the concrete reinforcement steel (Fig. 11-23),

Fig. 11-23 Installing reinforcements in the hollow section before the concrete filling

- The minimum concrete cover (gap between reinforcement bar and internal wall) depends on the maximum aggregate size D and lies between $1.5\,D$ and $2\,D$ (Fig. 11-24). The largest cover should be between 2 and 5 cm.

- For the cross sections with reinforcement, the maximum aggregate size D shall be smaller than the fictitious radius r for the reinforcement as defined for the smallest mesh. The calculation of r is to be done using the following equation (see also Fig. 11-25):

$$r = \frac{a' \cdot b'}{2\,(a' + b')}$$

Further D shall be $< \dfrac{b'}{2}$ and $< \dfrac{1}{3}(b - 2t - b')$.

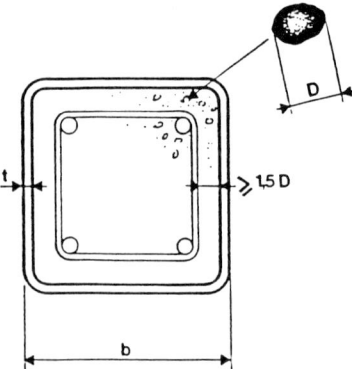

Fig. 11-24 Concrete cover of the reinforcement and maximum aggregate size

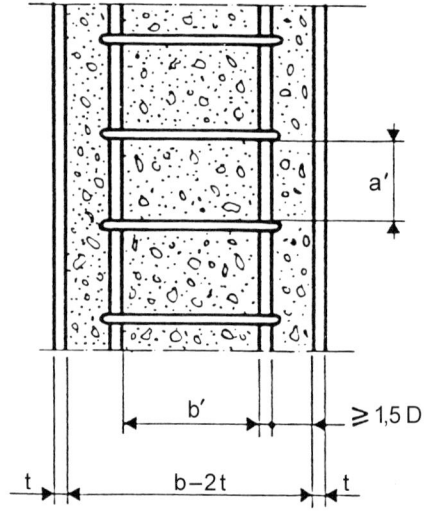

Fig. 11-25 Arrangement of reinforcement bars and stirrups in a concrete-filled hollow section

it is recommended to use a higher sand and cement content (water to cement ratio as low as possible) and a smaller aggregate size (particularly if the concrete is pumped into the column from the bottom). In some cases, plastifiers or liquidisers are added for the better preparation of concrete. Additives, which are likely to cause corrosion (e.g. calcium chloride), have to be excluded.

Maximum aggregate size D as determined by sieve shall be as follows:

- Smaller than 1/8 of the internal dimension of the hollow section for a composite column without reinforcement.

11.3.1.3 Reinforcements

In order to determine the design resistance of concrete-filled hollow section columns, the reinforcement amounting to maximum 3% of the cross-sectional area of the concrete should be taken into account at the room temperature. However this percentage is often exceeded, when the design strength has to be proved for columns exposed to fire. Then it is

11.3 Fabrication of concrete-filled hollow section columns

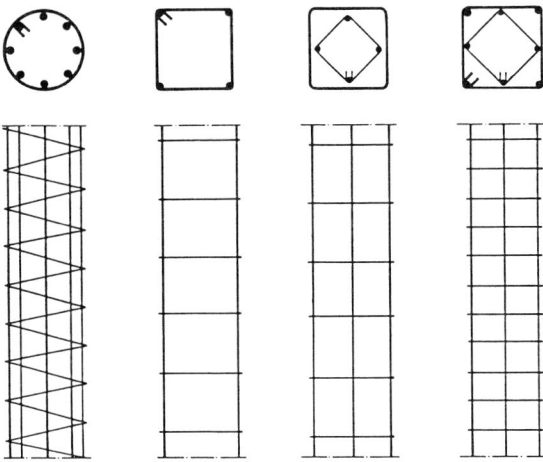

Fig. 11-26 Various reinforcement bar and stirrup arrangements

admissible to take larger share of reinforcement than 3% for the calculation of the design resistance of a concrete-filled hollow section column. Fig. 11-26 shows a number of reinforcement arrangements.

Due to practical reasons, it is not recommended to use reinforcement for hollow sections with sizes (width or diameter) smaller than 200 mm for site filling and 160 mm for workshop filling.

11.3.2 Concrete filling operation methods

Before concrete filling, it has to be assured that the inside of the vertically standing hollow section is free from water and impurities. The reinforcement cage, if any, is than introduced into the hollow section and wedged into its corner positions. The following concrete filling methods are applied:

Gravity filling

This can be carried out in the workshop (Fig. 11-27) as well as on site (Fig. 11-28).

Columns with a height up to 4 m, concrete filling is done with the help of a funnel (Fig. 11-29), for large sizes of hollow sections (column diameter > 500 mm) with a bottom emptying hopper (Fig. 11-30).

The fresh concrete is cast in layers of only 30 cm to 50 cm, each of which is vibrated immediately after being laid. Fig. 11-31 demonstrates this operation by means of an internal poker.

Further, it is recommended to use a funnel with a variable neck length in order to avoid the segregation of the concrete mix (Fig. 11-32).

Concrete pumping

Concrete is pumped into the hollow section column from above through a flexible pipe in this procedure. The end of the pipe is held either closely above or lower than the concrete surface. A better mixing of the concrete can be achieved in the latter case.

Fig. 11-27 Concrete filling in the workshop

Fig. 11-28 Concrete filling of a column on site

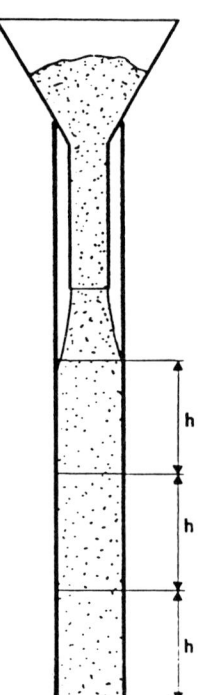

Fig. 11-29 Concrete filling by means of a funnel

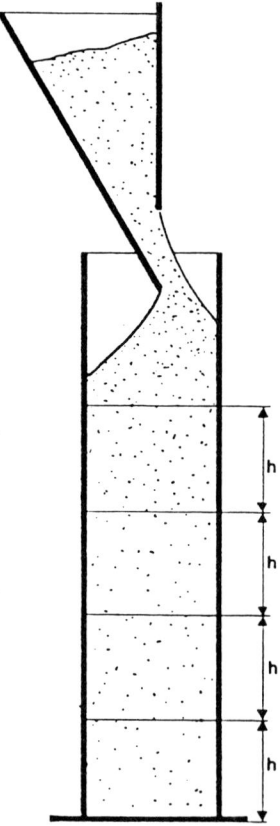

Fig. 11-30 Concrete filling with a bottom emptying hopper

11.3 Fabrication of concrete-filled hollow section columns

In some cases, concrete mix is also pumped into the column from the base through a hole in the hollow section.

Vibration belongs to the most important operations for concrete filling. Vibrators from outside held against the hollow section wall as well as pokers inside are applied in general. However, it is not advisable to use the external vibrator, because there is a risk of breaking the bonds between the columns and the components connected to them.

In order to ensure perfect concrete filling, a concrete-filled hollow section column is hit at close intervals by a hammer from the outside. The ensuing sounds reveal possible defects in the filling.

The defects by inadequate concrete filling can be eliminated by injecting cement through the pierced holes and then closing up the openings.

11.3.2.1 Connections of concrete-filled hollow section columns from floor to floor in a building

The connection of two concrete-filled column sections without reinforcement can be made by means of plates inserted between them (Fig. 11-33). Steel plates are welded to the open ends of the column sections and then bolted together.

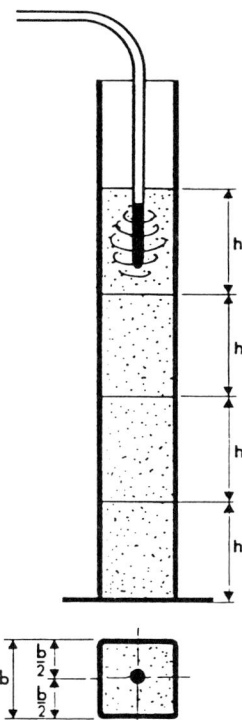

Fig. 11-31 Vibration of the concrete fill with an internal poker

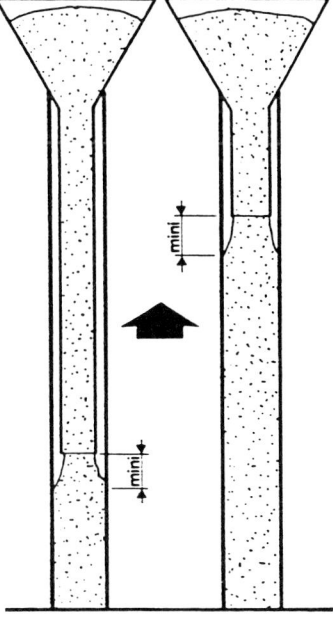

Fig. 11-32 Concrete filling using a funnel with a variable neck length

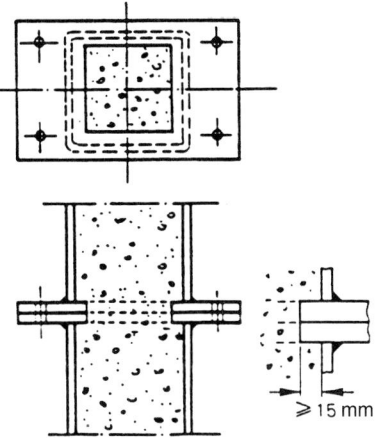

Fig. 11-33 Connection with steel plates inserted between the ends of concrete-filled hollow section columns without reinforcement

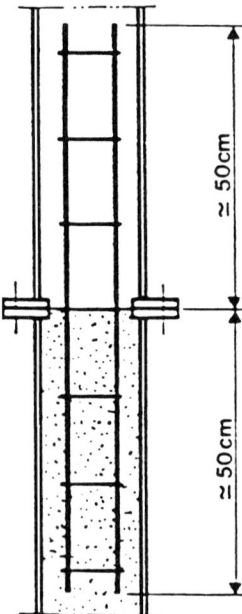

Fig. 11-34 Connection with light reinforcement cage set between two column sections

For reinforcements through column sections, the connection can be achieved by setting a light reinforcement cage through the connection between two column sections (Fig. 11-34). Local insertion of a light reinforcement cage can also be applied in the case of the connection between the unreinforced column sections.

11.4 List of symbols

CHS	circular hollow section
RHS	rectangular hollow section (also square)
A	cross-sectional area
A_v	shear area
E	modulus of elasticity
I	second moment of area
M	moment
N	axial force
V	shear force
W	section modulus
b, b_a, h, h_a	outer side lengths of RHS
d	outer diameter of CHS
f	axial stress
f_y	yield strength of hollow section material
l	length
l_K	buckling length
τ	shear stress
t	wall thickness
t_P	plate thickness

Indices:

a	steel hollow section
b	concrete
k	characteristic
K	buckling
o	without reinforcement or above
pl	plastic
Rd	resistance
s	reinforcement steel
Sd	applied load (action)
u	below

12 Corrosive behaviour and protection against corrosion of steel hollow sections and their structures

12.1 General

As structural elements, steel hollow sections are mostly subjected to mechanical damage and corrosion when they are exposed to atmosphere.

In steel structures, the corrosion occurs due to an electro-chemical reaction between steel and condensed humidity (above 70% relative humidity) forming oxides or rusts. This definition of rust signifies that the rust is not a homogenous oxide, but this requires for its formation the water in its fluid form besides an oxidising reagent. A certain importance falls to sulphur dioxide (SO_2), which is always present in the atmospheric pollutions.

Chlorides act in a similar manner in a sea climate or coastal areas. Steel surface begins to rust with about 70% relative humidity at a temperature between $-20\,°C$ and $+60\,°C$. Dust deposits on the steel sections promote corrosion particularly, as they absorb humidity (CHS is specially advantageous). Similar to the open rolled sections, the hollow sections in steel constructions have to satisfy the following requirements:

a) No reduction of wall thickness by forming rust
b) No discharge of fluid metal solution in the vicinage
c) No undesired change of appearance

With regard to a), an adequately small reduction of wall thickness can be tolerated when the load bearing strength does not relatively decrease. The requirements b) and c) depend on the corresponding applications and the local conditions.

The questions related to the damages due to corrosion, e.g. the detriments of the structural components described in the above mentioned points a) through c) and the required protection measures, cannot be answered in general, which means, each case of application has to be dealt with separately.

The requirements are therefore to be defined already in the planning stadium of a structural project.

The configuration of the steel structures has to be made in a manner, so that they offer less possibilities of attack by the corrosion agents. This requirement takes account of the hollow sections and their welded joints with gusset plates particularly (Fig. 1-13) (without hollow space, water or snow traps, re-entrant corners and the like). Fig. 12-1 illustrates this aspect clearly.

Due to the closed form of hollow sections, their external and internal surfaces are subjected to corrosion differently. Tests have shown that no rust is created in the inside of an airtight welded hollow section and therefore protection measures are not necessary there [14]. For external surfaces of hollow sections, this is not possible by nature. It is therefore to deal with internal and external surfaces of hollow sections separately.

12.2 Internal corrosion of hollow sections and their structures

According to the "state of the art", no internal corrosion takes place, when the hollow sections are airtight sealed by welding cap plates at the ends or by using other sealing devices e.g. washers, gasket rings or plugs for holes. The corrosion process is quickly stopped by the lack of renewed oxygen after the initial closed-in oxygen is exhausted.

Extensive and practical long-term tests had been carried out by the German Railways to

Fig. 12-1 Construction details. a) Neat joint without gusset plate, appropriate for corrosion prevention, b) favours corrosion by cluttered joints

determine the internal corrosive behaviour of steel hollow sections [16–19]. Lit. [14] contains a summary in abbreviated form.

Besides, a comprehensive list of the investigations on the realized structures is contained by [20]. American, English, French, Italian and Japanese investigations have been reported there. Fig. 12-2a shows an example of a truss in CHS, after it had been exposed to industry atmosphere for fifteen years. As the hollow sections are airtight welded, corrosion has not been detected on the inside (Fig. 12-2b).

Based on the results of the long-term corrosion tests by the German Railways on the hollow sections (RHS, CHS) and the CHS trusses, which were partly airtight, partly with openings for the introduction of humidity or partly not at all closed, the following conclusions have been reached:

- On the inside of airtight hollow sections, no condensation of humidity takes place and consequently no corrosion occurs.
  No protection against corrosion is therefore necessitated for the structural elements in hollow sections made airtight, the design and fabrication of which are without any problem technically. Today, welding, making the internal space airtight, is performed for most structures, as long as the selected corrosion protection procedure does not demand any other requirement.

- Condensed water may not collect also in hollow sections in a structure, which are not sealed. Through the openings, which are not perfectly sealed but does not allow surface water (rain, melted snow, dew) to get in, humid air might however enter; no condensation could be observed on the inside in this case. This could be determined that the humidity was transported outside with the air through the opening.

- Humid air can develop rust in a very restricted small area around the opening.

- If in any case surface water is sucked into the hollow section and drains down the inner wall, corrosion is restricted to the wetted locality only. This can be stopped by drilling a hole at a lower spot to drain any water away.

- Freezing in a hollow section is dangerous in the first place, because repeated freezing can lead to rupture of the wall through successive expansion and contraction until the elongation limit is reached. In order to avoid this, a hole with a drain plug is re-

12.3 External corrosion of hollow sections and their structures

a)

b)

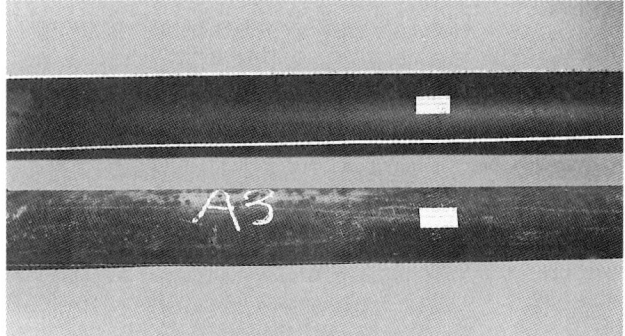

Fig. 12-2 a) Dismantling a truss in CHS after being exposed to industry atmosphere for 15 years, b) internal and external surfaces of a CHS member in the truss shown in a)

commended to be drilled at the lower end to remove the collected water from time to time. A discharge hole without a closing plug shall not be used.

- When a hollow section cannot be sealed due to constructional reasons, care has to be taken that air can pass through and water does not stagnate in the hollow section. Further, other protection measures have to be undertaken to prevent corrosion on the inside. DIN 55928 [1–10] contains further details.

- Bolted end-to-end connections of hollow sections with welded butt straps require special corrosion protection measures. The butt straps do not cover the existing gap airtight in the butt end. This zone has therefore to be partitioned from other hollow spaces. The rest space between the partitions has to be protected against corrosion.

Hand holes to fix the bolts have to be sealed with covers and rubber gaskets [21].

12.3 External corrosion of hollow sections and their structures

The corrosion behaviour of the external surface of the hollow sections does not differ fundamentally from that of the other rolled sections. The following special aspects are mentioned below:

- The surface areas of structures in hollow sections are significantly smaller than those in conventional rolled sections in regular cases. In certain cases, it can go up to 40%. Appreciable saving in spraying and coating materials as well as of labour is the consequence.

- Repeatations and repairs of the coatings for the structures in hollow sections with smooth surfaces and without any sharp edges and corners (Fig. 12-1 a) have to be carried out at longer intervals than those usual for conventional steel structures. In principle, the application of hollow sections instead of open rolled profiles represents already a protection measure against corrosion. The reasons are mainly the lack of sharp edges and small areas for the attack by the corrosive agents. The costs for obtaining uniform paint or spray thickness are low. This results also in a marked better long-time behaviour of corrosion protection.

As regards maintenance work, hollow section constructions are significantly more economical than structures in other profiles.

- Smooth surfaces of hollow sections enable a quick and uniform distribution of coating materials.

12.4 Protection measures against corrosion

Protection measures against corrosion for steel can be subdivided into two main groups: active and passive measures.

These measures can be interpreted so that in the first case, the protection measure takes part directly in the corrosion process itself; this is steered in one direction and does not act as damaging any more. In the second case, the unstable material under corrosive action is fully coated by a significantly more stable one, so that a direct contact of the corrosion agent with the material to be protected is prevented.

The coatings for the passive protection against corrosion can be classified as follows according to the nature of the protection materials:

- Organic coatings (paints and sprays in layers)
- Metallic coatings
- Inorganic non-metallic coatings (oxides, cements, silicates)

The following procedures belong to the active protection measures:

- Removal and neutralization of the aggressive corrosion agents
- Constructional measures
- Hot dip galvanization
- Electro-chemical polarisation (cathodic or anodic protection)

Further, hollow sections are nowadays delivered by the mills to the fabricators already coated with a "shop primer", which guarantees a timely limited protection against corrosion.

Organic and metallic coatings as well as metal platings by hot dip galvanisation have a prime position for the protection of hollow sections against corrosion. On the other hand, cathodic or anodic protection measures are applied only in the presence of a hydrous medium (e.g. for offshore-structures and underground pipelines).

12.4.1 Coatings for protection against corrosion

These consist of prime and top coatings with organic bonding agents and inorganic pigments (see Table 12-1) mostly. The prime coatings contain preferably pigments, which inhibit corrosion reactions on the steel surface. They have an indirect protective effect against corrosion by the increase of the resistance to permeability and above all have the task to lend a better resistance to weather and light to the top coatings. The top coats also contain pigments.

The prime coats have to be applied directly to the surface of hollow section as quick as possible. This is vital, as they not only prevent corrosion, but also ensure a good bond to the subsequent coatings.

The improvement of the adhesion between the prime and top coats belongs to the tasks of the intermediate coats. They are chemically inert and impervious to external atmosphere.

The final top coat is also impervious and resistant to chemical and mechanical actions.

In order to prepare the steel surface for the application of coatings against corrosion (mostly two prime and two top coating layers), the following measures have to be undertaken: thorough cleaning, degreasing, rusting off and descaling.

Cleaning and degreasing are carried out by washing to remove any foreign matter stuck

12.4 Protection measures against corrosion

Table 12-1 Coatings for protection against corrosion, fields for application of bonding agents and pigments

Coatings for atmospheric load			Coatings for chemical load			Coatings for thermal load		
Bonding agent	Pigment		Bonding agent	Pigment		Bonding agent	Pigment	
linseed oil	aluminium	DB	chlorinated rubber	zinc oxide	GB DB	phenolic resin	ferric oxide	GB DB
alkyd resin	red lead	GB	cycloidal rubber	titanic oxide	GB	silicious resin	hematite	DB
bituminous material	white lead	DB	vinyl resin	silicon carbide	GB DB	cumarinic resin	zinc oxide	GB DB
tar-bitumen-asphalt	calcium-plumbic-compound	DB	polyurethane	graphite	GB DB	alkaline silicate	graphite	GB DB
	ferric oxide	GB DB	epoxyd resin	ferric oxide	GB DB		aluminium	DB
	hematite	DB	polyester (unsaturated)	hematite	DB	ethylene silicate	zinc powder	GB
	zinc powder	GB	polychloroprene	lead powder	GB			
	zinc oxide	GB DB	chlorosulfonic polyethylene	stainless steel powder	DB			
	zinc chromate	GB	chlorinated polypropylene					
	titanic oxide	GB DB						
	lead silicone-chromate	GB DB						
	zinc phosphate	GB DB						

GB: Application in prime coat
DB: Application in top coat
} the bonding agents can mostly be used for prime and top coats

to the wall e.g. dust, mud etc., boiling off and spraying off with alkaline agents as well as with organic solutions, which are harmless physiologically.

Removing rusts and descaling are done applying the following procedures:

- Mechanical: manual or mechanical procedures e.g. shot blasting, grinding (with abrasive discs), hammering, needle gun scouring
- Thermal: flame radiation
- Chemical: pickling

Brushing and spraying (pneumatic by pulverisation or electrostatic) techniques, specially with high pressure nozzles, are most suitable for applying paints. Brushing is strongly recommended for the first coat. Zinc rich paints, with 34% metallic zinc content, are specially recommended for their excellent protective quality and easy application, as they dry and harden quickly.

The required specified thicknesses of the coating layers depend on the expected corrosive loads (climate, location, maintenance etc.). The number of the necessary layers is determined by the properties of the coatings.

12.4.1.1 Corrosion protection with a "shop primer"

Nowadays, pre-conserved steels are used frequently. For this purpose, hollow sections are descaled and cleaned in the mill and subsequently coated with a "shop primer" at once. Such a coating thickness is between 15 and 20 μm; this must be able to be overwelded directly and has the function of protecting the steel surface against corrosion during transport, storage and fabrication on site. Further, this acts partly as a prime coat. After the fabrication of the structural elements, the damages, which may occur due to transport, storage and fabrication, have to be repaired without delay. The first prime coat is then applied subsequently. Care has to be taken that the "shop primer" and the prime coat are compatible with each other [32, 33].

12.4.2 Metallic coating by spraying [29–31]

These processes are getting more and more importance, particularly because of the positive experiences with the modern application equipments. As metallic coating sprays, zinc or aluminium or a combination of both are used predominantly [12].

Prior to spraying, the steel surface has to be shot blasted by corundum, sharpedged, granulated material or wire grits. They create the required roughness (specified value about 20–30 µm) of the surface, which assures a safe adhesion of the sprayed layer. Two different metallic spray techniques are preferred: spraying metal melted with flame or electric arc.

In both processes, the spray material in the form of an wire is introduced into a metallising gun. The melting heat is generated by an oxyacetylene or oxypropane flame or by an electric arc. Metal coating is sprayed in an ionized form from the gun with a compressed air jet on the steel surface.

12.4.2.1 Hot dip galvanization

Galvanizing is a reliable, economical and technically advanced process, which protects both the **internal** and **external** surfaces of hollow sections against corrosion. Even in the worst outdoor atmospheres, the standard galvanized coating will last ten years without any maintenance.

Prior to hot dip galvanization, the steel surface has to be descaled, pickled in hydrochloric acid to obtain a chemically clean surface and activated by wetting with fluxing agents. Acid will find its way into tiny crevices. The best way to avoid the problem is to design it out, because zinc used for hot dip galvanization dissolves in acids of almost any strength.

Hollow sections or subassemblies or finished assemblies (depending on the dimension of the available bath) are immersed in a bath of molten zinc; the service temperature lies between 440 °C and 460 °C (at this temperature, iron and zinc react quickly), in special cases also higher.

Zinc is number 30 in the periodic table with a melting point of 419 °C and a vaporisation point of 907 °C. Working with liquid zinc can be dangerous.

In the molten bath, a non-porous metallic layer develops on the steel surface, which consists of layers of iron-zinc-alloy produced by the reciprocal diffusion of the molten zinc with the steel surface and a layer of pure zinc at the top adhered to the intermediate ones. Thereafter, the object is cooled in a water bath or air depending on iron or steel parts and their material compositions.

The duration of protection against corrosion by zinc coating is determined primarily by its thickness, which is measured and indicated in terms of µm (1 µm = 1/1000 mm). DIN EN ISO 1461 [35] recommends the minimum thicknesses of the zinc coatings, which also depend on the environmental conditions i.e. rural, maritime, urban, industrial and tropical. In practice, almost all galvanized steel fabrications in atmosphere have an average coating thickness of at least 85 µm on steel thicker than 6 mm.

The rate of the formation of the iron-zinc-alloy layer depends on the chemical composition of steel, particularly on its silicone content. There is a non-linear dependence between the Si-content and the thickness of the Zn-coating.

A steel work in hollow section must be provided with drilled holes of the correct size and position for zinc to enter and leave and for air to be expelled as the zinc enters. The molten zinc entering in the hollow section leads to an additional protection against internal corrosion.

If hollow sections do not have vent holes, two things may happen. Firstly, the section may explode in the galvanizing kettle as internal moisture flashes to steam. Secondly, steel is not protected on the inside, if the zinc cannot reach it. Normally it is not needed to plug these holes, if they are drilled after galvanization, because airflow through the hollow section provides the best conditions for long life for the internal galvanized coating. If it is needed to plug the vent holes due to safety reasons, i.a. to stop small fingers getting caught in a school playground, plastic or tapered aluminium plugs can be used.

While designing a construction with hollow sections, one should take care of providing every member with proper drainage (Fig. 12-3). Also some thought has to be given to the orientation of the drain holes. All holes should drain in the same direction. "V" notches make a good alternative to drilled holes.

When an welded construction is dipped in molten zinc, the steel is heated and the welding residual stresses in the construction are released consequently; this causes deformations or distortions [22, 23]. They have to be taken into account while designing a welded structure and subsequent galvanization. The constructional measure is to reduce welding to a minimum, because the residual stresses due to shrinkage becomes higher with increasing welding operations.

Heating of an object in the molten zinc bath does not affect the mechanical properties of the structural steels. This has been investigated in detail for S 235, S 355 and St E690 [23].

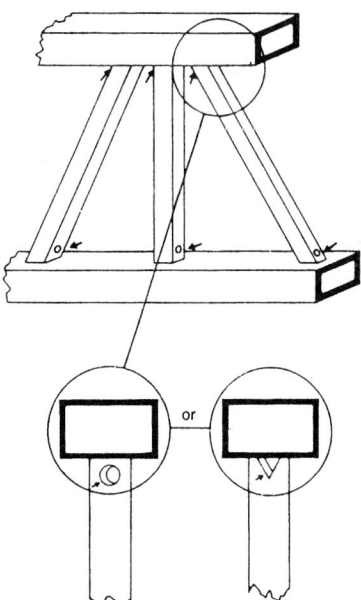

Fig. 12-3 Vent holes for lattice girders in hollow sections

In order to determine the economy of hot dip galvanization, the following influencing parameters have to be checked:

– Distance of galvanizing workshop
– Transport facilities
– Tonnage of the object
– Surface area of the object
– Material
– Pickling
– Form and dimension

An additional coating to the hot dip galvanized steel prolongs the duration of the protection against corrosion considerably (about 1.8 to 2.5 times larger than the sum of the duration of a painted or sprayed coating and the galvanized zinc layer). The synergism of this so-called "Duplex system" consists of the prevention of corrosion of the steel by the galvanized zinc layer and that of the zinc layer against the atmospheric corrosion by the painted or sprayed coatings. Both systems protect themselves mutually [27]. The thickness of the multiple-layer coatings lie between 70 and 360 μm.

Summarizing, there are three reasons for painting a galvanized object:

– To change the appearance
– To achieve an even longer life
– To protect part of a galvanized structure against acid splash

One can obtain all the benefits of the long life of galvanizing together with the good appearance of a coloured finish by powder coating or painting. The galvanized surface will not erode and will stay at the same thickness as long as the paint cover exists. Once the paint cover is disrupted, by age or damage, the galvanized coating thickness will gradually reduce at a rate determined by the conditions. There is no need ever to remove the paint completely from the surface of painted galvanizing. The maintenance requirement is a quick wire brushing followed by re-application of the finish coat.

Welding before and after galvanization

When welding is done before galvanizing, the galvanized coating "takes" to the weld in just the same way as it takes to the rest of the steel. Therefore, all the welding and flame cutting should be completed before galvanizing.

The sizes of the existing molten zinc baths have increased more and more in the last years; but however there are still cases, where it is not possible to galvanize a complete construction in the bath at a time. Therefore single sections or smaller subassemblies are hot dip galvanized, which are then welded together.

Besides welded constructions, bolted structures of galvanized elements are used frequently. It is recommended to use bolted connections wherever it is possible. Galvanized nuts and bolts are available as stock items and the galvanized coating thickness is specified in the European standard. It can be made sure that the threads are free from excess zinc by "spinning", which might stop the nut running up the bolt.

For welding after galvanizing, the problem of overwelding or repair of the damaged galvanization has to be discussed. Zinc melts at 420 °C and boils at 907 °C to form zinc vapour. Both these temperatures are well below the temperature in the welding arc or cutting torch flame. With the increase of the temperature beyond 900 °C, the pure zinc and the iron-zinc-alloy below it are oxidised and even go to smoke. The formation of gray-white toxic zinc oxide vapours hinders the welding by visual obstruction, spatter and unsteady welding course and causes danger to health. They can also generate worm holes and gas cavities in the weld deposits, which interrupt the continuity of the micro-structure. As regards health care, fume has to be removed from the weld site and the workplace has to be ventilated thoroughly.

A series of tests have been carried out in the last years in the Netherlands, France, Germany and England to investigate into the behaviour of welding of the hot dip galvanized structural parts. They show that the effect of the weldability on the zinc coating is small for the usual coating thickness up to about 100 µm. This effect (incorporation of zinc into the weld) may be significant for thicker coatings.

Demerits regarding the quality of the weld can be avoided by:

– Reduction of weld rate by about 20% (zinc vapour cannot escape completely from the weld and enter into the welding puddle, when the weld rate is fast)
– Distance of head plate depending on the material thickness of 1 through 3 mm
– Oscillating movement of the electrode (easy vaporisation and escape of zinc)
– Insignificant increase of current voltage
– Application of suitable electrodes
– Sucking off zinc oxide vapour

In general, today the zinc coating is burnt off in the welded zone (up to about 50 mm from the weld seam) and then removed by grit blasting or grinding and subsequent wire brushing, so that the coating material does not affect the welding. Welding is then conducted followed by cleaning the weld area and repairing the coating as soon as possible with painting or spraying reconditioning material e.g. zinc rich paint or metallic zinc. Zinc rich paints are preferred for its low cost, easy applications, good adhesion and adequate wear resistance.

Arc welding under controlled atmosphere, as performed by MAG method, does not require any step to burn off the zinc coating, which is done during welding itself without any loss of the mechanical properties of the weld. Reconditioning of the weld is done as described above.

Zinc coating by hot dip galvanization is enough to keep corrosion away from small areas of accidental damage; but more serious damages have to be repaired. The following methods can be used in these cases:

• The damaged area is to be wire brushed and then several coats of zinc rich paint are to be applied until 30 µm more than the original coating thickness is reached.

• The damaged area is to be wire brushed and then heated with a blow torch also applying one of the proprietary alloy rod compounds to the required thickness.

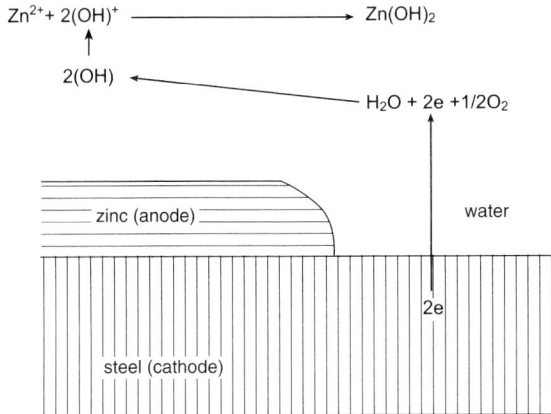

Fig. 12-4 Cathodic protection with zinc as sacrificing anode

- The damaged area is to be grit blasted and then sprayed by zinc coating until 30 μm more than the original coating thickness is obtained.

12.4.2.2 Zinc coating in galvanic bath

In this process, zinc coating is generated by immersing the steel in a zinc galvanization bath and then applying an appropriate voltage under direct current to it. Electrolytical zinc coatings of thickness between 5 and 25 μm according to the field of application can be put on only in the galvanizing workshops. Galvanized zinc coatings are often chromium plated to improve the protection against corrosion.

12.4.3 Electrochemical polarisation

The electrochemical protection against corrosion represents an active protection measure, in which the direct current to the steel surface to be protected is polarised in the potential zone, where the rate of corrosion is of an acceptable magnitude. This is called a "cathodic" protection (see Fig. 12-4), when the object to be protected acts as a cathode. The cathodic polarisation can also be obtained by the contact with metals, which have a more negative (base metal) corrosive potential in the observed medium than that of the object to be protected. The electrochemical protection can principally be applied in the permanent presence of an electrolyte e.g. water.

12.4.4 Application of hollow sections of weathering steels

The weathering steels show a stable behaviour against the atmospheric corrosion, where adhesive and impervious oxide layers are generated on the steel surface under normal weather conditions improving the corrosion resistance. This is specially activated by the alloy elements e.g. chromium, copper and nickel and partly by an increased content of phosphorus.

13 Structural hollow sections exposed to fire

13.1 General

The structures in steel are handicapped against concrete buildings, as their economy is often impaired by the additional requirements for fire protection. The extra costs can reach even about 30% of the total manufacturing cost. In order to obtain a technically expedient and economical solution, the fire behaviour of a steel construction must be taken into account already when the planning of a project starts.

With regard to fire protection, steel hollow sections offer more economical advantages than the conventional open profiles e.g. I, L and U forms, as they present a large range of facilities to assure protection against fire. These protection methods contain not only external insulation, where materials with lower thermal conductivity e.g. insulating clamp, sprayed plaster, heat protection shield etc. are applied on the outside, but also include other procedures, which are applicable to hollow sections only. Filling the hollow section with concrete or with stationary or circulating water belongs to them.

In contrast with the open profiles, the plate elements of hollow sections are only exposed one-sided to fire. Besides, the length of the perimeter of a hollow section is smaller than an open profile with the same ratio of the perimeter length to the cross-sectional area. This results in a smaller quantity of fire protection material and less labour for hollow sections. In other words, hollow sections have a significantly smaller surface area exposed to fire for a certain steel volume (A_m/V) than open profiles, which means an improved economy with their use.

The requirements for the fire protection of structures or structural elements are characterized by fire resistance time, which the objects can withstand under fire load. Fundamentally, the fire resistance of a structural element is determined by the development of temperature by the fire, which is represented by the amount of heat given off as the material burns (in general, it is expressed by the amount of burnable material to convert into equivalent amount of wood which would evolve the same amount of heat when burnt e.g. in kg (wood) per m² surface area: specific fire load) under the given ventilation conditions. As this may be considerably different from case to case, a time vs. temperature curve is taken as the basis for practical fire protection calculation, which is designated by the so-called "standard fire curve" defined in ISO 834 [4] (see Fig. 13-1). Alternative standard fire curves are also in use i.e. in the U.S.A. [20] or in maritime applications [21]; but they deviate little from the ISO-curve and are without any practical significance.

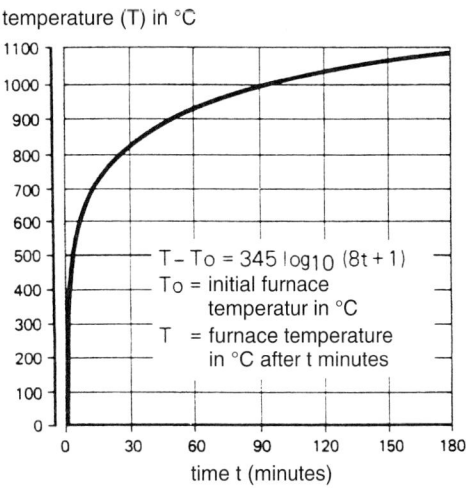

Fig. 13-1 The standard fire curve according to ISO

Table 13-1 Fire resistance classes R [22, 23]

Fire resistance class	Fire resistance (time) in minutes	Remarks
R 30 R 60	30 60	Fire retardant
R 90 R 120	90 120	Fire resistant
R 180	180	Highly fire resistant

Table 13-2 Variations in required fire resistance [26]

Type of building	Requirements	Fire resistance class
One storey	• None or low	• Possibly up to R 30
2 to 3 storey	• None up to medium	• Possibly up to R 3
More than 3 storey	• Medium	• R 60 to R 120
Tall buildings	• High	• R 90 and more

The time period of the heat exposure, which a building component can withstand according to the standard fire curve, is designated as "fire resistance". Basically, there are two possibilities to determine the fire resistance:

– Experimental approach based on standard fire tests
– Analytical approach based on fire engineering

The latter is a relatively modern development, which has become possible due to the advanced computer technology. The fire resistance of a structural element in a fire test can be determined by measuring the time in accordance with two alternatives:

1. Duration to failure of a structural element for specified fire exposure in the furnace under serviceability load.
2. Time required for a critical part of a structure or a structural element to reach the critical temperature for failure (about 550 °C for steel).

Required fire resistances are classified in various national codes [3, 19]; they depend on the following factors:

– Function of the structural element (i.e. load bearing, non-load bearing, space enclosing)
– Type of utilization
– Height and size of a building
– Effectiveness of fire brigade action
– Active measures such as vents and sprinklers (not in all countries)

In most countries of the world, the required fire resistance is not higher than about 90 to 120 minutes. If such requirements are set at all, the minimum value is usually 30 minutes (some countries however prescribe a minimum value of 15 or 20 minutes).

In general, the following rules apply:

– No special fire resistance is required for buildings with limited specific fire load (about 15 to 20 kg/m^2) or where the consequence of collapse of a structural element or a part of a structure is acceptable.
– In order to make a safe evacuation of the occupants and a quick fire brigade intervention possible, a certain but limited fire resistance is necessary.
– The required fire resistance of the main structure must be enlarged to ensure that the structure can survive a full burn-out of combustible materials in the building or a specified part of it.

The fire design of structures is normally carried out for equivalent static boundary conditions as for that under room temperature. In a multiple-storey, non-sway, braced structure, the buckling length of a column is usually set to be the height between two floors. However such structures are divided into compartments and fire is likely to be restricted to one storey. Therefore any column exposed to fire will lose its rigidity, while the adjacent members will remain relatively cold. Accordingly, if the column is rigidly connected to the adjacent members, built-in end conditions can be assumed in the case of fire. In this case, a smaller buckling length of the column may be taken into account depending on the boundary conditions. If the column heated on one level continues rigidly to the upper and lower levels, the minimum buckling length should be $L_{cr,\theta} = 0.5 \cdot L$ (L = height between two floors) (see Fig. 13-2). This value covers the majority of practical situations in multi-storey buildings.

For a column on the top floor, this condition will not generally be met as far as the connection of the column with the roof girders is concerned. At the top, the column itself and all the adjacent members are normally affected by the fire in a similar way. Consequently, at the top of the column, a hinge must be assumed. If this column continues rigidly to the lower level, the minimum buckling length should be $L_{cr,\theta} = 0.7 \cdot L$.

Structural fire design is defined by four levels of assessments:

Level 1: Design tables and diagrams
Level 2: Simple calculation models
Level 3: Advanced calculation models
Level 4: Testing

Level 1

The design tables and diagrams provide solutions on the safe side and allow a quick design, although for a limited range of application. This is clearly the lowest level of design solution.

Level 2

Simple calculation models provide simplified design methods applied to individual members using conventional, conservative design procedure, which however yield adequate accuracy normally.

In the following, main attention will be given to the simple calculation method.

Classification of cross sections:

In a fire design situation, the classification of cross section of a structural member is to be defined as for normal temperature design as given in Section 5.2.2.

Strength and deformation properties of steel at elevated temperature [22]:

For heating rate between 2 and 50 K/minute, the stress vs. strain relationship at elevated temperature can be obtained from Fig. 13-3. Table 13-3 gives the reduction factors $K_{y,\theta}$, $k_{p,\theta}$ and $k_{E,\theta}$ relative to the appropriate value

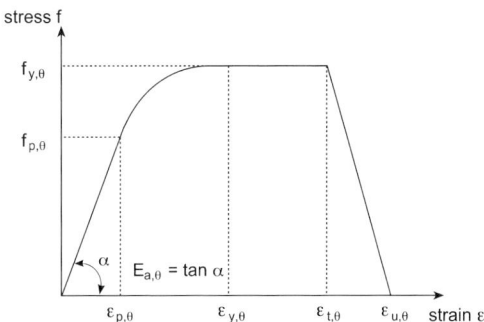

$f_{y,\theta}$ = the effective yield strength
$f_{p,\theta}$ = the proportional limit
$E_{a,\theta}$ = the slope of the linear elastic range
$\varepsilon_{p,\theta}$ = the strain at the proportional limit
$\varepsilon_{y,\theta}$ = the yield strain
$\varepsilon_{t,\theta}$ = the limiting strain for yield strength
$\varepsilon_{u,\theta}$ = the ultimate strain

Fig. 13-3 Stress vs. strain relationship for steel at elevated temperatures

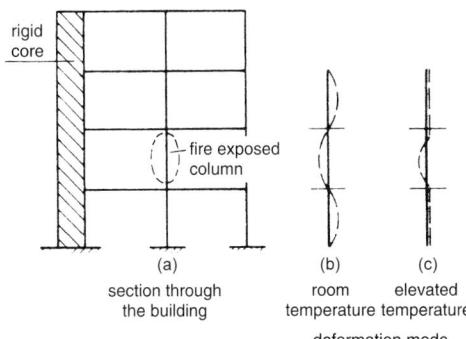

Fig. 13-2 Structural behaviour of columns in a braced structure

Table 13-3 Reduction factors $k_{y,\theta}$, $k_{p,\theta}$ and $k_{E,\theta}$ for stress vs. strain relationship of steel at elevated temperatures

Steel temperature θ_a (°C)	Reduction factor for effective yield strength $k_{y,\theta} = f_{y,\theta}/f_y$	Reduction factor for proportional limit $k_{p,\theta} = f_{p,\theta}/f_y$	Reduction factor for the slope of the linear elastic range $k_{E,\theta} = E_{a,\theta}/E_a$
20	1.0	1.0	1.0
100	1.0	1.0	1.0
200	1.0	0.807	0.900
300	1.0	0.613	0.800
400	1.0	0.420	0.700
500	0.780	0.360	0.600
600	0.470	0.180	0.310
700	0.230	0.075	0.130
800	0.110	0.050	0.090
900	0.060	0.0375	0.0675
1000	0.040	0.0250	0.0450
1100	0.020	0.0125	0.0225
1200	0.000	0.000	0.000

Note: For intermediate values of the steel temperature, linear interpolation may be used.

at 20 °C for the stress vs. strain curve in Fig. 13-3:

- Effective yield strength, relative to yield strength at 20 °C, $k_{y,0} = f_{y,\theta}/f_y$
- Proportional limit, relative to yield strength at 20 °C, $k_{p,\theta} = f_{p,\theta}/f_y$
- Slope of linear elastic range, relative to slope at 20 °C, $k_{E,\theta} = E_{a,\theta}/E_a$

According to [22], the partial safety factor for the fire situation $\gamma_{M,\theta}$ is recommended to be equal to 1.0 for both thermal and mechanical properties of steel.

Level 3

Advanced calculation models may be applied to individual members, to subassemblies or to entire structures. They represent the most sophisticated level providing a realistic analysis of structures exposed to fire including a complete thermal and mechanical analysis and the given material properties for the relevant temperature range. This method enables real boundary conditions including combined effects of mechanical actions, geometrical imperfections and thermal actions to be considered and takes into account the influence of non-uniform temperature distribution over the section based on the acknowledged principles and assumptions of the theory of heat transfer, which leads to more realistic design. The application of this method requires however expert knowledge and a number of computer programs [39–31] have been developed to solve problems regarding structures exposed to fire.

Level 4

For special cases, fire resistance of individual members, subassemblies or seldom also large parts of a structure is determined by actual fire tests in testing centres. The concept of fire testing is by and large the same in various countries. National testing procedures, at least in the European Union, have been harmonized [22–24].

13.2 Design resistance of unprotected and unfilled steel hollow sections

Steel belongs to the non-inflammable materials. But, as has been already mentioned, the structural element of steel loses its ability to function at about beyond the critical temperature of 550 °C.

For steel sections, the development of temperature based on the standard fire curve [4] depends on the "section factor" A_m/V, where A_m = envelope surface area of the member per unit length and V = volume of the member per unit length (see Fig. 13-4). The values for the section factor of the standardized sizes of hot formed hollow sections lie between 45 and 465 m^{-1}, which can be calculated as follows:

- Circular hollow section exposed to fire on all sides: $\dfrac{A_m}{V} = \dfrac{1}{s}$
- Rectangular hollow section (or welded box section of uniform thickness) exposed to fire on all sides for $s \ll b$: $\dfrac{A_m}{V} = \dfrac{1}{s}$, where s = wall thickness of the hollow section

Thickwalled sections (small A_m/V) are heated more slowly; hence they have higher fire resistance than thinwalled sections. Unprotected steel sections show in general only a limited fire resistance of 15 to 20 minutes depending on the load level and the section factor. They reach a fire resistance less than 30 minutes under standard fire load as a rule and therefore cannot be classified in the regular fire resistance classes (R).

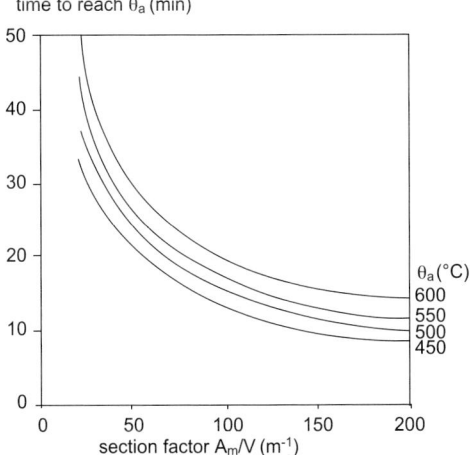

Fig. 13-4 Time for an unprotected steel section to reach a given mean temperature based on the standard fire curve according to ISO [4] as a function of the section factor

It is assumed while designing structural hollow sections using the simplified method that all four sides of the cross section are uniformly exposed to fire. If however a hollow section is built adjacent to a concrete wall, heat cannot be transferred to the flange of the hollow section through the concrete wall due to its very low conductivity. In this case, the hollow section is heated only on three sides. A lower section factor is recommended for the design accordingly.

Example:

Square hollow section 200 × 200 × 6.3 mm
A_m/V (exposed to fire on three sides) = 125 m^{-1}
A_m/V (exposed to fire on all four sides) = 165 m^{-1}

13.2.1 Design resistance of hollow section members under tension [25]

The design resistance $N_{\theta,Rd}$ of a tension member with a uniform temperature θ_a is determined from:

$$N_{\theta,Rd} = k_{y,\theta} \cdot N_{pl,Rd} [\gamma_{M,0}/\gamma_{M,\theta}] \quad (13\text{-}1)$$

where:
$k_{y,\theta}$ = see Table 13-3
$N_{pl,Rd}$ = plastic axial tension resistance at room temperature = $A_a \cdot f_y$
$\gamma_{M,0}$ = partial safety factor at room temperature
$\gamma_{M,\theta}$ = partial safety factor at elevated temperature

13.2.2 Design resistance of hollow section members with a uniform temperature θ_a under combined axial tension and bending moment

The degree of utilization η for hollow sections at elevated temperature is calculated as follows:

$$\eta = \frac{N_{Sd,\theta}}{N_{pl,Rd}} + \frac{M_{Sd,x,\theta}}{M_{pl,Rd,x}} + \frac{M_{Sd,y,\theta}}{M_{pl,Rd,y}} \quad (13\text{-}2)$$

where:
$N_{Sd,\theta}$ = design axial tension in fire case
$N_{pl,Rd}$ = plastic axial tension resistance of the effective cross section at room temperature

$M_{Sd,x,\theta}$ = design bending moment about x-axis in fire case

$M_{pl,Rd,x}$ = plastic moment resistance of the total cross section about x-axis at room temperature

$M_{Sd,y,\theta}$ = design bending moment about y-axis in fire case

$M_{pl,Rd,y}$ = plastic moment resistance of the total cross section about y-axis at room temperature

For members in bending moment only, the following relationship is valid:

a) $\eta = M_{Sd,\theta}/M_{pl,Rd}$ for laterally restrained beams
b) $\eta = M_{Sd,\theta}/M_b$ for laterally unstrained beam

where:
M_b = design moment resistance for lateral-torsional buckling

13.2.3 Design buckling resistance of hollow section members with class 1, class 2 or class 3 cross-sections with a uniform temperature θ_a under axial compression [25]

The design buckling resistance $N_{\theta,Rd}$ of a compression member with a uniform temperature θ_a is determined from:

$$N_{\theta,Rd} = \chi_\theta \cdot A \cdot k_{y,\theta} \cdot f_y / \gamma_{M,\theta} \quad (13\text{-}3)$$

where:
$k_{y,\theta}$ = see Table 13-3
$\gamma_{M,\theta}$ = partial safety factor at elevated temperature
χ_θ = reduction factor for flexural buckling at elevated temperature

The value of χ_θ should be taken as the lesser of the values of $\chi_{\theta,y}$ and $\chi_{\theta,z}$ determined according to:

$$\chi_\theta = \frac{1}{\varphi_\theta + \sqrt{\varphi_\theta^2 - \bar{\lambda}_\theta^2}} \quad (13\text{-}4)$$

where:

$$\varphi_\theta = \frac{1}{2}\left[1 + \alpha \cdot \bar{\lambda}_\theta + \bar{\lambda}_\theta^2\right] \quad (13\text{-}5)$$

$$\alpha = 0.65\sqrt{235/f_y} \quad (13\text{-}6)$$

$$\bar{\lambda}_\theta = \bar{\lambda}_k [k_{y,\theta}/k_{E,\theta}]^{0.5} \quad (13\text{-}7)$$

$\bar{\lambda}_k$ = see Eq. 5-53

$k_{y,\theta}, k_{E,\theta}$ = see Table 13-3

13.2.4 Design buckling resistance of hollow section members with class 1 and class 2 cross sections with a uniform temperature θ_a under axial compression and bending moment [25]

The design load is to be verified basing on the following condition:

$$\frac{N_{Sd,\theta}}{\chi_{\theta,min} \cdot A \cdot k_{y,\theta} \cdot \dfrac{f_y}{\gamma_{M,\theta}}} + \frac{\kappa_x \cdot M_{Sd,x,\theta}}{W_{pl,x} \cdot k_{y,\theta} \cdot \dfrac{f_y}{\gamma_{M,\theta}}} +$$

$$+ \frac{\kappa_y \cdot M_{Sd,y,\theta}}{W_{pl,y} \cdot k_{y,\theta} \cdot \dfrac{f_y}{\gamma_{M,\theta}}} \leq 1 \quad (13\text{-}8)$$

where:
$\chi_{\theta,min} = \min(\chi_{\theta,x}, \chi_{\theta,y})$ (see Eq. 13-4)

$$\kappa_x = 1 - \frac{\mu_x \cdot N_{Sd,\theta}}{\chi_{\theta,x} \cdot A \cdot k_{y,\theta} \cdot \dfrac{f_y}{\gamma_{M,\theta}}} \leq 3 \quad (13\text{-}9)$$

$$\mu_x = (1.2 \cdot \beta_{M,x} - 3)\bar{\lambda}_{x,\theta} + 0.44\beta_{M,x} - 0.29$$
$$\leq 0.8 \quad (13\text{-}10)$$

$$\kappa_y = 1 - \frac{\mu_y \cdot N_{Sd,\theta}}{\chi_{\theta,y} \cdot A \cdot k_{y,\theta} \cdot \dfrac{f_y}{\gamma_{M,\theta}}} \leq 3 \quad (13\text{-}11)$$

$$\mu_y = (2\beta_{M,y} - 5)\bar{\lambda}_{y,\theta} + 0.44\beta_{M,y} - 0.29 \leq 0.8 \quad (13\text{-}12)$$

$$\bar{\lambda}_{y,\theta} \leq 1.1 \quad (13\text{-}13)$$

$\beta_{M,x}$ and $\beta_{M,y}$ shall be obtained from Table 5-26.

For cross section class 3, W_{pl} shall be substituted by W_{el}.

13.2.5 Simplified design of a hollow section column exposed to fire under concentric and eccentric axial compressive load

The simple calculation is based on the mechanical response of axially loaded hollow section columns to the critical temperature of steel, which depends on the ratio between the load (action) which is present during a fire outbreak and the minimum collapse load at room temperature. This ratio is called the degree of utilization (η). This design method holds for class 1, 2 and 3 cross sections only and can be applied both to protected and unprotected columns. For columns with a class 4 cross section, a default value for the critical temperature of 350 °C is to be used.

For eccentric loading, not only the axial force but also the moment distribution has to be taken into account.

For a concentrically loaded column in axial compression:

$$\eta = \frac{N_{Sd,\theta}}{\chi_{min} \cdot N_{pl,Rd}} \quad (13\text{-}14)$$

$N_{pl,Rd}$ is to be calculated based on the gross cross section and χ_{min} is the reduction factor to be read from the curve "c" given in Fig. 5-4 (European buckling curves) irrespective of the type of hollow section or its material. According to the Eurocodes [22, 25], in the fire situation, generally only about 60% of the design load for normal conditions of use has to be taken into consideration. The partial safety factor $\gamma_{M,\theta}$ is to be equal to unity.

For an eccentrically compressed column, the interaction curve describing the combination of axial force and applied moment is given by the following relationship.

$$\eta = \frac{N_{Sd,\theta}}{\chi_{min} \cdot N_{pl,Rd}} + \frac{\kappa \cdot M_{Sd,\theta}}{M_{pl,Rd}} \leq 1 \quad (13\text{-}15)$$

where:
κ = reduction factor according to Eq. (5-88) [22], see Section 5.3.3.3
$M_{Sd,\theta}$ = design bending moment in fire case
$M_{pl,Rd}$ = plastic moment resistance of the total cross section at room temperature

In this case, it is also possible to define an equivalent degree of utilization η_{eq}, which can be described as follows:

$$\eta_{eq} = \frac{N_{eq,Sd}}{\chi_{min} \cdot N_{pl,Rd}} = \frac{N_{Sd,\theta}}{\chi_{min} \cdot N_{pl,Rd}} + \frac{\kappa \cdot M_{Sd,\theta}}{M_{pl,Rd}}$$

$$N_{eq,Sd} = N_{Sd,\theta} + \chi_{min} \cdot N_{pl,Rd} \cdot \frac{\kappa \cdot M_{Sd,\theta}}{M_{pl,Rd}}$$

$$(13\text{-}16)$$

By means of the equivalent axial force $N_{eq,Sd}$, the critical temperature of an eccentrically loaded columns can be determined using the simple calculation model, valid for concentrically loaded column in axial compression (see Eq. 13-14).

13.3 External protection against fire

13.3.1 External fire insulations

Fig. 13-5 shows the conventional methods for external fire protection by enveloping the object to be protected with insulating materials.

The design of insulation materials for steel structures depends on the following parameters:

– Type of the insulating material (mainly based on thermal conductivity)
– Thickness d_i of the insulating material
– Required fire resistance (R)
– Section factor of the steel profile (A_m/V)

In general, any desired fire resistance can be achieved by applying the required thickness of the insulating material. Informations regarding the thermal properties and the suitable fabrication conditions (stickability or adhesiveness, condition of the steel surface or any special surface preparation, design of joints or fixing systems, type and distance of the board connectors etc.) of an insulation material should be obtained from the manufacturers.

Fig. 13-5a shows a spray coating or plaster based mainly on mineral fibre or lightweight aggregates such as perlite and vermiculite, which is applied directly to the external surface of the hollow section.

In Fig. 13-5b, a metal or rockwool or glasswool mat is at first fixed to the outside wall of the hollow section and a fire proof coating is then sprayed on it.

Insulations by means of plaster boards made of gypsum, vermiculite or mineral fibre are shown in Fig. 13-5c and d. The boards are

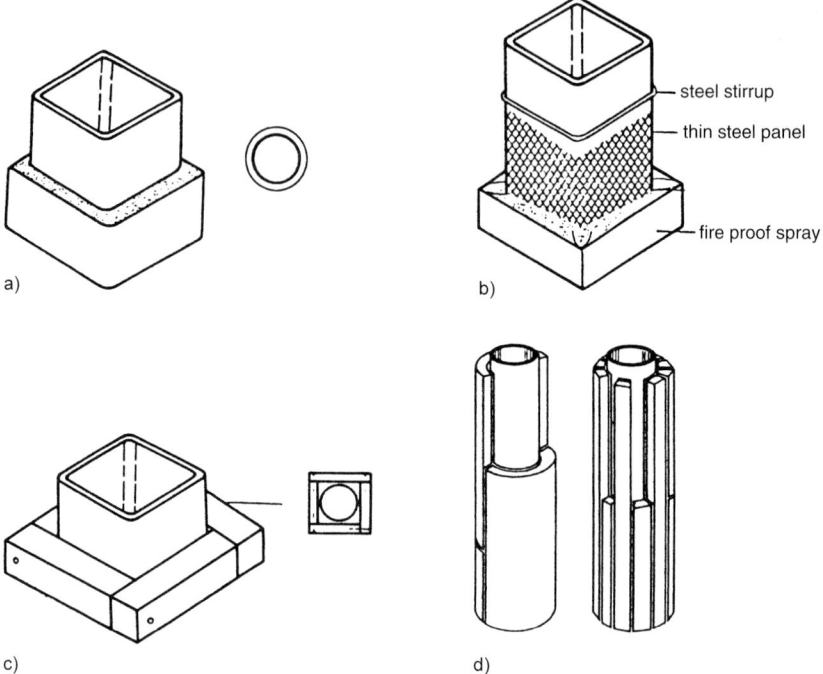

Fig. 13-5 Fire protection envelopes.
a) Plaster by in situ spraying on the external surface of RHS and CHS, b) fire proof material sprayed on metal mat, c) nailed plaster board of fire proof material on RHS and CHS, sometimes additionally sprayed or painted with fire proof materials, d) adhesive bonded fire poof plaster boards

nailed together in the former and are glued together in the latter. Finally they can be sprayed or shielded by thin steel panel on the outside in addition.

13.3.2 Fire protection paints

A special position with respect to the fire protection is occupied by the so-called intumescent coatings. These paints are mixtures, where the thin film coatings or mastics foam and swell under influence of heat and flames to a multiple of their original thickness (producing an insulating char layer sometimes 50 times thicker). They can provide up to 2 hours fire resistance. They can be applied by brush or spray.

A number of intumescent paint mixtures is available, which belong to the fire class R30. The products capable of achieving higher fire resistance are also offered in the market (see [1]). In most countries, the intumescent paints have to be approved by the governmental authorities.

13.3.3 Design of external insulation

The presentation of the required data for the fire design differ from country to country, which is demonstrated by the curves given in Fig. 13-6. In a design, where two of the parameters shown in Fig. 13-6 (e.g. required fire resistance, critical steel temperature and section factor) are known, the third parameter (e.g. thickness of insulation d_i) can be read from the diagrams.

Further, as mentioned in Section 13.1, there are also modern design procedures available, which take help of a number of computer programs [29–32] in order to determine the thickness of the insulation under actual boundary conditions (e.g. conductivity, temperature loading etc.).

13.4 Fire protection of steel hollow sections by water cooling

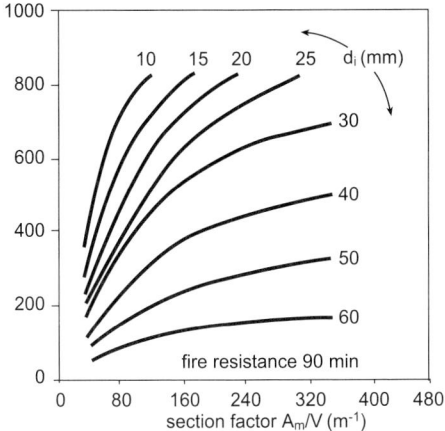

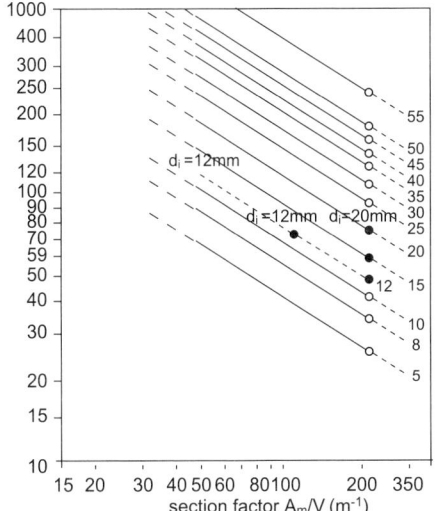

Fig. 13-6 Presentation of the thermal characteristics of a typical fire insulation material in a) France and b) Germany

13.4 Fire protection of steel hollow sections by water cooling

13.4.1 Fundamentals to water cooling systems

Water cooling of steel hollow sections using permanent water filling is an economical alternative to the conventional fire protection by external insulation. Contrary to the external insulation, which prevents the heat to reach the steel wall, the heat in this case is removed continuously by the natural circulation of water in the hollow section. This takes place most effectively by using water filled hollow sections, which are connected to one another by pipes in a closed circulating system. The natural water circulation in a properly designed system is activated, when the hollow sections (columns) are locally heated by the outbreak of fire. The density of hot water is lower than that of cold water, which produces pressure differentials activating natural water circulation. The effect is still more intensive, if the water boils locally and develops steam, as the mixture of water and steam has a significantly lower density than hot water. The production of steam increases with the further development of fire, which promotes the cooling effect by the activated natural water circulation. This behaviour is self controlling, as the cooling effect grows stronger with rising fire load.

Fire protection by water cooling in buildings can be applied both to overhead beams and columns in hollow sections; but the advantages for the beams are not adequate. However the application of hollow sections as water filled columns is comparatively frequent.

Figs. 13-7 and 13-8 illustrate the principle of water cooling schematically. The columns are connected at their tops and bottoms to one another by horizontal pipe lines and then filled up with water. In a water storage tank, which acts as surge, steam-out and reserve vessels at the same time, the water level is kept constant. In case of fire outbreak at one or more locations, the water is set in motion by the thermal convection phenomenon due to the different water densities in the heated and cold regions and evacuates the surplus heat in the form of steam by the vent.

Heat transfer can be described as follows: The heat from the fire flows to the column mainly by radiation and to a lesser extent by the convection of the hot gases. It then passes through the column plate by conduction. Boiling occurs on the inside of the column plate and heat is transferred to the water by conduction and convection. This heat is then absorbed chiefly as latent heat when the water is converted to steam and removed from the column by the vent.

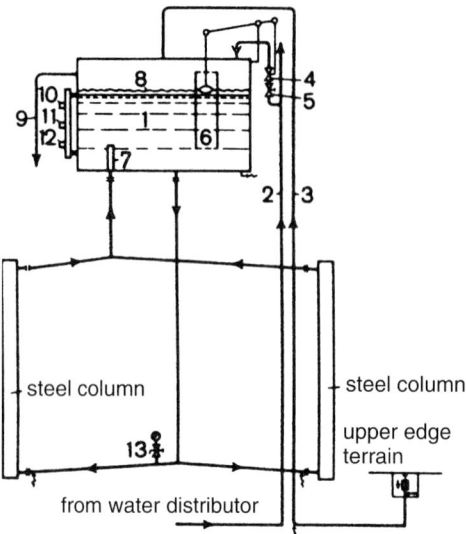

Fig. 13-7 Water circulating system (sketch) with reserve tank
1. reserve tank
2. feeding pipe from water distributor
3. dry riser
4. float valve with test equipment
5. shutoff valve with sealed wheel (open)
6. steadying pipe
7. gauge pipe
8. oil film
9. overflow
10. alarm overflow ⎫
11. normal service   ⎬ Filling inspection with movable contacts
12. alarm water shortage ⎭
13. contact manometer

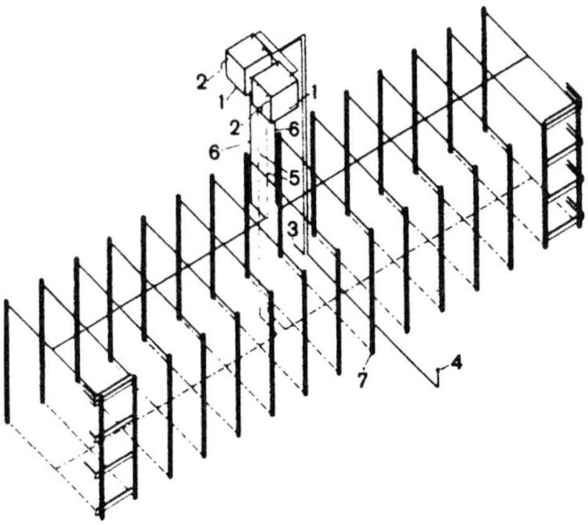

Fig. 13-8 Total view of the water cooling system for a building. 1. storage tank, 2. overflow, 3. feeder, 4. dry riser, 5. pipe loop below, 6. pipe loop above, 7. columns

The water circuit can be extended also to the floor (overhead) beams. Natural convection may no longer be sufficient in this case and pumping is then required to boost and regulate the flow. The circuit can be used to feed sprinklers or serve as a central heating system. However, it is not advisable to combine natural circulation and forced circulation by pumps, as there is danger of the pumps acting against natural circulation, which may lead to the collapse of the cooling system.

High thermal conductivity of steel as well as adequate heat transmission between steel and circulating water care for keeping the average steel temperature far below the critical temperature of about 550 °C also in case of long time fire. It is to be specially mentioned, unlimited fire resistance of the water cooled steel structure is possible by continuous refilling the system with new water.

In the tests, which were carried out during the construction of the three-storey office building for the Research Institute of the German Steel Industry Association (Fig. 13-9), the steel temperature oscillated finally to 200 °C after about 60 minutes and did not rise any more even by further heating [8] (Fig. 13-10).

Principally, the following three water cooling systems with permanent water filling are applied:

Unreplenished columns

In this system, the columns are simply filled with water with no provision for replacing the water lost by steam production. Due to the limited fire resistance (not more than about 60 minutes), this column type should be used only for lower fire resistance requirements.

Columns with external pipe connections
(see Figs. 13-11a and 13-12a)

This system is provided with a connecting down pipe between the top and bottom of the columns. The light mixture of water and steam flowing upwards has to be separated from each other at the top, so that water can

13.4 Fire protection of steel hollow sections by water cooling

Fig. 13-9 Office building of the Research Institute of the German Steel Industry Association

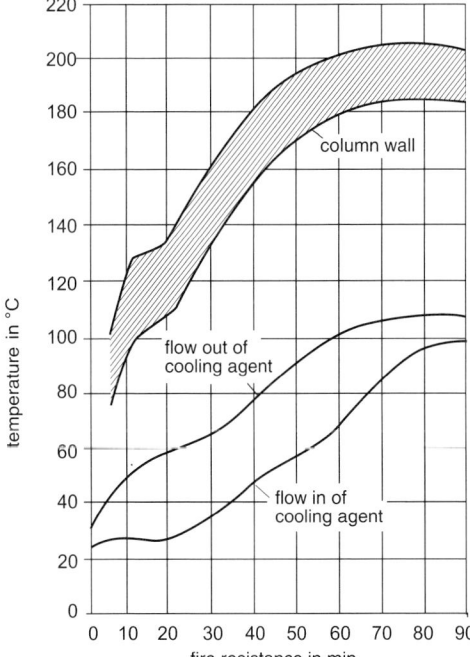

Fig. 13-10 Measured temperatures on the steel wall and in the cooling medium; heating of the furnace according to the standard fire curve, DIN 4102, part 2 [3] (about 1000 °C after 90 minutes)

flow down to the bottom of the columns through the pipe. A natural circulation of this type is hereby activated. In addition, the pipe is connected to a water storage tank at the top of the building in order to be able to compensate for the loss of water by steam production and also act as a chamber for the separation of steam and water. A group of individual columns is connected at their bottom to a shared connecting pipe; the same is done at the top of the columns. Only one down pipe is thereby necessary joining the top and bottom of the group of the columns.

Columns with internal pipe connections
(see Figs. 13-11 b and 13-12 b)

An internal down pipe within each column is the characteristic feature of this system. This provides the supply of cool water to the bottom of each column, which promotes the natural circulation of the mixture of water and steam flowing upwards and the water flowing downwards after the steam is separated and evacuated through the vent. Thus, each column can act as an individual member without any connection to the other columns. However, in order to minimize the number of the storage tank, the tops of several columns can be connected by a common pipe leading to one storage tank for the whole group.

An example of the practical application of this system is the hanging, five-storey office building in Hannover (see Fig. 13-13). The ground floor is suspended at a distance of about one meter above ground level. Each of the four upper storeys is suspended on a water cooled hanger fixed to the main girder at the

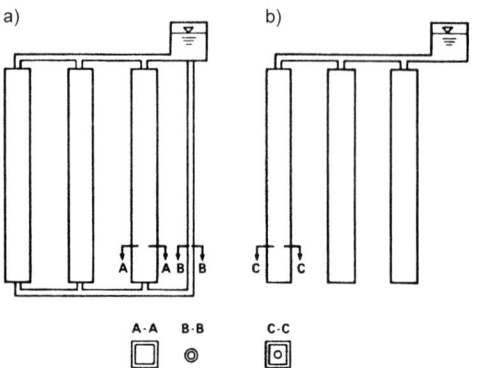

Fig. 13-11 Options for hollow section columns with a) external pipe connection, b) internal pipe connection

top of the building. The girder itself hangs under the top of the main load bearing column group. Each of the tubular columns is a water cooled system with a water storage tank for each group of four columns.

Mixed systems

The systems with internal or external pipe connections can be combined in a building as a mixed integrated system. This is advantageous for structures not only with water filled columns but also with hollow section braces filled with water. For horizontal members, a natural water circulation can be assured by a minimum inclination of about 45°.

The water in a cooling system must be protected against corrosion, frost, biological growth and scaling. The fluid formulation has to be done by the parts given by the weight, which can be different for particular projects. Based on the experiences obtained from the existing buildings, the following specification is recommended.

Desalinated, deoxygenated water	100 parts
Potassium carbonate K_2CO_3) additive as antifreeze	25 to 60 parts
Potassium nitrite (KNO_2) or sodium nitrite ($NaNO_2$) additive as a corrosion inhibitor	1 part

A thin oil film should be maintained on the free surface of the water in the storage tank to inhibit evaporation and to reduce the gradual

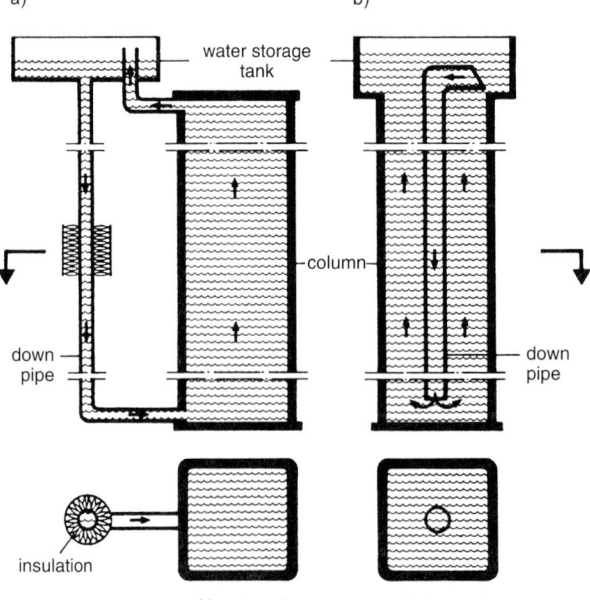

Fig. 13-12 Water cooling of hollow section column by natural circulation with a) external connection pipe, b) internal connection pipe

13.4 Fire protection of steel hollow sections by water cooling

Fig. 13-13 Norcon Building in Hannover with water cooled, tubular columns and hangars (Architects: Schuwirth and Erman, Hannover)

contamination of the coolant by oxygen, algae etc.

Examples of buildings with water cooled columns can be obtained from [5] and [7].

The principal advantages of water cooling for hollow section columns are summarized as follows:

- Slender columns without external protection provide more floor area in multi-storey buildings.
- The columns can in many cases be serviceable even after fire outbreaks (favoured cases by the insurance companies).
- Comparative cost estimations to determine the economy of a water cooling system lead to the conclusion that the water cooled external columns in hollow sections are more economical than columns in conventional rolled steel sections with fire protection envelopes or fire proof panels used in buildings with more than 6 storeys. The increasing economic viability for large number of floors is based on the fact that the expenditures involving water pipe circuits and storage tanks are nearly independent of the height of a building and are of less consequence.
- Exposed steelwork is possible, so that the structure of a building can be more clearly expressed architecturally.

13.4.2 Design of water cooling installations

In particular, two main criteria have to be fulfilled for the design:

- Maintenance of the natural circulation of the water
- Replacement of water loss due to steam production

Regarding the principles of water cooling, see Section 13.4.1 for the fundamentals.

The cooling water absorbs heat in two phases. In the first phase, the boiling point is reached by a rise from the initial temperature (1 kg of water raised 1 °C absorbs 4.187 kJ of heat). In the second phase, the boiling water is converted to steam (latent heat of vaporisation is about 2150 kJ/kg), which subsequently leaves the cooling system through the vent pipes. In principle, the design calculation consists of the solutions for the three following questions:

- How large is the volume of the required cooling water in the zone under fire load?
- How large is the required volume of the water storage tank?
- How large is the maximum water-steam flow rate in the pipework?

The flow rates for various pipe lengths within the water-steam network have to be determined at first. These are used in conjunction with standard steam and water friction coefficients, pressure head losses (obtained from formulae or tables) and the pipe work geometry (i.e. steam and water pipe diameters, lengths, number of bends etc.) to calculate the pressure head losses in the separate steam and water circuits. The pressure head needed to circulate the water and steam during any fire is then given by: Total water head losses + total steam head losses

This circulating pressure head is produced as a static head by ensuring that the base of the water tank is situated above the highest portion of the column subjected to flame impingement by that amount.

It is a very extensive work to present a design example for a water cooling installation with hollow section columns. Therefore, it has been dispensed with in this book and instead a number of reference literature [26, 33] is named.

13.5 Fire protection of steel hollow section columns by filling with concrete

13.5.1 Fundamentals

Steel hollow section is transformed into a composite structural element by filling with concrete. This enlarges initially the load bearing capacity of the steel section without increasing its external dimensions (see Chapter 11). Another feature is the improvement of fire resistance. In a concrete filled hollow section column under ambient conditions, the steel and concrete are constrained to move together and hence longitudinal steel and concrete strains are equal. Accordingly, the stress in each material is proportional to the ratio of the moduli of the elasticity of the two materials. Fig. 13-14 shows the behaviour of the steel and concrete when they are loaded by fire separately. This demonstrates that the loss of the yield strength of structural steel and that of compressive strength of concrete with the increase of temperature do not differ from each other substantially.

However, on heating, the steel hollow section which is directly exposed to fire, tries to expand more rapidly than the concrete and therefore begins to resist a greater proportion of the applied load. At the same time, the steel

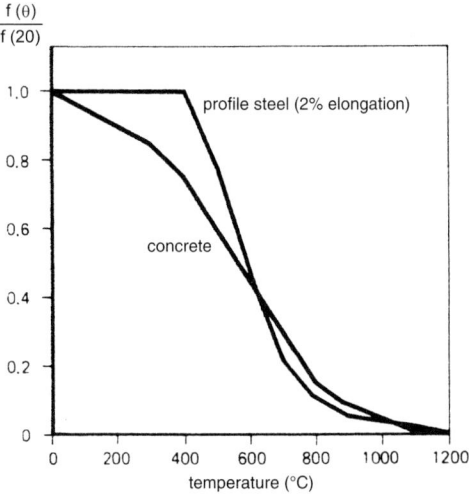

Fig. 13-14 Reduction of material strength for structural steel and concrete under fire load [22–24]

yield stress and modulus of elasticity begin to reduce and eventually, the steel sheds load into the concrete. Heat from the steel shell is transferred to the outer layers of the concrete causing their temperature to rise. Due to the low thermal conductivity and the large massivity of concrete, the rate of flow of heat through the concrete core is slow. As a consequence, the concrete core retains a large part of its strength longer. This explains in principle why a concrete filled hollow section yields a higher fire resistance than the unfilled ones.

Further, the degradation of concrete includes the driving off of the water present both as free moisture and from the hydrated constituents of the mix. A considerable amount of heat is hereby absorbed in converting the moisture to steam. The steel shell contains the concrete and prevents direct flame impingement so preventing progressive spalling and reducing the rate of degradation of the concrete core.

Failure in the column occurs when the combined strength of the steel and concrete is reduced to the level of the applied load. The lower the level of load applied, the lower the stresses produced and the longer the period of fire stability of the column.

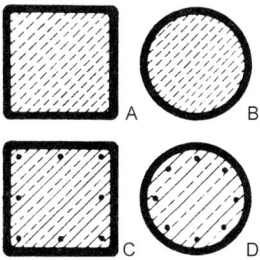

Fig. 13-15 Regular cross sections of composite hollow section columns

Column size is another important parameter, as the concrete core of a larger sized column supports a greater proportion of the total load compared to a smaller one.

Summarizing, the conclusion can be made that composite hollow section columns without external insulations can reach any required fire resistance by selecting appropriate column cross sections and the reduction of the applied load related to the allowable service load at room temperature.

The load bearing capacity (resistance) of composite hollow section columns consists of three components, namely that of:

- Enveloping steel (= hollow section)
- Concrete
- Reinforcement steel (if any) covered by concrete

Fig. 13-15 shows the regular cross sections of the composite circular and rectangular hollow section columns with and without reinforcement bars.

Due to the different arrangements in the cross section, each of the components demonstrates different reductions of design resistance as a function of time:

- Large reduction of resistance of steel hollow section within a short time due to direct exposure to fire
- Smaller reduction of resistance of concrete, mostly in the core area rather than near the internal surface area of the hollow section, fire resistance longer
- Retarded reduction of resistance of the reinforcement bars (if any), which are normally near the surface but is protected by 25–30 mm of concrete cover

The fire performance of concrete filled hollow section column is described by the basic diagram given in Fig. 13-16 demonstrating the characteristic behaviour of each component.

The design resistance R of a cross section results from the sum of the resistances of the single components r_j. In case of fire, all resistances are dependent on the time of fire exposure t:

$$R(t) = \sum r_j(t)$$

The combination of materials with markedly different thermal conductivities produces extreme transient heating behaviour and highly complex temperature differentials across the cross section. That is why simple calculation models cannot be applied for the fire design, which is based on the section factor A_m/V. Instead the special fire design method has to be used, which takes the thermal characteristics of the component materials and the resulting heat transfer into account [26].

The decisive part of the total design resistance of the composite column under room temperature is contributed by the steel hollow section. This can be attributed to the high strength of steel and the accumulation of steel on the outside. However, after a certain duration of fire exposure, only a small part of the initial resistance can be activated. In the case

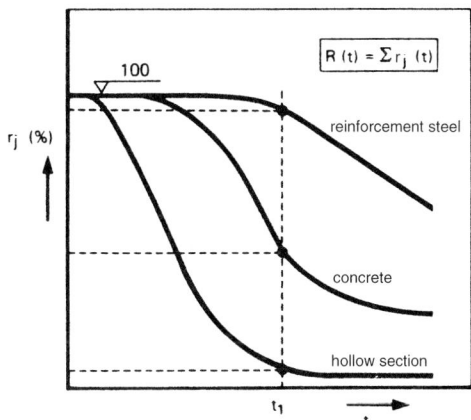

Fig. 13-16 Typical resistance reduction characteristics of the components of a concrete filled hollow section column

of fire, the predominant part of the load in the steel hollow section is transferred to the concrete, as this loses its strength and rigidity more slowly than the steel mantle. In case the concrete core is overloaded by the transferred load, a failure of the column may occur within a short time. From the aspects mentioned above, the following recommendations for the fire design of the concrete filled hollow section columns can be derived:

- Low design resistance of the steel mantle, i.a. small wall thickness and low strength of steel (side length of the hollow section a/wall thickness of the hollow section $s \geq 25$ and steel grade S 235 recommended).

- High design resistance of the concrete, i.e. use of high strength concrete (the water to cement factor has to be small and the application of liquidisers or plastifiers is of advantage).

- Significant improvement of fire resistance by using reinforcement bars (reinforcement steel at least S 400).

- In the vicinity of the head and the foot of a column (100–120 mm from the ends), small drain holes of 10 to 15 mm diameter are required in the walls (usually in pairs) as steam outlets [3, 4].

The reduction of the strengths of the single cross sections depends directly on the corresponding temperature development in the cross section. This makes a minimum size of the hollow section necessary in order to reach a required fire resistance. Further a minimum size of the hollow section is also required for proper positioning of the reinforcement bars and stirrups in the hollow section. The section should not be less than 140 mm square and 100 × 200 mm rectangular hollow section for plain concrete filled columns and not less than 200 mm square and 150 × 250 mm rectangular for bar reinforced ones.

In the diagrams obtained from investigations shown in Fig. 13-17, the fire resistance is plotted against the side length "a" of hollow section (for rectangular hollow sections, the shorter side length a_y) for the frequently used column length $l = 3$ m. Tests were carried out on eccentrically loaded columns ($e_0 = 5$ mm) with caged reinforcements. It is demonstrated that the degree of utilization η decreases with increasing fire resistance, which is evident.

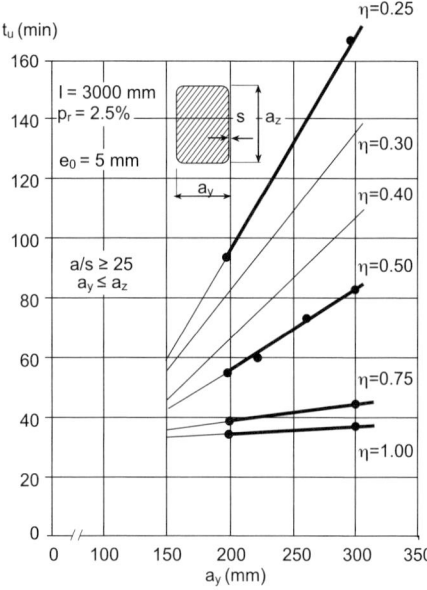

Fig. 13-17 Effect of cross section size on the fire resistance of concrete filled rectangular hollow section for varying degree of utilization η

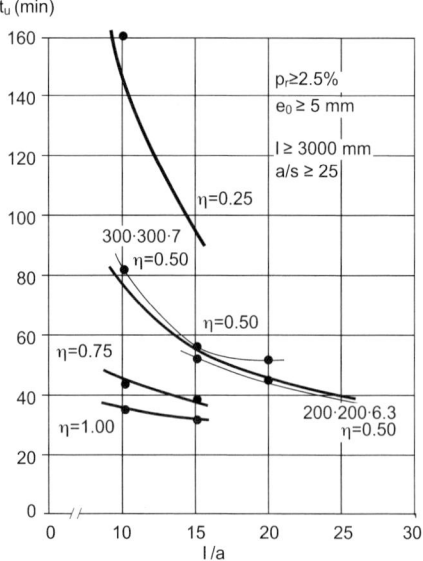

Fig. 13-18 Effect of slenderness l/a on the fire resistance

$$\eta = \frac{\text{Design axial force in fire case}}{\text{buckling force design at room temperature (20°C)}} = \frac{N_{Sd,\theta}}{\chi \cdot N_{pl,Rd}}$$

For the practical design, the influence of the slenderness of the column (represented by l/a) must also be taken into account. This has been done in Fig. 13-18, which is valid for a buckling length ≥ 3.0 m. The illustration is of general nature and leads to similar results as Fig. 13-17 for a buckling length = 3.0 m.

Mention should be made here, that the effect of accidental eccentricity and out of straightness in a column subjected to nominally axial loads is not significant for short columns. However, this can be significant for slender columns in fire.

13.5.2 Fire design of concrete filled hollow section columns without external insulation

The fundamentals for the fire design and construction of concrete filled hollow section columns were determined in the seventies and the eighties by means of extensive fire tests [10, 11, 14]. Fig. 13-19 shows a testing installation for fire tests of columns.

Fig. 13-19 Fire test on a concrete filled column

The results of these tests give important informations explaining the fire behaviour with regard to the following parameters:

– Degree of utilization η
– Minimum size of hollow section (hollow section, concrete, reinforcement)
– Steel grade of hollow section
– Grade of reinforcement steel
– Concrete grade
– Amount of reinforcement (%) (pr = $[A_s/(A_b + A_s)] \cdot 100$)
– Concrete cover on reinforcement bar (a_s, see Fig. 13-20)
– Critical buckling length of columns under fire condition $L_{cr,\theta}$
– Ultimate buckling resistance under fire condition $N_{cr,\theta}$

The evaluation of these research works led to the design rules and recommendations, which have been adopted by Eurocode 3 [22] and 4 [23] in three levels:

Level 1: Tabulated data
Level 2: Simple design diagrams
Level 3: General calculation models

13.5.2.1 Level 1 design: tabulated data

Table 13-4 is valid with the following limitations:

– Yield strength f_y of the steel hollow section ≤ 235 N/mm^2

$$\frac{\text{Main size of hollow section (width or diameter)}}{\text{wall thickness of hollow section } s} \geq 25$$

– Amount of reinforcement pr = 0; 1.5%; 3.0%; 6.0%

13.5.2.2 Level 2 design: simple design diagrams

Based on the test results described in Section 13.5.2, a computer program has been developed, which forms the basis for the buckling curves for the concrete filled hollow section columns at elevated temperature. In the following, a complete set of design charts (see

Table 13-4 Minimum cross-sectional dimensions (b or d), minimum amount of reinforcement (pr in %), minimum concrete cover for the reinforcement bar (a_s) for fire resistance classification depending on various degrees of utilization (η)

Hollow section cross section $b/s > 25$ or $d/s > 25$	Fire resistance class				
	R30	R60	R90	R120	R180
minimum cross-sectional dimensions for $\eta = 0.3$					
minimum width (b) or diameter (d)	160	200	220	260	400
minimum amount of reinforcement (pr in %)	0	1.5	3.0	6.0	6.0
minimum concrete cover for the reinforcement bar (a_s)	–	30	40	50	60
minimum cross-sectional dimensions for $\eta = 0.5$					
minimum width (b) or diameter (d)	260	260	400	450	500
minimum amount of reinforcement (pr in %)	0.0	3.0	6.0	6.0	6.0
minimum concrete cover for the reinforcement bar (a_s)	–	30	40	50	60
minimum cross-sectional dimensions for $\eta = 0.7$					
minimum width (b) or diameter (d)	260	450	500	–	–
minimum amount of reinforcement (pr in %)	3.0	6.0	6.0	–	–
minimum concrete cover for the reinforcement bar (a_s)	25	30	40	–	–

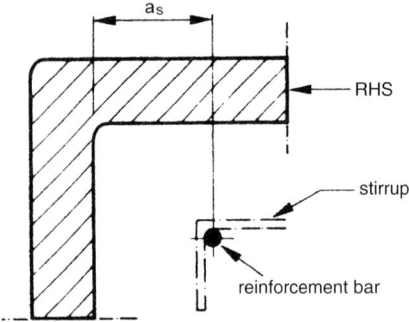

Fig. 13-20 Concrete cover a_s on the reinforcement bar

Table 13-5) is given, in which, for a standard fire exposure of (30), 60, 90 and 120 minutes, the ultimate axial buckling resistance $N_{cr,\theta}$ is presented as a function of the critical buckling length $L_{cr,\theta}$ of the column.

For the concrete cover a_s on the reinforcement bars (Fig. 13-20), the largest of the following values has been taken:

$a_s = 30$ mm

$a_s = \dfrac{b \text{ or } d}{8}$

Stirrups have been provided over the full length of the (reinforced) column and properly positioned (fixed) within the column. They do not have any further static function.

For the proposed design diagrams, only concentric axial force is valid. However, the design concept described in Section 13.1 considering the effect of the axial force eccentricity on the fire resistance of composite hollow section column is also applicable.

In the following, further geometrical properties valid for the diagrams (Table 13-5) are given:

Reinforcement bars:

Outer diameter $\varnothing = 8, 10, 12, 14, 16, 20$ mm

Amount of reinforcement pr = 1 or 2.5 or 4%

Number of reinforcement bars:

CHS – systematically 8 bars

Square hollow section – 8 bars for $b \geq 300$ mm; 4 bars if bar $\varnothing = 0.05\,b$ for $160 \leq b \leq 300$ mm, otherwise 8 bars

Concrete cover: $a_s = \max(30 \text{ mm}; b/8 \text{ or } d/8)$

Buckling length: $1 \text{ m} \leq L_{cr,\theta} \leq 4.5 \text{ m}$

13.5 Fire protection of steel hollow section columns by filling with concrete

Table 13-5 Design diagrams for externally unprotected concrete filled hollow section columns of the steel grade S 355

Fire resistance class	Concrete grade	Hollow section size [mm]	Diagram no.
R 60[a] R 90 R 120	C 20[b]	⌀ 219.1 × 4.5	I 1 I 2 I 3
R 60 R 90 R 120		⌀ 244.5 × 5.0	I 4 I 5 I 6
R 60 R 90 R 120	C 30	⌀ 273.0 × 5.0	I 7 I 8 I 9
R 60 R 90 R 120		⌀ 323.9 × 5.6	I 10 I 11 I 12
R 60 R 90 R 120	C 40	⌀ 355.6 × 5.6	I 13 I 14 I 15
R 60 R 90 R 120		⌀ 406.4 × 6.3	I 16 I 17 I 18
R 30 R 60 R 90	C 20	□ 180.1 × 6.3	I 19 I 20 I 21
R 30 R 60 R 90		□ 200 × 6.3	I 22 I 23 I 24
R 30 R 60 R 90		□ 220 × 6.3	I 25 I 26 I 27
R 60 R 90 R 120	C 30	□ 250 × 6.3	I 28 I 29 I 30
R 60 R 90 R 120		□ 260 × 6.3	I 31 I 32 I 33
R 60 R 90 R 120	C 40	□ 300 × 7.1	I 34 I 35 I 36
R 60 R 90 R 120		□ 350 × 8.0	I 37 I 38 I 39
R 60 R 90 R 120		□ 400 × 10.0	I 40 I 41 I 42

[a] R according to Eurocode 3 or 4 [22, 23]
[b] C according to Eurocode 2 or 4 [23, 24]

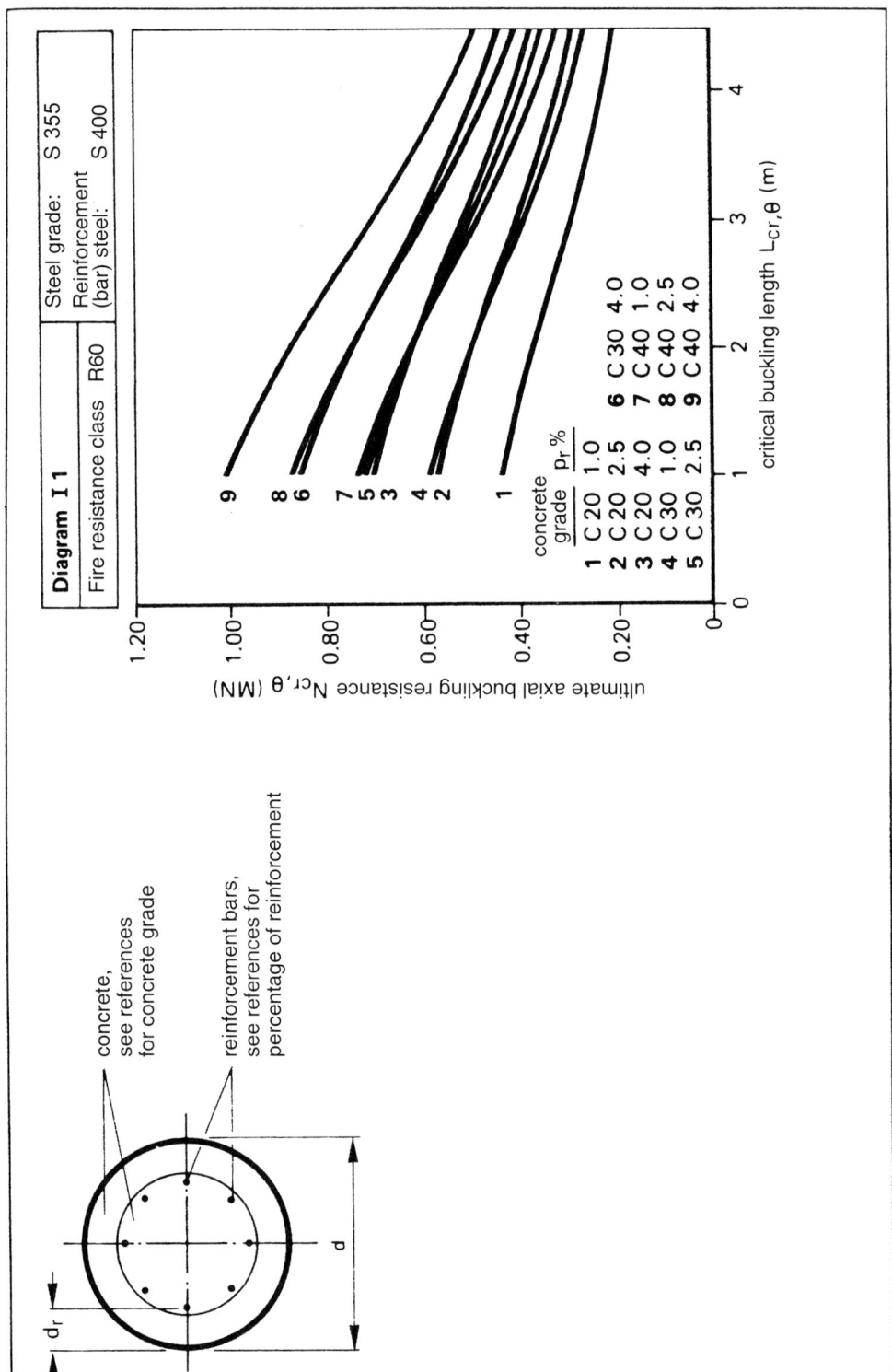

Buckling resistance diagrams for columns in CHS Ø 219.1 × 4.5

13.5 Fire protection of steel hollow section columns by filling with concrete

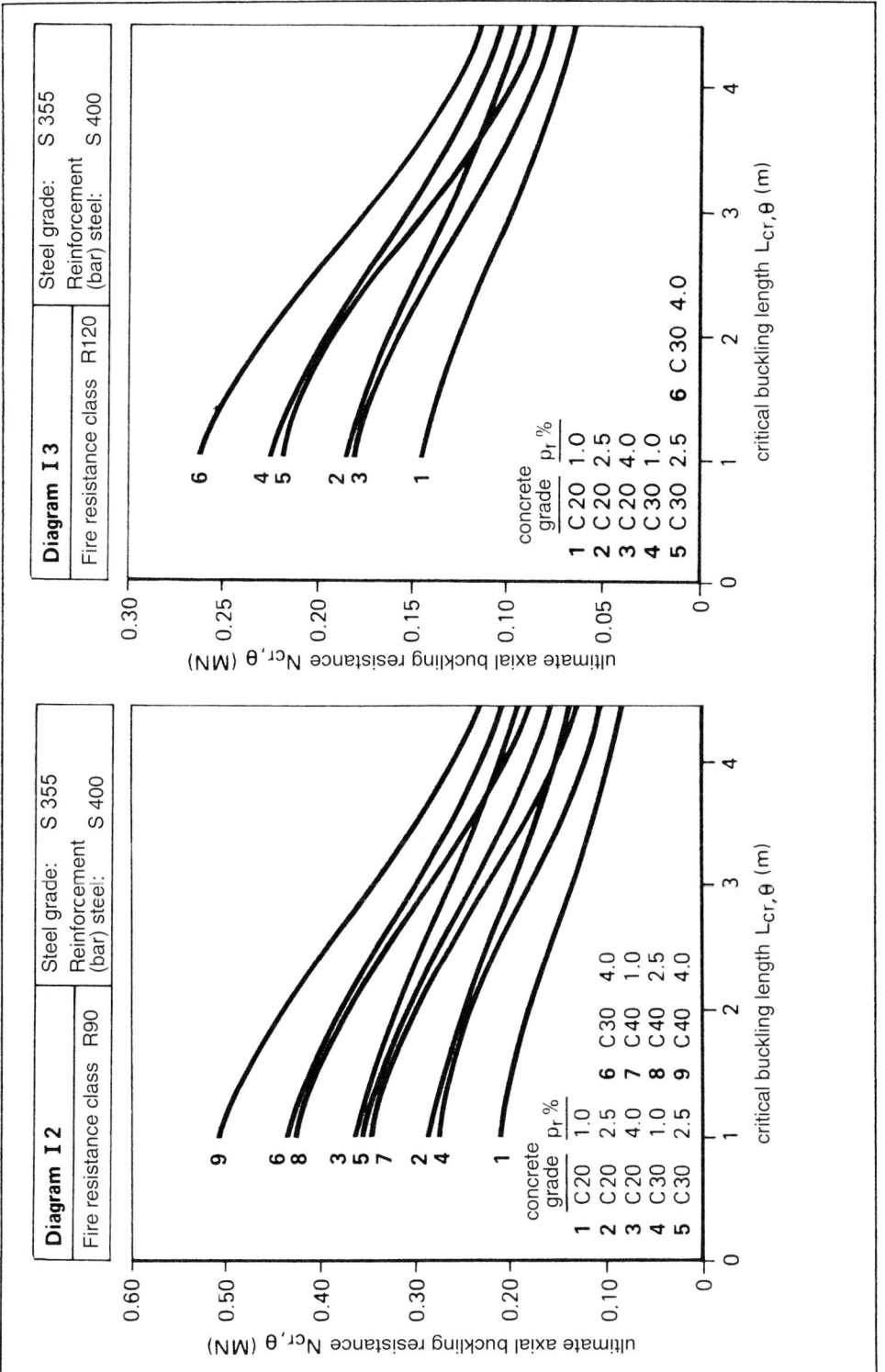

Buckling resistance diagrams for columns in CHS ⌀ 219.1 × 4.5

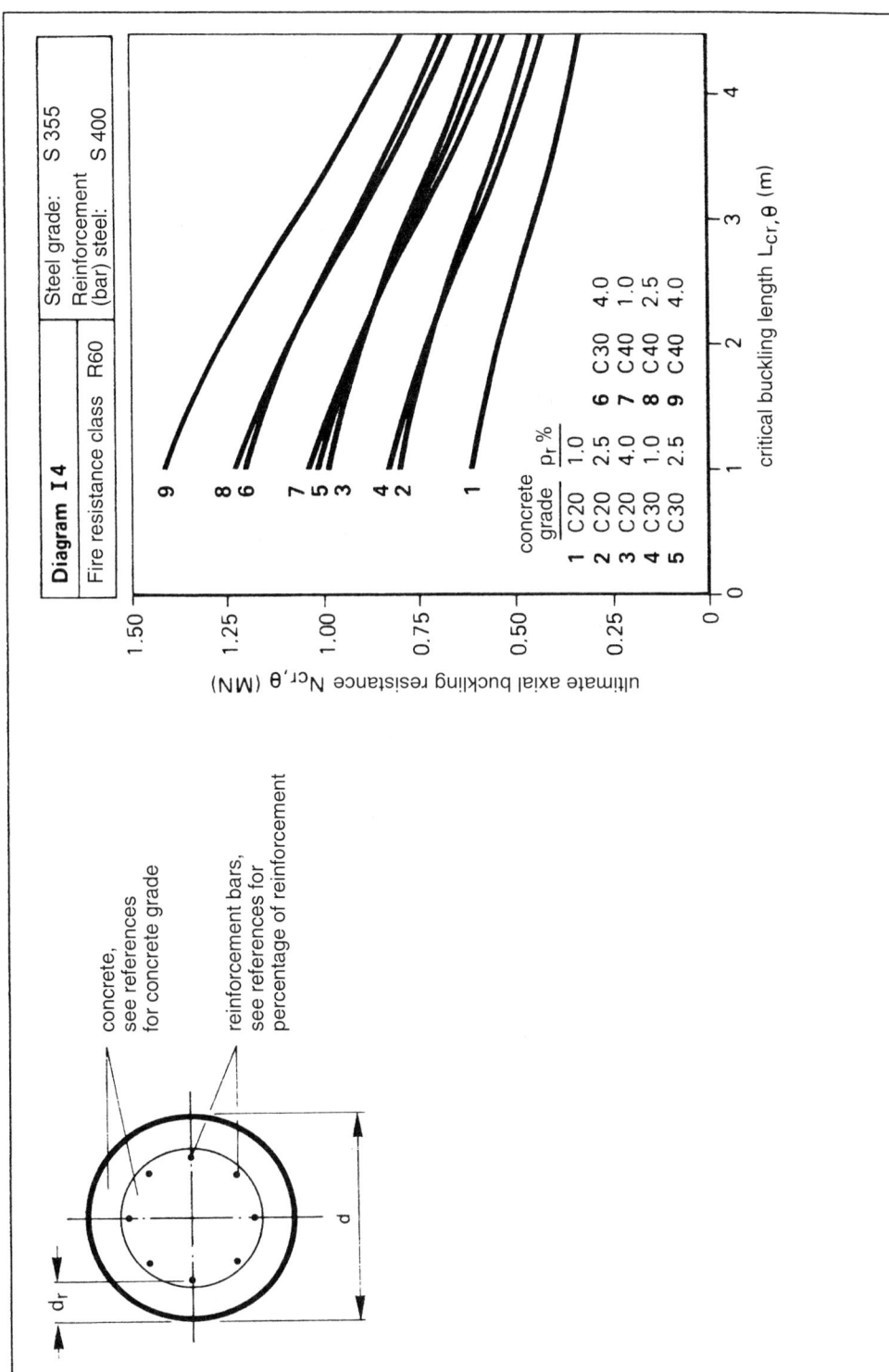

Buckling resistance diagrams for columns in CHS Ø 244.5 × 5.0

13.5 Fire protection of steel hollow section columns by filling with concrete

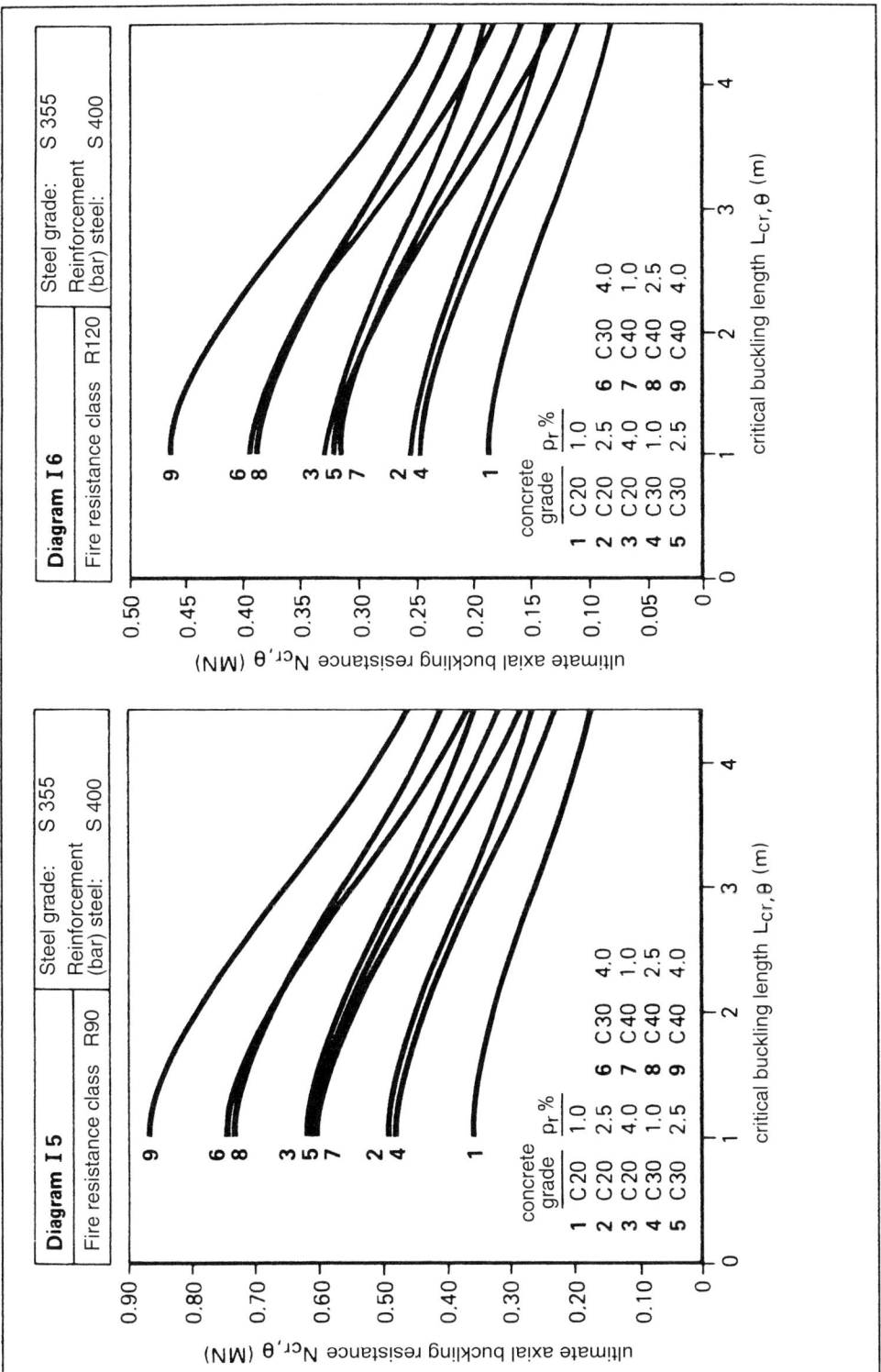

Buckling resistance diagrams for columns in CHS Ø 244.5 × 5.0

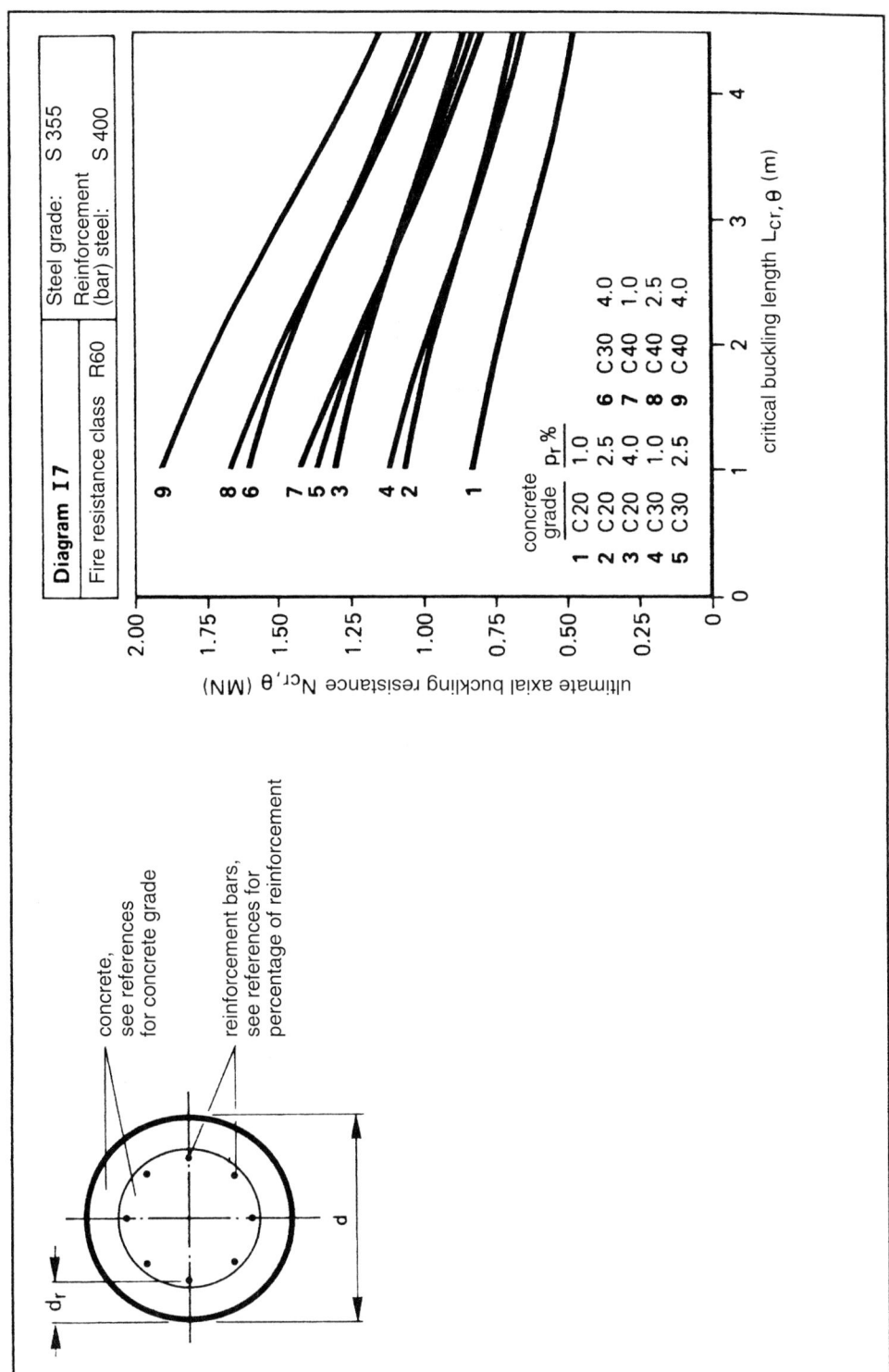

Buckling resistance diagrams for columns in CHS Ø 273.0 × 5.0

13.5 Fire protection of steel hollow section columns by filling with concrete

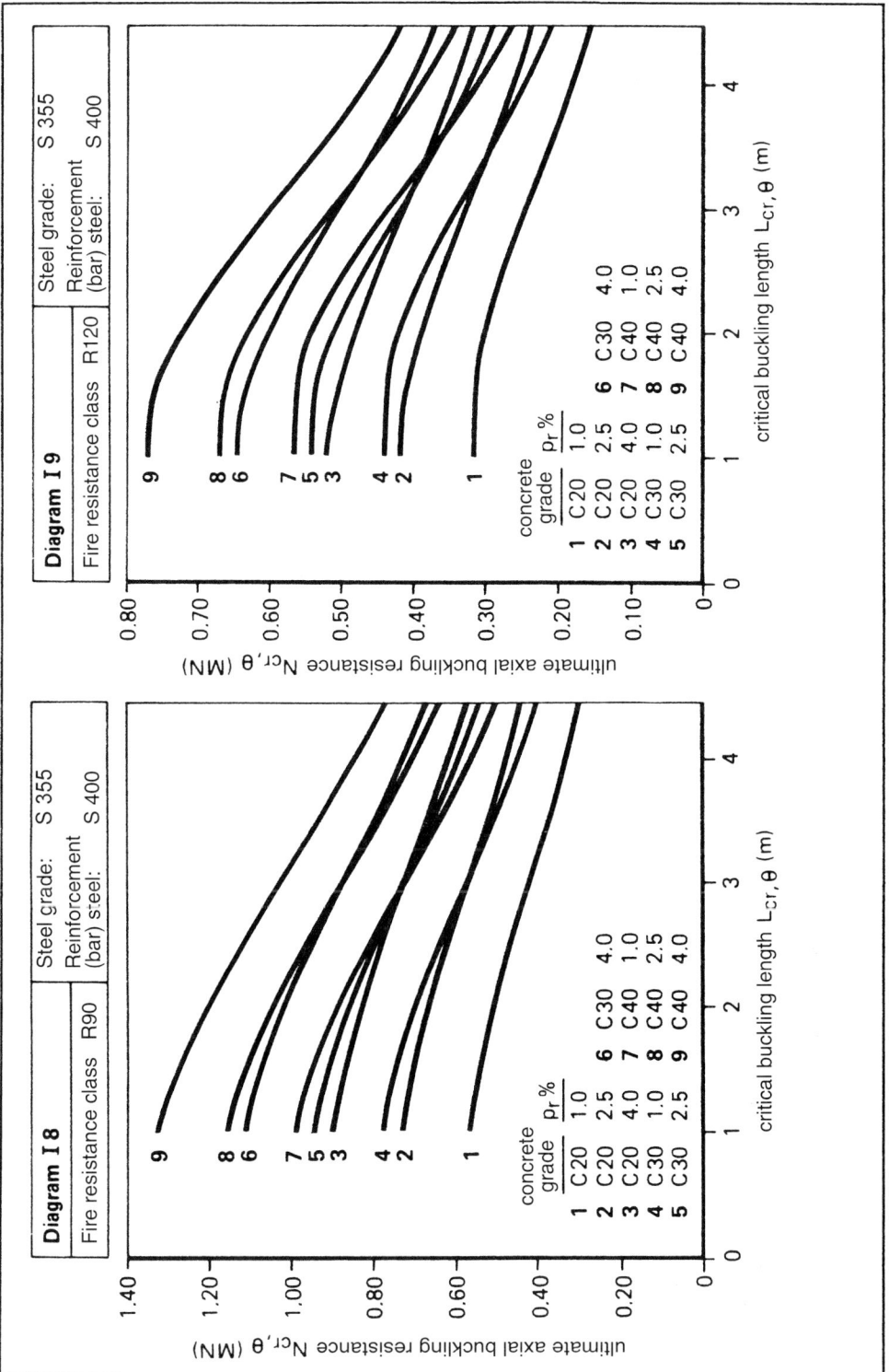

Buckling resistance diagrams for columns in CHS Ø 273.0 × 5.0

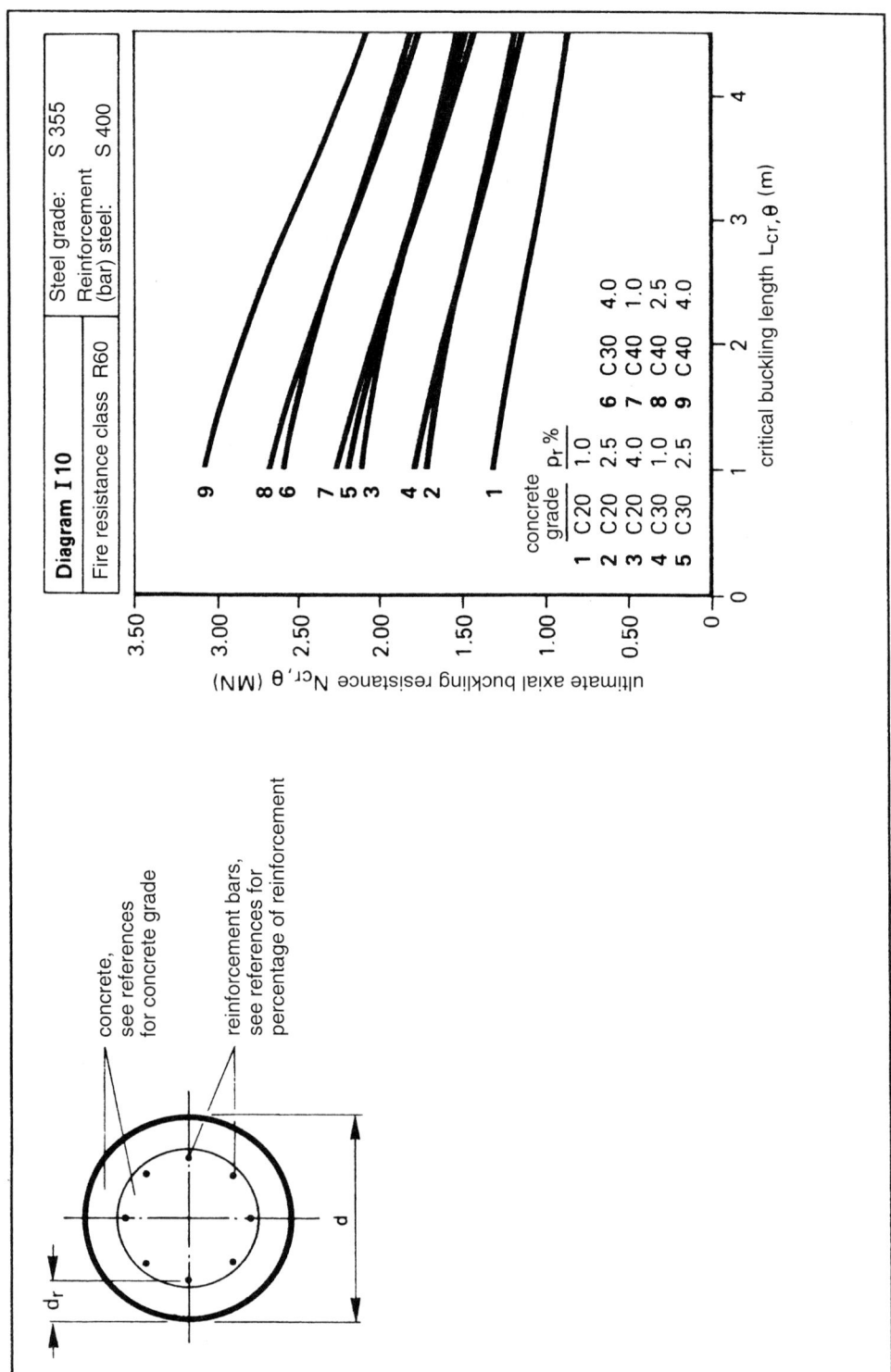

Buckling resistance diagrams for columns in CHS ⌀ 323.9 × 5.6

13.5 Fire protection of steel hollow section columns by filling with concrete

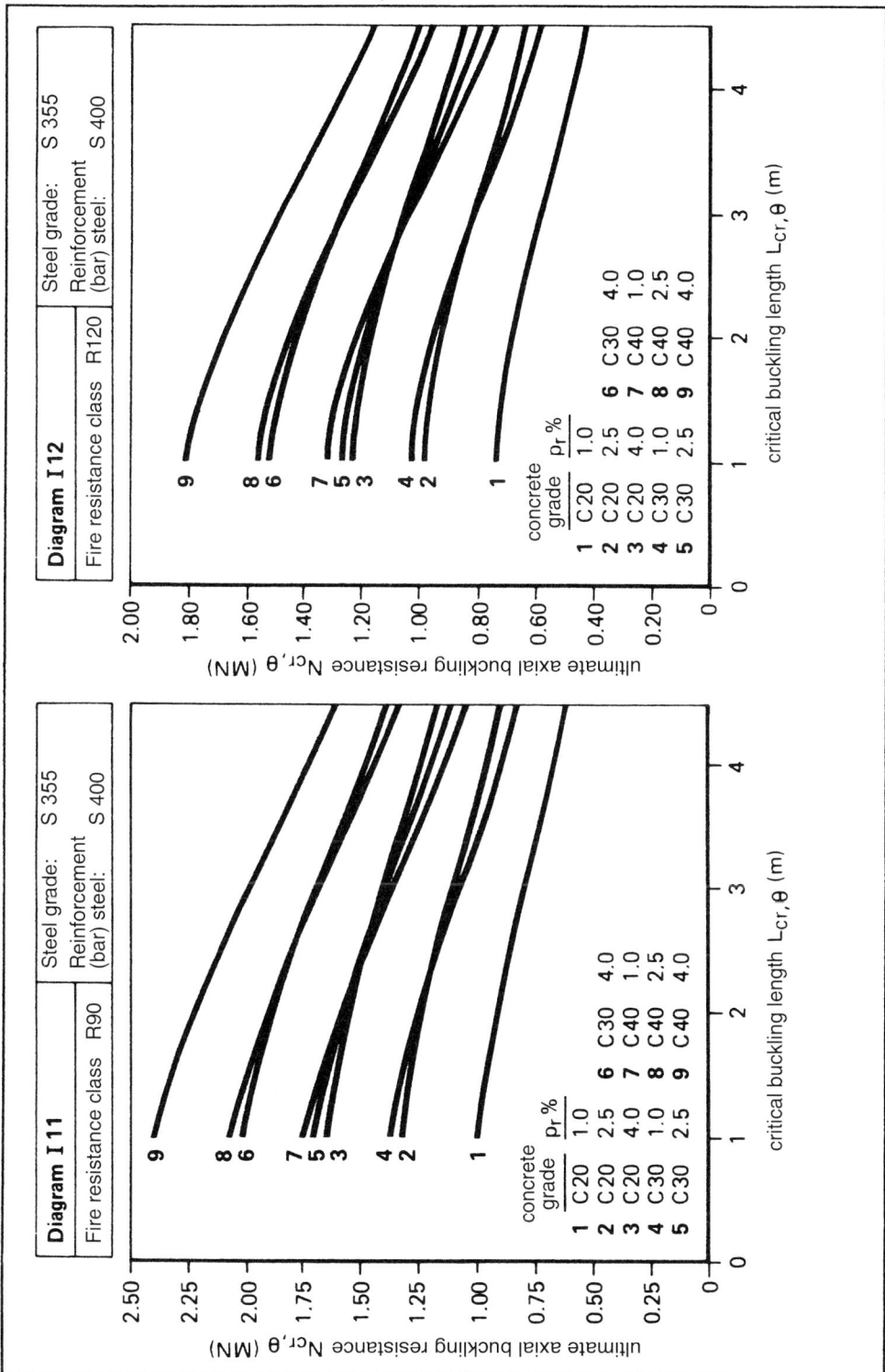

Buckling resistance diagrams for columns in CHS Ø 323.9 × 5.6

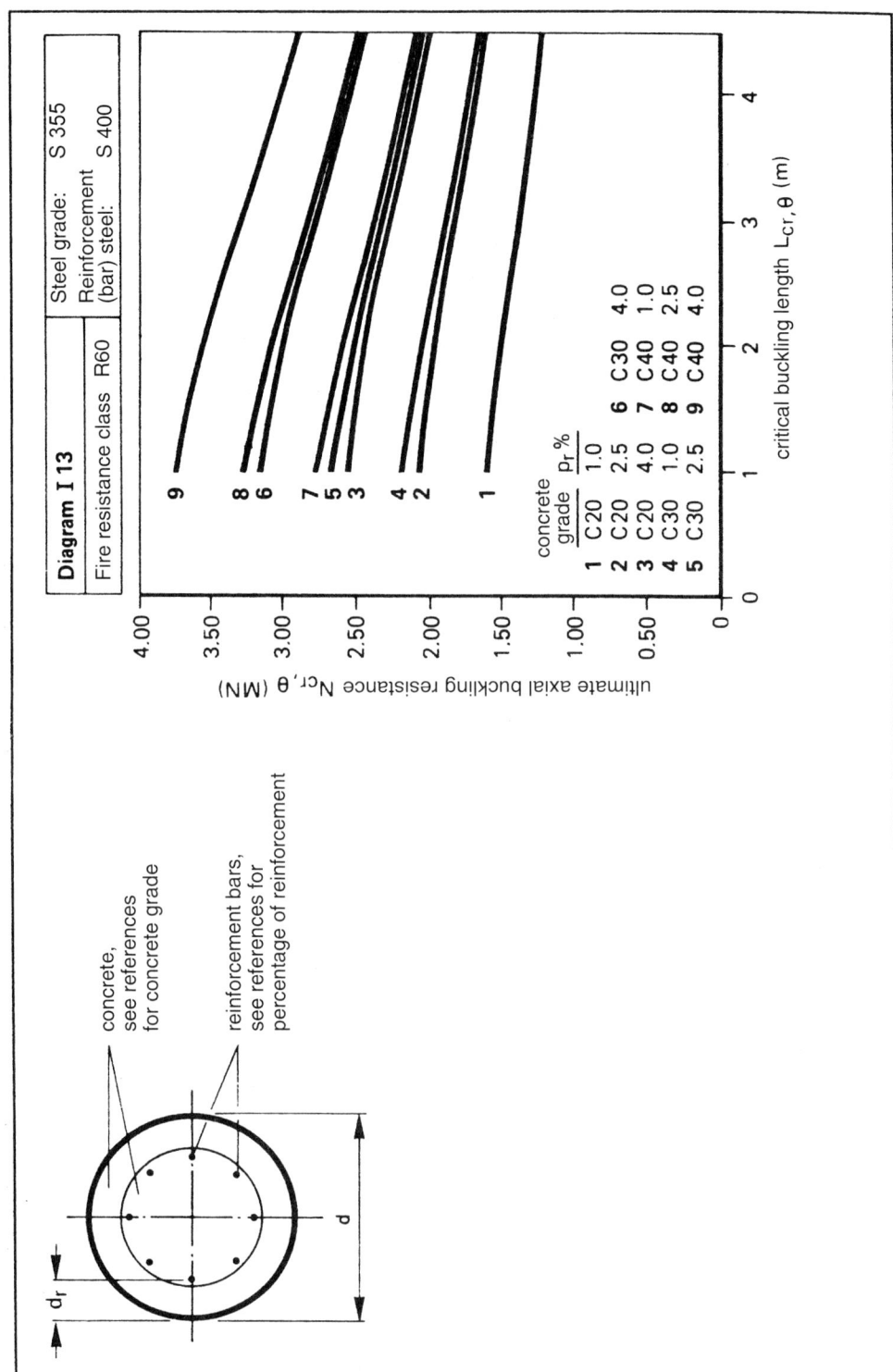

Buckling resistance diagrams for columns in CHS ⌀ 355.6 × 5.6

13.5 Fire protection of steel hollow section columns by filling with concrete

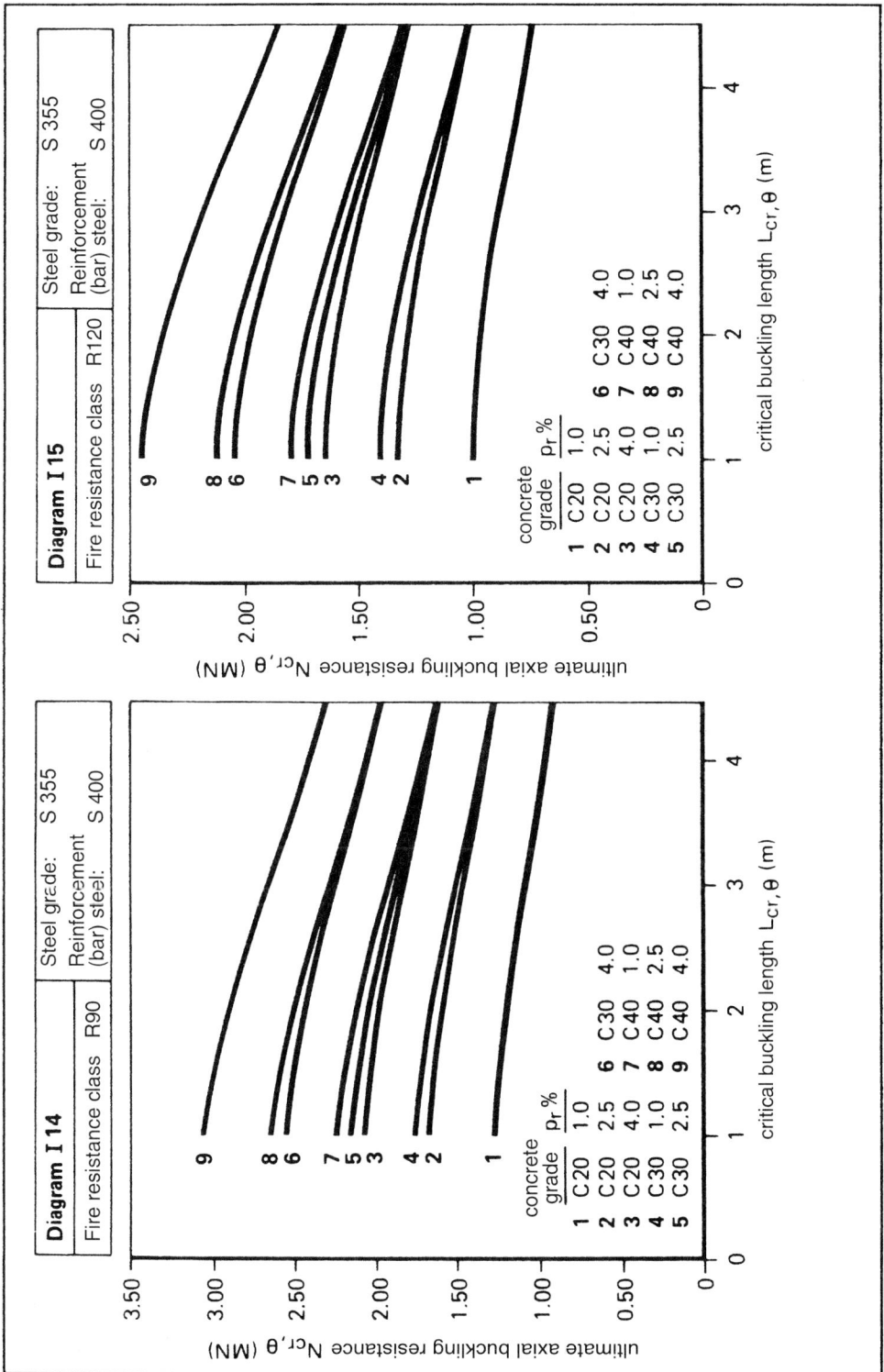

Buckling resistance diagrams for columns in CHS ⌀ 355.6 × 5.6

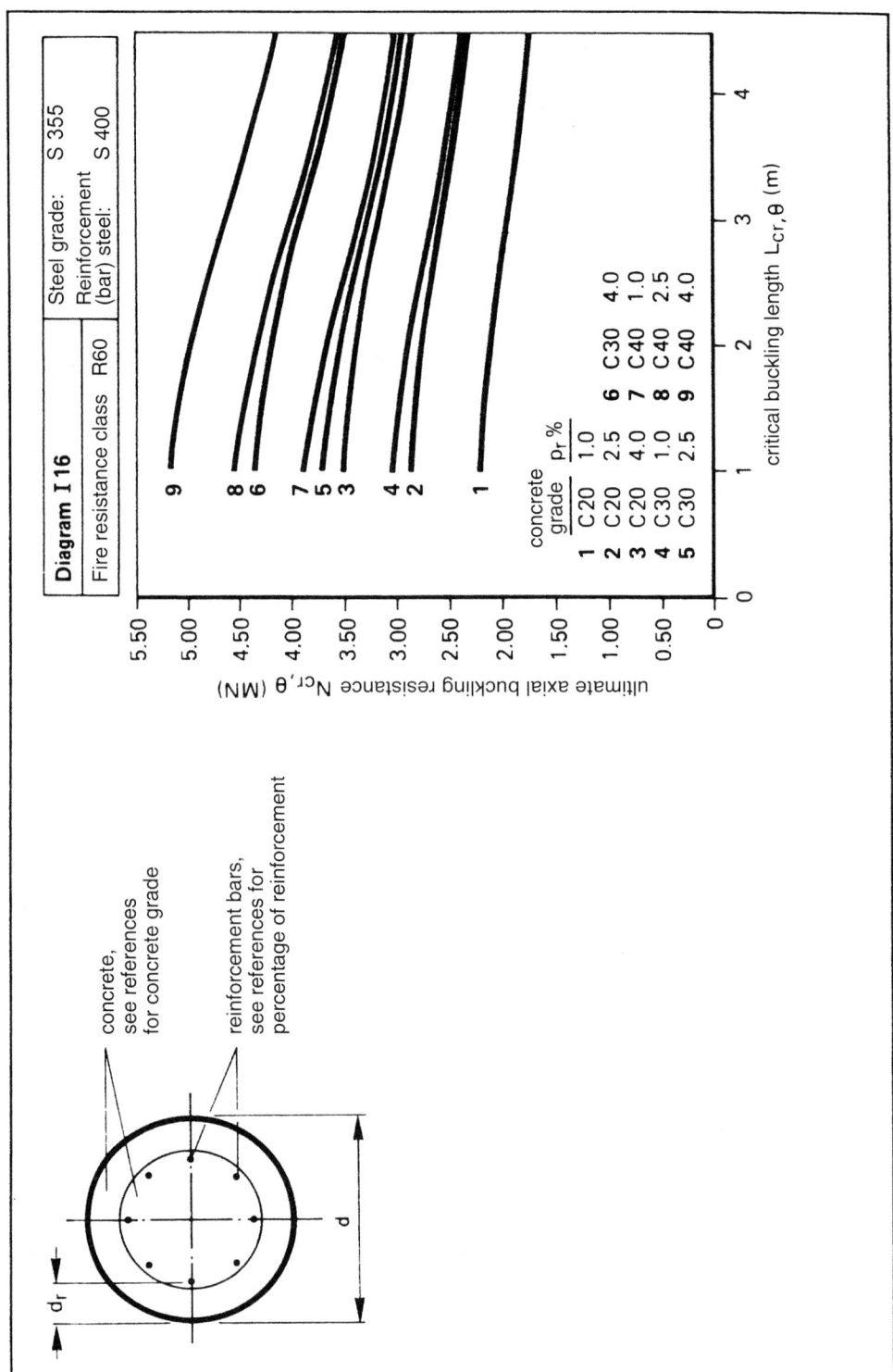

Buckling resistance diagrams for columns in CHS ⌀ 406.4 × 6.3

13.5 Fire protection of steel hollow section columns by filling with concrete

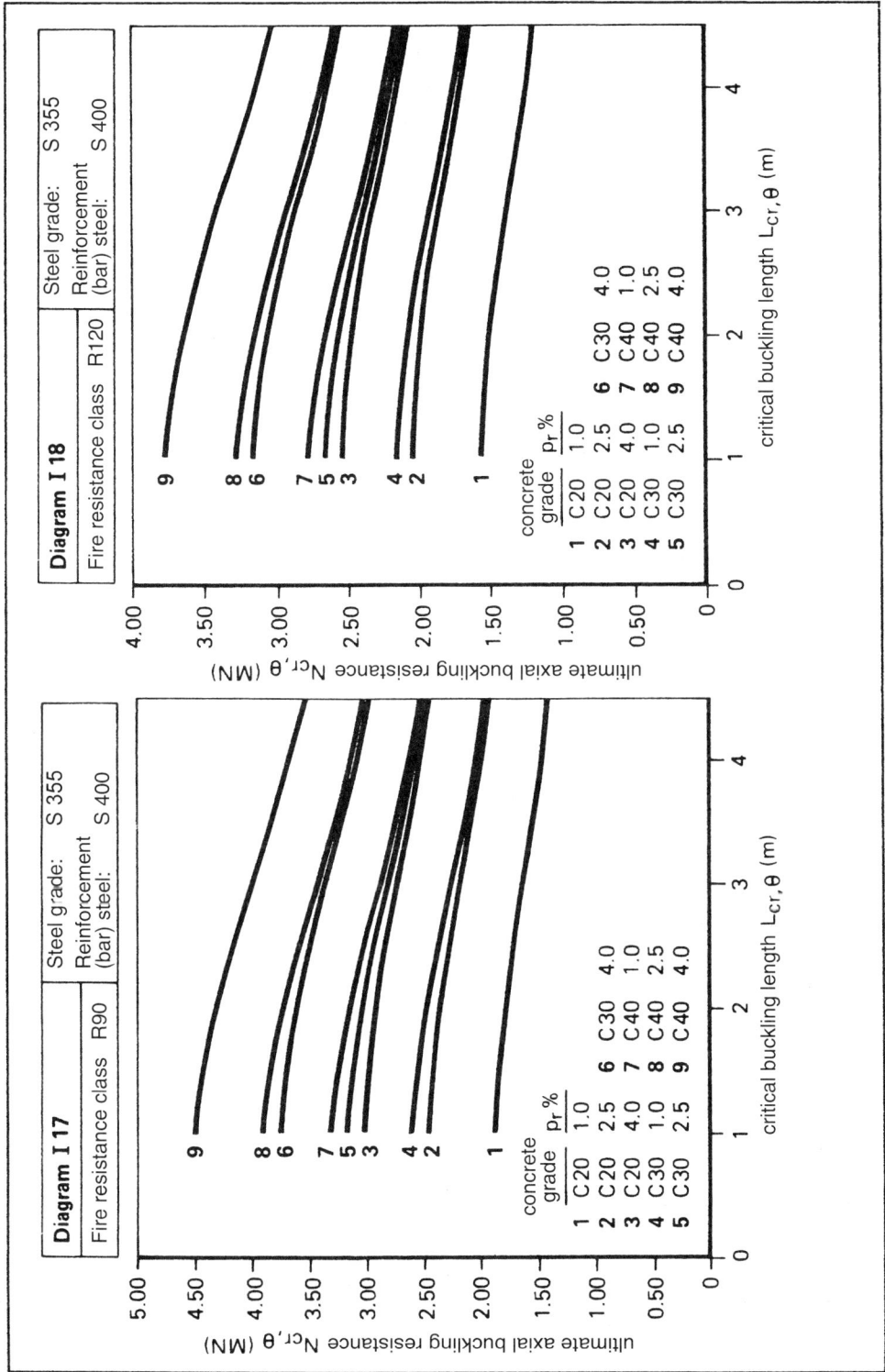

Buckling resistance diagrams for columns in CHS ⌀ 406.4 × 6.3

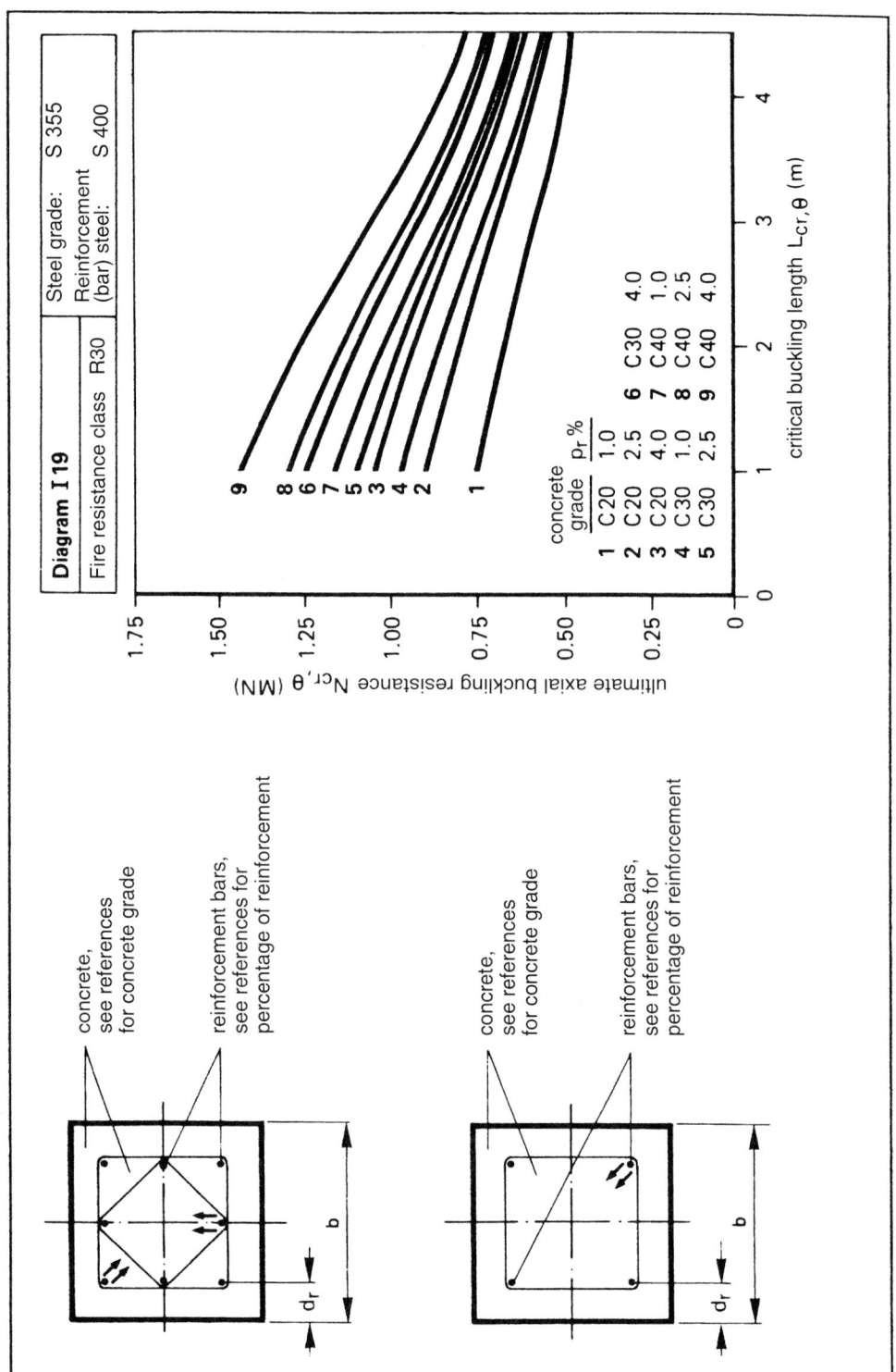

Buckling resistance diagrams for columns in RHS (square) □ 180 × 6.3

13.5 Fire protection of steel hollow section columns by filling with concrete

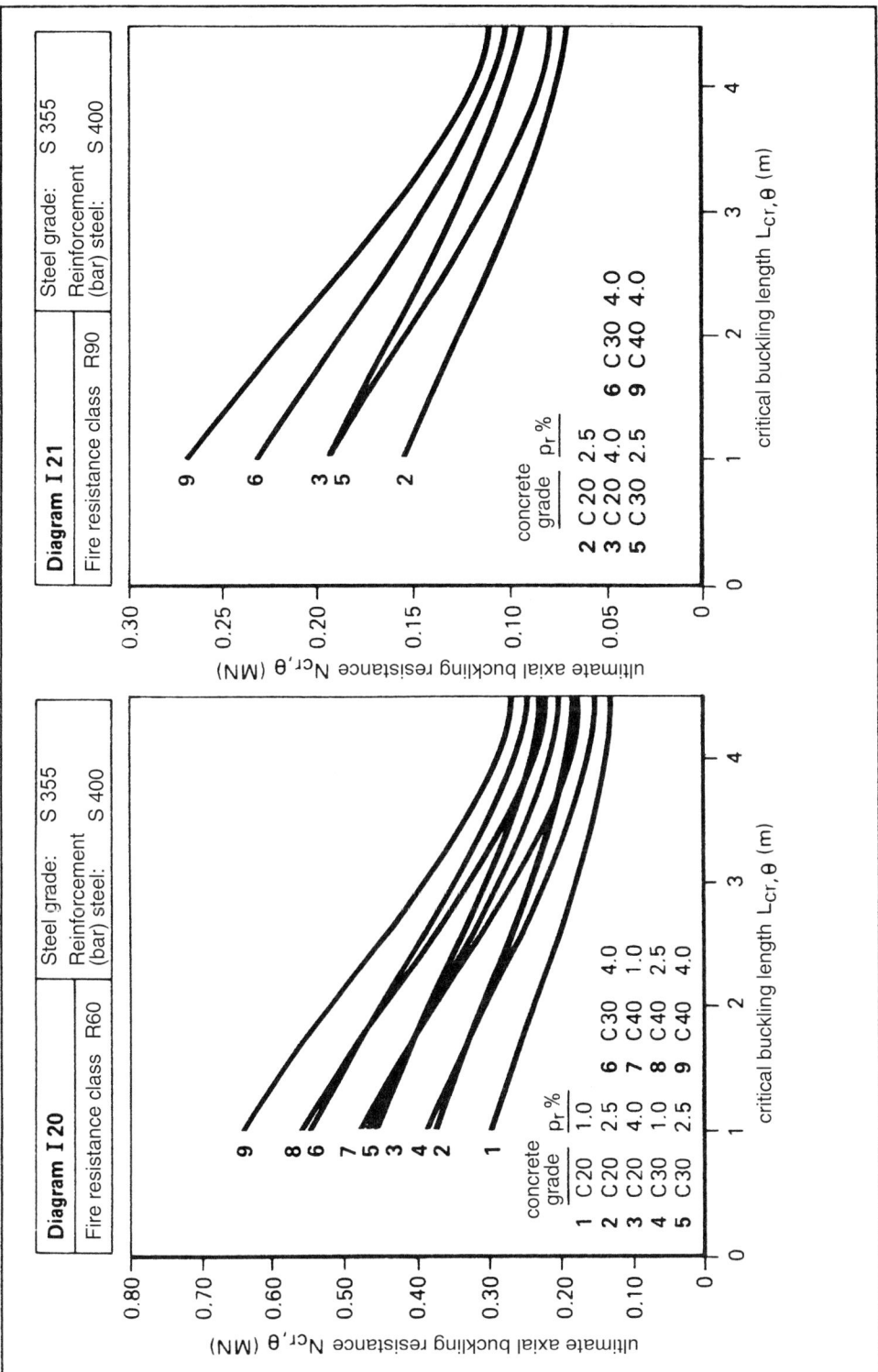

Buckling resistance diagrams for columns in RHS (square) □ 180 × 6.3

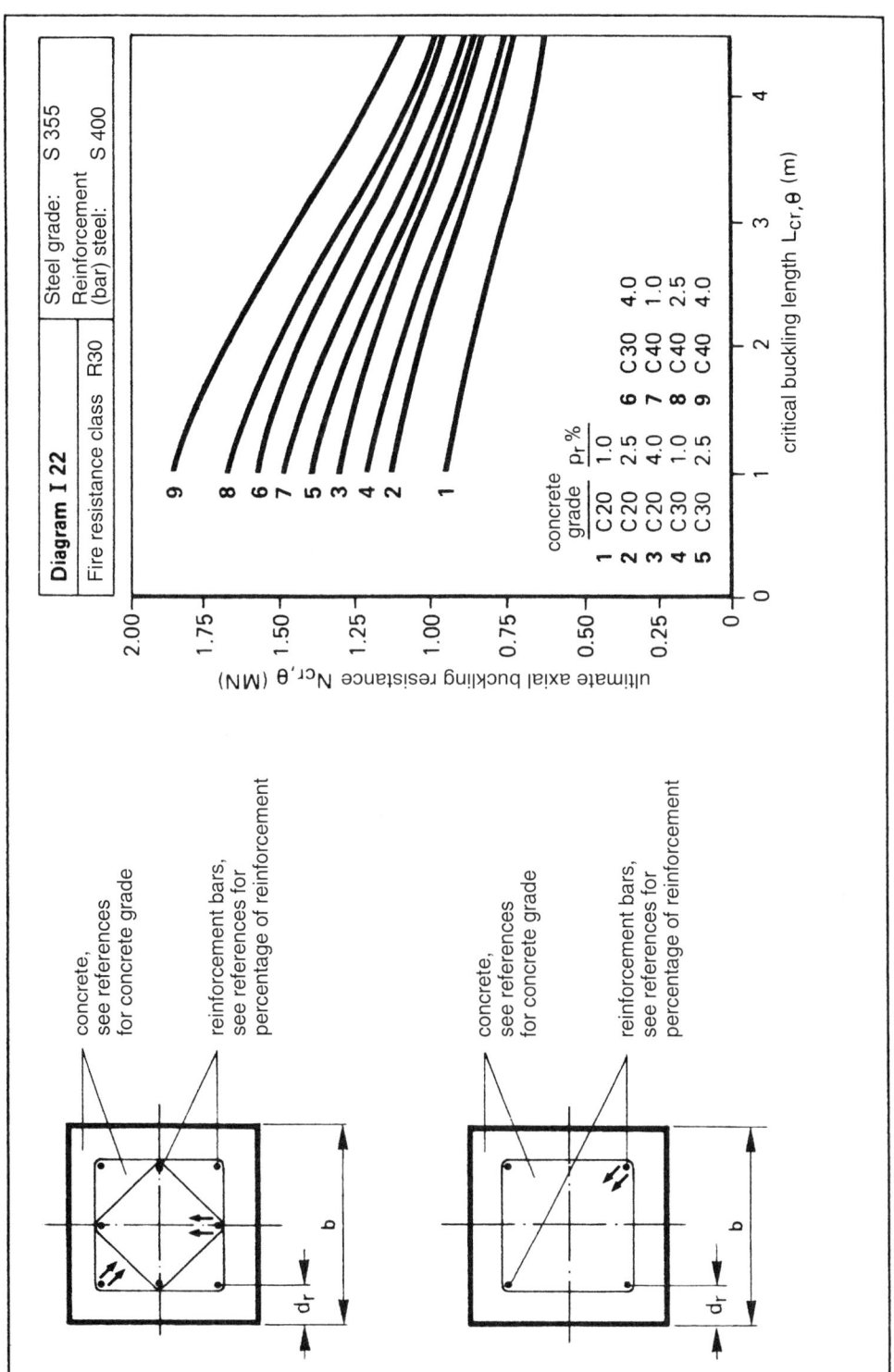

Buckling resistance diagrams for columns in RHS (square) □ 200 × 6.3

13.5 Fire protection of steel hollow section columns by filling with concrete

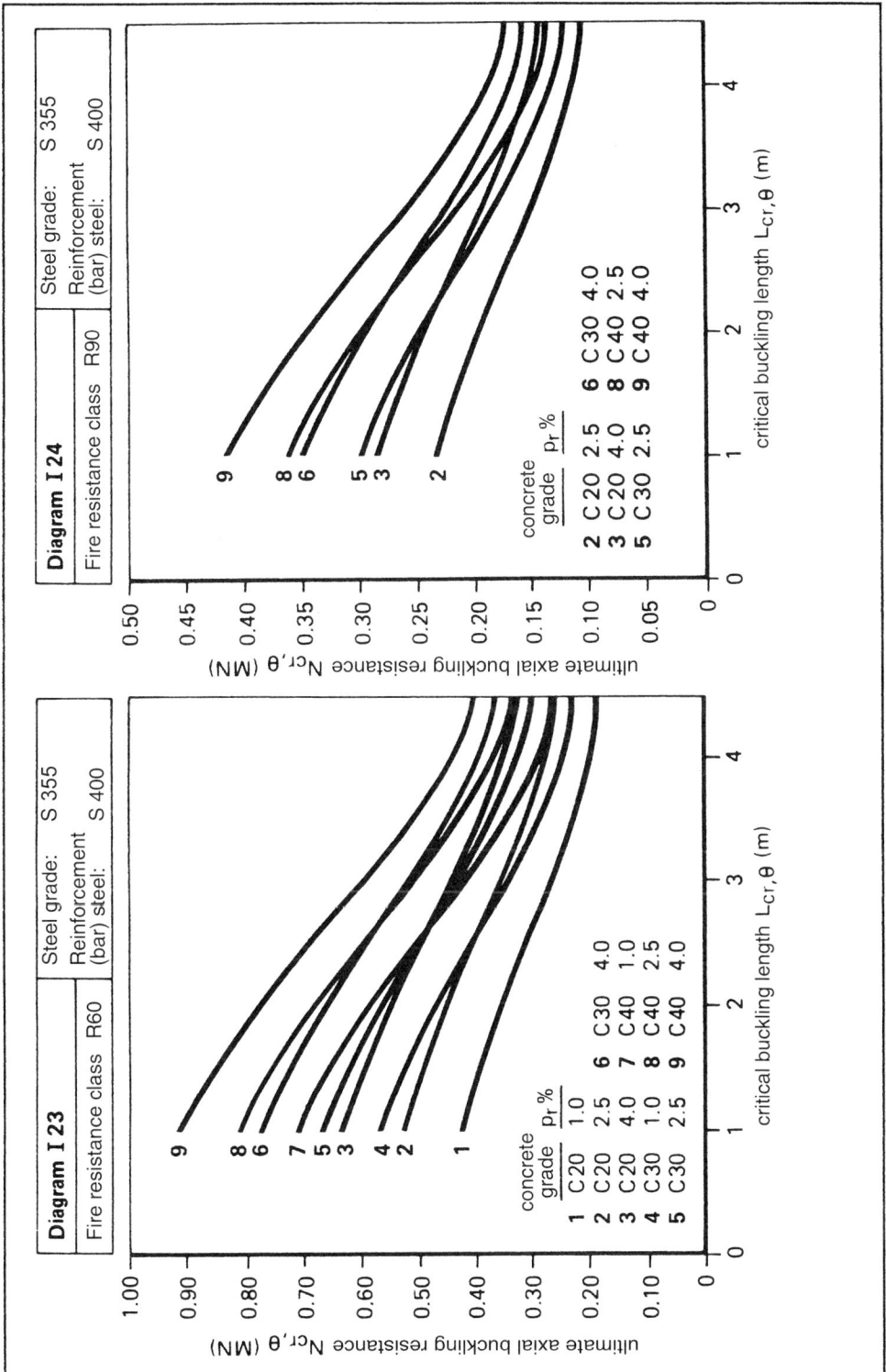

Buckling resistance diagrams for columns in RHS (square) □ 200 × 6.3

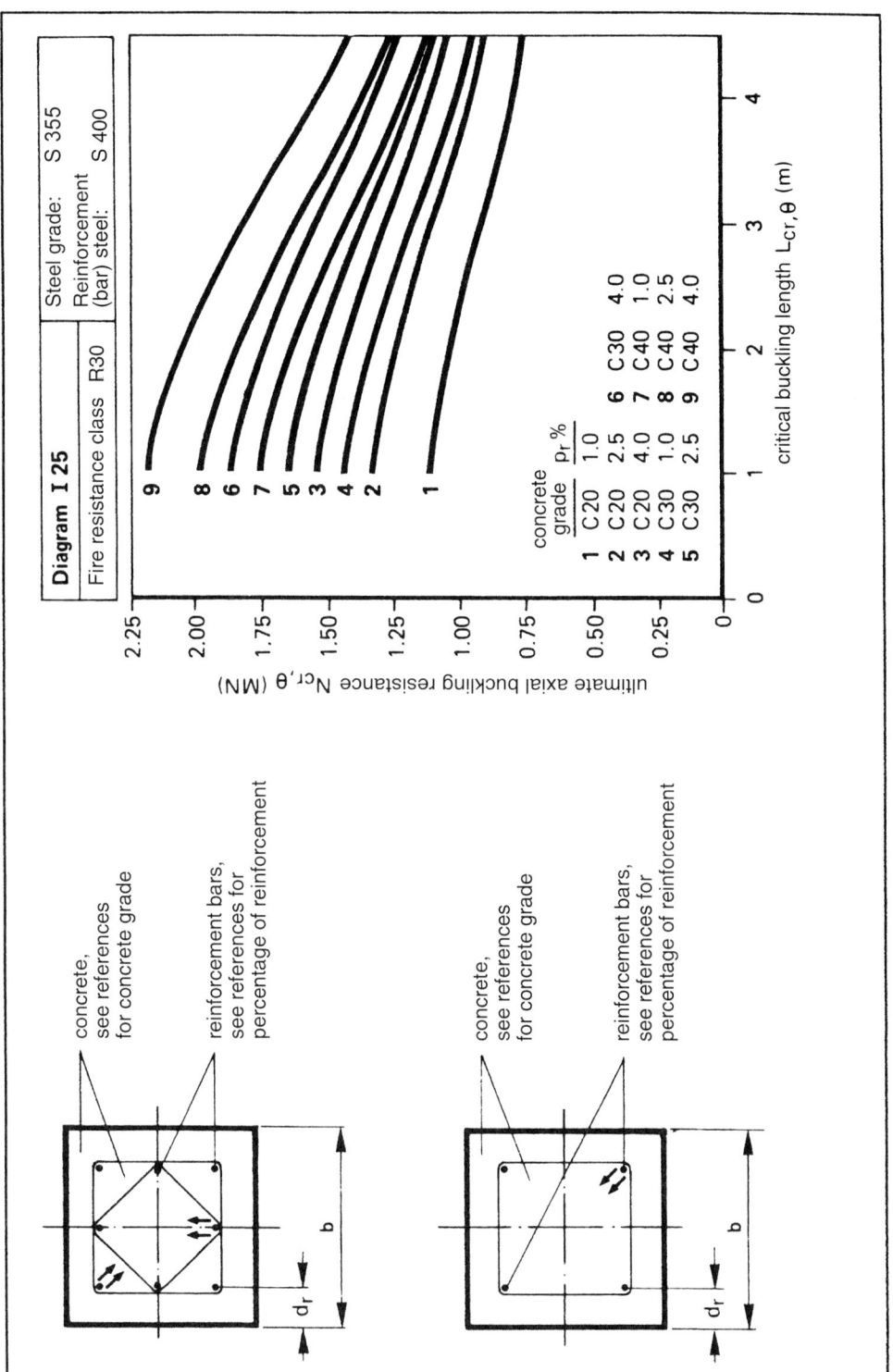

Buckling resistance diagrams for columns in RHS (square) □ 220 × 6.3

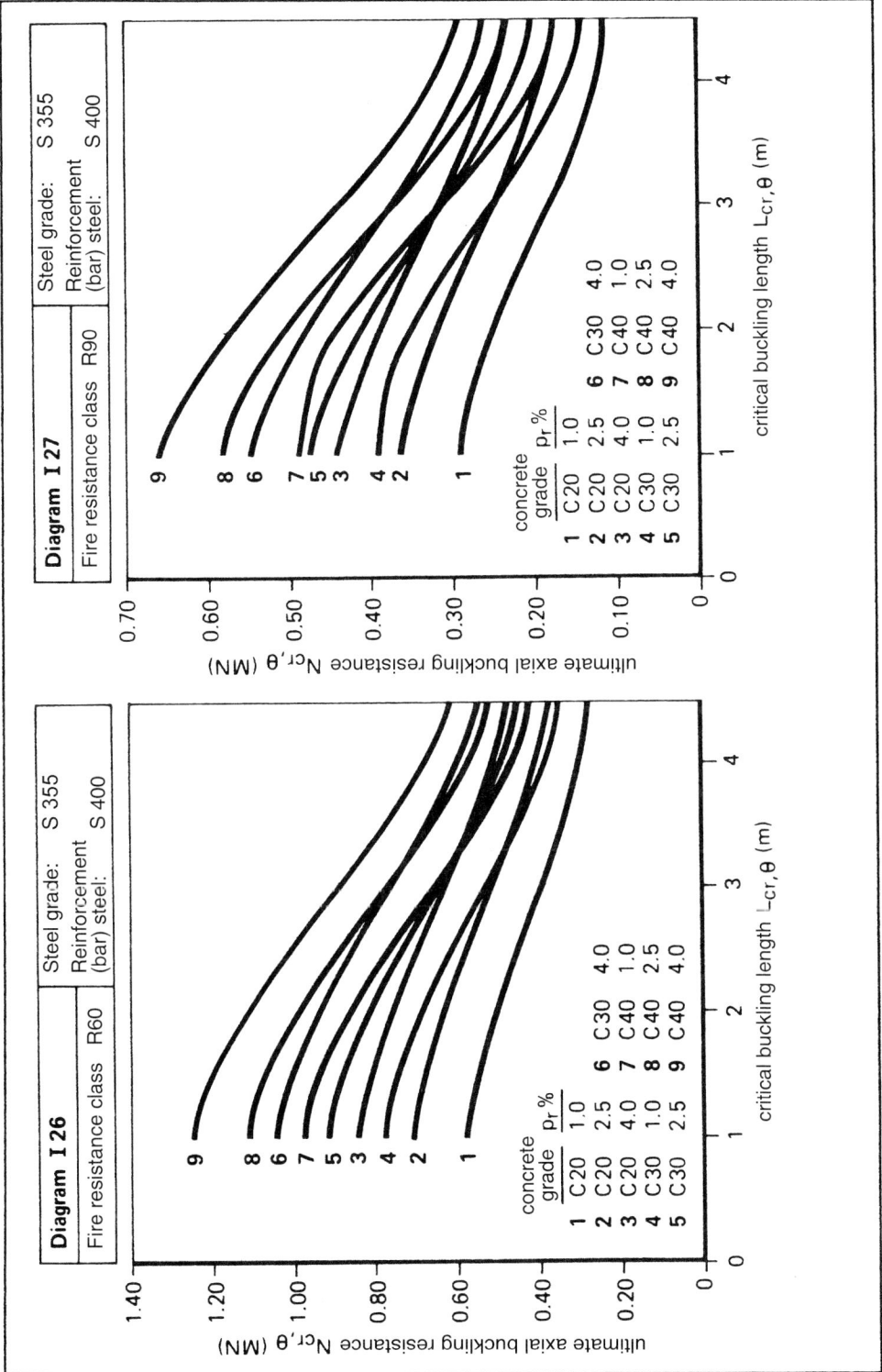

Buckling resistance diagrams for columns in RHS (scuare) □ 220 × 6.3

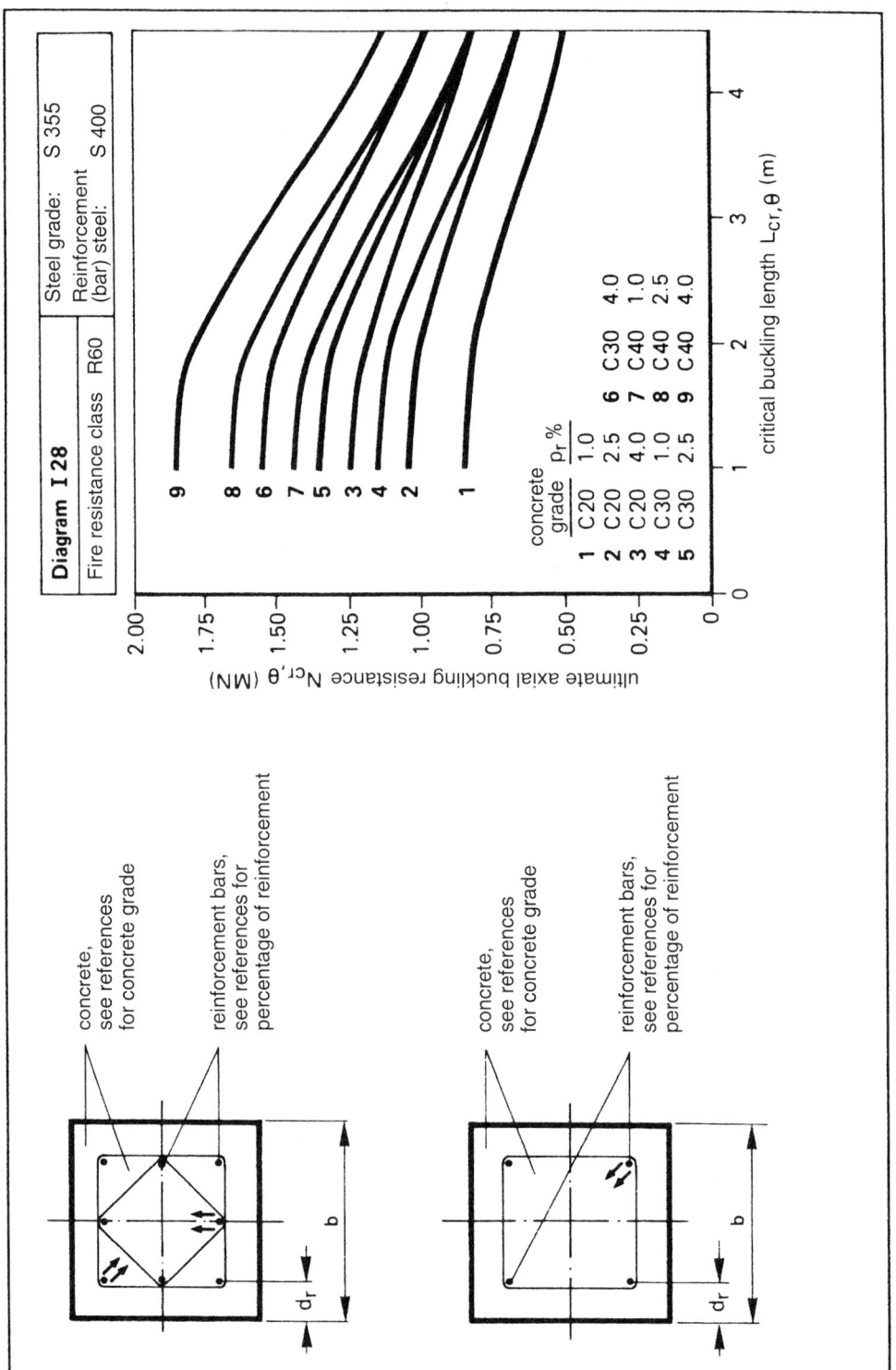

Buckling resistance diagrams for columns in RHS (square) □ 250 × 6.3

13.5 Fire protection of steel hollow section columns by filling with concrete

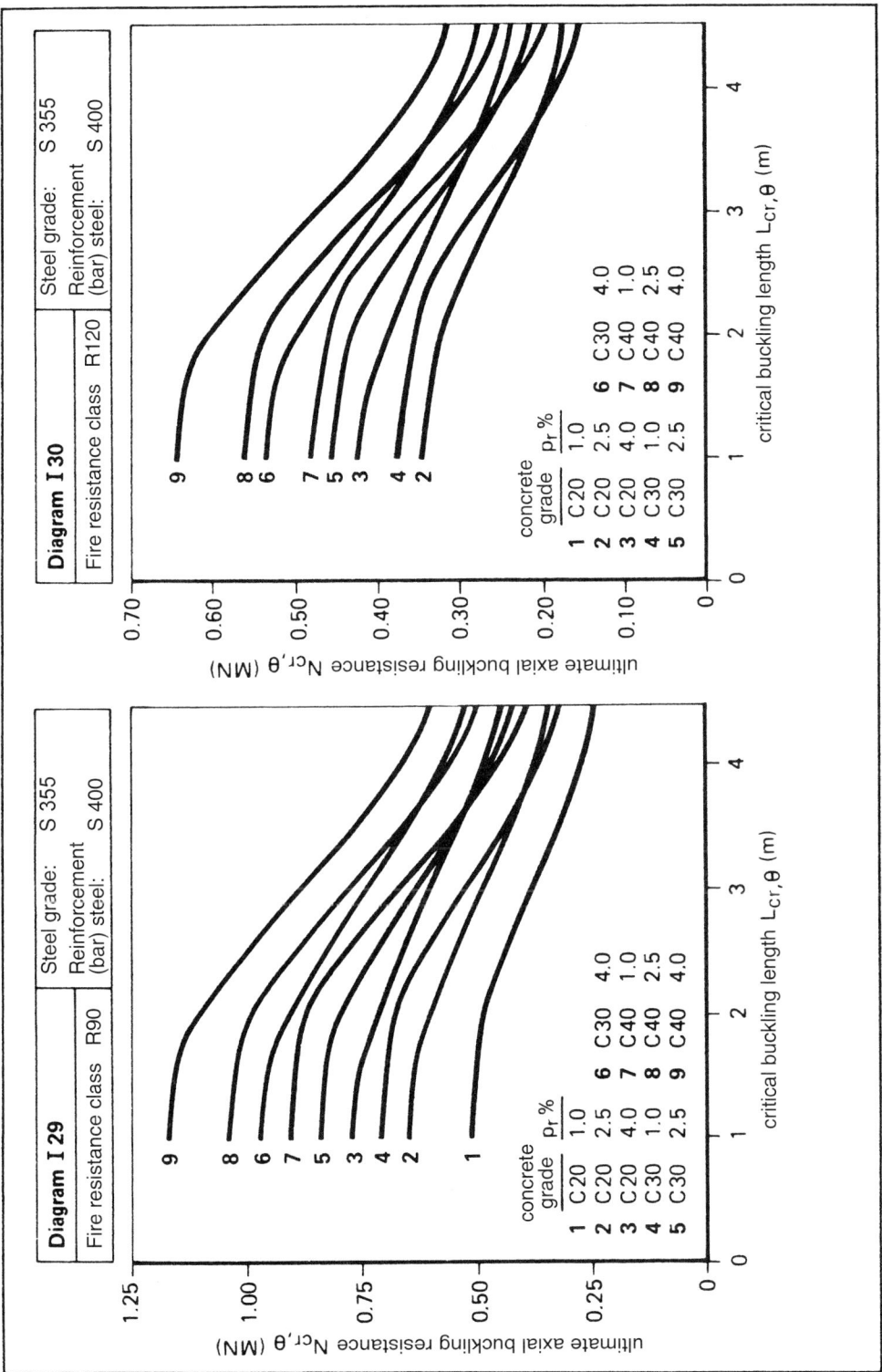

Buckling resistance diagrams for columns in RHS (square) □ 250 × 6.3

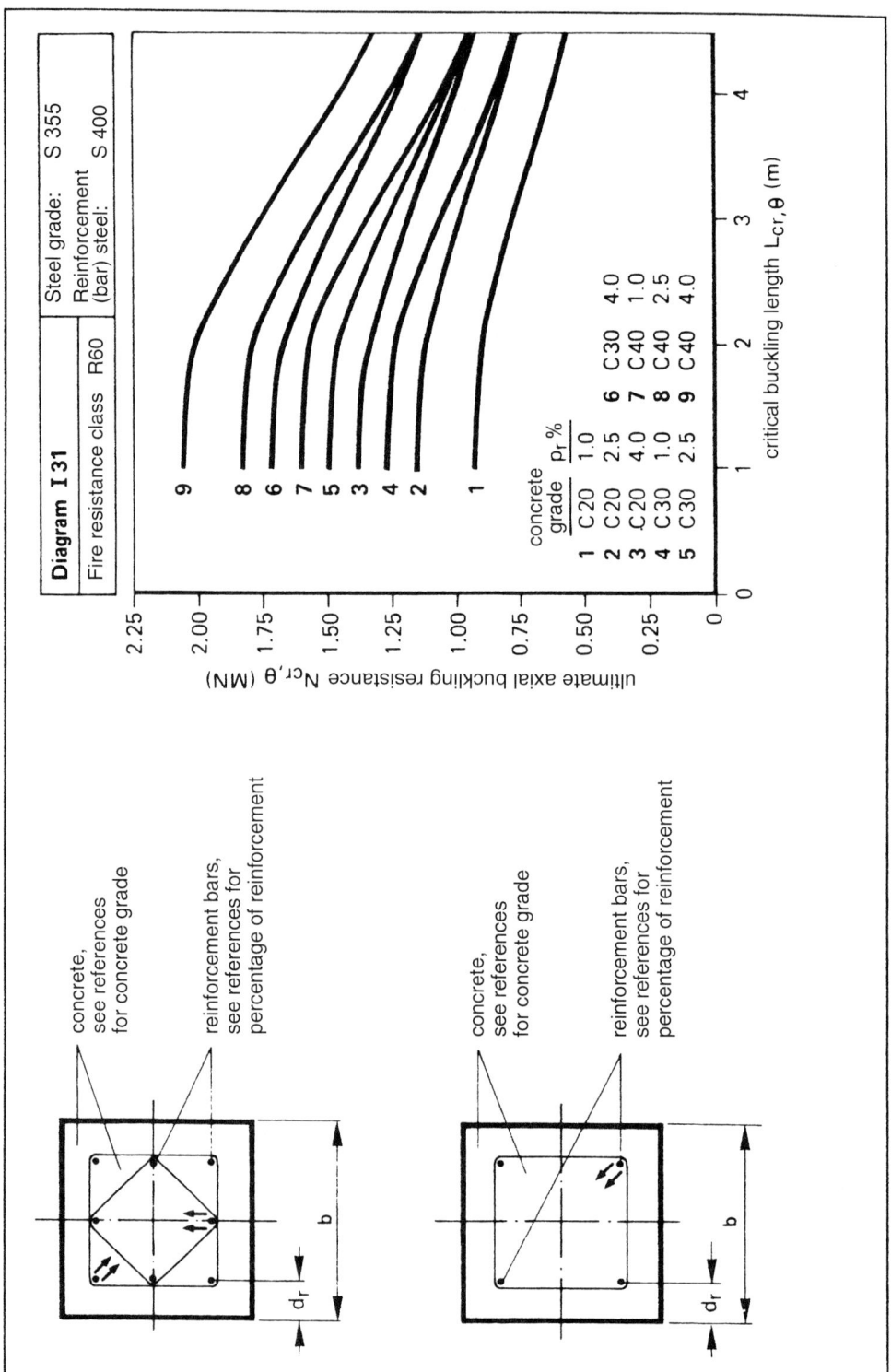

Buckling resistance diagrams for columns in RHS (square) □ 260 × 6.3

13.5 Fire protection of steel hollow section columns by filling with concrete

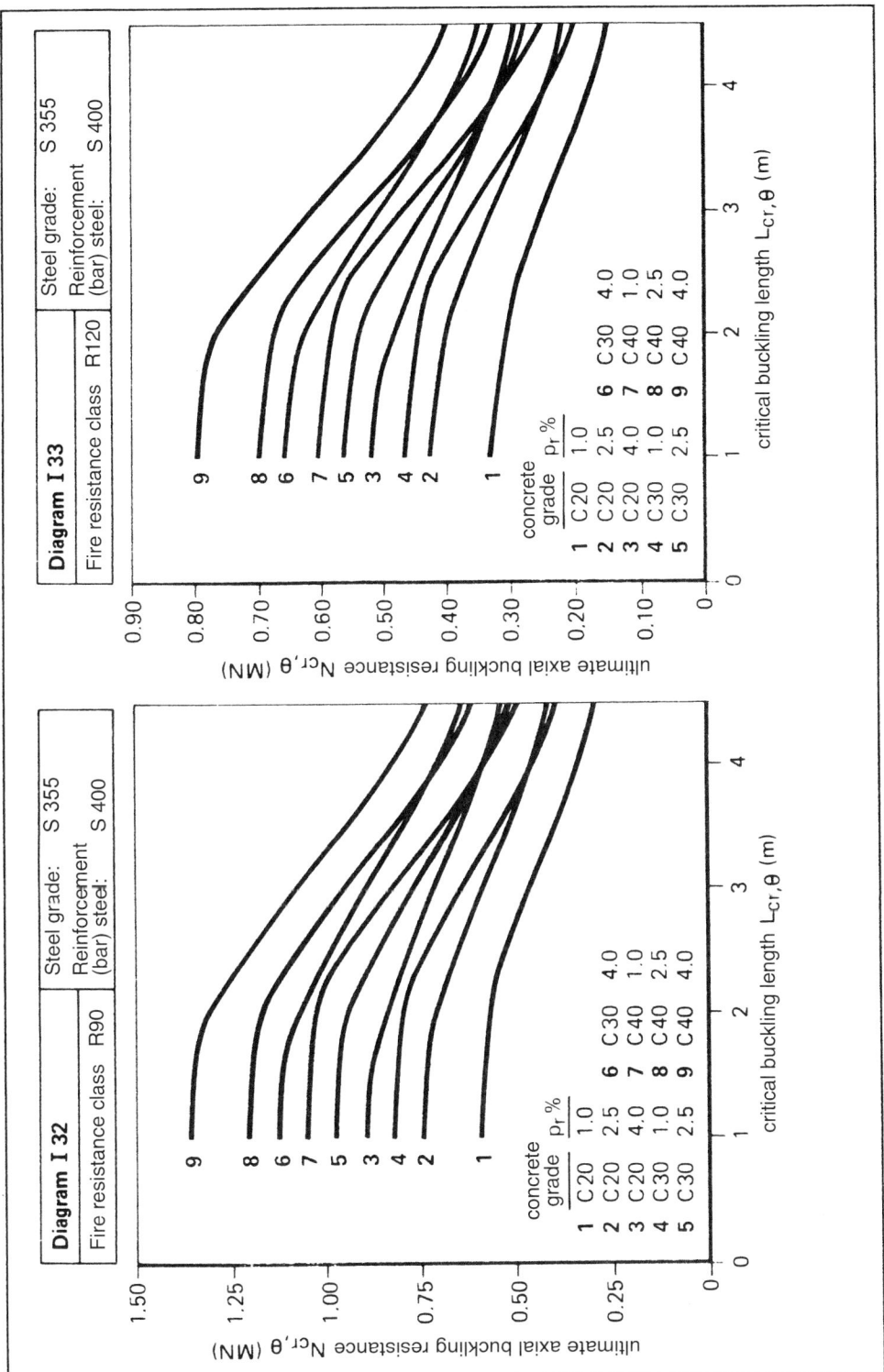

Buckling resistance diagrams for columns in RHS (square) □ 260 × 6.3

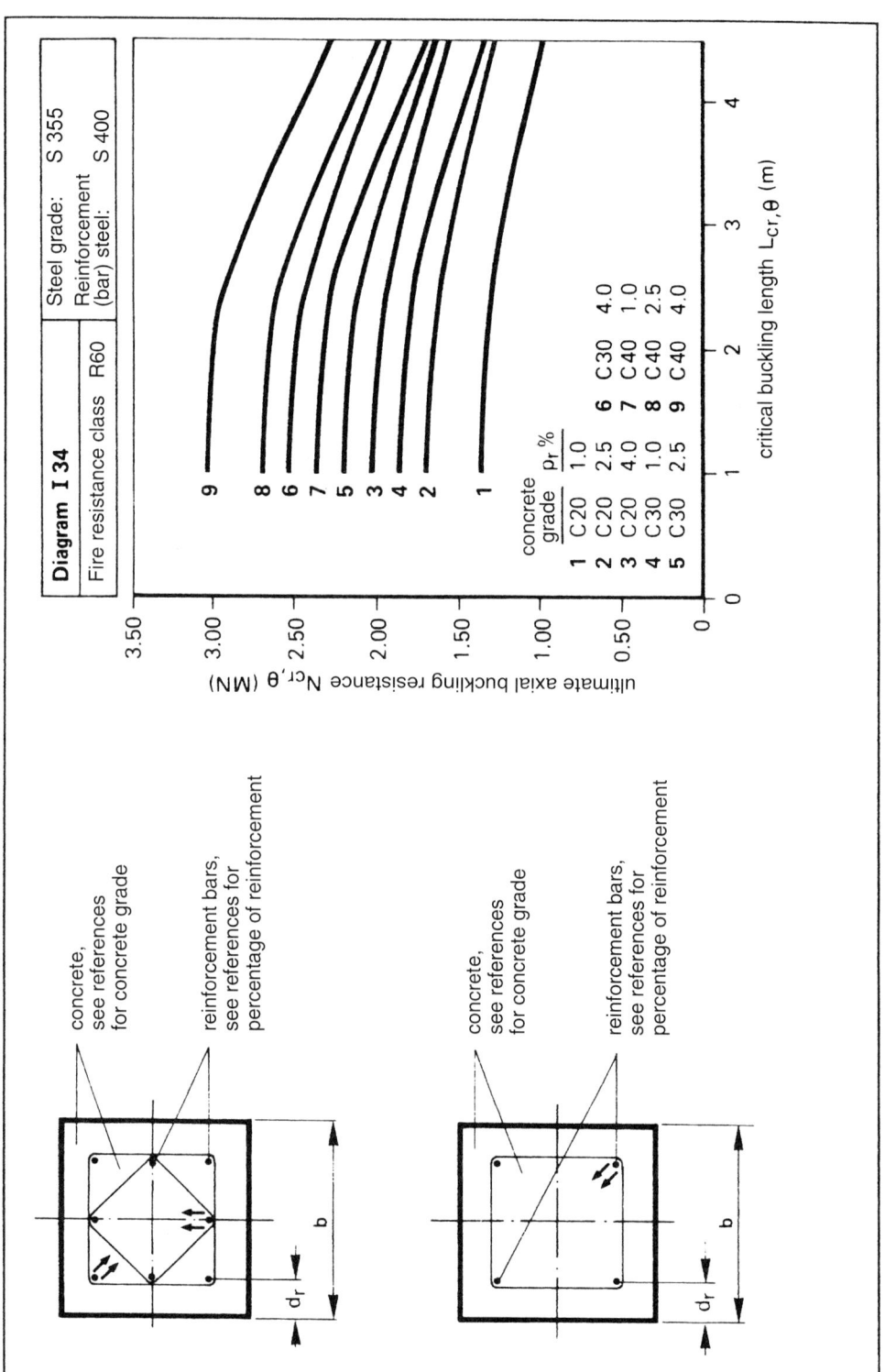

Buckling resistance diagrams for columns in RHS (square) □ 300 × 7.1

13.5 Fire protection of steel hollow section columns by filling with concrete

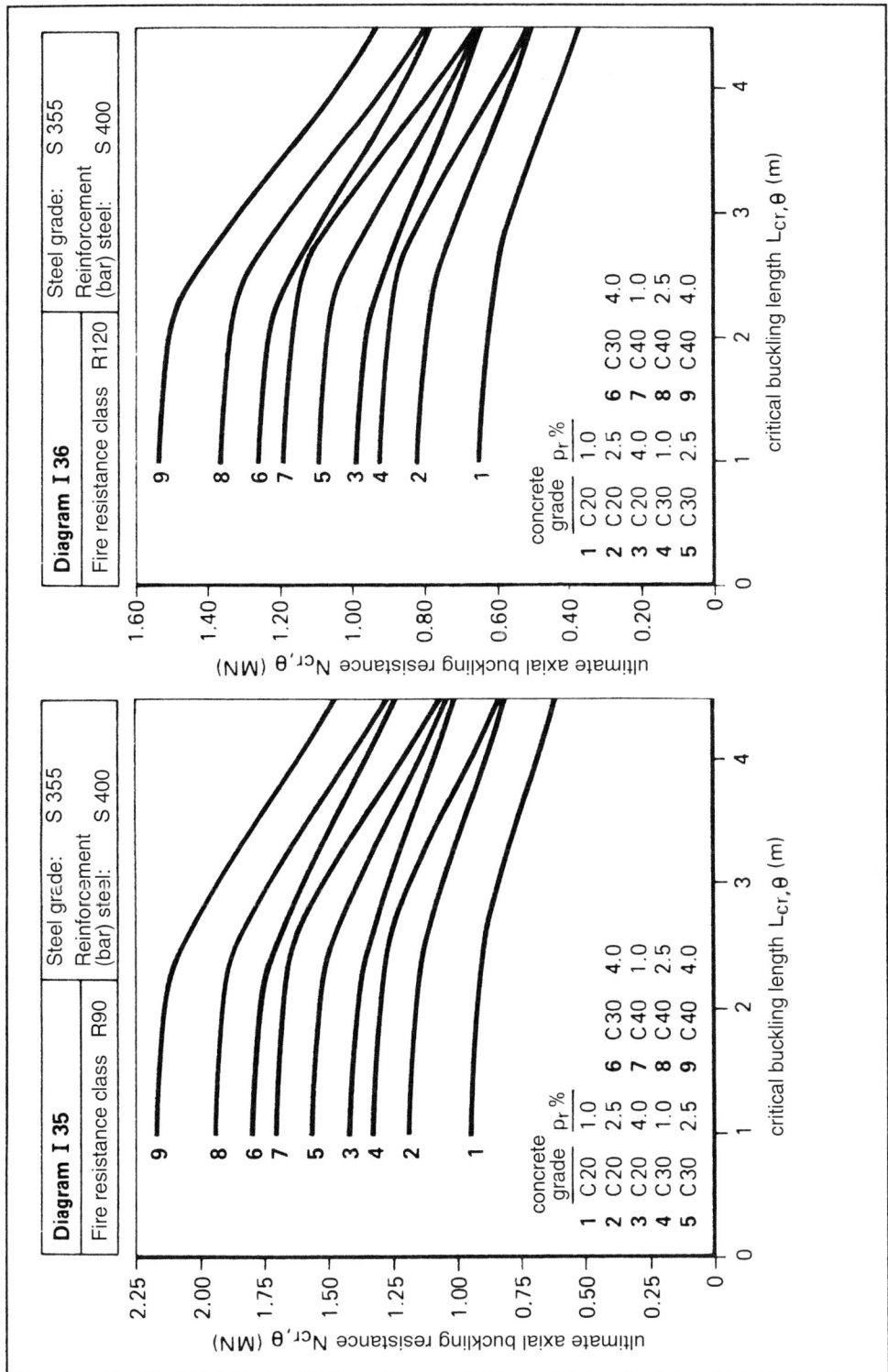

Buckling resistance diagrams for columns in RHS (square) □ 300 × 7.1

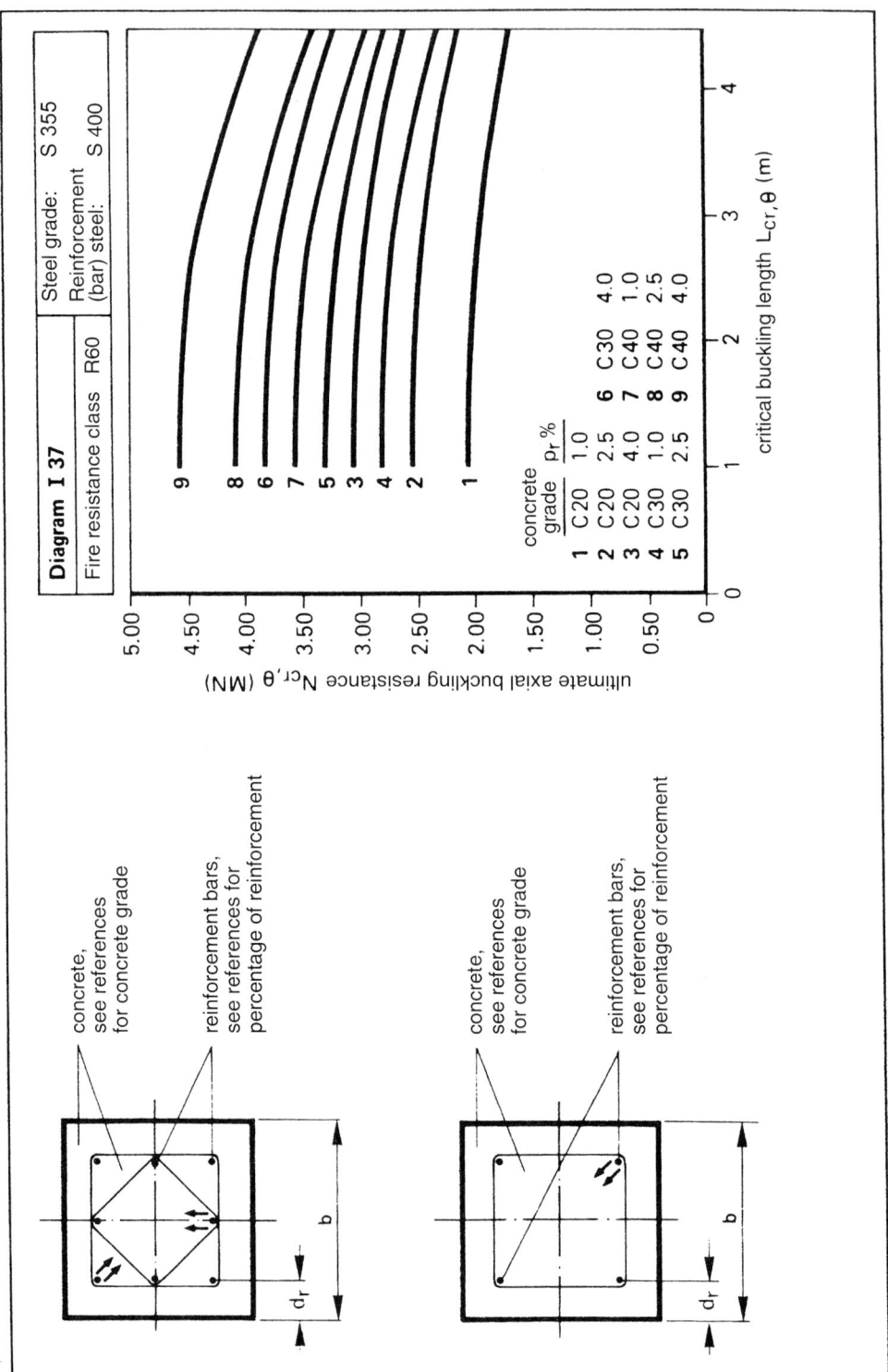

Buckling resistance diagrams for columns in RHS (square) □ 350 × 8.0

13.5 Fire protection of steel hollow section columns by filling with concrete

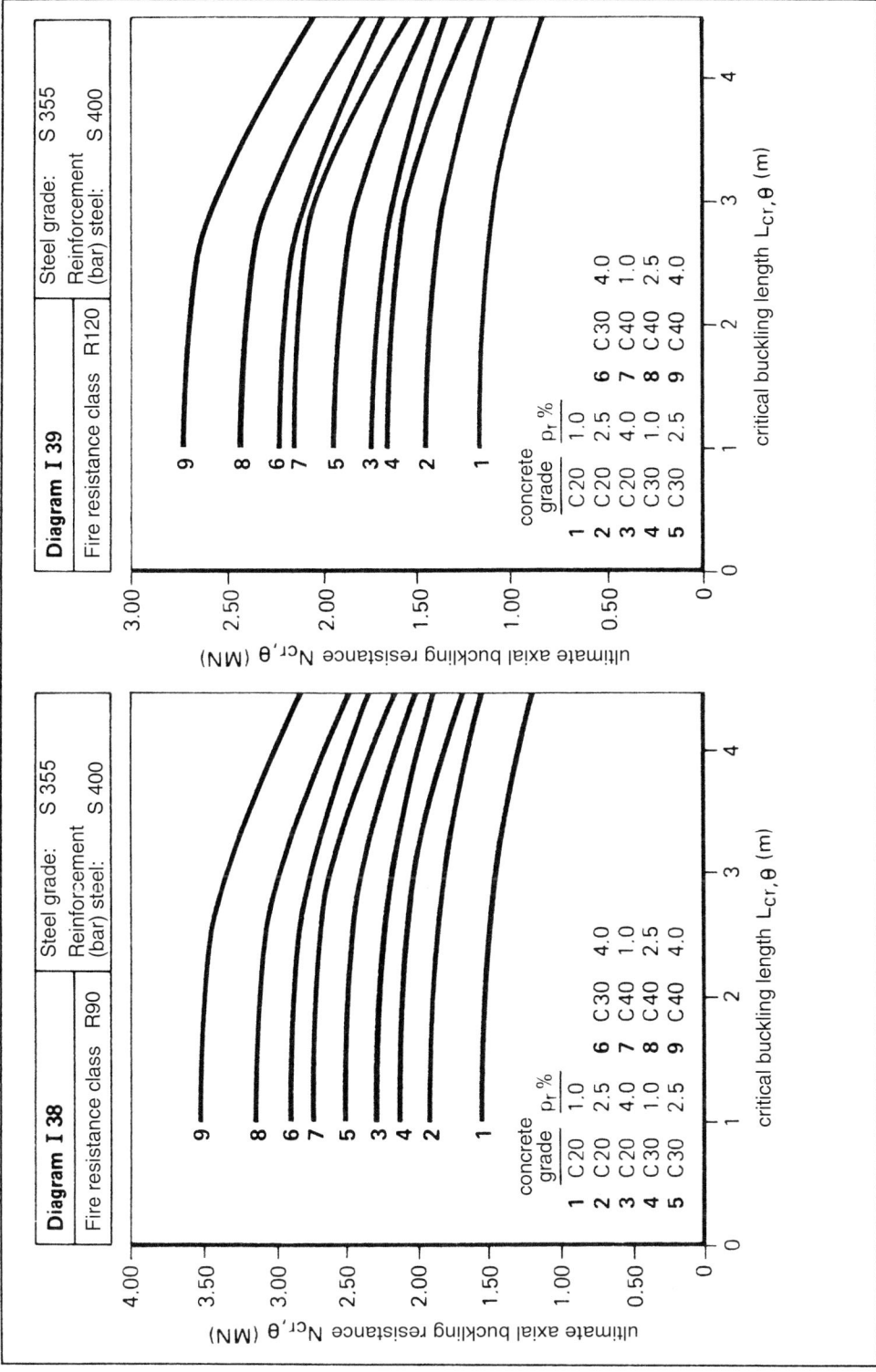

Buckling resistance diagrams for columns in RHS (square)   □ 350 × 8.0

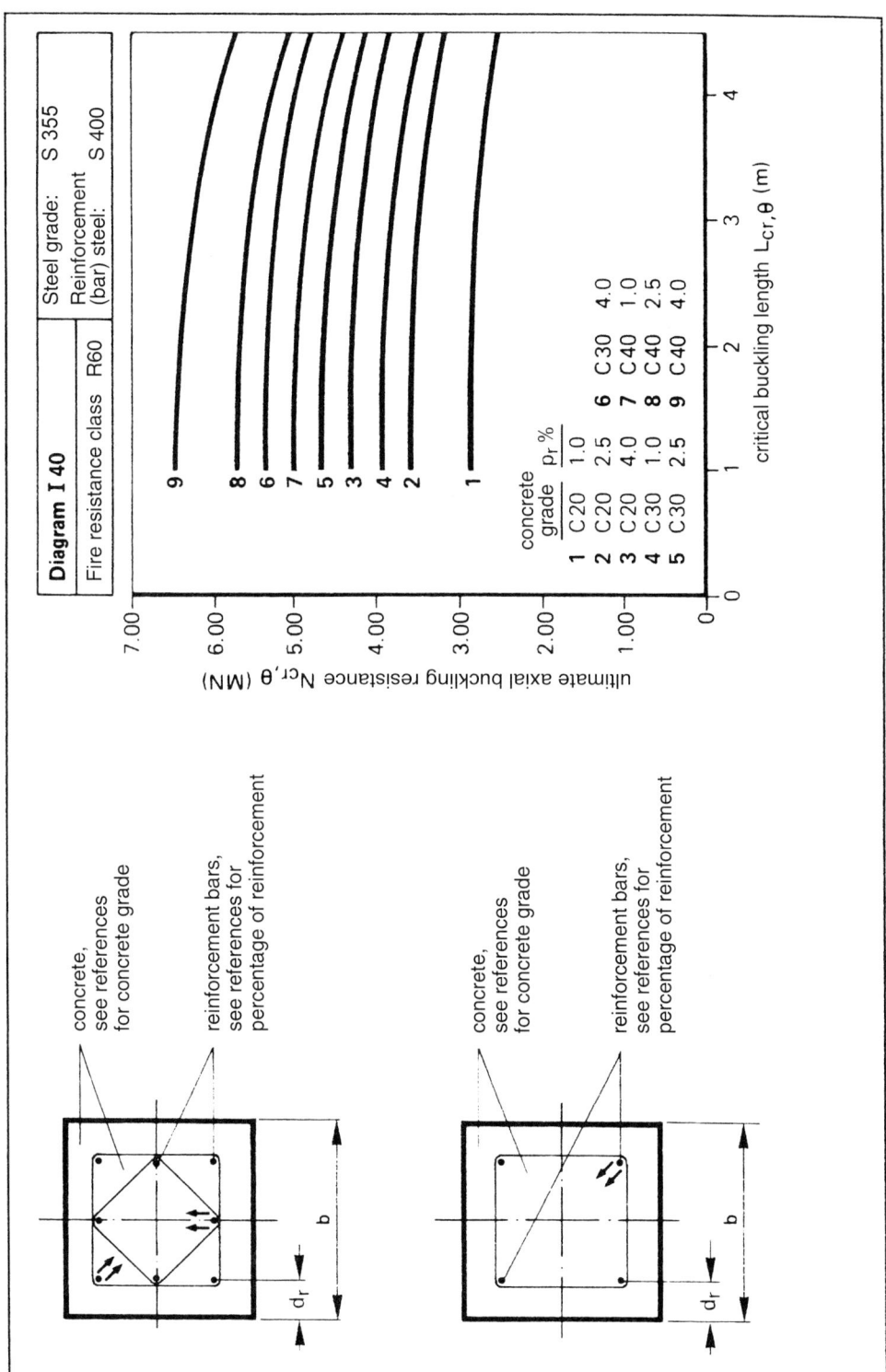

Buckling resistance diagrams for columns in RHS (square) □ 400 × 10.0

13.5 Fire protection of steel hollow section columns by filling with concrete

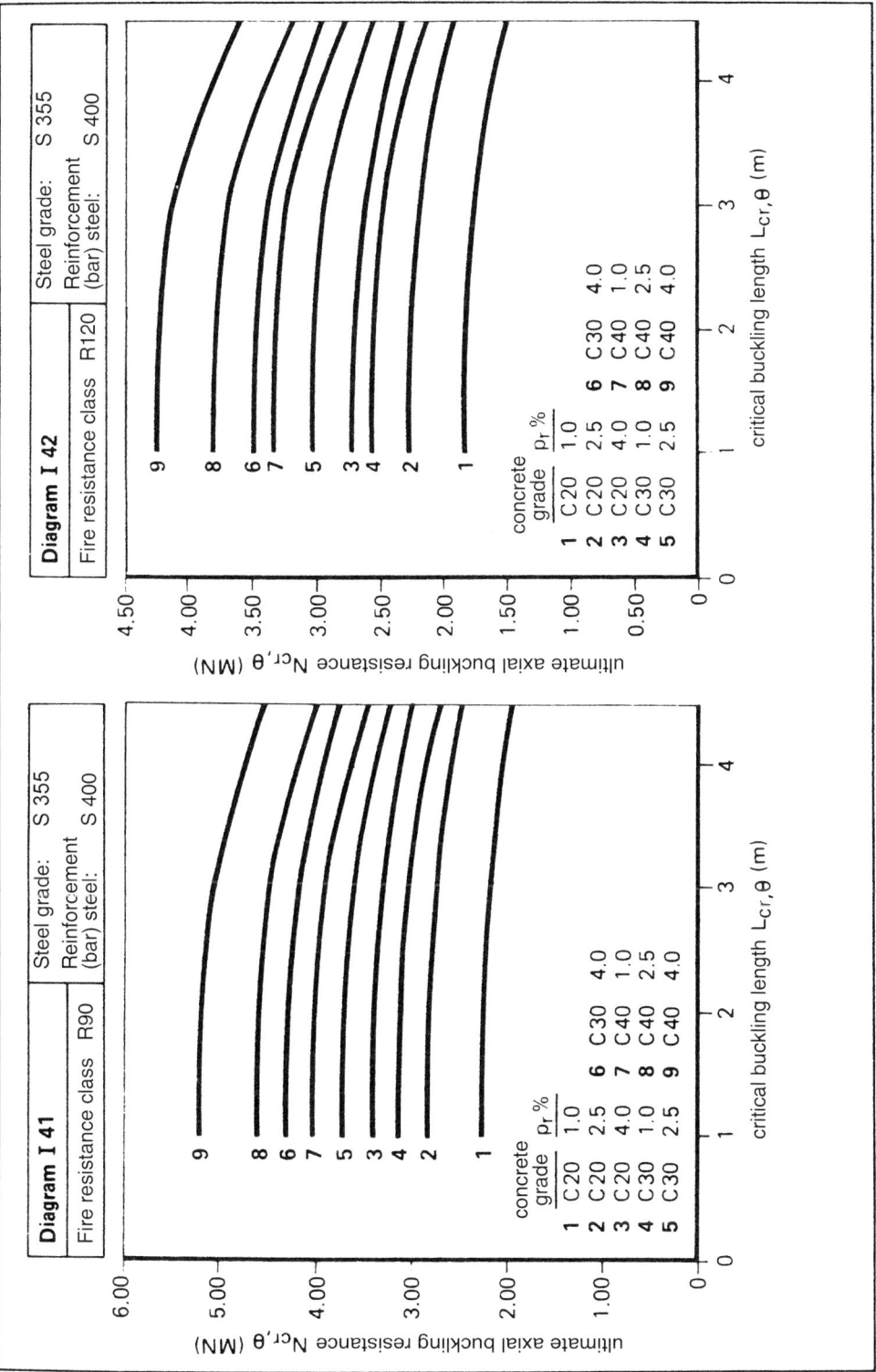

Buckling resistance diagrams for columns in RHS (square)   □ 400 × 10.0

The results obtained from the diagrams do not practically change and are on the safe side, whether a higher steel grade or a larger wall thickness is chosen for the hollow section.

Example:

Given:
CHS 273.0 ∅ × 5.0 mm
Amount of reinforcement (pr) 3%
Concrete grade C 30
Hollow section steel grade S 355
Reinforcement steel grade S 400
Column length 5.0 m (in the top floor)

Required:

Fire resistance class R 90
Axial buckling resistance 600 kN

Buckling length in the top floor =
0.7 × 5.0 = 3.5 m

Relevant design diagram: I 8

Values for the curves 5 (pr = 2.5%) and 6 (pr = 4.0%) are interpolated.

For 2.5% reinforcement, ultimate axial buckling load = 630 kN

For 4.0% reinforcement, ultimate axial buckling load = 756 kN

For 3% reinforcement,

ultimate axial buckling load =

$$630 + (756 - 630) \cdot \frac{3.0 - 2.5}{4.0 - 2.5} = 671.9 > 600 \text{ kN}$$

13.5.2.3 Level 3 design: general calculation method

This process calculates the fire resistance assessing the real fire, material and structural conditions as well as the non-linear temperature distribution over the cross section of the composite column. Due to the complicated and time-consuming calculation, the application of special computer programs is necessary, which have been developed in the last years [29–32].

13.5.3 Fire protection of hollow section columns filled by concrete with steel fibre reinforcement without external insulation

The addition of chopped steel fibres to the concrete mix has shown to produce a significant increase in the fire resistance of axially loaded concrete filled columns when compared to the equivalent column with plain concrete. It is assumed that the fibres act during heating as crack arrestors and so prevent premature crack propagation through the concrete matrix from the outer degraded layers to the sound cooler inner area of the core. In addition, the presence of fibres improves the flexural behaviour of the core and hence gives additional stability during heating.

The fire behaviour of this type of columns has not been adequately investigated until now. An evaluation of the available research results is given in Table 13-6, where the fire resistances are recorded for the various degrees of utilization η. For calculation purposes, the modulus of the concrete can be considered to be unchanged, so that the values of the non-dimensional slenderness $\bar{\lambda}$ of the fibre reinforced concrete filled columns are calculated exactly as for plain concrete.

For hollow section columns filled up with plain or fibre reinforced concrete:

$$\bar{\lambda}_b = \sqrt{\frac{N_{pl,Rd}}{N_{ki}}} \text{ (for plain as well as fibre reinforced concrete)}$$

where:

$$N_{pl,Rd} = A_a \cdot \frac{f_y}{\gamma_a} + A_b \cdot \frac{f_b}{\gamma_b} \quad (\gamma_a = 1.1; \gamma_b = 1.5)$$

$$N_{ki} = \frac{\pi^2}{l_k^2}(E_a \cdot I_a + E_b \cdot I_b)$$

The degree of utilization η of the plain concrete filled hollow section column is determined as follows:

$$\eta = \frac{N_{Sd,\theta}}{\chi_{min} \cdot N_{pl,Rd}}$$

where:
$N_{Sd,\theta}$ = applied axial force in fire case
$N_{pl,Rd}$ = see above

χ_{min} = reduction factor read from the buckling curve "c" (Fig. 5-4) based on the value of $\tilde{\lambda}_\theta$ calculated above

Further test results are shown by the $N_{cr,\theta}$ vs. $L_{cr,\theta}$ diagrams in Figs. 13-21 to 13-23 [39].

Table 13-6 Fire resistance of concrete filled hollow section columns depending on the degree of utilization

Fire resistance (minutes)	Degree of utilization η	
	Plain concrete	5% fibre concrete
30	No check needed	
60	0.51	0.67
90	0.40	0.53
120	0.36	0.47

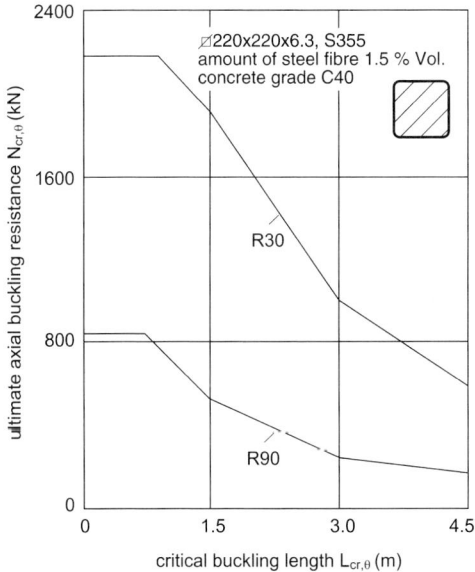

Fig. 13-21 Buckling resistance diagrams for hollow section columns filled with steel fibre reinforced concrete [39]

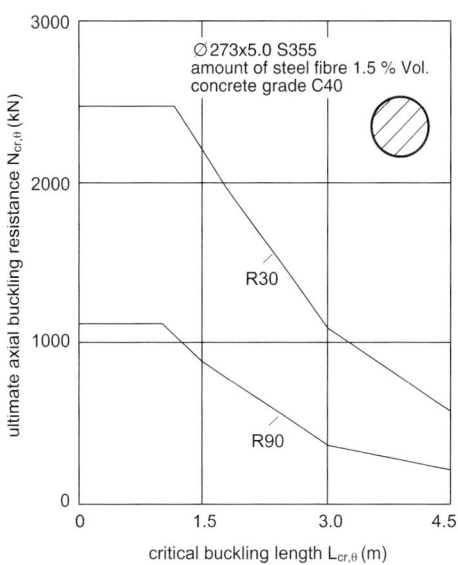

Fig. 13-22 Buckling resistance diagrams for hollow section columns filled with steel fibre reinforced concrete [39]

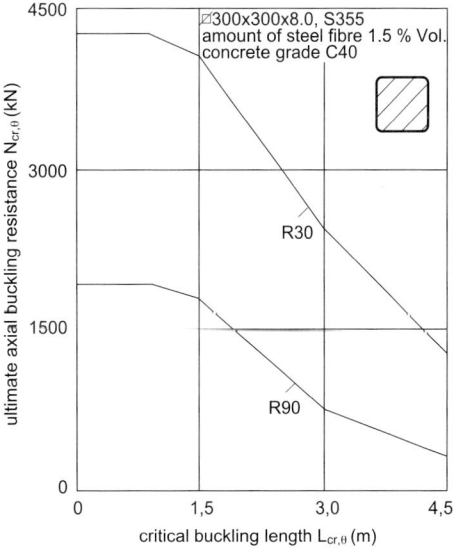

Fig. 13-23 Buckling resistance diagrams for hollow section columns filled with steel fibre reinforced concrete [39]

13.6 Fire protection of hollow section column to beam connections

13.6.1 Unfilled hollow section columns with or without external insulation

These connections are designed in practice based on the design codes for room temperature. For bolted connections, the bolt heads and nuts have to be fire protected as the cleats by means of insulations.

13.6.2 Concrete filled hollow section columns

These connections are designed similarly as for pure steel constructions, so that the loads can be transferred from the beams to the columns in such a way that all components – steel hollow section, concrete and reinforcement – contribute to the design resistance according to their strengths. An adequate fire resistance has to be assured without disturbing external protective cladding.

Fig. 13-24 illustrates beam to concrete filled column connections to transfer shear load in a braced structure (central core). For continuous beams (Fig. 13-24a), the connection can be classified in the same fire resistance class as the composite column and the beam without taking any special measure. For continuous columns (Fig. 13-24b), the pinned joint with connection plate to the column can be either fire protected, overdesigned or specially designed, so that the forces can be transferred even though the material loses its strength when exposed to fire.

Although connection plates can be welded to the hollow sections without special measures, the following two connection types are recommended to improve the mechanism to transfer the loads from steel to concrete:

– Connection with a cleat passed through the column (Fig. 13-25)
– Connection with a support bracket for the beam (Fig. 13-26)

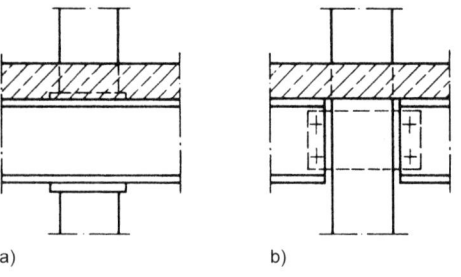

Fig. 13-24 Column to beam connection
a) with continuous beam, b) with continuous column

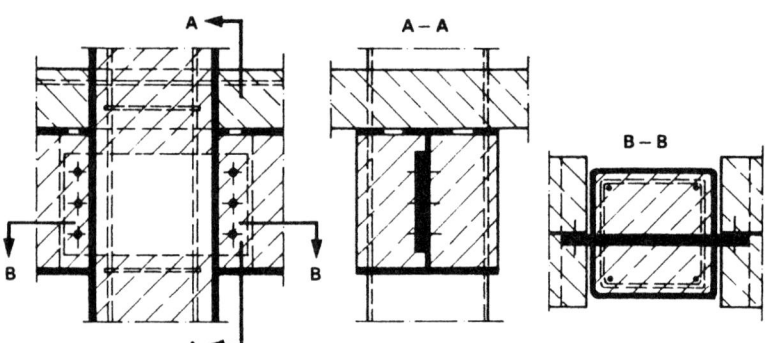

Fig. 13-25 Beam to column connection with a cleat through the column

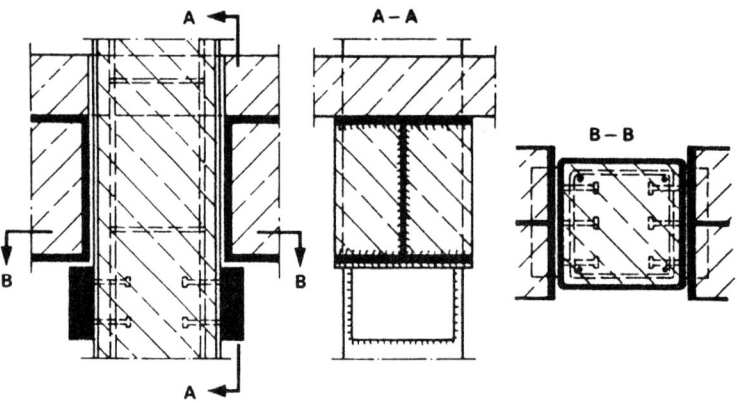

Fig. 13-26 Beam to column connection with a support bracket for the beam

13.6.3 Water cooled hollow section columns

Although no experimental evidence is available, it can be assumed that the same recommendations for the fire protection are valid for these connections as for those with unfilled hollow sections.

13.7 List of symbols

CHS	circular hollow section
RHS	rectangular hollow section (also square)
CEN	Comité Européen de Normalisation (European Committee for Standardisation)
EC	Eurocode
EN	Euronorm
ENV	European prestandard
ISO	International Standards Organization
A	cross-sectional area
A_m	envelope surface area of a structural member
a, a_y, a_z	side length of a concrete filled hollow section column (RHS)
a_s, d_r	concrete cover of a reinforcement bar
b	width of a rectangular hollow section
C	concrete grade
d	outer diameter of a circular hollow section
d_i	thickness of insulation
E	modulus of elasticity
e_0	distance of the eccentric load from the central axis of the RHS column
f	stress
f_b	28 day cylinder strength of concrete
f_y	yield strength of steel at room temperature
h	depth of a rectangular hollow section
I	second moment of area
K	absolute temperature (Kelvin)
$L_{cr,\theta}$	critical buckling length in fire case
L, l	length
l_K	buckling length
M	moment
$M_{pl,Rd}$	plastic moment design resistance of a cross section at room temperature
$M_{Sd,\theta}$	moment design load (action) in fire case
N	axial force
N_{eq}	equivalent axial force
$N_{pl,Rd}$	plastic axial force design resistance at room temperature
$N_{Sd,\theta}$	axial force design load (action) in fire case
R	fire resistance class
r_j	design resistance of an individual component of a composite cross section
s	wall thickness of a hollow section
T, θ	temperature
t	time in fire exposure
t_u	ultimate duration to failure in fire case
V	volume of a structural member per unit length
W	section modulus
ε	strain
η	degree of utilization
χ_{min}	minimum buckling coefficient according to the European buckling curve "c" (see Fig. 5–4) of EC 3, part 1.1 [22] or any equivalent national buckling curve
κ	moment reduction factor according to EC 3, part 1.1 [22], see Section 5.3.3.3
$\bar{\lambda}$	non-dimensional slenderness
γ	partial safety factor

Indices:

a	steel hollow section
b	concrete
cr	critical
E	slope of the linear elastic range
s	reinforcement
θ	elevated temperature
K	buckling

14 Wind loads on circular and rectangular hollow sections and their lattice structures

14.1 General

The performance of circular hollow sections is by far superior to that of the conventional open profiles (U, L, T, I and Z forms), when they are exposed to the forces induced by the oncoming stream of wind and water. That is why they are frequently applied in the open air or free water environments, where they are loaded by wind forces (as for example in masts, cranes, conveyor bridges etc.) or by water streams (as for example in dolphins, offshore platforms etc.). They are distinguished by their extraordinarily low drag coefficient C_W (wind fixed reference system).

This chapter deals mainly with the wind loads and the response of the hollow section structures to them. In this case, the calculation of wind loads has to cover not only the usual type of structures, which are relatively rigid, but also slender and flexible structures, which are very sensitive to the fluctuating effect of wind. It is a highly complex operation, as it involves the study of both local and general effects. The local effect is related to the properties of the envelope such as roughness, while the general effect is due to the general shape of the structure.

In order to determine the drag behaviour of cylindrical structural elements and their lattice structures, extensive tests were carried out in the seventies and the eighties on national as well as on international basis. The informations containing the analysis of the wind tunnel tests, the formulation of a theory, a calculation method and the preparation of working diagrams are available in the literature [1, 2, 4, 6, 9–14, 16].

The major part of the experiments and the related evaluations were done in the Sub-sonic Wind Tunnel at Porz-Wahn of the German Research Centre for Aeronautics and Astronautics (DFVLR) [1, 2, 9–11] and in the Compressed Air Tunnel of the National Physical Laboratories at Teddington, United Kingdom [6].

A number of theoretical and experimental research works had also been carried out in Porz-Wahn [4] and Karlsruhe [5] to investigate into the flow phenomena of the wind stream on cylinders of square hollow sections with varying corner radii. They show lower drag coefficients C_W for the square hollow sections depending on corner radii in comparison with the sharpedged structural bodies and open rolled profiles. Lit. [4, 5] explain the measured results of C_W-values for square hollow sections. The corner radii of the slender, rectangular bodies have also been taken into account in the European codes [15].

14.2 Wind loads on single circular cylinders

The drag coefficients C_W for all profiles with strongly rounded corners, especially for CHS without any corner, depend significantly on

$$\text{Reynolds Number Re} = \frac{V \cdot D}{v}$$

(V = windvelocity; D = outer diameter of CHS; v = kinematic viscosity of air (m²/sec)).

Below a certain Reynolds Number Re (subcritical range), the drag coefficient C_W remains constant at a high level. When this critical value for Re is exceeded, a dramatic decrease of C_W occurs. C_W rises slowly again, when Re increases further, but does not reach its old level. The aerodynamical superiority of the CHS cylinders to other edged profiles is based on this fall of the drag coefficient in the supercritical Re-range.

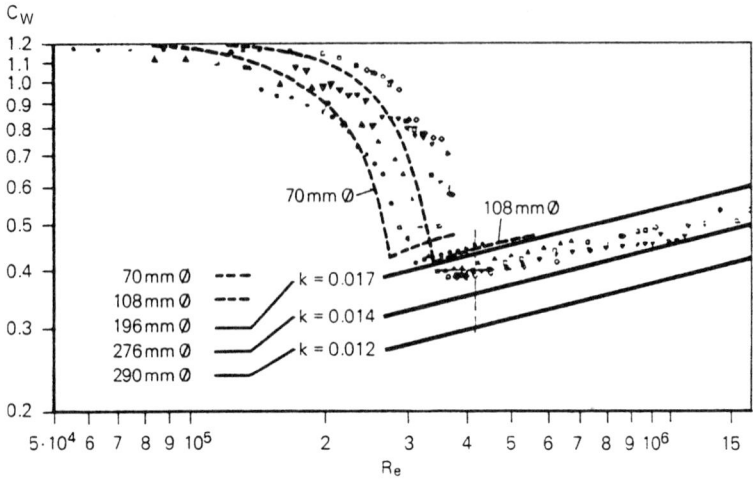

Fig. 14-1 Drag coefficient C_W for circular cylinders with mean roughness [1, 11]

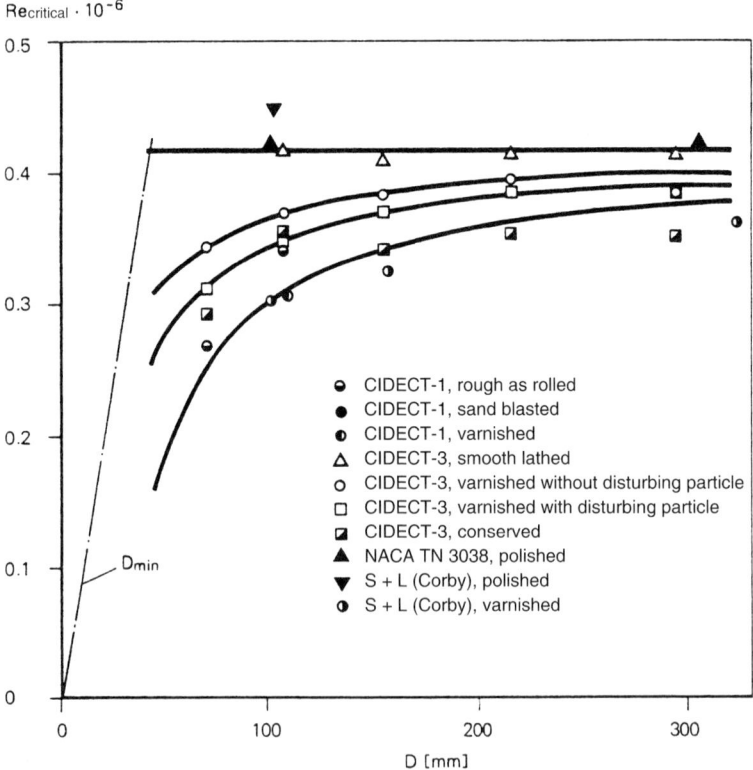

Fig. 14-2 Critical Reynolds Number $Re_{critical}$ of circular cylinders as a function of diameter D and surface roughness k [9, 11–14]

14.3 Wind loads on single square cylinders

Further to the knowledge of the critical Reynolds Number Re, the grade of decrease of C_W and the slope $\dfrac{dC_W}{d\text{Re}}$ are important characteristics. Both of them do not have constant values and depend on the surface roughness k and the cylinder diameter D.

The turbulence of wind plays also its role. Fig. 14-1 shows the test measurements for cylinders with mean roughness (see [1, 9]).

The critical Re for circular cylinders as a function of the diameter D and the surface roughness k is illustrated in Fig. 14-2. For polished surface, no dependence on diameter D exists. $\text{Re}_{\text{critical}}$ decreases with increasing surface roughness and reducing diameter.

For the supercritical range (see Fig. 14-1), the following interpolation formula is recommended to describe the increase of C_W [1]:

$$C_{W,\text{supercritical}} = k \cdot \text{Re}^{1/4}$$

In the subcritical range, $C_{w,\text{subcritical}} = 1.20$ (= constant).

14.3 Wind loads on single square cylinders

The following general statement can be made regarding the aerodynamical load on the cross-sectional shapes:

Wind loads on the sharpedged profiles or on the sections with edges of small sharpness ($R/D < 0.025$) (it is without any significance whether they are open or closed) are independent of Reynolds Number Re (R = corner radius). This independence is valid for all wind loading directions.

Figs. 14-3 and 14-4 show the drag and transversal force coefficients for the range:

$$10^3 < \frac{V \cdot D}{v} = \text{Re} < 3 \cdot 10^5$$

where:
V = mean wind velocity (m/sec)
D = width of the square section (m)
v = kinematic viscosity of air (m²/sec)

They describe the drag coefficients for the resistance W (wind fixed reference system) for loading in line with the wind direction (Fig.

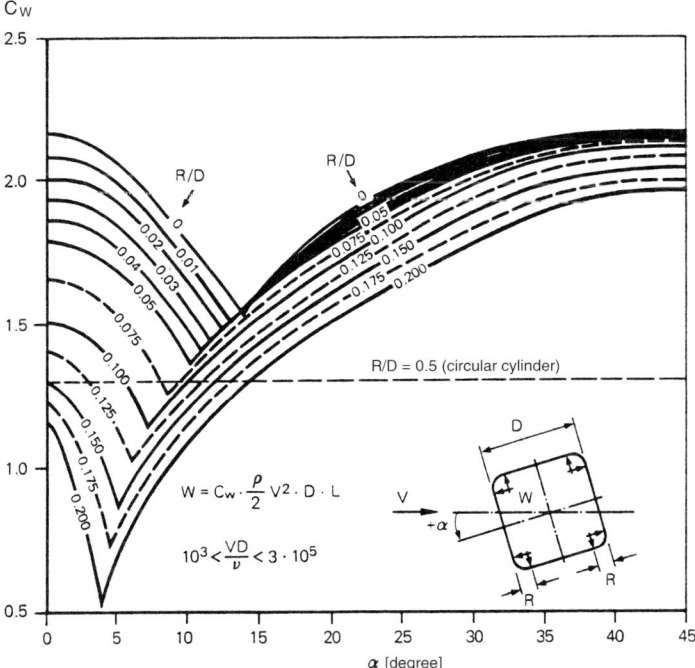

Fig. 14-3 Drag coefficient C_W for square sections with corner radius versus wind direction angle α (W = mean drag) [4]

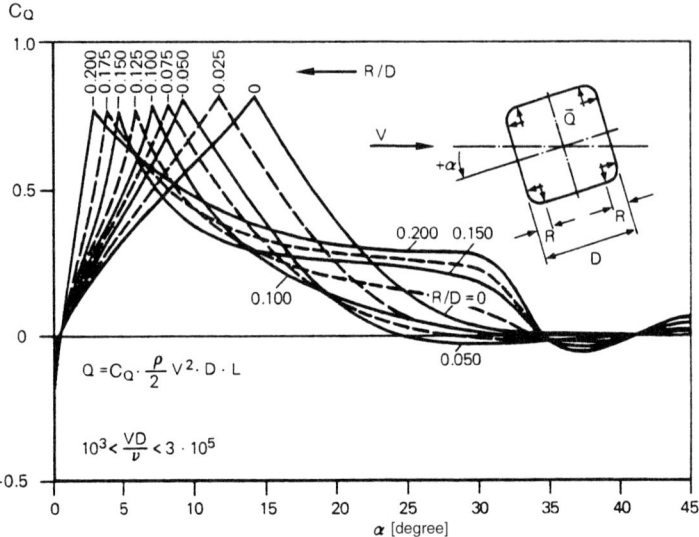

Fig. 14-4 Transversal force coefficient C_Q for square sections with corner radius versus wind direction angle α (Q = mean transversal force across wind)

14-3) and those (C_Q) for the transversal forces Q (wind fixed reference system) for loading across the wind (Fig. 14-4) versus the wind direction angle α. The other parameter is the ratio $\dfrac{\text{corner radius } R}{\text{width } D}$.

$$W = C_W \cdot \frac{\rho}{2} \cdot V^2 \cdot D \cdot L \qquad (14\text{-}1)$$

where:
W = mean drag
C_W = drag coefficient
ρ = air density $\left(\dfrac{N \cdot \sec^2}{m^4}\right)$
V = mean wind velocity
D = width of the square section
L = Length of the square section

$$Q = C_Q \cdot \frac{\rho}{2} \cdot V^2 \cdot D \cdot L \qquad (14\text{-}2)$$

where:
Q = mean transversal force
C_Q = transversal force coefficient

The measurements by DFVLR [4] lead to the conclusion that larger surface roughness causes smaller force coefficient. The diagrams, which were established for smooth surfaces with $k/D = 6.25 \cdot 10^{-4}$, are therefore also valid for larger roughness.

14.4 Wind loads on lattice structures in circular hollow sections

In [2], the procedure to calculate the wind loads on lattice structures of circular hollow sections has been described, where the wind loads consist of the loads on the single members and the interference resistances. Decisive for the interference resistance is the solidity ratio φ given by the effective frontal area of the lattice frame work (so-called reference area A_{ref}) to the enveloping area of the effective frontal area (A_u). Also the distance between the single lattice structures or of single structural elements e.g. chords is of vital importance regarding the interference behaviour (wind shielding).

However, only lattice structures, which are manufactured in a large number, are usually tested in a wind tunnel. On the other hand, the measured values for a lattice structure are in general based on the sum of the wind-loaded single profiles. They take further the following effects into account:

- Displacement effects, which increase or decrease the wind load of a single member (the deformation of wind stream about a single member caused by a neighbouring member changes its direction and magnitude locally).

- Wind shielding effects, which decrease the wind load of a single member (one element lies in the dead water of another element).
- Wake effects, where a backward-lying single member gets the impact of the fluctuating flow separation of a frontal member and is flowed around with higher turbulence. For rounded profiles, higher turbulence can lead to lower loads.

The drag coefficients C_W (related to the resistance W in line with the wind direction, wind fixed reference system) and the force coefficient C_N (related to the normal force N, body fixed reference system) and C_T (related to the tangential force T, body fixed reference system) had been determined by wind tunnel tests [1] on two-dimensional lattice frames and triangular or square lattice masts in CHS. In Figs. 14-5 to 14-10, they are plotted against Reynolds Number Re varying the solidity ratio $\varphi = \left(\dfrac{A_{\text{ref}}}{A_u}\right)$. Another parameter is the wind direction angle β measured against the perpendicular to the reference plane A_{ref}.

For special lattice structures, it is to be decided whether the wind loads as the sum of the loads on the single members can be corrected with the help of the known interference theories [1, 7, 8] or more clarity can be attained with the wind tunnel tests.

14.4.1 Approximate calculation of wind loads on lattice masts

The calculation is to be carried out in the following steps:

1. Calculation of Reynolds Number Re and dynamic pressure q

$$\text{Re} = \dfrac{V \cdot d_c}{v}$$

$$q = \dfrac{\rho \cdot V^2}{2}$$

2. Calculation of geometric solidity ratio φ

$$\varphi = \dfrac{A_{\text{ref}}}{A_u} =$$

$$= \dfrac{\text{effective frontal area or reference area}}{\text{enveloping area of the effective frontal area}}$$

These two areas can be determined using the following equation:

$$\varphi = \dfrac{2 d_c}{b}\left[1 + \dfrac{n}{2}\left(1 - \dfrac{d_c}{b}\right)\left(1 + \tan^2 \gamma_d\right)^{0.5}\right]$$

where:
d_c = outer diameter of the chord member
d_d = outer diameter of the bracing member
n = d_d/d_c
b = distance between the central axes of the two chords
γ_d = angle of the bracing members to the vertical
$\tan \gamma_d = b/l_c$ = inclination of the bracings
l_c = half panel length of mast section

3. Determination of the drag (C_W) and force (C_N, C_T) coefficients from the Figs. 14-5 to 14-10

4. Calculation of wind forces and drag

Normal force $N = C_N \cdot q \cdot A_{\text{ref}}$
Tangential force $T = C_T \cdot q \cdot A_{\text{ref}}$
Drag $W = C_W \cdot q \cdot A_{\text{ref}}$

5. Load component of the individual lattice planes for triangular or square lattice masts

Table 14-1 is to be used for this calculation. The shielding factor is given in Fig 14-11. The proportion of the total force is thus obtained, which the individual planes of a lattice frame have to carry.

Example:
Triangular lattice mast (Fig. 14-12)
Wind direction $\beta = 0°$
$d_c = 0.0337$ m; $d_d = 0.0214$ m; $b = 0.300$ m; $l_c = 0.356$
$\tan \gamma_d = 0.3/0.356 = 0.8426$
Wind velocity $V = 30$ m/sec

Reynolds Number $\text{Re} = \dfrac{V \cdot d_c}{v} =$
$= 2.05 \cdot 0.0337 \cdot 10^6 = 0.06915 \cdot 10^6$

where:
$v = 1.462 \cdot 10^{-5}$ m²/sec

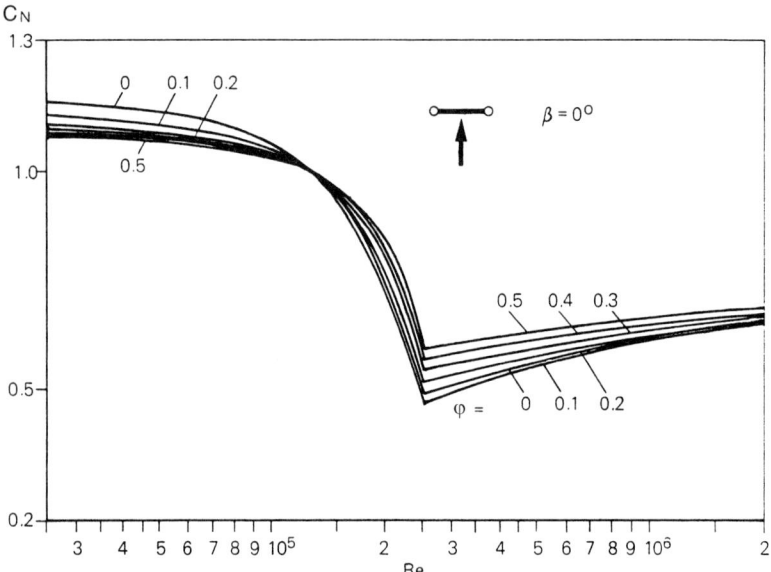

Fig. 14-5 Normal force coefficient C_N of a single plane lattice in CHS for wind perpendicular to plane ($\beta = 0°$) as a function of Reynolds Number Re and solidity ratio φ [2]

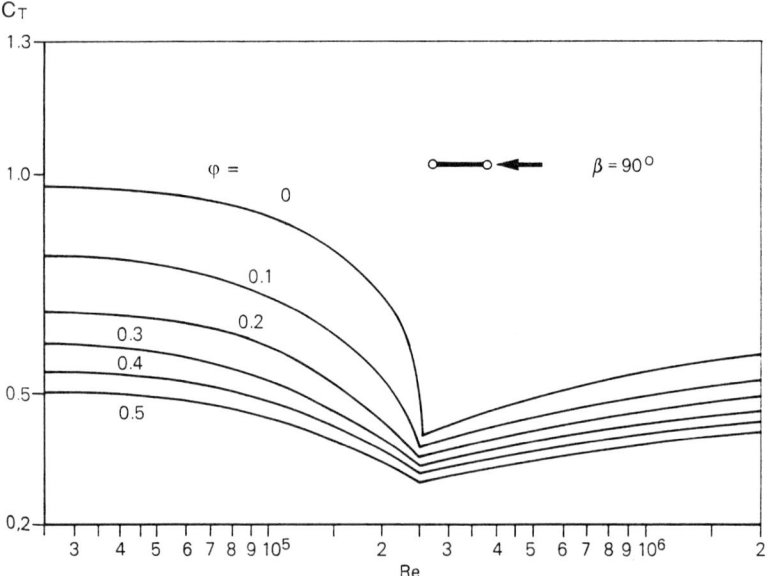

Fig. 14-6 Tangential force coefficient C_T of a single plane lattice in CHS for wind parallel to plane ($\beta = 90°$) as a function of Reynolds Number Re and solidity ratio φ [2]

14.4 Wind loads on lattice structures in circular hollow sections

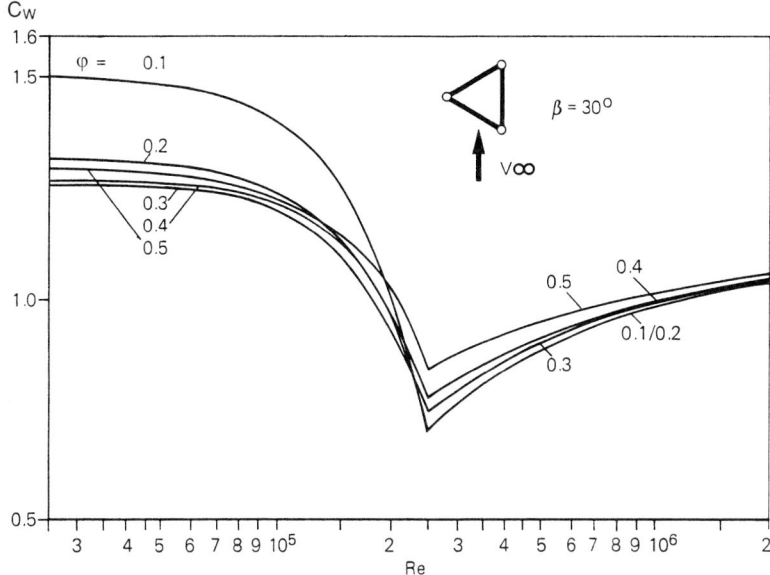

Fig. 14-7 Drag coefficient C_W for a triangular lattice mast in CHS ($\beta = 30°$) as a function of Re and φ [2]

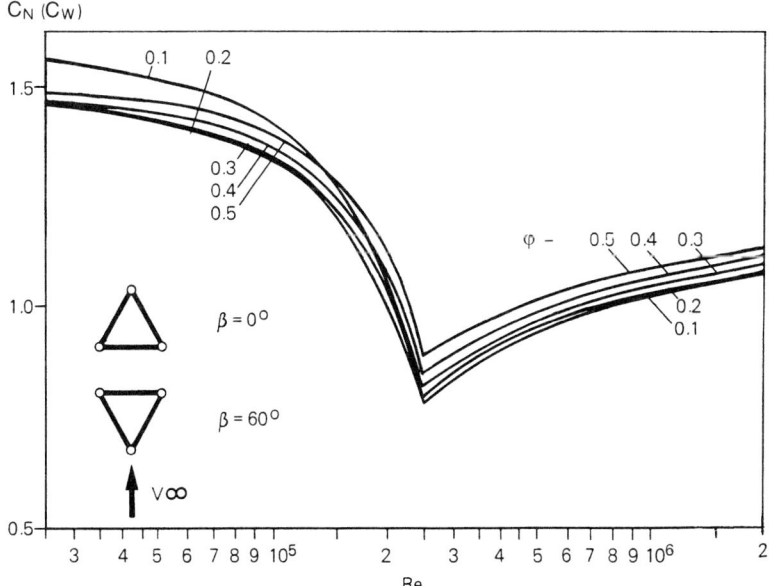

Fig. 14-8 Drag coefficient C_W (= Normal force coefficient C_N) for a triangular lattice mast in CHS (for $\beta = 0°$ or $60°$) as functions of Re and φ [2]

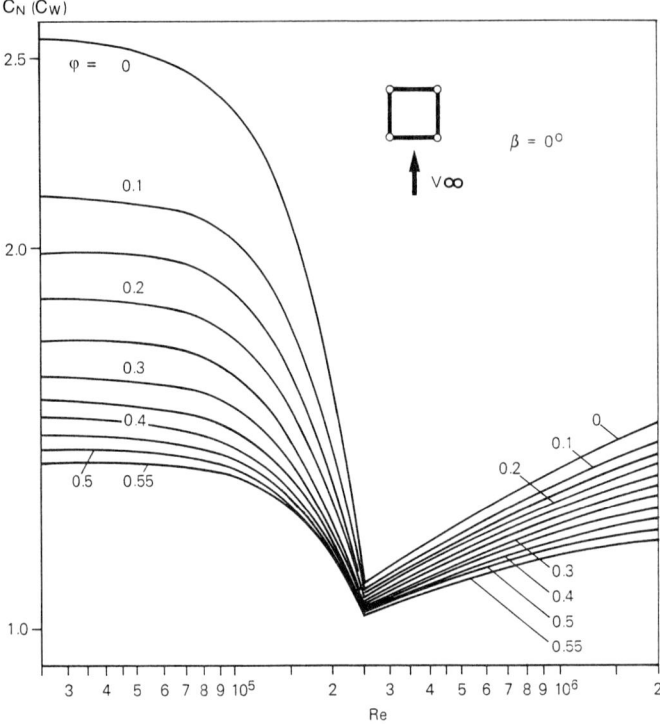

Fig. 14-9 Drag coefficient C_W (= Normal force coefficient C_N) for a square lattice mast in CHS (for $\beta = 0$) as a function of Re and φ [2]

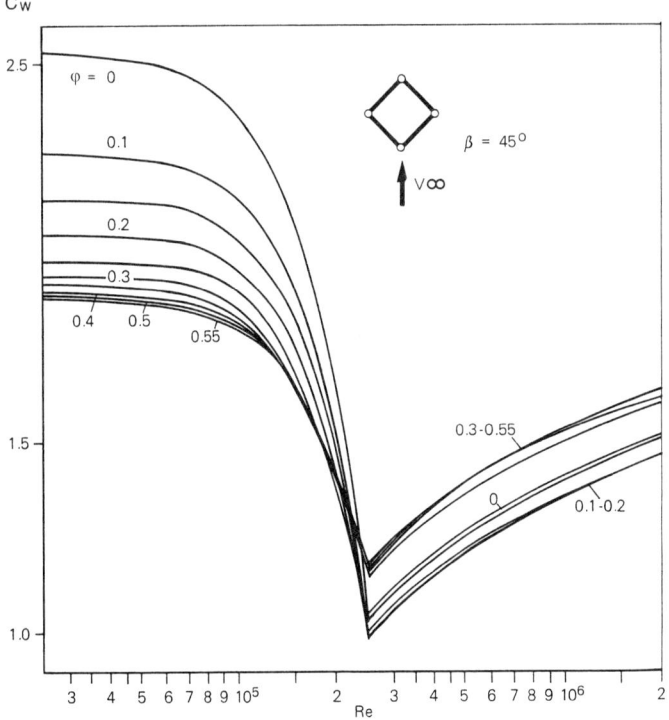

Fig. 14-10 Drag coefficient C_W for a square lattice mast in CHS ($\beta = 45°$) as a function of Re and φ [2]

14.4 Wind loads on lattice structures in circular hollow sections 475

Table 14-1 Load distribution in triangular and square masts using shielding factor η (see Fig. 14-11)

Wind direction	Direction to Plane	Proportion of the total wind load W_{total} for Plane No.			
		I	II	III	IV
(square, normal wind)	normal	$\dfrac{1}{1+\eta}$	0	$\dfrac{\eta}{1+\eta}$	0
	tangential	0	0	0	0
(square, diagonal wind)	normal	$\dfrac{0.7071}{2(1+\eta)}$	$\dfrac{0.7071}{2(1+\eta)}$	$\dfrac{0.7071\,\eta}{2(1+\eta)}$	$\dfrac{0.7071\,\eta}{2(1+\eta)}$
	tangential	$\dfrac{0.7071}{2(1+\eta)}$	$\dfrac{0.7071}{2(1+\eta)}$	$\dfrac{0.7071\,\eta}{2(1+\eta)}$	$\dfrac{0.7071\,\eta}{2(1+\eta)}$
(triangular, normal)	normal	$\dfrac{1}{1+0.5\eta}$	$\dfrac{0.25\,\eta}{1+0.5\eta}$	$\dfrac{0.25\,\eta}{1+0.5\eta}$	–
	tangential	0	$\dfrac{0.3\,\eta}{1+0.5\eta}$	$\dfrac{0.3\,\eta}{1+0.5\eta}$	–
(triangular, angled)	normal	$\dfrac{0.9}{1+0.8\eta}$	0	$\dfrac{0.5\,\eta}{1+0.8\eta}$	–
	tangential	$\dfrac{0.45}{1+0.8\eta}$	$\dfrac{0.3\,\eta}{1+0.8\eta}$	$\dfrac{0.18\,\eta}{1+0.8\eta}$	–
(triangular, other)	normal	$\dfrac{0.25}{\eta+0.5}$	$\dfrac{0.25}{\eta+0.5}$	$\dfrac{\eta}{\eta+0.5}$	–
	tangential	$\dfrac{0.28}{\eta+0.5}$	$\dfrac{0.28}{\eta+0.5}$	0	–

The geometric solidity ratio φ is:

$$\varphi = \frac{2 \cdot 0.0337}{0.3}\left[1+\frac{0.0214}{2\cdot 0.0337}\left(1-\frac{0.0337}{0.3}\right)\right.$$
$$\left.\cdot (1+0.8426^2)^{0.5}\right] = 0.3075$$

The value of normal force coefficient C_N can be read for Re = $0.06915 \cdot 10^6$ and $\varphi = 0.3075$ from the diagram in Fig. 14-8: $C_N = 1.39$ (subcritical)

Calculation of the normal force N acting on the entire triangular mast

Dynamic pressure

$$q = \frac{\rho \cdot V^2}{2} = \frac{1.226 \cdot 30^2}{2} = 552 \text{ N/m}^2$$

where:
$\rho = 1.226\ N\cdot\sec^2/m^4$

Effective frontal area $A_{ref} = \varphi \cdot A_u = \varphi \cdot b \cdot h$, where the height h of the mast has been assumed to be 7.12 m:

$A_{ref} = 0.3075 \cdot 0.3 \cdot 7.12 = 0.6568$ m^2

Normal force $N = q \cdot C_N \cdot A_{ref} = 552 \cdot 1.39 \cdot 0.6568 = 503$ N

Load distribution in 3 planes:

Shielding factor η can be read from the diagram in Fig. 14-11 based on the aerodynamic solidity ratio $(C_N \cdot \varphi) = (1.07 \cdot 0.3075) = 0.329$ (C_N has to be taken for the front face): $\eta = 0.705$

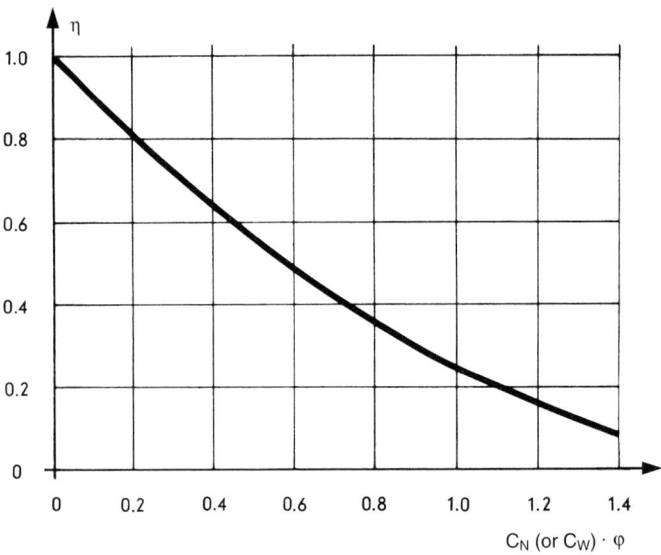

Fig. 14-11 Shielding factor η (for CHS)

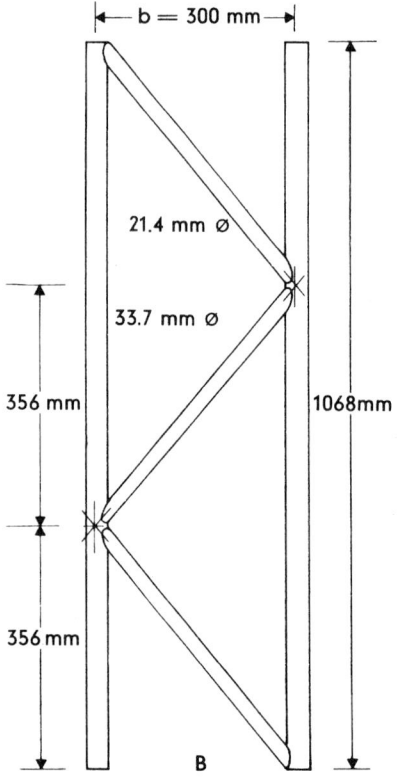

Fig. 14-12 Triangular lattice mast of CHS ($\beta = 0°$)

Subsequently the percentage load on the three planes is determined using Table 14-1:

Plane I (front):

$$N_\mathrm{I} = N \cdot \frac{1}{1+0.5\,\eta} = 503 \cdot \frac{1}{1+0.5 \cdot 0.705} =$$
$$= 372 \text{ N}$$

$T_\mathrm{I} = 0$

Plane II and III:

$$N_\mathrm{II} = N_\mathrm{III} = N \cdot \frac{0.25\,\eta}{1+0.5\,\eta} = 503 \cdot 0.1305 =$$
$$= 65.7 \text{ N}$$

$$T_\mathrm{II} = T_\mathrm{III} = N \cdot \frac{0.3\,\eta}{1+0.5\,\eta} = 503 \cdot 0.156 =$$
$$= 78.7 \text{ N}$$

14.5 Wind force coefficients for circular hollow section cylinders and their lattice structures according to DIN 1055-4, Issue 8/86 [3] and EC 1 [15]

The above mentioned standards propose the wind load coefficients for single circular cylinders and those for the lattice structures made of them. They are based on the results of the investigations described in [1].

14.6 Wind loads on uni-planar and multi-planar lattice structures with square hollow sections

Table 14-2 Effective slenderness λ for circular cylinders, rectangular sections and lattice structures [3, 15]

Position of the structure, wind normal to the plane	Effective slenderness λ
for $1 \geq 4d$	$\dfrac{1}{d}$
for $d \leq 1$; for $d \leq 1$	$1.4\,\dfrac{1}{d} \leq 70$ for $1 \geq 50$ m $2\,\dfrac{1}{d} \leq 70$ for $1 \leq 15$ m for CHS cylinder: $0.7\,\dfrac{1}{d} \leq 70$ for $1 \geq 50$ m $\dfrac{1}{d} \leq 70$ for $1 \leq 15$ m
(rectangular section on wall)	linear interpolation is to be done for intermediate values
(lattice frame)	$0.7\,\dfrac{1}{d} \geq 70$ for $1 \geq 50$ m $\dfrac{1}{d} \geq 70$ for $1 \leq 15$ m linear interpolation is to be done for intermediate values

In the following, only those parts of the standards are compiled, which concern the CHS and their lattice frames.

Table 14-2 deals with the position of the structure, while Tables 14-3 and 14-4 as well as Figs. 14-13 and 14-14 involve structural bodies.

14.6 Wind loads on uni-planar and multi-planar lattice structures with square hollow sections [16]

The total windload on a lattice structure is originated by the sum of the windloads on the single members. The mutual influence (interference) of the single members on one another

Table 14-3 Force coefficients c_f of CHS cylinders, closed on all sides [3]

1 Structural body	2 Slenderness	3 Wind direction angle β	4 Reference area A	5 Force coefficient C_f
Vertical and horizontal cylinder	$l/d \leq \infty$	Perpendicular to the cylinder axis	$l \cdot d$	$C_f = C_{f,0} \cdot \Psi$ where: $C_{f,0}$ = force coefficient for $\lambda = \infty$ equal to 1.2 or more accurately according to Fig. 14-13 Ψ = reduction factor according to Fig. 14-14 for solidity ratio $\varphi = 1$

Table 14-4 Force coefficient C_f of uni-planar and multi-planar lattice frames with CHS members [3]

1 Form and length of lattice frame	2 Reference area A_{ref}	3 Force coefficient C_f
Cross section of lattice frame without gusset plate [a]	Total area of the members or a lattice wall (normal to the wall)	$C_{f,0} \cdot \Psi$ where: $C_{f,0}$ = force coefficient [c] for $\lambda = \infty$ according to Fig. 14-15 a–f versus Reynolds Number Re and solidity ratio φ [b] Ψ = reduction factor according to Fig. 14-14

[a] In lattice frames with gusset plates, the wind load on the gusset plate has to be taken into account additionally. It can be determined by taking the total joint area and a force coefficient $C_f = 1.6$ into consideration. While calculating $C_{f,0}$ for lattice frames with gusset plates, the calculation is to be based on the solidity ratio φ obtained from the total area of CHS members and the gusset plates of the lattice wall.

[b] Solidity $\varphi = A_{ref}/A_u$ with A_{ref} according to column 2 and $A_u = d \cdot l$ (enveloping area); column 1 shows d and l is the length of the lattice frame.

[c] For $C_{f,0}$, the usual surface roughnesses (varnish, rust) are taken into account. It is assumed that the enveloping stream about the CHS cross section is not disturbed over a long length (i. e. by electric cables).

14.6 Wind loads on uni-planar and multi-planar lattice structures with square hollow sections

Table 14-5 Surface roughness values k [15]

Surface	Equivalent roughness k [mm]
Brickwork	3.0
Smooth concrete	0.2
Rough concrete	1.0
Bright steel	0.05
Cast iron	0.2
Galvanised steel	0.2
Rust	2.0
Polished metal	0.002
Fine paint	0.006
Spray paint	0.02

has therefore to be taken into consideration. The influencing factors are:

– Joints
– Grade of blockage
– Arrangement of structural members

The drag W and the transversal force Q on the lattice structure are to be determined as a function of the wind direction angle α by computation taking the above mentioned factors into account:

$$W(\alpha) = \frac{\rho}{2} \cdot V^2 \sum C_{W,i} \cdot A_i \qquad (14\text{-}3)$$

$$Q(\alpha) = \frac{\rho}{2} \cdot V^2 \sum C_{Q,i} \cdot A_i \qquad (14\text{-}4)$$

where:
ρ, V, C_W, C_Q = see Eqs. (14-1) and (14-2)
i = index for the single member ($i = 1, 2, 3 \ldots$)
A_i = width of a member D_i × length of a member l_i
α = wind direction angle

The drag coefficient as well as the transversal force coefficient depend on α and the slenderness $\lambda = l/D$ besides the geometry of the lattice frame. They can be calculated as follows:

$$C_W(\alpha) \approx C_{W_o}(\alpha) - 9 \cdot \frac{D}{l} \qquad (14\text{-}5)$$

$$C_Q(\alpha) \approx C_{Q_o}(\alpha) - 9 \cdot \frac{D}{l} \qquad (14\text{-}6)$$

with

$C_{W_o}(\alpha)$ and $C_{Q_o}(\alpha)$ for $\lambda = \dfrac{l}{D} \geq 150$

The values for $C_{W_o}(\alpha)$ and $C_{Q_o}(\alpha)$ can be read from Figs. 14-3 and 14-4, which have been established by means of tests.

Further, the slenderness λ and the form of the lattice frame are decisive for the solidity ratio $\varphi = \dfrac{\sum A_i}{A_u}$, where A_u is the enveloping area of the effective frontal area depending on the wind direction (Fig. 14-24).

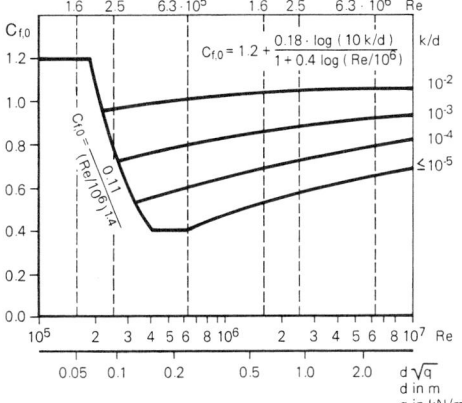

Fig. 14-13 Force coefficient $C_{f,0}$ versus Re with varying relative surface roughness k/d (values of k according to Table 14-5), Re $= \dfrac{V \cdot d}{1.5 \cdot 10^{-5}}$ with $V = 40\sqrt{q}$ in m/sec, q in kN/m², d in m [3, 15]

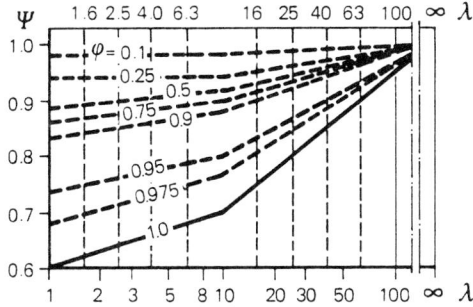

Fig. 14-14 Reduction factor ψ versus effective slenderness λ (according to Table 14-3) with varying solidity ratio φ [3, 15]

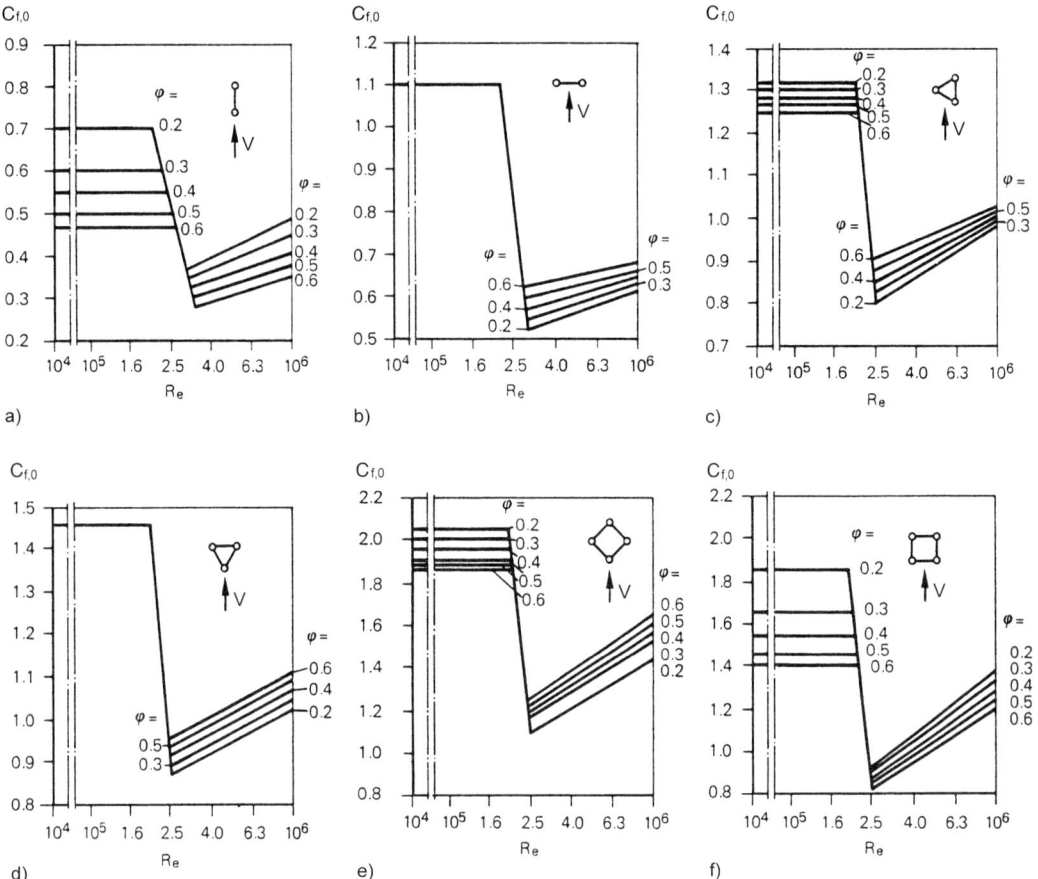

Fig. 14-15 Force coefficient $C_{f,0}$ for lattice frame of CHS versus Reynolds Number Re for varying solidity ratio φ [3, 15]

$$Re = \frac{V \cdot d_c}{1.5 \cdot 10^{-5}} \text{ with } V = 40 \sqrt{q} \text{ in m/sec, } q \text{ in kN/m}^2 \text{ and chord diameter } d_c \text{ in m}$$

a), b) force coefficient $C_{f,0}$ for uni-planar lattice frame
c), d) force coefficient $C_{f,0}$ for triangular (isosceles) lattice frame
e), f) force coefficient $C_{f,0}$ for quadrangular (right-angled) lattice frame

14.6.1 Uni-planar lattice structures

The wind loads on a uni-planar lattice frame are defined by Eqs. (14-3) and (14-4). The effect of the interference of the joints on the loads can be considered by the following relationship according to Biriulin [18]:

$$C_W(\alpha) = K \cdot \frac{\sum C_{W,i}(\alpha) \cdot A_i}{\sum A_i} \quad (14-7)$$

$$C_Q(\alpha) = K \cdot \frac{\sum C_{Q,i}(\alpha) \cdot A_i}{\sum A_i} \quad (14-8)$$

The factor K represents the effect of joint interference depending on the solidity ratio $\varphi = \frac{\sum A_i}{A_u}$ and can be read from Fig. 14-16.

14.6.2 Multi-planar (space) lattice structures with square or rectangular plan

A space structure is treated as an addition of a number of uni-planar structures. The profiles on the sides are more or less shielded by the chord members depending on the wind direction angle α. The leeward side is subjected to

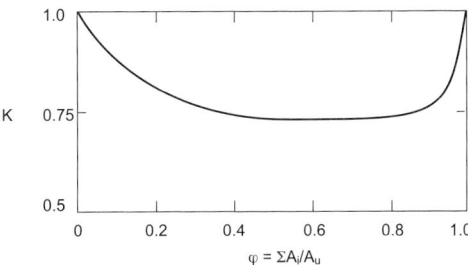

Fig. 14-16 Coefficient for joint interference K versus solidity ratio $\varphi = \dfrac{\sum A_i}{A_u}$ [16]

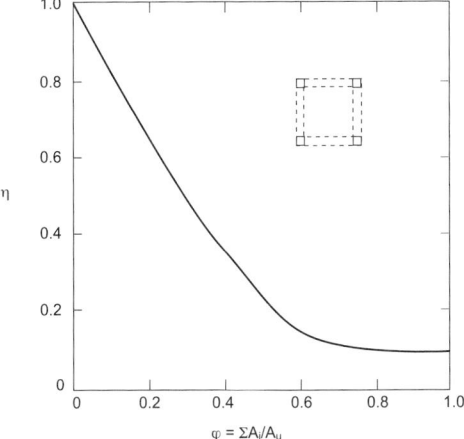

Fig. 14-17 Wind shielding factor η versus solidity ratio φ for the RHS (square) lattice frame with a square plan [16]

smaller wind loads, as it is within the wind shield of the windward side.

The side, which is directly loaded by wind, can be treated as a plane lattice frame (see Section 14.6.1). A reduction factor η (shielding factor) is to be used to calculate the windload on the leeward side. This factor is a function of the solidity ratio φ and the plan geometry of the lattice structure. Fig. 14-17 shows the "η vs. φ" diagram for a square plan of a multi-planar square lattice frame.

The coefficients for the total drag $W(\alpha)$ and for the total transversal force $Q(\alpha)$ can be determined using the following equations:

$$C_W(\alpha) = \frac{\sum C_{W,i}(\alpha) \cdot A_i}{\sum A_i} \cdot K(1+\eta) \qquad (14\text{-}9)$$

$$C_Q(\alpha) = \frac{\sum C_{Q,i}(\alpha) \cdot A_i}{\sum A_i} \cdot K(1+\eta) \qquad (14\text{-}10)$$

14.6.3 Design example

Fig. 14-18 illustrates a square lattice mast consisting of square hollow sections with rounded corners.

Vertical chords: 100 × 100 mm
Horizontal chords: 80 × 80 mm
Diagonals: 60 × 60 mm

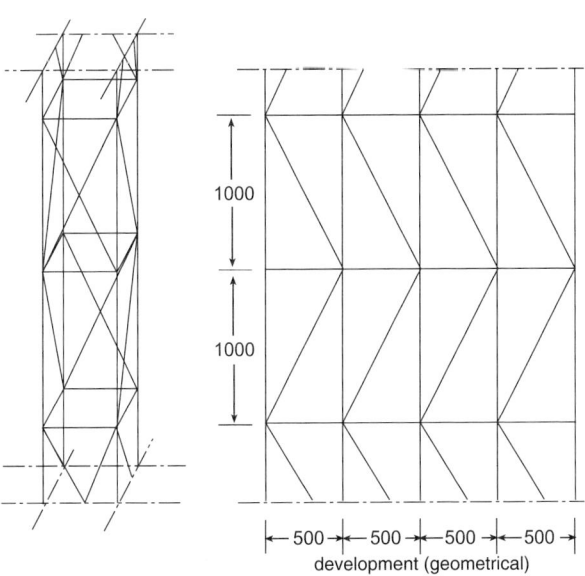

development (geometrical)

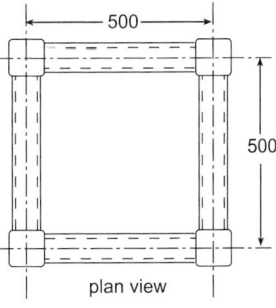

Fig. 14-18 Geometry of the calculation example (see Section 14.6.3)

Solidity ratio $\varphi = \dfrac{\sum A_i}{A_u} = 0.32$

The following values can be read from the Figs. 14-16 and 14-17:

Joint interference factor $K = 0.76$

Wind shielding factor $\eta = 0.47$
$K(1+\eta) = 1.117$

Eqs. (14-9) and (14-10) lead to the $C_W(\alpha)$ and $C_Q(\alpha)$ results.

Figs. 14-19 and 14-20 show the results for the corner radius ratio $R/D = 0.1$.

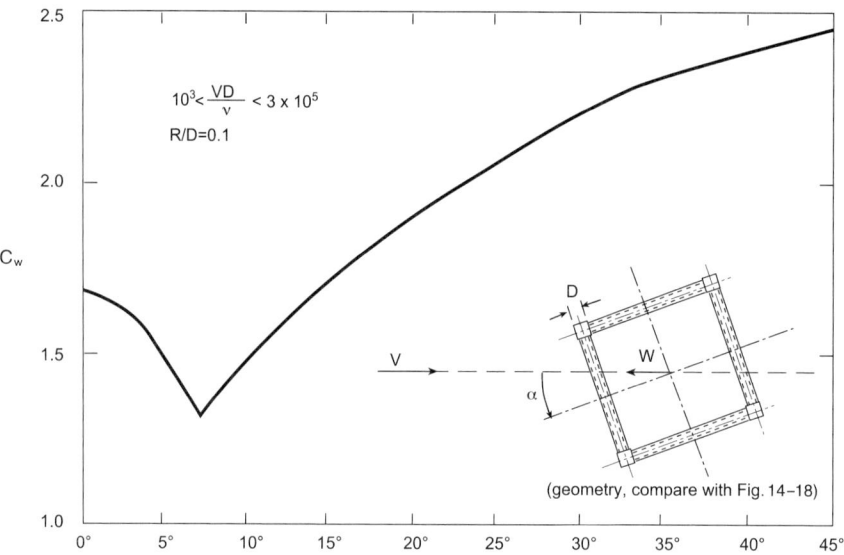

Fig. 14-19 Coefficient of the total drag $C_W(\alpha)$ for the square lattice mast [16]

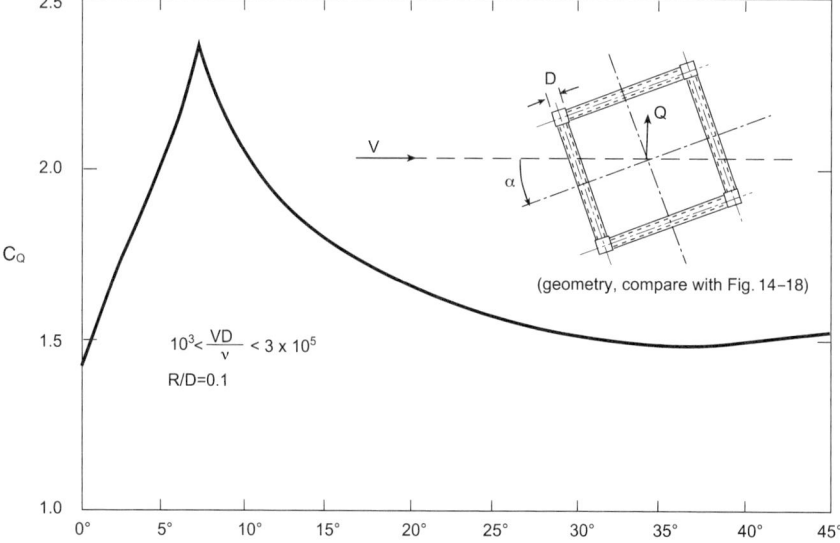

Fig. 14-20 Coefficient of the total transversal force $C_Q(\alpha)$ for the square lattice mast [16]

14.7 Wind loads according to Eurocode 1 [15]

Based on the available studies and the national standards, which had been described in the previous Sections, the recommendations to calculate the wind loads on circular and rectangular hollow sections as well as on uni-planar or multi-planar lattice frames made of them have been worked out in Eurocode 1 [15].

14.7.1 Wind loads

Wind loads acting on the effective frontal area of a single member or of a lattice framework is calculated as follows:

$$W = G \cdot C_f \cdot q \qquad (14\text{-}11)$$

where:
G = gust reaction factor (simple assumption: 2.0 for total or part structures, 2.5 for components of a structure or cladding materials)
$q = \frac{1}{2} \cdot \rho \cdot V_{ref}^2$
ρ = density of air ($= \frac{1}{800}$ kN · sec²/m⁴)
V_{ref} = 10 minutes mean wind velocity at 10 m above ground with a probability of 50 years, also dependent on the topography and the terrain roughness
C_f = aerodynamical force coefficient

14.7.1.1 Wind force coefficient C_f for rectangular hollow sections with rounded-off corners

The force coefficient C_f for the wind direction perpendicular to the surface of RHS is calculated as follows:

$$C_f = C_{f,0} \cdot \psi_r \cdot \psi_\lambda \qquad (14\text{-}12)$$

where:
$C_{f,0}$ = force coefficient for $\lambda = \infty$ (see Fig. 14-21)
ψ_r = reduction factor for RHS with rounded-off corners (see Fig. 14-22)
ψ_λ = reduction factor depending on the finite slenderness λ with the solidity ratio $\varphi = A_i/A_u$ (see Fig. 14-23)

Table 14-6 lists the $C_{f,0}$ values for various d/b-ratios based on Fig. 14-21.

Further, the turbulence intensity I is described by the following equation:

$$I(Z_e) = \frac{1}{C_t(Z_e) \cdot \ln(Z_e/Z_o)} \qquad (14\text{-}13)$$

where:
$C_t(Z_e)$ = topography factor at a height of Z_e
Z_0 = roughness length

$C_t(Z_e)$ and Z_0 can be obtained from a table or with the aid of equations in [15].

14.7.1.2 Wind force coefficient C_f for CHS members with $\lambda = l/d$

For finite CHS cylinder,

$$C_f = C_{f,0} \cdot \psi_\lambda \qquad (14\text{-}14)$$

where:
$C_{f,0}$ = force coefficient (see Fig. 14-13) for $\lambda = l/d = \infty$ and various relative roughness k/d
ψ_λ = slenderness reduction factor (see Fig. 14-23)

14.7.1.3 Wind force coefficient C_f for lattice structures and scaffolding

In this case, C_f is calculated as follows:

$$C_f = C_{f,0} \cdot \psi_\lambda \cdot \psi_{SC} \qquad (14\text{-}15)$$

where:
$C_{f,0}$ = force coefficient for $\lambda = \infty$ ($\lambda = l/d$), l = length, d = width or diameter, see Fig. 14-24)

Fig. 14-25 and 14-26 show the $C_{f,0}$ values for uni-planar and spatial lattice frames of angle steel depending on the solidity ratio φ. Fig. 14-27 illustrates the dependence of $C_{f,0}$ on the Reynolds Number Re and the solidity ratio φ for the uni-planar and spatial frames of CHS.

For the scaffolding, $C_{f,0}$ can be assumed to be equal to 1.3, in case all members are CHS. The Reynolds Number Re can be determined with the aid of the diameter d_i of the structural element.

ψ_λ – slenderness reduction factor depending on the solidity ratio $\varphi = \frac{\sum A_i}{A_u}$ (see Fig. 14-23)

484 14 Wind loads on circular and rectangular hollow sections and their lattice structures

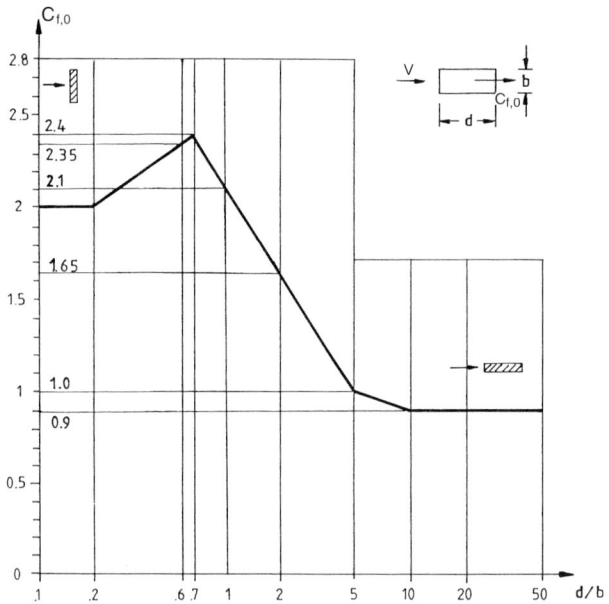

Fig. 14-21 Force coefficient $C_{f,0}$ of RHS with sharp corners and slenderness $\lambda = b/l = \infty$ and turbulence intensity $I \geq 6\%$ [15]

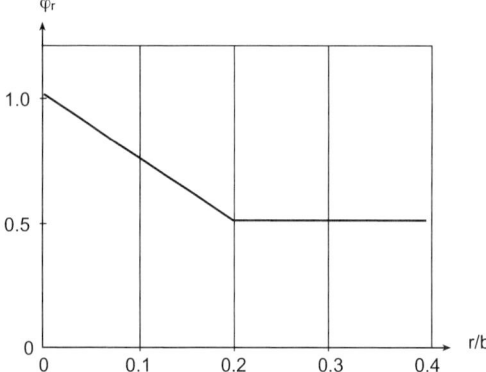

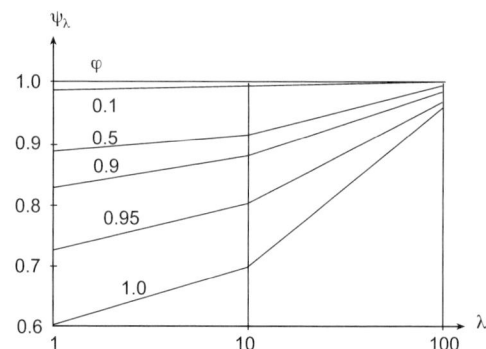

Fig. 14-23 Slenderness reduction factor ψ_λ as a function of solidity ratio φ versus slenderness λ [15]

$\lambda = \infty$
$A_{ref} = l \times b$

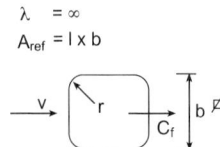

Fig. 14-22 Reduction factor ψ_r for the wind force coefficient of RHS with rounded-off corners [15]

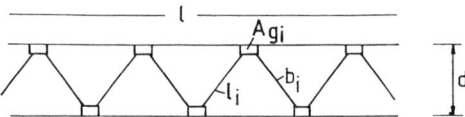

Fig. 14-24 Lattice structures or scaffolding.
Solidity ratio $= \sum A_i / A_u$

$\sum A_i$ = sum of the areas of the members and the gusset plates: $\sum b_i \cdot l_i + \sum A_{gi}$; for multi-planar lattice structures, one plane area (in wind plane) shall be used
A_u = envelope area $d \cdot l$
l = length of the lattice frame
b_i, l_i = width or diameter and length of single member i
A_{gi} = surface area of the gusset plates

Table 14-6 $C_{f,0}$-values (compare with Fig. 14-21)

d/b	0.1	0.2	0.7	5.0	≥ 10
$C_{f,0}$	2.0	2.0	2.4	1.0	0.9

14.7 Wind loads according to Eurocode 1

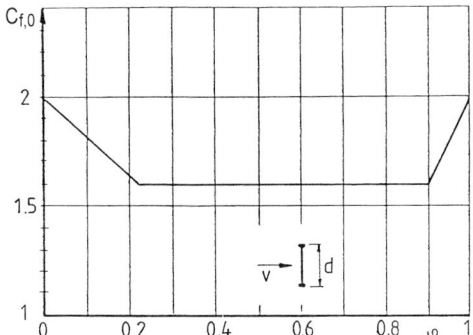

Fig. 14-25 Force coefficient $C_{f,0}$ for a plane lattice structure with right-angled members (angles) as a function of solidity ratio φ [15]

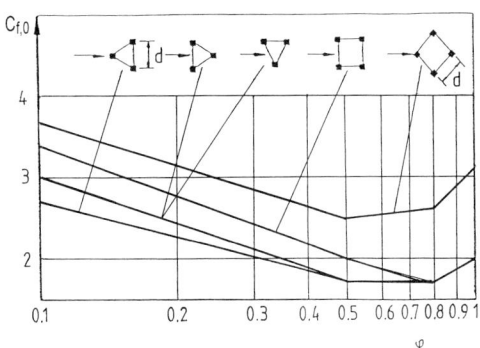

Fig. 14-26 Force coefficient $C_{f,0}$ for a multi-planar lattice structure with right-angled members (angles) as a function of solidity ratio φ [15]

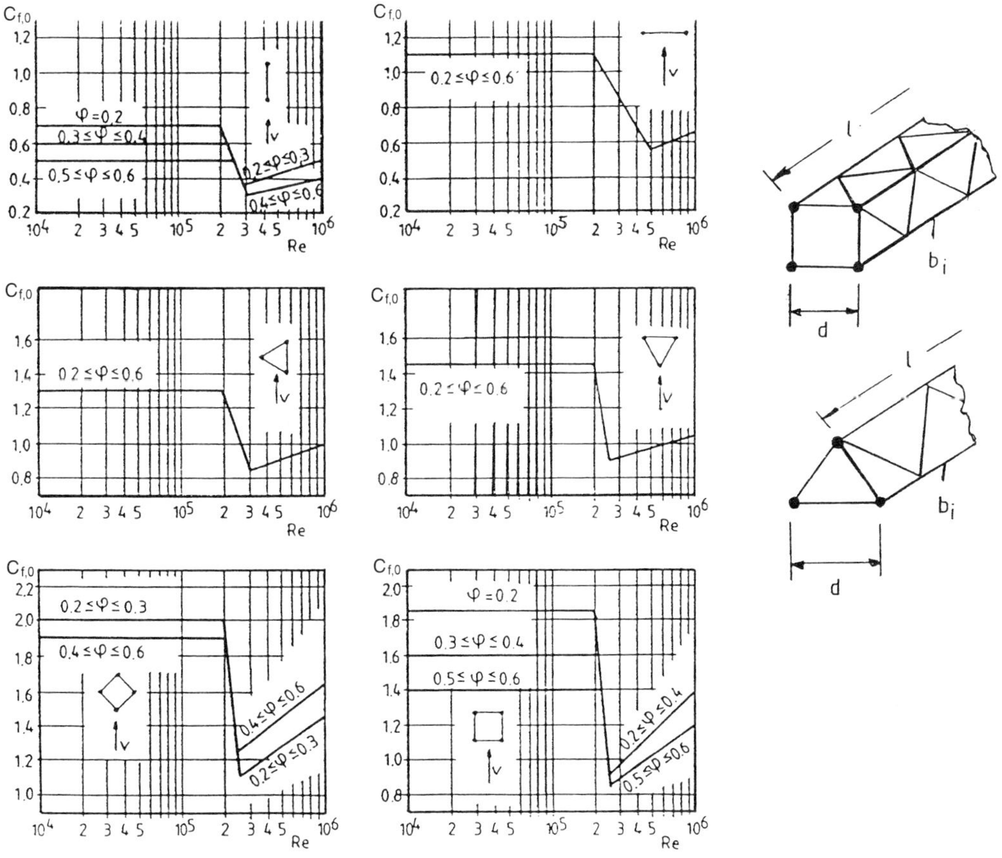

Fig. 14-27 Force coefficient $C_{f,0}$ for uni-planar and multi-planar lattice structures with CHS members versus Reynolds Number Re for varying solidity ratio φ [15]

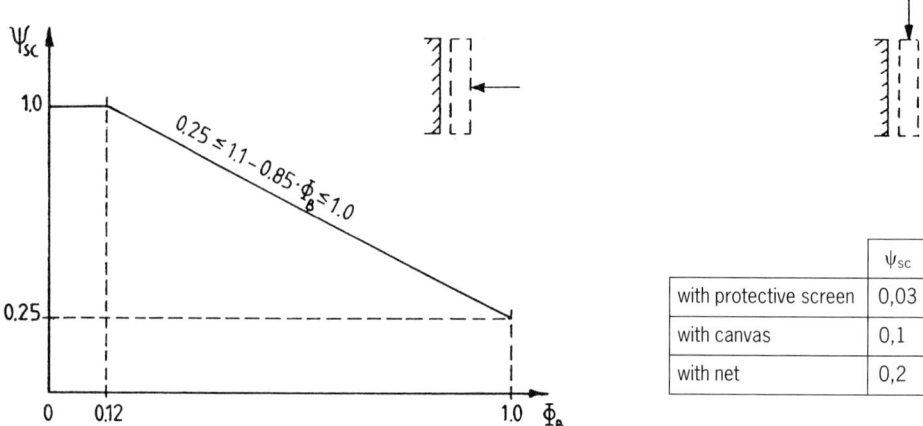

Fig. 14-28 Reduction factor ψ_{sc} for wind force coefficient of a scaffolding without air-tightness devices, affected by solid building-faces

ψ_{sc} – reduction factor for $C_{f,0}$ of scaffolding depending on the obstruction factor Φ (see Fig. 14-28)

$$\Phi_B = \frac{A_{B,II}}{A_{B,g}}$$

where:
$A_{B,II}$ = net surface area of façade
$A_{B,g}$ = transversal surface area of façade

14.8 List of symbols

CHS	circular hollow section
RHS	rectangular (also square) hollow section
CEN	Comité Européen de Normalisation (European Committee for Standardisation)
EC	Eurocode
A	surface area
A_i	surface area of a structural element i ($i = 0, 1, 2, \ldots$)
A_u	enveloping area of the effective frontal area
A_{ref}	effective frontal area of a lattice framework – also reference area
C_f	force coefficient
$C_{f,0}$	force coefficient for $\lambda = \infty$
C_N	normal force coefficient
C_Q	transversal force coefficient
C_T	tangential force coefficient
C_t	topography factor
C_W	drag coefficient
D, d	outer diameter of CHS or width of RHS (also square) (represents also the height of a lattice girder)
I	intensity of turbulence
k	surface roughness
K	interference factor (RHS lattice frame, see Fig. 14-16)
l	length of a structural body
Q	transversal force
q	dynamic pressure $\left(= \rho \cdot \dfrac{V^2}{2}\right)$
R, r	corner radius of RHS
V	wind velocity
W	drag
Z_e	height of a body above the ground
Z_0	length of the surface roughness
α, β	wind direction angle
η	shielding factor
λ	slenderness
ν	kinematic viscosity of air
ρ	density of air
φ	solidity ratio $\left(= \dfrac{A_{ref}}{A_u}\right)$
ψ_r	reduction factor for corner radius (RHS)
ψ_λ	slenderness reduction factor
ψ_{sc}	reduction factor for $C_{f,0}$ of a scaffolding

Appendices I–VI

Nominal sizes and geometric properties of cold formed circular hollow sections according to EN 10219-2 (see [2], Chapter 4)

Outside diameter	Wall thickness	Mass per unit length	Cross sectional area	Second moment of area	Radius of gyration	Elastic section modulus	Plastic section modulus	Torsional inertia constant	Torsional modulus constant	Superficial area per meter	Nominal length per metric ton
D	T	M	A	I	i	W_{el}	W_{pl}	I_t	C_t	A_s	
mm	mm	kg/m	cm^2	cm^4	cm	cm^3	cm^3	cm^4	cm^3	m^2/m	m
21.3	2.0	0.95	1.21	0.571	0.686	0.536	0.748	1.14	1.07	0.0669	1050
21.3	2.5	1.16	1.48	0.664	0.671	0.623	0.889	1.33	1.25	0.0669	863
21.3	3.0	1.35	1.72	0.741	0.656	0.696	1.01	1.48	1.39	0.0669	739
26.9	2.0	1.23	1.56	1.22	0.883	0.907	1.24	2.44	1.81	0.0845	814
26.9	2.5	1.50	1.92	1.44	0.867	1.07	1.49	2.88	2.14	0.0845	665
26.9	3.0	1.77	2.25	1.63	0.852	1.21	1.72	3.27	2.43	0.0845	566
33.7	2.0	1.56	1.99	2.51	1.12	1.49	2.01	5.02	2.98	0.106	640
33.7	2.5	1.92	2.45	3.00	1.11	1.78	2.44	6.00	3.56	0.106	520
33.7	3.0	2.27	2.89	3.44	1.09	2.04	2.84	6.88	4.08	0.106	440
42.4	2.0	1.99	2.54	5.19	1.43	2.45	3.27	10.4	4.90	0.133	502
42.4	2.5	2.46	3.13	6.26	1.41	2.95	3.99	12.5	5.91	0.133	407
42.4	3.0	2.91	3.71	7.25	1.40	3.42	4.67	14.5	6.84	0.133	343
42.4	4.0	3.79	4.83	8.99	1.36	4.24	5.92	18.0	8.48	0.133	264
48.3	2.0	2.28	2.91	7.81	1.64	3.23	4.29	15.6	6.47	0.152	438
48.3	2.5	2.82	3.60	9.46	1.62	3.92	5.25	18.9	7.83	0.152	354
48.3	3.0	3.35	4.27	11.0	1.61	4.55	6.17	22.0	9.11	0.152	298
48.3	4.0	4.37	5.57	13.8	1.57	5.70	7.87	27.5	11.4	0.152	229
48.3	5.0	5.34	6.80	16.2	1.54	6.69	9.42	32.3	13.4	0.152	187
60.3	2.0	2.88	3.66	15.6	2.06	5.17	6.80	31.2	10.3	0.189	348
60.3	2.5	3.56	4.54	19.0	2.05	6.30	8.36	38.0	12.6	0.189	281
60.3	3.0	4.24	5.40	22.2	2.03	7.37	9.86	44.4	14.7	0.189	236
60.3	4.0	5.55	7.07	28.2	2.00	9.34	12.7	56.3	18.7	0.189	180
60.3	5.0	6.82	8.69	33.5	1.96	11.1	15.3	67.0	22.2	0.189	147
76.1	2.0	3.65	4.66	32.0	2.62	8.40	11.0	64.0	16.8	0.239	274
76.1	2.5	4.54	5.78	39.2	2.60	10.3	13.5	78.4	20.6	0.239	220
76.1	3.0	5.41	6.89	46.1	2.59	12.1	16.0	92.2	24.2	0.239	185
76.1	4.0	7.11	9.06	59.1	2.55	15.5	20.8	118	31.0	0.239	141
76.1	5.0	8.77	11.2	70.9	2.52	18.6	25.3	142	37.3	0.239	114
76.1	6.0	10.4	13.2	81.8	2.49	21.5	29.6	164	43.0	0.239	96.4
76.1	6.3	10.8	13.8	84.8	2.48	22.3	30.8	170	44.6	0.239	92.2
88.9	2.0	4.29	5.46	51.6	3.07	11.6	15.1	103	23.2	0.279	233
88.9	2.5	5.33	6.79	63.4	3.06	14.3	18.7	127	28.5	0.279	188
88.9	3.0	6.36	8.10	74.8	3.04	16.8	22.1	150	33.6	0.279	157

EN 10219-2 (continued)

Outside diameter	Wall thickness	Mass per unit length	Cross sectional area	Second moment of area	Radius of gyration	Elastic section modulus	Plastic section modulus	Torsional inertia constant	Torsional modulus constant	Superficial area per meter	Nominal length per metric ton
D	T	M	A	I	i	W_{el}	W_{pl}	I_t	C_t	A_s	
mm	mm	kg/m	cm²	cm⁴	cm	cm³	cm³	cm⁴	cm³	m²/m	m
88.9	4.0	8.38	10.7	96.3	3.00	21.7	28.9	193	43.3	0.279	119
88.9	5.0	10.3	13.2	116	2.97	26.2	35.2	233	52.4	0.279	96.7
88.9	6.0	12.3	15.6	135	2.94	30.4	41.3	270	60.7	0.279	81.5
88.9	6.3	12.8	16.3	140	2.93	31.5	43.1	280	63.1	0.279	77.9
101.6	2.0	4.91	6.26	77.6	3.52	15.3	19.8	155	30.6	0.319	204
101.6	2.5	6.11	7.78	95.6	3.50	18.8	24.6	191	37.6	0.319	164
101.6	3.0	7.29	9.29	113	3.49	22.3	29.2	226	44.5	0.319	137
101.6	4.0	9.63	12.3	146	3.45	28.8	38.1	293	57.6	0.319	104
101.6	5.0	11.9	15.2	177	3.42	34.9	46.7	355	69.9	0.319	84.0
101.6	6.0	14.1	18.0	207	3.39	40.7	54.9	413	81.4	0.319	70.7
101.6	6.3	14.8	18.9	215	3.38	42.3	57.3	430	84.7	0.319	67.5
114.3	2.5	6.89	8.78	137	3.95	24.0	31.3	275	48.0	0.359	145
114.3	3.0	8.23	10.5	163	3.94	28.4	37.2	325	56.9	0.359	121
114.3	4.0	10.9	13.9	211	3.90	36.9	48.7	422	73.9	0.359	91.9
114.3	5.0	13.5	17.2	257	3.87	45.0	59.8	514	89.9	0.359	74.2
114.3	6.0	16.0	20.4	300	3.83	52.5	70.4	600	105	0.359	62.4
114.3	6.3	16.8	21.4	313	3.82	54.7	73.6	625	109	0.359	59.6
114.3	8.0	21.0	26.7	379	3.77	66.4	90.6	759	133	0.359	47.7
139.7	3.0	10.1	12.9	301	4.83	43.1	56.1	602	86.2	0.439	98.9
139.7	4.0	13.4	17.1	393	4.80	56.2	73.7	786	112	0.439	74.7
139.7	5.0	16.6	21.2	481	4.77	68.8	90.8	961	138	0.439	60.2
139.7	6.0	19.8	25.2	564	4.73	80.8	107	1129	162	0.439	50.5
139.7	6.3	20.7	26.4	589	4.72	84.3	112	1177	169	0.439	48.2
139.7	8.0	26.0	33.1	720	4.66	103	139	1441	206	0.439	38.5
139.7	10.0	32.0	40.7	862	4.60	123	169	1724	247	0.439	31.3
168.3	3.0	12.2	15.6	532	5.85	63.3	82.0	1065	127	0.529	81.8
168.3	4.0	16.2	20.6	697	5.81	82.8	108	1394	166	0.529	61.7
168.3	5.0	20.1	25.7	856	5.78	102	133	1712	203	0.529	49.7
168.3	6.0	24.0	30.6	1009	5.74	120	158	2017	240	0.529	41.6
168.3	6.3	25.2	32.1	1053	5.73	125	165	2107	250	0.529	39.7
168.3	8.0	31.6	40.3	1297	5.67	154	206	2595	308	0.529	31.6
168.3	10.0	39.0	49.7	1564	5.61	186	251	3128	372	0.529	25.6
177.8	4.0	17.1	21.8	825	6.15	92.8	121	1650	186	0.559	58.3

Nominal sizes and geometric properties of cold formed circular hollow sections according to EN 10219-2

EN 10219-2 (continued)

Outside diameter	Wall thickness	Mass per unit length	Cross sectional area	Second moment of area	Radius of gyration	Elastic section modulus	Plastic section modulus	Torsional inertia constant	Torsional modulus constant	Superficial area per meter	Nominal length per metric ton
D	T	M	A	I	i	W_{el}	W_{pl}	I_t	C_t	A_s	
mm	mm	kg/m	cm²	cm⁴	cm	cm³	cm³	cm⁴	cm³	m²/m	m
177.8	5.0	21.3	27.1	1014	6.11	114	149	2028	228	0.559	46.9
177.8	6.0	25.4	32.4	1196	6.08	135	177	2392	269	0.559	39.3
177.8	6.3	26.6	33.9	1250	6.07	141	185	2499	281	0.559	37.5
177.8	8.0	33.5	42.7	1541	6.01	173	231	3083	347	0.559	29.9
177.8	10.0	41.4	52.7	1862	5.94	209	282	3724	419	0.559	24.2
177.8	12.0	49.1	62.5	2159	5.88	243	330	4318	486	0.559	20.4
177.8	12.5	51.0	64.9	2230	5.86	251	342	4460	502	0.559	19.6
193.7	4.0	18.7	23.8	1073	6.71	111	144	2146	222	0.609	53.4
193.7	5.0	23.3	29.6	1320	6.67	136	178	2640	273	0.609	43.0
193.7	6.0	27.8	35.4	1560	6.64	161	211	3119	322	0.609	36.0
193.7	6.3	29.1	37.1	1630	6.63	168	221	3260	337	0.609	34.3
193.7	8.0	36.6	46.7	2016	6.57	208	276	4031	416	0.609	27.3
193.7	10.0	45.3	57.7	2442	6.50	252	338	4883	504	0.609	22.1
193.7	12.0	53.8	68.5	2839	6.44	293	397	5678	586	0.609	18.6
193.7	12.5	55.9	71.2	2934	6.42	303	411	5869	606	0.609	17.9
219.1	4.0	21.2	27.0	1564	7.61	143	185	3128	286	0.688	47.1
219.1	5.0	26.4	33.6	1928	7.57	176	229	3856	352	0.688	37.9
219.1	6.0	31.5	40.2	2282	7.54	208	273	4564	417	0.688	31.7
219.1	6.3	33.1	42.1	2386	7.53	218	285	4772	436	0.688	30.2
219.1	8.0	41.6	53.1	2960	7.47	270	357	5919	540	0.688	24.0
219.1	10.0	51.6	65.7	3598	7.40	328	438	7197	657	0.688	19.4
219.1	12.0	61.3	78.1	4200	7.33	383	515	8400	767	0.688	16.3
219.1	12.5	63.7	81.1	4345	7.32	397	534	8689	793	0.688	15.7
244.5	5.0	29.5	37.6	2699	8.47	221	287	5397	441	0.768	33.9
244.5	6.0	35.3	45.0	3199	8.43	262	341	6397	523	0.768	28.3
244.5	6.3	37.0	47.1	3346	8.42	274	358	6692	547	0.768	27.0
244.5	8.0	46.7	59.4	4160	8.37	340	448	8321	681	0.768	21.4
244.5	10.0	57.8	73.7	5073	8.30	415	550	10146	830	0.768	17.3
244.5	12.0	68.8	87.7	5938	8.23	486	649	11877	972	0.768	14.5
244.5	12.5	71.5	91.1	6147	8.21	503	673	12295	1006	0.768	14.0
273.0	5.0	33.0	42.1	3781	9.48	277	359	7562	554	0.858	30.3
273.0	6.0	39.5	50.3	4487	9.44	329	428	8974	657	0.858	25.3
273.0	6.3	41.4	52.8	4696	9.43	344	448	9392	688	0.858	24.1

EN 10219-2 (continued)

Outside diameter	Wall thickness	Mass per unit length	Cross sectional area	Second moment of area	Radius of gyration	Elastic section modulus	Plastic section modulus	Torsional inertia constant	Torsional modulus constant	Superficial area per meter	Nominal length per metric ton
D	T	M	A	I	i	W_{el}	W_{pl}	I_t	C_t	A_s	
mm	mm	kg/m	cm²	cm⁴	cm	cm³	cm³	cm⁴	cm³	m²/m	m
273.0	8.0	52.3	66.6	5852	9.37	429	562	11703	857	0.858	19.1
273.0	10.0	64.9	82.6	7154	9.31	524	692	14308	1048	0.858	15.4
273.0	12.0	77.2	98.4	8396	9.24	615	818	16792	1230	0.858	12.9
273.0	12.5	80.3	102	8697	9.22	637	849	17395	1274	0.858	12.5
323.9	5.0	39.3	50.1	6369	11.3	393	509	12739	787	1.02	25.4
323.9	6.0	47.0	59.9	7572	11.2	468	606	15145	935	1.02	21.3
323.9	6.3	49.3	62.9	7929	11.2	490	636	15858	979	1.02	20.3
323.9	8.0	62.3	79.4	9910	11.2	612	799	19820	1224	1.02	16.0
323.9	10.0	77.4	98.6	12158	11.1	751	986	24317	1501	1.02	12.9
323.9	12.0	92.3	118	14320	11.0	884	1168	28639	1768	1.02	10.8
323.9	12.5	96.0	122	14847	11.0	917	1213	29693	1833	1.02	10.4
355.6	5.0	43.2	55.1	8464	12.4	476	615	16927	952	1.12	23.1
355.6	6.0	51.7	65.9	10071	12.4	566	733	20141	1133	1.12	19.3
355.6	6.3	54.3	69.1	10547	12.4	593	769	21094	1186	1.12	18.4
355.6	8.0	68.6	87.4	13201	12.3	742	967	26403	1485	1.12	14.6
355.6	10.0	85.2	109	16223	12.2	912	1195	32447	1825	1.12	11.7
355.6	12.0	102	130	19139	12.2	1076	1417	38279	2153	1.12	9.83
355.6	12.5	106	135	19852	12.1	1117	1472	39704	2233	1.12	9.45
355.6	16.0	134	171	24663	12.0	1387	1847	49326	2774	1.12	7.46
355.6	20.0	166	211	29792	11.9	1676	2255	59583	3351	1.12	6.04
406.4	6.0	59.2	75.5	15128	14.2	745	962	30257	1489	1.28	16.9
406.4	6.3	62.2	79.2	15849	14.1	780	1009	31699	1560	1.28	16.1
406.4	8.0	78.6	100	19874	14.1	978	1270	39748	1956	1.28	12.7
406.4	10.0	97.8	125	24476	14.0	1205	1572	48952	2409	1.28	10.2
406.4	12.0	117	149	28937	14.0	1424	1867	57874	2848	1.28	8.57
406.4	12.5	121	155	30031	13.9	1478	1940	60061	2956	1.28	8.24
406.4	16.0	154	196	37449	13.8	1843	2440	74898	3686	1.28	6.49
406.4	20.0	191	243	45432	13.7	2236	2989	90864	4472	1.28	5.25
406.4	25.0	235	300	54702	13.5	2692	3642	109404	5384	1.28	4.25
457.0	6.0	66.7	85.0	21618	15.9	946	1220	43236	1892	1.44	15.0
457.0	6.3	70.0	89.2	22654	15.9	991	1280	45308	1983	1.44	14.3
457.0	8.0	88.6	113	28446	15.9	1245	1613	56893	2490	1.44	11.3
457.0	10.0	110	140	35091	15.8	1536	1998	70183	3071	1.44	9.07

Nominal sizes and geometric properties of cold formed circular hollow sections according to EN 10219-2

EN 10219-2 (continued)

Outside diameter	Wall thickness	Mass per unit length	Cross sectional area	Second moment of area	Radius of gyration	Elastic section modulus	Plastic section modulus	Torsional inertia constant	Torsional modulus constant	Superficial area per meter	Nominal length per metric ton
D	T	M	A	I	i	W_{el}	W_{pl}	I_t	C_t	A_s	
mm	mm	kg/m	cm^2	cm^4	cm	cm^3	cm^3	cm^4	cm^3	m^2/m	m
457.0	12.0	132	168	41556	15.7	1819	2377	83113	3637	1.44	7.59
457.0	12.5	137	175	43145	15.7	1888	2470	86290	3776	1.44	7.30
457.0	16.0	174	222	53959	15.6	2361	3113	107919	4723	1.44	5.75
457.0	20.0	216	275	65681	15.5	2874	3822	131363	5749	1.44	4.64
457.0	25.0	266	339	79415	15.3	3475	4671	158830	6951	1.44	3.75
457.0	30.0	316	402	92173	15.1	4034	5479	184346	8068	1.44	3.17
508.0	6.0	74.3	94.6	29812	17.7	1174	1512	59623	2347	1.60	13.5
508.0	6.3	77.9	99.3	31246	17.7	1230	1586	62493	2460	1.60	12.8
508.0	8.0	98.6	126	39280	17.7	1546	2000	78560	3093	1.60	10.1
508.0	10.0	123	156	48520	17.6	1910	2480	97040	3820	1.60	8.14
508.0	12.0	147	187	57536	17.5	2265	2953	115072	4530	1.60	6.81
508.0	12.5	153	195	59755	17.5	2353	3070	119511	4705	1.60	6.55
508.0	16.0	194	247	74909	17.4	2949	3874	149818	5898	1.60	5.15
508.0	20.0	241	307	91428	17.3	3600	4766	182856	7199	1.60	4.15
508.0	25.0	298	379	110918	17.1	4367	5837	221837	8734	1.60	3.36
508.0	30.0	354	451	129173	16.9	5086	6864	258346	10171	1.60	2.83
610.0	6.0	89.4	114	51924	21.4	1702	2189	103847	3405	1.92	11.2
610.0	6.3	93.8	119	54439	21.3	1785	2296	108878	3570	1.92	10.7
610.0	8.0	119	151	68551	21.3	2248	2899	137103	4495	1.92	8.42
610.0	10.0	148	188	84847	21.2	2782	3600	169693	5564	1.92	6.76
610.0	12.0	177	225	100814	21.1	3305	4292	201627	6611	1.92	5.65
610.0	12.5	184	235	104755	21.1	3435	4463	209509	6869	1.92	5.43
610.0	16.0	234	299	131781	21.0	4321	5647	263563	8641	1.92	4.27
610.0	20.0	291	371	161490	20.9	5295	6965	322979	10589	1.92	3.44
610.0	25.0	361	459	196906	20.7	6456	8561	393813	12912	1.92	2.77
610.0	30.0	429	547	230476	20.5	7557	10101	460952	15113	1.92	2.33
711.0	6.0	104	133	82568	24.9	2323	2982	165135	4645	2.23	9.59
711.0	6.3	109	139	86586	24.9	2436	3129	173172	4871	2.23	9.13
711.0	8.0	139	177	109162	24.9	3071	3954	218324	6141	2.23	7.21
711.0	10.0	173	220	135301	24.8	3806	4914	270603	7612	2.23	5.78
711.0	12.0	207	264	160991	24.7	4529	5864	321981	9057	2.23	4.83
711.0	12.5	215	274	167343	24.7	4707	6099	334686	9415	2.23	4.64
711.0	16.0	274	349	211040	24.6	5936	7730	422080	11873	2.23	3.65

EN 10219-2 (continued)

Outside diameter	Wall thickness	Mass per unit length	Cross sectional area	Second moment of area	Radius of gyration	Elastic section modulus	Plastic section modulus	Torsional inertia constant	Torsional modulus constant	Superficial area per meter	Nominal length per metric ton
D	T	M	A	I	i	W_{el}	W_{pl}	I_t	C_t	A_s	
mm	mm	kg/m	cm²	cm⁴	cm	cm³	cm³	cm⁴	cm³	m²/m	m
711.0	20.0	341	434	259351	24.4	7295	9552	518702	14591	2.23	2.93
711.0	25.0	423	539	317357	24.3	8927	11770	634715	17854	2.23	2.36
711.0	30.0	504	642	372790	24.1	10486	13922	745580	20973	2.23	1.98
762.0	6.0	112	143	101813	26.7	2672	3429	203626	5345	2.39	8.94
762.0	6.3	117	150	106777	26.7	2803	3598	213555	5605	2.39	8.52
762.0	8.0	149	190	134683	26.7	3535	4548	269366	7070	2.39	6.72
762.0	10.0	185	236	167028	26.6	4384	5655	334057	8768	2.39	5.39
762.0	12.0	222	283	198855	26.5	5219	6751	397710	10439	2.39	4.51
762.0	12.5	231	294	206731	26.5	5426	7023	413462	10852	2.39	4.33
762.0	16.0	294	375	260973	26.4	6850	8906	521947	13699	2.39	3.40
762.0	20.0	366	466	321083	26.2	8427	11014	642166	16855	2.39	2.73
762.0	25.0	454	579	393461	26.1	10327	13584	786922	20654	2.39	2.20
762.0	30.0	542	690	462853	25.9	12148	16084	925706	24297	2.39	1.85
813.0	8.0	159	202	163901	28.5	4032	5184	327801	8064	2.55	6.30
813.0	10.0	198	252	203364	28.4	5003	6448	406728	10006	2.55	5.05
813.0	12.0	237	302	242235	28.3	5959	7700	484469	11918	2.55	4.22
813.0	12.5	247	314	251860	28.3	6196	8011	503721	12392	2.55	4.05
813.0	16.0	314	401	318222	28.2	7828	10165	636443	15657	2.55	3.18
813.0	20.0	391	498	391909	28.0	9641	12580	783819	19282	2.55	2.56
813.0	25.0	486	619	480856	27.9	11829	15529	961713	23658	2.55	2.06
813.0	30.0	579	738	566374	27.7	13933	18402	1132748	27866	2.55	1.73
914.0	8.0	179	228	233651	32.0	5113	6567	467303	10225	2.87	5.59
914.0	10.0	223	284	290147	32.0	6349	8172	580294	12698	2.87	4.49
914.0	12.0	267	340	345890	31.9	7569	9764	691779	15137	2.87	3.75
914.0	12.5	278	354	359708	31.9	7871	10159	719417	15742	2.87	3.60
914.0	16.0	354	451	455142	31.8	9959	12904	910284	19919	2.87	2.82
914.0	20.0	441	562	561461	31.6	12286	15987	1122922	24572	2.87	2.27
914.0	25.0	548	698	690317	31.4	15105	19763	1380634	30211	2.87	1.82
914.0	30.0	654	833	814775	31.3	17829	23453	1629550	35658	2.87	1.53
1016.0	8.0	199	253	321780	35.6	6334	8129	643560	12668	3.19	5.03
1016.0	10.0	248	316	399850	35.6	7871	10121	799699	15742	3.19	4.03
1016.0	12.0	297	378	476985	35.5	9389	12097	953969	18779	3.19	3.37
1016.0	12.5	309	394	496123	35.5	9766	12588	992246	19532	3.19	3.23

EN 10219-2 (continued)

Outside diameter	Wall thickness	Mass per unit length	Cross sectional area	Second moment of area	Radius of gyration	Elastic section modulus	Plastic section modulus	Torsional inertia constant	Torsional modulus constant	Superficial area per meter	Nominal length per metric ton
D	T	M	A	I	i	W_{el}	W_{pl}	I_t	C_t	A_s	
mm	mm	kg/m	cm²	cm⁴	cm	cm³	cm³	cm⁴	cm³	m²/m	m
1016.0	16.0	395	503	628479	35.4	12372	16001	1256959	24743	3.19	2.53
1016.0	20.0	491	626	776324	35.2	15282	19843	1552648	30564	3.19	2.04
1016.0	25.0	611	778	956086	35.0	18821	24557	1912173	37641	3.19	1.64
1016.0	30.0	729	929	1130352	34.9	22251	29175	2260704	44502	3.19	1.37
1067.0	10.0	261	332	463792	37.4	8693	11173	927585	17387	3.35	3.84
1067.0	12.0	312	398	553420	37.3	10373	13357	1106840	20747	3.35	3.20
1067.0	12.5	325	414	575666	37.3	10790	13900	1151332	21581	3.35	3.08
1067.0	16.0	415	528	729606	37.2	13676	17675	1459213	27352	3.35	2.41
1067.0	20.0	516	658	901755	37.0	16903	21927	1803509	33805	3.35	1.94
1067.0	25.0	642	818	1111355	36.9	20831	27149	2222711	41663	3.35	1.56
1067.0	30.0	767	977	1314864	36.7	24646	32270	2629727	49292	3.35	1.30
1168.0	10.0	286	364	609843	40.9	10443	13410	1219686	20885	3.67	3.50
1168.0	12.0	342	436	728050	40.9	12467	16037	1456101	24933	3.67	2.92
1168.0	12.5	356	454	757409	40.9	12969	16690	1514818	25939	3.67	2.81
1168.0	16.0	455	579	960774	40.7	16452	21235	1921547	32903	3.67	2.20
1168.0	20.0	566	721	1188632	40.6	20353	26361	2377264	40707	3.67	1.77
1168.0	25.0	705	898	1466717	40.4	25115	32666	2933434	50230	3.67	1.42
1219.0	10.0	298	380	694014	42.7	11387	14617	1388029	22773	3.83	3.35
1219.0	12.0	357	455	828716	42.7	13597	17483	1657433	27193	3.83	2.80
1219.0	12.5	372	474	862181	42.7	14146	18196	1724362	28291	3.83	2.69
1219.0	16.0	475	605	1094091	42.5	17951	23157	2188183	35901	3.83	2.11
1219.0	20.0	591	753	1354155	42.4	22217	28755	2708309	44435	3.83	1.69
1219.0	25.0	736	938	1671873	42.2	27430	35646	3343746	54860	3.83	1.36

Nominal sizes and geometric properties of hot finished circular hollow sections according to EN 10210-2 (see [1], Chapter 4)

Outside diameter	Wall thickness	Mass per unit length	Cross sectional area	Second moment of area	Radius of gyration	Elastic section modulus	Plastic section modulus	Torsional inertia constant	Torsional modulus constant	Superficial area per meter	Nominal length per metric ton
D	T	M	A	I	i	W_{el}	W_{pl}	I_t	C_t	A_s	
mm	mm	kg/m	cm²	cm⁴	cm	cm³	cm³	cm⁴	cm³	m²/m	m
21.3	2.3	1.08	1.37	0.629	0.677	0.590	0.834	1.26	1.18	0.0669	928
21.3	2.6	1.20	1.53	0.681	0.668	0.639	0.915	1.36	1.28	0.0669	834
21.3	3.2	1.43	1.82	0.768	0.650	0.722	1.06	1.54	1.44	0.0669	700
26.9	2.3	1.40	1.78	1.36	0.874	1.01	1.40	2.71	2.02	0.0845	717
26.9	2.6	1.56	1.98	1.48	0.864	1.10	1.54	2.96	2.20	0.0845	642
26.9	3.2	1.87	2.38	1.70	0.846	1.27	1.81	3.41	2.53	0.0845	535
33.7	2.6	1.99	2.54	3.09	1.10	1.84	2.52	6.19	3.67	0.106	501
33.7	3.2	2.41	3.07	3.60	1.08	2.14	2.99	7.21	4.28	0.106	415
33.7	4.0	2.93	3.73	4.19	1.06	2.49	3.55	8.38	4.97	0.106	341
42.4	2.6	2.55	3.25	6.46	1.41	3.05	4.12	12.9	6.10	0.133	392
42.4	3.2	3.09	3.94	7.62	1.39	3.59	4.93	15.2	7.19	0.133	323
42.4	4.0	3.79	4.83	8.99	1.36	4.24	5.92	18.0	8.48	0.133	264
48.3	2.6	2.93	3.73	9.78	1.62	4.05	5.44	19.6	8.10	0.152	341
48.3	3.2	3.56	4.53	11.6	1.60	4.80	6.52	23.2	9.59	0.152	281
48.3	4.0	4.37	5.57	13.8	1.57	5.70	7.87	27.5	11.4	0.152	229
48.3	5.0	5.34	6.80	16.2	1.54	6.69	9.42	32.3	13.4	0.152	187
60.3	2.6	3.70	4.71	19.7	2.04	6.52	8.66	39.3	13.0	0.189	270
60.3	3.2	4.51	5.74	23.5	2.02	7.78	10.4	46.9	15.6	0.189	222
60.3	4.0	5.55	7.07	28.2	2.00	9.34	12.7	56.3	18.7	0.189	180
60.3	5.0	6.82	8.69	33.5	1.96	11.1	15.3	67.0	22.2	0.189	147
76.1	2.6	4.71	6.00	40.6	2.60	10.7	14.1	81.2	21.3	0.239	212
76.1	3.2	5.75	7.33	48.8	2.58	12.8	17.0	97.6	25.6	0.239	174
76.1	4.0	7.11	9.06	59.1	2.55	15.5	20.8	118	31.0	0.239	141
76.1	5.0	8.77	11.2	70.9	2.52	18.6	25.3	142	37.3	0.239	114
88.9	3.2	6.76	8.62	79.2	3.03	17.8	23.5	158	35.6	0.279	148
88.9	4.0	8.38	10.7	96.3	3.00	21.7	28.9	193	43.3	0.279	119
88.9	5.0	10.3	13.2	116	2.97	26.2	35.2	233	52.4	0.279	96.7
88.9	6.0	12.3	15.6	135	2.94	30.4	41.3	270	60.7	0.279	81.5
88.9	6.3	12.8	16.3	140	2.93	31.5	43.1	280	63.1	0.279	77.9
101.6	3.2	7.77	9.89	120	3.48	23.6	31.0	240	47.2	0.319	129
101.6	4.0	9.63	12.3	146	3.45	28.8	38.1	293	57.6	0.319	104
101.6	5.0	11.9	15.2	177	3.42	34.9	46.7	355	69.9	0.319	84.0
101.6	6.0	14.1	18.0	207	3.39	40.7	54.9	413	81.4	0.319	70.7

Nominal sizes and geometric properties of hot finished circular hollow sections according to EN 10210-2

EN 10210-2 (continued)

Outside diameter	Wall thickness	Mass per unit length	Cross sectional area	Second moment of area	Radius of gyration	Elastic section modulus	Plastic section modulus	Torsional inertia constant	Torsional modulus constant	Superficial area per meter	Nominal length per metric ton
D	T	M	A	I	i	W_{el}	W_{pl}	I_t	C_t	A_s	
mm	mm	kg/m	cm²	cm⁴	cm	cm³	cm³	cm⁴	cm³	m²/m	m
101.6	6.3	14.8	18.9	215	3.38	42.3	57.3	430	84.7	0.319	67.5
101.6	8.0	18.5	23.5	260	3.32	51.1	70.3	519	102	0.319	54.2
101.6	10.0	22.6	28.8	305	3.26	60.1	84.2	611	120	0.319	44.3
114.3	3.2	8.77	11.2	172	3.93	30.2	39.5	345	60.4	0.359	114
114.3	4.0	10.9	13.9	211	3.90	36.9	48.7	422	73.9	0.359	91.9
114.3	5.0	13.5	17.2	257	3.87	45.0	59.8	514	89.9	0.359	74.2
114.3	6.0	16.0	20.4	300	3.83	52.5	70.4	600	105	0.359	62.4
114.3	6.3	16.8	21.4	313	3.82	54.7	73.6	625	109	0.359	59.6
114.3	8.0	21.0	26.7	379	3.77	66.4	90.6	759	133	0.359	47.7
114.3	10.0	25.7	32.8	450	3.70	78.7	109	899	157	0.359	38.9
139.7	4.0	13.4	17.1	393	4.80	56.2	73.7	786	112	0.439	74.7
139.7	5.0	16.6	21.2	481	4.77	68.8	90.8	961	138	0.439	60.2
139.7	6.0	19.8	25.2	564	4.73	80.8	107	1129	162	0.439	50.5
139.7	6.3	20.7	26.4	589	4.72	84.3	112	1177	169	0.439	48.2
139.7	8.0	26.0	33.1	720	4.66	103	139	1441	206	0.439	38.5
139.7	10.0	32.0	40.7	862	4.60	123	169	1724	247	0.439	31.3
139.7	12.0	37.8	48.1	990	4.53	142	196	1980	283	0.439	26.5
139.7	12.5	39.2	50.0	1020	4.52	146	203	2040	292	0.439	25.5
168.3	4.0	16.2	20.6	697	5.81	82.8	108	1394	166	0.529	61.7
168.3	5.0	20.1	25.7	856	5.78	102	133	1712	203	0.529	49.7
168.3	6.0	24.0	30.6	1009	5.74	120	158	2017	240	0.529	41.6
168.3	6.3	25.2	32.1	1053	5.73	125	165	2107	250	0.529	39.7
168.3	8.0	31.6	40.3	1297	5.67	154	206	2595	308	0.529	31.6
168.3	10.0	39.0	49.7	1564	5.61	186	251	3128	372	0.529	25.6
168.3	12.0	46.3	58.9	1810	5.54	215	294	3620	430	0.529	21.6
168.3	12.5	48.0	61.2	1868	5.53	222	304	3737	444	0.529	20.8
177.8	5.0	21.3	27.1	1014	6.11	114	149	2028	228	0.559	46.9
177.8	6.0	25.4	32.4	1196	6.08	135	177	2392	269	0.559	39.3
177.8	6.3	26.6	33.9	1250	6.07	141	185	2499	281	0.559	37.5
177.8	8.0	33.5	42.7	1541	6.01	173	231	3083	347	0.559	29.9
177.8	10.0	41.4	52.7	1862	5.94	209	282	3724	419	0.559	24.2
177.8	12.0	49.1	62.5	2159	5.88	243	330	4318	486	0.559	20.4
177.8	12.5	51.0	64.9	2230	5.86	251	342	4460	502	0.559	19.6

EN 10210-2 (continued)

Outside diameter	Wall thickness	Mass per unit length	Cross sectional area	Second moment of area	Radius of gyration	Elastic section modulus	Plastic section modulus	Torsional inertia constant	Torsional modulus constant	Superficial area per meter	Nominal length per metric ton
D	T	M	A	I	i	W_{el}	W_{pl}	I_t	C_t	A_s	
mm	mm	kg/m	cm²	cm⁴	cm	cm³	cm³	cm⁴	cm³	m²/m	m
193.7	5.0	23.3	29.6	1320	6.67	136	178	2640	273	0.609	43.0
193.7	6.0	27.8	35.4	1560	6.64	161	211	3119	322	0.609	36.0
193.7	6.3	29.1	37.1	1630	6.63	168	221	3260	337	0.609	34.3
193.7	8.0	36.6	46.7	2016	6.57	208	276	4031	416	0.609	27.3
193.7	10.0	45.3	57.7	2442	6.50	252	338	4883	504	0.609	22.1
193.7	12.0	53.8	68.5	2839	6.44	293	397	5678	586	0.609	18.6
193.7	12.5	55.9	71.2	2934	6.42	303	411	5869	606	0.609	17.9
193.7	16.0	70.1	89.3	3554	6.31	367	507	7109	734	0.609	14.3
219.1	5.0	26.4	33.6	1928	7.57	176	229	3856	352	0.688	37.9
219.1	6.0	31.5	40.2	2282	7.54	208	273	4564	417	0.688	31.7
219.1	6.3	33.1	42.1	2386	7.53	218	285	4772	436	0.688	30.2
219.1	8.0	41.6	53.1	2960	7.47	270	357	5919	540	0.688	24.0
219.1	10.0	51.6	65.7	3598	7.40	328	438	7197	657	0.688	19.4
219.1	12.0	61.3	78.1	4200	7.33	383	515	8400	767	0.688	16.3
219.1	12.5	63.7	81.1	4345	7.32	397	534	8689	793	0.688	15.7
219.1	16.0	80.1	102	5297	7.20	483	661	10590	967	0.688	12.5
219.1	20.0	98.2	125	6261	7.07	572	795	12520	1143	0.688	10.2
244.5	5.0	29.5	37.6	2699	8.47	221	287	5397	441	0.768	33.9
244.5	6.0	35.3	45.0	3199	8.43	262	341	6397	523	0.768	28.3
244.5	6.3	37.0	47.1	3346	8.42	274	358	6692	547	0.768	27.0
244.5	8.0	46.7	59.4	4160	8.37	340	448	8321	681	0.768	21.4
244.5	10.0	57.8	73.7	5073	8.30	415	550	10146	830	0.768	17.3
244.5	12.0	68.8	87.7	5938	8.23	486	649	11877	972	0.768	14.5
244.5	12.5	71.5	91.1	6147	8.21	503	673	12295	1006	0.768	14.0
244.5	16.0	90.2	115	7533	8.10	616	837	15066	1232	0.768	11.1
244.5	20.0	111	141	8957	7.97	733	1011	17914	1465	0.768	9.03
244.5	25.0	135	172	10517	7.81	860	1210	21034	1721	0.768	7.39
273.0	5.0	33.0	42.1	3781	9.48	277	359	7562	554	0.858	30.3
273.0	6.0	39.5	50.3	4487	9.44	329	428	8974	657	0.858	25.3
273.0	6.3	41.4	52.8	4696	9.43	344	448	9392	688	0.858	24.1
273.0	8.0	52.3	66.6	5852	9.37	429	562	11703	857	0.858	19.1
273.0	10.0	64.9	82.6	7154	9.31	524	692	14308	1048	0.858	15.4
273.0	12.0	77.2	98.4	8396	9.24	615	818	16792	1230	0.858	12.9

Nominal sizes and geometric properties of hot finished circular hollow sections according to EN 10210-2

 EN 10210-2 (continued)

Outside diameter	Wall thickness	Mass per unit length	Cross sectional area	Second moment of area	Radius of gyration	Elastic section modulus	Plastic section modulus	Torsional inertia constant	Torsional modulus constant	Superficial area per meter	Nominal length per metric ton
D	T	M	A	I	i	W_{el}	W_{pl}	I_t	C_t	A_s	
mm	mm	kg/m	cm^2	cm^4	cm	cm^3	cm^3	cm^4	cm^3	m^2/m	m
273.0	12.5	80.3	102	8697	9.22	637	849	17395	1274	0.858	12.5
273.0	16.0	101	129	10707	9.10	784	1058	21414	1569	0.858	9.86
273.0	20.0	125	159	12798	8.97	938	1283	25597	1875	0.858	8.01
273.0	25.0	153	195	15127	8.81	1108	1543	30254	2216	0.858	6.54
323.9	5.0	39.3	50.1	6369	11.3	393	509	12739	787	1.02	25.4
323.9	6.0	47.0	59.9	7572	11.2	468	606	15145	935	1.02	21.3
323.9	6.3	49.3	62.9	7929	11.2	490	636	15858	979	1.02	20.3
323.9	8.0	62.3	79.4	9910	11.2	612	799	19820	1224	1.02	16.0
323.9	10.0	77.4	98.6	12158	11.1	751	986	24317	1501	1.02	12.9
323.9	12.0	92.3	118	14320	11.0	884	1168	28639	1768	1.02	10.8
323.9	12.5	96.0	122	14847	11.0	917	1213	29693	1833	1.02	10.4
323.9	16.0	121	155	18390	10.9	1136	1518	36780	2271	1.02	8.23
323.9	20.0	150	191	22139	10.8	1367	1850	44278	2734	1.02	6.67
323.9	25.0	184	235	26400	10.6	1630	2239	52800	3260	1.02	5.43
355.6	6.0	51.7	65.9	10071	12.4	566	733	20141	1133	1.12	19.3
355.6	6.3	54.3	69.1	10547	12.4	593	769	21094	1186	1.12	18.4
355.6	8.0	68.6	87.4	13201	12.3	742	967	26403	1485	1.12	14.6
355.6	10.0	85.2	109	16223	12.2	912	1195	32447	1825	1.12	11.7
355.6	12.0	102	130	19139	12.2	1076	1417	38279	2153	1.12	9.83
355.6	12.5	106	135	19852	12.1	1117	1472	39704	2233	1.12	9.45
355.6	16.0	134	171	24663	12.0	1387	1847	49326	2774	1.12	7.46
355.6	20.0	166	211	29792	11.9	1676	2255	59583	3351	1.12	6.04
355.6	25.0	204	260	35677	11.7	2007	2738	71353	4013	1.12	4.91
406.4	6.0	59.2	75.5	15128	14.2	745	962	30257	1489	1.28	16.9
406.4	6.3	62.2	79.2	15849	14.1	780	1009	31699	1560	1.28	16.1
406.4	8.0	78.6	100	19874	14.1	978	1270	39748	1956	1.28	12.7
406.4	10.0	97.8	125	24476	14.0	1205	1572	48952	2409	1.28	10.2
406.4	12.0	117	149	28937	14.0	1424	1867	57874	2848	1.28	8.57
406.4	12.5	121	155	30031	13.9	1478	1940	60061	2956	1.28	8.24
406.4	16.0	154	196	37449	13.8	1843	2440	74898	3686	1.28	6.49
406.4	20.0	191	243	45432	13.7	2236	2989	90864	4472	1.28	5.25
406.4	25.0	235	300	54702	13.5	2692	3642	109404	5384	1.28	4.25
406.4	30.0	278	355	63224	13.3	3111	4259	126447	6223	1.28	3.59

EN 10210-2 (continued)

Outside diameter	Wall thickness	Mass per unit length	Cross sectional area	Second moment of area	Radius of gyration	Elastic section modulus	Plastic section modulus	Torsional inertia constant	Torsional modulus constant	Superficial area per meter	Nominal length per metric ton
D	T	M	A	I	i	W_{el}	W_{pl}	I_t	C_t	A_s	
mm	mm	kg/m	cm²	cm⁴	cm	cm³	cm³	cm⁴	cm³	m²/m	m
406.4	40.0	361	460	78186	13.0	3848	5391	156373	7696	1.28	2.77
457.0	6.0	66.7	85.0	21618	15.9	946	1220	43236	1892	1.44	15.0
457.0	6.3	70.0	89.2	22654	15.9	991	1280	45308	1983	1.44	14.3
457.0	8.0	88.6	113	28446	15.9	1245	1613	56893	2490	1.44	11.3
457.0	10.0	110	140	35091	15.8	1536	1998	70183	3071	1.44	9.07
457.0	12.0	132	168	41556	15.7	1819	2377	83113	3637	1.44	7.59
457.0	12.5	137	175	43145	15.7	1888	2470	86290	3776	1.44	7.30
457.0	16.0	174	222	53959	15.6	2361	3113	107919	4723	1.44	5.75
457.0	20.0	216	275	65681	15.5	2874	3822	131363	5749	1.44	4.64
457.0	25.0	266	339	79415	15.3	3475	4671	158830	6951	1.44	3.75
457.0	30.0	316	402	92173	15.1	4034	5479	184346	8068	1.44	3.17
457.0	40.0	411	524	114949	14.8	5031	6977	229898	10061	1.44	2.43
508.0	6.0	74.3	94.6	29812	17.7	1174	1512	59623	2347	1.60	13.5
508.0	6.3	77.9	99.3	31246	17.7	1230	1586	62493	2460	1.60	12.8
508.0	8.0	98.6	126	39280	17.7	1546	2000	78560	3093	1.60	10.1
508.0	10.0	123	156	48520	17.6	1910	2480	97040	3820	1.60	8.14
508.0	12.0	147	187	57536	17.5	2265	2953	115072	4530	1.60	6.81
508.0	12.5	153	195	59755	17.5	2353	3070	119511	4705	1.60	6.55
508.0	16.0	194	247	74909	17.4	2949	3874	149818	5898	1.60	5.15
508.0	20.0	241	307	91428	17.3	3600	4766	182856	7199	1.60	4.15
508.0	25.0	298	379	110918	17.1	4367	5837	221837	8734	1.60	3.36
508.0	30.0	354	451	129173	16.9	5086	6864	258346	10171	1.60	2.83
508.0	40.0	462	588	162188	16.6	6385	8782	324376	12771	1.60	2.17
508.0	50.0	565	719	190885	16.3	7515	10530	381770	15030	1.60	1.77
610.0	6.0	89.4	114	51924	21.4	1702	2189	103847	3405	1.92	11.2
610.0	6.3	93.8	119	54439	21.3	1785	2296	108878	3570	1.92	10.7
610.0	8.0	119	151	68551	21.3	2248	2899	137103	4495	1.92	8.42
610.0	10.0	148	188	84847	21.2	2782	3600	169693	5564	1.92	6.76
610.0	12.0	177	225	100814	21.1	3305	4292	201627	6611	1.92	5.65
610.0	12.5	184	235	104755	21.1	3435	4463	209509	6869	1.92	5.43
610.0	16.0	234	299	131781	21.0	4321	5647	263563	8641	1.92	4.27
610.0	20.0	291	371	161490	20.9	5295	6965	322979	10589	1.92	3.44
610.0	25.0	361	459	196906	20.7	6456	8561	393813	12912	1.92	2.77

Nominal sizes and geometric properties of hot finished circular hollow sections according to EN 10210-2

EN 10210-2 (continued)

Outside diameter	Wall thickness	Mass per unit length	Cross sectional area	Second moment of area	Radius of gyration	Elastic section modulus	Plastic section modulus	Torsional inertia constant	Torsional modulus constant	Superficial area per meter	Nominal length per metric ton
D	T	M	A	I	i	W_{el}	W_{pl}	I_t	C_t	A_s	
mm	mm	kg/m	cm²	cm⁴	cm	cm³	cm³	cm⁴	cm³	m²/m	m
610.0	30.0	429	547	230476	20.5	7557	10101	460952	15113	1.92	2.33
610.0	40.0	562	716	292333	20.2	9585	13017	584666	19169	1.92	1.78
610.0	50.0	691	880	347570	19.9	11396	15722	695140	22791	1.92	1.45
711.0	6.0	104	133	82568	24.9	2323	2982	165135	4645	2.23	9.59
711.0	6.3	109	139	86586	24.9	2436	3129	173172	4871	2.23	9.13
711.0	8.0	139	177	109162	24.9	3071	3954	218324	6141	2.23	7.21
711.0	10.0	173	220	135301	24.8	3806	4914	270603	7612	2.23	5.78
711.0	12.0	207	264	160991	24.7	4529	5864	321981	9057	2.23	4.83
711.0	12.5	215	274	167343	24.7	4707	6099	334686	9415	2.23	4.64
711.0	16.0	274	349	211040	24.6	5936	7730	422080	11873	2.23	3.65
711.0	20.0	341	434	259351	24.4	7295	9552	518702	14591	2.23	2.93
711.0	25.0	423	539	317357	24.3	8927	11770	634715	17854	2.23	2.36
711.0	30.0	504	642	372790	24.1	10486	13922	745580	20973	2.23	1.98
711.0	40.0	662	843	476242	23.8	13396	18031	952485	26793	2.23	1.51
711.0	50.0	815	1038	570312	23.4	16043	21888	1140623	32085	2.23	1.23
711.0	60.0	963	1127	655583	23.1	18441	25500	1311166	36882	2.23	1.04
762.0	6.0	112	143	101813	26.7	2672	3429	203626	5345	2.39	8.94
762.0	6.3	117	150	106777	26.7	2803	3598	213555	5605	2.39	8.52
762.0	8.0	149	190	134683	26.7	3535	4548	269366	7070	2.39	6.72
762.0	10.0	185	236	167028	26.6	4384	5655	334057	8768	2.39	5.39
762.0	12.0	222	283	198855	26.5	5219	6751	397710	10439	2.39	4.51
762.0	12.5	231	294	206731	26.5	5426	7023	413462	10852	2.39	4.33
762.0	16.0	294	375	260973	26.4	6850	8906	521947	13699	2.39	3.40
762.0	20.0	366	466	321083	26.2	8427	11014	642166	16855	2.39	2.73
762.0	25.0	454	579	393461	26.1	10327	13584	786922	20654	2.39	2.20
762.0	30.0	542	690	462853	25.9	12148	16084	925706	24297	2.39	1.85
762.0	40.0	712	907	593011	25.6	15565	20873	1186021	31129	2.39	1.40
762.0	50.0	878	1118	712207	25.2	18693	25389	1424414	37386	2.39	1.14
813.0	8.0	159	202	163901	28.5	4032	5184	327801	8064	2.55	6.30
813.0	10.0	198	252	203364	28.4	5003	6448	406728	10006	2.55	5.05
813.0	12.0	237	302	242235	28.3	5959	7700	484469	11918	2.55	4.22
813.0	12.5	247	314	251860	28.3	6196	8011	503721	12392	2.55	4.05
813.0	16.0	314	401	318222	28.2	7828	10165	636443	15657	2.55	3.18

EN 10210-2 (continued)

Outside diameter	Wall thickness	Mass per unit length	Cross sectional area	Second moment of area	Radius of gyration	Elastic section modulus	Plastic section modulus	Torsional inertia constant	Torsional modulus constant	Superficial area per meter	Nominal length per metric ton
D	T	M	A	I	i	W_{el}	W_{pl}	I_t	C_t	A_s	
mm	mm	kg/m	cm^2	cm^4	cm	cm^3	cm^3	cm^4	cm^3	m^2/m	m
813.0	20.0	391	498	391909	28.0	9641	12580	783819	19282	2.55	2.56
813.0	25.0	486	619	480856	27.9	11829	15529	961713	23658	2.55	2.06
813.0	30.0	579	738	566374	27.7	13933	18402	1132748	27866	2.55	1.73
914.0	8.0	179	228	233651	32.0	5113	6567	467303	10225	2.87	5.59
914.0	10.0	223	284	290147	32.0	6349	8172	580294	12698	2.87	4.49
914.0	12.0	267	340	345890	31.9	7569	9764	691779	15137	2.87	3.75
914.0	12.5	278	354	359708	31.9	7871	10159	719417	15742	2.87	3.60
914.0	16.0	354	451	455142	31.8	9959	12904	910284	19919	2.87	2.82
914.0	20.0	441	562	561461	31.6	12286	15987	1122922	24572	2.87	2.27
914.0	25.0	548	698	690317	31.4	15105	19763	1380634	30211	2.87	1.82
914.0	30.0	654	833	814775	31.3	17829	23453	1629550	35658	2.87	1.53
1016.0	8.0	199	253	321780	35.6	6334	8129	643560	12668	3.19	5.03
1016.0	10.0	248	316	399850	35.6	7871	10121	799699	15742	3.19	4.03
1016.0	12.0	297	378	476985	35.5	9389	12097	953969	18779	3.19	3.37
1016.0	12.5	309	394	496123	35.5	9766	12588	992246	19532	3.19	3.23
1016.0	16.0	395	503	628479	35.4	12372	16001	1256959	24743	3.19	2.53
1016.0	20.0	491	626	776324	35.2	15282	19843	1552648	30564	3.19	2.04
1016.0	25.0	611	778	956086	35.0	18821	24557	1912173	37641	3.19	1.64
1016.0	30.0	729	929	1130352	34.9	22251	29175	2260704	44502	3.19	1.37
1067.0	10.0	261	332	463792	37.4	8693	11173	927585	17387	3.35	3.84
1067.0	12.0	312	398	553420	37.3	10373	13357	1106840	20747	3.35	3.20
1067.0	12.5	325	414	575666	37.3	10790	13900	1151332	21581	3.35	3.08
1067.0	16.0	415	528	729606	37.2	13676	17675	1459213	27352	3.35	2.41
1067.0	20.0	516	658	901755	37.0	16903	21927	1803509	33805	3.35	1.94
1067.0	25.0	642	818	1111355	36.9	20831	27149	2222711	41663	3.35	1.56
1067.0	30.0	767	977	1314864	36.7	24646	32270	2629727	49292	3.35	1.30
1168.0	10.0	286	364	609843	40.9	10443	13410	1219686	20885	3.67	3.50
1168.0	12.0	342	436	728050	40.9	12467	16037	1456101	24933	3.67	2.92
1168.0	12.5	356	454	757409	40.9	12969	16690	1514818	25939	3.67	2.81
1168.0	16.0	455	579	960774	40.7	16452	21235	1921547	32903	3.67	2.20
1168.0	20.0	566	721	1188632	40.6	20353	26361	2377264	40707	3.67	1.77
1168.0	25.0	705	898	1466717	40.4	25115	32666	2933434	50230	3.67	1.42
1219.0	10.0	298	380	694014	42.7	11387	14617	1388029	22773	3.83	3.35

EN 10210-2 (continued)

Outside diameter	Wall thickness	Mass per unit length	Cross sectional area	Second moment of area	Radius of gyration	Elastic section modulus	Plastic section modulus	Torsional inertia constant	Torsional modulus constant	Superficial area per meter	Nominal length per metric ton
D	T	M	A	I	i	W_{el}	W_{pl}	I_t	C_t	A_s	
mm	mm	kg/m	cm^2	cm^4	cm	cm^3	cm^3	cm^4	cm^3	m^2/m	m
1219.0	12.0	357	455	828716	42.7	13597	17483	1657433	27193	3.83	2.80
1219.0	12.5	372	474	862181	42.7	14146	18196	1724362	28291	3.83	2.69
1219.0	16.0	475	605	1094091	42.5	17951	23157	2188183	35901	3.83	2.11
1219.0	20.0	591	753	1354155	42.4	22217	28755	2708309	44435	3.83	1.69
1219.0	25.0	736	938	1671873	42.2	27430	35646	3343746	54860	3.83	1.36

Nominal sizes and geometric properties of cold formed square hollow sections according to EN 10219-2 (see [2], Chapter 4)

Size	Wall thickness	Mass per unit length	Cross sectional area	Second moment of area	Radius of gyration	Elastic section modulus	Plastic section modulus	Torsional inertia constant	Torsional modulus constant	Superficial area per meter	Nominal length per metric ton
B = H	T	M	A	I	i	W_{el}	W_{pl}	I_t	C_t	A_s	
mm	mm	kg/m	cm^2	cm^4	cm	cm^3	cm^3	cm^4	cm^3	m^2/m	m
20	2.0	1.05	1.34	0.692	0.720	0.692	0.877	1.21	1.06	0.0731	953
25	2.0	1.36	1.74	1.48	0.924	1.19	1.47	2.53	1.80	0.0931	733
25	2.5	1.64	2.09	1.69	0.899	1.35	1.71	2.97	2.07	0.0914	610
25	3.0	1.89	2.41	1.84	0.874	1.47	1.91	3.33	2.27	0.0897	529
30	2.0	1.68	2.14	2.72	1.13	1.81	2.21	4.54	2.75	0.113	596
30	2.5	2.03	2.59	3.16	1.10	2.10	2.61	5.40	3.20	0.111	492
30	3.0	2.36	3.01	3.50	1.08	2.34	2.96	6.15	3.58	0.110	423
40	2.0	2.31	2.94	6.94	1.54	3.47	4.13	11.3	5.23	0.153	434
40	2.5	2.82	3.59	8.22	1.51	4.11	4.97	13.6	6.21	0.151	355
40	3.0	3.30	4.21	9.32	1.49	4.66	5.72	15.8	7.07	0.150	303
40	4.0	4.20	5.35	11.1	1.44	5.54	7.01	19.4	8.48	0.146	238
50	2.0	2.93	3.74	14.1	1.95	5.66	6.66	22.6	8.51	0.193	341
50	2.5	3.60	4.59	16.9	1.92	6.78	8.07	27.5	10.2	0.191	278
50	3.0	4.25	5.41	19.5	1.90	7.79	9.39	32.1	11.8	0.190	236
50	4.0	5.45	6.95	23.7	1.85	9.49	11.7	40.4	14.4	0.186	183
50	5.0	6.56	8.36	27.0	1.80	10.8	13.7	47.5	16.6	0.183	152
60	2.0	3.56	4.54	25.1	2.35	8.38	9.79	39.8	12.6	0.233	281
60	2.5	4.39	5.59	30.3	2.33	10.1	11.9	48.7	15.2	0.231	228
60	3.0	5.19	6.61	35.1	2.31	11.7	14.0	57.1	17.7	0.230	193
60	4.0	6.71	8.55	43.6	2.26	14.5	17.6	72.6	22.0	0.226	149
60	5.0	8.13	10.4	50.5	2.21	16.8	20.9	86.4	25.6	0.223	123
60	6.0	9.45	12.0	56.1	2.16	18.7	23.7	98.4	28.6	0.219	106
60	6.3	9.55	12.2	54.4	2.11	18.1	23.4	100	28.8	0.213	105
70	2.5	5.17	6.59	49.4	2.74	14.1	16.5	78.5	21.2	0.271	193
70	3.0	6.13	7.81	57.5	2.71	16.4	19.4	92.4	24.7	0.270	163
70	4.0	7.97	10.1	72.1	2.67	20.6	24.8	119	31.1	0.266	126
70	5.0	9.70	12.4	84.6	2.62	24.2	29.6	142	36.7	0.263	103
70	6.0	11.3	14.4	95.2	2.57	27.2	33.8	163	41.4	0.259	88.3
70	6.3	11.5	14.7	93.8	2.53	26.8	33.8	168	42.1	0.253	86.7
80	3.0	7.07	9.01	87.8	3.12	22.0	25.8	140	33.0	0.310	141
80	4.0	9.22	11.7	111	3.07	27.8	33.1	180	41.8	0.306	108
80	5.0	11.3	14.4	131	3.03	32.9	39.7	218	49.7	0.303	88.7
80	6.0	13.2	16.8	149	2.98	37.3	45.8	252	56.6	0.299	75.7

Nominal sizes and geometric properties of cold formed square hollow sections according to EN 10219-2

 EN 10219-2 (continued)

Size	Wall thickness	Mass per unit length	Cross sectional area	Second moment of area	Radius of gyration	Elastic section modulus	Plastic section modulus	Torsional inertia constant	Torsional modulus constant	Superficial area per meter	Nominal length per metric ton
B = H	T	M	A	I	i	W_{el}	W_{pl}	I_t	C_t	A_s	
mm	mm	kg/m	cm^2	cm^4	cm	cm^3	cm^3	cm^4	cm^3	m^2/m	m
80	6.3	13.5	17.2	149	2.94	37.1	46.1	261	57.9	0.293	74.0
80	8.0	16.4	20.8	168	2.84	42.1	53.9	307	66.6	0.286	61.1
90	3.0	8.01	10.2	127	3.53	28.3	33.0	201	42.5	0.350	125
90	4.0	10.5	13.3	162	3.48	36.0	42.6	261	54.2	0.346	95.4
90	5.0	12.8	16.4	193	3.43	42.9	51.4	316	64.7	0.343	77.9
90	6.0	15.1	19.2	220	3.39	49.0	59.5	368	74.2	0.339	66.2
90	6.3	15.5	19.7	221	3.35	49.1	60.3	382	76.2	0.333	64.6
90	8.0	18.9	24.0	255	3.25	56.6	71.3	456	88.8	0.326	53.0
100	3.0	8.96	11.4	177	3.94	35.4	41.2	279	53.2	0.390	112
100	4.0	11.7	14.9	226	3.89	45.3	53.3	362	68.1	0.386	85.2
100	5.0	14.4	18.4	271	3.84	54.2	64.6	441	81.7	0.383	69.4
100	6.0	17.0	21.6	311	3.79	62.3	75.1	514	94.1	0.379	58.9
100	6.3	17.5	22.2	314	3.76	62.8	76.4	536	97.0	0.373	57.3
100	8.0	21.4	27.2	366	3.67	73.2	91.1	645	114	0.366	46.8
100	10.0	25.6	32.6	411	3.55	82.2	105	750	130	0.357	39.1
100	12.0	28.3	36.1	408	3.36	81.6	110	794	136	0.338	35.3
100	12.5	29.1	37.0	410	3.33	82.1	111	804	137	0.336	34.4
120	3.0	10.8	13.8	312	4.76	52.1	60.2	488	78.2	0.470	92.3
120	4.0	14.2	18.1	402	4.71	67.0	78.3	637	101	0.466	70.2
120	5.0	17.5	22.4	485	4.66	80.9	95.4	778	122	0.463	57.0
120	6.0	20.7	26.4	562	4.61	93.7	112	913	141	0.459	48.2
120	6.3	21.4	27.3	572	4.58	95.3	114	955	146	0.453	46.7
120	8.0	26.4	33.6	677	4.49	113	138	1163	175	0.446	37.9
120	10.0	31.8	40.6	777	4.38	129	162	1376	203	0.437	31.4
120	12.0	35.8	45.7	806	4.20	134	174	1518	219	0.418	27.9
120	12.5	36.9	47.0	817	4.17	136	178	1551	223	0.416	27.1
140	4.0	16.8	21.3	652	5.52	93.1	108	1023	140	0.546	59.7
140	5.0	20.7	26.4	791	5.48	113	132	1256	170	0.543	48.3
140	6.0	24.5	31.2	920	5.43	131	155	1479	198	0.539	40.8
140	6.3	25.4	32.3	941	5.39	134	160	1550	205	0.533	39.4
140	8.0	31.4	40.0	1127	5.30	161	194	1901	248	0.526	31.8
140	10.0	38.1	48.6	1312	5.20	187	230	2274	291	0.517	26.2
140	12.0	43.4	55.3	1398	5.03	200	253	2567	322	0.498	23.1

EN 10219-2 (continued)

Size	Wall thickness	Mass per unit length	Cross sectional area	Second moment of area	Radius of gyration	Elastic section modulus	Plastic section modulus	Torsional inertia constant	Torsional modulus constant	Superficial area per meter	Nominal length per metric ton
B=H	T	M	A	I	i	W_{el}	W_{pl}	I_t	C_t	A_s	
mm	mm	kg/m	cm²	cm⁴	cm	cm³	cm³	cm⁴	cm³	m²/m	m
140	12.5	44.8	57.0	1425	5.00	204	259	2634	329	0.496	22.3
150	4.0	18.0	22.9	808	5.93	108	125	1265	162	0.586	55.5
150	5.0	22.3	28.4	982	5.89	131	153	1554	197	0.583	44.9
150	6.0	26.4	33.6	1146	5.84	153	180	1833	230	0.579	37.9
150	6.3	27.4	34.8	1174	5.80	156	185	1922	239	0.573	36.6
150	8.0	33.9	43.2	1412	5.71	188	226	2364	289	0.566	29.5
150	10.0	41.3	52.6	1653	5.61	220	269	2839	341	0.557	24.2
150	12.0	47.1	60.1	1780	5.44	237	298	3231	380	0.538	21.2
150	12.5	48.7	62.0	1817	5.41	242	306	3321	389	0.536	20.5
150	16.0	58.7	74.8	2009	5.18	268	351	3830	440	0.518	17.0
160	4.0	19.3	24.5	987	6.34	123	143	1541	185	0.626	51.9
160	5.0	23.8	30.4	1202	6.29	150	175	1896	226	0.623	42.0
160	6.0	28.3	36.0	1405	6.25	176	206	2239	264	0.619	35.4
160	6.3	29.3	37.4	1442	6.21	180	213	2349	275	0.613	34.1
160	8.0	36.5	46.4	1741	6.12	218	260	2897	334	0.606	27.4
160	10.0	44.4	56.6	2048	6.02	256	311	3490	395	0.597	22.5
160	12.0	50.9	64.9	2224	5.86	278	346	3997	443	0.578	19.6
160	12.5	52.6	67.0	2275	5.83	284	356	4114	455	0.576	19.0
160	16.0	63.7	81.2	2546	5.60	318	413	4799	520	0.558	15.7
180	4.0	21.8	27.7	1422	7.16	158	182	2210	237	0.706	45.9
180	5.0	27.0	34.4	1737	7.11	193	224	2724	290	0.703	37.1
180	6.0	32.1	40.8	2037	7.06	226	264	3223	340	0.699	31.2
180	6.3	33.3	42.4	2096	7.03	233	273	3383	354	0.693	30.0
180	8.0	41.5	52.8	2546	6.94	283	336	4189	432	0.686	24.1
180	10.0	50.7	64.6	3017	6.84	335	404	5074	515	0.677	19.7
180	12.0	58.5	74.5	3322	6.68	369	454	5865	584	0.658	17.1
180	12.5	60.5	77.0	3406	6.65	378	467	6050	600	0.656	16.5
180	16.0	73.8	94.0	3887	6.43	432	550	7178	698	0.638	13.6
200	4.0	24.3	30.9	1968	7.97	197	226	3049	295	0.786	41.2
200	5.0	30.1	38.4	2410	7.93	241	279	3763	362	0.783	33.2
200	6.0	35.8	45.6	2833	7.88	283	330	4459	426	0.779	27.9
200	6.3	37.2	47.4	2922	7.85	292	341	4682	444	0.773	26.8
200	8.0	46.5	59.2	3566	7.76	357	421	5815	544	0.766	21.5

Nominal sizes and geometric properties of cold formed square hollow sections according to EN 10219-2

 EN 10219-2 (continued)

Size	Wall thickness	Mass per unit length	Cross sectional area	Second moment of area	Radius of gyration	Elastic section modulus	Plastic section modulus	Torsional inertia constant	Torsional modulus constant	Superficial area per meter	Nominal length per metric ton
B = H	T	M	A	I	i	W_{el}	W_{pl}	I_t	C_t	A_s	
mm	mm	kg/m	cm²	cm⁴	cm	cm³	cm³	cm⁴	cm³	m²/m	m
200	10.0	57.0	72.6	4251	7.65	425	508	7072	651	0.757	17.6
200	12.0	66.0	84.1	4730	7.50	473	576	8230	743	0.738	15.2
200	12.5	68.3	87.0	4859	7.47	486	594	8502	765	0.736	14.6
200	16.0	83.8	107	5625	7.26	562	706	10210	901	0.718	11.9
220	5.0	33.2	42.4	3238	8.74	294	340	5038	442	0.863	30.1
220	6.0	39.6	50.4	3813	8.70	347	402	5976	521	0.859	25.3
220	6.3	41.2	52.5	3940	8.66	358	417	6277	543	0.853	24.3
220	8.0	51.5	65.6	4828	8.58	439	516	7815	668	0.846	19.4
220	10.0	63.2	80.6	5782	8.47	526	625	9533	804	0.837	15.8
220	12.0	73.5	93.7	6487	8.32	590	712	11149	922	0.818	13.6
220	12.5	76.2	97.0	6674	8.29	607	735	11530	951	0.816	13.1
220	16.0	93.9	120	7812	8.08	710	881	13971	1129	0.798	10.7
250	5.0	38.0	48.4	4805	9.97	384	442	7443	577	0.983	26.3
250	6.0	45.2	57.6	5672	9.92	454	524	8843	681	0.979	22.1
250	6.3	47.1	60.0	5873	9.89	470	544	9290	711	0.973	21.2
250	8.0	59.1	75.2	7229	9.80	578	676	11598	878	0.966	16.9
250	10.0	72.7	92.6	8707	9.70	697	822	14197	1062	0.957	13.8
250	12.0	84.8	108	9859	9.55	789	944	16691	1226	0.938	11.8
250	12.5	88.0	112	10161	9.52	813	975	17283	1266	0.936	11.4
250	16.0	109	139	12047	9.32	964	1180	21146	1520	0.918	9.18
260	6.0	47.1	60.0	6405	10.3	493	569	9970	739	1.02	21.2
260	6.3	49.1	62.6	6635	10.3	510	591	10475	772	1.01	20.4
260	8.0	61.6	78.4	8178	10.2	629	734	13087	955	1.01	16.2
260	10.0	75.8	96.6	9865	10.1	759	894	16035	1156	0.997	13.2
260	12.0	88.6	113	11200	9.96	862	1028	18878	1337	0.978	11.3
260	12.5	91.9	117	11548	9.93	888	1063	19553	1381	0.976	10.9
260	16.0	114	145	13739	9.73	1057	1289	23986	1663	0.958	8.77
300	6.0	54.7	69.6	9964	12.0	664	764	15434	997	1.18	18.3
300	6.3	57.0	72.6	10342	11.9	689	795	16218	1042	1.17	17.5
300	8.0	71.6	91.2	12801	11.8	853	991	20312	1293	1.17	14.0
300	10.0	88.4	113	15519	11.7	1035	1211	24966	1572	1.16	11.3
300	12.0	104	132	17767	11.6	1184	1402	29514	1829	1.14	9.65
300	12.5	108	137	18348	11.6	1223	1451	30601	1892	1.14	9.30

EN 10219-2 (continued)

Size	Wall thickness	Mass per unit length	Cross sectional area	Second moment of area	Radius of gyration	Elastic section modulus	Plastic section modulus	Torsional inertia constant	Torsional modulus constant	Superficial area per meter	Nominal length per metric ton
B=H	T	M	A	I	i	W_{el}	W_{pl}	I_t	C_t	A_s	
mm	mm	kg/m	cm^2	cm^4	cm	cm^3	cm^3	cm^4	cm^3	m^2/m	m
300	16.0	134	171	22076	11.4	1472	1774	37837	2299	1.12	7.46
350	8.0	84.2	107	20681	13.9	1182	1366	32557	1787	1.37	11.9
350	10.0	104	133	25189	13.8	1439	1675	40127	2182	1.36	9.61
350	12.0	123	156	29054	13.6	1660	1949	47598	2552	1.34	8.16
350	12.5	127	162	30045	13.6	1717	2020	49393	2642	1.34	7.86
350	16.0	159	203	36511	13.4	2086	2488	61481	3238	1.32	6.28
400	10.0	120	153	38216	15.8	1911	2214	60431	2892	1.56	8.35
400	12.0	141	180	44319	15.7	2216	2587	71843	3395	1.54	7.07
400	12.5	147	187	45877	15.7	2294	2683	74598	3518	1.54	6.81
400	16.0	184	235	56154	15.5	2808	3322	93279	4336	1.52	5.43

Nominal sizes and geometric properties of hot finished square hollow sections according to EN 10210-2 (see [1], Chapter 4)

Size	Wall thickness	Mass per unit length	Cross sectional area	Second moment of area	Radius of gyration	Elastic section modulus	Plastic section modulus	Torsional inertia constant	Torsional modulus constant	Superficial area per meter	Nominal length per metric ton
B=H	T	M	A	I	i	W_{el}	W_{pl}	I_t	C_t	A_s	
mm	mm	kg/m	cm²	cm⁴	cm	cm³	cm³	cm⁴	cm³	m²/m	m
20	2.0	1.10	1.40	0.739	0.727	0.739	0.930	1.22	1.07	0.0748	912
20	2.5	1.32	1.68	0.835	0.705	0.835	1.08	1.41	1.20	0.0736	757
25	2.0	1.41	1.80	1.56	0.932	1.25	1.53	2.52	1.81	0.0948	709
25	2.5	1.71	2.18	1.81	0.909	1.44	1.82	2.97	2.08	0.0936	584
25	3.0	2.00	2.54	2.00	0.886	1.60	2.06	3.35	2.30	0.0923	501
30	2.0	1.72	2.20	2.84	1.14	1.89	2.29	4.53	2.75	0.115	580
30	2.5	2.11	2.68	3.33	1.11	2.22	2.74	5.40	3.22	0.114	475
30	3.0	2.47	3.14	3.74	1.09	2.50	3.14	6.16	3.60	0.112	405
40	2.5	2.89	3.68	8.54	1.52	4.27	5.14	13.6	6.22	0.154	346
40	3.0	3.41	4.34	9.78	1.50	4.89	5.97	15.7	7.10	0.152	293
40	4.0	4.39	5.59	11.8	1.45	5.91	7.44	19.5	8.54	0.150	228
40	5.0	5.28	6.73	13.4	1.41	6.68	8.66	22.5	9.60	0.147	189
50	2.5	3.68	4.68	17.5	1.93	6.99	8.29	27.5	10.2	0.194	272
50	3.0	4.35	5.54	20.2	1.91	8.08	9.70	32.1	11.8	0.192	230
50	4.0	5.64	7.19	25.0	1.86	9.99	12.3	40.4	14.5	0.190	177
50	5.0	6.85	8.73	28.9	1.82	11.6	14.5	47.6	16.7	0.187	146
50	6.0	7.99	10.2	32.0	1.77	12.8	16.5	53.6	18.4	0.185	125
50	6.3	8.31	10.6	32.8	1.76	13.1	17.0	55.2	18.8	0.184	120
60	2.5	4.46	5.68	31.1	2.34	10.4	12.2	48.5	15.2	0.234	224
60	3.0	5.29	6.74	36.2	2.32	12.1	14.3	56.9	17.7	0.232	189
60	4.0	6.90	8.79	45.4	2.27	15.1	18.3	72.5	22.0	0.230	145
60	5.0	8.42	10.7	53.3	2.23	17.8	21.9	86.4	25.7	0.227	119
60	6.0	9.87	12.6	59.9	2.18	20.0	25.1	98.6	28.8	0.225	101
60	6.3	10.3	13.1	61.6	2.17	20.5	26.0	102	29.6	0.224	97.2
60	8.0	12.5	16.0	69.7	2.09	23.2	30.4	118	33.4	0.219	79.9
70	3.0	6.24	7.94	59.0	2.73	16.9	19.9	92.2	24.8	0.272	160
70	4.0	8.15	10.4	74.7	2.68	21.3	25.5	118	31.2	0.270	123
70	5.0	9.99	12.7	88.5	2.64	25.3	30.8	142	36.8	0.267	100
70	6.0	11.8	15.0	101	2.59	28.7	35.5	163	41.6	0.265	85.1
70	6.3	12.3	15.6	104	2.58	29.7	36.9	169	42.9	0.264	81.5
70	8.0	15.0	19.2	120	2.50	34.2	43.8	200	49.2	0.259	66.5
80	3.0	7.18	9.14	89.8	3.13	22.5	26.3	140	33.0	0.312	139
80	4.0	9.41	12.0	114	3.09	28.6	34.0	180	41.9	0.310	106

EN 10210-2 (continued)

Size	Wall thickness	Mass per unit length	Cross sectional area	Second moment of area	Radius of gyration	Elastic section modulus	Plastic section modulus	Torsional inertia constant	Torsional modulus constant	Superficial area per meter	Nominal length per metric ton
B = H	T	M	A	I	i	W_{el}	W_{pl}	I_t	C_t	A_s	
mm	mm	kg/m	cm^2	cm^4	cm	cm^3	cm^3	cm^4	cm^3	m^2/m	m
80	5.0	11.6	14.7	137	3.05	34.2	41.1	217	49.8	0.307	86.5
80	6.0	13.6	17.4	156	3.00	39.1	47.8	252	56.8	0.305	73.3
80	6.3	14.2	18.1	162	2.99	40.5	49.7	262	58.7	0.304	70.2
80	8.0	17.5	22.4	189	2.91	47.3	59.5	312	68.3	0.299	57.0
90	4.0	10.7	13.6	166	3.50	37.0	43.6	260	54.2	0.350	93.7
90	5.0	13.1	16.7	200	3.45	44.4	53.0	316	64.8	0.347	76.1
90	6.0	15.5	19.8	230	3.41	51.1	61.8	367	74.3	0.345	64.4
90	6.3	16.2	20.7	238	3.40	53.0	64.3	382	77.0	0.344	61.6
90	8.0	20.1	25.6	281	3.32	62.6	77.6	459	90.5	0.339	49.9
100	4.0	11.9	15.2	232	3.91	46.4	54.4	361	68.2	0.390	83.9
100	5.0	14.7	18.7	279	3.86	55.9	66.4	439	81.8	0.387	68.0
100	6.0	17.4	22.2	323	3.82	64.6	77.6	513	94.3	0.385	57.5
100	6.3	18.2	23.2	336	3.80	67.1	80.9	534	97.8	0.384	54.9
100	8.0	22.6	28.8	400	3.73	79.9	98.2	646	116	0.379	44.3
100	10.0	27.4	34.9	462	3.64	92.4	116	761	133	0.374	36.5
120	5.0	17.8	22.7	498	4.68	83.0	97.6	777	122	0.467	56.0
120	6.0	21.2	27.0	579	4.63	96.6	115	911	141	0.465	47.2
120	6.3	22.2	28.2	603	4.62	100	120	950	147	0.464	45.1
120	8.0	27.6	35.2	726	4.55	121	146	1160	176	0.459	36.2
120	10.0	33.7	42.9	852	4.46	142	175	1382	206	0.454	29.7
120	12.0	39.5	50.3	958	4.36	160	201	1578	230	0.449	25.3
120	12.5	40.9	52.1	982	4.34	164	207	1623	236	0.448	24.5
140	5.0	21.0	26.7	807	5.50	115	135	1253	170	0.547	47.7
140	6.0	24.9	31.8	944	5.45	135	159	1475	198	0.545	40.1
140	6.3	26.1	33.3	984	5.44	141	166	1540	206	0.544	38.3
140	8.0	32.6	41.6	1195	5.36	171	204	1892	249	0.539	30.7
140	10.0	40.0	50.9	1416	5.27	202	246	2272	294	0.534	25.0
140	12.0	47.0	59.9	1609	5.18	230	284	2616	333	0.529	21.3
140	12.5	48.7	62.1	1653	5.16	236	293	2696	342	0.528	20.5
150	5.0	22.6	28.7	1002	5.90	134	156	1550	197	0.587	44.3
150	6.0	26.8	34.2	1174	5.86	156	184	1828	230	0.585	37.3
150	6.3	28.1	35.8	1223	5.85	163	192	1909	240	0.584	35.6
150	8.0	35.1	44.8	1491	5.77	199	237	2351	291	0.579	28.5

EN 10210-2 (continued)

Size	Wall thick-ness	Mass per unit length	Cross sectional area	Second moment of area	Radius of gyration	Elastic section modulus	Plastic section modulus	Torsional inertia constant	Torsional modulus constant	Superficial area per meter	Nominal length per metric ton
B=H	T	M	A	I	i	W_{el}	W_{pl}	I_t	C_t	A_s	
mm	mm	kg/m	cm²	cm⁴	cm	cm³	cm³	cm⁴	cm³	m²/m	m
150	10.0	43.1	54.9	1773	5.68	236	286	2832	344	0.574	23.2
150	12.0	50.8	64.7	2023	5.59	270	331	3272	391	0.569	19.7
150	12.5	52.7	67.1	2080	5.57	277	342	3375	402	0.568	19.0
150	16.0	65.2	83.0	2430	5.41	324	411	4026	467	0.559	15.3
160	5.0	24.1	30.7	1225	6.31	153	178	1892	226	0.627	41.5
160	6.0	28.7	36.6	1437	6.27	180	210	2233	264	0.625	34.8
160	6.3	30.1	38.3	1499	6.26	187	220	2333	275	0.624	33.3
160	8.0	37.6	48.0	1831	6.18	229	272	2880	335	0.619	26.6
160	10.0	46.3	58.9	2186	6.09	273	329	3478	398	0.614	21.6
160	12.0	54.6	69.5	2502	6.00	313	382	4028	454	0.609	18.3
160	12.5	56.6	72.1	2576	5.98	322	395	4158	467	0.608	17.7
160	16.0	70.2	89.4	3028	5.82	379	476	4988	546	0.599	14.2
180	5.0	27.3	34.7	1765	7.13	196	227	2718	290	0.707	36.7
180	6.0	32.5	41.4	2077	7.09	231	269	3215	340	0.705	30.8
180	6.3	34.0	43.3	2168	7.07	241	281	3361	355	0.704	29.4
180	8.0	42.7	54.4	2661	7.00	296	349	4162	434	0.699	23.4
180	10.0	52.5	66.9	3193	6.91	355	424	5048	518	0.694	19.0
180	12.0	62.1	79.1	3677	6.82	409	494	5873	595	0.689	16.1
180	12.5	64.4	82.1	3790	6.80	421	511	6070	613	0.688	15.5
180	16.0	80.2	102	4504	6.64	500	621	7343	724	0.679	12.5
200	5.0	30.4	38.7	2445	7.95	245	283	3756	362	0.787	32.9
200	6.0	36.2	46.2	2883	7.90	288	335	4449	426	0.785	27.6
200	6.3	38.0	48.4	3011	7.89	301	350	4653	444	0.784	26.3
200	8.0	47.7	60.8	3709	7.81	371	436	5778	545	0.779	21.0
200	10.0	58.8	74.9	4471	7.72	447	531	7031	655	0.774	17.0
200	12.0	69.6	88.7	5171	7.64	517	621	8208	754	0.769	14.4
200	12.5	72.3	92.1	5336	7.61	534	643	8491	778	0.768	13.8
200	16.0	90.3	115	6394	7.46	639	785	10340	927	0.759	11.1
220	6.0	40.0	51.0	3875	8.72	352	408	5963	521	0.865	25.0
220	6.3	41.9	53.4	4049	8.71	368	427	6240	544	0.864	23.8
220	8.0	52.7	67.2	5002	8.63	455	532	7765	669	0.859	19.0
220	10.0	65.1	82.9	6050	8.54	550	650	9473	807	0.854	15.4
220	12.0	77.2	98.3	7023	8.45	638	762	11091	933	0.849	13.0

EN 10210-2 (continued)

Size	Wall thickness	Mass per unit length	Cross sectional area	Second moment of area	Radius of gyration	Elastic section modulus	Plastic section modulus	Torsional inertia constant	Torsional modulus constant	Superficial area per meter	Nominal length per metric ton
B=H	T	M	A	I	i	W_{el}	W_{pl}	I_t	C_t	A_s	
mm	mm	kg/m	cm^2	cm^4	cm	cm^3	cm^3	cm^4	cm^3	m^2/m	m
220	12.5	80.1	102	7254	8.43	659	789	11481	963	0.848	12.5
220	16.0	100	128	8749	8.27	795	969	14054	1156	0.839	10.0
250	6.0	45.7	58.2	5752	9.94	460	531	8825	681	0.985	21.9
250	6.3	47.9	61.0	6014	9.93	481	556	9238	712	0.984	20.9
250	8.0	60.3	76.8	7455	9.86	596	694	11525	880	0.979	16.6
250	10.0	74.5	94.9	9055	9.77	724	851	14106	1065	0.974	13.4
250	12.0	88.5	113	10556	9.68	844	1000	16567	1237	0.969	11.3
250	12.5	91.9	117	10915	9.66	873	1037	17164	1279	0.968	10.9
250	16.0	115	147	13267	9.50	1061	1280	21138	1546	0.959	8.67
260	6.0	47.6	60.6	6491	10.4	499	576	9951	740	1.02	21.0
260	6.3	49.9	63.5	6788	10.3	522	603	10417	773	1.02	20.1
260	8.0	62.8	80.0	8423	10.3	648	753	13006	956	1.02	15.9
260	10.0	77.7	98.9	10242	10.2	788	924	15932	1159	1.01	12.9
260	12.0	92.2	117	11954	10.1	920	1087	18729	1348	1.01	10.8
260	12.5	95.8	122	12365	10.1	951	1127	19409	1394	1.01	10.4
260	16.0	120	153	15061	9.91	1159	1394	23942	1689	0.999	8.30
300	6.0	55.1	70.2	10080	12.0	672	772	15407	997	1.18	18.2
300	6.3	57.8	73.6	10547	12.0	703	809	16136	1043	1.18	17.3
300	8.0	72.8	92.8	13128	11.9	875	1013	20194	1294	1.18	13.7
300	10.0	90.2	115	16026	11.8	1068	1246	24807	1575	1.17	11.1
300	12.0	107	137	18777	11.7	1252	1470	29249	1840	1.17	9.32
300	12.5	112	142	19442	11.7	1296	1525	30333	1904	1.17	8.97
300	16.0	141	179	23850	11.5	1590	1895	37622	2325	1.16	7.12
350	8.0	85.4	109	21129	13.9	1207	1392	32384	1789	1.38	11.7
350	10.0	106	135	25884	13.9	1479	1715	39886	2185	1.37	9.44
350	12.0	126	161	30435	13.8	1739	2030	47154	2563	1.37	7.93
350	12.5	131	167	31541	13.7	1802	2107	48934	2654	1.37	7.62
350	16.0	166	211	38942	13.6	2225	2630	60990	3264	1.36	6.04
400	10.0	122	155	39128	15.9	1956	2260	60092	2895	1.57	8.22
400	12.0	145	185	46130	15.8	2306	2679	71181	3405	1.57	6.90
400	12.5	151	192	47839	15.8	2392	2782	73906	3530	1.57	6.63
400	16.0	191	243	59344	15.6	2967	3484	92442	4362	1.56	5.24
400	20.0	235	300	71535	15.4	3577	4247	112489	5237	1.55	4.25

Nominal sizes and geometric properties of cold formed rectangular hollow sections according to EN 10219-2 (see [2], Chapter 4)

Size		Wall thickness	Mass per unit length	Cross sectional area	Second moment of area		Radius of gyration		Elastic section modulus		Plastic section modulus		Torsional inertia constant	Torsional modulus constant	Superficial area per meter	Nominal length per metric ton
H × B		T	M	A	I_{xx}	I_{yy}	i_{xx}	i_{yy}	$W_{el,xx}$	$W_{el,yy}$	$W_{pl,xx}$	$W_{pl,yy}$	I_t	C_t	A_s	
mm	mm	mm	kg/m	cm²	cm⁴	cm⁴	cm	cm	cm³	cm³	cm³	cm³	cm⁴	cm³	m²/m	m
40	20	2.0	1.68	2.14	4.05	1.34	1.38	0.793	2.02	1.34	2.61	1.60	3.45	2.36	0.113	596
40	20	2.5	2.03	2.59	4.69	1.54	1.35	0.770	2.35	1..54	3.09	1.88	4.06	2.72	0.111	492
40	20	3.0	2.36	3.01	5.21	1.68	1.32	0.748	2.60	1.68	3.50	2.12	4.57	3.00	0.110	423
50	30	2.0	2.31	2.94	9.54	4.29	1.80	1.21	3.81	2.86	4.74	3.33	9.77	4.84	0.153	434
50	30	2.5	2.82	3.59	11.3	5.05	1.77	1.19	4.52	3.37	5.70	3.98	11.7	5.72	0.151	355
50	30	3.0	3.30	4.21	12.8	5.70	1.75	1.16	5.13	3.80	6.57	4.58	13.5	6.49	0.150	303
50	30	4.0	4.20	5.35	15.3	6.69	1.69	1.12	6.10	4.46	8.05	5.58	16.5	7.71	0.146	238
60	40	2.0	2.93	3.74	18.4	9.83	2.22	1.62	6.14	4.92	7.47	5.65	20.7	8.12	0.193	341
60	40	2.5	3.60	4.59	22.1	11.7	2.19	1.60	7.36	5.87	9.06	6.84	25.1	9.72	0.191	278
60	40	3.0	4.25	5.41	25.4	13.4	2.17	1.58	8.46	6.72	10.5	7.94	29.3	11.2	0.190	236
60	40	4.0	5.45	6.95	31.0	16.3	2.11	1.53	10.3	8.14	13.2	9.89	36.7	13.7	0.186	183
60	40	5.0	6.56	8.36	35.3	18.4	2.06	1.48	11.8	9.21	15.4	11.5	42.8	15.6	0.183	152
70	50	2.0	3.56	4.54	31.5	18.8	2.63	2.03	8.99	7.50	10.8	8.58	37.5	12.2	0.233	281
70	50	2.5	4.39	5.59	38.0	22.6	2.61	2.01	10.9	9.04	13.2	10.4	45.8	14.7	0.231	228
70	50	3.0	5.19	6.61	44.1	26.1	2.58	1.99	12.6	10.4	15.4	12.2	53.6	17.1	0.230	193
70	50	4.0	6.71	8.55	54.7	32.2	2.53	1.94	15.6	12.9	19.5	15.4	68.1	21.2	0.226	149
70	50	5.0	8.13	10.4	63.5	37.2	2.48	1.90	18.1	14.9	23.1	18.2	80.8	24.6	0.223	123
80	40	2.0	3.56	4.54	37.4	12.7	2.87	1.67	9.34	6.36	11.6	7.17	30.9	11.0	0.233	281
80	40	2.5	4.39	5.59	45.1	15.3	2.84	1.65	11.3	7.63	14.1	8.72	37.6	13.2	0.231	228

EN 10219-2 (continued)

Size H × B		Wall thickness T	Mass per unit length M	Cross sectional area A	Second moment of area		Radius of gyration		Elastic section modulus		Plastic section modulus		Torsional inertia constant	Torsional modulus constant	Superficial area per meter	Nominal length per metric ton
					I_{xx}	I_{yy}	i_{xx}	i_{yy}	Wel_{xx}	Wel_{yy}	Wpl_{xx}	Wpl_{yy}	I_t	C_t	A_s	
mm	mm	mm	kg/m	cm²	cm⁴	cm⁴	cm	cm	cm³	cm³	cm³	cm³	cm⁴	cm³	m²/m	m
80	40	3.0	5.19	6.61	52.3	17.6	2.81	1.63	13.1	8.78	16.5	10.2	43.9	15.3	0.230	193
80	40	4.0	6.71	8.55	64.8	21.5	2.75	1.59	16.2	10.7	20.9	12.8	55.2	18.8	0.226	149
80	40	5.0	8.13	10.4	75.1	24.6	2.69	1.54	18.8	12.3	24.7	15.0	65.0	21.7	0.223	123
80	60	2.0	4.19	5.34	49.5	31.9	3.05	2.44	12.4	10.6	14.7	12.1	61.2	17.1	0.273	239
80	60	2.5	5.17	6.59	60.1	38.6	3.02	2.42	15.0	12.9	18.0	14.8	75.1	20.7	0.271	193
80	60	3.0	6.13	7.81	70.0	44.9	3.00	2.40	17.5	15.0	21.2	17.4	88.3	24.1	0.270	163
80	60	4.0	7.97	10.1	87.9	56.1	2.94	2.35	22.0	18.7	27.0	22.1	113	30.3	0.266	126
80	60	5.0	9.70	12.4	103	65.7	2.89	2.31	25.8	21.9	32.2	26.4	136	35.7	0.263	103
90	50	2.0	4.19	5.34	57.9	23.4	3.29	2.09	12.9	9.35	15.7	10.5	53.4	15.9	0.273	239
90	50	2.5	5.17	6.59	70.3	28.2	3.27	2.07	15.6	11.3	19.3	12.8	65.3	19.2	0.271	193
90	50	3.0	6.13	7.81	81.9	32.7	3.24	2.05	18.2	13.1	22.6	15.0	76.7	22.4	0.270	163
90	50	4.0	7.97	10.1	103	40.7	3.18	2.00	22.8	16.3	28.8	19.1	97.7	28.0	0.266	126
90	50	5.0	9.70	12.4	121	47.4	3.12	1.96	26.8	18.9	34.4	22.7	116	32.7	0.263	103
100	40	2.5	5.17	6.59	79.3	18.8	3.47	1.69	15.9	9.39	20.2	10.6	50.5	16.8	0.271	193
100	40	3.0	6.13	7.81	92.3	21.7	3.44	1.67	18.5	10.8	23.7	12.4	59.0	19.4	0.270	163
100	40	4.0	7.97	10.1	116	26.7	3.38	1.62	23.1	13.3	30.3	15.7	74.5	24.0	0.266	126
100	40	5.0	9.70	12.4	136	30.8	3.31	1.58	27.1	15.4	36.1	18.5	87.9	27.9	0.263	103
100	50	2.5	5.56	7.09	91.2	31.1	3.59	2.09	18.2	12.4	22.7	14.0	75.4	21.5	0.291	180
100	50	3.0	6.60	8.41	106	36.1	3.56	2.07	21.3	14.4	26.7	16.4	88.6	25.0	0.290	152

Nominal sizes and geometric properties of cold formed rectangular hollow sections according to EN 10219-2

EN 10219-2 (continued)

Size H × B	Wall thickness T	Mass per unit length M	Cross sectional area A	Second moment of area		Radius of gyration		Elastic section modulus		Plastic section modulus		Torsional inertia constant I_t	Torsional modulus constant C_t	Superficial area per meter A_s	Nominal length per metric ton
				I_{xx}	I_{yy}	i_{xx}	i_{yy}	We_{lxx}	We_{lyy}	Wpl_{xx}	Wpl_{yy}				
mm	mm	kg/m	cm²	cm⁴	cm⁴	cm	cm	cm³	cm³	cm³	cm³	cm⁴	cm³	m²/m	m
100 50	4.0	8.59	10.9	134	44.9	3.50	2.03	26.8	18.0	34.1	20.9	113	31.3	0.286	116
100 50	5.0	10.5	13.4	158	52.5	3.44	1.98	31.6	21.0	40.8	25.0	135	36.8	0.283	95.4
100 50	6.0	12.3	15.6	179	58.7	3.38	1.94	35.8	23.5	46.9	28.5	154	41.4	0.279	81.5
100 50	6.3	12.5	15.9	176	58.2	3.32	1.91	35.1	23.3	46.9	28.6	158	42.1	0.271	79.9
100 60	2.5	5.96	7.59	103	46.9	3.69	2.49	20.6	15.6	25.1	17.7	103	26.2	0.311	168
100 60	3.0	7.07	9.01	121	54.6	3.66	2.46	24.1	18.2	29.6	20.8	122	30.6	0.310	141
100 60	4.0	9.22	11.7	153	68.7	3.60	2.42	30.5	22.9	37.9	26.6	156	38.7	0.306	108
100 60	5.0	11.3	14.4	181	80.8	3.55	2.37	36.2	26.9	45.6	31.9	188	45.8	0.303	88.7
100 60	6.0	13.2	16.8	205	91.2	3.49	2.33	41.1	30.4	52.5	36.6	216	51.9	0.299	75.7
100 60	6.3	13.5	17.2	203	90.9	3.44	2.30	40.7	30.3	52.8	36.9	223	53.0	0.293	74.0
100 80	2.5	6.74	8.59	127	90.2	3.84	3.24	25.4	22.5	30.0	25.8	166	35.7	0.351	148
100 80	3.0	8.01	10.2	149	106	3.82	3.22	29.8	26.4	35.4	30.4	196	41.9	0.350	125
100 80	4.0	10.5	13.3	189	134	3.77	3.17	37.9	33.5	45.6	39.2	254	53.4	0.346	95.4
100 80	5.0	12.8	16.4	226	160	3.72	3.12	45.2	39.9	55.1	47.2	308	63.7	0.343	77.9
100 80	6.0	15.1	19.2	258	182	3.67	3.08	51.7	45.5	63.8	54.7	357	73.0	0.339	66.2
100 80	6.3	15.5	19.7	259	183	3.62	3.04	51.8	45.7	64.6	55.4	371	75.0	0.333	64.6
120 60	2.5	6.74	8.59	161	55.2	4.33	2.53	26.9	18.4	33.2	20.6	133	31.7	0.351	148
120 60	3.0	8.01	10.2	189	64.4	4.30	2.51	31.5	21.5	39.2	24.2	156	37.1	0.350	125
120 60	4.0	10.5	13.3	241	81.2	4.25	2.47	40.1	27.1	50.5	31.1	201	47.0	0.346	95.4

EN 10219-2 (continued)

Size H × B		Wall thickness T	Mass per unit length M	Cross sectional area A	Second moment of area		Radius of gyration		Elastic section modulus		Plastic section modulus		Torsional inertia constant I_t	Torsional modulus constant C_t	Superficial area per meter A_s	Nominal length per metric ton
					I_{xx}	I_{yy}	i_{xx}	i_{yy}	Wel_{xx}	Wel_{yy}	Wpl_{xx}	Wpl_{yy}				
mm	mm	mm	kg/m	cm²	cm⁴	cm⁴	cm	cm	cm³	cm³	cm³	cm³	cm⁴	cm³	m²/m	m
120	60	5.0	12.8	16.4	287	96.0	4.19	2.42	47.8	32.0	60.9	37.4	242	55.8	0.343	77.9
120	60	6.0	15.1	19.2	328	109	4.13	2.38	54.7	36.3	70.6	43.1	280	63.6	0.339	66.2
120	60	6.3	15.5	19.7	327	109	4.07	2.35	54.5	36.4	71.2	43.7	289	65.1	0.333	64.6
120	60	8.0	18.9	24.0	375	124	3.95	2.27	62.6	41.3	84.1	51.3	340	75.0	0.326	53.0
120	80	3.0	8.96	11.4	230	123	4.49	3.29	38.4	30.9	46.2	35.0	255	50.8	0.390	112
120	80	4.0	11.7	14.9	295	157	4.44	3.24	49.1	39.3	59.8	45.2	331	64.9	0.386	85.2
120	80	5.0	14.4	18.4	353	188	4.39	3.20	58.9	46.9	72.4	54.7	402	77.8	0.383	69.4
120	80	6.0	17.0	21.6	406	215	4.33	3.15	67.7	53.8	84.3	63.5	469	89.4	0.379	58.9
120	80	6.3	17.5	22.2	408	217	4.28	3.12	68.1	54.3	85.6	64.7	488	92.1	0.373	57.3
120	80	8.0	21.4	27.2	476	252	4.18	3.04	79.3	62.9	102	76.9	584	108	0.366	46.8
140	80	4.0	13.0	16.5	430	180	5.10	3.30	61.4	45.1	75.5	51.3	412	76.5	0.426	77.0
140	80	5.0	16.0	20.4	517	216	5.04	3.26	73.9	54.0	91.8	62.2	501	91.8	0.423	62.6
140	80	6.0	18.9	24.0	597	248	4.98	3.21	85.3	62.0	107	72.4	584	106	0.419	53.0
140	80	6.3	19.4	24.8	603	251	4.93	3.19	86.1	62.9	109	74.0	609	109	0.413	51.4
140	80	8.0	23.9	30.4	708	293	4.82	3.10	101	73.3	131	88.4	731	129	0.406	41.8
150	100	4.0	14.9	18.9	595	319	5.60	4.10	79.3	63.7	95.7	72.5	662	105	0.486	67.2
150	100	5.0	18.3	23.4	719	384	5.55	4.05	95.9	76.8	117	88.3	809	127	0.483	54.5
150	100	6.0	21.7	27.6	835	444	5.50	4.01	111	88.8	137	103	948	147	0.479	46.1
150	100	6.3	22.4	28.5	848	453	5.45	3.98	113	90.5	140	106	992	152	0.473	44.6

Nominal sizes and geometric properties of cold formed rectangular hollow sections according to EN 10219-2

EN 10219-2 (continued)

Size H × B		Wall thickness T	Mass per unit length M	Cross sectional area A	Second moment of area		Radius of gyration		Elastic section modulus		Plastic section modulus		Torsional inertia constant I_t	Torsional modulus constant C_t	Superficial area per meter A_s	Nominal length per metric ton
					I_{xx}	I_{yy}	i_{xx}	i_{yy}	Wel_{xx}	Wel_{yy}	Wpl_{xx}	Wpl_{yy}				
mm	mm	mm	kg/m	cm²	cm⁴	cm⁴	cm	cm	cm³	cm³	cm³	cm³	cm⁴	cm³	m²/m	m
150	100	8.0	27.7	35.2	1008	536	5.35	3.90	134	107	169	128	1206	182	0.466	36.1
150	100	10.0	33.4	42.6	1162	614	5.22	3.80	155	123	199	150	1426	211	0.457	29.9
150	100	12.0	37.7	48.1	1207	642	5.01	3.65	161	128	215	163	1573	229	0.438	26.5
150	100	12.5	38.9	49.5	1225	651	4.97	3.63	163	130	220	166	1606	233	0.436	25.7
160	80	4.0	14.2	18.1	598	204	5.74	3.35	74.7	50.9	92.9	57.4	494	88.0	0.466	70.2
160	80	5.0	17.5	22.4	722	244	5.68	3.30	90.2	61.0	113	69.7	601	106	0.463	57.0
160	80	6.0	20.7	26.4	836	281	5.62	3.26	105	70.2	132	81.3	702	122	0.459	48.2
160	80	6.3	21.4	27.3	846	286	5.57	3.24	106	71.4	135	83.3	732	126	0.453	46.7
160	80	8.0	26.4	33.6	1001	335	5.46	3.16	125	83.7	163	100	882	150	0.446	37.9
160	80	10.0	31.8	40.6	1146	380	5.32	3.06	143	95.0	191	117	1031	172	0.437	31.4
160	80	12.0	35.8	45.7	1171	391	5.06	2.93	146	97.8	204	125	1111	183	0.418	27.9
160	80	12.5	36.9	47.0	1185	396	5.02	2.90	148	98.9	208	127	1129	185	0.416	27.1
180	100	4.0	16.8	21.3	926	374	6.59	4.18	103	74.8	126	84.0	854	127	0.546	59.7
180	100	5.0	20.7	26.4	1124	452	6.53	4.14	125	90.4	154	103	1045	154	0.543	48.3
180	100	6.0	24.5	31.2	1310	524	6.48	4.10	146	105	181	120	1227	179	0.539	40.8
180	100	6.3	25.4	32.3	1335	536	6.43	4.07	148	107	186	124	1283	185	0.533	39.4
180	100	8.0	31.4	40.0	1598	637	6.32	3.99	178	127	226	150	1565	222	0.526	31.8
180	100	10.0	38.1	48.6	1859	736	6.19	3.89	207	147	268	177	1859	260	0.517	26.2
180	100	12.0	43.4	55.3	1965	782	5.96	3.76	218	156	292	194	2073	285	0.498	23.1

EN 10219-2 (continued)

Size		Wall thickness	Mass per unit length	Cross sectional area	Second moment of area		Radius of gyration		Elastic section modulus		Plastic section modulus		Torsional inertia constant	Torsional modulus constant	Superficial area per meter	Nominal length per metric ton
H × B		T	M	A	I_{xx}	I_{yy}	i_{xx}	i_{yy}	We_{lxx}	We_{lyy}	Wpl_{xx}	Wpl_{yy}	I_t	C_t	A_s	
mm	mm	mm	kg/m	cm²	cm⁴	cm⁴	cm	cm	cm³	cm³	cm³	cm³	cm⁴	cm³	m²/m	m
180	100	12.5	44.8	57.0	2001	796	5.92	3.74	222	159	300	199	2122	290	0.496	22.3
200	100	4.0	18.0	22.9	1200	411	7.23	4.23	120	82.2	148	91.7	985	142	0.586	55.5
200	100	5.0	22.3	28.4	1459	497	7.17	4.19	146	99.4	181	112	1206	172	0.583	44.9
200	100	6.0	26.4	33.6	1703	577	7.12	4.14	170	115	213	132	1417	200	0.579	37.9
200	100	6.3	27.4	34.8	1739	591	7.06	4.12	174	118	219	135	1483	208	0.573	36.6
200	100	8.0	33.9	43.2	2091	705	6.95	4.04	209	141	267	165	1811	250	0.566	29.5
200	100	10.0	41.3	52.6	2444	818	6.82	3.94	244	164	318	195	2154	292	0.557	24.2
200	100	12.0	47.1	60.1	2607	876	6.59	3.82	261	175	350	215	2414	322	0.538	21.2
200	100	12.5	48.7	62.0	2659	892	6.55	3.79	266	178	359	221	2474	329	0.536	20.5
200	120	4.0	19.3	24.5	1353	618	7.43	5.02	135	103	164	115	1345	172	0.626	51.9
200	120	5.0	23.8	30.4	1649	750	7.37	4.97	165	125	201	141	1652	210	0.623	42.0
200	120	6.0	28.3	36.0	1929	874	7.32	4.93	193	146	237	166	1947	245	0.619	35.4
200	120	6.3	29.3	37.4	1976	898	7.27	4.90	198	150	244	172	2040	255	0.613	34.1
200	120	8.0	36.5	46.4	2386	1079	7.17	4.82	239	180	298	209	2507	308	0.606	27.4
200	120	10.0	44.4	56.6	2806	1262	7.04	4.72	281	210	356	250	3007	364	0.597	22.5
200	120	12.0	50.9	64.9	3031	1368	6.84	4.59	303	228	395	278	3419	406	0.578	19.6
200	120	12.5	52.6	67.0	3099	1397	6.80	4.57	310	233	406	285	3514	416	0.576	19.0
250	150	5.0	30.1	38.4	3304	1508	9.28	6.27	264	201	320	225	3285	337	0.783	33.2
250	150	6.0	35.8	45.6	3886	1768	9.23	6.23	311	236	378	266	3886	396	0.779	27.9

Nominal sizes and geometric properties of cold formed rectangular hollow sections according to EN 10219-2

EN 10219-2 (continued)

Size H × B		Wall thickness T	Mass per unit length M	Cross sectional area A	Second moment of area I_{xx}	Second moment of area I_{yy}	Radius of gyration i_{xx}	Radius of gyration i_{yy}	Elastic section modulus $W_{el,xx}$	Elastic section modulus $W_{el,yy}$	Plastic section modulus $W_{pl,xx}$	Plastic section modulus $W_{pl,yy}$	Torsional inertia constant I_t	Torsional modulus constant C_t	Superficial area per meter A_s	Nominal length per metric ton
mm	mm	mm	kg/m	cm²	cm⁴	cm⁴	cm	cm	cm³	cm³	cm³	cm³	cm⁴	cm³	m²/m	m
250	150	6.3	37.2	47.4	4001	1825	9.18	6.20	320	243	391	276	4078	412	0.773	26.8
250	150	8.0	46.5	59.2	4886	2219	9.08	6.12	391	296	482	340	5050	504	0.766	21.5
250	150	10.0	57.0	72.6	5825	2634	8.96	6.02	466	351	582	409	6121	602	0.757	17.6
250	150	12.0	66.0	84.1	6458	2925	8.77	5.90	517	390	658	463	7088	684	0.738	15.2
250	150	12.5	68.3	87.0	6633	3002	8.73	5.87	531	400	678	477	7315	704	0.736	14.6
250	150	16.0	83.8	106.8	7660	3453	8.47	5.69	613	460	805	566	8713	823	0.718	11.9
260	180	5.0	33.2	42.4	4121	2350	9.86	7.45	317	261	377	294	4695	426	0.863	30.1
260	180	6.0	39.6	50.4	4856	2763	9.81	7.40	374	307	447	348	5566	501	0.859	25.3
260	180	6.3	41.2	52.5	5013	2856	9.77	7.38	386	317	463	361	5844	523	0.853	24.3
260	180	8.0	51.5	65.6	6145	3493	9.68	7.29	473	388	573	446	7267	642	0.846	19.4
260	180	10.0	63.2	80.6	7363	4174	9.56	7.20	566	464	694	540	8850	772	0.837	15.8
260	180	12.0	73.5	93.7	8245	4679	9.38	7.07	634	520	790	615	10328	884	0.818	13.6
260	180	12.5	76.2	97.0	8482	4812	9.35	7.04	652	535	815	635	10676	911	0.816	13.1
260	180	16.0	93.9	120	9923	5614	9.11	6.85	763	624	977	759	12890	1079	0.798	10.7
300	100	6.0	35.8	45.6	4777	842	10.2	4.30	318	168	411	188	2403	306	0.779	27.9
300	100	6.3	37.2	47.4	4907	868	10.2	4.28	327	174	425	194	2515	318	0.773	26.8
300	100	8.0	46.5	59.2	5978	1045	10.0	4.20	399	209	523	238	3080	385	0.766	21.5
300	100	10.0	57.0	72.6	7106	1224	9.90	4.11	474	245	631	285	3681	455	0.757	17.6
300	100	12.0	66.0	84.1	7808	1343	9.64	4.00	521	269	710	321	4177	508	0.738	15.2

EN 10219-2 (continued)

Size		Wall thickness	Mass per unit length	Cross sectional area	Second moment of area		Radius of gyration		Elastic section modulus		Plastic section modulus		Torsional inertia constant	Torsional modulus constant	Superficial area per meter	Nominal length per metric ton
H × B		T	M	A	I_{xx}	I_{yy}	i_{xx}	i_{yy}	Wel_{xx}	Wel_{yy}	Wpl_{xx}	Wpl_{yy}	I_t	C_t	A_s	
mm	mm	mm	kg/m	cm²	cm⁴	cm⁴	cm	cm	cm³	cm³	cm³	cm³	cm⁴	cm³	m²/m	m
300	100	12.5	68.3	87.0	8010	1374	9.59	3.97	534	275	732	330	4292	521	0.736	14.6
300	100	16.0	83.8	107	9157	1543	9.26	3.80	610	309	865	386	4939	592	0.718	11.9
300	150	6.0	40.5	51.6	6074	2080	10.8	6.35	405	277	500	309	4988	479	0.879	24.7
300	150	6.3	42.2	53.7	6266	2150	10.8	6.32	418	287	517	321	5234	499	0.873	23.7
300	150	8.0	52.8	67.2	7684	2623	10.7	6.25	512	350	640	396	6491	612	0.866	18.9
300	150	10.0	64.8	82.6	9209	3125	10.6	6.15	614	417	776	479	7879	733	0.857	15.4
300	150	12.0	75.4	96.1	10298	3498	10.4	6.03	687	466	883	546	9153	837	0.838	13.3
300	150	12.5	78.1	99.5	10594	3595	10.3	6.01	706	479	912	563	9452	862	0.836	12.8
300	150	16.0	96.4	123	12387	4174	10.0	5.83	826	557	1092	673	11328	1015	0.818	10.4
300	200	6.0	45.2	57.6	7370	3962	11.3	8.29	491	396	588	446	8115	651	0.979	22.1
300	200	6.3	47.1	60.0	7624	4104	11.3	8.27	508	410	610	463	8524	680	0.973	21.2
300	200	8.0	59.1	75.2	9389	5042	11.2	8.19	626	504	757	574	10627	838	0.966	16.9
300	200	10.0	72.7	92.6	11313	6058	11.1	8.09	754	606	921	698	12987	1012	0.957	13.8
300	200	12.0	84.8	108	12788	6854	10.9	7.96	853	685	1056	801	15236	1167	0.938	11.8
300	200	12.5	88.0	112	13179	7060	10.8	7.94	879	706	1091	828	15768	1204	0.936	11.4
300	200	16.0	109	139	15617	8340	10.6	7.75	1041	834	1319	1000	19223	1442	0.918	9.18
350	250	6.0	54.7	69.6	12457	7458	13.4	10.3	712	597	843	671	14554	967	1.18	18.3
350	250	6.3	57.0	72.6	12923	7744	13.3	10.3	738	620	876	698	15291	1010	1.17	17.5
350	250	8.0	71.6	91.2	16001	9573	13.2	10.2	914	766	1092	869	19136	1253	1.17	14.0

Nominal sizes and geometric properties of cold formed rectangular hollow sections according to EN 10219-2   521

EN 10219-2 (continued)

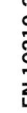

Size H × B		Wall thickness T	Mass per unit length M	Cross sectional area A	Second moment of area		Radius of gyration		Elastic section modulus		Plastic section modulus		Torsional inertia constant	Torsional modulus constant	Superficial area per meter	Nominal length per metric ton
					I_{xx}	I_{yy}	i_{xx}	i_{yy}	We_{xx}	We_{yy}	Wpl_{xx}	Wpl_{yy}	I_t	C_t	A_s	
mm	mm	mm	kg/m	cm²	cm⁴	cm⁴	cm	cm	cm³	cm³	cm³	cm³	cm⁴	cm³	m²/m	m
350	250	10.0	88.4	113	19407	11583	13.1	10.1	1109	927	1335	1062	23500	1522	1.16	11.3
350	250	12.0	104	132	22197	13261	13.0	10.0	1268	1061	1544	1229	27749	1770	1.14	9.65
350	250	12.5	108	137	22922	13690	12.9	9.99	1310	1095	1598	1272	28764	1830	1.14	9.30
350	250	16.0	134	171	27580	16434	12.7	9.81	1576	1315	1954	1554	35497	2220	1.12	7.46
400	200	8.0	71.6	91.2	18974	6517	14.4	8.45	949	652	1173	728	15820	1133	1.17	14.0
400	200	12.5	108	137	27100	9260	14.1	8.22	1355	926	1714	1062	23594	1644	1.14	9.30
400	200	16.0	134	171	32547	11056	13.8	8.05	1627	1106	2093	1294	28928	1984	1.12	7.46
400	300	10.0	104	133	30609	19726	15.2	12.2	1530	1315	1824	1501	38407	2132	1.36	9.61
400	300	12.0	123	156	35284	22747	15.0	12.1	1764	1516	2122	1747	45527	2492	1.34	8.16
400	300	12.5	127	162	36489	23517	15.0	12.0	1824	1568	2198	1810	47237	2580	1.34	7.86
400	300	16.0	159	203	44350	28535	14.8	11.9	2218	1902	2708	2228	58730	3159	1.32	6.28

Nominal sizes and geometric properties of hot finished rectangular hollow sections according to EN 10210-2 (see [1], Chapter 4)

Size		Wall thickness	Mass per unit length	Cross sectional area	Second moment of area		Radius of gyration		Elastic section modulus		Plastic section modulus		Torsional inertia constant	Torsional modulus constant	Superficial area per meter	Nominal length per metric ton
H × B		T	M	A	I_{xx}	I_{yy}	i_{xx}	i_{yy}	Wel_{xx}	Wel_{yy}	Wpl_{xx}	Wpl_{yy}	I_t	C_t	A_s	
mm	mm	mm	kg/m	cm²	cm⁴	cm⁴	cm	cm	cm³	cm³	cm³	cm³	cm⁴	cm³	m²/m	m
50	25	2.5	2.69	3.43	10.4	3.39	1.74	0.994	4.16	2.71	5.33	3.22	8.42	4.61	0.144	371
50	25	3.0	3.17	4.04	11.9	3.83	1.72	0.973	4.76	3.06	6.18	3.71	9.64	5.20	0.142	315
50	30	2.5	2.89	3.68	11.8	5.22	1.79	1.19	4.73	3.48	5.92	4.11	11.7	5.73	0.154	346
50	30	3.0	3.41	4.34	13.6	5.94	1.77	1.17	5.43	3.96	6.88	4.76	13.5	6.51	0.152	293
50	30	4.0	4.39	5.59	16.5	7.08	1.72	1.13	6.60	4.72	8.59	5.88	16.6	7.77	0.150	228
50	30	5.0	5.28	6.73	18.7	7.89	1.67	1.08	7.49	5.26	10.0	6.80	19.0	8.67	0.147	189
60	40	2.5	3.68	4.68	22.8	12.1	2.21	1.60	7.61	6.03	9.32	7.02	25.1	9.73	0.194	272
60	40	3.0	4.35	5.54	26.5	13.9	2.18	1.58	8.82	6.95	10.9	8.19	29.2	11.2	0.192	230
60	40	4.0	5.64	7.19	32.8	17.0	2.14	1.54	10.9	8.52	13.8	10.3	36.7	13.7	0.190	177
60	40	5.0	6.85	8.73	38.1	19.5	2.09	1.50	12.7	9.77	16.4	12.2	43.0	15.7	0.187	146
60	40	6.0	7.99	10.2	42.3	21.4	2.04	1.45	14.1	10.7	18.6	13.7	48.2	17.3	0.185	125
60	40	6.3	8.31	10.6	43.4	21.9	2.02	1.44	14.5	11.0	19.2	14.2	49.5	17.6	0.184	120
80	40	3.0	5.29	6.74	54.2	18.0	2.84	1.63	13.6	9.00	17.1	10.4	43.8	15.3	0.232	189
80	40	4.0	6.90	8.79	68.2	22.2	2.79	1.59	17.1	11.1	21.8	13.2	55.2	18.9	0.230	145
80	40	5.0	8.42	10.7	80.3	25.7	2.74	1.55	20.1	12.9	26.1	15.7	65.1	21.9	0.227	119
80	40	6.0	9.87	12.6	90.5	28.5	2.68	1.50	22.6	14.2	30.0	17.8	73.4	24.2	0.225	101
80	40	6.3	10.3	13.1	93.3	29.2	2.67	1.49	23.3	14.6	31.1	18.4	75.6	24.8	0.224	97.2
80	40	8.0	12.5	16.0	106	32.1	2.58	1.42	26.5	16.1	36.5	21.2	85.8	27.4	0.219	79.9
90	50	3.0	6.24	7.94	84.4	33.5	3.26	2.05	18.8	13.4	23.2	15.3	76.5	22.4	0.272	160

Nominal sizes and geometric properties of hot finished rectangular hollow sections according to EN 10210-2

EN 10210-2 (continued)

Size H × B		Wall thickness T	Mass per unit length M	Cross sectional area A	Second moment of area		Radius of gyration		Elastic section modulus		Plastic section modulus		Torsional inertia constant	Torsional modulus constant	Superficial area per meter	Nominal length per metric ton
					I_{xx}	I_{yy}	i_{xx}	i_{yy}	$W_{el,xx}$	$W_{el,yy}$	$W_{pl,xx}$	$W_{pl,yy}$	I_t	C_t	A_s	
mm	mm	mm	kg/m	cm²	cm⁴	cm⁴	cm	cm	cm³	cm³	cm³	cm³	cm⁴	cm³	m²/m	m
90	50	4.0	8.15	10.4	107	41.9	3.21	2.01	23.8	16.8	29.8	19.6	97.5	28.0	0.270	123
90	50	5.0	9.99	12.7	127	49.2	3.16	1.97	28.3	19.7	36.0	23.5	116	32.9	0.267	100
90	50	6.0	11.8	15.0	145	55.4	3.11	1.92	32.2	22.1	41.6	27.0	133	37.0	0.265	85.1
90	50	6.3	12.3	15.6	150	57.0	3.10	1.91	33.3	22.8	43.2	28.0	138	38.1	0.264	81.5
90	50	8.0	15.0	19.2	174	64.6	3.01	1.84	38.6	25.8	51.4	32.9	160	43.2	0.259	66.5
100	50	3.0	6.71	8.54	110	36.8	3.58	2.08	21.9	14.7	27.3	16.8	88.4	25.0	0.292	149
100	50	4.0	8.78	11.2	140	46.2	3.53	2.03	27.9	18.5	35.2	21.5	113	31.4	0.290	114
100	50	5.0	10.8	13.7	167	54.3	3.48	1.99	33.3	21.7	42.6	25.8	135	36.9	0.287	92.8
100	50	6.0	12.7	16.2	190	61.2	3.43	1.95	38.1	24.5	49.4	29.7	154	41.6	0.285	78.8
100	50	6.3	13.3	16.9	197	63.0	3.42	1.93	39.4	25.2	51.3	30.8	160	42.9	0.284	75.4
100	50	8.0	16.3	20.8	230	71.7	3.33	1.86	46.0	28.7	61.4	36.3	186	48.9	0.279	61.4
100	60	3.0	7.18	9.14	124	55.7	3.68	2.47	24.7	18.6	30.2	21.2	121	30.7	0.312	139
100	60	4.0	9.41	12.0	158	70.5	3.63	2.43	31.6	23.5	39.1	27.3	156	38.7	0.310	106
100	60	5.0	11.6	14.7	189	83.6	3.58	2.38	37.8	27.9	47.4	32.9	188	45.9	0.307	86.5
100	60	6.0	13.6	17.4	217	95.0	3.53	2.34	43.4	31.7	55.1	38.1	216	52.1	0.305	73.3
100	60	6.3	14.2	18.1	225	98.1	3.52	2.33	45.0	32.7	57.3	39.5	224	53.8	0.304	70.2
100	60	8.0	17.5	22.4	264	113	3.44	2.25	52.8	37.8	68.7	47.1	265	62.2	0.299	57.0
120	60	4.0	10.7	13.6	249	83.1	4.28	2.47	41.5	27.7	51.9	31.7	201	47.1	0.350	93.7
120	60	5.0	13.1	16.7	299	98.8	4.23	2.43	49.9	32.9	63.1	38.4	242	56.0	0.347	76.1

EN 10210-2 (continued)

Size H × B		Wall thickness T	Mass per unit length M	Cross sectional area A	Second moment of area		Radius of gyration		Elastic section modulus		Plastic section modulus		Torsional inertia constant I_t	Torsional modulus constant C_t	Superficial area per meter A_s	Nominal length per metric ton
					I_{xx}	I_{yy}	i_{xx}	i_{yy}	Wel_{xx}	Wel_{yy}	Wpl_{xx}	Wpl_{yy}				
mm	mm	mm	kg/m	cm²	cm⁴	cm⁴	cm	cm	cm³	cm³	cm³	cm³	cm⁴	cm³	m²/m	m
120	60	6.0	15.5	19.8	345	113	4.18	2.39	57.5	37.5	73.6	44.5	279	63.8	0.345	64.4
120	60	6.3	16.2	20.7	358	116	4.16	2.37	59.7	38.8	76.7	46.3	290	65.9	0.344	61.6
120	60	8.0	20.1	25.6	425	135	4.08	2.30	70.8	45.0	92.7	55.4	344	76.6	0.339	49.9
120	60	10.0	24.3	30.9	488	152	3.97	2.21	81.4	50.5	109	64.4	396	86.1	0.334	41.2
120	80	4.0	11.9	15.2	303	161	4.46	3.25	50.4	40.2	61.2	46.1	330	65.0	0.390	83.9
120	80	5.0	14.7	18.7	365	193	4.42	3.21	60.9	48.2	74.6	56.1	401	77.9	0.387	68.0
120	80	6.0	17.4	22.2	423	222	4.37	3.17	70.6	55.6	87.3	65.5	468	89.6	0.385	57.5
120	80	6.3	18.2	23.2	440	230	4.36	3.15	73.3	57.6	91.0	68.2	487	92.9	0.384	54.9
120	80	8.0	22.6	28.8	525	273	4.27	3.08	87.5	68.1	111	82.6	587	110	0.379	44.3
120	80	10.0	27.4	34.9	609	313	4.18	2.99	102	78.1	131	97.3	688	126	0.374	36.5
140	80	4.0	13.2	16.8	441	184	5.12	3.31	62.9	46.0	77.1	52.2	411	76.5	0.430	75.9
140	80	5.0	16.3	20.7	534	221	5.08	3.27	76.3	55.3	94.3	63.6	499	91.9	0.427	61.4
140	80	6.0	19.3	24.6	621	255	5.03	3.22	88.7	63.8	111	74.4	583	106	0.425	51.8
140	80	6.3	20.2	25.7	646	265	5.01	3.21	92.3	66.2	115	77.5	607	110	0.424	49.6
140	80	8.0	25.1	32.0	776	314	4.93	3.14	111	78.5	141	94.1	733	130	0.419	39.9
140	80	10.0	30.6	38.9	908	362	4.83	3.05	130	90.5	168	111	862	150	0.414	32.7
150	100	4.0	15.1	19.2	607	324	5.63	4.11	81.0	64.8	97.4	73.6	660	105	0.490	66.4
150	100	5.0	18.6	23.7	739	392	5.58	4.07	98.5	78.5	119	90.1	807	127	0.487	53.7
150	100	6.0	22.1	28.2	862	456	5.53	4.02	115	91.2	141	106	946	147	0.485	45.2

Nominal sizes and geometric properties of hot finished rectangular hollow sections according to EN 10210-2

EN 10210-2 (continued)

Size H × B		Wall thickness T	Mass per unit length M	Cross sectional area A	Second moment of area		Radius of gyration		Elastic section modulus		Plastic section modulus		Torsional inertia constant	Torsional modulus constant	Superficial area per meter	Nominal length per metric ton
					I_{xx}	I_{yy}	i_{xx}	i_{yy}	Wel_{xx}	Wel_{yy}	Wpl_{xx}	Wpl_{yy}	I_t	C_t	A_s	
mm	mm	mm	kg/m	cm²	cm⁴	cm⁴	cm	cm	cm³	cm³	cm³	cm³	cm⁴	cm³	m²/m	m
150	100	6.3	23.1	29.5	898	474	5.52	4.01	120	94.8	147	110	986	153	0.484	43.2
150	100	8.0	28.9	36.8	1087	569	5.44	3.94	145	114	180	135	1203	183	0.479	34.7
150	100	10.0	35.3	44.9	1282	665	5.34	3.85	171	133	216	161	1432	214	0.474	28.4
150	100	12.0	41.4	52.7	1450	745	5.25	3.76	193	149	249	185	1633	240	0.469	24.2
150	100	12.5	42.8	54.6	1488	763	5.22	3.74	198	153	256	190	1679	246	0.468	23.3
160	80	4.0	14.4	18.4	612	207	5.77	3.35	76.5	51.7	94.7	58.3	493	88.1	0.470	69.3
160	80	5.0	17.8	22.7	744	249	5.72	3.31	93.0	62.3	116	71.1	600	106	0.467	56.0
160	80	6.0	21.2	27.0	868	288	5.67	3.27	108	72.0	136	83.3	701	122	0.465	47.2
160	80	6.3	22.2	28.2	903	299	5.66	3.26	113	74.8	142	86.8	730	127	0.464	45.1
160	80	8.0	27.6	35.2	1091	356	5.57	3.18	136	89.0	175	106	883	151	0.459	36.2
160	80	10.0	33.7	42.9	1284	411	5.47	3.10	161	103	209	125	1041	175	0.454	29.7
160	80	12.0	39.5	50.3	1449	455	5.37	3.01	181	114	240	142	1175	194	0.449	25.3
160	80	12.5	40.9	52.1	1485	465	5.34	2.99	186	116	247	146	1204	198	0.448	24.5
180	100	4.0	16.9	21.6	945	379	6.61	4.19	105	75.9	128	85.2	852	127	0.550	59.0
180	100	5.0	21.0	26.7	1153	460	6.57	4.15	128	92.0	157	104	1042	154	0.547	47.7
180	100	6.0	24.9	31.8	1350	536	6.52	4.11	150	107	186	123	1224	179	0.545	40.1
180	100	6.3	26.1	33.3	1407	557	6.50	4.09	156	111	194	128	1277	186	0.544	38.3
180	100	8.0	32.6	41.6	1713	671	6.42	4.02	190	134	239	157	1560	224	0.539	30.7
180	100	10.0	40.0	50.9	2036	787	6.32	3.93	226	157	288	188	1862	263	0.534	25.0

EN 10210-2 (continued)

Size		Wall thickness	Mass per unit length	Cross sectional area	Second moment of area		Radius of gyration		Elastic section modulus		Plastic section modulus		Torsional inertia constant	Torsional modulus constant	Superficial area per meter	Nominal length per metric ton
H × B		T	M	A	I_{xx}	I_{yy}	i_{xx}	i_{yy}	Wel_{xx}	Wel_{yy}	Wpl_{xx}	Wpl_{yy}	I_t	C_t	A_s	
mm	mm	mm	kg/m	cm²	cm⁴	cm⁴	cm	cm	cm³	cm³	cm³	cm³	cm⁴	cm³	m²/m	m
180	100	12.0	47.0	59.9	2320	886	6.22	3.85	258	177	333	216	2130	296	0.529	21.3
180	100	12.5	48.7	62.1	2385	908	6.20	3.82	265	182	344	223	2191	303	0.528	20.5
200	100	4.0	18.2	23.2	1223	416	7.26	4.24	122	83.2	150	92.8	983	142	0.590	54.9
200	100	5.0	22.6	28.7	1495	505	7.21	4.19	149	101	185	114	1204	172	0.587	44.3
200	100	6.0	26.8	34.2	1754	589	7.16	4.15	175	118	218	134	1414	200	0.585	37.3
200	100	6.3	28.1	35.8	1829	613	7.15	4.14	183	123	228	140	1475	208	0.584	35.6
200	100	8.0	35.1	44.8	2234	739	7.06	4.06	223	148	282	172	1804	251	0.579	28.5
200	100	10.0	43.1	54.9	2664	869	6.96	3.98	266	174	341	206	2156	295	0.574	23.2
200	100	12.0	50.8	64.7	3047	979	6.86	3.89	305	196	395	237	2469	333	0.569	19.7
200	100	12.5	52.7	67.1	3136	1004	6.84	3.87	314	201	408	245	2541	341	0.568	19.0
200	100	16.0	65.2	83.0	3678	1147	6.66	3.72	368	229	491	290	2982	391	0.559	15.3
200	120	6.0	28.7	36.6	1980	892	7.36	4.94	198	149	242	169	1942	245	0.625	34.8
200	120	6.3	30.1	38.3	2065	929	7.34	4.92	207	155	253	177	2028	255	0.624	33.3
200	120	8.0	37.6	48.0	2529	1128	7.26	4.85	253	188	313	218	2495	310	0.619	26.6
200	120	10.0	46.3	58.9	3026	1337	7.17	4.76	303	223	379	263	3001	367	0.614	21.6
200	120	12.0	54.6	69.5	3472	1520	7.07	4.68	347	253	440	305	3461	417	0.609	18.3
200	120	12.5	56.6	72.1	3576	1562	7.04	4.66	358	260	455	314	3569	428	0.608	17.7
250	150	6.0	36.2	46.2	3965	1796	9.27	6.24	317	239	385	270	3877	396	0.785	27.6
250	150	6.3	38.0	48.4	4143	1874	9.25	6.22	331	250	402	283	4054	413	0.784	26.3

Nominal sizes and geometric properties of hot finished rectangular hollow sections according to EN 10210-2

EN 10210-2 (continued)

Size H × B		Wall thickness T	Mass per unit length M	Cross sectional area A	Second moment of area		Radius of gyration		Elastic section modulus		Plastic section modulus		Torsional inertia constant	Torsional modulus constant	Superficial area per meter	Nominal length per metric ton
					I_{xx}	I_{yy}	i_{xx}	i_{yy}	Wel_{xx}	Wel_{yy}	Wpl_{xx}	Wpl_{yy}	I_t	C_t	A_s	
mm	mm	mm	kg/m	cm²	cm⁴	cm⁴	cm	cm	cm³	cm³	cm³	cm³	cm⁴	cm³	m²/m	m
250	150	8.0	47.7	60.8	5111	2298	9.17	6.15	409	306	501	350	5021	506	0.779	21.0
250	150	10.0	58.8	74.9	6174	2755	9.08	6.06	494	367	611	426	6090	605	0.774	17.0
250	150	12.0	69.6	88.7	7154	3168	8.98	5.98	572	422	715	497	7088	695	0.769	14.4
250	150	12.5	72.3	92.1	7387	3265	8.96	5.96	591	435	740	514	7326	717	0.768	13.8
250	150	16.0	90.3	115	8879	3873	8.79	5.80	710	516	906	625	8868	849	0.759	11.1
260	180	6.0	40.0	51.0	4942	2804	9.85	7.42	380	312	454	353	5554	502	0.865	25.0
260	180	6.3	41.9	53.4	5166	2929	9.83	7.40	397	325	475	369	5810	524	0.864	23.8
260	180	8.0	52.7	67.2	6390	3608	9.75	7.33	492	401	592	459	7221	644	0.859	19.0
260	180	10.0	65.1	82.9	7741	4351	9.66	7.24	595	483	724	560	8798	775	0.854	15.4
260	180	12.0	77.2	98.3	8999	5034	9.57	7.16	692	559	849	656	10285	895	0.849	13.0
260	180	12.5	80.1	102	9299	5196	9.54	7.13	715	577	879	679	10643	924	0.848	12.5
260	180	16.0	100	128	11245	6231	9.38	6.98	865	692	1081	831	12993	1106	0.839	10.0
300	200	6.0	45.7	58.2	7486	4013	11.3	8.31	499	401	596	451	8100	651	0.985	21.9
300	200	6.3	47.9	61.0	7829	4193	11.3	8.29	522	419	624	472	8476	681	0.984	20.9
300	200	8.0	60.3	76.8	9717	5184	11.3	8.22	648	518	779	589	10562	840	0.979	16.6
300	200	10.0	74.5	94.9	11819	6278	11.2	8.13	788	628	956	721	12908	1015	0.974	13.4
300	200	12.0	88.5	113	13797	7294	11.1	8.05	920	729	1124	847	15137	1178	0.969	11.3
300	200	12.5	91.9	117	14273	7537	11.0	8.02	952	754	1165	877	15677	1217	0.968	10.9
300	200	16.0	115	147	17390	9109	10.9	7.87	1159	911	1441	1080	19252	1468	0.959	8.67

EN 10210-2 (continued)

Size		Wall thickness	Mass per unit length	Cross sectional area	Second moment of area		Radius of gyration		Elastic section modulus		Plastic section modulus		Torsional inertia constant	Torsional modulus constant	Superficial area per meter	Nominal length per metric ton
H × B		T	M	A	I_{xx}	I_{yy}	i_{xx}	i_{yy}	Wel_{xx}	Wel_{yy}	Wpl_{xx}	Wpl_{yy}	I_t	C_t	A_s	
mm	mm	mm	kg/m	cm²	cm⁴	cm⁴	cm	cm	cm³	cm³	cm³	cm³	cm⁴	cm³	m²/m	m
350	250	6.0	55.1	70.2	12616	7538	13.4	10.4	721	603	852	677	14529	967	1.18	18.2
350	250	6.3	57.8	73.6	13203	7885	13.4	10.4	754	631	892	709	15215	1011	1.18	17.3
350	250	8.0	72.8	92.8	16449	9798	13.3	10.3	940	784	1118	888	19027	1254	1.18	13.7
350	250	10.0	90.2	115	20102	11937	13.2	10.2	1149	955	1375	1091	23354	1525	1.17	11.1
350	250	12.0	107	137	23577	13957	13.1	10.1	1347	1117	1624	1286	27513	1781	1.17	9.32
350	250	12.5	112	142	24419	14444	13.1	10.1	1395	1156	1685	1334	28526	1842	1.17	8.97
350	250	16.0	141	179	30011	17654	12.9	9.93	1715	1412	2095	1655	35325	2246	1.16	7.12
400	200	8.0	72.8	92.8	19562	6660	14.5	8.47	978	666	1203	743	15735	1135	1.18	13.7
400	200	10.0	90.2	115	23914	8094	14.4	8.39	1196	808	1480	911	19259	1376	1.17	11.1
400	200	12.0	107	137	28059	9418	14.3	8.30	1403	942	1748	1072	22622	1602	1.17	9.32
400	200	12.5	112	142	29063	9738	14.3	8.28	1453	974	1813	1111	23438	1656	1.17	8.97
400	200	16.0	141	179	35738	11824	14.1	8.13	1787	1182	2256	1374	28871	2010	1.16	7.12
450	250	8.0	85.4	109	30082	12142	16.6	10.6	1337	971	1622	1081	27083	1629	1.38	11.7
450	250	10.0	106	135	36895	14819	16.5	10.5	1640	1185	2000	1331	33284	1986	1.37	9.44
450	250	12.0	126	161	43434	17359	16.4	10.4	1930	1389	2367	1572	39260	2324	1.37	7.93
450	250	12.5	131	167	45026	17973	16.4	10.4	2001	1438	2458	1631	40719	2406	1.37	7.62
450	250	16.0	166	211	55705	22041	16.2	10.2	2476	1763	3070	2029	50545	2947	1.36	6.04
500	300	10.0	122	155	53762	24439	18.6	12.6	2150	1629	2595	1826	52450	2696	1.57	8.22
500	300	12.0	145	185	63446	28736	18.5	12.5	2538	1916	3077	2161	62039	3167	1.57	6.90

Nominal sizes and geometric properties of hot finished rectangular hollow sections according to EN 10210-2

EN 10210-2 (continued)

Size H × B		Wall thickness T	Mass per unit length M	Cross sectional area A	Second moment of area		Radius of gyration		Elastic section modulus		Plastic section modulus		Torsional inertia constant I_t	Torsional modulus constant C_t	Superficial area per meter A_s	Nominal length per metric ton
					I_{xx}	I_{yy}	i_{xx}	i_{yy}	Wel_{xx}	Wel_{yy}	Wpl_{xx}	Wpl_{yy}				
mm	mm	mm	kg/m	cm²	cm⁴	cm⁴	cm	cm	cm³	cm³	cm³	cm³	cm⁴	cm³	m²/m	m
500	300	12.5	151	192	65813	29780	18.5	12.5	2633	1985	3196	2244	64389	3281	1.57	6.63
500	300	16.0	191	243	81783	36768	18.3	12.3	3271	2451	4005	2804	80329	4044	1.56	5.24
500	300	20.0	235	300	98777	44078	18.2	12.1	3951	2939	4885	3408	97447	4842	1.55	4.25

New ultimate joint resistance formulae for uni-planar and multi-planar CHS T, X and XX joints
(see [116], Chapter 6)

1. Axially loaded uni-planar X joints ($\alpha = 11.5$), unrestrained chord ends

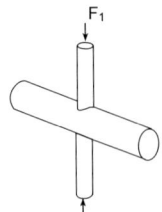

$$\frac{F_{1,u}}{f_{y,0} \cdot t_0^2} = \frac{8.7 \, \gamma^{0.5\beta - 0.5\beta^2}}{(1 - 0.9\beta) + \sqrt{(1 - 0.9\beta)^2 + \frac{2 - (0.9\beta)^2}{\gamma^2}}} \qquad \text{(IV 1)}$$

validity range:
$\alpha = 11.5$
$0.25 \leq \beta \leq 1.0$
$14.5 \leq 2\gamma \leq 50.8$

2. Axially loaded uni-planar X joints ($6 \leq \alpha \leq 18$), unrestrained chord ends

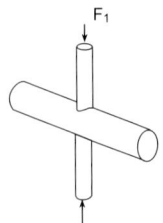

$$F_{1,u}(\alpha) = f(\alpha) \cdot F_{1,u}(\alpha = 11.5) \qquad \text{(IV 2)}$$

with the chord length function $f(\alpha)$ defined as:

$$f(\alpha) = \frac{12.5\alpha}{11.5(1 + \alpha)} \qquad \text{(IV 3)}$$

validity range:
$6.0 \leq \alpha \leq 18.0$
$0.25 \leq \beta \leq 1.0$
$2\gamma = 25.4$

3. Uni-planar X joints loaded by in-plane bending moment

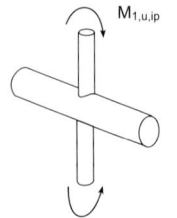

$$\frac{M_{1,u,ip}}{f_{y,0} \cdot t_0^2 \cdot d_1} = \frac{5.1 \, \gamma^{1.04\beta - 0.43\beta^2}}{(1 - 0.4\beta) + \sqrt{(1 - 0.4\beta)^2 + \frac{2 - (0.4\beta)^2}{\gamma^2}}} \qquad \text{(IV 4)}$$

validity range:
$\alpha = 12.0$
$0.25 \leq \beta \leq 1.0$
$14.5 \leq 2\gamma \leq 50.8$

4. Uni-planar X joints loaded by out-of-plane bending moment

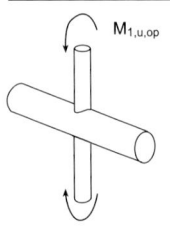

$$\frac{M_{1,u,op}}{f_{y,0} \cdot t_0^2 \cdot d_1} = 1.56 \, \gamma^{0.33\beta - 0.24\beta^2} \frac{C_1 \cdot 2\gamma^2 \left[C_2 + \sqrt{(1 - 0.9\beta)^2 + \frac{2 - (0.9\beta)^2}{\gamma^2}} \right] + C_3}{(0.9\beta + 0.55) \cdot (2 - (0.9\beta)^2)^2} \qquad \text{(IV 5)}$$

with:
$C_1 = 3 + 0.9\beta - 2 \cdot (0.9\beta)^2$
$C_2 = 0.9\beta - 1$
$C_3 = -(0.9\beta)^4 + 3 \cdot (0.9\beta)^2 - 2$

validity range:
$\alpha = 12.0$
$0.25 \leq \beta \leq 1.0$
$14.5 \leq 2\gamma \leq 50.8$

5. Axially loaded multi-planar XX joints ($\alpha = 16$)

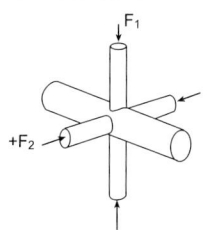

$J = 0.0$:

$$\frac{F_{1,u}}{f_{y,0} \cdot t_0^2} = \frac{8.0 \, \gamma^{0.7\beta - \beta^2}}{\sqrt{1 - (0.9\beta)^2} - 0.9\beta + \sqrt{\left(\sqrt{1 - (0.9\beta)^2} - 0.9\beta\right)^2 + \frac{2}{\gamma^2}}} \quad \text{(IV 6)}$$

$-0.6 \leq J \leq 1.0$:

$$F_{1,u}(J) = \frac{F_{1,u}(J = 0.0)}{1 - (1.6\beta - 1.2\beta^2)J + (1.5\beta - 2.5\beta^2)J^2} \quad \text{(IV 7)}$$

$F_{2,u} = J F_{1,u}$

validity range:
$\alpha = 16.0$
$0.22 \leq \beta \leq 0.60$
$14.5 \leq 2\gamma \leq 50.8$

J = load ratio between the load on the out-of-plane bracing and that on the in-plane bracing

6. Axially loaded multi-planar XX joints ($6 \leq \alpha \leq 16$)

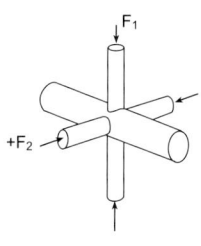

$$F_{1,u}(\alpha) = f(\alpha, J) \cdot F_{1,u}(\alpha = 16.0) \quad \text{(IV 8)}$$

with the chord length function $f(\alpha, J)$ defined as:

$$f(\alpha, J) = \frac{17.0 \, \alpha}{16.0(1 + \alpha)} \cdot (1 + 0.5 \, J e^{-0.3\alpha}) \quad \text{(IV 9)}$$

validity range:
$6.0 \leq \alpha \leq 16.0$
$0.22 \leq \beta \leq 0.60$
$2\gamma = 25.4$
$-0.60 \leq J \leq 1.0$

J = load ratio between the load on the out-of-plane bracing and that on the in-plane bracing

7. Multi-planar XX joints loaded by in-plane bending moment on the in-plane bracings and axial forces on the out-of-plane bracings

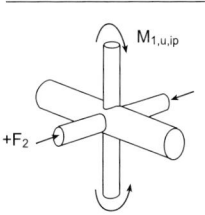

$$M_{1,u,ip}(F_2) = M_{1,u,ip}(F_2 = 0.0) + F_2 \cdot \left(-0.4 + 0.7\beta + 1.5 \frac{\gamma}{100}\right) \quad \text{(IV 10)}$$

validity range:
$\alpha = 12.0$
$0.22 \leq \beta \leq 0.60$
$14.5 \leq 2\gamma \leq 50.8$
$-0.6 F_u \text{ (tension)} \leq F_2 \leq 0.6 F_u \text{ (compression)} \quad \text{(IV 11)}$

$$F_u = \frac{7.46}{1 - 0.812\beta} (2\gamma)^{-0.05} \cdot \left(\frac{f_{u,0}}{f_{y,0}}\right)^{0.173} \cdot f_{y,0} \cdot t_0^2 \quad \text{(IV 12)}$$

8. Multi-planar XX joints loaded by in-plane bending moment on the in-plane and out-of-plane bracings

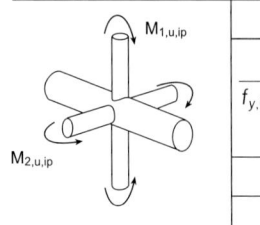

$J = 0.0$:

$$\frac{M_{1,u,ip}}{f_{y,0} \cdot t_0^2 \cdot d_1} = \frac{4.8\gamma^{1.2\beta-0.6\beta^2}}{\sqrt{1-(0.4\beta)^2} - 0.4\beta + \sqrt{\left(\sqrt{1-(0.4\beta)^2} - 0.4\beta\right)^2 + \frac{2}{\gamma^2}}} \quad \text{(IV 13)}$$

$-1.0 \leq J \leq 1.0$:

$$M_{1,u,ip}(J) = \frac{M_{1,u,ip}(J=0.0)}{1 - 0.5\beta^2 J - (0.16 - 0.9\beta + 0.42\beta^2)J^2} \quad \text{(IV 14)}$$

$$M_{2,u,ip} = J M_{1,u,ip} \quad \text{(IV 15)}$$

validity range:
$\alpha = 12.0$
$0.22 \leq \beta \leq 0.60$
$14.5 \leq 2\gamma \leq 50.8$

J = load ratio between the load on the out-of-plane bracing and that on the in-plane bracing

9. Axially loaded uni-planar T joints (the influence of overall chord bending has been excluded by applying compensating moments to the chord ends)

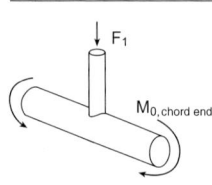

$$\frac{F_{1,u,loc}}{f_{y,0} \cdot t_0^2} = \frac{2.3\gamma^{0.66\beta-0.3\beta^2} \cdot (1+C_1)}{\left(1 - \frac{0.8\psi_2}{\pi}\right)\sin(0.8\psi_2)(1+C_1) - \left(1 - \frac{\arcsin 0.8\beta}{\pi}\right)0.8\beta(1+\cos 0.8\psi_2) + \frac{0.7}{\gamma^2}} \quad \text{(IV 16)}$$

with:

$C_1 = \sqrt{1 - (0.8\beta)^2}$

$\psi_2 = 1.2 + 0.8\beta^2$ rad.

validity range:
$0.22 \leq \beta \leq 1.0$
$14.5 \leq 2\gamma \leq 50.8$

$F_{1,u,loc}$ = ultimate axial force on member i ($i=1$) excluding the influence of overall chord bending

10. Axially loaded uni-planar T joints (the influence of overall chord bending is included)

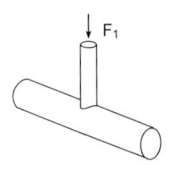

$$0.32 \left(\frac{M_{0,u}}{M_{0,pl,V}}\right) + \left(\frac{F_{1,u}}{F_{1,u,loc}}\right) = 1.0 \quad \text{(IV 17)}$$

with:

$M_{0,pl,V} = \frac{1}{4} F_{1,pl,V} (l_0 - d_1)$

$$\frac{F_{1,pl,V}}{f_{y,0} \cdot t_0^2} = \frac{8\gamma^2}{\sqrt{12}} \frac{\left(1 - \left(1 - \frac{1}{\gamma}\right)^2\right)\left(1 - \left(1 - \frac{1}{\gamma}\right)^3\right)}{\sqrt{\left(1 - \left(1 - \frac{1}{\gamma}\right)^3\right)^2 + \frac{3}{16}\alpha^2 \left(1 - \frac{2\beta}{\alpha}\right)^2 \left(1 - \left(1 - \frac{1}{\gamma}\right)^2\right)^2}} \quad \text{(IV 18)}$$

validity range:
$0.22 \leq \beta \leq 1.0$
$14.5 \leq 2\gamma \leq 50.8$
$M_{0,u}/M_{0,pl,V} \leq 1.0$

$M_{0,u}$ = bending moment in the chord as a result of ultimate axial force in the bracing
$M_{0,pl,V}$ = reduced plastic moment capacity of the chord due to the combination of bending moments and shear forces in the chord

11. Uni-planar T joints loaded by in-plane bending moment

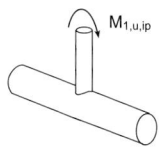

$$\frac{M_{1,u,ip}}{f_{y,0} \cdot t_0^2 \cdot d_1} = \frac{\gamma^{0.78\beta - 0.13\beta^2} \cdot (1 + C_1)}{\left(1 - \frac{0.4\psi_2}{\pi}\right)\sin(0.4\psi_2)(1 + C_1) - \left(1 - \frac{\arcsin 0.4\beta}{\pi}\right)0.4\beta(1 + \cos 0.4\psi_2)} + \frac{0.7}{\gamma^2}$$

(IV 19)

with:

$C_1 = \sqrt{1 - (0.4\beta)^2}$

$\psi = 1.2 + 0.8\beta^2$ rad.

validity range:
$\alpha = 12.0$
$0.25 \le \beta \le 1.0$
$14.5 \le 2\gamma \le 50.8$

12. Uni-planar T joints loaded by out-of-plane bending moment

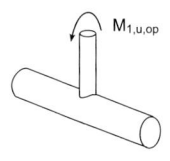

$$\frac{M_{1,u,op}}{f_{y,0} \cdot t_0^2 \cdot d_1} = \frac{2.5\gamma^{0.28\beta} \cdot (0.8\beta + \sin 0.8\psi_2)}{(0.8\beta + 1.0)\left[\left(1 - \frac{\arcsin 0.8\beta}{\pi}\right)\sin(0.8\psi_2) - \left(1 - \frac{0.8\psi_2}{\pi}\right)0.8\beta\right]} + \frac{0.5}{\gamma^2}$$

(IV 20)

with:
$\psi_2 = 1.8 + 0.5\beta^2$ rad.

validity range:
$\alpha = 12.0$
$0.25 \le \beta \le 1.0$
$14.5 \le 2\gamma \le 50.8$

13. Axially loaded multi-planar TT joints (the influence of overall chord bending has been excluded by applying compensating moments to the chord ends)

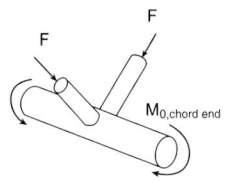

TT joints, which fail like a uni-planar T joint:

$$\beta_{eq.} = \left[0.93 + 0.07 \frac{\sin\left(\frac{\phi}{2} + \arcsin\beta\right)}{\beta}\left(2 - \frac{\sin\left(\frac{\phi}{2} + \arcsin\beta\right)}{\beta}\right)\right]\sin\left(\frac{\phi}{2} + \arcsin\beta\right)$$

(IV 21)

Ultimate resistance is calculated by setting the $\beta_{eq.}$-value in Eq. (IV 16)

TT joints, which fail like a uni-planar X joint:

$$\beta_{eq.} = \frac{\beta}{\sin\frac{1}{2}\phi}$$

(IV 22)

Ultimate resistance is calculated by setting the $\beta_{eq.}$-value in Eq. (IV 1)

validity range:
$0.22 \le \beta \le 0.60$
$2\gamma = 36.9$

New ultimate resistance formulae for welded, uni-planar and multiplanar I beam to RHS column connections (see [265], Chapter 6)

Symbols: β: width ratio between I beam flange and RHS column b_1/b_0
η: I-beam depth to RHS column width ratio h_1/b_0
2γ: width to thickness ratio of RHS column b_0/t_0
t_1: thickness of I-beam flange

		Range of validity
Axially loaded uni-planar connection without web at the connection	Chord face yielding $$N_{1,u} = f(\beta, \eta)(0.5 + 0.7\beta) \cdot \frac{4}{\sqrt{1-\beta}} \cdot f_{y0} \cdot t_0^2$$ for $\eta < 2\sqrt{1-\beta}$: $f(\beta, \eta) = 1 + \frac{\eta}{2\sqrt{1-\beta}}$ for $\eta \geq 2\sqrt{1-\beta}$: $f(\beta, \eta) = 2$	$0.2 \leq \beta \leq 1.0$
	Chord side wall failure	$15 \leq 2\gamma \leq 30$
	for $h_1 \geq 2t_1 + 5t_0$: $N_{1,u} = 4(t_1 + 5t_0) f_{y0} \cdot t_0$ for $h_1 < 2t_1 + 5t_0$: $N_{1,u} = 2(h_1 + 5t_0) f_{y0} \cdot t_0$	$0.8 \leq \eta \leq 2.5$
	Crack after initial plastification of the column face $N_{1,u}(+C) = 2N_{1,u}^{**}(+C)$ where $N_{1,u}^{**}(+C) = f(c)(0.5 + 0.7\beta) \cdot \frac{4}{\sqrt{1-0.9\beta}} \cdot f_{y0} \cdot t_0^2$ $f(c)$ = effect function of concrete filling in column = 1.3	$0.4 \leq \beta \leq 0.6$ $2\gamma = 30$ $0.8 \leq \eta \leq 1.2$
Column filled up with concrete		
Axially loaded uni-planar connection with a web at the connection	Chord face yielding $$N_{1,u} = f_i(\beta, \eta)(0.5 + 0.7\beta) \cdot \frac{4}{\sqrt{1-\beta}} \cdot f_{y0} \cdot t_0^2 \quad i = 1, 2$$	
	Chord side wall failure	
	for $h_1 \geq 2t_1 + 5t_0$: $N_{1,u} = 4(t_1 + 5t_0) \cdot f_{y0} \cdot t_0$ for $h_1 < 2t_1 + 5t_0$: $N_{1,u} = 2(h_1 + 5t_0) f_{y0} \cdot t_0$	$0.2 \leq \beta \leq 1.0$ $15 \leq 2\gamma \leq 37.5$
	for $\eta \geq 0.5$: $f_1(\beta, \eta) = \left\{ \frac{1}{1.12(1-0.9\beta)} + \frac{\eta}{(0.8+2.4\beta)\sqrt{(1-0.9\beta)}} \right\} \cdot \{1 - (0.9\beta)^2\}$ for $\eta < 0.5$: (linear interpolation between 1 and $f_1(\beta, \eta)$ for $\eta = 0.5$) i. e. $f_2(\beta, \eta) = 1 + \frac{\eta}{0.5}\{[f_1(\beta, \eta)/\eta = 0.5] - 1\}$	$0.3 \leq \eta \leq 2.5$

	Chord face yielding	Range of validity
Axially loaded multi-planar connection with a web at the connection $J = N_2/N_1$	$N_{1,u} = F(J) \cdot N_{1,u}^*$ where $N_{1,u}^* = f(\beta, \eta)(0.5 + 0.7\beta) \cdot \dfrac{4}{\sqrt{1-\beta}} \cdot f_{y0} \cdot t_0^2$ for $\eta < 2\sqrt{1-\beta}$: $f(\beta, \eta) = 1 + \dfrac{\eta}{2\sqrt{1-\beta}}$ for $\eta \geq 2\sqrt{1-\beta}$: $f(\beta, \eta) = 2$ for $J \geq 0$: $f(J) = 1$ for $J < 0$: $f(J) = 1 + 0.37 J$	$0.2 \leq \beta \leq 0.75$ $15 \leq 2\gamma \leq 37.5$ $0.3 \leq \eta \leq 2.0$
Uni-planar connection loaded by in-plane bending moment (excluding pre-loading in the column) with a web at the connection	Chord face yielding $M_{1.ip,u} = (0.5 + 0.7\beta) \cdot \dfrac{4}{\sqrt{1-0.9\beta}} \cdot f_{y0} \cdot t_0 \cdot h_m$ where $h_m = h_1 - t_1$ (h_1 = depth of I beam; t_1 = thickness of I-beam flange)	$0.2 \leq \beta \leq 1.0$ $15 \leq 2\gamma \leq 37.5$
	Chord side failure for $h_1 \geq 2t_1 + 5t_0$, $M_{1.ip,u} = 2(t_1 + 5t_0) \cdot f_{y0} \, t_0 \cdot (h_1 - t_1)$ for $h_1 < 2t_1 + 5t_0$, $M_{1.ip,u} = 0.5(h_1 + 5t_0)^2 \cdot f_{y0} \, t_0$	$0.3 \leq \eta \leq 2.0$
	Cracking of the flange tips after initial yielding $M_{1.ip,u} = f(c)(0.5 + 0.7\beta) \cdot \dfrac{4}{\sqrt{1-0.9\beta}} \cdot f_{y0} \cdot t_0^2 \cdot h_m$ where $f(c) = 1.7$	$\beta = 0.4$ $2\gamma = 30$ $\eta = 0.8$
Column filled up with concrete		
Un-iplanar connection loaded by in-plane bending moment (including pre-loading in the column) with a web at the connection	Chord face yielding (+ local buckling under large pre-compression) $M_{1.ip,u}(n \neq 0) = f(n)(0.5 + 0.7\beta) \cdot \dfrac{4}{\sqrt{1-0.9\beta}} \cdot f_{y0} \cdot t_0^2 \cdot h_m$ for $n \geq 0$ (tension), $f(n) = 1$ for $n < 0$ (compression), $f(n) = 1 + 1.48\,(2\gamma)^{-0.33} \cdot n - 0.46\,(2\gamma)^{(0.33-0.1\beta^2)} \cdot n^{1.5}$ but $f(n) \leq 1$ $n = N_0/N_{0,pl}$ $N_{0,pl} = A_0 \cdot F_{y0}$	$0.2 \leq \beta \leq 1.0$ $15 \leq 2\gamma \leq 37.5$ $0.3 \leq \eta \leq 2.0$

Multi-planar connection loaded by in-plane bending moment (excluding pre-loading in the column) with a web at the connection 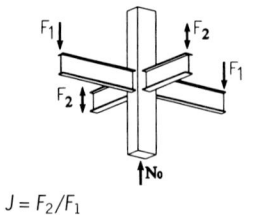 $J = F_2/F_1$	Chord face yielding $M_{1,ip,u} = f(J)(0.5 + 0.7\beta) \cdot \dfrac{4}{\sqrt{1 - 0.9\beta}} \cdot f_{y0} \cdot t_0 \cdot h_m$ for $J \geq 0 : f(J) = 1$ for $J < 0 : f(J) = 1 + 0.4 J$	Range of validity $0.2 \leq \beta \leq 0.75$ $15 \leq 2\gamma \leq 37.5$ $0.3 \leq \eta \leq 2.0$

New ultimate joint resistance formulae for uni-planar and multi-planar RHS X, XX and TX joints
(see [266], Chapter 6)

	Chord flange plastification		Validity range
a) Axially loaded uni-planar X joint (without chord axial pre-loading)	$N_{1,u} = 2 \dfrac{f_{y0} \cdot t_0^2}{(1-\beta)} \cdot \left(\eta + 2\sqrt{1-\beta}\right) \cdot f(\beta, \eta)$ $f(\beta, \eta) = 0.7 + 0.6\beta + 0.1\eta$	(VI 1) (VI 2)	$0.2 \leq \beta \leq 0.85$ $15 \leq 2\gamma \leq 35$ $0.5\beta \leq \eta \leq 2.0\beta$
	Chord side wall failure		
	$N_{1,u} = 4\chi\left(\sqrt{\gamma} + \gamma\eta\right) \cdot f_{y0}\, t_0^2$	(VI 3)	
	χ = according to EC 3 [3] curve "a"* using a column slenderness ratio $\sqrt{3}\left(\dfrac{h_0}{t_0} - 2\right)$, see also Eqs. (5-53) and (5-54) * see Fig. 5-4 $\lambda_K = \sqrt{3}\left(\dfrac{h_0}{t_0} - 2\right)$ $\lambda_E = \pi\sqrt{E/f_{y0}}$ $\bar{\lambda}_K = \dfrac{\lambda_K}{\lambda_E}$		$\beta = 1.0$ $15 \leq 2\gamma \leq 35$ $0.5\beta \leq \eta \leq 2.0\beta$
	Linear interpolation between (VI 1) and (VI 3) + check for brace effective width for chord punching shear according to Eqs. (6-80) and (6-81)		$0.85 < \beta < 1.0$
b) Axially loaded uni-planar X joint (with chord axial pre-loading)	**Chord flange plastification**		
	$N_{1,u}$ = Eq. (VI 1) $\times f(n)$	(VI 4)	$0.2 \leq \beta \leq 1.0$
	$f(n) = \left\{1 - n^{\frac{2}{(1+0.004\gamma^2)(1-0.85\beta^3)}}\right\}^{0.4}$		$15 \leq 2\gamma \leq 35$
	$n = \dfrac{N_0}{N_{0,Rd}} = \dfrac{N_0}{A_0 \cdot f_{y0}}$		$\beta = \eta$ $f_{y0} \leq 355\,\text{N/mm}^2$
c) Axially loaded uni-planar X joint (with chord bending moment)	**Chord flange plastification**		
	$N_{1,u}$ = Eq. (VI 1) $\times f(J_m)$	(VI 5)	
	Based on cross section classes 1, 2 or 3: $f(J_m) = \left\{1 - J_m^{\frac{1}{(1-0.85\beta^{1.6})}}\right\}^{1/3}$ where $J_m = \begin{cases} M_0/M_{pl,Rd} \text{ (class 1 or 2)} \\ M_0/M_{el,Rd} \text{ (class 3)} \end{cases}$ Based on the plastic moment capacity only: $f(J_m) = \left\{1 - \left(\dfrac{M_0}{M_{pl,Rd}}\right)^{\frac{2}{(1+0.004\gamma^2)(1-0.85\beta^3)}}\right\}^{0.4}$ Based on the elastic moment capacity only: $f(J_m) = 1 - 0.23(1 - 0.85\beta)\sqrt{\gamma}\left(\dfrac{M_0}{M_{el,Rd}}\right)^2$		$0.2 \leq \beta \leq 1.0$ $15 \leq 2\gamma \leq 35$ $\beta = \eta$ $f_{y0} \leq 355\,\text{N/mm}^2$

	Chord flange plastification	Validity range
Uni-planar X joint loaded with in-plane bending moment	$M_{1.ip.u} = \left(\dfrac{2}{\sqrt{1-\beta}} + \dfrac{\eta}{1-\beta} + \dfrac{1}{2\eta} \right) f_{y0} \cdot t_0^2 \cdot h_1 \cdot f(\beta, \eta)$ (VI 6) where $f(\beta, \eta) = 1 + 0.6\beta - 0.25\eta$	$0.2 \leq \beta \leq 0.85$ $15 \leq 2\gamma \leq 35$ $0.5\beta \leq \eta \leq 2.0\beta$
	Chord side wall failure	
	$M_{1.ip.u} = \chi \left(2\sqrt{\gamma} + \gamma\eta + \dfrac{1}{2\eta} \right) f_{y0} \cdot t_0^2 \cdot h_1$ (VI 7) where $\chi = \begin{cases} 1 \; (\eta \leq 1.0) \\ \dfrac{1}{\phi + \sqrt{\phi^2 + \bar{\lambda}_K^2}} \; (\eta = 2.0) \\ 1 + (\eta - 1)\left(\dfrac{1}{\phi + \sqrt{\phi^2 + \bar{\lambda}_K^2}} - 1 \right) (1.0 < \eta < 2.0) \end{cases}$ $\phi, \bar{\lambda}_K$: see Eq. (VI 3)	$\beta \leq 1.0$ $15 \leq 2\gamma \leq 35$ $0.5\beta \leq \eta \leq 2.0\beta$
	$M_{1.ip.u}$ = Linear interpolation between Eqs. (VI 6) and (VI 7) + check for brace effective width and for chord punching shear according to Eqs. (6-80) and (6-81)	$0.85 < \beta < 1.0$ $15 \leq 2\gamma \leq 35$ $0.5\beta \leq \eta \leq 2.0\beta$
Uni-planar X joint loaded with out-of-plane bending moment	Chord flange plastification	
	$M_{1.op.u} = f_{y0} \cdot t_0^2 \cdot b_1 \left(\dfrac{\eta(1+\beta)}{2\beta(1-\beta)} + \sqrt{\dfrac{2(1+\beta)}{\beta(1-\beta)}} \right) \cdot f(\beta, \eta)$ (VI 8) where $f(\beta, \eta) = 1 + 0.5\beta$	$0.2 \leq \beta \leq 0.85$ $15 \leq 2\gamma \leq 35$ $0.5\beta \leq \eta \leq 2.0\beta$
	Chord side wall failure	
	$M_{1.op.u} = \chi \cdot \left(\sqrt{2(1+2\gamma)} + 2\gamma\eta \right) \cdot f_{y0} \cdot t_0^2 \cdot b_1$ (VI 9) where χ: see Eq. (VI 3)	$\beta \leq 1.0$ $15 \leq 2\gamma \leq 35$ $0.5\beta \leq \eta \leq 2.0\beta$
	$M_{1.op.u}$ = Linear interpolation between Eqs. (VI 8) and (VI 9) + check for brace effective width and for chord punching shear according to Eqs. (6-80) and (6-81)	$0.85 < \beta < 1.0$ $15 \leq 2\gamma \leq 35$ $0.5\beta \leq \eta \leq 2.0\beta$
Axially loaded multi-planar XX joints	$N_{1.u}(J_{AA} = 0) = c_m \times$ Eq. (VI 1) (VI 10) $N_{1.u}(J_{AA}) = f(J_{AA}) \cdot N_{1.u}(J_{AA} = 0)$ (VI 11) where $J_{AA} = N_2/N_1$	
	$c_m = 1 + 0.4\beta^4$ (VI 12)	$0.2 \leq \beta \leq 1.0$ $15 \leq 2\gamma \leq 35$
	$f(J_{AA}) = 1 + 0.37 J_{AA} \; (J_{AA} \leq 0)$ (VI 13)	$0.2 \leq \beta \leq 0.85$
	$f(J_{AA}) = 1 + \left(0.03 + \dfrac{2\beta}{\gamma} + 0.3\beta^2 \right) J_{AA} \; (J_{AA} \geq 0)$ (VI 14)	$15 \leq 2\gamma \leq 35$
	$f(J_{AA}) = 1 - \dfrac{3}{\gamma} \cdot J_{AA}^2 \; (-1.0 \leq J_{AA} \leq 1.0)$ (VI 15)	$\beta \leq 1.0$ $15 \leq 2\gamma \leq 35$
	$f(J_{AA})$ = Linear interpolation between Eqs. (VI 13) and (VI 15) or between Eqs. (VI 14) and (VI 15)	$0.85 < \beta < 1.0$

Multi-planar XX joint loaded with in-plane bending moments	$M_{1,ip,u}(J_\parallel = 0) = c_m \times$ Eq. (VI 6)	(VI 16)	
	$M_{1,ip,u}(J_\parallel) = f(J_\parallel) \cdot M_{1,ip,u}(J_\parallel = 0)$	(VI 17)	
	where $J_\parallel = M_2/M_1$		
	$c_m = 1 + 0.4\,\beta^8$	(VI 18)	$0.2 \leq \beta \leq 1.0$ $15 \leq 2\gamma \leq 35$
	$f(J_\parallel) = 1 + 0.53\,\beta \cdot J_\parallel \quad (J_\parallel \leq 0)$	(VI 19)	$0.2 \leq \beta \leq 0.85$ $15 \leq 2\gamma \leq 35$
	$f(J_\parallel) = 1.0 \quad (J_\parallel \geq 0)$	(VI 20)	
	$f(J_\parallel) = 1 + 0.08\,J_\parallel - 0.18\,J_\parallel^2 \quad (-1 \leq J_\parallel \leq 1)$	(VI 21)	$\beta \leq 1.0$ $15 \leq 2\gamma \leq 35$
	$f(J_\parallel) =$ Linear interaction between Eqs. (VI 19) and (VI 21) or between Eqs. (VI 20) and (VI 21)		$0.85 < \beta < 1.0$
Multi-planar XX joint loaded with in-plane bending moment and axial force	$M_{1,ip,u}(J_{IA} = 0) = c_m \times$ Eq. (VI 6)	(VI 22)	
	$M_{1,ip,u}(J_{IA}) = f(J_{IA}) \cdot M_{1,ip,u}(J_{IA} = 0)$	(VI 23)	
	where		
	$J_{IA} = N_2 \cdot h_1/M_1$		
	$c_m = 1 + 0.4\,\beta^8$	(VI 24)	$0.2 \leq \beta \leq 1.0$ $15 \leq 2\gamma \leq 35$
	$f(J_{IA}) = 1 + 0.43(1-\beta) \cdot J_{IA} \quad (J_{IA} \leq 0$	(VI 25)	$0.2 \leq \beta \leq 1.0$ $15 \leq 2\gamma \leq 35$
	$f(J_{IA}) = 1 \quad (J_{IA} \geq 0)$	(VI 26)	
Axially loaded multi-planar TX joint excluding chord bending moment	$N_{1,u}(J_{AA} = 0) = c_m \times$ Eqn. (VI1)	(VI 27)	
	$N_{1,u}(J_{AA}) = f(J_{AA}) \cdot N_{1,u}(J_{AA} - 0)$	(VI 28)	
	where $J_{AA} = N_2/N_1$		
	$c_m = 1 + 0.4\,\beta^4$	(VI 29)	$0.2 \leq \beta \leq 1.0$ $15 \leq 2\gamma \leq 35$
	$f(J_{AA}) = 1 + 0.37\,J_{AA} \quad (J_{AA} \leq 0)$	(VI 30)	$\beta > 0.6$
	$f(J_{AA}) = 1.0 \quad (J_{AA} \geq 0)$	(VI 31)	
Moment M_0 is to compensate the moment in the chord due to $N_1 : N_1 \cdot (l_0 - h_1)/4$			

References

References to Chapter 2

[1] EN 10002-1: Metallische Werkstoffe – Zugversuch, Teil 1: Prüfverfahren (bei Raumtemperatur)
[2] EN 10045-1: Metallische Werkstoffe – Kerbschlagbiegeversuch nach Charpy, Teil 1: Prüfverfahren

References to Chapter 3

[1] ISO 630: Structural steels. International Organization of Standardization
[2] DIN 17100: Allgemeine Baustähle, Gütenorm
[3] DIN 17119: Kaltgefertigte, geschweißte, quadratische und rechteckige Stahlrohre (Hohlprofile) für den Stahlbau, Technische Lieferbedingungen
[4] DIN 17120: Geschweißte kreisförmige Rohre aus allgemeinen Baustählen für den Stahlbau, Technische Lieferbedingungen
[5] DIN 17121: Nahtlose kreisförmige Rohre aus allgemeinen Baustählen für den Stahlbau, Technische Lieferbedingungen
[6] BS 4360: Weldable structural steels. British Standards Institution
[7] NF 49-501: Profils creux finis a chaud pour construction. Norme Française
[8] NF 49-541: Profils creux finis a froid pour construction. Norme Française
[9] UNI 7806: Prodotti finiti di acciaio formati a caldo per construzioni metalliche, Profilati cavi, Qualita, prescrizioni e prove. Norma Italiana
[10] UNI 7810: Prodotti finiti di acciaio formati a freddo per construzioni metalliche. Profilati cavi, Qualita, prescrizioni e prove. Norma Italiana

American Society for Testing and Materials ASTM:

[11] A 36: Specification for structural steel
[12] A 441: Specification for high-strength low-alloy structural Manganese-Vanadium steel
[13] A 500: Specification for cold-formed welded and seamless carbon steel structural tubing in rounds and shapes
[14] A 501: Specification for hot-formed welded and seamless carbon steel structural tubing
[15] A 529: Specification for standard steel with 42 KSi minimum yield point (1/2 in. [13 mm] maximum thickness)
[16] A 572: Specification for high-strength low-alloy columbium-Vanadium steels of structural quality
[17] A 588: Specification for high-strength low-alloy structural steel with 50 KSi minimum yield point to 4 in. (100 mm) thick
[18] A 618: Grade II and III (Grade I if the properties are suitable for welding), Specification for hot-formed welded and seamless high-strength low-alloy structural tubing
[19] A 633: Specification for normalized high-strength low-alloy structural steel
[20] JISG 3444: Carbon steel tubes for general structural purposes. Ministry of International Trade and Industry, Japan
[21] JISG 3466: Carbon steel square pipes for general structural purposes. Ministry of International Trade and Industry, Japan
[22] EN 10210-1: Hot finished structural hollow sections of non-alloy and fine grain structural steels – Part 1: Technical delivery conditions, European Standard, 1994
[23] EN 10219-1: Cold formed welded structural hollow sections of non-alloy and fine grain structural steels – Part 1: Technical delivery conditions, European Standard, 1997
[24] DASt-Ri 011: Richtlinie 011 des Deutschen Ausschusses für Stahlbau, Hochfeste schweißgeeignete Feinkornbaustähle St E 460 und St E 690, Anwendung für Stahlbauten, Februar 1988
[25] DASt-Ri 007: Richtlinie 007 des Deutschen Ausschusses für Stahlbau, Richtlinien für die Lieferung, Verarbeitung und Anwendung wetterfester Baustähle
[26] EN 10155: Weathering steels. Eurostandard, CEN
[27] ENV 1993-1-1, Eurocode 3: Design of steel structures, part 1.1: General rules and rules for buildings, February 1992, CEN (European Committee of Standardisation)
[28] DASt-Ri 009: Richtlinie 009 des Deutschen Ausschusses für Stahlbau, Empfehlungen zur Wahl der Stahlgütegruppen für geschweißte Stahlbauten
[29] Bathke, W.: Schweißen in kaltverformten Bereichen an Winkelproben mit 6 mm Schenkeldicke. In: Schweißen und Schneiden 37 (1985), Heft 11

[30] Mang, F.; Bucak, Ö.; Herion, S.: Fatigue behaviour of hollow section joints of hight-tensile steel. Final report, CIDECT programme 7L, March 1997

References to Chapter 4

[1] EN 10210-2: Hot finished structural hollow sections of non-alloy and fine grain structural steels, Part 2: Tolerances, dimensions and sectional properties. European Standard, CEN, 1997
[2] EN 10219-2: Cold formed welded structural hollow sections of non-alloy and fine grain structural steels, Part 2: Tolerances, dimensions and sectional properties. European Standard, CEN, 1997
[3] ISO 657/14: Hot rolled steel sections, Part 14: Hot formed structural hollow sections – Dimensions and sectional properties. International Organization for Standardization
[4] ISO 4019: Cold finished steel structural hollow sections – Dimensions and sectional properties. International Organization for Standardization
[5] CAN/CSA-G312.3-M92: Metric dimensions of structural steel shapes and hollow structural sections. Canadian Standards Association, Toronto, Canada
[6] A 500: Standard specifications for cold formed welded and seamless carbon steel structural tubing in rounds and shapes. American Society for Testing and Materials
[7] A 501: Specification for hot formed welded and seamless carbon steel structural tubing. American Society for Testing and Materials
[8] A 618: Specification for hot formed welded and seamless low-alloy structural tubing. American Society for Testing and Materials

References to Chapter 5

[1] DIN 18800: Stahlbauten, Teil 1: Bemessung und Konstruktion, November 1990; Teil 2: Stabilitätsfälle, Knicken von Stäben und Stabwerken, November 1990; Teil 3: Stabilitätsfälle, Plattenbeulen, November 1990; Teil 4: Stabilitätsfälle, Schalenbeulen, November 1990
[2] ENV 1993-1-1, Eurocode 3: Design for steel structures, part 1.1: General rules and rules for buildings, February 1992, CEN (European Committee for Standardization)
[3] ENV 1991, Eurocode 1: Basis of design and actions on structures
[4] ISO 2394: General principles for the verifications of the safety of structures. International Organization of Standardization
[5] DIN: Grundlagen zur Festlegung von Sicherheitsforderungen für bauliche Anlagen, Beuth Verlag, Berlin
[6] DIN 18808: Stahlbauten, Tragwerke aus Hohlprofilen unter vorwiegend ruhender Beanspruchung, Oktober 1984
[7] Anpassungsrichtlinie Teil 2 zu DIN 18 800 – Stahlbauten – Teile 1–4/11.90, Abschnitt 4.4: DIN 18 808/10.84 – Stahlbauten, Tragwerke aus Hohlprofilen unter vorwiegend ruhender Beanspruchung, Mitteilungen des Deutschen Instituts für Bautechnik, 1994, Heft 4, S. 132–133
[8] Dutta, D.; Würker, K.-G.: Handbuch Hohlprofile in Stahlkonstruktionen, Verlag TÜV Rheinland GmbH, Köln, 1988
[9] Rondal, J.; Würker, K.-G.; Dutta, D.; Wardenier, J.; Yeomans, N.: Structural stability of hollow sections, CIDECT series "Construction with hollow steel sections", Verlag TÜV Rheinland, Cologne 1996
[10] Sfintesco, D.: Fondement experimental des courbes europeenes de flambement, Constructions métallique, Nr. 3, Paris, 1970
[11] Braham, M.; Grimault, J. P.; Rondal, J.: Flambement de profils creux á minces – Cas des Profils rectangulaires chargés axialement, Schlußbericht EGKS-Nr. 6210/SA/3/301-1981-EUR 6730 Fr
[12] Braham, M.; Grimault, J. P.; Massonet, C.; Mouty, J.; Rondal, J.: Das Knickverhalten dünnwandiger Hohlprofile, acier-Stahl-steel, 1/1980
[13] Beer, H.; Schulz, G.: Die Traglast des mittig gedrückten Stabes mit Imperfektionen, VDI-Zeitschrift, Bd. 111 (1969), Nr. 21, 23, 24
[14] Beer, H.: Aktuelle Probleme der Stabilitätsforschung, Sonderdruck 1971, Österreichischer Stahlbauverband, Wien
[15] SSRC: Stability of metal structures – a world view. Structural Stability Research Council, 2nd edition, 1991
[16] CIDECT: Buckling behaviour of hot finished SHS in high strength steel S460. Research project no. 2T
[17] DIN 4114: Stahlbau; Stabilitätsfälle (Knickung, Kippung, Beulung), Teil 1: Berechnungsgrundlagen, Vorschriften, Juli 1952; Teil 2: Berechnungsgrundlagen, Richtlinien, Februar 1953
[18] Klöppel, K.; Scheer, J.; Klöppel, K.; Möhler, J.: Beulwerte ausgesteifter Rechteckplatten, Band I und II, Verlag Wilhelm Ernst und Sohn, Berlin, München
[19] DASt-Ri 012: Beulsicherheitsnachweise für Platten, Deutscher Ausschuß für Stahlbau, Oktober 1978
[20] v. Karman, Th.; Tsien, H. S.: The buckling of thin cylindrical shells under axial compression. Journal of Aeronautical Science, 1941
[21] SIA 161: Stahlbauten, Norm Ausgabe 1979, Schweizerischer Ingenieur- und Architekten-Verein

[22] AISI: Specification for the design of cold formed steel structural members, 2nd edition, 1968–1973, Deutsche Übersetzung herausgegeben von der Beratungsstelle für Stahlverwendung, Düsseldorf, 1976
[23] BS 5950: Structural use of steelwork in building, Part 1. Code of practice for design in simple and continuous construction: hot rolled sections. British Standards Institution, London, 1985
[24] ECCS: European recommendations for steel structures. ECCS-CECM-EKS 77-2E, March 1978
[25] Lorenz, R.: Achsensymmetrische Verzerrungen in dünnwandigen Hohlzylindern, VDI-Zeitschrift 52 (1908), Heft 43, S. 1706
[26] Timoshenko, S.: Einige Elastizitätsprobleme aus der Elastizitätstheorie, Zeitschrift für Mathematik und Physik 58 (1910), H. 4, S. 378
[27] Robertson, A.: The strength of tubular struts. Proceedings of Royal Society, London, A 121 (1928), S. 553
[28] Wilson, W.; Newmark, N.: The strength of thin cylindrical shells as columns. Engineering Experiment Station, University of Illinois, Bulletin 255 (1933) und Bulletin (1941)
[29] Donnel, L. H.: A new theory for the buckling of thin cylinders under axial compression and bending. Transactions ASME 56 (1934), S. 795
[30] Kappus, R.: Druck-, Biege- und Torsionsversuche mit Holmrohren aus Stahl, Aero 70, Jahrbuch der Deutschen Luftfahrtforschung (1939), S. 173
[31] Donnel, L. H.; Wan: Effect of imperfections on buckling of thin cylinders and columns under axial compression. Journal of Applied Mechanics 17 (1950)
[32] Lindenberger, H.: Bericht über Druckversuche an Kreiszylindern (Fortschritte im Stahlbrückenbau), Stahlbau-Verlag, Köln, 1958
[33] Almroth, B. O.; Holmes, A. M. C.; Brush, D. O.: An experimental study of the buckling of cylinders under axial compression. Experimental Mechanics, September 1964, USA
[34] Steinhardt, O.; Schulz, V.: Zum Beulverhalten von Kreiszylinderschalen, Schweizerische Bauzeitung 89 (1971), Heft 1
[35] Plantema, J.: Collapsing stresses of circular cylinders and round tubes. Report 5280, Nat. Luchtvaartlaboratorium, Amsterdam, 1946
[36] DNV Det Norske Veritas: Rules for the design, construction and inspection of offshore structures. 2nd edition, Oslo, 1977
[37] DASt-Ri 013: Beulsicherheitsnachweise für Schalen, Deutscher Ausschuß für Stahlbau, Juli 1980
[38] ECCS-EKS: European recommendations for steel construction – Section 4.6 Buckling of shells. European Convention for Constructional Steelwork, 1981
[39] Roik, K.; Kindmann, R.: Das Ersatzstabverfahren – eine Nachweisform für den einfeldrigen Stab bei planmäßig einachsiger Biegung mit Druckkraft, Der Stahlbau 12, 1981
[40] Roik, K.; Kindmann, R.: Das Ersatzstabverfahren – Tragsicherheitsnachweise für Stabwerke bei einachsiger Biegung und Normalkraft, Der Stahlbau 5, 1982
[41] Rondal, J.; Maquoi, R.: Formulations d'Ayrton-Perry pour le flambement des barres métalliques, Construction metallique, Nr. 4, Paris, 1979
[42] Roik, K.; Kuhlmann, U.: Beitrag zur Bemessung von Stäben für zweiachsige Biegung mit Druckkraft, Der Stahlbau 9, 1985
[43] DIN 18808: Stahlbauten, Tragwerke aus Hohlprofilen unter vorwiegend ruhender Beanspruchung, Normenausschuß Bauwesen (NA Bau) im DIN Deutsches Institut für Normung, Beuth Verlag, Berlin, Oktober 1984
[44] API: API Recommended practice for planning, designing and constructing fixed offshore platforms. American Petroleum Institute, API-RP-2A, January 1980
[45] DNV: Rules for the design, construction and inspection of offshore structures. Det Norske Veritas, 1977
[46] de Jong, H.: The effect of the joint rigidity on the buckling behaviour of a member in compression. Technical University Delft, Stevinreport 6-86-6, September 1986
[47] Mouty, J.: Knicklängen-Ermittlung der Stäbe von Fachwerkträgern, CIDECT, Monografie 4, Paris, 1980
[48] Grimault, J. P.: Longueurs de flambement de treillis en profils creux sur membrures eu profils creux, programmes 3E et 3G, rapport CIDECT, mars 1977
[49] Mouty, J.: Calcul des longueurs de flambement par la méthode de rotules élastiques, Chambre Syndicale des Fabricants de Tubes d'Acier, Notice 1074, Paris, 1978
[50] Rondal, J.: Effective lengths of tubular lattice girder members, statistical tests. CIDECT-report 3K-88/9, University of Liége, August 1988
[51] Spira, E.; Pollner, E.: Le problème du flambage des poutres tubulaires soudées en treillis, Acier-Stahl-Steel Nr. 11, Brüssel, 1968
[52] Baar, S.: Étude théorique et expérimentale du déversement des poutres à membrures tubulaires, Collection des publications de la Faculté des Sciences Appliquées de Université de Liége, N° 10, 1968
[53] Kollbrunner, C. F.; Basler, K.: Torsion, Springer-Verlag, Berlin, 1966
[54] Timoshenko, S.: Theory of elasticity, 2nd edition. McGraw-Hill, New York, 1951

References to Chapter 6

[1] Kato, B.; Nishiyama, I.: Behaviour of rigid frame connections subjected to horizontal force. CIDECT programme 5Z, Final report, University of Tokyo, 1979

[2] DIN 4114: Stahlbau: Stabilitätsfälle (Knickung, Kippung, Beulung), Teil 1: Berechnungsgrundlagen, Vorschriften, Juli 1952; Teil 2: Berechnungsgrundlagen, Richtlinien, Februar 1953 (am 1.1.1997 ausgelaufen)

[3] ENV 1993-1-1, Eurocode 3: Design of steel structures, part 1.1: General rules and rules for buildings. February 1992, CEN (European Committee of Standardisation)

[4] Mang, F.; Steidl, G.; Bucak, Ö.: Design of welded lattice joints and moment resisting knee joints made of hollow sections. Doc.No. XV-463-80, International Institute of Welding, University of Karlsruhe, Germany, 1980

[5] DIN 18808: Stahlbauten, Tragwerke aus Hohlprofilen unter vorwiegend ruhender Beanspruchung, Deutsches Institut für Normung, Berlin, Deutschland, Oktober 1984

[6] Packer, J. A.; Wardenier, J.; Kurobane, Y.; Dutta, D.; Yeomans, N.: Design guide for rectangular hollow section (RHS) joints under predominantly static loading. CIDECT-series "Construction with hollow steel sections", published by TÜV Rheinland, Cologne, Germany, 1996

[7] Wardenier, J.: Hollow section joints. Delft University Press, Delft, The Netherlands, 1982

[8] Packer, J. A.; Hendersen, J. E.: Design guide for hollow structural section connections. Canadian Institute of Steel Construction, Ontario, Canada, 1992

[9] Dutta, D.; Würker, K.-G.: Handbuch Hohlprofile in Stahlkonstruktionen, Verlag TÜV Rheinland, Köln, 1988

[10] DIN 18800: Stahlbauten, Teil 1: Bemessung und Konstruktion, November 1990; Teil 2: Stabilitätsfälle, Knicken von Stäben und Stabwerken, November 1990; Teil 3: Stabilitätsfälle, Plattenbeulen, November 1990; Teil 4: Stabilitätsfälle, Schalenbeulen, November 1990

[11] CIDECT-Project 6E: L-type joint of circular hollow sections. Final report, Testing centre for steel, timber and stone, University of Karlsruhe (TH), August 1995

[12] Mang, F.; Dutta, D.: Static strength of plate connections to circular and rectangular hollow section. CIDECT-Programme 5AM, Final report, Testing centre for steel, timber and stone, University of Karlsruhe (TH), 1989

[13] AWS 1992: Structural welding code-steel. ANSI/AWS D 1.1-92, American Welding Society

[14] Rockey, K. C.; Griffiths, D. W.: The behaviour of bolted flanged joints in tension. University of Cardiff, 1970

[15] Rockey, K. C.; Griffiths, D. W.: The effect of initial bolt tension on the performance of bolted flanged joints in tension. University of Cardiff, 1971

[16] Griffiths, D. W.; Rockey, K. C.: The behaviour of ring flanges in bolted joints in tension. University of Cardiff, 1973

[17] Griffiths, D. W.: The behaviour of blank flanges in bolted joints in tension. University of Cardiff, 1973

[18] Griffiths, D. W.: Tensile test results of the flanges of bolted joints in tubular members. University of Cardiff, 1974

[19] Timoshenko, S.: Theory of plates and shells, 2nd edition. McGraw-Hill Book Company, 1959

[20] DASt-Ri 010: Anwendung hochfester Schrauben im Stahlbau, Deutscher Ausschuß für Stahlbau, 1974

[21] CSA 1989: Limit states design of steel structures. CAN/CSA-S 16.1-M89, Canadian Standards Association, Rexdale, Ontario

[22] Kato, B.; Hirose, A.: Bolted tension flanges joining circular hollow section members. CIDECT report 8C-84/24-E

[23] Igarashi, S.; Wakiyama, K.; Inoue, K.; Matsumoto, T.; Murase, Y.: Limit design of high strength tube flange joint, Part 1 and 2. Journal of Structural and Construction Engineering Transactions of AIJ, Department of Architecture reports, Osaka University, Japan, August 1985

[24] AIJ: Recommendations for the design and fabrication of tubular structures in steel, 3rd edition. Architectural Institute of Japan, 1990

[25] Mang, F.: Untersuchungen an geschraubten Stirnplatten – Regelanschlüssen für Rechteck- und Rundhohlprofile, Projekt 38, Studiengesellschaft für Anwendungstechnik von Eisen und Stahl e.V., Düsseldorf, Mai 1981

[26] Kato, B.; Mukai, A.: Bolted tension flanges joining square hollow section members. Final report, CIDECT programme 8B, March 1982

[27] Packer, J. A.; Bruno, L.; Birkemoe, P. C.: Limit analysis of bolted RHS flange plate joints. Journal of Structural Engineering, American Society of Civil Engineers, Vol. 115, No. 9, September 1989, pp. 2226–2242

[28] Birkemoe, P. C.; Packer, J. A.: Ultimate strength design of bolted tubular tension connections. Conference on Steel Structures – Recent Research Advances and their Applications to Design, Budva, Yugoslavia, 1986, Proceedings, pp. 153–168

[29] Struik, J. H. A.; de Back, J.: Tests on bolted T-Stubs with respect to a bolted beam-to-column

connection. Stevin Laboratory report 6-69-13, Delft University of Technology, The Netherlands, 1969
[30] Bouwmann, L. P.: Fatigue of bolted connections and bolts in tension. Stevin Laboratory report 6-79-9, Delft University of Technology, The Netherlands, 1979
[31] Kurobane, Y.: Design of plate-to-tube joints. International Institute of Welding, IIW- Doc. XV-E-90-158
[32] Mang, F.; Bucak, Ö.; Karcher, D.: Load-bearing behaviour of hollow sections with inserted plates. Proceedings of 5th International Symposium on Tubular Structures, Nottingham, United Kingdom 1993, E & FN Spon, London
[33] Lindner, J.; Bamm, D.; Müller W.: Geschraubte Anschlüsse an Hohlprofilen, Teil 1: Bestimmung der Lochleibungstragfähigkeit; Teil 2: Querkraftbeanspruchte Verbindungen, Schlußbericht zum DFG Forschungsvorhaben Li 351/5-1, Bericht VR 2101, Technische Universität Berlin, August 1990
[34] Lindner, J.: Bolted connections to hollow sections with through bolts. Proceedings of the fifth International Symposium on Tubular Structures, Nottingham, United Kingdom, 1993, E & FN Spon, London
[35] Scheer, J.; Maier, W.; Klarhold, M.; Vajem, K.: Bestimmung der reinen Lochleibungsfestigkeiten und des Lochleibungspressungs-Verformungsverhaltens, Bericht Nr. 6066, Institut für Stahlbau, Technische Universität Braunschweig, 1985
[36] Scheer, J.; Maier, W.; Klarhold, M.; Vajem, K.: Zur Lochleibungsbeanspruchung in Schraubenverbindungen, Der Stahlbau 56 (1987), S. 129–136
[37] Mang, F.: Geschraubte Hohlprofil-Laschen-Verbindungen, Abschlußbericht, CIDECT-Programm 6E, Versuchsanstalt für Stahl, Holz und Steine, Universität Karlsruhe (TH), August 1993
[38] ENV 1090-4: Execution of steel structures, part 4: supplementary rules for hollow section lattice structure. CEN, May 1995
[39] DIN 2559: Schweißnahtvorbereitung, Teil 1: Richtlinien für Fugenformen, Schmelzschweißen von Stumpfstößen an Stahlrohren, Ausgabe Mai 1973; Teil 2: Anpassen der Innendurchmesser für Rundnähte an nahtlosen Rohren, Ausgabe Februar 1984
[40] DIN 8551, Teil 1: Schweißnahtvorbereitung, Fugenformen an Stahl, Gasschweißen, Lichtbogenhandschweißen und Schutzgasschweißen, Ausgabe Juni 1976
[41] Dutta, D.; Wardenier, J.; Yeomans, N.; Sakae, K.; Bucak, Ö.; Packer, J. A.: Design guide for fabrication, assembly and erection of hollow section structures. CIDECT-Series "Construction with hollow steel sections", Verlag TÜV Rheinland, Cologne, 1997
[42] CSA: Welded steel construction (metal arc welding). W 59-M 1989, Canadian Standards Association, 1989
[43] AS 4100: Steel structures. Standards Australia, 1990
[44] Cramer, K.: Einiges zum Messen von Stumpf- und Kehlnähten, Der Praktiker 10 (1980)
[45] DIN EN 10025: Warmgewalzte Erzeugnisse aus unlegierten Baustählen, Technische Lieferbedingungen, 1990
[46] prEN 10113, Part 1–3: Hot-rolled products in weldable fine grain structural steels. 1990
[47] EN 10210-1: Hot finished structural hollow sections of non-alloy and fine grain structural steels, Part 1: Technical delivery conditions. European Standard, 1994
[48] EN 10219-1: Cold formed welded structural hollow sections of non-alloy and fine grain structural steels, Part 1: Technical delivery conditions. European Standard, 1997
[49] IIW 1989: Design recommendations for hollow section joints – predominantly statically loading. Doc. XV-701–89, International Institute of Welding
[50] Jamm, W.: Gestaltfestigkeit geschweißter Rohrverbindungen und Rohrkonstruktionen unter statischer Belastung, Schweißen und Schneiden, Jahrgang 3 (1951), Sonderheft
[51] Toprac, A. A.; Natarajan, M.: Studies of tubular joints in Japan, Part I. Report no. R682, Technology Laboratories, Austin, Texas, prepared for Welding Research Council, New York, September 1968
[52] Togo, T.: Experimental study on mechanical behaviour of tubular joints (in Japanese). Doctorial dissertation, Osaka University, January 1967
[53] Kurobane, Y.: New developments and practices in tubular joint design. IIW-Doc. XV-488-81 + Addendum
[54] Kurobane, Y.: Welded truss joints of tubular structural members. Memoirs of the Faculty of Engineering, Kumamoto University, Vol. XII, No. 2, December 1964
[55] Kanatani, H.: Experimental study on welded tubular connections. Memoirs of the Faculty of Engineering, Kobe University, No. 12, 1966
[56] Naka, T.; Kato, B.; Kanatani, H.: Experimental study on welded tubular connections. Research Institute of Welding, University of Tokyo, 1964
[57] Washio, K.; Togo, T.; Mitsui, N.: Experimental study on local failure of chords in tubular truss joints, Part I+II. Technology reports of Osaka University, Vol. 18, No. 850 (1968) and Vol. 19, No. 874 (1969)

[58] Washio, K.; Togo, T.; Mitsui, N.: Cross joints of tubular members (in Japanese). Report Kinki Branch of AIJ, May 1966

[59] Kurobane, Y.; Makino, Y.; Honda, T.; Mitsui, Y.: Additional tests on tubular K-joints with CHS members under static loads. IIW-Doc. XV-460-80

[60] Kurobane, Y.; Makino, Y.; Mitsui, Y.: Re-analysis of ultimate strength data for truss connections in circular hollow sections. IIW-Doc. XV-461-80

[61] Kurobane, Y.; Makino, Y.; Mitsui, Y.: Ultimate strength formulae for simple tubular joints. IIW-Doc. XV-385-76, Department of Architecture, Kumamoto University, May 1976

[62] Makino, Y.; Kurobane, Y.; Minoda, Y.: Design of CHS X- and T-joints under tensile brace loading. IIW-Doc. XV-487-81

[63] Kurobane, Y.; Natarajan, M.; Toprac, A. A.: Studies on tubular joints in Japan, Part II. Specifications on tubular steel-structures and list of literature, June 1968

[64] Marshall, P. W.; Toprac, A. A.: Basis for tubular design. ASCE preprint 2008, April 1973, also Welding Journal, May 1974

[65] Marshall, P. W.: Design of simple tubular joints. CDG Report 12, Shell Oil Co., January 1967

[66] Reber, J. B.: Ultimate strength design of tubular joints. OTC 1964, Proceedings of Offshore Technology Conference, 1972, Houston, Texas

[67] Pan, B. P.; Plummer, F. B.; Kuang, J. G.: Ultimate strength design of tubular joints. OTC 2644 (1976)

[68] Lee, M. S.; Cheng, M. P.: Plastic consideration on punching shear strength of tubular joints. OTC 2641 (1976)

[69] Marshall, P. W.: A review of American criteria for tubular structures and proposed revisions. IIW-Doc. XV-405-77

[70] Graff, W. F.; Marshall, P. W.; Minas, A. N.: Review of design considerations for tubular joints. ASCE, Int. Conv. and Exp., New York, May 1981

[71] Beale, L. A.; Toprac, A. A.: Analysis of in-plane, T, Y and K welded tubular connections. Bulletin no. 125, Welding Research Council, October 1967

[72] Bouwkamp, J. G.: Behaviour of tubular truss joints under static loads. Structure and Material Research, Report no. SESM-65-4, College of Engineering, University of California, Berkeley, July 1965

[73] Beale, L. A.; Toprac, A. A.; Natarajan, M.: Experiment in tubular joints: elastic stresses. IIW-Doc. XV-215-66, 1966

[74] Bouwkamp, J. G.: Research on tubular connections in structural work. WRC-Bul. 71, 1961

[75] Bouwkamp, J. G.: Concept on tubular joints design. Proceedings of ASCE, Vol. 90, No. ST2, 1964

[76] Dundrova, V.: Stress and strain investigations of a cylindrical shell loaded along a curve. Structures fatigue research lab., Dept. of Civil Engineering, Univ. of Texas, Austin, SFRL, Techn. Report, pp. 550–4, July 1965

[77] Dundrova, V.: Stresses at intersection of tubes cross and T-joints. Structures fatigue research lab., Techn. Report P550-5, Univ. of Texas, Austin, 1965

[78] Bijlaard, P. P.: Stresses from radial loads and external moments in cylindrical pressure vessels. The Welding Journal 34, Research Suppl., pp. 608–617, 1955

[79] Bijlaard, P. P.: Additional data on stresses in cylindrical shells under local loading. Welding Research Council, Bul.-no. 50, pp. 10–50, May 1959

[80] Brown, R. C.; Toprac, A. A.: An experimental investigation of tubular T-joints. Structures fatigue research lab., Report no. P550-8, Univ. of Texas, Austin, Texas 1966

[81] Andrian, I. E.; Sewell, K. A.; Womack, W. R.: Partial investigation of directly loaded pipe T-joints. Thesis for the Civil Engineering Dept., Southern Methodist Univ., Dallas, 1958

[82] Boone, T. J.; Yura, J. A.; Hoadley, P. W.: Chord stress effects on the ultimate strength of tubular joints. PMFSEL, Report no. 82-1, Univ. of Texas, Austin, 1982

[83] Brown and Root Company: An investigation of welded tubular joints loaded by axial and moment loads. Offshore Structure Dept., Job no. ER-0169, Houston, Texas, 1976

[84] Grigory, S. C.: Experimental determination of the ultimate strength of tubular joints. Southwest Research Institute, Project no. 03-3054, Report to Humble Oil and Refining Co., San Antonio, 1971

[85] Healy, B. E.; Zettlemoyer, N.: In-plane bending strength of circular tubular joints. Proceedings of 5th International Symposium on Tubular Structures, Nottingham, U.K., pp. 325–344, 1993

[86] Hoadley, P. W.; Yura, J. A.: Ultimate strength of tubular joints subjected to combined loads. PMFSEL, Report no. 83-3, Univ. of Texas, Austin, 1983

[87] Kurobane, Y.; Makino, Y.; Ogawa, K.; Maruyama, Y.: Capacity of CHS T-joints under combined OPB and axial loads and its interactions with frame behaviour. Proceedings of 4th International Symposium on Tubular Structures, Delft, The Netherlands, pp. 412–423, 1991

[88] Makino, Y.; Kurobane, Y.; Ochi, K.: Reliability-based design criteria for tubular X-joints under tension. Summary papers, Annual Conference of Architectural Institute of Japan, pp. 1781–1782, 1982

[89] Makino, Y.; Kurobane, Y.; Takizawa, S.; Yamamoto, N.: Behaviour of tubular T- and K-joints under combined loads. Proceedings of Offshore Technology Conference, Paper no. 5133, pp. 429–438, 1986

[90] Makino, Y.; Kurobane, Y.; Ochi, K.: Ultimate capacity of tubular double K-joints. Proceedings of International Conference on Welding of Tubular Structures, Boston, pp. 451–458, 1984

[91] Makino, Y.; Kurobane, Y.; Paul, J. C.; Orita, Y.; Hiraishi, K.: Ultimate capacity of gusset plate-to-tube joints under axial and in-plane bending loads. Proceedings of 4th International Symposium on Tubular Structures, Delft, pp. 424–434, 1991

[92] Makino, Y.; Kurobane, Y.; Paul, J. C.: Ultimate behaviour of diaphragm stiffened tubular KK-joints. Proceedings of 5th International Symposium on Tubular Structures, Nottingham, U.K., pp. 465–472, 1993

[93] Makino, Y.; Kurobane, Y.; Paul, J. C.: Further tests on unstiffened tubular KK-joints. Proceedings of 5th International Conference on Steel Structures, Jakarta, Indonesia, pp. 183–190, 1994

[94] Makino, Y.; Kurobane, Y.: Tests on CHS KK-joints under anti-symmetrical loads. Proceedings of 6th International Symposium on tubular Structures, Melbourne, Australia, pp. 446–449, 1994

[95] Makino, Y.; Kurobane, Y.; Wilmhurst, S. R.; Lee, M. M. K.: Proposed ultimate capacity equations for CHS KK-joints under anti-symmetrical loads. Proceedings of 5th International Offshore and Polar Engineering Conference, The Hague, The Netherlands, Vol. IV, pp. 6–11, 1995

[96] Ochi, K.; Kurobane, Y.; Makino, Y.: Further tests on CHS K-joints with various d/D ratios. Technical reports, Vol. 30, No. 3, Kumamoto University, Kumamoto, Japan, pp. 119–124, 1981

[97] Ohtake, A.; Sakamoto, S.; Tanaka, T.; Kai, T.; Nakazato, T.; Takizawa, T.: Static and fatigue strength of steel tubular joints for offshore structures. Proceedings of Offshore Technology Conference, Paper no. 3254, pp. 1747–1755, 1978

[98] Paul, J. C.; Ueno, T.; Makino, Y.; Kurobane, Y.: The ultimate behaviour of circular multiplanar TT-joints. Proceedings of 4th International Symposium on Tubular Structures, Delft, pp. 448–460, 1991

[99] Paul, J. C.; Ueno, T.; Makino, Y.; Kurobane, Y.: Ultimate behaviour of multiplanar double K-joints of circular hollow section members. Proceedings of 2nd International Offshore and Polar Engineering Conference, San Francisco, pp. 377–383, 1992

[100] Paul, J. C.; Makino, Y.; Kurobane, Y.: Ultimate resistance of tubular double T-joints under axial brace loading. Journal of Constructional Steel Research, Vol. 24, No. 3, pp. 205–228, 1993

[101] Paul, J. C.; Ueno, T.; Makino, Y.; Kurobane, Y.: Ultimate behaviour of multiplanar double K-joints. International Journal of Offshore and Polar Engineering, Vol. 3, No. 1, pp. 43–50, 1993

[102] Sanders, D. H.; Yura, J. A.: Ultimate strength and behaviour of double-tee tubular joints in tension. PMFSEL, Report 86-2, Univ. of Texas, Austin, 1989

[103] Takizawa, S.; Yamamoto, N.; Mihara, J.; Ohkata, S.: Full scale experiments of tee and cross type tubular joints under static and cyclic loading. Technical reports, Kawasaki Steel Technical Review, Vol. 11, No. 2, Kawasaki Steel Corp., pp. 115–125, 1979

[104] Toprac, A. A.; Natarajan, M.; Erzurumulu, H.; Kanoo, A. L. J.: Research in tubular joints static and fatigue loads. Proceedings of Offshore Technology Conference, Paper no. 1062, 1969

[105] Yamasaki, T.; Takizawa, S.; Komatsu, M.: Static and fatigue tests on large-size tubular T-joints. Proceedings of Offshore Technology Conference, Paper no. 3424, pp. 583–591, 1979

[106] Yonemura, H.; Makino, Y.; Kurobane, Y.; van der Vegte, G. J.: Tests on CHS planar KK-joints under antisymmetrical loads. Proceedings of 7th International Symposium on Tubular Structures, Miscolc, Hungary, 1996

[107] Yura, J. A.; Howell, L. E.; Frank, K. H.: Ultimate load tests on tubular connections. Civil Engineering Structures Research Laboratory, Univ. of Texas, Report no. 78-1, Austin, 1978

[108] Yura, J. A.; Zettlemoyer, N.; Edwards, I. F.: Ultimate capacity equations for tubular joints. Proceedings of Offshore Technology Conference, Paper no. 3690, Houston, 1980

[109] Paul, J. C.; Valk, C. A. C.; Wardenier, J.: The static strength of multiplanar X-joints. Proceedings of 3rd International Symposium on Tubular Structures, pp. 73–80, Lappeenranta, Finland, 1989

[110] Koning, C. H. M.; Wardnier, J.: The static strength of welded CHS K-joints. Stevin Report 6-81-13, Stevin Laboratory, Delft University of Technology, Delft, The Netherlands, 1981

[111] Stol, H. G. A.; Puthli, R. S.; Bijlaard, F. S. K.: Static strength of welded tubular T-joints under combined loading. TNO-IBBC, Report B-84-561/63.6.0829, Delft, The Netherlands, 1984

[112] Stol, H. G. A.; Puthli, R. S.; Bijlaard, F. S. K.: Experimental research on tubular T-joints under proportionally applied combined static loading. Proceedings of the Conference on the

Behaviour of Offshore Structures (BOSS), Delft, The Netherlands, 1985

[113] van der Vegte, G. J.; Koning, C. H. M.; Puthli, R. S.; Wardenier, J.: Static behaviour of multiplanar welded joints in circular hollow sections. Stevin Report 25.6.90.13/A1/11.03, TNO-IBBC Report BI-90-106/63.5.3860, Delft, The Netherlands, 1990, Rev. February 1991

[114] van der Vegte, G. J.; Puthli, R. S.; Wardenier, J.: The influence of the chord and can length on the static strength of uniplanar tubular steel X-joints. Proceedings of 4th International Symposium on Tubular Structures, pp. 435–447, Delft, The Netherlands, 1991

[115] van der Vegte, G. J.; Lu, L. H.; Puthli, R. S.; Wardenier, J.: The non-linear behaviour of uniplanar tubular steel X-joints under out-of-plane loading. Proceedings of 2nd International Offshore and Polar Engineering Conference (ISOPE), Vol. IV, pp. 361–368, San Francisco, 1992

[116] van der Vegte, G. J.: The static strength of uniplanar and multiplanar tubular T- and X-joints. Doctoral Thesis, Delft University of Technology, Delft, The Netherlands, 1995

[117] Wardenier, J.; Koning, C. H. M.: Investigation into the static strength of welded Warren type joints made of circular hollow sections. Stevin Report no. BI-77-19, Stevin Laboratory, Delft University of Technology, Delft, The Netherlands, 1977

[118] Dexter, E. M.: Overlapped tubular joints in CHS, validation and the effect of overlap on strength. Civil Engineering Research Report no. CR/825/94, Univ. College of Swansea, U.K., 1994

[119] Dexter, E. M.; Haswell, J. V.; Lee, M. M. K.: A comparative study on out-of-plane moment capacity of tubular T/Y-joints. Proceedings of 5th International Symposium on Tubular, Structures, Nottingham, U.K., pp. 675–682, 1993

[120] Eimanis, A.; Grundy, P.: Load capacity of innovative tubular X-joints. Proceedings of 5th International Symposium on Tubular Structures, Nottingham, U.K., pp. 485–494, 1993

[121] Lee, M. M. K.; Dexter, E. M.: A parametric study on the out-of-plane bending strength of T/Y-joints. Proceedings of 6th International Symposium on tubular Structures, Melbourne, Australia, pp. 433–440, 1994

[122] Ma, S. Y. A.: A test programme on the static ultimate strength of welded fabricated tubular. Offshore Tubular Joints Conference (OTJ '88), Surrey, U.K.

[123] Tebbett, I. E.; Becket, D. C.; Billington, C. J.: The punching shear strength of tubular joints reinforced with a grouted pile. Proceedings of Offshore Technology Conference, Paper no. 3463, 1979

[124] Wilmhurst, S. R.; Lee, M. M. K.: Finite element analysis of KK-joints – an assessment of the test data of Mouty and Rondal. University College of Swansea, Report no. CR/781/93, Swansea, U.K., 1993

[125] Stammenkovic, A.; Sparrow, K. D.: Load interaction in T-joints of steel circular hollow sections. Journal of Structural Engineering, ASCE, Vol. 109, No. 9, 1983

[126] Stammenkovic, A.; Sparrow, K. D.: In-plane bending and interaction of CHS T- and X-joints. Proceedings of 1st International Symposium on Tubular Structures, Boston, pp. 543–554, 1984

[127] Mouty, J.; Rondal, J.: Etude du Comportement sous Charge Statique des Assemblages Soudes de Profils Creux Circulaires dans les Poutres de Sections Triangulaires et Quadrangulaires, Valexy, CECA 7210.SA.310, Bruxelles, 1990

[128] Brandi, R.: Behaviour of unstiffened and stiffened tubular joints. Commission of the European Community, Proceedings of International Conference in Marine Structures, Paris, Paper no. 6.1, 1981

[129] Zimmermann, W.: Tests on panel point type joints for large diameter tubes. Otto Graf Institute, Technische Hochschule Stuttgart, Report to CIDECT, Germany, 1965

[130] Davarpanah, P.: Load transmission in cruciform joints. Construction Metallique, No. 2, 1972

[131] Gibstein, M. B.: The static strength of T-joints subjected to in-plane bending moments. Det Norske Veritas, Report no. 76-137, Oslo, Norway, 1976

[132] Gibstein, M. B.: Static strength of tubular joints. Det Norske Veritas, Report no. 73-86C, Oslo, Norway, 1973

[133] Hlavacek, V.: Strength of welded tubular joints in lattice girders. Construzioni Metalliche, No. 6, 1970

[134] Scola, S.: Behaviour of axially loaded tubular V-joints. M. Sc. Thesis, McGill University, Montreal, Canada, 1989

[135] Scola, S.; Redwood, R. G.; Mitri, H. S.: Behaviour of axially loaded V-joints. Journal of Constructional Steel Research, No. 16, pp. 89–109, 1990

[136] Makino, Y.; Kurobane, Y.; Ochi, K.; van der Vegte, G. J.; Wilmhurst, S. R.: Database of test and numerical analysis results for unstiffened tubular joints. Kumamoto University, IIW-Doc. XV E-96-220, 1996

References

[137] Sammet, H.: Die Festigkeit knotenblechloser Verbindungen im Stahlbau, Schweißtechnik, Berlin, November 1963

[138] Bader, W.: Stahlrohrkonstruktion für statische und dynamische Beanspruchung, Schweißtechnik, Berlin, Dezember 1962

[139] Brodka, J.: Stahlrohrkonstruktionen, Verlagsgesellschaft Rudolf Müller, Köln, Braunsfeld, 1968

[140] Wanke, J.: Stahlrohrkonstruktionen, Springer-Verlag, Wien, New York, 1968

[141] Stammenkovic, A.; Sparrow, K. D.: Existing methods for calculating the static strength of welded T, Y, N, K joints in CHS, Part 1 and 2. Kingston Polytechnic, England, Juni 1977

[142] Wardenier, J.: The static strength of welded lattice girder joints in structural hollow sections, Part 3: Joints made of circular hollow sections. Stevin Report no. 6-78-4, ECSC Report-Eur 6428 e, MF, 1980

[143] Wardenier, J.: The ultimate static strength of tubular cross joints. Stevin Report no. 6-77-22, Delft University of Technology, Delft, 1977

[144] Wardenier, J.: The ultimate static strength of tubular T- and Y-joints. Stevin Report no. 6-77-23, Delft University of Technology, Delft, 1977

[145] Yura, J. A.: Ultimate capacity equations for tubular joints. Proceedings of Offshore Technology Conference, Paper no. 3690, 1980

[146] Sparrow, K. D.: Ultimate strengths of welded joints in tubular steel structures. Thesis, School of civil Engineering, Kingston Polytechnik, England, 1979

[147] Fessler, H.; Spooner, H.: Experimental determination of stiffness of tubular joints. Conference Integrity of Offshore Structures, Glasgow, July 1981

[148] American Petroleum Institute API RP2A: Recommended practice for planning, designing and constructing fixed platforms. 1991

[149] Det Norske Veritas: Rules for the design, construction and inspection of fixed offshore structures

[150] Wardenier, J.; Giddings, T. W.: The strength and behaviour of statically loaded welded connections in structural hollow sections. Monograph no. 6, CIDECT, Corby, England, 1986

[151] Giddings, T. W.: The development of recommendations for the design of welded joints between steel structural hollow sections or between steel structural hollow sections and H-sections. ECSC Final Report no. 7210-SA/814, February 1984

[152] Giddings, T. W.; Yeomans, N. F.; Wardenier, J.: The development of recommendations for the design of welded joints between steel structural hollow sections or between steel structural hollow sections and H-sections. Summary Report for ECSC Project 7210-SA/814, September 1985

[153] de Koning, C. H. M.; Wardenier, J.: The static strength of welded joints between structural hollow sections or between structural hollow sections and H-sections Part 1: Joints between circular hollow sections; Part 2: Joints between rectangular hollow sections; Part 3: Joints between structural hollow sections bracings and H-section chord. Stevin Report no. 6-84-18, 19, 20, Delft University of Technology, December 1984, March/April 1985

[154] Wardenier, J.; de Koning, C. H. M.: Investigation into the static strength of welded lattice girder joints in structural hollow sections; Part 1: Joints with rectangular hollow sections; Part 2: Joints with circular hollow sections and a rectangular boom. Stevin Report no. 6-76-6, Delft University of Technology, Delft

[155] Wardenier, J.; Stark, J. W. B.: The static strength of lattice girder joints. Stevin Report no. 6-78-4, Delft, also ECSC Report EUR 6428 E, MF-1980, European Community of Steel and Coal

[156] de Koning, C. H. M.; Wardenier, J.: Tests on welded joints in complete girders made of square hollow sections. Stevin Report 6-79-4, also TNO-IBBC Report no. 79-19/0063.4.3471, Delft

[157] Wardenier, J.; de Koning, C. H. M.: Investigation into the static strength of welded joints with SHS bracings and an I profile as chord. Stevin Report no. 6-76-19, also TNO-IBBC Report no. BI-76-89/35.3.51210, Delft

[158] Wardenier, J.; Davies, G.: The strength of predominantly statically loaded joints with a square or rectangular hollow section chord. IIW-Doc. XV-492-81

[159] Davies, G.; Wardenier, J.; Stolle, P.: The effective width of branch cross walls for RR cross joints in tension. Stevin Report no. 6-81-7, Delft University of Technology, Delft

[160] Wardenier, J.; de Koning, C. H. M.; van Douwen, A. A.: Behaviour of axially loaded K- and N-type gap joints with bracings of structural hollow sections and an I-profile as chord. IIW-Doc. XV-401-72

[161] Wardenier, J.: Comparison of various investigations into the static strength of tubular joints, Part 1. Cross joints. Stevin Report no. 6-76-3, also TNO-IBBC Report no. BI-76-33/35.3.51210, Delft

[162] Wardenier, J.; de Koning, C. H. M.: Supplement with test results of welded joints in structural hollow sections with rectangular boom. Stevin Report no. 6-76-5, also TNO-IBBC Report no. BI-76-122/35.3.51210, Delft

[163] Wardenier, J.; de Koning, C. H. M.; van Douwen, A. A.: Investigation into the static strength of welded Warren and Pratt type joints of rectangular hollow sections. Stevin Report no. 6-76-9, also TNO-IBBC Report no. BI-76-65/35.3.51210, Delft

[164] Wardenier, J.; de Koning, C. H. M.: Investigation into the static strength of welded joints with RHS bracings and a channel profile as chord. Stevin Report no. BI-77-4/35.3.51210, CIDECT Final Report, 1977

[165] Wardenier, J.; de Koning, C. H. M.: Comparison of static strength of welded joints made of RHS with different steel qualities. Stevin Report no. 6-77-20, also TNO-IBBC Report no. 77-109/05.3.31310, Delft

[166] Wardenier, J.; de Koning, C. H. M.: Investigations into the static strength of welded TK joints with three bracings made of RHS or CHS. TNO-IBBC Report no. BI-77-37/35.3.51210, also Stevin Report no. 6-77-6, Delft

[167] Wardenier, J.; Mouty, J.: Design rules for predominantly statically loaded welded joints with hollow sections as bracings and an I- or H-section as chord. Welding in the World, Vol. 17, No. 9/10, 1979

[168] Bettzieche, P.: Konstruktive Gestaltung von Knotenpunkten aus Vierkanthohlprofilen, Studienhefte zum Fertigbau, No. 12, Vulkan Verlag, Essen, 1969

[169] Mang, F.; Bucak, Ö.; Striebel, A.: The load carrying behaviour of unstiffened K-joints of large size thinwalled rectangular hollow sections of steel grade St 42 and St 52. IIW-Doc. XV-417-78

[170] Mang, F.; Bucak, Ö.: Investigations into the behaviour of high tensile steel joints of rectangular hollow sections. IIW-Doc. XV-416-78

[171] Mang, F.; Bucak, Ö.; Wolfmüller, F.: Bemessungsverfahren für T-Knoten aus Rechteckhohlprofilen, Foschungsbericht, Projekt-Nr. 82 der Studiengesellschaft für Anwendungstechnik von Eisen und Stahl e.V., Düsseldorf, Mai 1981

[172] Mang, F.; Bucak, Ö.; Knödel, P.: Ermittlung des Tragverhaltens von biegesteifen Rahmenecken aus Rechteckhohlprofilen (St 37, St 52) unter statischer Belastung, Forschungsbericht, Projekt-Nr. 70 der Studiengesellschaft für Anwendungstechnik von Eisen und Stahl e.V., Düsseldorf, Juni 1981

[173] Mang, F.; Bucak, Ö.; Steidl, G.: Untersuchungen an Verbindungen von geschlossenen und offenen Profilen aus hochfesten Stählen, Forschungsbericht, Projekt-Nr. 71 der Studiengesellschaft für Anwendungstechnik von Eisen und Stahl e.V., Düsseldorf, Juni 1981

[174] Mang, F.; Bucak, Ö.: Hohlprofilkonstruktion, Stahlbau Handbuch, Band 1, Stahlbau-Verlag, Köln, 1983

[175] Mee, B. L.: The structural behaviour of joints in rectangular hollow sections. Ph. D. Thesis, University of Sheffield, 1969

[176] Eastwood, A.; Wood, A. A.: The static strength of welded joints in structural hollow sections. Constructional Steelwork, January 1981

[177] Davie, J.; Giddings, T. W.: Research into the strength of welded lattice girder joints in structural hollow sections. CIDECT Report 5EC/71/7E, April 1971

[178] Anonymous: The behaviour of welded joints in complete lattice girders with RHS chords. CIDECT Report 5FC-77/31, prepared by British Steel Corporation, Corby, U.K.

[179] Packer, J. A.: Theoretical behaviour and analysis of welded steel joints with RHS chord sections. Ph. D. Thesis, University of Nottingham, 1978

[180] Haleem, R.: Determination of ultimate joint strength for statically loaded SHS welded lattice girder joints with RHS chords. CIDECT Report no. 77/37, October 1977 and addendum, March 1978

[181] Davies, G.: Estimating the strength of some welded lap joints formed from RHS members. Proceedings of International Conference on Joints in Structural Steelwork, Teeside, England, April 1981

[182] Coutie, M. G.; Davies, G.: The strength of welded gap joints with RHS members. Proceedings of International Conference on Joints in Structural Steelwork, Teeside, England, April 1981

[183] Korol, R. M.; El-Zanaty, M.; Brady, F. J.: Unequal width connections of square hollow sections in Vierendeel trusses. Canadian Journal of Civil Engineering, Vol. 4, No. 2, 1977

[190] Korol, R. M.; Mirza, F. A.; Elhifnawy, L.: Elastic-plastic finite element analysis of rectangular hollow section T-joints. McMaster University, Canada, CIDECT Report 5Jt-81/8

[191] Redwood, R. G.; Harries, P. J.: Welded joints for triangular trusses. McGill University, Montreal, Canada, Dept. of Civil Engineering and Applied Mechanics, Report no. 81-2, 1981

[192] Chidiac, M.; Korol, R. M.: Rectangular hollow section double chord T-joints. ASCE Journal of the Structural Division, Vol. 105, No. St. 8, August 1979

[193] Mehrotra, L.; Redwood, R. G.: Load transfer through connections between box sections. Canadian Engineering Institute, Report no. C-70-BR and Str. 10, Aug.–Sept. 1970

[194] Loo, Y.: Moment connections for Vierendeel trusses of square hollow structural sections. Project Thesis, McMaster University, Canada, 1973

[195] Patel, N. M.; Graff, W. J.; White, A.: Punching shear characteristics of RHS joints. ASCE National Structural Engineering Meeting, San Francisco, April 1980

[196] Mouty, J.: Calcul des charges ultimes des assemblages soudes de profils creux carrés et rectangulaires, Construction Metallique, No. 2, Juin 1976

[197] Strating, J.: The interpretation of test results for a level I code. IIW-Doc. XV-462-80

[198] Brockenbrough, R. L.: Strength of square-tube connections under combined loads. Proceedings of ASCE, Journal of Structural Division, St. 12, December 1972

[199] Bailly, R.; Mouty, J.: Experimental research on K and N welded joints composed of web members of hollow sections and of chords of HE and IPE sections. IIW-Doc. XV-425-78

[200] Bailly, R.: Etude des assemblages soudes-profiles creux sur profils ouverts (I et H), CIDECT-Programm 5N, CIDECT Report 77/15/5N, Juin 1977

[201] Roloos, A.: The effective weld length of beam to column connections without stiffening plates. IIW-Doc. XV-276-69

[202] Korol, R. M.; Mansour, M. H.: Theoretical analysis of haunch reinforced T-joints in square hollow sections. Canadian Journal of Civil Engineering, Vol. 6, No. 4, 1979

[203] Morris, G. A.: Designing HSS trusses with end cropped webs. CIDECT Reference SK-83/2, University of Manitoba, Winnipeg, Canada, February 1983

[204] Eastwood, W.; Osgerby, C.; Wood, A.; Blockley, D. I.: An experimental investigation into the behaviour of joints between structural hollow sections. Dept. of Civil and Structural Engineering, University of Sheffield, 1967

[205] Thiensiripipat, N.; Morris, G. A.; Pinkney, R. B.: Statical behaviour of cropped-web tubular truss joints. Canadian Journal for Civil Engineering, Vol. 7, No. 3

[206] Morris, G. A.; Frolich, L. E.; Thiensiripipat, N.: An experimental investigation of flattened-end tubular truss joints. Dept. of Civil Engineering, University of Manitoba, Canada, 1974

[207] Thiensiripipat, N.: Statical behaviour of cropped web joints in tubular trusses. Ph. D. Thesis, University of Manitoba, Canada, 1979

[208] Ciwko, B. J.: Statical behaviour of cropped-web joints for trusses with round tubular members. M. Sc. Thesis, University of Manitoba, Canada, 1980

[209] Davies, G.; Panjeshahi, E.: TEE joints in rectangular hollow sections (RHS) under axial loading and bending. Dept. of Civil Engineering, University of Nottingham, 1983

[210] Jubb, J. E. M.; Redwood, R. G.: Design of joints to box sections. Institute of Structural Engineers, Conference on Industrial Buildings and Structural Engineering, U.K., May 1966

[211] Lazar, B. E.; Fang, P. J.: T-type moment connections between rectangular tubular sections. Res. Rep., Sir George Williams University, Faculty of Engineering, Montreal, Canada, 1971

[212] Duff, G.: Joint behaviour of a welded beamcolumn connection in rectangular hollow sections. Ph. D. Thesis, The College of Aeronautics, Cranfield, U.K., 1963

[213] Cute, D.; Camo, S.; Rumpf, J. L.: Welded connections for square and rectangular structural steel tubing. Res. Report no. 292-10, Drexel Institute of Technology, Philadelphia, USA, 1968

[214] Korol, R. M.: The behaviour of HSS double chord Warren trusses and aspects of design. CIDECT Reference 5V-83/3, McMaster University, Hamilton, Canada, February 1983

[215] Lalani, M.: Static strength – design of simple and complex welded joints. Offshore Tubular Joints Conference (OTJ), London, 1985

[216] Lalani, M.; Bolt, H. M.: Strength of multiplanar joints on offshore platforms. Proceedings of 3rd International Symposium on Tubular Structures, Lappeenranta, Finland, pp. 90–102, 1989

[217] Ma, S. Y. A.; Tebbett, I. E.: New data on the ultimate strength of tubular welded K-joints under moment loads. Proceedings of Offshore Technology Conference, Paper No. 5831, Houston, USA, 1988

[218] Mäkeläinen, P. K.; Puthli, R. S.: Semi-analytical models for the static behaviour of T and DT tubular joints. Proceedings of International Conference on Behaviour of Offshore Structures (BOSS), Trondheim, Norway, pp. 1285–1300, 1988

[219] Wilmshurst, S. R.; Lee, M. M. K.: Ultimate capacity of axially loaded multiplanar double K-joints. Proceedings of 5th International Symposium on Tubular Structures, Nottingham, U.K., pp. 712–719, 1993

[220] Weinstein, R. M.; Yura, J. A.: The effect of chord stresses on the static strength of DT tubular connections. Proceedings of Offshore Technology Conference, Paper no. 5135, Houston, USA, 1986

[221] Boone, T. J.: Ultimate strength of tubular joints – chord stress effects. Proceedings of Offshore Technology Conference, Paper no. 4828, Houston, USA, 1984

[222] Davies, G.; Morita, K.: Three dimensional cross joints under combined axial branch loading. Proceedings of 4th International Symposium on Tubular Structures, Delft, The Netherlands, pp. 324–333, 1991

[223] Lau, B. L.; Morris, G. A.; Pinkney, R. B.: Testing of Warrentype cropped-web tubular truss joints. Proceedings of CSCE Annual Conference, Saskatoon, Saskatchewan, Canada, 1985

[224] Morris, G. A.; Packer, J. A.: Yield line analysis of cropped-web Warren truss joints. Proceedings of CSCE Annual Conference, Calgary, Alberta, Canada, 1988

[225] Frater, G. S.; Packer, J. A.: Design of fillet weldments for hollow structural section trusses. CIDECT Report no. 5AN/2-90/7, University of Toronto, Ontario, Canada, 1990

[226] Wardenier, J.; Kurobane, Y.; Packer, J. A.; Dutta, D.; Yeomans, N.: Design guide for circular hollow section (CHS) joints under predominantly static loading. CIDECT-Series "Construction with hollow steel sections", Published by TÜV Rheinland, Cologne, Germany, 1991

[227] Packer, J. A.; Birkemoe, P. C.; Tucker, W. J.: Canadian implementation of CIDECT Monograph 6. University of Toronto, Dept. of Civil Engineering, Publ. 84-04, 1984

[228] Davies, G.; Packer, J. A.; Coutie, M. G.: The behaviour of full width RHS cross joints. In welding of tubular structures, Pergamon Press, Oxford, pp. 411–418, 1984

[229] Anpassungsrichtlinie Teil 2 zu DIN 18800 – Stahlbauten – Teile 1 bis 4/11.90, Abschnitt 4.4: DIN 18808/10.84 – Stahlbauten, Tragwerke aus Hohlprofilen unter vorwiegend ruhender Beanspruchung, Mitteilungen des Deutschen Instituts für Bautechnik, Heft 4, 1994, S. 132–133

[230] ENV 1993-1-1: 1992/A1:1994, Eurocode 3: Design of steel structures, part 1.1: General rules and rules for buildings, Draft, Complement A1, November 1994

[231] Shinouda, M. R.: Stiffened tubular joints. Ph. D. Thesis, University of Sheffield, England, 1967

[232] Packer, J. A.: Cranked-chord HSS connections. Journal of Structural Engineering, American Society of Civil Engineers, Vol. 117, No. 8, August 1991, pp. 2224–2240

[233] Wardenier, J.; de Koning, C. H. M.; de Back, J.: Behaviour of axially loaded K- and N-type joints with bracings of rectangular hollow sections with a channel profile as chord. IIW-Doc. XV-402-77

[234] Roik, K.; Wagenknecht, G.: Traglastdiagramme zur Bemessung von Druckstäben mit doppelsymmetrischem Querschnitt aus Baustahl, Mitteilungen des Instituts für konstruktiven Ingenieurbau, Universität Bochum, Heft 27, Januar 1977

[235] Philiastides, A.: Fully overlapped rolled hollow section welded joints in trusses. Ph. D. Thesis, University of Nottingham, England, 1988

[236] Coutie, M. G.; Davies, G.; Philiastides, A.; Yeomans, N.: Testing of full-scale lattice girders fabricated with RHS members. Conference on Structural Assessment based on Full and Large Scale Testing, Building Research Station, Watford, England, April 1987

[237] Frater, G. S.: Performance of welded rectangular hollow structural section trusses. Ph. D. Thesis, University of Toronto, Canada, 1991

[238] Czeckowski, A.; Gasparski, T.; Zycinski, J.; Brodka, J.: Investigation into the static behaviour and strength of lattice girders made of RHS. IIW-Doc. XV-562-84, Poland, 1984

[239] Wilmhurst, S. R.; Lee, M. M. K.: Nonlinear FEM study of ultimate strength of tubular multiplanar double K-joints. 12th International Conference on Mechanics and Arctic Engineering (OMAE), ASME, Vol. III-B, Matrials Engineering, 1993

[240] Davies, G.; Crockett, P.: Interaction diagrams for CHS T-DT multiplanar joints under axial loads. 4th International Offshore Polar Engineering Conference, Osaka, Japan, Vol. IV, pp. 15–20, also IIW-Doc. XVE-94-204

[241] Mitri, H. S.; Scola, S.; Redwood, R. G.: Experimental investigation into the behaviour of axially loaded tubular V-joints. Proceedings of the 1987 CSCE Centennial Conference, Montreal, May 1987

[242] Project Team 1A, Dutta, D.; Grotmann, D.; Wardenier, J.: Eurocode 3 Annex KK, Hollow section connections. Preliminary draft, Aix-la-Chapelle, Germany, February 1993

[243] Paul, J. C.: The ultimate behaviour of multiplanar TT- and KK-joints made of circular hollow sections. Doctoral Dissertation, Kumamoto University, Kumamoto, Japan, 1992

[244] Ng, C. F.: Influence of chord preload on behaviour of tubular truss joints with cropped webs. M. Sc. Thesis, University of Manitoba, 1980

[245] Rondal, J.: Study of maximum permissible weld gaps in connections with plane end cuttings (5AH2); Simplification of circular hollow section welded joints (5AP). CIDECT Report 5AH2/5AP-20/90

[246] Morris, G. A.: Influence of cropping on effective length factors of tubular steel struts. Canadian Journal for Civil Engineering, Vol. 6, No. 2, pp. 260–267, 1979

[247] Mouty, J.: Buckling lengths for tubular members in welded lattice girders. CIDECT Report 69/22/E, September 1969

[248] Korol, R. M.; Chidiac, M. A.: K-joints of double chord square hollow sections. Canadian Journal of Civil Engineering, 7(3), pp. 523–539, 1980

[249] Kanatani, H.; Fujiwara, K.; Tabuchi, M.; Kamba, T.: Bending tests on T-joints of RHS chord and RHS or H-shape branch. CIDECT Report 5AF-80/15, Croydon, England, 1980

[250] Kanatani, H.; Kamba, T.; Tabuchi, M.: Effect of the local deformation of the joints on RHS Vierendeel trusses. Proceedings International Meeting on Safety Criteria in Design of Tubular Structures, Tokyo, Japan, pp. 127–137, 1986

[251] Szlendak, J.; Brodka, J.: Design of strengthened frame RHS joints. Proceedings International Meeting on Safety Criteria in Design of Tubular Structures, Tokyo, Japan, pp. 159–168, 1986

[252] Szlendak, J.: Interaction curves for M-N loaded T RHS joints. Proceedings International Meeting on Safety Criteria in Design of Tubular Structures, Tokyo, Japan, 1986

[253] Staples, C. J. L.; Harrison, C. C.: Test results of 24 rightangled branches fabricated from rectangular hollow sections. University of Manchester, Institute of Science and Technology (UMIST) Report, England

[254] Zhao, X. L.; Hancock, G. J.: Tubular T-joints subject to combined actions. Proceedings 10th International Specialty Conference on Cold-formed Steel Structures, St. Louis, Missouri, USA, pp. 545–573, October 1990

[255] EGKS-Bericht „Entwicklung von Konstruktionsempfehlungen für geschweißte Knoten aus Stahlhohlprofilen (T- und X-Knoten)", Forschungsprojekt-Nr. 7210.SA.109, Mannesmannröhren-Werke AG, Dezember 1983

[256] Det Norske Veritas: Rules for the design, construction and inspection of fixed offshore structures. Oslo, 1977

[257] Efthymiou, M.: Local rotational stiffness of unstiffened tubular joints. KSEPL Report, RKER 85.199

[258] Davies, G.; Packer, J. A.: Analysis of web crippling in a rectangular hollow section. Proceedings of the Institution of Civil Engineers, part 2, 83785-798, 1987

[259] Makino, Y.; Kurobane, Y.: Recent research in Kumamoto University in tubular joint design. International Institute of Welding, IIW-Doc. XV-615-86, July 1986

[260] Wardenier, J.; Davies, G.; Stolle, P.: The effective width of branch plate to RHS chord connections in cross joints. Stevin Report 6-81-6, Stevin Laboratory, Delft University of Technology, Delft, The Netherlands, 1981

[261] Dawe, J. L.; Grondin, G. Y.: W-shape beam to RHS column connections. Canadian Journal of Civil Engineering, 17(5), pp. 788–797

[262] Guravich, S. J.: Reinforced branch plate-to-RHS connections in tension and compression. M. Sc. Thesis, Department of Civil Engineering, The University of New Brunswick, Fredericton, Canada, 1992

[263] Ono, T.; Iwata, M.; Ishida, K.: An experimental study on joints of new truss system using rectangular hollow sections. 4th International Symposium on Tubular Structures, Delft, The Netherlands, June 1991, Proceedings, pp. 344–353

[264] Akiyama, N.; Yajima, M.; Akiyama, H.; Otake, F.: Experimental study of strength of joints in steel tubular structures, ISSC, Vol. 10, No. 102, 1974

[265] Lu, L. H.: The static strength of I-beam to rectangular hollow section column connections, Doctorate thesis, ISBN 90-407-1603 X, Delft University Press, The Netherlands, 1997

[266] Yu, Y.: The static strength of uni-planar and multi-planar connections in rectangular hollow sections, Doctorate thesis, ISBN 90-407-1615-3, Delft University Press, The Netherlands, 1997

[267] Lu, L. H. et al.: Deformation limit for the ultimate strength of hollow section joints. Proceedings Sixth International Symposium on Tubular Structures, Melbourne/Australia, 14–16 December 1994, ISBN 90 5410 5208, A. A. Balkema, Rotterdam/Netherlands

[268] International Institute of Welding: Design recommendations for hollow section joints – predominantly statically loaded. IIW Doc. XV-701-89, 2nd edition, 1989

References to Chapter 7

[1] Kuang, F. C.; Potvin, A. E.; Leick, R. D.: Stress concentrations in tubular joints. OTC 2205, Offshore Techn. Conf. 1975, Houston, Texas

[2] Beale, L. A.; Toprac, A. A.: Analysis of in-plane, T, Y and K welded tubular connections. Welding Research Council 125/October 1967

[3] Gibstein, M. E.: Parametrical stress analysis of T joints, Select Seminar. European Offshore Steels Research, November 1978 at the Welding Institute, Abington Hall, Cambridge, U.K.

[4] Wordsworth, A. C.; Smedley, G. P.: Stress concentrations at unstiffened tubular joints. Select Seminar, Offshore Steels Research, November 1978 at the Welding Institute, Abington Hall, Cambridge, U.K.

[5] Zirn, R.: Schwingfestigkeitsverhalten geschweißter Rohrknotenpunkte und Rohrlaschenverbindungen, Techn. Wiss. Bericht MPA Stuttgart (1975), Heft 75-01

[6] Maeda, T.; Uchino, K.; Sakurai, H.: Experimental study on the fatigue strength of welded tubular K joints. IIW-Doc. XV-260-69
[7] DIN 15018, Teil 1: Krane, Grundsätze für Stahltragwerke, Berechnung, Ausg. Nov. 1984
[8] Leitfaden für eine Betriebsfestigkeitsrechnung, Empfehlungen zur Lebensdauerabschätzung von Bauteilen in Hüttenwerksanlagen, Bericht-Nr. ABF 01, Arbeitsgemeinschaft Betriebsfestigkeit, Verein Deutscher Eisenhüttenleute, Düsseldorf, Februar 1977
[9] Noordhoek, C.; Wardenier, J.; Dutta, D.: The fatigue behaviour of welded joints in square hollow sections. ECSC Research 6210-KD/1/103, May 1980
[10] Dutta, D.; Mang, F.; Wardenier, J.: Fatigue behaviour of welded hollow section joints, CIDECT Monograph no. 7, Constrado, Croyolon, April 1982
[11] Wardenier, J.; Dutta, D.: The fatigue behaviour of lattice girder joints in square hollow sections. IIW-Doc. XV-493-81, XIII-1005-81
[12] API RP2A: Planning, designing and constructing fixed offshore platforms
[13] AWS D 1.1–94 Structural Welding Code. American Welding Society, Miami, USA, 1994
[14] Toprac, A. A.; Louis, B. G.: The fatigue behaviour of tubular connections. Structures Fatigue Research Laboratory, University of Texas, Austin, Texas, May 1970
[15] Kurobane, Y.; Konomi, M.: Fatigue strength of tubular K joints – S-N releationship proposed as tentative design criteria. IIW-Doc. XV-370-73
[16] Uchino, K.; Sakurai, H.; Sugiyama, S.: Experimental study on the fatigue strength of welded tubular K joints. IIW-Doc. XV-690-73, IIW-Doc. XV-344-73
[17] Kurobane, Y.: Effects of low-cycle alternating loads on tubular K- joints. IIW-Doc. XV-271-69
[18] Toprac, A. A.; Louis, B. G.: The fatigue behaviour of tubular constructions. IIW-Doc. XV-293-70, May 1970
[19] CIDECT Programme 7 B: Fatigue strength of overlapping CHS N-joints. Final report, Sept. 1981
[20] Dijkstra, O. D.; Hartog, J.: Dutch part of the large scale welded tubular T-joints. Select Seminar, European Offshore Steels Research, Preprints Vol. 2, At the Welding Institute, Abington Hall, Cambridge, U.K., 27–29 November 1978
[21] Kwan, C.; Graff, W. F.: Analysis of tubular T-Connections by the finite element method, comparison with experiments. Preprint OTC 1669, Offshore Technology Conference, Houston, Texas, 1969
[22] Visser, W.: On the structural design of tubular joints. Preprint OTC 2117, Offshore Technology Conference, Houston, Texas, 1974
[23] Reber, J. B., Jr.: Ultimate strength design of tubular joints. Preprint OTC 1664, Offshore Technology Conference, Houston, Texas, 1972
[24] Haibach, E.: Schwingfestigkeitsverhalten von Schweißverbindungen, VDI Darmstadt, VDI-Bericht Nr. 268, 1976
[25] Haibach, E.: Einfluß von Eigenspannungen auf den Schwingbruch, LBF, Vortrag beim schweißtechnischen Kolloquium am 2. und 6. Mai 1977, „Eigenspannungen in geschweißten Konstruktionen" in der schweißtechnischen Lehr- und Versuchsanstalt Duisburg
[26] Bouwman, L. P.: Fatigue of bolted connections and bolt loaded in tension. Report 6-79-9, Delft University of Technology, Dept. of Civil Engineering
[27] Fisher, J. W.; Struik, J. H. A.: Guide to design criteria for bolted and riveted joints. John Wiley & Sons, New York, London, Sydney, 1973
[28] Austen, I. M.: Factors affecting corrosion fatigue crack growth in steels. Select Seminar, European Offshore Steels Research, Preprints Vol. 1, November 1978, Welding Institute, Abington Hall, Cambridge, U.K.
[29] Wildschut, H.; de Back, J.; Dortland, W.; von Leeuween, J. L.: Fatigue behaviour of welded joints in air and sea water, Select Seminar, European Offshore Steels. Research, Preprints, Vol. 1, November 1978, Welding Institute, Abington Hall, Cambridge, U.K.
[30] Haibach, E.: Die Schwingfestigkeit von Schweißverbindungen aus der Sicht einer örtlichen Beanspruchungsmessung, Laboratorium für Betriebsfestigkeit, Darmstadt, Bericht Nr. FB-77 (1968)
[31] Dutta, D.; Wardenier, J.; Noordhoek, C.: The fatigue behaviour of welded joints in circular hollow sections. ECSC Research 6210-KD/1-103, April 1980
[32] Dutta, D.; Hauk, V.; Mang, F.: A Survey of investigation into the behaviour of unstiffened structures. Lectures in Offshore Engineering 1978, Institute of Building Technology and Structural Engineering, Aalborg University Centre, Denmark
[33] Lauresen, A. A.; Dijkstra, O. D.: Fatigue tests on large post-weld heat-treated and as-welded tubular T-joints. Preprint OTC 4405, Offshore Technology Conference, Houston, Texas, May 1982
[34] Wardenier, J.: Hollow Section joints. Delft University Press, 1982
[35] Dijkstra, O. D.; Hartog, J.: Dutch part of the large scale welded tubular T-joints. Select Seminar, European Offshore Steels Research, Preprints Vol. 2, Abington Hall, Cambridge, U.K., November 1978

[36] de Back, J.: Testing tubular joints. Session developer's report, International Conference on Steel in Marine Structures, Paris, October 1981

[37] Mang, F.; Bucak, Ö.: Hohlprofilkonstruktionen, Stahlbau-Handbuch, Bd. I, Stahlbau-Verlag, Köln, 1983

[38] Snedden, N. W.: Offshore Installations: Guidance on design and construction, Proposed new fatigue design procedures for steel welded joints in offshore structures. Recommendations of the Department of Energy, "Guidance Notice" Revision Drafting Panel AERE Harwell, Oxfordshire, U.K., June 1981

[39] Snedden, N. W.: Background to proposed new fatigue design rules for steel welded joints in offshore structures. Report of the Department of Energy "Guidance Notes" Revision Drafting Panel, AERE Harwell, Oxfordshire, U.K., May 1981

[40] IIW-Subcomm. XV E: Recommended fatigue design procedure for hollow section joints, Part 1 – Hot spot stress method for nodal joints. Doc. XV-582-85, International Institute of Welding Annual Assembly, Strasbourg, France, 1985

[41] Committee TC6 "Fatigue", European Convention for Constructional Steelwork: Recommendations for the fatigue design of steel structures, No. 43, 1985

[42] ENV 1993-1-1; Eurocode 3: Design of steel structures, part 1.1: General rules and rules for buildings, February 1992, CEN (European Committee of Standardisation)

[43] Schütz, W.: Lebensdauervorhersage an geschweißten Rohrknoten

[44] Schütz, W.; Zenner, H.: Schadensakkumulationshypothesen zur Lebensdauervorhersage bei schwingender Beanspruchung – ein kritischer Überblick, Zeitschrift für Werkstofftechnik, Teil 1, S. 25–33; Teil 2, S. 97–102, 1973

[45] Mang, F.; Bucak, Ö.; Steidl, G.: Über den Einfluß von Eigenspannungen auf die Schwingfestigkeit von Hohlprofilknoten, Heft 5 der Schriftenreihe des Institutes für Fördertechnik an der Universität Karlsruhe, 1983

[46] de Back, J.; Vaessen, G. H. G.: Effect of plate thickness, temperature and weld toe profile on the fatigue and corrosion fatigue behaviour of welded offshore structures, Part I & II. Final report, Project no. 7210-KG/601, Commission of the European Communities, Luxembourg, EUR 10309 EN, 1986

[47] de Back, J.: Strength of tubular joints. International Conference "Steel in Marine Structures", Paris, 1981, Commission of the European Community

[48] Wardenier, J.; Mang, F.; Dutta, D.: Fatigue strength of welded joints in latticed structures and Vierendeel girders. Final report, ECSC research programme 7210-SA/111, August 1989

[49] Research team: Fatigue behaviour of multiplanar welded hollow section joints and reinforcement measures for repair. Final report, ECSC research programme 7210-SA/114, September 1992

[50] van Delft, D. R. V.; Noordhoek, C.; Da Re, M. L.: The results of the European fatigue tests on welded tubular joints compared with SCF formulas and design lines. Steel in Marine Structures (SIM '87), Delft, The Netherlands, Elsevier applied science publishers ltd., Amsterdam/London/New York, Tokyo, June 1987

[51] Romeijn, A.: Stress and strain concentration factors of welded multiplanar tubular joints. Delft University Press, The Netherlands, 1994

[52] Frater, G.: Performance of welded rectangular hollow structural section trusses. Ph. D. thesis, University of Toronto, Canada, 1991

[53] van Wingerde, A. M.: The fatigue behaviour of T- and X-joints made of square hollow sections. Ph. D. thesis, Delft University of Technology, The Netherlands, 1992

[54] Panjeh Shahi, E.: Stress and strain concentration factors of welded multiplanar joints between square hollow sections. Ph. D. thesis, Delft University Press, The Netherlands, 1994

[55] Dover, W. D.; Petrie, J. R.: In-plane bending fatigue of a tubular welded T-joint. Proceedings S.E.E. Conference "Fatigue Testing and Design", 1976

[56] Brink, F. I. A.; Krogt, A. H. van de: Stress analysis of a tubular cross joint without internal stiffening for offshore structures. Proceedings W.I. Conference, Paper 5, Newcastle, November 1974

[57] Wordsworth, A. C.: Stress concentration factors at K and KT-joints. Proceedings Conference "Fatigue in Offshore Structural Steel", ICE, London, 1981

[58] Efthymiou, M.; Durkin, S.: Stress concentrations in T/Y and gap/overlap K-joints. Proceedings Conference BOSS 1985, pp. 429–440, Delft, July 1985

[59] Efthymiou, M.: Development of SCF formulae and generalised influence functions for use in fatigue analysis. Proceedings OTJ '88 "Recent developments in tubular joints technology", Eaglefield Green, Surrey, U.K., October 1988

[60] Underwater Engineering Group: Design of tubular joints for offshore structures, Vol. 3, Part F, 1985

[61] Lalani, M.: Developments in tubular joints technology for offshore structures. Proceedings Second International Offshore and Polar Engineering Conference, San Francisco, June 1992

[62] Tyler, E.; Gibstein, M. B. et al.: Parametrical stress analysis of T-joints. Technical report 77–253, Det Norske Veritas, Oslo, November 1977

[63] Gibstein, M. B.: Parametrical stress analysis of T joints. European Offshore Steel Research Seminar, Cambridge, November 1978

[64] de Back, J.; Vaessen, G. H. G.: Fatigue behaviour and corrosion fatigue behaviour of offshore structures. Final report ECSC programme 7210-KB/6/602 (J.7.1 f/76), Foundation for Materials Research in the Sea, Delft/Apeldoorn, The Netherlands, April 1981

[65] Marshall, P. W.: Design of welded tubular connections: basis and use of AWS code provisions. Report, Civil Engineering Consultant, Shell Oil Company, Houston, USA, 1989

[66] Romeijn, A.; Puthli, R. S.; de Koning, C.; Wardenier, J.: Stress and strain concentration factors of multiplanar joints made of circular hollow sections. Proceedings 2nd International Offshore and Engineering Conference, San Francisco, USA, International Society of Offshore and Polar Engineers, June 1992

[67] Romeijn, A.; Puthli, R. S.; de Koning, C.; Wardenier, J.; Dutta, D.: Fatigue behaviour and influence of repair on multiplanar K joints made of circular hollow sections. Proceedings 3rd International Offshore and Engineering Conference, Singapore, International Society of Offshore and Polar Engineers, June 1993

[68] Research team: Fatigue strength of welded, unstiffened RHS joints in latticed structures and Vierendeel girders. Final reports, ECSC research programme 7210-SA/111, August 1989

[69] van Wingerde, A. M.; Packer, J. A.; Wardenier, J.; Dutta, D.: The fatigue behaviour of K-joints made of square hollow sections. Final report, CIDECT Programme 7P, University of Toronto, September 1995

[70] Herion, S.: Räumliche K-Knoten aus Rechteck-Hohlprofilen, Dissertation, Universität Fredriciana zu Karlsruhe (TH), Karlsruhe, 1994

[71] van Wingerde, A. M.; Yu, Y.; Puthli, R. S.; Wardenier, J.; Dutta, D.: Influence of corner radius and weld dimensions on the stress concentration factors of SHS T- and X-joints. Proceedings 4th International Symposium "Tubular Structures", Delft, The Netherlands, Delft University Press, June 1991

[72] Herion, S.; Mang, F.: Parametric study on multiplanar K-joints made of RHS regarding axial force, in-plane-bending and out-of-plane-bending moments. Proceedings 4th International Offshore and Polar Engineering Conference, April 10–15, Osaka, Japan, 1994

[73] Department of Energy: Offshore installation, guidance on design and construction. London, United Kingdom, 1990

[74] Thorpe, T. W.; Sharp, J. V.: The fatigue performance of tubular joints in air and seawater. MaTSU report, Harwell Laboratory, Oxfordshire, United Kingdom, 1989

[75] van Wingerde, A. M.; Packer, J. A.; Wardenier, J.: New guidelines for fatigue design of HSS connection. IIW XV-E-96-221, International Institute of Welding, 1996

[76] Herion, S.; Mang, F.: Comparison of uniplanar and multiplanar K-joints with gap made of rectangular hollow sections. Proceedings 6th International Offshore and Polar Engineering Conference, May 26–31, 1996, Los Angeles, USA

[77] Mang, F.; Bucak, Ö.; Klinger, J.: Wöhlerlinienkatalog für Hohlprofilverbindungen, Studiengesellschaft für Anwendungstechnik von Eisen und Stahl e.V., Düsseldorf, Juni 1987

[78] DIN 18808: Stahlbauten, Tragwerke aus Hohlprofilen unter vorwiegend ruhender Beanspruchung, Deutsches Institut für Normung, Berlin, Oktober 1984

[79] DASt-Richtlinie 011: Hochfeste schweißgeeignete Feinkornbaustähle mit Mindeststreckgrenzenwerten von 460 und 690 N/mm^2, Anwendung für Stahlbauten, Deutscher Ausschuß für Stahlbau, Berlin, Februar 1988

[80] Mang, F.; Bucak, Ö.; Stauff, K.: Fatigue behaviour of rectangular hollow section joints made of high strength steels. Proceedings 4th International Symposium on Tubular Structures, Delft, June 1991

[81] Mang, F.; Bucak, Ö.; Herion, S.: Fatigue behaviour of hollow section joints of high tensile steel. CIDECT-programme 7L, Final report, March 1997

[82] Van Wingerde, A. M.; Packer, J. A.; Wardenier, J.; Dutta, D.: Simplified design graphs and comparison of fatigue design guidelines for uniplanar RHS K-joints. CIDECT Programme 7R, Report 7R-01/97, January 1997, Univ. Toronto, Canada

[83] Soh, A. K.; Soh, C. K.: Stress concentrations in T/Y and K square-to-square and square-to-round tubular joints. Journal of Offshore Mechanics and Arctic Engineering (OMAE), 1992, Vol. 114, No. 3, pp. 220–230

[84] Niemi, E. J.: Fatigue resistance predictions for RHS K-joints using two alternative methods. Proceedings 7th International Symposium on Tubular Structures, Miscolc, Hungary, 1996

[85] Puthli, R. S.; van Foeken, R. J.; Romeijn, A.: Guidelines on the determination of stress concentration factors of circular and rectangular hollow section joints. Fatigue in offshore structures, Vol. I, ISBN 81-204-1101-3, Oxford & IBH Publishing Co. PVT. LTD., New Delhi, 1996

[86] Dutta, D.: Parameters influencing the stress concentration factors in joints in offshore structures. Fatigue in offshore structures, Vol. I,

ISBN 81-204-2202-3, Oxford & IBH Publishing Co. PVT. LTD., New Delhi, 1996

[87] Berge, S.: Fatigue strength of tubular joints. Fatigue in offshore structures, Vol. 2, ISBN 81-204-1102-1, Oxford & IBH Publishing Co. PVT. LTD., New Delhi, 1996

[88] IIW Subcommission XV-E: Recommended fatigue design procedure for welded hollow section joints. Part 1: Recommendation, Part 2: Commentary, Second edition, IIW Doc. XV-1035-99, International Institute of Welding, October, 1999

[89] Herion, S.; Weyand, K.: Ermüdungsgerechte Bemessung von Hohlprofilknoten nach dem CIDECT Fatigue Design Guide, Stahlbau, 69. Jahrgang, Heft 4, April 2000

[90] Efthymiou, M.; Durkin, S.: Stress concentration in T/Y and gap/overlap K-joints. Behaviour of offshore Structures, pp. 429–440, Elsevier Science Publishers, Amsterdam, 1985

[91] van Wingerde, A. M.; Packer, J. A.; Wardenier, J.: IIW fatigue rules for tubular joints. IIW International Conference on "Performance of dynamically loaded welded structures", July 14–15, 1997, San Francisco, USA

[92] Packer, J. A.; Wardenier, J.: Stress concentration factors for non-90° X-connections made of square hollow sections, Can. J. Civ. Eng., 25(2), pp. 370–375

[93] van Wingerde, A. M.: The fatigue behaviour of T and X joints made of square hollow sections, Heron, The Netherlands, 37(2), pp. 1–180

[94] van Wingerde, A. M.; Wardenier, J.; Packer, J. A.: Simplified design graphs for the fatigue design of multi-planar K joints with gap, CIDECT final report 7R-23/98, Delft Univ. of Techn., The Netherlands, September, 1998

[95] Zhao, X. L.; Puthli, R. S.: Cpmparison of SCF formulae and fatigue strength of uni-planar RHS K-joints with gap, IIW Doc. XVE-98-235, International Institute of Welding

[96] Zhao, X. L. et al.: Design guide for circular and rectangular hollow section welded joints under fatigue loading, CIDECT series "Construction with hollow steel sections", Serial no. 8, published by TÜV Verlag, Cologne, Germany

[97] Kobayashi, K. et al.: Improvement in the fatigue strength of fillet welded joint by use of the new welding electrode. IIW-Doc. XIII-828-77, International Institute of Welding, 1977

[98] Haagensen, P. J. et al.: Prediction of the improvement in fatigue life of welded joints due to grinding, TIG dressing, weld shape control and shot peening. TS 35, ibid

[99] Verheul, A.; Wardenier, J.: The low cycle fatigue behaviour of axially loaded T-joints between rectangular hollow sections. CIDECT report 7H-89/1-E, TNO-IBBC report BI-89-060/63.5.3820, Delft, The Netherlands

[100] Recommendations on fatigue of welded components. IIW-DOC.XIII-1539-94/XV-845-94, September 1994

References to Chapter 8

[1] British Steel Tubes & Pipes: Flowdrill jointing system, Part I. Mechanical integrity tests, Part II. Structural hollow section connections. CIDECT report no. 6F-13A+B/96, 1996

[2] British Steel Tubes & Pipes: Flowdrill jointing system for hollow section connections. CIDECT report no. 6D-16/94, 1994

[3] Sidercad: Hollow section connections using (Hollo-Fast) Hollo-Bolt expansion bolting. CIDECT report no. 6G-19/94 and 6G-16/95, 1994/95

[4] British Steel Tubes & Pipes: Hollo-Fast and Hollo-Bolt system for hollow section connections. CIDECT report no. 6G-14(a)/96, 1996

[5] Rondal, J.: Study of maximum permissible weld gaps in connections with plane end cuttings (5AH2); Simplification of circular hollow section welded joints (5AP). CIDECT report no. 5AH2/4AP-90/20, 1990

[6] Wardenier, J.; Dutta, D.; Yeomans, N.; Packer, J. A.; Bucak, Ö.: Design guide for structural hollow sections in mechanical applications. CIDECT series "Construction with hollow steel sections", No. 6, published by TÜV Rheinland, Cologne, Germany 1995

[7] Stahlrohr-Handbuch, 9. Aufl. 1982, Vulkan-Verlag, Essen

[8] Kennedy, J. B.: Minimum bending radii for square and rectangular hollow section (3 roller cold bending). CIDECT report 11C-88/14-E, 1988

[9] Brady, F. J.: Determination of minimum radii for cold bending of square and rectangular hollow structural sections. Final report, CIDECT-Programme 11B, May 1978

[10] CAN/CSA-S16.1-M89: Limit states design of steel structures. Canadian Standards Association

[11] AS 4100-1990: Steel structures. Standards Australia

[12] AISC 1978: Recommendations for the design, fabrication and erection of building structures in steel. American Institute of Steel Construction

[13] ENV 1993-1-1: 1992, Eurocode 3: Design of steel structures – Part 1.1: General rules and rules for buildings. European Commitee for Standardisation (CEN), 1992

[14] EN 10210-1: Hot finished structural hollow sections of non-alloy and fine grain structural steels, Part 1: Technical delivery conditions. European Standard (CEN)

[15] EN 10219-1: Cold formed welded structural hollow sections of non-alloy and fine grain structural steels, Part 1: Technical delivery conditions. European Standard (CEN)

[16] AWS 1992: Structural welding code – Steel. ANSI/AWS D.10.1-92, American Welding Society, USA

[17] AS 1163: Structural steel hollow sections. Standards Australia

[18] JIS G3444: Carbon steel tubes for general structural purposes. Ministry of Interntional Trade and Industry, Japan

[19] JIS G3466: Carbon steel square pipes for general structural purposes. Ministry of International Trade and Industry, Japan

[20] A 500: Standard specifications for cold-formed welded and seamless carbon steel structural tubing in rounds and shapes. American Society for Testing and Materials

[21] A 501: Specification for hot formed welded and seamless carbon steel structural tubing. American Society for Testing and Materials

[22] A 242: Specifications for high strength low-alloy structural steel. American Society for Testing and Materials

[23] A 588-91a: Standard specifications for high strength low-alloy structural steel with 50 KSi (345 MPa) minimum yield point to 4 in. (100 mm) thick. American Society for Testing and Materials

[24] A 618: Specification for hot formed welded and seamless high strength low-alloy structural tubing. American Society for Testing and Materials

[25] CAN/CSA-G40.20-M92: General requirements for rolled or welded structural quality steel. Canadian Standards Association

[26] CAN/CSA-G40.21-M92: Structural quality steels. Canadian Standards Association

[27] ISO 630: Structural steels. International Organisation of Standardisation

[28] Stahl-Eisen-Werkstoffblatt 088-87: Schweißgeeignete Feinkornbaustähle, Richtlinien für die Weiterverarbeitung, besonders für das Schmelzschweißen, Verlag Stahleisen, Düsseldorf, 1987

[29] DIN EN 288, Teil 1: Anforderung und Anerkennung von Schweißverfahren für metallische Werkstoffe, Allgemeine Regeln für das Schmelzschweißen, Deutsche Fassung EN 288-1: 1992, Ausgabe April 1992

[30] DIN EN 288-2, Teil 2: Anforderungen und Anerkennung von Schweißverfahren für metallische Werkstoffe, Schweißanweisung für das Lichtbogenschweißen, Deutsche Fassung EN 288-2: 1992, Ausgabe April 1992

[31] DIN EN 288, Teil 3: Anforderungen und Anerkennung von Schweißverfahren für metallische Werkstoffe, Schweißverfahrensprüfungen für das Lichtbogenschweißen von Stählen, Deutsche Fassung EN 288-3: 1992, Ausgabe April 1992

[32] Wasserstoff- und Bauteilverhalten unter Schwingbeanspruchung, VDI-Berichte 268, 1976

[33] ENV 1090-1: Execution of steel structures, Part 1: General rules and rules for buildings. Draft, CEN

[34] CSA 1983: Certification of companies for fusion welding of steel structures. W47.1-1983, Canadian Standards Association

[35] DIN EN 287, Teil 1: Prüfung von Schweißern, Schmelzschweißen, Stähle, CEN, April 1992

[36] DIN 18800, Teil 7: Stahlbauten; Herstellen, Eignungsnachweise zum Schweißen, Mai 1993

[37] Anpassungsrichtlinie Teil 2 zu DIN 18800 – Stahlbauten – Teile 1–4/11.90, Abschnitt 4.4: DIN 18 808/10.84 – Stahlbauten, Tragwerke aus Hohlprofilen unter vorwiegend ruhender Beanspruchung, Mitteilungen des Deutschen Instituts für Bautechnik, Heft 4, 1994, S. 132–133

[38] Packer, J. A.; Krutzler, R. T.: Nailing of steel tubes. Proceedings of 6th International Symposium on Tubular Structures, Delft, Balkema, Rotterdam/Brookfield, 1994

[39] Shakir-Khalil, H.: Resistance of concrete-filled steel tubes to pushout forces. The Structural Engineer 71(13), 1993

[40] Dutta, D.; Wardenier, J.; Yeomans, N.; Sakae, K.; Bucak, Ö.; Packer, J. A.: Design guide for fabrication, assembly and erection of hollow section structures. CIDECT series "Construction with hollow steel sections", No. 7, published by TÜV Rheinland, Cologne, Germany 1998

[41] Hilti Corporation: Fastening technology manual. Issue 5/93, Schaan, Liechtenstein, 1993

[42] AISC 1993: Load and resistance factor design specification for structural steel buildings. 2nd edition, American Institute of Steel Construction, Chicago

[43] ENV 1993-1-1, Eurocode 3: Design of steel structures, part 1-1: General rules and rules for buildings. February 1992, CEN (European Committee of Standardisation)

[44] Schlaich, J.; Schober, H.: Rohrknoten aus Stahlguß, Stahlbau Heft 8, 1999, Ernst & Sohn

[45] Mang, F.; Herion, S.; Koch, E.: On the fatigue behaviour of welded cast steel–steel connections. Proceedings of International Conference on Current and Future Trends in Bridge Design, Construction and Maintenance, Singapore, 1999

[46] Fröhlich, J.: Oktaplatte in Rohrkonstruktion, Der Stahlbau, 28. Jg., Heft 9, September, 1959, Ernst & Sohn

[47] Mang, F.; Herion, S.: Guss im Bauwesen, Sonderdruck aus Stahlbau Kalender, 2001, Ernst & Sohn, Berlin
[48] Schlaich, J.; Seidel, J.: Die Eislaufhalle im Olympiapark München, Bauingenieur 60, 1985
[49] Leonhardt, F.; Schlaich, J. & associates: Vorgespannte Seilnetzkonstruktionen, Das Olympiadach in München, Heft 9/10/12, 1972, Hefte 2/3/4/6, 1973, Der Stahlbau, Ernst & Sohn
[50] Pichler, G.; Guggisberg, R.: Deutsches Technikmuseum Berlin, Heft 7, 1998, Heft 4, 1999, Der Stahlbau, Ernst & Sohn
[51] Hartmann, R.; Jahn, H.; Riese, D.; Provoost, W.: Besonderheiten der Fertigung und Einschwimmmontage des Stahlüberbaues der Kronprinzenbrücke in Berlin, Heft 10, 1996, Der Stahlbau, Ernst & Sohn
[52] Mecse, L.: Bericht zur Berechnung des Stahlgussknotens für die Nesenbachtalbrücke in Stuttgart, 1998
[53] Schlaich, J.; Schober, H.: Bahnbrücken am Lehrter Bahnhof in Berlin, Die Humboldthafenbrücke, Heft 6, 1999, Der Stahlbau, Ernst & Sohn
[54] Bergmann, J.; Biswas, K.; Hoff, H.; Seeger, T.-H.: Fatigue behaviour of HOESCH GsArk 10, part 1: Experimental results. Publications of the department of material mechanics in Technical University Darmstadt, FI-13/1981
[55] Seeger, T.: Fatigue behaviour of HOESCH GsArk 10, part 2: Fatigue design concept. Publications of the department of material mechanics in Technical University Darmstadt, FI-14/1981
[56] Sonsino, C. M.; Lipp, K.: Übertragbarkeit des an Winkelproben ermittelten Betriebsfestigkeitsverhaltens auf große Rohrknoten für die Offshoretechnik, Fraunhofer Institute for Serviceability Strength (LBF), LBF-Report no. TB 180, Darmstadt, 1988
[57] UEG: Design of tubular joints for offshore structures. Vol. 3, part 6: Cast joints, Publication UR 33, United Kingdom, 1985

References to Chapter 9

[1] Makowski, Z. S.: Räumliche Tragwerke aus Stahl, Verlag Stahleisen, Düsseldorf, 1963
[2] Makowski, Z. S.: Approximate Methods of Analysis of Grid Frameworks. Survey 1979 (Übersetzung von Witte, H.: Raumfachwerke, Beratungsstelle für Stahlverwendung, Düsseldorf, 1980)
[3] Witte, H.: Einfache Regeln zur Vorbemessung von Raumfachwerken, Merkblatt 110, Beratungsstelle für Stahlverwendung, Düsseldorf, 1981
[4] Scheer, J.; Koep, H.: Wirtschaftlich optimierte Raumfachwerke, Forschungsbericht, Projekt 43, Studiengesellschaft für Anwendungstechnik von Eisen und Stahl e.V., Düsseldorf, November 1980
[5] Mengeringhausen, M.: Raumfachwerke aus Stäben und Knoten, Komposition im Raum, Bd. 1, Bauverlag GmbH, Wiesbaden und Berlin, 1975
[6] Emde, H.: Geometrie der Knoten-Stab-Tragwerke, Hrsg.: Strukturforschungszentrum e.V., Würzburg 1979
[7] Girkmann, K.: Flächentragwerke, Springer Verlag, Wien, New York, 1956
[8] Eberlein, H.: Räumliche Fachwerksstrukturen: Versuch einer Zusammenstellung und Wertung hinsichtlich Konstruktion, Statik und Montage, Acier-Stahl-Steel, Heft 2, 1975, S. 50–66
[9] Lacher, G.: Ein neues Raumtragwerk zur Überbrückung großer Spannweiten im Hochbau, Der Stahlbau 7/1977, S. 205–212
[10] Dauner, H.-G.: Einige Gedanken zur Wahl, Konstruktion und Berechnung ebener Raumfachwerke, Acier-Stahl-Steel, Heft 3, 1977, S. 107–112
[11] Demidov, N.; Klimke, H.: Optimierung einiger Parameter bei vorgespannten Raumstabwerken, Bauingenieur 55 (1980), Heft 4, S. 137–140
[12] Massonet, C.; Deprez, G.; Maquoi, R.; Müller, R.; Fonder, G.: Calcul des Structures sur ordinateur. Eyrolles et Masson Éditeurs, Paris, 1972

References to Chapter 10

[1] Gregor, A.: Der praktische Stahlbau, Bd. I, Verlagsgesellschaft R. Müller, Köln-Braunsfeld, 1972
[2] Widenroth, M.: Einspanntiefe und zulässige Belastung eines in einen Betonkörper eingespannten Stabes, Die Bautechnik 12, 1971
[3] Sherif, G.: Elastisch eingespannte Bauwerke, Verlag W. Ernst & Sohn, Berlin/München/Düsseldorf, 1974
[4] Wölfer: Elastisch gebettete Balken, Bauverlag Wiesbaden, Berlin, 1970
[5] Mang, F.: Stützen aus Rechteck-Hohlprofilen mit Einspannung in Betonfundamenten, Schlußbericht, CIDECT-Forschungsprogramm 2J, Versuchsanstalt für Stahl, Holz und Steine, Universität Karlsruhe, September 1979
[6] Beton Kalender, Verlag W. Ernst & Sohn, Berlin/München/Düsseldorf

References to Chapter 11

[1] Klöppel, K.; Goder, W.: Traglastversuche mit ausbetonierten Stahlrohren und Aufstellung einer Bemessungsformel, Der Stahlbau 26 (1957)
[2] British Code BS 5400: Steel, concrete and composite bridges, Part 5, Section 11: Composite columns, 1979
[3] ACI-Building Code 318-77/USA, Section 10.14: Composite compression members, 1977

[4] DIN 18806 Teil 1: Verbundkonstruktionen; Verbundstützen, Ausg. März 1984
[5] Eurocode 4: Design of composite steel and concrete Structures, Part 1.1: General rules and rules for buildinges. ENV 1994-1-1: 1992
[6] Guiaux, P.; Janss, J.: Comportement an flambement de colonnes constituées de tubes en acier remplis de beton, CRIF-Report MT 65, Brüssel, 1970
[7] Janss, J.: Charges ultimes des profils creux remplis de beton charges axialement, CRIF-Report MT 101, Brüssel, 1974
[8] Asourian, P., San, H. K.: Rigid frame connections to concrete-filled tubular steel columns. Triaxial stresses in short concrete-filled tubular steel columns. CRIF-Report MT 86, Brüssel, Jan. 1974
[9] Guiaux, P.; Janss, J.: Noeuds d'ossature comprenant des colonnes tubulaires en acier remplis de beton et des pontes en acier, CRIF-Report MT 103, Brüssel, 1975
[10] CIDECT, Monographie Nr. 5: Calcul des poteaux en profils creux remplis de beton, 1979, Abaques de calcul
[11] Basu, A. K.; Sommerville, W.: Derivation of formula for the design of rectangular composite columns. Proceedings of the Institution of civil Engineers, Supplement volume, paper 7206 S, London, 1969
[12] Neogi, P. K.; Sen, H. K.; Chapman, J.G.: Concrete-filled tubular steel columns under excentric loading. The Structural Engineer, Vol. 47, Nr. 5, London, Mai 1969
[13] Virdi, K. S.; Dowling, P. J.: The ultimate strength of composite columns in biaxial bending. Proceedings of the Institution of Civil Engineers, Vol. 55, Part 2, London, 1973
[14] Virdi, K. S.; Dowling, P. J.: A unified design method for composite columns. IABSE, Mémoires, Vol. 36-II, Zürich, 1976
[15] Bridge, R. O.: Concrete-filled steel tubular columns. Civil Engineering transaction, 1976
[16] Dowling, P. J.; Janss, J.; Virdi, K. S.: The design of composite-steel concrete columns. II. Int. Coll. on Stability, 1977, Introductory Report
[17] Roik, K.; Bergmann, R.; Bode, H.; Wagenknecht, G.: Tragfähigkeit von ausbetonierten Hohlprofilen aus Baustahl, Ruhr-Universität Bochum, Institut für konstruktiven Ingenieurbau, TWM-Heft Nr. 75-4, Mai 1975
[18] Roik, K.; Wagenknecht, G.: Ermittlung der Grenztragfähigkeit von ausbetonierten Hohlprofilstützen aus Baustahl, Bauingenieur, 51. Jahrgang, Heft 5, Mai 1976
[19] Bergmann, R.; Bode, H.; Grube, R.: Die Bemessung von Verbundstützen mit Hilfe von Tabellen und Interaktionsdiagrammen, Universität Bochum, Institut für konstruktiven Ingenieurbau, Heft 32, Vulkan-Verlag, Essen, 1978
[20] Roik, K.; Bergmann, R.; Bode, H.: Einfluß von Kriechen und Schwinden des Betons auf die Tragfähigkeit von ausbetonierten Hohlprofilen, Herausg. Studiengesellschaft für Anwendungstechnik von Eisen und Stahl e.V., Projekt 27, Düsseldorf, 1979
[21] Bergmann, R.: Traglastberechnung von Verbundstützen, Mitteilung Nr. 81-2, Institut für konstruktiven Ingenieurbau, Ruhr-Universität Bochum, Februar 1981
[22] Roik, K.; Bergmann, R.: Berechnung der Traglast von dickwandigen Verbundstützen mit zentrischer und geringer exzentrischer Normalkraft mit beliebig über den Querschnitt verteilten Werkstoffeigenschaften, Bericht Nr. 7801, Institut für konstruktiven Ingenieurbau, Ruhr-Universität Bochum
[23] DIN 18800, Teil 2: Stahlbauten, Stabilitätsfälle, Knicken von Stäben und Stabwerken, November 1990
[24] Merkblatt Nr. 167, Betongefüllte Stahlhohlprofilstützen, Beratungsstelle für Stahlverwendung, Düsseldorf, 1981
[25] Stahlbauhandbuch, Band 1, Stahlbau-Verlags GmbH, Köln, 1982
[26] DIN 4102, Teil 4: Brandverhalten von Baustoffen und Bauteilen; Zusammenstellung und Anwendung klassifizierter Baustoffe, Bauteile und Sonderbauteile, Ausgabe März 1981
[27] Roik, K.; Breit, M.; Schwalbenhofer, K.: Untersuchung der Verbundwirkung zwischen Stahlprofil und Beton bei Stützenkonstruktionen, Projekt P51 der Studiengesellschaft für Anwendungstechnik von Eisen und Stahl, Düsseldorf, Mai 1984
[28] Roik, K.; Breit, M.: Momentfreier Anschluß betongefüllter Hohlprofilstützen, Projekt P52 der Studiengesellschaft für Anwendungstechnik von Eisen und Stahl, Düsseldorf, Oktober 1981
[29] Roik, K.; Schwalbenhofer, K.: Experimentelle Untersuchungen zum plastischen Verhalten von Verbundstützen, Projekt P125 der Studiengesellschaft für Anwendungstechnik von Eisen und Stahl, Düsseldorf
[30] Merkblatt Verbundträger, Beratungsstelle für Stahlverwendung, Düsseldorf, 1987
[31] Bergmann, R.; Matsui, C.; Mainsma, C.; Dutta, D.: Bemessung von betongefüllten Hohlprofil-Verbundstützen unter statischer und seismischer Beanspruchung, CIDECT-Reihe „Konstruieren mit Stahlhohlprofilen", Nr. 5, Verlag TÜV Rheinland, Köln, 1995
[32] ENV 1993-1-1; Eurocode 3: Design of steel structures, part 1.1: General rules and rules for buildings, February 1992, CEN (European Committee of Standardisation)

[33] Packer, J. A.; Fear, C. E.: Concrete-filled rectangular hollow section X and T connections. Proceedings of 4th International Symposium on Tubular Structures, Delft, The Netherlands, pp. 382–391, June 1991

References to Chapter 12

[1] DIN 55928 Teil 1: Korrosionsschutz von Stahlbauten durch Beschichtungen und Überzüge; Allgemeines, Ausg. Nov. 1976
[2] DIN 55928 Teil 2: Korrosionsschutz von Stahlbauten durch Beschichtungen und Überzüge; Korrosionsschutzgerechte Gestaltung, Ausg. Okt. 1979
[3] DIN 55928 Teil 3: Korrosionsschutz von Stahlbauten durch Beschichtungen und Überzüge; Planung der Korrosionsschutzarbeiten, Ausg. Nov. 1978
[4] DIN 55928 Teil 4: Korrosionsschutz von Stahlbauten durch Beschichtungen und Überzüge; Vorbereitung und Prüfung der Oberfläche, Ausg. Jan. 1977
[5] DIN 55928 Teil 4 Beiblatt 1: Korrosionsschutz von Stahlbauten durch Beschichtungen und Überzüge; Vorbereitung und Prüfung der Oberflächen, Photographische Vergleichsmuster, Ausg. Aug. 1978
[6] DIN 55928 Teil 5: Korrosionsschutz von Stahlbauten durch Beschichtungen und Überzüge; Beschichtungsstoffe und Schutzsysteme, Ausg. März 1980
[7] DIN 55928 Teil 6: Korrosionsschutz von Stahlbauten durch Beschichtungen und Überzüge; Ausführung und Überwachung der Korrosionsschutzarbeiten, Ausg. Nov. 1978
[8] DIN 55928 Teil 7: Korrosionsschutz von Stahlbauten durch Beschichtungen und Überzüge; Technische Regeln für Kontrollflächen, Ausg. Febr. 1980
[9] DIN 55928 Teil 8: Korrosionsschutz von Stahlbauten durch Beschichtungen und Überzüge; Korrosionsschutz von tragenden, dünnwandigen Bauteilen (Stahlleichtbau), Ausg. März 1980
[10] DIN 55928 Teil 9: Korrosionsschutz von Stahlbauten durch Beschichtungen und Überzüge; Bindemittel und Pigmente für Beschichtungsstoffe, Ausg. Aug. 1982
[11] DASt-Richtlinie 007 – für die Lieferung, Verarbeitung und Anwendung wetterfester Baustähle, Stahlbau-Verlags-GmbH, Köln, Februar 1970
[12] Friehe, W.; van Oeteren, K. A.; Schwenk, W.: Korrosionsschutz im Stahlbau-Leistungsbereich DIN 55928, Merkblatt 259, Beratungsstelle für Stahlverwendung, Düsseldorf, 1982
[13] Buchholz, H.; Kleingarn, J.-P.; Schier, K. H.: Feuerverzinkungsgerechtes Konstruieren im Stahlbau, Merkblatt 359, Beratungsstelle für Stahlverwendung, Düsseldorf
[14] Kranitzky, W.: Das Verhalten der Innenflächen geschlossener Hohlbauteile aus Stahl hinsichtlich Feuchtigkeitsniederschlag und Korrosion, Der Stahlbau, Heft 7, 1983
[15] DIN 18808: Stahlbauten, Tragwerke aus Hohlprofilen unter vorwiegend ruhender Belastung, Ausg. Okt. 1984
[16] Seils, A.; Kranitzky, W.: Sind Stahlbauwerke, bei denen allseits geschlossene Hohlkörper verwendet wurden, durch Wassersammlung und Innenkorrosion gefährdet? Der Stahlbau 22 (1953), Nr. 4, S. 80–84 und Nr. 5, S. 113–118
[17] Seils, A.; Kranitzky, W.: Die Verwendung geschlossener Hohlquerschnitte im Stahlbau und ihr Korrosionsschutz, Eisenbahntechnische Rundschau (ETR), Sonderausgabe 4 „Brückenbau und Ingenieurhochbauten", Juli 1954, S. 119–135
[18] Seils, A.; Kranitzky, W.: Der Korrosionsschutz im Inneren geschlossener Hohlkästen, Der Stahlbau 28 (1959), Nr. 2, S. 46–53
[19] Kranitzky, W.: Untersuchung der Innenkorrosion bei Stahlrohrkonstruktionen und geschlossenen Kastenquerschnitten aus Stahlblechen, Bericht im Auftrag der Studiengesellschaft für Anwendungstechnik von Eisen und Stahl e.V., Düsseldorf
[20] Tournay, M.: La résistance à corrosion de l'interieur des profiles creux en acier (Korrossionsfestigkeit im Inneren geschlossener Hohlprofile), Studie im Auftrag von CSFTA und CIDECT, Paris, 1978. Auszugsweise veröffentlicht in Acier-Stahl-Steel 43 (1978), Nr. 2, S. 67–75
[21] Seils, A.: Optimaler Korrosionsschutz der Stahlbauwerke, Der Stahlbau 37 (1968), Nr. 3, S. 72–81
[22] Marberg, J.: Verminderung der Verzugsgefahr beim Feuerverzinken, Verzinken 6 (1977), S. 17–19
[23] Horstmann, D.: Das Verhalten mikrolegierter Baustähle mit höherer Festigkeit beim Feuerverzinken, Archiv des Eisenhüttenwesens 46 (1975), S. 137–141
[24] Peterson, Ch.: Dauerfestigkeit von Schweißverbindungen nach Überschweißung der Feuerverzinkung, Der Stahlbau 46 (1977), S. 277–282
[25] Kranitzky, W.: Klimatische Bedingungen und Korrosion im Inneren großer Hohlkästen aus Stahl, Der Stahlbau, Heft 2, 1983
[26] Kleingarn, J.-P.: Feuerverzinken von Einzelteilen aus Stahl, Stückverzinken, Merkblatt 293, Beratungsstelle für Stahlverwendung, Düsseldorf, 1980
[27] Van Oeteren, K. A.: Feuerverzinken und Beschichten = Duplex-System, Merkblatt 329, Beratungsstelle für Stahlverwendung, Düsseldorf, 1981
[28] Horstmann, D.: Das Feuerverzinken siliziumhaltiger Stähle, Gemeinschaftsausschuß Verzin-

ken e.V. Düsseldorf, Vortrags- und Diskussionsveranstaltung 1974, Vortragstexte, S. 9–33

[29] DIN 8565: Korrosionsschutz von Stahlbauten durch thermisches Spritzen von Zink und Aluminium; Allgemeine Grundsätze, Ausg. März 1977

[30] DIN 8566 Teil 1: Zusätze für das thermische Spritzen; Massivdrähte zum Flammspritzen, Ausg. März 1979. Teil 2: Zusätze für das thermische Spritzen; Massivdrähte zum Lichtbogenspritzen, Technische Lieferbedingungen, Ausg. Dez. 1984

[31] DIN 8567: Vorbereitung von Oberflächen metallischer Werkstücke und Bauteile für das thermische Spritzen, Ausg. Aug. 1984

[32] Friehe, W.; Schwenk, W.: Entwicklungsarbeiten zum Korrosionsschutz von Stahlbauten durch Beschichtungen, Der Stahlbau 50 (1981), S. 115–120

[33] von Oeteren, K. A.: Werkstättengrundanstriche – Ablieferungsanstriche und Walzstahlkonservierung mit Shop-Primer (eine technische Bestandsaufnahme), Werkstoffe und Korrosion 16 (1965), Heft 7

[34] DASt-Richtlinie 006: Vorläufige Richtlinien für die Auswahl von Fertigungstrichen bei der Walzstahlkonservierung im Stahlbau, Stahlbau-Verlags GmbH, Köln, Juni 1968

[35] DIN EN ISO 1461: Durch Feuerverzinken auf Stahl aufgebrachte Zinküberzüge (Stückverzinken)

References to Chapter 13

[1] Stahlbau-Kalender, jährliche Neuausgabe, Herausgeber: Deutscher Stahlbauverband, Ebertplatz 1, Köln

[2] Bauaufsichtliche Brandschutzanforderungen an Bauteile aus Stahl, eine zusammenfassende Darstellung der bauaufsichtlichen Vorschriften, Verordnungen und Normen, Stahlbau-Verlag, Köln

[3] DIN 4102, Beiblatt 1: Brandverhalten von Baustoffen und Bauteilen; Inhaltsverzeichnisse, Ausg. Mai 1981; Teil 1: Brandverhalten von Baustoffen und Bauteilen; Baustoffe; Begriffe, Anforderungen und Prüfungen, Ausg. Mai 1981; Teil 2: Brandverhalten von Baustoffen und Bauteilen; Bauteile, Begriffe, Anforderungen und Prüfungen, Ausg. Sept. 1977; Teil 3: Brandverhalten von Baustoffen und Bauteilen; Brandwände und nichttragende Außenwände, Begriffe, Anforderungen und Prüfungen, Ausg. Sept. 1977; Teil 4: Brandverhalten von Baustoffen und Bauteilen; Zusammenstellung und Anwendung klassifizierter Baustoffe, Bauteile und Sonderbauteile, Ausg. März 1981; Teil 5: Brandverhalten von Baustoffen und Bauteilen; Feuerschutzabschlüsse in Fahrschachtwänden und gegen feuerwiderstandsfähige Verglasungen, Begriffe, Anforderungen und Prüfungen, Ausg. Sept. 1977; Teil 6: Brandverhalten von Baustoffen und Bauteilen; Lüftungsleitungen, Begriffe, Anforderungen und Prüfungen, Ausg. Sept. 1977; Teil 7: Brandverhalten von Baustoffen und Bauteilen; Bedachungen, Begriffe, Anforderungen und Prüfungen, Ausg. Sept. 1977; Teil 8: Brandverhalten von Baustoffen und Bauteilen; Kleinprüfstand, Ausg. Mai 1986

[4] Fire Resistance Tests of Structures. ISO Recommendation R834, International Organization for Standards, 1968

[5] Wassergefüllte Tragwerke, Merkblatt 467, Beratungsstelle für Stahlverwendung, Düsseldorf, 1981

[6] Giddings, T. W.: Fire resistent constructions using HSS. International Symposium on Hollow Structural Sections, Toronto, May 1977

[7] Instruct Ing.: Brandverhalten von Stahl- und Stahlverbundkonstruktionen, Stahlkonstruktionen mit Wasserkühlung, Stud. Ges. P86, Studiengesellschaft für Anwendungstechnik von Eisen und Stahl e.V., Düsseldorf, 1981

[8] Polthier, K.: Entwicklung und Anwendung wassergefüllter Stützen im Hochbau, Forschungsbericht EGKS, EUR 5317d 1975, zu beziehen durch: Verlag Bundesanzeiger, Köln

[9] Klingsch, W.; Würker, K. G.: Hohlprofil-Verbundstützen, sichtbarer Stahl für feuerwiderstandsfähige Konstruktionen, DBZ 11/82

[10] Quast, U.; Rudolf, K.: Brandverhalten von Stahl und Stahlverbundkonstruktionen – Bemessungshilfe für Verbund-Stützen mit definierten Feuerwiderstandsklassen, Band 1, 2, 3, Studiengesellschaft für Anwendungstechnik von Eisen und Stahl e.V., Düsseldorf, 1985, Projekt BMFT Bau 6004/Stud.Ges. P86/Akt. 2.3

[11] Kordina, K.; Klingsch, W.: Brandverhalten von Stahlstützen im Verbund mit Beton und massiven Stahlstützen ohne Beton, Institut für Baustoffe, Massivbau und Brandschutz, Technische Universität Braunschweig, Abschlußbericht zum Forschungsprojekt P35 der Studiengesellschaft für Anwendungstechnik (EGKS 6210/SA/1-108), 1983

[12] Klingsch, W.: Grundlagen für die rechnerische Ermittlung des Tragverhaltens von Bauteilen im Brandfall, Bauphysik 1 (1979), Heft 1, S. 29–33

[13] Klingsch, W.: Grundlagen der brandschutztechnischen Auslegung und Beurteilung von Verbundstützen, Bauphysik 3 (1981), Heft 4, S. 129–133

[14] Grandjean, G.; Grimault, J. P.; Petit, L.: Determination de la durée au feu des Profils creux remplis de Beton, Report Cometube, Paris (CIDECT 15B/80-10, CECA 7210/SA/3/302), 1980

[15] Kordina, K.; Klingsch, W.: Fire resistance of composite columns of concrete filled hollow sections, Report CIDECT 15C/83-27, 1983
[16] DIN 1045: Beton und Stahlbeton; Bemessung und Ausführung, Ausgabe Dez. 1978
[17] Witte, H.; Schwenk, W.: Stahlhohlprofile mit Wasserkühlung für den baulichen Brandschutz, VDI-Sonderdruck, Vortrag anläßlich einer Tagung der VDI-Gesellschaft Werkstofftechnik „Korrosion und Korrosionsschutz metallischer Bau- und Installationsstelle innerhalb Gebäuden" – Außenkorrosion der Bau- und Installationsstelle – 29. und 30. November 1984 in Mannheim
[18] Haß, R.; Quast, U.: Brandverhalten von Verbundstützen mit Berücksichtigung der unterschiedlichen Stützen/Riegel-Verbindung, Studiengesellschaft für Anwendungstechnik von Eisen und Stahl e.V., Düsseldorf, 1985, Projekt BMFT Bau 6004/Stud. Ges. P86/Akt 2.2
[19] BS 5950: The structural use of steelwork in buildings, Part 8: Code of practice for fire resistant design. British Standards Institution BSI, 1990
[20] Underwriters Laboratory: Fire tests of building construction and matrials. UL263, USA, 1991
[21] IMO: Recommendation on fire test procedure for "A", "B" and "F" class divisions. IMO Resolution A.517(13), November 1983
[22] ENV 1993-1-2, Eurocode 3: Design of steel structures, Part 1.2: Structural fire design. CEN, 1994
[23] ENV 1994-1-2, Eurocode 4: Design of composite steel and concrete structures, Part 1.2: Structural fire design. CEN, 1994
[24] ENV 1992, Eurocode 2: Design of concrete structures, Part 10: Structural fire design of concrete structures. CEN, 1994
[25] EN 1991-2-2, Eurocode 1: Basis of design and actions on structures, Part 2.2: Actions on structures exposed to fire. CEN
[26] Twilt, L.; Hass, R.; Klingsch, W.; Edward, M.; Dutta, D.: Design guide for structural hollow section columns exposed to fire. CIDECT series "Construction with steel hollow sections", No. 4, published by TÜV Rheinland, Cologne, 1994
[27] Steel Promotion Committee of Eurofer: Steel and fire safety, a global approach. Eurofer, Brussels, 1990
[28] Twilt, L.; Both, C.: Technical notes on the realistic behaviour and design of fire exposed steel and composite structures. Final report ECSC 7210SA112, Activity D: Basis for technical notes, TNO Building and Construction Research, BL-91-069, 1991
[29] CEFICOSS: Computer engineering of the fire resistance for composite and steel structures; Computer code for both thermal and mechanical response of steel and composite structures exposed to fire. ARBED, Luxembourg
[30] COMSYS-T: Computer code for the determination of the ultimate load capacity in fire case. Wuppertal University, Institute for Structural Engineering and Fire Safety, Germany
[31] STABA-F: Computer code for the determination of load bearing and deformation behaviour of uni-axial structural elements (beams, columns) under fire action. Technical University of Brunwick, Brunswick, Germany
[32] DIANA: DIsplacement method ANAlyser, a general purpose finite element programme suitable for the calculation of geometrical and physical non-linear problems. TNO Building and Construction Research, Rijswijk, The Netherlands
[33] Bond, G. V. L.: Fire and steel construction; water cooled hollow columns. The Steel Construction Institute, Ascot, United Kingdom, 1975
[34] Klingsch, W.: Optimization of cross sections of steel composite columns. Proceedings of the 3rd International Conference on Steel-Concrete Composite Structures, Fukuoka, Japan, 1991, pp. 99–105
[35] Twilt, L.; Haar, v.d. P. W.: Harmonization of the calculation rules for the fire resistance of concrete filled SHS columns. CIDECT report 15F-86/7-0, IBBC-TNO report B-86-461, August 1986
[36] Twilt, L.: Design charts for the fire resistance of concrete filled HSS columns under centric loading. Final report, CIDECT project 15J, TNO report BL-88-134, August 1988
[37] Design for SHS fire resistance to BS 5950, Part 8. TD 361/5E, British Steel plc, Tubes and Pipes, Corby, Northants, May 1993
[38] Roik, K.; Bergmann, R.; Haensel, J.; Hanswille, G.: Verbundkonstruktionen, Bemessung auf der Grundlage des Eurocode 4, Teil 1, Betonkalender, Verlag Ernst & Sohn, Berlin, 1993
[39] Hass, R.: Reinforcement of concrete by steel fibres in composite columns; simplified manufacture and defined fire-resistance. Final report, CIDECT research project 15L, 1991

References to Chapter 14

[1] Schulz, G.: Der Windwiderstand von Fachwerken aus zylindrischen Stäben und seine Berechnung, Interntionaler Normenvergleich für die Windlasten auf Fachwerken, Monographie Nr. 3, CIDECT (Comité International pour le Developpement et l'Etude de la Construction Tubulaire), Düsseldorf, 1970
[2] Constrado: Wind forces on unclad tubular structures. Croydon, England, January 1975

[3] DIN 1055, Teil 4: Lastannahmen für Bauten, Verkehrslasten, Windlasten bei nicht schwingungsanfälligen Bauwerken, Ausgabe August 1986

[4] Wichmann, K.: Windkraftmessungen an Quadratprofilen mit verschiedenem Eckradius, Bericht der DFVLR, IB 157-79C 28

[5] Richter, A.: Wind forces on square sections with various corner radii, investigations and evaluations. Institute for hydromechanics, University of Karlsruhe, March, 1984

[6] Gould, R. W. F.; Raymer, W. G.: Measurements over a wide range of Reynolds numbers of the wind forces on models of lattice frameworks with tubular members. National Physical Laboratory (NPL), Sci. Rep. No. 5-72 (Department of Trade and Industry)

[7] Zuranski, J. A.: Windbelastung von Bauwerken und Konstruktionen, Verlagsgesellschaft R. Müller, Köln, 1969

[8] Rosemeier, G.: Winddruckprobleme bei Bauwerken, Springer-Verlag, Berlin, Heidelberg, New-York, 1976

[9] Schulz, G.; Hayn, F.: Widerstandsmessungen an Systemteilen von Rohrkonstruktionen, Teil I: Rohre und ebener Rohrknoten, Bericht AM 506 (1-NK-I-66-19), Nov. 1966

[10] Hayn, F.: Widerstandsmessungen an Systemteilen von Rohrkonstruktionen, Teil II: Dreidimensionaler Mastschuß und Ebene eines Mastschusses, Bericht AM 507 (1-NK-I-67-30), August 1967

[11] Schulz, G.; Hayn, F.: Widerstandsmessungen an Systemteilen von Rohrkonstruktionen, Teil III: Einfluß des Durchmessers und der Oberflächenstruktur auf den Widerstand von Zylinder, Bericht 1-NK-I-68-34, April 1968

[12] Delany, N. K.; Sorensen, N. E.: Low speed drag of cylinders of various shapes. NACA Technical Note 3038, 1953

[13] Roshko, A.: Experiments on the flow past a circular cylinder at very high Reynolds number. Journal of Fluid Mechanics 10, S. 345–356, 1961

[14] Fa. Stewarts & Lloyds: Aerodynamic drag and shielding. Stewarts & Lloyds, Department of Research and Technical Development, Corby, Report No. E. 56/22, July 1960

[15] ENV 1991, Eurocode 1: Basis of design and actions on structures, Part 2.4: Wind actions. CEN, 1994

[16] Richter, A.: Wind forces on square sections with various corner radii. CIDECT report 9D, Supplement, Institute for hydromechanics, University of Karlsruhe, August 1986

[17] Hoerner, S. F.: Fluid-dynamic drag. published by the author, New York, 1964

[18] Biriuliu, A. P.: Aerodynamiczeskuje charakteristiki stierzniej: rieszetczatych fierm, 1974

[19] Richter, A.: Wind forces on square sections with various corner radii. Final report, CIDECT project 9D, August 1986

Subject index

A
abutment   102
acoustic transmitter   30, 346
actions   53
– accidental   53
– permanent   53
– variable   53
active gas   27
additives   400
adhesion   408
adjustable support   97
aesthetic reasons   184
air conditioning   7
airtight   406
alloys   27
alternating load   232
analytical model   150, 180
angle of inclination   133 f., 140 f., 206
annealing   29
anode   413
arched girders   102
arched lattice girder   11, 130
arches   102
arching   327
assembly   317, 361
– frame   361
– preassembly   361
– subassembly   361
asymmetrical joint   134
auxiliary construction   365
axial compression   57, 62, 86
axial force   56, 58, 61, 105
axial load   180
axial tension   56

B
backing ring   122, 125
barrel-type piercer   22
base steel   35
bath   410

beams   94
– castellated   94
– single section   94
bearing length, effective   178
bearing pressure resistance   117, 120
bell method   2
bending   327
– bi-axial   58, 128
– cold   329
– hot   331
– uni-axial   57 f., 128
bending moments   105, 180, 183 ff.
– in-plane   208, 213, 215
– out-of-plane   215, 219
– primary   145, 185, 283
– secondary   145, 183, 207, 283, 293, 295
bending stiffness   54
bevel   26
bi-axial bending   58, 78, 80, 128, 388,
bi-axial bending moment   86
billet   22
bill-shaped   226
bird mouth   226
blind bolting   333
blind bolts   93
blockage   479
bloom   21, 25
bolts
– arrangement   119
– contact   113
– high strength   111
– pitch   114 f.
– shaft diameter   119
– shear   117
– tightening   311
– torque   110
bolted joints, detachable   204
bolting   332
– blind   333
bondage   395

bonding agents 408
bracing 82 f., 131
– effective width 151
– efficiency 176
– end preparation 194
– periphery 146
– pin-ended 186
– plate stiffener 218
brittle fracture 41 f.
brushing 409
buckling
– curve 62 ff., 74, 162
– in-plane 85
– local 155, 209
– out-of-plane 85
– resistance 62
– slenderness 72
– stiffness 54
– stress 75, 161
buckling length
– coefficient 82
– effective 66, 82 f., 84 ff., 87, 184, 203
buckling slenderness 72
– flexural 67
– local 67 f.
– plate 68
buckling stress
– critical 153, 162
– ideal 74
butt straps 407
butt weld 122

C
camber 136
cambering 332
Carbon contents 336
Carbon Equivalent Value 336
carry-out effect 250
carry-over brace 272
cast steel 352
castellated beams 94
cathode 413
CEN 8, 45
certificates of competency 347
characteristic resistance 154
characteristic values 52
characteristic yield strength 79, 82
chard side wall plate 174
Charpy-V 30
– test 42

chemical composition 33, 39
chord 82 f., 131
– distortional failure 211
– intersection 164
– plastification 157, 159, 215, 222
– punching shear 215
chord flange 180
– plastification 155, 161 f., 164, 167 f., 171, 173 f., 211, 224
– shear 180
– stiffener 218
chord joint, separated 205 f.
chord shear 164, 176, 181, 190
– failure 209
– yield model 153 f.
chord side wall
– buckling 173
– failure 171
– shear 153, 173
chord web
– bearing model 153
– buckling 162
– failure 158, 161, 209 f., 211 f., 224
– shear 206
– stability 176
– yielding 176, 226
clamping 311
classification 295
– method 293
cleaning 408
coating 9, 408, 410
– primer 318
– top 318
coils 25
cold bending 329
cold drawing 29
cold formed 36, 42, 45
column axis 61
column bases 96
columns 94
– flexural 94
– lateral torsion 94
– lattice 95
– local buckling 94
– plane-cut 94
– single leg 95 f.
compact 55
composite sections 381 f.
compression resistance 57
concavity 124
concavity/convexity 47

concentric 61, 383
concrete
– core 396
– filling 381
– grade 433
– fundament 377
conduction 424
conductive electric current 28
configuration 185
connection plate 205
connections 127
– direct 127
– indirect 127
– mixed 127
– symmetrical 118
– unsymmetrical 118
console 98 f.
constant amplitude load 235
contact bolts 113
contact face 313
continuous pressure welding 28
continuous rolling 21, 25
controlled torque 311
convection 434
conversion factor 249
cooling bed 25
corner area 90
corner radii 274
corner radius 42, 140
– external 48
– internal 48
– outside 46
corrosion 128, 175, 184, 238
– external 407
– internal 405
– protection 183
corrosive environment 238
couplant 31
cover plate 117
crack 155, 164, 178, 303
– initiation 148, 179, 194
– stopper 306
cradles 361
cranked chord 175
crash barrier 121
crimping press 26
crippling 100
critical buckling stress 153
critical elastic stress 57
critical plate buckling stress 72
cropped end 135, 194 f.

cropping 203
– machine 196
cross section classes 70 f., 76, 77, 80
cross section classification 55
cross-sectional area 48
crown 142, 242, 246 f.
cube compressive strength 383
cumulative load 235
curvature 367
cut-off limit 295
cuts
– double 323
– mitre 330
– plane 323
– profile 323
cutting 318
– laser 325
– plasma 325
cyclic loading 238
cylinder compressive strength 383

D
deflection 99, 128
deformation 100, 146, 152, 164
– capacity 154, 217
– local 147
degreasing 408
density 54
deoxidation 35
depth of fixity 378
descaling 409
design
– bearing pressure resistance 119
– criterion 187 f.
– effect 51
– loading 51
– moment resistance 211
– resistance 51, 56, 145, 154, 157, 161, 166, 173 f., 176, 181, 187 f., 215, 221 f., 224, 226
destructive tests 30
detachable connections 103 f., 108
detail categories 294, 298
diagonal member 301
dial gauge 208
diaphragms 100
dime test 310
disc 310
disc test 310
discontinuity in shape 238
dislocation 133

displacement effects   470
distance pieces   99, 194, 198
double cut   323
double plane cut   133
double profile cut   133
double-chord truss   204 f.
double-tongued forks   109
dowel   396 f.
drag coefficients   467
drawing die   23
ductility   41 f., 149
Duplex system   411
dye penetration test   31, 345
dynamic pressure   475
dynamometer   208

E
eccentricity   100, 117, 168, 185, 192
– joint   144
– moment   145, 185
edge miller   27
effective (reduced) width   69 f.
effective area   57, 178
effective area coefficient   227
effective length   178
effective periphery   179
effective width   57, 69, 72, 150, 159, 161, 165 f., 173 f., 176, 181, 211 f., 226
efficiency   212 ff.
elastic   75
– analysis   293
– calculation   57, 80
– design   59
– design bending moment resistance   80
– limit   148 f.
– resistance pressure welding   28
– section modulus   48 f., 77, 212
– stress   70 f.
elastic-plastic   75
electrochemical polarisation   413
electrode   27, 306, 337
electrolyte   413
elongation   30, 34, 38 ff., 149
– to failure   41
embedment modulus   378
end fixity   183, 377
end-to-end connection   108, 115, 122
endurance limit   233
equilibrium of loads   144
equivalent stress   61, 138

equivalent uniform moment factor   78 f., 80 ff., 87
Euler curve   68
Eulerian buckling load   66
Eulerian buckling stress   66
Eulerian slenderness   66
Eurocode   8
European buckling curve   66, 73, 84
external corrosion   407
extrusion
– mandrel   22, 25
– process   21 f., 25

F
fabrication   317, 399
failure
– criterion   168
– critical   154
– mode   153, 155, 171, 180, 182, 208, 210, 214, 219, 303
– model   209
– probability   234
fatigue
– damage   235
– life   233
– limit   295
– lines   286
– loading   122, 291
– resistance   233
FCAW   336, 339
field assembly   94
field weld   94
field welding   143
filler metal   26, 28
fillet weld   122, 135 ff., 139, 274
– concave   175
fin plate   220
fine grain   36
finite elements   190, 215, 217, 243, 245
fire
– design   417
– insulation   421
– protection   101
– resistance   416
– resistance class   433
fish plate   117, 120
fitting   308
– tolerance   133
fixity parameter   258
flame cut   194

flame cutting   318
– machine   128
flank   142
flattened ends   84, 193 f., 203
flattened edge   195
flattening   203, 327
– cold   196
– hot   195
– non-symmetrical   195
– partial   195, 197 f.
flexibility   99
flexural (overall)   61
flexural buckling   61, 67 f., 78
floor slab   101
flowdrill   333
fork ends   108 f.
fork plate   116 f.
forming rolls   28
freezing   407
Fretz-Moon   28
friction welding   337
full penetration groove weld   186
fully killed steel   35
funnel   402 f.
fusion, full   123
fusion welding   27, 337, 354

G

galvanization   410
gamma rays   346
gap   130, 132 f., 140, 144, 157 ff., 165, 301
gas-shielded arc welding   27
gauge   135 f., 149
– weld measuring   137
– self-made   124, 136
geometrical imperfection   62, 75
geometrical parameters   251
girders
– arched lattice   130
– deflection   217
– double-chord   206
– Howe type   129
– quadrangular   186, 189, 203 f.
– Pratt type   129
– triangular   186, 189, 203 f.
– Warren type   129
gliding modulus   89
GMAW   336, 338, 363
gouging   306 f.
gravity   401

grid   367
grinding   242, 306, 324
grits   410
groove weld   143, 274
– complete penetration   123, 139 ff., 203
– partial penetration   126
groove welding   122
grout   398
gusset plate   9, 93, 103, 220 ff.

H

hammer peening   308 f.
hardening   42
hardness   77
haunch   106 f.
haunch stiffener   218 f.
head plate   98
heat affected zone   326, 336
heat analysis   39
heel   186, 242, 246
helix   26
high strength bolt   111
high-cycle fatigue   43
high-frequency induction welding   28
hinge   100
hinged supports   82, 97
holing   333
hollow shell   22
hopper   402
hot bending   331
hot finished   36, 45
hot flattening installation   194 f.
hot rolling   21
hot spot strain   243
hot spot stress   239, 243
hot spot stress range   247
housing   7
Howe type girder   129
Huck Ultra Twist   334
humidity   405

I

identification labels   318
IIW   8
impact bend test   30
impact energy   40
impact text   34
imperfection   67, 74, 77 f.
– factor   66
impurities   42
inaccuracy   133

inclination 295
- angle of 138
indentation 310
inert gas 27
influence function 250, 256
ingot 21, 23
in-plane buckling 103
inside diameter 48
inspection
- ultrasonic 346
- visual 345
interaction 67, 179, 212
- curves 385 f.
- linear 214
- surface 393
interference 481
internal corrosion 405
interpolation 161 f.
intersection length 162
intumescent coatings 422
ISO 8, 45

J
jig 140, 148 f.
joints
- asymmetrical 134
- centricity 185
- configuration 198
- eccentricity 144, 163, 180, 191, 295
- geometry 185
- mixed combination 134
- multi-planar 141, 189 ff.
- parameters 160, 163, 177, 183
- rigidity 83, 215
- strength 156
- uni-planar 141

K
kinematic viscosity 469
knee joints 103 ff.

L
ladle analysis 30, 34
lamellar tearing 42, 112, 154
laser cutting 325
laser welding 349
lateral stability 85
lateral support 186
lateral-torsional buckling 57 f., 61, 78, 82
lateral-torsional stability 94, 128, 169

lattice column 95
lattice geometry 128
lattice girder 82 ff., 85, 126, 158
- multi-planar 126
- quadrangular 126
- single plane 128
- triangular 126
- uni-planar 126
leg depth 123
levelling 398
limit states 51
Lindapter Hollo-Fast 334
linear expansion, coefficient of 54
linear extrapolation 245
linear scale 234
liquidiser 400, 430
loads
- introduction region 396
- random loading 231
- regular loading 231
- reorientation 354
- sine-curve loading 231
- spectrum 235, 293
- transmission 151, 354
loading cycles 233
- to failure 234
local buckling 57, 61, 66 f., 74 ff., 153, 155, 209
- curve 70
logarithmic scale 234
long term effect 384
longitudinal plate 224
low cycle fatigue 248
low temperature steel 35

M
macrography 30
MAG welding 26, 336, 338, 353
magnetic particle test 31, 345
magnetic stray flux test 31
major axis 48 f.
mandrel bar with plug 22
Mannesmann, Max 2
Mannesmann, Reinhard 2
mantle 27
mashes 367
mechanical expanding 27
mechanical properties 34
membrane 367
- action 151, 156
- effect 210

microstructure 33, 40
MIG welding 336, 338, 353
minor axis 48
mitre cuts 330
– inductive heating 331 f.
mixed combination 139
– joints 134
mixed torsion 89
MMA 353
modules 369
modulus of elasticity 54, 66 f., 89
moment of inertia 3
moment resistance 207, 216
mortar 397 f.
multi-axial bending 128
multi-axial stress 30
multi-planar 247, 273
– joints 250, 278
– lattice structure 12
Murdoch, William 2

N
nail shear failure 352
nailing 351
neutral axis 387
– shift 73
nodal connectors 370
nominal stress 239
nominal stress range 295
non-alloy mild steels 43
non-alloy structural steels 33
non-destructive tests 30
non-linearity 77
non-rimming 33
non-rimming steel 34
normalizing rolling 35, 37
notch effect 43
nozzle 27
numerical simulation 62

O
obtuse angle 105 f.
offcut 106 f.
optimisation 373
oscilloscope 346
out-let hole 7
out-of-plane 83
outside diameter 45, 48
overlap 130, 132, 144, 157 ff., 301
– full 133
– grade 132

– partial 132 f., 134 f., 140, 142
overloading 308 f.

P
paint 184
Palmgren-Miner rule 235
panel points 145
parametric formulae 251
partial safety factors 52, 56, 65, 77, 154
peak load 154
peak stress 147, 164, 239
peening hammer 308 f.
peening shot 308 f.
penetration groove weld 138
performance qualification 347
pierced bloom 24
piercing mandrel 21 f., 24
piercing mill 25
piercing press 23
pigments 408
pilger mandrel 22
pilger rolling 21, 23
pilger rolls 22
pin joint 127, 203, 207
plane cut 128, 132 f., 193, 322
– double 134
plane trusses 156
plasma 309, 311
– cutting 325
plastic
– bending moment 52
– bracing moment, transmission 218
– calculation 57
– deformation 147
– design 146
– design resistance 86
– failure load 150
– hinge 114
plastic moment 164
– plateau 41
– resistance 58, 60, 61, 216, 218
– section modulus 48 f., 105, 212 f., 218 f.
– shear 52
– stress 70 f., 210
plastification, local 145
plastifier 400, 430
plate buckling 67
– factor 67, 72
– reduction factor 69
– slenderness 68

– stress 68
– zone 68
plates
– connection 205
– cover 117
– fin 220
– gusset 220 ff.
– longitudinal 224
– splice 117
– stiffening 205
– tie 205
– tip 116
– transversal 224
plug 22
– rolling 21 f., 24
– welding 121, 341
pneumatic pusher 25
Poensgen, Albert 2
point stiffness, distribution 146
Poisson's coefficient 54
Poisson's ratio 67, 89
poker 403
polar moment of inertia 57, 89
pollution 405
post-heat treatment 342
post-weld heat treatment 308 f.
Pratt truss 175
Pratt type girder 129
preassembly 361
pre-deformation 342
prefabrication 371
pre-heat treatment 336, 350
prequalified joint 139
preset shoe 98
pressing 329
pressure distribution 378
pressure rollers 28
pressure welding 337
prestressing load 200 f.
primer coating 318
principal strains 243
product analysis 39
profile 325
profile cut 133, 193, 318, 323
– double 133
– end 203
protection 408
prying force 111, 113 f.
pulsating load 232
pumping 401
punching 158

– shear 155, 157 f., 161, 164, 173, 209, 221 f.
– shear area 151
– shear in chord 168
– shear model 150 ff.
– shear strength 151
purlin 169 f., 220
– continuous 170 f.
– lattice 170 f.
push bench 21 f., 24
pyramid stiffener, truncated 219

Q
quadratic extrapolation 245
quality control 185
quenched and tempered 40

R
radiographic inspection 346
radius of gyration 48, 66
– effective 73
rail transport 363
rain water down pipe 97
random loading 231
range spacers 115
real buckling stress 76 f.
recessed die 196 f.
reconstruction 303
redistribution of stresses 145
reducing mill 24
reducing process 28
reduction factor 63 ff., 77, 80, 105 ff., 191 f., 201, 214
reeling 26
reference brace 272
reference effect 250
reflector 30
regression analysis 199, 251
regular loading 231
reinforcement 9, 107, 196, 217, 303 ff., 381, 383, 400
– cage 404
repair 303
residual stress 62, 77, 141, 236, 342, 411
– range 295
– technique 308
resistances 54
rewelding 307
Reynolds Number 467
rigidity 210, 217
rimming steel 34

Subject index

ring inductor   28
ring model   150
road transport   363
rolling process   22
root depth   124
rotary burr   310
rotary hearth   23
rotary hearth furnace   25
rotation capacity   41, 54, 99, 105, 146
rotation frames   362
rotational capacity   207
rotational rigidity   215 f.
rotational stiffness   207
roughness   468, 479
rust   405

S
saddle   142, 245, 247, 318
– location   150
safety factor   200
– partial   105
Saint-Venant torsional moment   88
sand blasting   311
saw   194
SAW   336, 339
sawing   322
scatter   162, 235
screwed coupling   121
seam legs   136
seamless tubes   21
section modulus, effective   57, 70, 73
second moment of area   48
second moment of inertia   59, 66
section factor   419
sectional properties   2, 45, 47, 54
– effective   69, 73
semi-compact   55
semi-rigid   210, 217
– joints   145
serviceability   154, 186
– deformation limit   154, 158
– limit   162
– limit state   51
shaft diameter   119 f.
– of bolt   119
shaping   29
– press   26
sharp edges   7
shear   26, 58, 61, 105
– area   60, 162, 165
– failure   224

– force   59 f.
– load   180
– modulus   54
– resistance   120, 223
– stress   151
shearing off   324
shell buckling   77
shell slenderness   77
shielding factor   475
shop primer   409
shot blasting   318
shot peening   308 f.
shrink   141
shrinkage   411
side length ratio   67 f.
simulation calculation   66
sine-curve loading   231
single leg column   95 f.
single section beams   94
size rolling   23
sizing
– mill   24
– tool   27
skelp   28
skew roll piercing   21 ff.
sleeve   121
slenderness   65, 66, 73 f., 79, 153, 383, 477
slip-resistant   56
slit cylinder   26
slots   115
slotting   326
SMAW   336
snow trap   7, 128
socket wrench   110
solid bar   108
solidification simulation   357
solidity ratio   470
space structure   189, 367
space truss   205
space tube   118
span   85, 128
spatial variation   53
specific fire load   415
spiral weld   26
splice plate   117
– connection   299 f.
spray   184
spraying   409 f.
squareness   46
stability   83
– cases   61

573

standard fire curve   415
stand-off height   351
static loading   233
static moment   59
statistical analysis   156
steel
– fibres   462
– fine grain   138
– forging   108
– grade   33 f.
– mantle   396
– non-alloy   138
stickle   397
stiffening plate   205
stiffening ribs   111
stirrups   379
stocky sections   127
straightening   26
straightness   45 f.
strain
– concentration factor   243, 247
– gauge   84, 128, 243
– gauge rosette   245, 249
– hardening   151, 210
– hardening effect   29
– ratio   243
stretch-reducing mill   23
stress
– amplitude   232
– concentration   90, 162, 164
– concentration factor   243, 247
– hot spot   239
– maximum   232
– mean   232
– minimum   232
– nominal   239
– nominal range   295
– peak   239
– plastic   210
– range   232, 295
– ratio   232
– residual   295
– redistribution   146
– relief annealing   43
– relieving heat treatment   237
– relieving vibration   308, 310
– yield   210
stretch rolls   22
stretch-reducing process   23 ff., 28 f.
strip   25
– strain gauge   245, 249

structural imperfection   62
stud welding   348 f.
subassembly   361
subcritical   469
submerged-arc welding   26
supercritical   469
superficial area   48
superposition   256
symmetrical connection   118
system length   82

T
tack welding   26, 140 f., 340
taper cone   196
tapering   195
technical delivery requirements   30
tee section   98 f.
template   321
– weld gauge   137
tensile strength   30, 34, 38 ff., 54
tensile test   41
tension resistance   56
thermo-mechanical rolling   43, 288
thermo-mechanical milled steels
    43 f.
throat angle   136
through bolts   118
through welding   123
tie beam   103
tie plate   205
tie-strut   102 f.
TIG   309, 311, 338, 353
tip of the plate   116
toe   130, 142, 242
tolerance properties   45
top coating   318
torque bolts, controlled   110
torsional angle   89
torsional inertia constant   48 f., 89
torsional modulus constant   48 f.
torsional moment of inertia   3
torsional resistance   88
torsional rigidity   61, 186
torsional section modulus   89
torsional shear stress   88, 90
torsional stiffness   54, 128
transition length   196 f.
transmissibility   148
transport   317, 363
– rail   363
– road   363

Subject index

– water 363
transversal plate 224
triangular 126 f.
truss 126, 146
– deflection 186
– depth 128
– forces 128
truss joint
– partial safety factor 138
– uni-polar 176
tubular jacket 103
tungsten electrode 27
twist 47
twisting moment 89

U

ultimate fatigue resistance 234
ultimate load 148, 152
ultimate resistance 154, 199
ultimate strength 149
ultimate tensile strength 41
ultrasonic inspection 346
ultrasonic test 30
undercut 239
uni-axial bending 57 f., 77, 385
uni-planar 246
– lattice girders 11
unsymmetrical connection 118
utilization 421

V

validity range 158, 160, 163, 177, 181 f., 214, 221
vent holes 399, 410
vertical 133 ff.
– of the main truss 171
– plate 174
vibration 403
Vierendeel girders 207, 216 ff.
Vierendeel trusses 207, 211
visual inspection 345

W

wake effects 471
wall thickness 45, 48
– ratio 213
warping 88
– torsion 89
Warren type girder 129
water
– circulation 7, 423

– cooling 423
– transport 363
Watt, James 1
weather resistant 41
weathering steels 413
weld effects, repairs 343
welds
– execution 291
– execution sequence 219
– fillet 274
– failure 155
– full penetration 139
– groove 274
– inspection 343
– joints, uni-planar 156 f.
– legs 136
– length 116, 142 f., 239
– metal 140, 142
– penetration 350
– preparation 125
– prequalification 186
– profiles 344
– roof 129
– root 122, 137 f., 141
– rupture 164
– seam 116, 135, 138, 164
– stress 124
– thickness 123, 126, 135, 138, 140, 239
– thickness, concave 135 f.
– thickness, convex 135 f.
– thickness, flat 135 f.
– throat 172
– toe 239
– toe transition 311
– transition 310
– volume 186
weldability 335
welded joint
– rigidity 210
– uni-planar 157, 159, 181 f., 187 f.
welded top hat section 101
welded truss joints, multi-planar 186
welded tubes 25
welding 335
– chamfer 320
– flux 26
– friction 337
– fusion 337, 354

- laser 349
- plug 341
- position 337, 339 f.
- pressure 337
- rolls 28
- sequence 141 f., 340
- stud 348
Whitehouse, Cornelius 2
wind
- shielding effects 471
- wind tunnel 470
wiring 7
Woehler curve 234

X
X-rays 346

Y
yield hinge theory 78
yield limit 149, 152, 162
yield line 150
yield line mode 210
yield line model 151 f., 172, 202
yield line theory 113, 172
yield strain 248
yield stress 210
yielding 147
yield strength 30, 34, 38 ff., 54, 65, 80, 163, 178, 186, 218
- average 87
- inhomogeneity 77

Z
zinc baths 326